W0106933

Thin Film Growth
Techniques for
Low-Dimensional Structures

NATO ASI Series

Advanced Science Institutes Series

A series presenting the results of activities sponsored by the NATO Science Committee, which aims at the dissemination of advanced scientific and technological knowledge, with a view to strengthening links between scientific communities.

The series is published by an international board of publishers in conjunction with the NATO Scientific Affairs Division

A	**Life Sciences**	Plenum Publishing Corporation
B	**Physics**	New York and London
C	**Mathematical and Physical Sciences**	D. Reidel Publishing Company Dordrecht, Boston, and Lancaster
D	**Behavioral and Social Sciences**	Martinus Nijhoff Publishers
E	**Engineering and Materials Sciences**	The Hague, Boston, Dordrecht, and Lancaster
F	**Computer and Systems Sciences**	Springer-Verlag
G	**Ecological Sciences**	Berlin, Heidelberg, New York, London,
H	**Cell Biology**	Paris, and Tokyo

Recent Volumes in this Series

Series B: Physics

Thin Film Growth Techniques for Low-Dimensional Structures

Edited by

R. F. C. Farrow and
S. S. P. Parkin

IBM Almaden Research Center
San Jose, California

P. J. Dobson and
J. H. Neave

Philips Research Laboratories
Surrey, United Kingdom

and

A. S. Arrott

Simon Fraser University
Burnaby, Canada

Springer Science+Business Media, LLC

Proceedings of a NATO Advanced Research Workshop on
Thin Film Growth Techniques for Low-Dimensional Structures,
held September 15–19, 1986,
at the University of Sussex, Brighton, United Kingdom

This book contains the proceedings of a NATO Advanced Reserach Workshop
held within the program of activities of the NATO Special Program on Con-
densed Systems of Low Dimensionality, running from 1983 to 1988 as part of
the activities of the NATO Science Committee.

Other books previously published as a result of the activities of the Special
Program are:

Volume 148   INTERCALATION IN LAYERED MATERIALS
              edited by M. S. Dresselhaus

Volume 152   OPTICAL PROPERTIES OF NARROW-GAP LOW-DIMENSIONAL
              STRUCTURES
              edited by C. M. Sotomayor Torres, J. C. Portal, J. C. Maan, and
              R. A. Stradling

Library of Congress Cataloging in Publication Data

NATO Advanced Research Workshop on Thin Film Growth Techniques for
    Low-Dimensional Structures (1986: Brighton, East Sussex)
    Thin film growth techniques for low-dimensional structures / edited by
R. F. C. Farrow . . . [et al.].
        p.   cm.—(NATO advanced science institutes series. Series B, Physics; v.
163)
    "Proceedings of a NATO Advanced Research Workshop on Thin Film
Growth Techniques for Low-Dimensional Structures, held Sept. 15–19, 1986,
. . . in Brighton, United Kingdom."
    "Published in cooperation with NATO Scientific Affairs Division."
    Bibliography: p.
    Includes index.

    ISBN 978-1-4684-9147-0   ISBN 978-1-4684-9145-6 (eBook)
    DOI 10.1007/978-1-4684-9145-6

    1. Semiconductors—Surfaces—Congresses. 2. Crystals—Growth—Con-
gresses. 4. Layer structure (Solids)—Congresses. I. Farrow, R. F. C. II. Title. III.
Series.
QC611.6.S9N38   1986
537.6′22—dc19   87-22766
                                                                           CIP

© 1987 Springer Science+Business Media New York
Originally published by Plenum Press, New York in 1987

A Division of Plenum Publishing Corporation
233 Spring Street, New York, N.Y. 10013

All rights reserved
No part of this book may be reproduced, stored in a retrieval system, or transmitted
in any form or by any means, electronic, mechanical, photocopying, microfilming,
recording, or otherwise, without written permission from the Publisher

PREFACE

This work represents the account of a NATO Advanced Research
Workshop on "Thin Film Growth Techniques for Low Dimensional
Structures", held at the University of Sussex, Brighton, England from
15-19 Sept. 1986. The objective of the workshop was to review the
problems of the growth and characterisation of thin semiconductor and
metal layers. Recent advances in deposition techniques have made it
possible to design new material which is based on ultra-thin layers and
this is now posing challenges for scientists, technologists and
engineers in the assessment and utilisation of such new material.

Molecular beam epitaxy (MBE) has become well established as a
method for growing thin single crystal layers of semiconductors. Until
recently, MBE was confined to the growth of III-V compounds and alloys,
but now it is being used for group IV semiconductors and II-VI
compounds. Examples of such work are given in this volume. MBE has one
major advantage over other crystal growth techniques in that the
structure of the growing layer can be continuously monitored using
reflection high energy electron diffraction (RHEED). This technique has
offered a rare bonus in that the time dependent intensity variations of
RHEED can be used to determine growth rates and alloy composition rather
precisely. Indeed, a great deal of new information about the kinetics
of crystal growth from the vapour phase is beginning to emerge. This
subject is not without some controversy and the reader may identify
differing views of the interpretation of RHEED intensity variations in
this book. Metal-organic chemical vapour deposition (MOCVD) is another
technique well suited to growing semiconductors of atomic layer
thickness and many believe it will be more applicable to the large scale
production of device structures. As will be evident here, the results
are very impressive. A similar remark applies to what can be achieved
using liquid phase epitaxy, which although not originally intended for
the growth of layers of a few atoms thickness, can now achieve this for
some materials systems.

Future trends are identified in the extension of these growth
techniques to materials systems outside of the III-V compounds and to
layers which are amorphous. In the latter case this also removes or
relaxes the constraint of seeking combinations of materials which

lattice match and surely must open up many new areas and scope for
combining layers of metals, dielectrics and semiconductors. Metallic
multilayer systems also emerge as a very promising area of research.
There are many examples of the control that can be now exercised on the
design of new magnetic materials. Of course this development also
highlights the need to explore new ways of characterising the magnetic
properties at the interfacial and atomic level.

The organizers of the workshop were aware of the need to question
how appropriate are the contemporary characterisation techniques.
Optical measurements are clearly very useful and important for the
assessment of semiconductor material, and for some magnetic
measurements. Structural investigations must rely on careful x-ray
diffraction and transmission electron microscopy and diffraction.
Several examples of the utility of these techniques applied to low
dimensional structures are given in this book. It was a feeling of
everyone at this research workshop that we are only just starting to
realise the enormous possibilities of growing material with control of
thickness at the atomic level. During the next few years, growth and
assessment techniques will be clearly pushed to new limits.

P.J. Dobson
Philips Research Labs
Redhill
England

CONTENTS

INTRODUCTION

This volume contains the contributions to a NATO Advanced Research Workshop on Thin Film Growth Techniques for Low Dimensional Structures which was held at the University of Sussex between 15-19 September 1986.

In recent years there has been intense activity in the semiconductor field for the manufacture of structures in which at least one dimension is restricted. This has resulted in the study and use of quantum wells and superlattices for lasers, photodetectors, optical modulators and other optoelectronic devices. There is now a worldwide activity in the development of many new electronic and optical devices which depend on the growth of accurately controlled semiconductor layers of only a few atomic layers thickness. The demands being placed on growth technology to realize such structures with the required high degree of crystalline perfection are severe, and scientists and technologists are rising to these challenges. There is an interesting interplay between the science and technology here. As new structures are designed to investigate the nature of material in low-dimensional form, so problems are created for the technology to make the structure. In solving the problems, new insight is gained into the basics physics and chemistry of the growth processes. Hopefully, and indeed it is already evident, we can use some of this insight to improve the technology for making structures which might eventually reach the market-place. This workshop concentrated specifically on aspects of growth, but it should be remembered that there are many other problems in achieving small dimensions in a device structure.

In some respects such a workshop is timely since the field is just beginning to expand and encompass many more materials. For the past decade most work on two-dimensional structures, i.e, quantum wells and superlattices, has concentrated on GaAs, (AlGa)As and other III-V alloy systems. For example layers of GaAs of thickness from about $13\overset{\circ}{A}$ to $250\overset{\circ}{A}$ sandwiched between (AlGa)As alloy layers have been extensively investigated with regard to optical and electrical properties and have been made into quantum well lasers. Now, group IV and II-VI semiconductor superlattices are being grown and the field is rapidly expanding to include metal and dielectric layers. Much of the motivation in studying metal multilayers is the prospect that new magnetic material can be designed, in much the same way that semiconductor technologists are seeking to design new semiconductors from thin layers of dissimilar material. We are only at the threshold of a new era of "designed material". However actually achieving the design parameters presents many challenges.

The different growth techniques being used to grow low dimensional structures were well represented at the workshop. Molecular beam epitaxy (MBE) perhaps received rather more emphasis because it is more adaptable for different materials and more easily used in combination with modern surface analytical techniques, particularly reflection high energy electron diffraction (RHEED). This technique is capable of yielding rather precise measurements of the growth rate via oscillations of the diffracted intensity with time which occur during growth. At this workshop there were examples of these so-called

RHEED oscillations for the growth of growth IV, III-V and II-VI semiconductors and, for the first time, metals. These changes in the intensity which accompany uniform layer type growth modes will contribute, to our understanding of surface kinetics. Vapor phase epitaxy and particularly the variant which uses metal-organic vapors for the source of the group III component (MOVPE) has been shown to be capable of growing very high quality device material. Much of these achievements in MOVPE have been made without access to the detailed atomistic understanding which exists for MBE. Light beams are the only realistic tools to monitor crystal growth at high gas pressures and their exploitation was well illustrated in the course of plasma deposition of amorphous superlattices by the Exxon group (Yang and Abeles). This new class of structures opens up a wide range of possibilities where the lattice parameter matching criterion is less significant. Liquid phase epitaxy (LPE), which has been used for some time to grow device structures and high quality "bulk" semiconductor material, has also been used successfully to grow quantum well structures.

Techniques for the the growth of single crystalline metallic superlattice structures are still in their infancy compared to those developed for comparable semiconductor structures. In the past a tremendous amount of work has been carried out on metallic thin film structures grown by sputtering and to a lesser extent electron beam evaporation in moderately- high vacuum systems. Such superlattice structures have shown novel properties linked to the imposed super-lattice period. However without ultra high vacuum, high pumping speeds and detailed in-situ surface analytical capabilities the quality of such structures is poor. Since the influence of the interface on the magnetic properties of adjacent layers goes to zero within a few atomic planes, particularly for the 3d transition metals, this has led to considerable controversy concerning the magnetic properties of such systems. A very limited discussion of some of this work on textured multi-layer systems is included in these proceedings.

Recently the techniques of Molecular Beam Epitaxy in ultra high vacuum systems have been used to grow a number of very carefully characterized magnetic metal film and superlattice systems. Many of the groups involved in this work are represented in these proceedings. The successful growth of single-crystal rare earth metal superlattices ,first carried out by Kwo et al, opens the door to the growth of novel artificial magnetic structures. The existence of magnetic coupling of Gd and Dy layers through nonmagnetic Y layers has provided a beautiful demonstration of the RKKY interaction. The understanding of these data has only been possible by the use of powerful techniques to establish and quantify the structural perfection across the interface, including for example the use of magnetic x-ray scattering.

The rare earth superlattices contain components with identical lattice structures and very similar lattice parameters. Very different superlattice systems have recently been studied in which the two components have different crystal structures. This can lead to novel but theoretically predicted long-period stacking fault configurations as observed in Ru-Ir "bicrystal" superlattices. Coherency strains can have important effects on the magnetic properties of magnetic films and superlattice, in particular with respect to magnetic ordering temperatures and magnetic anisotropies. Some of these effects are described in these proceedings. Of particular interest are the dependence of these properties on film thickness as described for iron films grown epitaxially on GaAs and for Dy films grown on yttrium.

A second important type of metallic film whose growth has been made possible by the use of MBE techniques is that of crystallographic phases of various 3d transition metals which are not normally stable. Metastable phases of Co and Mn stabilized by their epitaxial growth onto substrates of GaAs(100) and Ru(0001) respectively are described.

The characterization of ultra thin films is difficult, both with respect to crystal and magnetic structure. Some of these techniques can effect the properties of the films, making their interpretation ambiguous. A large number of sophisticated analytical probes are presented in these proceedings, including a variety of relatively novel techniques for examining the magnetic behavior of thin films. These include SMOKE (surface magneto-optic Kerr effect), a number of spin-polarized electron spectroscopies, polarized neutron reflectometry, and magnetic neutron scattering. The same type of oscillations in RHEED intensity, recently found during the growth of semiconductor films, has similarly been observed for metal films. Oscillations in other experimental measurements, for example, the resistivity of the film are described. These types of in-situ measurements may prove invaluable in the growth of more perfect metallic film structures.

In summary, the growth of highly ordered metal thin- film structures has been demonstrated. This has, in large part, become possible by using many of the same methods developed for the growth of semiconductor film structures. In the future it seems likely that the application of more of these methods can improve the quality of metal film structures and will make possible the growth of artificial metal materials with novel physical properties. The understanding of the properties of such films, in particular magnetic properties, may depend upon the further development of appropriate characterization probes for examining the magnetic moment and susceptibility as a function of temperature of oligatomic films.

All the materials systems are on common ground when it comes to microstructural characterization. Here, high resolution transmission electron microscopy and X-ray diffraction are likely to provide most of the information for many years. At this workshop we were given new insights into the ways in which these techniques are developing to meet the challenges set by low-dimensional structures.

EFFECT OF BARRIER CONFIGURATION AND INTERFACE QUALITY ON STRUCTURAL AND ELECTRONIC PROPERTIES OF MBE-GROWN $Al_xGa_{1-x}As$ / GaAs, $Al_xGa_{1-x}Sb$ / GaSb AND $Al_xIn_{1-x}As$ / $Ga_xIn_{1-x}As$ SUPERLATTICES

K. Ploog, W. Stolz, and L. Tapfer

Max-Planck-Institut für Festkörperforschung
D-7000 Stuttgart-80, FR-Germany

ABSTRACT

The challenge for the design and growth of quantum wells and super-lattices made of III-V semiconductors is to minimize scattering from impurities, alloy clusters or interface irregularities so that the confined carriers can move freely along the interfaces. We present a few examples for the influence of interface quality and barrier configuration on the structural as well as electrical and optical properties of III-V semiconductor quantum wells and superlattices grown by molecular beam epitaxy (MBE).

1. INTRODUCTION

The fabrication of ultrathin semiconductor layers and multiple layers with abrupt junctions and precise dopant control plays an important role for the development of new photonic and electronic devices. The technique of molecular beam epitaxy (MBE) provides atomic abruptness and smoothness between layers of different lattice-matched and lattice-mismatched III-V semiconductors at their interfaces or heterojunctions. The heterointerfaces between epitaxial layers of different composition($Al_xGa_{1-x}As$/GaAs, $Al_xIn_{1-x}As$ /$Ga_xIn_{1-x}As$, $Al_xGa_{1-x}Sb$/GaSb, etc.) are used to confine electrons or holes to two-dimensional (2D) motion. The challenge for the design and growth of materials is to minimize scattering from impurities, alloy clusters or interfaces. In this paper we present a few selected examples for the influence of barrier configuration and interface quality on the structural and electronic properties of quantum wells and superlattices made of III-V semiconductors. The material systems AlAs/GaAs, $Al_xIn_{1-x}As$/$Ga_xIn_{1-x}As$, and AlSb/ GaSb are attractive for application in advanced photonic devices, as they cover the 0.6 - 2.0 μm wavelength range in emission and absorption, and in high-speed devices because of the formation of high-mobility 2D electron and hole systems.

2. DETERMINATION OF SUPERLATTICE PERIOD AND INTERFACE DISORDER BY DOUBLE-CRYSTAL X-RAY DIFFRACTION

High-angle X-ray diffraction is a powerful non-destructive technique for investigation of interface disorder effects in superlattices and multi quantum well heterostructures, if a dynamical analysis of the diffraction curves is performed. We have recently developed a semi-kinematical approach of the dynamical theory of X-ray diffraction to determine the strain profile, the composition (periodicity), and the interface quality of $Al_xGa_{1-x}As/GaAs$ and of $Al_{0.48}In_{0.52}As/Ga_{0.47}In_{0.53}As$ heterostructures and superlattices / 1 / . In the following we briefly outline those parts of the theory that are relevant to our measurements.

The scattering of X-rays from strained single crystals is described by Taupin's formalism of the dynamical theory of X-rays / 2 / . The semi-kinematical approximation of Petrashen / 3 / uses the first iteration of Taupin's equation for the amplitude ratio of the diffracted and incident waves and is valid if the thickness of the deformed layer is small compared with the X-ray extinction length. The reflectivity is then given in integral form which reduces the calculation time for a diffraction curve. For our calculation the epitaxial layer is devided into n lamellae of equal thickness $\Delta = z_0/n$, where z_0 is the total thickness of the layer measured in units of the extinction length devided by π. The reflectivity $R(y)$ is given by the product of the diffraction curve for an undeformed crystal $R_p(y) = |g - y|^2$ and a deformation dependent factor

$$R(y) = R_p(y) \left| 1-2iy \sum_{j=1}^{n} \exp(i\phi_j) \frac{\sin(y-s_j)\Delta}{y-s_j} \right|^2 \tag{1}$$

For a crystal without a centre of inversion we have

$$y = \frac{-2b \sin(2\theta_B)\Delta\theta - (1-b)(\chi_0^r + i\chi_0^i)}{2C (|b|(\chi_h^r\chi_{\bar{h}}^r - \chi_h^i\chi_{\bar{h}}^i))^{1/2}} \tag{2}$$

$$g = \text{sign}(y) \ (y^2 - 1 - 2i \frac{\chi_{\bar{h}}^r\chi_h^i + \chi_h^r\chi_{\bar{h}}^i}{\chi_h^r\chi_{\bar{h}}^r + \chi_h^i\chi_{\bar{h}}^i})^{1/2} \tag{3}$$

and

$$\phi_j = -y\Delta \ (2(n-j)+1) + \Delta s_j + 2\Delta + \sum_{j=j+1}^{n} s_j \tag{4}$$

where $\chi_h^{r,i} = r_e \frac{\lambda^2}{\pi V} F_h^{r,i}$ are the h-th Fourier coefficients of the polariza-bility multiplied by 4π, λ is the X-ray wavelength, r_e is the classical electron radius, V is the unit cell volume and F^r, F^i are the the real and imaginary part of the structure factor. $b = \gamma_0/\gamma_h$ is the asymmetry factor, and $\gamma_0 = \sin(\theta_B - \alpha)$ and $\gamma_h = \sin(\theta + \alpha)$ are the direction cosines of the incident and diffracted waves respectively, θ_B is the kinematical Bragg angle and α the angle between crystal surface and reflecting lattice plane. Δ is the deviation from the Bragg angle. C is the polarization factor, which is equal to unity for σ-polarization and $\cos 2\theta_B$ for π-polarization. s_j is the normalized strain for the j-th lamella. This constant value within each lamella is given by the equation

$$s = \frac{\lambda |b|^{1/2}}{2\pi C (\chi_h^r\chi_{\bar{h}}^r - \chi_h^i\chi_{\bar{h}}^i)^{1/2}} \frac{\partial}{\partial s_h} (\vec{h}\vec{u}) \tag{5}$$

where $\bar{h}$ is the diffraction vector, $\bar{u}$ is the displacement field, and s_h is the direction of the diffracted wave. If there is no shear strain, as in epitaxial layers grown in (001) and (111) orientation, and if the crystal is bound by the xy plane and the diffraction plane is the xz plane, Eq.(5) takes the form

6

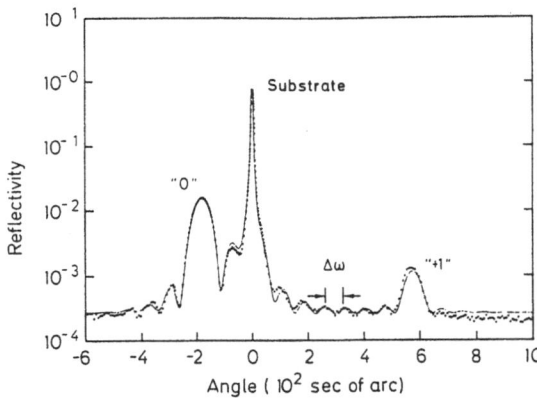

Fig. 1. Experimental (dotted curve) and theoretical (solid line) X-ray
diffraction curve of an $(AlAs)_{44}$ $(GaAs)_{46}$ superlattice in the
vicinity of the (004) reflection using $CuK\alpha_1$ radiation

$$s(z) = \frac{2|b|^{1/2} \sin^2\theta_B}{(\chi_h^r \chi_{\bar{h}}^r - \chi_h^i \chi_{\bar{h}}^i)^{1/2}} (\varepsilon_{zz}(z)\cos^2\alpha + \varepsilon_{xx}(z)\sin^2\alpha + (\varepsilon_{zz}(z) - \varepsilon_{xx}(z))\frac{\sin2\alpha}{2\tan\theta_B}) \quad (6)$$

Here the depth dependence of the normalized strain is taken into account,
and $\varepsilon_{zz}(z)$ and $\varepsilon_{xx}(z)$ are the strains at depth z perpendicular and parallel
to the crystal surface, respectively.

The sinusoidal term in Eq. (1) describes the amplitude of a single
lamella, whereas ϕ_j in the exponential term describes the phase of a single
lamella. One of the interesting consequences of Eq. (1) is the possibility
to observe oscillations on the X-ray diffraction curve, i.e., Pendellösung
fringes, as a result of interferences of waves scattered at different depths
in the epitaxial layer (see $\Delta\omega$ in Fig. 1). If we assume that s(z) is constant in the whole epilayer of the thickness D, the sinusoidal term of the
amplitude in Eq. (1) oscillates with the period π. For the argument of the
sinusoidal term we then obtain the relation

$$yD = \pi. \quad (7)$$

Using Eq. (2) and considering the fact that D must be multiplied by l_{ex}
we find the known relation between the angular spacing of the Pendellösung
fringes $\Delta\omega$ and the thickness D of the strained surface layer, i.e.

$$D = \frac{\lambda |\gamma_h|}{\Delta\omega \sin(2\theta_B)}. \quad (8)$$

If we apply the same evaluation to a superlattice, i.e., a one-dimensional
periodic structure with a period length T, we find the identical relation
as given in Eq. (8). The whole epilayer thickness is replaced by the superlattice period T, and the oscillation spacing is replaced by the angular
distance between the satellite peaks which occur in X-ray diffraction curves
from superlattices. The amplitude of the deformation factor in Eq. (1),

7

which describes the scattering of the deformed layer, has a maximum if $y = \bar{s}$, there $\bar{s}$ is the average value of $s(z)$ in the strained layer. This consideration leads to the relation between (i) the angular spacing of the main diffraction peaks of the strained layer and of the unstrained crystal, $\Delta\theta^o$, and (ii) the average strain value $\bar{\varepsilon}_{zz}$, which exists in the strained layer.

Semiconductor superlattices are one-dimensional periodic structures consisting of thin layers of alternating composition. Therefore, the depth variation of the strain in Eq. (6) and the depth variation of the structure factor in the deformation dependent term of Eq. (1) must be included. The depth distribution of the strain and of the chemical composition are determined by a comparison of the calculated X-ray diffraction curve with the experimental data. A certain strain and an approximate composition distribution is initially presumed from the employed growth conditions. The best coincidence of the theoretical and experimental X-ray diffraction curve is then found by varying successively the values for the structural parameters, i.e. the strain and composition profiles as well as the epilayer thickness.

The X-ray diffraction measurements were performed with a computer-controlled double-crystal X-ray diffractometer in non-dispersive (+, -) Bragg geometry. An asymmetrically cut (100) Ge crystal was used for monochromizing and collimating the X-rays / 1 / . As the lattice parameter and the scattering factors of superlattices are subject to a one-dimensional modulation in growth direction, the diffraction patterns consist of satellite reflections located symmetrically around the Bragg reflections, as shown in Fig. 1 for an AlAs/GaAs superlattice. From the position of the satellite peaks the superlattice periodicity can be deduced. Detailed information about thickness fluctuations of the constituent layers, inhomogeneity of composition, and interface quality can be extracted from the halfwidths and

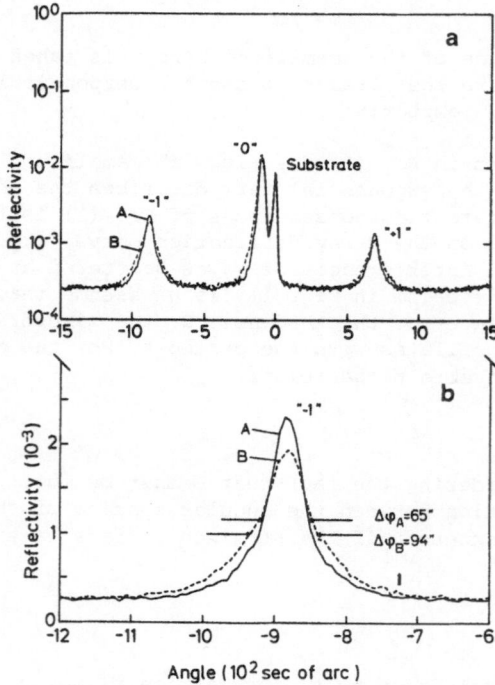

Fig. 2. X-ray diffraction curves of two AlAs / GaAs superlattices grown (A) with and (B) without growth interruption at the heterointerfaces, recorded with CuKα₁ radiation in the vicinity of the quasiforbidden (002) reflection.

intensities of the satellite peaks. The excellent agreement between experimental and theoretical diffraction curve in Fig. 1 indicates extremely abrupt AlAs/GaAs interfaces to within one monolayer.

Since MBE growth occurs predominantly in a two-dimensional layer-by-layer growth mode, the compositional changes at heterointerfaces of closely lattice-matched materials, like AlAs/GaAs, should occur over no more than one monolayer. However, for the widely used $Al_xGa_{1-x}As$ / GaAs heterojunction it is well established that the sequence of layer growth is critical for compositional gradients and crystal perfection, which in turn strongly affect the excitonic and the transport properties of quantum wells. While the $GaAs/Al_xGa_{1-x}As$ heterointerface is abrupt to within one monolayer when the ternary alloy is grown on the binary compound, this is not the case for the inverse growth sequence under typical MBE growth conditions / 4 / . This phenomenon is probably caused by the difference in the relative surface diffusion lengths of Ga and Al on (100) surfaces / 5 / . Various attempts have been made to minimize the interface roughness (or disorder) by modifying the MBE growth conditions. The most successful modification is the method of growth interruption at each interface, which allows the small terraces between monolayer steps on the $Al_xGa_{1-x}As$ surface to relax into larger terraces via diffusion of the surface atoms, and thus reduces the step density. The time of closing both the Al and the Ga shutters (while the As shutter is left open) depends on the actual growth conditions, and values ranging from a few seconds to several minutes have been reported. In X-ray diffraction the existence of interface disorder manifests itself in an increase of the halfwidths and a decrease of the intensities of the satellite peaks, as shown in Fig. 2. During MBE growth of these two $(AlAs)_{42}$ $(GaAs)_{34}$ superlattices, the adjustment of the shutter motion at the transition from AlAs to GaAs and vice versa was changed in the two growth runs. Sample A was grown with growth interruption at each AlAs/GaAs and GaAs/AlAs interface, whereas sample B was grown continuously. While the positions of all the diffraction peaks of sample A coincide with those of sample B, the halfwidths of the satellite peaks from sample A are narrower and their reflected intensities are higher. A growth interruption of 10 s was sufficient to smooth the growing surface which then provides sharp heterointerfaces. When the heterojunctions are grown continuously, the monolayer roughness of the growth surface leads to a disorder and thus broadening of the interface. In X-ray diffraction this broadening manifests itself as a random variation of the superlattice period of about one lattice constant (~ 5.6 Å) for sample B.

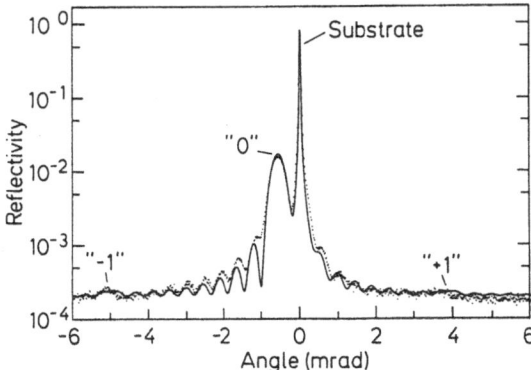

Fig.3. $CuK\alpha_1$ (400) diffraction pattern of $Al_xIn_{1-x}As/Ga_xIn_{1-x}As$ superlattice on ($\bar{1}$00) InP with $L_Z = L_B = 10.2$ nm (..... experiment, ——— theory).

The quantitative evaluation of the interface quality by X-ray diffraction becomes even more important if the lattice parameters of the epilayer have to be matched to those of the substrate by appropriate choice of the layer composition, as for $Al_{0.48}In_{0.52}As/Ga_{0.47}In_{0.53}As$ superlattices lattice-matched to InP substrates. In Fig. 3 we show the X-ray diffraction pattern of such a superlattice with a periodicity of 20.4 nm. For the investigation of this all-ternary material system the X-ray diffraction patterns were recorded in the vicinity of the symmetric (200) and (400) reflections and of the asymmetric (422) and (440) reflections using $CuK\alpha_1$ radiation. Both symmetric and asymmetric diffraction data yield the average lattice strain perpendicular, $\bar{\epsilon}_{zz}$, and parallel, $\bar{\epsilon}_{xx}$, to the (100) substrate surface. The lattice strains $\bar{\epsilon}_{zz}$ and $\bar{\epsilon}_{xx}$ are correlated with the angular distances $\Delta\theta_I$ and $\Delta\theta_{II}$ between the substrate diffraction maximum and the main epitaxial layer peak ("O"-peak) by the equation

$$\begin{pmatrix} \bar{\epsilon}_{zz} \\ \bar{\epsilon}_{xx} \end{pmatrix} = \begin{pmatrix} A_I & B_I \\ A_{II} & B_{II} \end{pmatrix}^{-1} \begin{pmatrix} \Delta\theta_I \\ \Delta\theta_{II} \end{pmatrix} \qquad (9)$$

with

$$A_{I,II} = \cos\alpha_{I,II} * [\cos\alpha_{I,II} * \tan\theta_{I,II} + \sin\alpha_{I,II}]$$
$$B_{I,II} = \sin\alpha_{I,II} * [\sin\alpha_{I,II} * \tan\theta_{I,II} - \cos\alpha_{I,II}] \qquad (10)$$

and

$$\bar{\epsilon}_{zz} = (\bar{d}_e^{\perp} - d_s) / d_s$$
$$\bar{\epsilon}_{xx} = (\bar{d}_e^{\parallel} - d_s) / d_s \qquad (11)$$

The indices I and II hold for symmetric and asymmetric reflections, respectively. Here, $\bar{d}_e^{\perp}$ and $\bar{d}_e^{\parallel}$ are the average interplanar spacings of the epitaxial layer perpendicular and parallel to the crystal surface, while d_s is the interplanar spacing of the substrate crystal, respectively. θ_I and θ_{II} are the kinematic Bragg angles, and α_I and α_{II} the angles between crystal surface and reflection planes I and II. The evaluation of our experimental diffraction data reveals that the superlattice is not misoriented with respect to the substrate crystal and the $\bar{\epsilon}_{xx} = 0$ for all samples, i.e.

TABLE 1 Structural parameters of four 10-period $Al_xIn_{1-x}As/Ga_xIn_{1-x}As$ superlattices grown by MBE with different temperature settings of the group-III-element effusion cells, as determined by X-ray diffraction.

Sample No.	Thickness of $Ga_xIn_{1-x}As$ layers (nm)	Thickness of $Al_xIn_{1-x}As$ layers (nm)	Average lattice mismatch of superlattice $\bar{\epsilon}_{zz}$ (x 10^{-4})	Lattice strain in $Ga_xIn_{1-x}As$ layers $\bar{\epsilon}_{zz}$ (x 10^{-4})	Lattice strain in $Al_xIn_{1-x}As$ layers $\bar{\epsilon}_{zz}$ (x 10^{-4})	Mole fraction x of $Ga_xIn_{1-x}As$ layers	Mole fraction x of $Al_xIn_{1-x}As$ layers
5649	10.6	10.6	− 4.5	+ 6.0	− 15.0	0.464	0.487
5653	10.2	10.2	+ 9.4	9.5	9.3	0.462	0.470
5654	10.2	10.2	− 4.0	0.0	− 8.0	0.468	0.482
5655	10.2	10.2	− 3.5	7.0	− 14.0	0.464	0.486

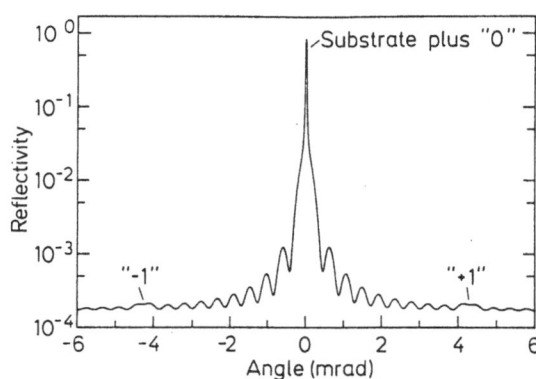

Fig. 4. Calculated diffraction pattern of a perfectly lattice-matched $Ga_xIn_{1-x}As/Al_xIn_{1-x}As$ superlattice on (100) InP [$CuK\alpha_1$ (400) reflection].

the lattice spacing parallel to the crystal surface in the constituent $Ga_xIn_{1-x}As$ and $Al_xIn_{1-x}As$ layers and in the InP substrate crystal are the same. Hence it follows that the lattice strain in the $Ga_xIn_{1-x}As$ and $Al_xIn_{1-x}As$ epilayers perpendicular to the crystal surface is a measure of their mole fraction x. The relation between elastic strain and chemical composition in the $Ga_xIn_{1-x}As$ layers is given by

$$x = \frac{1}{a_{GaAs} - a_{InAs}} *[a_{InP}(1+ \frac{c_{11}}{c_{11}+2c_{12}} \varepsilon_{zz}) - a_{InAs}] \qquad (12)$$

where c_{11} and c_{12} are the elastic stiffness constants of the epilayer material. For the $Al_xIn_{1-x}As$ layers the lattice constant a_{GaAs} in Eq. (12) must be replaced by that of a_{AlAs}.

In Fig. 3 we show the experimental (dotted line) and the theoretically fitted (solid line) diffraction patterns obtained from a representative $Al_xIn_{1-x}As$ / $Ga_xIn_{1-x}As$ superlattice in the vicinity of the symmetrical (400) $CuK\alpha_1$ reflection. From the theoretical diffraction pattern we obtain the thickness as well as the lattice strain of the individual layers. The chemical compositions are determined by using Eq. (12). In Table 1 we summarize the measured and the calculated structural data from four $Al_xIn_{1-x}As/Ga_xIn_{1-x}As$ superlattices grown under different conditions. The growth conditions were adjusted such that for all samples the barrier width L_B equals the well width L_z. The lattice mismatch perpendicular to the (100) growth face for the tetragonally distorted epilayers on InP is less than / 2.8×10^{-3} / , i.e. less than the lattice mismatch in the AlAs/GaAs system. The pendellösung fringes observed between the main diffraction peaks and the satellite peaks "-1" and "+1" demonstrate the excellent thickness and composition homogeneity perpendicular and parallel to the crystal surface.

In Fig. 4 we show the theoretical diffraction pattern for a perfectly lattice-matched $Ga_{0.468}In_{0.532}As/Al_{0.477}In_{0.523}As$ superlattice with $L_B = L_z = 10.6$ nm. It should be noted that the satellite peaks "-1" and "+1"

11

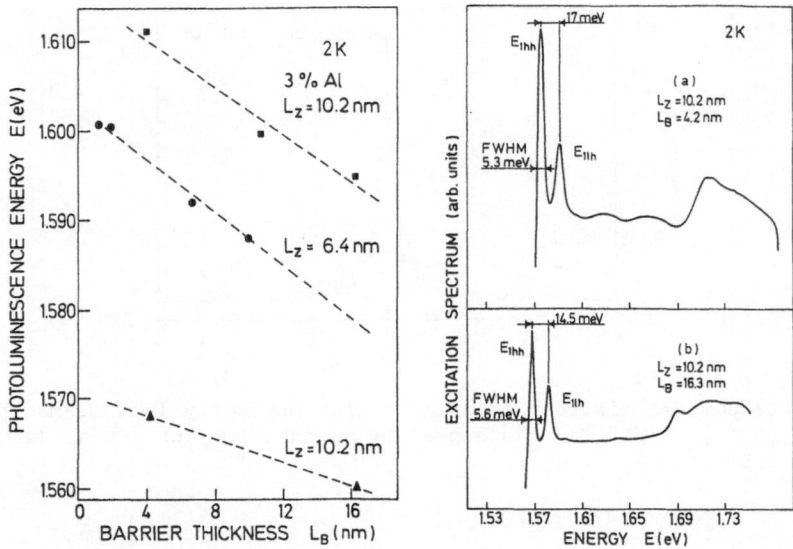

Fig. 5. Variation of luminescence peak-energy (left) and splitting between
E_{1h} and E_{11} free-exciton resonances (right) as a function of barrier
thickness in AlAs/GaAs superlattices for three series of samples
with constant well widths.

have almost disappeared. This finding is in contrast to that observed in
strained layer superlattices, where a strain periodicity produces strong
satellite peaks. The low intensity and the Pendellösung fringes in Fig. 4
are caused only by the periodicity of the structure factors in the superlattice. A weak intensity of the satellite peaks is also observed if the
lattice strains of the $Ga_xIn_{1-x}As$ and $Al_xIn_{1-x}As$ layers are of the same
magnitude, which occurs in the sample displayed in Fig. 3 (see also Table 1).

3. EFFECT OF BARRIER THICKNESS ON LUMINESCENCE PROPERTIES OF AlAs/GaAs
SUPERLATTICES AND MULTI QUANTUM WELL HETEROSTRUCTURES

When the barrier thickness L_B in $Al_xGa_{1-x}As$/GaAs superlattices is
reduced to below 3 nm, the wavefunctions of the GaAs wells couple through
the barriers and subbands of finite width parallel to the layer plane are
formed. At this transition from a multi quantum well heterostructure with
isolated GaAs wells to a real Esaki-Tsu superlattice the luminescence peak
energy decreases for a constant well width L_z due to the broadening of
the subbands. We have recently found, however, that even for isolated
GaAs wells with thick barriers in AlAs/GaAs superlattices the barrier
thickness has an unexpected influence on the excitonic transitions / 6 / .
In Fig. 5 we show that for constant GaAs well widths of L_z = 10.2 nm and
L_z = 6.4 nm the excitonic peaks shift to higher energies and the splitting
between heavy- (E_{1h}) and light-hole (E_{11}) free excitons becomes larger
when the AlAs barrier thickness is reduced from L_B = 16 to L_B = 2nm. The
same high-energy shift exists when 3 mole percent Al is added to the well.
This phenomenon is in contrast to the expectation from a simple coupling
between adjacent wells, and a conclusive explanation has not yet been
found. For interpretation we have to take into account the complex band
structure of GaAs quantum wells arising from (i) the valence band mixing,
(ii) the nonparabolicity of the conduction band, and (iii) the indirect
nature of the barrier material. For practical application it is important
that our results demonstrate the inadequacy of luminescence spectroscopy

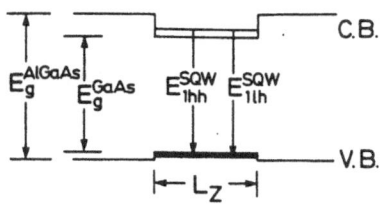

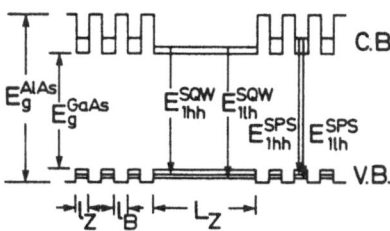

Fig. 6. Real-space energy band diagram of GaAs quantum well confined by
ternary $Al_xGa_{1-x}As$ alloy (top) or by all-binary AlAs/GaAs SPS
(bottom).

to determine the well widths of superlattices accurately. For this purpose
additional techniques like double-crystal X-ray diffraction are required.
We have used this technique to determine the thickness of the constituent
layers of the three series of samples precisely, since the knowledge of
the actual values is crucial for interpretation of the luminescence data.

4. SHORT-PERIOD SUPERLATTICE BARRIERS

The concept of short-period superlattice (SPS) barriers was originally
developed to replace the ternary alloy $Al_xGa_{1-x}As$ by all-binary AlAs/GaAs
superlattices / 7 / , because (i) $Al_xGa_{1-x}As$ is an indirect semiconductor
for x > 0.4, (ii) deep donors exist in n-doped $Al_xGa_{1-x}As$ for x > 0.2, and
(iii) the growth sequence $Al_xGa_{1-x}As$/GaAs exhibits considerable disorder
at the interface. In Fig. 6 we show schematically the energy band edges of
a GaAs QW confined either by homogeneous ternary $Al_xGa_{1-x}As$ barriers or by
GaAs/AlAs SPS barriers. The effective barrier height for carrier confinement in the QW is adjusted by appropriate choice of the layer thickness
of the lower-gap material in the SPS.

Several years before we studied the improvement of the optical properties of GaAs, GaSb, and $Ga_{0.47}In_{0.53}As$ QW by using SPS barriers systematically, we had already unintentionally produced periodic compositional
oscillations in the growth direction of $Al_xGa_{1-x}As$ barriers, as shown in
Fig. 7, which are related to the substrate rotation. The regular fringes
observed in the dark-field transmission electron micrograph (TEM) through
the cross-section of a GaAs double quantum well heterostructure were
caused by compositional oscillations with a period of 5 nm, due to variations of the Ga and Al flux profiles over the substrate area. The observed
period was consistent with the growth rate of 1.2 μm/hr and the substrate
rotation frequency of 4 rpm. Although not intentionally introduced and
often undesired, in that particular case the periodic variation of the
$Al_xGa_{1-x}As$ alloy composition normal to the layers resulted in highly
improved luminescence properties of GaAs quantum wells as shown by the
photoluminescence excitation (PLE) spectrum in Fig. 7. In addition to the
parity-allowed n = 1 and n = 2 electron heavy- and light-hole exciton
resonances, also a forbidden transition (E_{21h}) was observed at 1.62 eV.

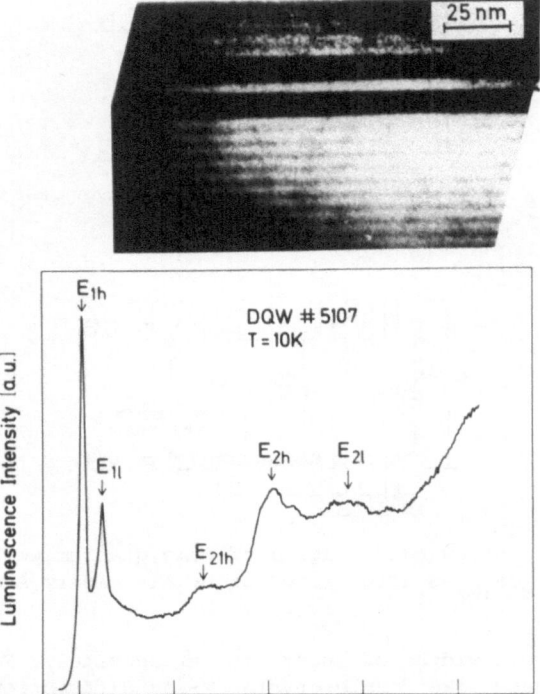

Fig. 7. (100) cross-sectional TEM (top) and photoluminescence excitation
spectrum (bottom) of a 9-nm GaAs double quantum well heterostructure confined by $Al_xGa_{1-x}As$ SPS (average x = 0.24) due to compositional oscillations induced by substrate rotation.

The very sharp exciton resonances and the large intensity ratio of the E_{1h}
and E_{1l} transition (about 3 to 1) indicate the excellent quality of this
GaAs QW, in particular the smoothness of the heterointerfaces that were
grown four years ago without any growth interruption. We assume that the
observed improvement of the optical properties of SPS confined QW is due
(i) to a removal of substrate defects by the SPS layers, (ii) to an ame-
lioration of the interface between QW and barrier, and (iii) to a modification of the dynamics of injected carriers in the SPS barrier.

The application of SPS barriers has an even more dramatic effect on
the improvement of excitonic recombination processes in GaSb / 8 / and in
$Ga_{0.47}In_{0.53}As$ quantum wells / 9 / . In Fig. 8 we show the results of photoluminescence (PL) and PLE measurements. In GaSb quantum wells the effect
of strain and the formation of defect centers at the interface, due to
the considerable lattice mismatch of $\Delta a/a_0$ = 0.65% between GaSb and AlSb,
is significantly reduced by all-binary AlSb/GaSb SPS confinement layers.
As a result the luminescence of SPS confined GaSb quantum wells is dominated by free-exciton emission in the temperature range 4 - 200 K. The
small value of 7.5 meV at 5K for the Stokes shift between PL and PLE heavyhole excitonic peak indicates the absence of impurity related trapping of
excitons in GaSb quantum wells. The application of all-ternary $Al_{0.48}In_{0.52}As/Ga_{0.47}In_{0.53}As$ SPS to confine $Ga_{0.47}In_{0.53}As$ quantum wells resulted in
the first experimental evidence for intrinsic free-exciton recombination

14

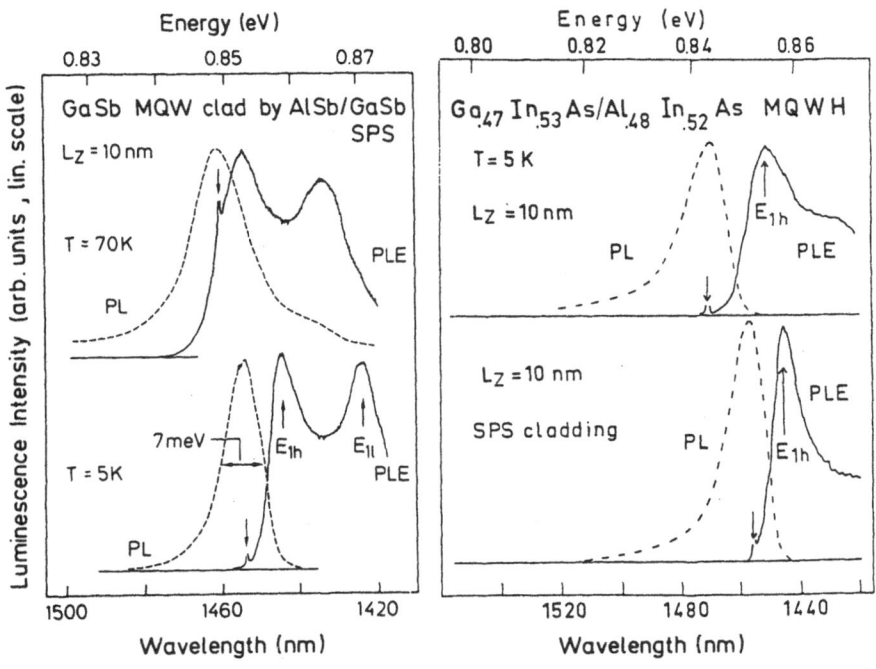

Fig. 8. Low-temperature PL and PLE spectra of GaSb quantum wells confined by
all-binary AlSb/GaSb SPS (left) and of $Ga_{0.47}In_{0.53}As$ quantum wells
confined by ternary $Al_{0.48}In_{0.52}As$ and by $Ga_{0.47}In_{0.53}As/Al_{0.48}In_{0.52}As$ SPS (right).

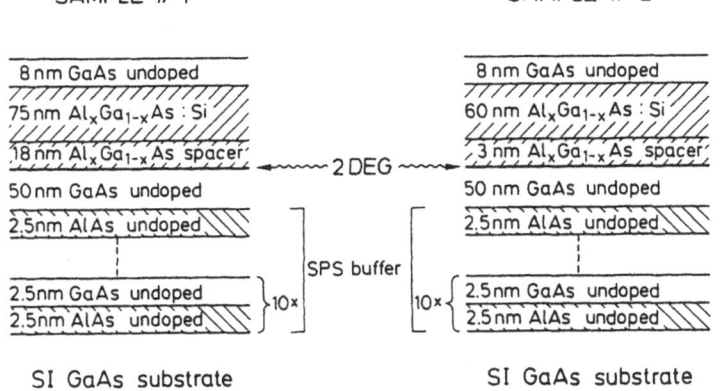

Fig. 9. Schematic layer sequence of two selectively doped $n-Al_xGa_{1-x}As$ /
GaAs heterostructures having the same AlAs / GaAs SPS buffer
layer but different spacer widths.

in this material system. The narrow linewidth of 3.1 meV for the E_{1h} exciton peak in the PLE spectrum and the small Stokes shift of 6.5 meV between PL and PLE exciton peaks manifest the excellent quality of the SPS confined $Ga_{0.47}In_{0.53}As$ quantum wells lattice-matched to InP substrates which were grown without any growth interruption at the heterointerfaces.

A distinct example for the removal of substrate defects by SPS buffer layers and for the improvement of the interface between quantum well and SPS barrier is given by a modified selectively doped $Al_xGa_{1-x}As$ / GaAs heterostructure with high-mobility 2D electron gas (2DEG) / 9 / , whose layer sequence is schematically shown in Fig. 9. The 10-period GaAs/AlAs SPS prevents propagation of dislocations from the substrate so that the thickness of the "active" GaAs layer containing the 2DEG can be reduced to 50 nm. The results of Hall effect measurements (Fig. 10) demonstrate the excellent mobilities of the 2DEG obtained for two samples with different spacer widths. The SPS-confined narrow "active" GaAs layer of the heterostructure is of distinct importance for transistor operation, because the electrons cannot escape too far from the two-dimensional channel during pinchoff. This implies a higher transconductance for the high electron mobility transistor (HEMT). Finally, the growth time of the complete heterostructure is reduced to less than 15 min. An additional 15 min for wafer exchange and heat and cool time makes a total of 30 min throughput time per high-quality heterostructure wafer grown by MBE.

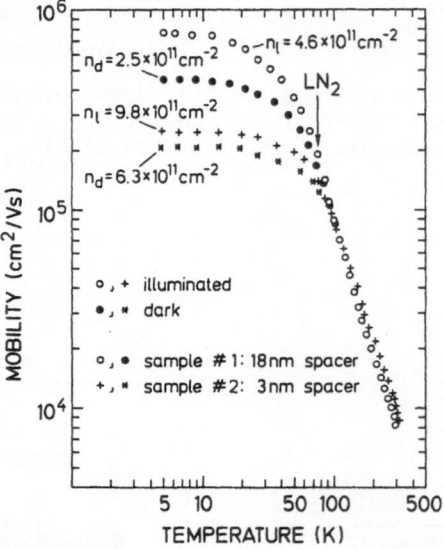

Fig. 10. Hall electron mobilities versus temperature obtained from two selectively doped $Al_xGa_{1-x}As$ / GaAs heterostructures with different spacer widths whose configurations are schematically shown in Fig. 9.

5. APPLICATION OF AlAs SPACER IN SELECTIVELY DOPED n-$Al_xGa_{1-x}As$ / GaAs HETEROSTRUCTURES

It is generally accepted that the peak electron mobility of selectively doped single-interface n-$Al_xGa_{1-x}As$/GaAs heterostructures increases with the width of the undoped $Al_xGa_{1-x}As$ spacer layer / 11 / . This mobility increase is due to the reduction of Coulomb scattering from the remote ionized Si donors. When we replace the ternary $Al_xGa_{1-x}As$ spacer by an

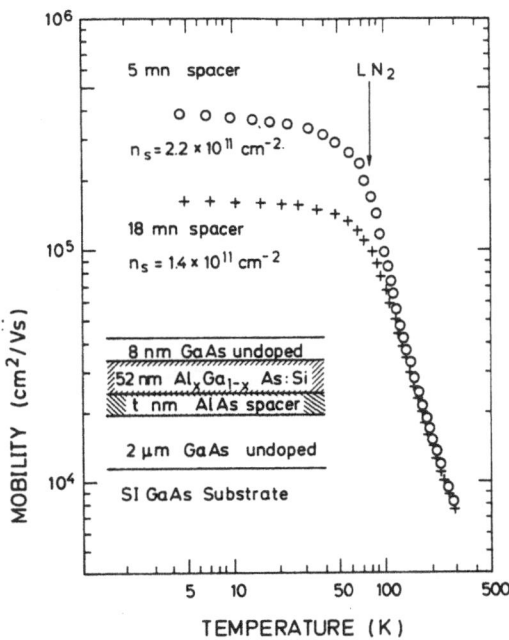

Fig. 11. Hall electron mobilities versus temperature obtained from two
selectively doped n-Al$_x$Ga$_{1-x}$As/AlAs/GaAs heterostructures with
different spacer width t whose configuration is schematically
shown in the inset.

undoped binary AlAs spacer in order to improve the heterointerface, the
dependence of the peak mobility on spacer width changes significantly.
In these n-Al$_x$Ga$_{1-x}$As/AlAs/GaAs heterostructures the height of the tunnel-
ing barrier is increased. For AlAs spacer widths t > 10 nm, the 4 K mobility of the 2 DEG was observed to saturate at about 2 x 10^5 cm^2/Vs. In addition, the mobility enhancement in the temperature range 100 K to 4 K is
only marginal, as indicated in Fig. 11 for a spacer width of t = 18 nm. A
significant increase of the 2 DEG mobility is achieved if the AlAs spacer
widths are reduced to t < 10 nm. In Fig. 11 we show that a mobility enhancement of nearly 4 x 10^5 cm^2/Vs was obtained for a width of the AlAs spacer
as narrow as 5 nm. These results indicate that the assumption of a simple
dependence of mobility enhancement on the real-space distance between
2 DEG and remote ionized impurities has to be modified for selectively
doped n-Al$_x$Ga$_{1-x}$As/AlAs/GaAs heterostructures.

6. CONCLUSION

The heterointerfaces between epitaxial layers of different III-V semiconductors are used to confine electrons and holes to two-dimensional motion.
We have presented a few examples for the effect of barrier configuration
and interface quality on the structural and electronic properties of quantum
wells and superlattices formed by the material systems GaAs/AlAs, Ga$_x$In$_{1-x}$As
/Al$_x$In$_{1-x}$As, and GaSb/AlSb. A careful analysis of the X-ray diffraction
data obtained from GaAs/AlAs superlattices revealed that the interface
disorder can be minimized by growth interruption at each heterointerface.
In contrast to previous assumptions for uncoupled GaAs quantum wells, the
excitonic peaks shift to higher energies when the AlAs barrier thickness
is reduced from 16 to 2 nm, due to the complex band structure of this
quantum well system. The introduction of short period superlattices as
barriers resulted in a substantial improvement of the optical and electrical properties of GaAs, GaSb, and Ga$_{0.47}$In$_{0.53}$As quantum wells and heterostructures. The application of AlAs spacer layers in selectively doped

17

n-Al$_x$Ga$_{1-x}$As/GaAs heterostructures modifies the simple dependence of 2 DEG mobility enhancement on the distance between 2 DEG and remote ionized impurities.

ACKNOWLEDGEMENTS

The active contribution of A. Fischer, K. Fujiwara, J. Knecht, and J.L. de Miguel to the reviewed results is gratefully acknowledged. This work was sponsored by the Stiftung Volkswagenwerk and by the Bundesministerium für Forschung und Technologie of the Federal Republic of Germany.

REFERENCES

/ 1 / L. Tapfer and K. Ploog, Phys. Rev. B 33 (1986) 5565

/ 2 / D. Taupin, Bull. Soc. Fr. Minéral. Cristallogr. 87 (1964) 496

/ 3 / P.V.Petrashen, Fiz. Tverd. Tela (Leningrad) 16 (1974) 2168; ibd. 17 (1975) 2814

/ 4 / Y. Suzuki and H. Okamoto, J. Appl. Phys. 58 (1985) 3456

/ 5 / B.A. Joyce, P.J. Dobson, J.H. Neave, K. Woodbridge, J. Zhang, P.K. Larsen, and B. Bôlger, Surf. Sci. 168 (1986) 423

/ 6 / J.L. de Miguel, K. Fujiwara, L. Tapfer, and K. Ploog, Appl. Phys. Lett. 47 (1985) 836

/ 7 / K. Fujiwara and K. Ploog, Appl. Phys. Lett. 45 (1984) 1222

/ 8 / K. Ploog, Y. Ohmori, H. Okamoto, W. Stolz, and J. Wagner, Appl. Phys. Lett. 47 (1985) 384

/ 9 / J. Wagner, W. Stolz, J. Knecht, and K. Ploog, Solid State Commun. 57 (1986) 781

/ 10 / K. Ploog and A. Fischer, Appl. Phys. Lett. 48 (1986) 1392

/ 11 / G. Weimann and W. Schlapp, Appl. Phys. Lett. 46 (1985) 411.

DYNAMIC RHEED TECHNIQUES AND INTERFACE QUALITY IN MBE-GROWN

GaAs/(Al,Ga)As STRUCTURES

B.A. Joyce, J.H. Neave, J. Zhang*, P.J. Dobson, P. Dawson,
K.J. Moore and C.T. Foxon

Philips Research Laboratories
Cross Oak Lane, Redhill, Surrey, U.K.

* Department of Physics, Imperial College of Science and
  Technology, South Kensington, London SW7 2AZ, U.K.

ABSTRACT

The RHEED intensity oscillation technique has received widespread
attention in the study of MBE growth and is, in principle, capable of
providing detailed information on growth dynamics. There are, however, some
rather serious complications associated with diffraction behaviour which
must be resolved before its full potential can be realised.

We first describe our present understanding of diffraction effects and
then demonstrate the application of the technique to the assessment of
surface migration during growth. This is extended to the evaluation of the
morphological and compositional structure of heterojunctions. Finally, we
present photoluminescence and photoluminescence excitation spectroscopy
measurements designed to probe the interface structure of quantum wells and
indicate the extent to which these results can be correlated with RHEED
data. We derive interface models which are apparently consistent with the
information obtained from both approaches.

INTRODUCTION

Reflection high energy electron diffraction (RHEED) is very widely used
for in-situ monitoring of the structure and growth of films prepared by
molecular beam epitaxy (MBE). The forward scattering geometry of RHEED is
fully compatible with the conventional MBE arrangement of almost normally
incident atomic and molecular beams. This makes possible not only the
evaluation of surface crystallography and morphology (1-4), but also the
investigation of growth dynamics. It has been firmly established[5][6] that
the intensity of all diffraction features oscillates with a period
corresponding precisely to monolayer growth rate. The technique is now used
routinely to calibrate beam fluxes and alloy composition and to control the
thicknesses of quantum wells and superlattice layers[7][8]. Aspects of surface
diffusion and growth mechanisms have also been investigated by the same
method[9-12].

In this paper we first illustrate the influence of diffraction parameters on the waveform and phase of the intensity oscillations, with the aim of demonstrating that the diffraction processes involved must first be established before the technique can be reliably applied to dynamic and mechanistic growth studies.

Models for the formation of growing surfaces of GaAs, AlAs and (Al,Ga)As are then developed, based principally on studies of surface diffusion using RHEED techniques. Difficulties are introduced, however, by surface reconstruction changes associated with the higher growth temperatures which are appropriate for producing material with good optical properties.

These surface measurements have been extended to interface formation, but a detailed interpretation of the RHEED results in terms of interface perfection is not yet available. Our present understanding will be discussed in the context of the level at which the real-time, in-situ RHEED intensity oscillation results can be correlated with ex-situ low temperature ($\approx$ 4K) photoluminescence (PL) and photoluminescence excitation spectroscopy (PLE) measurements on multi-quantum wells (MQWs). We will be addressing the question of how an exciton probe relates to an electron beam probe in this system.

DIFFRACTION EFFECTS ON RHEED INTENSITY OSCILLATIONS

It appears to be well established and fully accepted that a two-dimensional, layer-by-layer growth mode is responsible for the periodic intensity variations observed in RHEED features during MBE. The oscillatory behaviour is not simple, however, and to be able to evaluate waveform, period and damping effects it is important to establish first the nature of the diffraction processes. We believe it is essential to invoke a dynamical scattering treatment and our evidence starts with diffraction from a static surface. We have previously discussed[13] rocking curves and the associated diffraction patterns for various azimuths and surface reconstructions obtained from GaAs (001) surfaces maintained at similar temperatures and As_2 fluxes to those used for growth, but with no Ga flux. Rocking curves from the 00 rod for the 2x4 reconstructed surface in the [110], [$\bar{1}$10], [010] and [130] azimuths are shown in Fig. 1. The substrate temperature was 580°C and J_{As_2} = 2×10^{14} molecules $cm^{-2}s^{-1}$.

The most important factor is the amount of structure in the curves, indicating the strong dynamical nature of diffraction under these conditions. The principal elastic scattering effects are due to primary Bragg diffraction (assumed to be a single scattering event) and multiple diffraction processes. These include multiple excitations between allowed reflections and surface resonances. In the latter case, beams excited inside the surface potential region travel at low angles relative to the surface but are prevented from emerging by the inner potential of the crystal. Refraction effects introduced by the inner potential in fact need to be incorporated generally, since they strongly influence the angular positions at which beams emerge. We have used a value of -14.5 eV for GaAs[3]. Detailed assignments of the various features in the rocking curves have been discussed elsewhere[13], and although the curves shown in Fig. 1 were obtained on accurately oriented material (better than 0.03°) in precise azimuths, it is important to recognise that the structure is not eliminated by moving off-orientation or off-azimuth. It is also important to note the large intensity range of the specular spot, which spans approximately three decades for the angular range used, and the wide variations between different azimuths. Finally, rocking curves are very sensitive to surface structure, in that large differences exist between differently reconstructed surfaces[13].

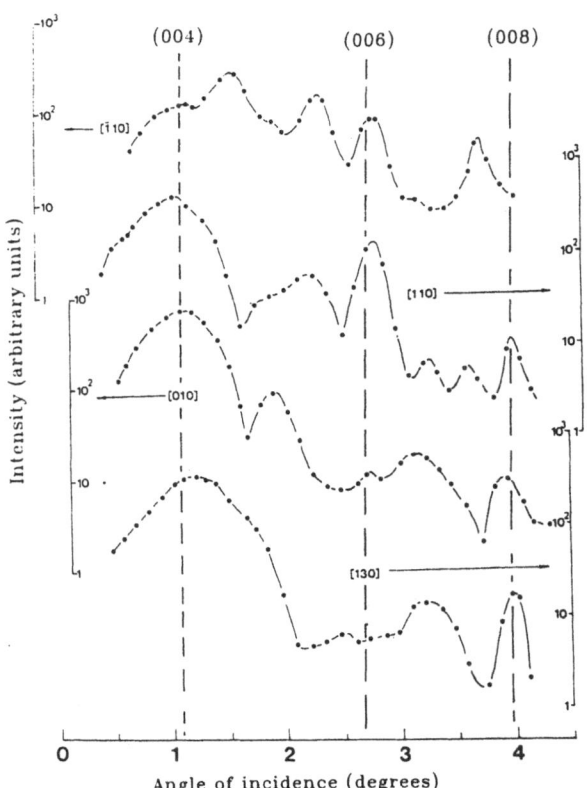

Fig. 1   Rocking curves for the 00 rod in [110], [$\bar{1}$10], [010] amd [130] azimuths from a GaAs (001)-2x4 reconstructed surface maintained at a substrate temperature of 580°C and As$_2$ flux of 2x10^{14} molecules cm^{-2}s^{-1}. The calculated positions of the primary Bragg peaks for a −14.5 eV inner potential are indicated. Primary beam energy = 12.5 keV.

We may summarise the important features of the measured rocking curves as follows:

1.   They contain a large amount of structure (related principally to surface resonances and beam emergence effects).

2.   They cover a large intensity range.

3.   They are very sensitive to surface structure.

4.   Structure is still present on slightly misoriented surfaces.

If now we turn to growth induced intensity oscillations, we show in Fig. 2(a,b) the form of the oscillations for the specular spot on the 00 rod as a function of the incidence angle of the primary beam for the [110] and [010] azimuths respectively. The growth conditions were Ga flux∼ 1x10^{14} atoms cm^{-2}s^{-1}; As$_2$ flux ∼ 2x10^{14} molecules cm^{-2}s^{-1}; substrate temperature ∼ 580°C, producing a GaAs (001) 2x4 reconstructed surface.

It is immediately evident from Fig. 2 that the form of the oscillations is a very sensitive function of the angle of incidence of the primary beam and that there are also very strong azimuthal effects. Three main features require explanation: (i) the intensity response to growth initiation, which

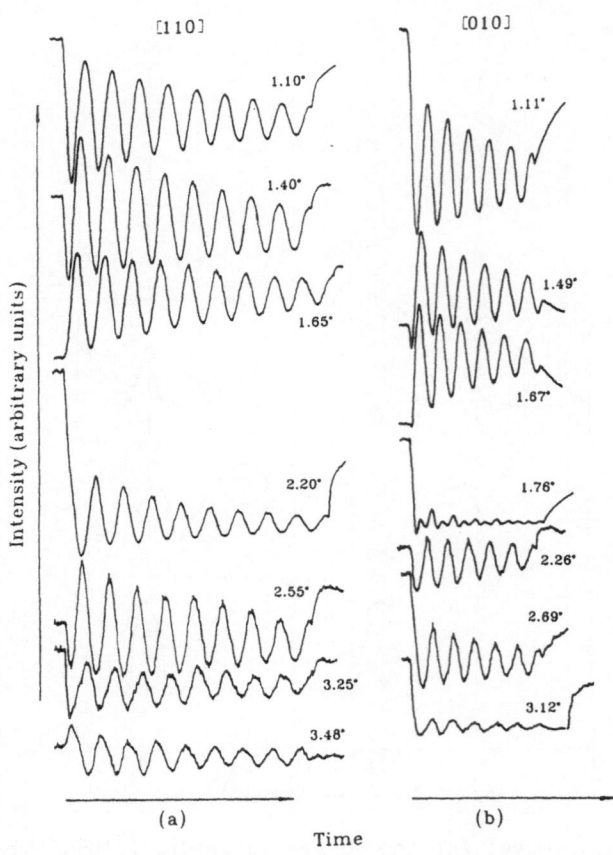

Fig. 2 a,b   RHEED intensity oscillations of the specular spot on the 00 rod
in (a) [110]; (b) [010] azimuths from a GaAs (001)-2x4
reconstructed surface at different primary beam incident
angles; primary beam energy = 12.5 keV. T_S = 580°C;
$J_{Ga} \approx 1 \times 10^{14}$ atoms $cm^{-2}s^{-1}$; $J_{As_2} \approx 2 \times 10^{14}$ molecules
$cm^{-2}s^{-1}$.

can be an increase or a decrease, (ii) the double periodicity which occurs
under certain limited diffraction conditions, e.g. 1.76° incidence angle in
the [010] azimuth for the data shown here, but this is by no means a unique
condition[14], (iii) the phase of the oscillations, which is defined in this
context as the time taken to reach a defined point in the oscillations,
normalised by the steady state growth period. The normalisation serves to
eliminate small variations in growth rate and we have arbitrarily chosen the
'defined point' to be the second minimum in the oscillation trace.

A fourth effect is illustrated in Fig. 3, which shows oscillations
recorded in the [$\bar{1}$10] azimuth at low angles of incidence. Under these
diffraction conditions there is a very marked initial transient response,
such that for the lowest angle there is a single short period oscillation
with an initial intensity decrease, but with increasing incidence angle a
second short period appears at the expense of the first.

Finally, it is essential to be aware that while oscillations are
measured at one position only, intensity variations occur over the whole
diffraction pattern during growth. This can most readily be observed by
recording patterns at various stages of layer growth, using a TV system
incorporating intensity to colour conversion[14].

22

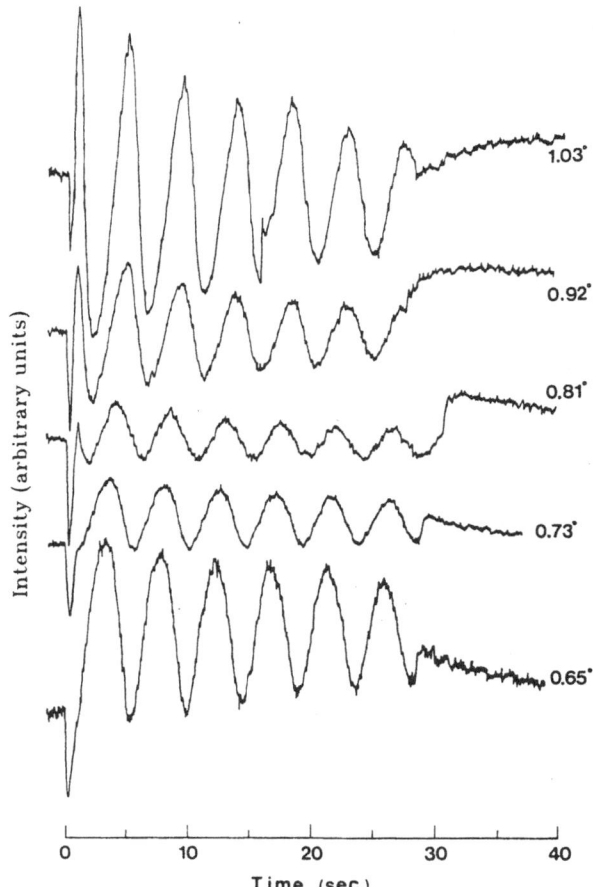

Fig. 3   RHEED intensity oscillations of the specular spot on the 00 rod in the [1̄10] azimuth from a GaAs (001)-2x4 reconstructed surface at low incidence angles (θ = 0.65° - 1.17°); primary beam energy = 12.5 keV. Growth conditions as for Fig. 2.

To understand the origin of these effects, observed under _fixed_ growth conditions, we must consider all of the contributing processes which can produce diffracted intensity. We have already indicated the principal elastic effects, i.e. primary Bragg diffraction, surface resonances and multiple excitations between allowed reflections, in discussing the rocking curve structure. Inelastic (diffuse scattering) processes are of equal, if not greater importance, however. We are concerned with emergent electrons which have undergone one or more scattering events involving phonons or plasmons and which have therefore lost energy and are incoherent with respect to the primary beam. These emergent electrons can be considered to be derived from an array of point sources within the substrate produced by the inelastic events from which subsequent elastic diffraction gives rise to Kikuchi lines related either to Bragg diffraction or surface resonances[15]. Kikuchi lines are clearly visible in the diffraction patterns, but the experimental indication of a large diffuse scattered contribution to the measured intensity during growth is the observation of strong features in the diffraction pattern on or very near the 00 rod which are fixed in position with respect to the shadow edge (or crystal) as the angle of incidence of the primary beam is varied. The electrons in the corresponding plane wave must therefore be incoherent with the incident beam.

Finally, we need to consider beam coherence effects. The high sensitivity of RHEED to changes in surface morphology is a consequence of its unique geometry, but an equally important factor is the coherence length of the electrons parallel and perpendicular to the beam direction. In diffraction patterns from stable, well-ordered surfaces, all features, including those in fractional order Laue zones, are well defined[3]. Under these circumstances the coherence lengths in the two orthogonal directions must be significantly greater than the lattice parameters of the reconstructed surface unit cell. The width of diffraction features in zero and higher order Laue zones suggests the actual value could be at least 100Å. This implies that substantial long range order exists and that coherent scattering will tend to dominate over incoherent (diffuse) scattering. At partial monolayer coverages during growth, the long range order in relation to the coherence lengths is disrupted by the additional steps being created and incoherent scattering is enhanced at the expense of coherent scattering. Experimental justification for this statement is the enhancement of those diffraction features which are fixed with respect to the crystal (shadow edge) at fractional monolayer coverages.

The significance of these various diffraction processes on intensity oscillations during growth is best illustrated by combining an intensity scan along the 00 rod (normal to the shadow edge) with oscillation measurements at specific points on the rod as the angle of incidence is varied by a small amount. In Fig. 4 we show results for the [010] azimuth for incidence angles close to 1.7°, which clearly demonstrate the influence of separate diffraction processes on the double periodicity effect. The even numbered maxima in the oscillations labelled b, c and d are all associated with one feature (x) in the profile, whereas the odd numbered maxima are associated with a different feature (y). The slightly increasing angle of incidence from oscillation trace a to d changes the relative contribution of x and y and hence the relative amplitudes of the odd and

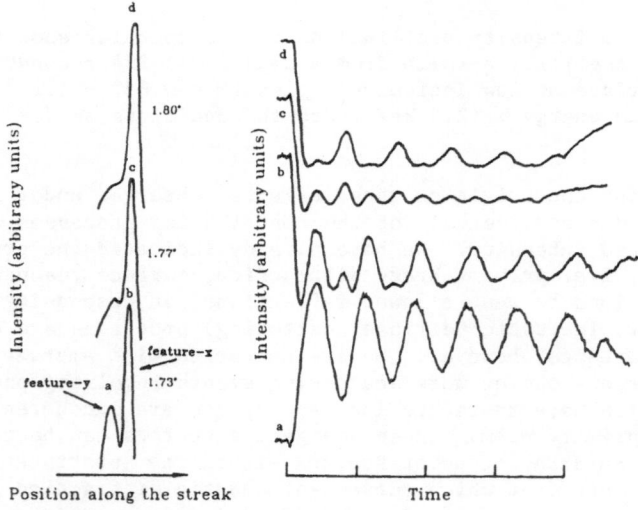

Fig. 4   RHEED intensity oscillations for on- and off-specular beam positions on the 00 rod in the [010] azimuth from a GaAs (001)-2x4 reconstructed surface for incidence angles close to 1.7°; primary beam energy = 12.5 keV. Intensity profiles along the 00 rod are shown alongside the oscillations. The position of the specular beam and, if different, the position at which oscillations were recorded, are indicated. Growth conditions as for Fig. 2.

24

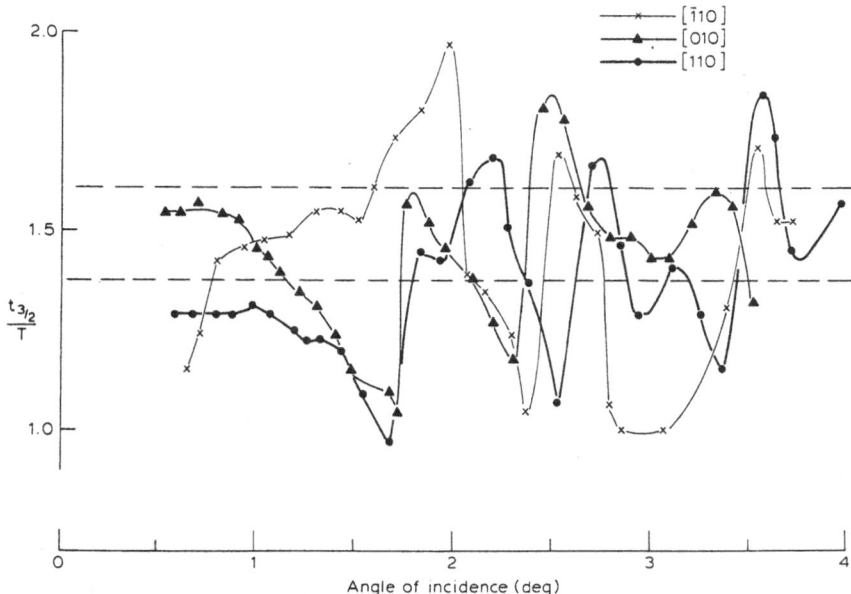

Fig. 5   Phase relationships of RHEED oscillations as a function of the
          angle of incidence for [110], [010] and [Ī10] azimuths from a
          GaAs (001)-2x4 reconstructed surface. The growth conditions
          were as for Fig. 2. Phase is defined as the time taken to
          reach the second oscillation minimum normalised by the time of
          a complete period. The horizontal lines indicate the maximum
          experimental error possible in the time measurements.

even numbered oscillations. x is strictly the specular beam, whilst y is
off-specular, but it is apparent that the dynamical interaction between them
is very strong. Some contribution from each of them will therefore always
be present in oscillations nominally recorded from the specular beam close
to this angle of incidence whatever the resolution of the detector.

      The same combined diffraction feature effect is also responsible for
the changing phase of the oscillations with changing angle of incidence (and
the direction of the initial intensity change), for which we have summarised
results in Fig. 5. Data points were obtained by measuring the time to the
second minimum, $t_{3/2}$, and normalising with respect to the period at steady
state, T, to allow for any minor changes in growth rate. We emphasise that
the steady state period is effectively independent of diffraction conditions
and dependent only on growth rate. It is therefore only when the ordinate
value is 1.5 that oscillation maxima correspond to layer completion. It is
evident from Fig. 5, even allowing a significant error range, that this is
not usually the case.

      We can demonstrate the origin of this phase behaviour most easily with
a specific example and we will take the set of oscillations in the [110]
azimuth obtained for angles of incidence between 1.0° and 1.7°, which span a
phase change of π (Fig. 2(a)). Over this angular range the specular beam
intensity from the non-growing surface goes from its maximum value (at 1°)
to its minimum value (at 1.7°) as shown in Fig. 1. Examination of the
streak profile of the 00 rod (Fig. 6) shows more than one feature to be
present and in particular there is a component at a take-off angle of 1.6°

which remains fixed with respect to the shadow edge as the angle of incidence varies; i.e. it is Kikuchi line related. Oscillations from it are π out of phase with those from the specular beam. At 1.0° incident angle it is comparatively weak from a non-growing surface, but of comparable intensity to the specular beam at half-monolayer coverage from a growing surface. By 1.7° angle of incidence, however, it is the dominant feature and is coincident with the specular beam position. As a consequence, the oscillations show a phase change of π from those obtained at 1.0° incident angle, where the specular beam itself dominates the intensity. We would emphasise that the phase change of the steady state oscillations calls for careful examination of this RHEED technique as applied to studies of growth optimisation. Most of the reported work[12][13] is based on the assumption that the maximum specular beam intensity in the oscillations corresponds to the completion of a monolayer. This interpretation becomes suspect when it is realised that the phase of the oscillations varies with diffraction conditions.

The final point we need to answer is the reason for the transient effects which occur when growth is initiated, a condition seen most prominently in the [$\bar{1}$10] azimuth at low incidence angle, but which can be observed to some extent at all angles. We have previously suggested[17] that this phenomenon is not solely related to diffraction effects, but that it results from a transient reconstruction change from the As-stable 2x4 to the more Ga-rich 3x1 as the Ga flux first impinges. This can lead to a large change in the specular beam intensity, as demonstrated by the rocking curves shown in Fig. 1. Furthermore, the [$\bar{1}$10] azimuth is the most sensitive to changes in reconstruction since the strength of the $\frac{1}{4}$ order features is greatly reduced during growth and the multiple scattering contribution to the specular beam intensity is also changed. Nevertheless the effect should be present to some extent in all azimuths at certain incidence angles, which is consistent with our observations.

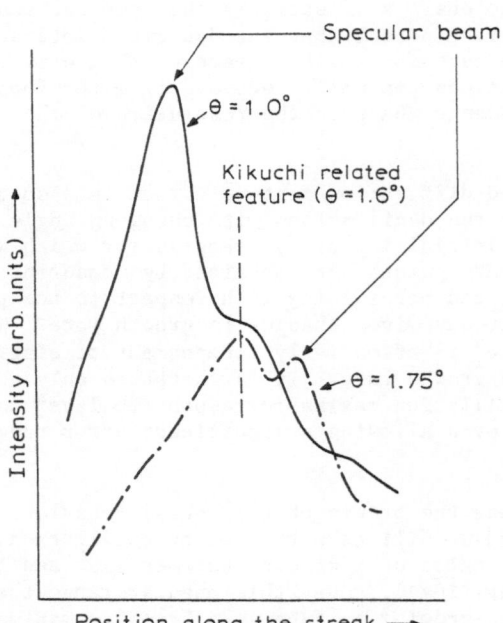

Fig. 6   Intensity profiles along the 00 rod for different incidence angles for the beam directed along the [110] azimuth. Note the changing contribution of the feature which is fixed with respect to the shadow edge as the angle of incidence varies.

GROWTH DYNAMICS FROM RHEED MEASUREMENTS

Having established some of the basic diffraction effects involved in the generation of RHEED intensity oscillations, we are in a position to consider application of the technique to the determination of growth models and the study of interface formation. We will not discuss here the straightforward measurement of growth and evaporation ratio and alloy composition based on the oscillation period, but rather we will summarise our results on surface migration, two-dimensional growth and interface effects.

Surface Migration

Surface migration in this context refers to the motion on the substrate surface of atoms which have been adsorbed from the vapour phase but which have still to be incorporated in crystal lattice sites. It can be investigated using a substrate whose surface is slightly misoriented from a precise low index plane (a vicinal surface), which is assumed to break up into monatomic steps with precise low index terraces. An average terrace width, W, can therefore be defined which is free of steps. For III-V compounds it is the surface migration of the Group III atoms which determines the growth mode and if the mean migration distance is λ, for $\lambda \geqslant W$ the step edge acts as the major sink for migrating adatoms and there is no 2-D growth on the terraces, i.e. no new steps are created. Since intensity oscillations occur as the result of 2-D growth, they would be expected to be absent when step propagation is the dominant growth mode. If growth conditions are varied to the point where $\lambda \leqslant W$, which can be achieved by lowering the temperature or increasing the Ga flux, 2-D growth occurs on the terraces and oscillations should be observed. That this prediction can be realised is shown by the data set for GaAs in Fig. 7, where for a constant Ga flux, the intensity oscillations are seen to decrease and then disappear as the temperature is raised from 540°C to 600°C. A GaAs (001)-2x4 surface reconstruction was maintained throughout and the mean terrace width was ~ 70Å.

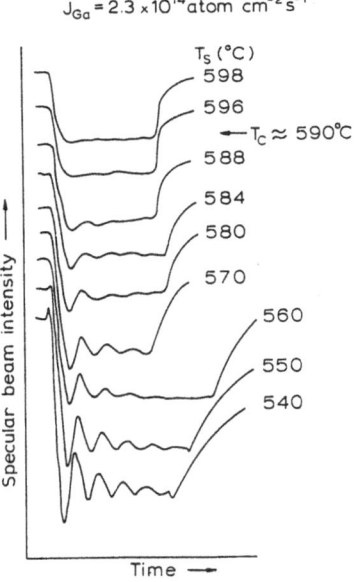

$J_{Ga} = 2.3 \times 10^{14}$ atom cm^{-2}s^{-1}

Fig. 7   Typical data set showing the transition from oscillations to constant response as a function of substrate temperature, with constant J_{Ga}.

It is possible to obtain quantitative values of migration parameters, or at least limiting values, with two assumptions. Firstly, we assume this type of migration can be treated as an isotropic surface diffusion process, for which:

$$x^2 = 2D\tau \qquad (1)$$

where x is the mean displacement and D a diffusion coefficient. Secondly, to evaluate τ, we assume, following Madhukar[18], that we are concerned with configurations sampled by individual atom migrations, not with the maximum range which can be sampled within the total residence time. As a consequence, $\tau = N_s/J_{Ga}$, where N_s is the number of surface sites and J_{Ga} the Ga flux. The activation energy can be calculated from the usual Arrhenius relationship:

$$D = D_o \exp(-E_D/kT) \qquad (2)$$

For Ga on a GaAs (001) 2x4 surface, $E_D \approx 1.3$ eV and $D_o \approx 10^{-5} cm^2 s^{-1}$ [19].

It is important to understand that these migration parameters are not intrinsic values for the surface diffusion of Ga on GaAs. The measurement is made during growth and is only valid for the particular conditions obtaining. It relates specifically to the migration of an adatom (Ga in this case) prior to its incorporation in a lattice site. For GaAs (001), provided the surface reconstruction does not change over the temperature range required to perform the experiment, only small differences are observed with As_2 flux and Ga/As_2 flux ratio variations. A problem occurs with AlAs, however, because the reconstruction does not remain constant over the relevant temperature and Al/As_2 flux ratio range. Intensity oscillations obtained from AlAs grown on a singular (001) surface as a function of T_s are shown in Fig. 8. It is apparent that the amplitude and damping rate change very significantly, indicative of a possible growth mode change with reconstruction or temperature. This means that it is difficult to obtain reliable migration data from vicinal plane measurements in which diffusion distances are varied by a temperature change. There is no problem if measurements are made at a single temperature using the extent of misorientation (i.e. terrace width) as the variable.

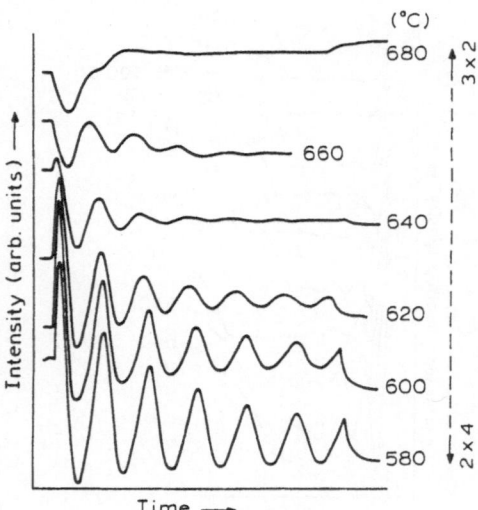

Fig. 8   Intensity oscillations during AlAs growth at indicated temperatures. Note the change of reconstruction; (001) surface, [110] azimuth, 00 rod, specular spot, $\theta_i \approx 1.5°$.

Such experiments have not yet been performed, but an E_D value of
~ 1.5 eV for Al can be estimated by assuming that it scales with that for Ga
in proportion to the relative cohesive energies of GaAs and AlAs. We would
expect the Al migration parameters to determine the growth dynamics of
(Al,Ga)As alloys, except for very low ($\lesssim$ 10%) Al concentrations, principally
because of its shorter mean migration length.

   <u>Oscillation Damping</u>. In addition to enabling migration parameters to
be determined, vicinal plane measurements, which involve a pre-determined
step density, clearly demonstrate the intrinsic process responsible for
oscillation damping. This simply reflects the tendency to reach a steady
state step-terrace distribution during growth, as determined by the surface
migration length. As steady state develops, the step density increases to
the point where no new steps form on a statistical basis, growth occurs
predominantly from existing steps and RHEED intensity oscillations cease.

   There are other factors which affect the damping, however, the simplest
being a growth rate variation across the part of the substrate which is
sampled by the incident RHEED beam, which can give rise to a "beating" of
oscillations[20] resulting from phase (i.e. growth rate) differences.

   More complex is the effect of As_2 flux on the oscillation intensity and
damping at a fixed growth temperature. This is shown in Fig. 9 for GaAs and
AlAs at T_s = 580°C, where in both cases the reconstruction was (001)-2x4.
Two points can immediately be made: (i) the As_2 flux affects the damping
rate (i.e. the steady state step density and migration length) but (ii) it
is in the opposite sense for AlAs and GaAs – more rapid damping with
increasing As_2 flux for GaAs, the inverse for AlAs. However AlAs at higher
temperatures (620°C) behaves the same as GaAs at 580°C, implying strongly
that the migration length of the group III atoms is the crucial parameter in
determining the influence of As_2 but the mechanism is not clear. It may be

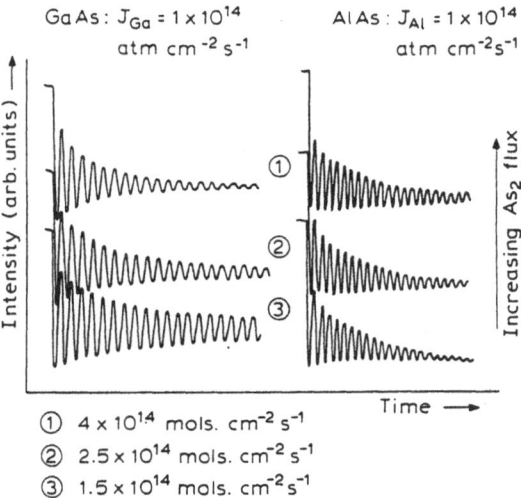

Fig. 9   Effects of As_2 flux on intensity oscillations during GaAs and
            AlAs growth at 580°C; (001) surface, [010] azimuth, 00 rod,
            specular spot, θ_i ~ 1°.

related to the group V incorporation models discussed by Madhukar and co-workers[21]; reaction limited incorporation rate (RLI), where the limitation is the dissociation rate of the group V molecule on a particular surface configuration of group III atoms or configuration limited reactive incorporation (CLRI), where the slow step is the formation of a particular group III configuration on the surface. This would imply that with GaAs at 580°C we are in the RLI regime, while for the slower diffusing Al at the same temperature it is CLRI. As the temperature is increased, however, the formation rate of appropriate Al configurations increases to the point where it is no longer rate limiting and rate control switches to the RLI situation. It is however, premature to attempt any quantitative correlation.

TWO-DIMENSIONAL GROWTH AND INTERFACE FORMATION

A fundamental requirement for the preparation of MQW and superlattice structures is a growth mode which is able to produce compositionally abrupt interfaces. It is worth emphasising that interface morphology is essentially concerned with compositional effects, the lattice planes are contiguous across the heterojunction. The occurrence of RHEED intensity oscillations establishes that the MBE growth of III–V compounds and alloys occurs in a two dimensional layer-by-layer fashion and is therefore an ideal method for controlled interface formation. It is a simple matter to record oscillations across interfaces between GaAs/(Al,Ga)As and (Al,Ga)As/GaAs, and we can distinguish two cases: (i) each region is sufficiently thin that oscillations are still present when an interface is reached, i.e. a steady state terrace width distribution has not been attained in either material, (ii) thicker structures, where the oscillation amplitude has damped effectively to zero before a heterojunction is formed.

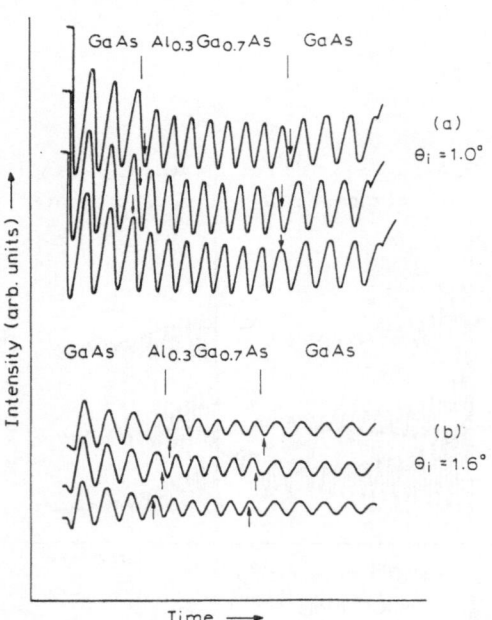

Fig. 10 a,b Intensity oscillations across GaAs–$Al_{0.3}Ga_{0.7}As$–GaAs interfaces, growth temperature = 580°C; (001) surfaces, [110] azimuth, 00 rod, specular spot, (a) $\theta_i \approx 1°$, (b) $\theta_i \approx 1.6°$. Interfaces formed before oscillations have damped.

Case (i) is illustrated in Fig. 10(a and b) for specular beam positions on the 00 rod for incident angles of ~ 1° and 1.6° respectively in the [110] azimuth. The latter angle corresponds to a strong Kikuchi line crossing and is consequently less surface sensitive than the former. The growth temperature was constant throughout at 580°C and a 2x4 reconstruction was maintained for both GaAs and (Al,Ga)As. The alloy composition was approximately $Al_{0.3}Ga_{0.7}As$. Two important points can immediately be made. Firstly, the RHEED intensity behaviour across both interfaces is independent of the point in the oscillation where the Al flux was switched on or off (indicated by the arrows) for both sets of diffraction conditions. Secondly, apart from the change in period attributed to the increased group III flux in the (Al,Ga)As regions, there is no detectable change at either interface for the Kikuchi crossing dominated diffraction conditions, and only a small intensity change when the more surface sensitive condition is used. In general, therefore, it can be stated that RHEED intensity is not very sensitive to composition changes when the terrace width distribution is not at steady state.

The situation is quite different when the oscillations in each composition region have effectively damped out. Here, at a GaAs-(Al,Ga)As interface, oscillations immediately restart, consistent with the shorter migration distance of Al producing a new, shorter terrace width distribution. At an (Al,Ga)As-GaAs interface there is usually a change (increase) in intensity, but no definite recommencement of oscillations. These effects are illustrated in Fig. 11.

In summary, we may conclude that RHEED clearly demonstrates that growth in these systems follows a two-dimensional layer-by-layer mode, with steady state terrace widths being determined by the mean migration length of the group III atoms on the surface. These migration distances are exponentially dependent on growth temperature and rather more weakly dependent on absolute fluxes and group III/group V flux ratios. They determine the compositional "roughness" of interfaces.

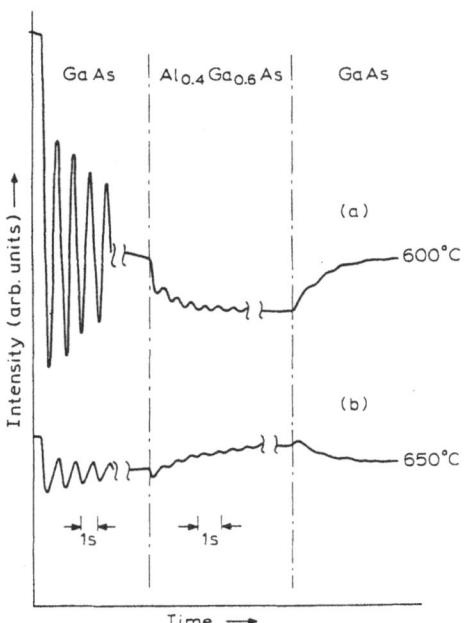

Fig. 11 a,b Intensity oscillations across GaAs-$Al_{0.4}Ga_{0.6}As$-GaAs interfaces, growth temperature (a) = 600°C, (b) 650°C, off [110] azimuth, 00 rod, specular spot, θ_i ~ 1°. Interfaces formed when oscillations have almost fully damped.

Finally, we may consider to what extent it is possible to relate our apparent understanding of interface formation with the optical properties of MQWs. We have studied a number of GaAs-$(Al_xGa_{1-x})As$ MQW structures with well widths in the range 25-150 Å and x = 0.35. All the structures were grown in a Varian Gen II MBE system. The layers were deposited on (001)-orientated semi-insulating GaAs substrates at a temperature of 630°C. The growth sequence was as follows: (a) 1.0 μm of GaAs buffer material, (b) 0.13 μm of (Al,Ga)As cladding, (c) 60 periods of GaAs wells and (Al,Ga)As barriers, and (d) 0.13 μm of (Al,Ga)As cladding. None of the layers was intentionally doped and the barriers were thick enough to ensure that the mean terrace widths were approaching their steady-state values and also that the GaAs wells were electronically decoupled. The well width for each sample was determined by fitting the features observed in detailed studies of the low temperature photoluminescence excitation spectra to a theoretical model[22], using our measured values of the exciton binding energies[23].

In this study we combine high resolution PL measurements with detailed PLE spectroscopy to probe the interface quality of our quantum well structures. In samples where the n=1 (e-hh) exciton peak observed in the PLE spectrum is coincident with the PL peak we interpret the emission as due to the recombination of free excitons[24]. In this case we infer that the quantum well interfaces are essentially smooth over lateral regions larger than the exciton diameter. If instead the n=1 (e-hh) exciton peak in the PLE spectrum is at a higher energy than the emission peak then we assign the emission process to the recombination of localised excitons[25]. This energy difference (the "Stokes shift") arises because in creation the excitons sample an average well width, while in emission the excitons become bound at potential fluctuations arising from steps at the heterojunction interface. In such samples we infer that there are lateral variations in well width on a scale smaller than the exciton diameter. It should be stressed that in the limit of very small terrace widths there should be no Stokes shift.

Again we emphasise that the samples we have studied are in two groups. The first has comparatively wide GaAs wells ($\gtrsim$ 100Å). Fig. 12 shows the PL spectra of a MQW structure with 112Å wells at 6K and 77K. The 6K spectrum has a narrow total linewidth (FWHM = 2.7 meV), but there is a significant high energy shoulder. As the measurement temperature is increased to 77K the shoulder becomes increasingly prominent until it is comparable in strength to the main peak. The peaks correspond to n=1 (e-hh) exciton transitions and the energy separation between the main peak and the shoulder (or between the two peaks at 77K) is exactly equivalent to a change of 1 ML ($\equiv a_o/2$) in the well width. Excitation spectra from this sample show n=1 (e-hh) exciton peaks at exactly the same energy as the two emission peaks when the detection energy is moved from one to the other; i.e. there is no Stokes shift associated with either peak.

These results are fully consistent with a model of the well, indicated in the inset of Fig. 12, in which one interface has terraces which are long compared with the exciton diameter (say $\approx$ 150Å), while the other has much shorter terraces. There are therefore two exciton states, corresponding to thickness L_z and $(L_z - a_o/2)$ becoming more prominent at higher temperatures as a result of thermalization effects. This temperature dependence would not be observed if the 2 peaks were due to recombination from well-to-well changes in thickness through the stucture. This relates at least qualitatively to the RHEED evaluated cation surface migration values, which indicate much larger terrace widths for a GaAs-(Al,Ga)As interface than for the inverse structure. Each layer (GaAs and (Al,Ga)As) is sufficiently thick to have achieved values close to steady state for their terrace width distribution.

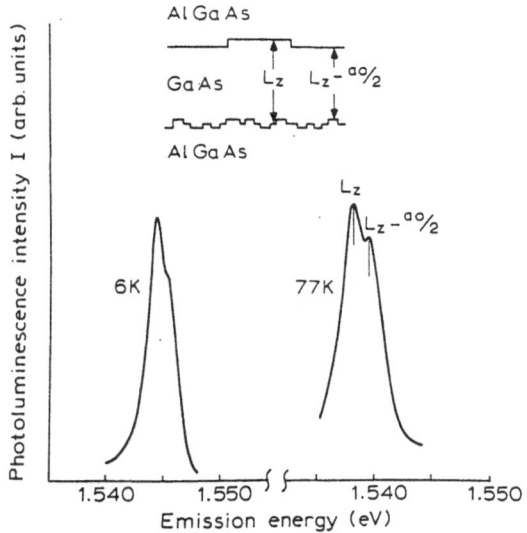

Fig. 12   PL spectra at 6K and 77K from an MQW with 112Å well width.
           Inset shows proposed interface model.

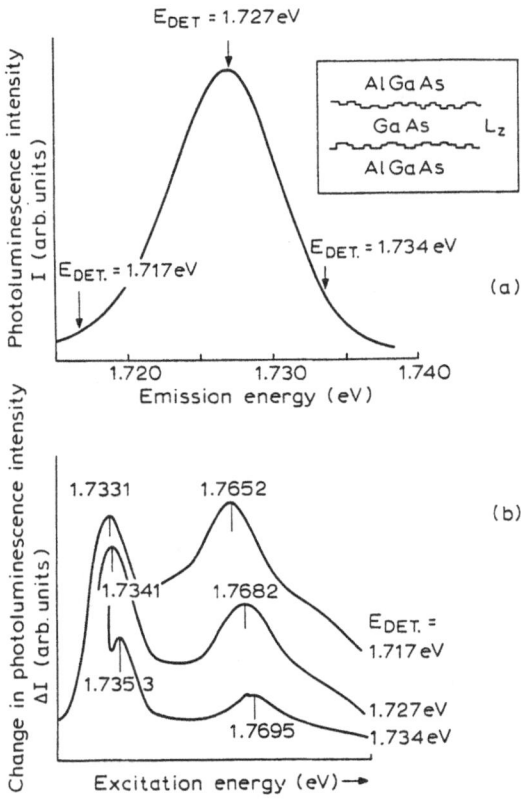

Fig. 13 a,b (a) 6K PL spectrum from an MQW with 25Å well width.
             (b) 6K PLE spectra from the same sample at different detection
                  energies.
                  Inset shows proposed interface model.

The second class of structure we have considered are those MQW samples with narrow well widths (< 60Å). The optical properties are quite different from the previous sample, as illustrated in Fig. 13(a,b) for a structure with 25Å wells. The 6K PL spectrum (Fig. 13(a)) is a single line of FWHM = 9 meV. Emission associated with a well different in width by 1 ML would easily be resolved ($\sim$ 18 meV splitting) from this line; no such feature is observed in the spectrum. Fig. 13(b) shows the 6K PLE spectrum for this sample, recorded as a function of detection energy. There is clearly a Stokes shift associated with this sample which increases from 6 meV to 8 meV as the detection energy is moved through the emission to higher energy. It is important to note that these shifts are significantly less than would correspond to $a_0/2$ well width variations.

A possible model which accounts for these effects has both interfaces with mean terrace widths shorter than the exciton diameter, as shown in the inset of Fig. 13. In this structure, excitons are bound with slightly different energies at different sites along the interfaces, corresponding to potential fluctuations on a scale small in relation to the exciton diameter. The measured PL peak then represents an effective envelope of a series of peaks whose gradually varying energy positions result from the different exciton localisation energies.

Again, there is a qualitative correspondence with RHEED observations, which indicate that the (Al,Ga)As terraces would be short compared to an exciton diameter and the GaAs wells are too narrow to produce equilibrium terrace lengths.

We may conclude that it is possible to relate RHEED observations to the interface structure of QWs deduced from optical measurements, but much more work in this area is necessary before quantitative evaluation can be achieved.

REFERENCES

1. A.Y. Cho, J. Appl. Phys. 47:2841 (1976).
2. J.H. Neave and B.A. Joyce, J. Crystal Growth 44:387 (1978).
3. B.A. Joyce, J.H. Neave, P.J. Dobson and P.K. Larsen, Phys. Rev. B29:814 (1984).
4. C.S. Lent and P.I. Cohen, Phys. Rev. B33:8329 (1986).
5. J.H. Neave, B.A. Joyce, P.J. Dobson and N. Norton, Appl. Phys. A31:1 (1983).
6. J.M. Van Hove, C.S. Lent, P.R. Pukite and P.I. Cohen, J. Vac. Sci. Technol. B1:741 (1983).
7. T. Sakamoto, H. Funabashi, K. Ohta, T. Nakagawa, N.J. Kawai and T. Kojima, Japan. J. Appl. Phys. 23:L657 (1984).
8. B.A. Joyce, P.J. Dobson, J.H. Neave, K. Woodbridge, J. Zhang, P.K. Larsen B. Bölger, Surface Sci. 168:423 (1986).
9. J.H. Neave, P.J. Dobson, B.A. Joyce and J. Zhang, Appl. Phys. Lett. 47:400 (1985).
10. B.F. Lewis, F.J. Grunthaner, A. Madhukar, T.C. Lee and R. Fernandez, J. Vac. Sci. Technol. B3:1317 (1985).
11. P. Cohen, A. Madhukar, J.Y. Kim and T.C. Lee, Appl. Phys. Lett. 48:650 (1986).
12. A. Madhukar, S.V. Ghaisas, T.C. Lee, M.Y. Chen, P. Chen, J.Y. Kim and P.G. Newman, Proc. SPIE 524:78 (1985).
13. P.K. Larsen, P.J. Dobson, J.H. Neave, B.A. Joyce, B. Bölger and J. Zhang, Surface Sci. 169:176 (1986).
14. J. Zhang, J.H. Neave, P.J. Dobson and B.A. Joyce, to be published.

15. A. Ichimiya, K. Kambe and G. Lehmpfuhl, J. Phys. Soc. Japan, 49:684 (1980).
16. H. Sakaki, M. Tanaka and J. Yoshino, Japan. J. Appl. Phys. 224:L417 (1985).
17. B.A. Joyce, P.J. Dobson, J.H. Neave and J. Zhang, in "Two-Dimensional Systems: Physics and New Devices", G. Bauer, F. Kuchar and H. Heinrich, eds., Springer Series in Solid State Sciences No. 67 p.42 (1986).
18. A. Madhukar, Surface Sci. 132:344 (1983).
19. J.H. Neave, P.J. Dobson, J. Zhang and B.A. Joyce, Appl. Phys. Letts. 47:100 (1985).
20. J.M. van Hove, P.R. Pukite and P.I. Cohen, J. Vac. Sci. Technol. B3:563 (1985).
21. S.V. Ghaisas and A. Madhukar, J. Vac. Sci. Technol. B3:540 (1985).
22. G. Duggan, H.I. Ralph and K.J. Moore, Phys. Rev. B32:8395 (1985).
23. K.J. Moore, P. Dawson and C.T. Foxon, Phys. Rev. B34:6022 (1986).
24. C. Weisbuch, R.C. Miller, R. Dingle, A.C. Gossard and W. Wiegmann, Solid State Commun. 37:219 (1981).
25. G. Bastard, C. Delalande, M.H. Meynadier, P.M. Frijlink and M. Voos, Phys. Rev. B29:7042 (1984).

MOLECULAR BEAM EPITAXIAL GROWTH KINETICS, MECHANISM(S) AND

INTERFACE FORMATION: COMPUTER SIMULATIONS AND EXPERIMENTS

Anupam Madhukar

Departments of Materials Science and Physics
University of Southern California
Los Angeles, California 90089-0241

I. INTRODUCTION

When, during the early phase of organizing this research workshop, Dr. Robin Farrow approached me, not only was I supportive of the idea and its timeliness in bringing together a disparate community of researchers in the area of thin film epitaxial growth, but also was pleased to have the opportunity to present some of the work undertaken at University of Southern California in the area of molecular beam epitaxial (MBE) growth of III-V compound semiconductors. Judging by the level of interest and discussions, often quite spirited, that occurred during the workshop, it is quite clear that the organizers succeeded in their objective quite well. In writing this article I am therefore mindful of capturing some of the spirit, excitement, and controversies, real or imagined.

The flow chart shown as fig. 1 illustrates the overall approach to examining the nature of MBE growth and its consquences that we at USC have been pursuing for some time. It represents a multipronged approach essential for arriving at even a reasonable level of understanding of the subject. The major ingredients are:

(1) Computer simulations of the MBE growth process based upon a model derived from information available on the relevant kinetic processes,

(2) Predicting the consequences of the resulting structural and chemical nature of the surfaces and interfaces for properties that might be amenable to measurement, such as the behavior of the reflection high energy electron diffraction (RHEED) intensities,

(3) Systematic measurements of the RHEED behavior, being mindful of differentiating between growth condition related and diffraction measurement related effects,

(4) Comparison of (2) and (3) to examine expected and unexpected correlations between these and consequently the connection between RHEED observations and the growth

condition dependent growth kinetics and mechanisms,

(5) Systematic growth of quantum well structures under conditions determined via RHEED and motivated by the expected consequences for the interfaces implied by (4),

(6) Examination of quantum wells via appropriate optical and electrical means as an independent (i.e. of RHEED behavior) check on the inferred relation between growth kinetics and the behavior of RHEED (i.e. point (4) above).

As should readily be apparent, it is a sizeable undertaking and no definitive conclusions ought necessarily to be expected in any or all areas (1) through (6) at this, still too early, stage of the development of the subject. It is nevertheless this author's view that only through such a comprehensive approach, whether undertaken by individual research groups or, as is more common, collectively by independent researchers interested only in parts of the subject, can an eventual understanding evolve. Inferences derived from only limited information on any partial ingredient of the flow chart, as appears to be the case with most papers in the literature, not only have little chance of surviving the test of time, but also tend to cause distractions and confusion along the way. This word of caution is of particular significance since although most of the effort represented by the flow chart is concentrated on the lattice matched $GaAs/Al_xGa_{1-x}As(100)$ system, a clear and reliable understanding evolved through study of this system will serve as a basis for understanding the more complicated systems such lattice mismatched $GaAs/In_xGa_{1-x}As$, the IV-VI magnetic semiconductor superlattices, and the metallic superlattices - all discussed at this workshop and covered in other chapters of this volume.

The individual sections of this chapter are divided according to the ingredients noted in the flow chart, fig. 1. The emphasis in each category is to capture the essentials of the present status and to make precise what the issues are, if not what some of the possible answers may be.

II. MBE GROWTH AND COMPUTER SIMULATIONS

The experimental aspects of MBE growth and the variety of low dimensional structures involving semiconductor-semiconductor, semiconductor-dielectric, semiconductor-metal and metal-metal combinations are well covered in several review articles[1-3], at least to the extent that such attempts can remain current with this rapidly advancing area. Consequently we do not cover this aspect here. Similarly, the details of the computer simulation procedures and the variety of situations explored are also fairly well covered in the recent review[4] by Madhukar and Ghaisas. The emphasis in this section is thus placed on providing some illustrative examples of the types of results obtained which provide considerable insight into the relationship between the structural and chemical nature of the growing surface and the underlying growth kinetics and mechanisms attendant to the chosen growth conditions.

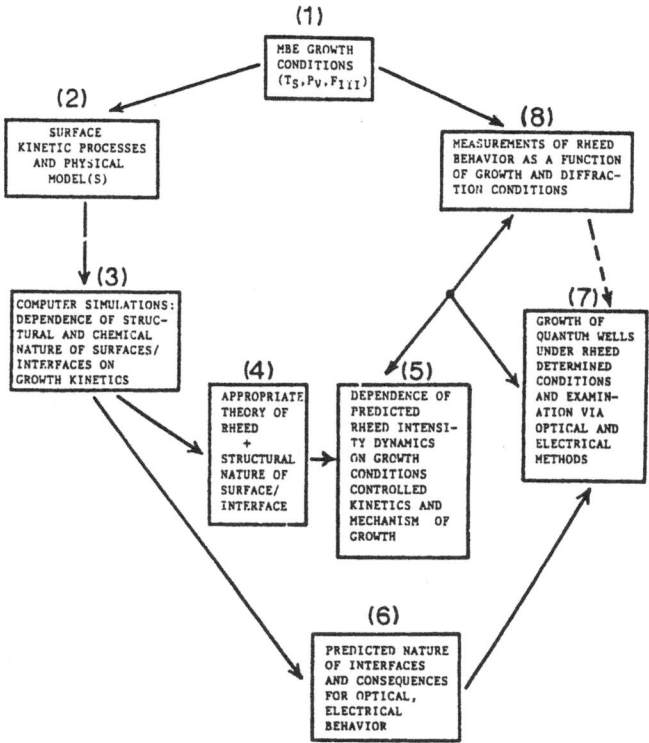

Fig. 1. Flow chart of the combined theoretical and
experimental approach being pursued in the
author's research group.

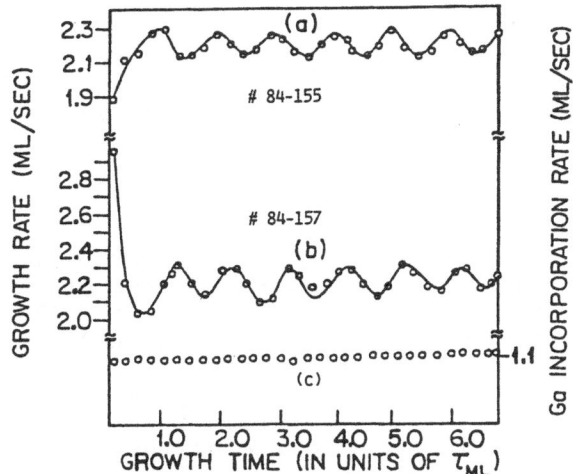

Fig. 2. The time dependent GaAs(100) growth rate obtained
from computer simulations. Curves (a) and (b)
correspond to the reaction-limited-incorporation
(RLI) and configuration-limited-reactive-
incorporation (CLRI) growth mechanism regimes,
respectively. Curve (c) shows the Ga incorporation
rate in both cases.

II.A. Growth Kinetics, Mechanism and Surface Morphology

To appreciate the atomistic nature of the growth process in MBE a few simple aspects need to be clearly recognized. We note these via the example of homo-epitaxy of GaAs(100), although their relevance to other material systems should be self-evident. First, the starting GaAs(100) substrate in the usual range of substrate temperatures (400°C to 700°C) employed for good epitaxial growth requires that it be maintained under an overpressure of arsenic (typically an As_4 beam) to compensate for the preferential loss of arsenic (in the form of As_2) and thus maintain stoichiometry. The sticking coefficient of arsenic on arsenic being very low at these temperatures, no growth occurs. To initiate growth, a beam of atomic Ga is allowed to impinge and stoichiometric epitaxial growth begins, provided the As_4 overpressure is sufficient. The second feature thus to be remembered is that for the usual growth situation, the time averaged growth rate over times longer than monolayer growth time is controlled by the group III flux. However, on time scales of the addition of an atomic layer of Ga and one of arsenic (i.e. a bilayer of Ga-As which, unfortunately, has come to be referred to as a "monolayer" (ML) in the literature), the growth rate is not controlled by the incorporation kinetics of Ga alone. The dissociative reaction and incorporation kinetics of As_4 (or As_2) plays a critical role in the growth of GaAs, as will be seen in the computer simulation results to follow. This is the third important feature to be recognized.

The dissociative reaction and incorporation of As_4 (or As_2) occurs from a precursor, weakly adsorbed, state which we denote as As_2^*. In this highly mobile state, whose population $N_{As_2^*}$ is maintained via a steady state balance between various adsorption, reaction and desorption steps involving As_4 (or As_2) and As_2^*, the As_2^* can react and dissociatively chemisorb at a variety of Ga atom configurations consisting of a single Ga atom to several which may be either intraplanar or interplanar (i.e. step) configurations, each with a different reaction rate. The total incorporation rate for a given configuration labelled i is then

$$N_{As_2}^*(P_{As_4},\ T)\ \cdot\ R_i(T)\ \cdot\ P_i(P_{As_4},\ \{h_\alpha(T)\};\ t)$$

where P_i is the time dependent probability of occurrence of the ith Ga configuration during the course of growth and is itself dependent upon the P_{As4} and $\{h_i(T)\}$, the latter being the set of surface hopping rates for Ga atoms in different configurations. While details of this can be found in reference (4), the simple underlying physics is that the creation of a given Ga configuration itself depends upon the ability of the Ga atoms to migrate on the surface. This migration ability is set not only by the individual hopping (or jump) rates determined solely by the chosen substrate temperature and intrinsic properties such as the prefactors and activation energies in the usual Arhenius form of rate expressions, but also by another user controlled growth parameter, the arsenic pressure, and the intrinsic As_2^* reaction rates, $R_i(t)$. Thus an "effective" surface migration ability during the course of the growth of a monolayer is in operation and will itself be a time dependent entity. This is the fourth significant aspect of MBE growth that needs to be appreciated if an atomistic understanding of the growth kinetics and mechanism is to be sought.

The features noted above manifest themselves most clearly in Monte-Carlo computer simulations of Ghaisas and Madhukar. The underlying physical and chemical model incorporating the various kinetic processes has been called by these authors the configuration-dependent-reactive-incorporation (CDRI) model of MBE growth. The name emphasises the significance of the dissociative reactive incorporation of the molecular group V species at various group III configurations in controlling both, the mechanism and the resulting structural and chemical nature of the dynamic growth front. This is well illustrated by the nature of the time dependent GaAs(100) growth rate in two limiting regimes of the CDRI model, shown in fig. 2. Curve (a) corresponds to the situation of slow As_2^* incorporation rates at low Ga configurations and is called the reaction limited incorporation (RLI) regime. By contrast, curve (b) corresponds to rapid incorporation of As_2^* at low Ga configurations and is called the configuration limited reactive incorporation (CLRI) regime since, after initial rapid incorporation of, on the average, two As atoms for each Ga upon initiation of a layer, As_2^* incorporation beyond slightly less than 0.5 ML of Ga is limited by unavailability of Ga configurations for growth of the next layer until the first is nearly complete.

The single most important fact to recognize about the atomistic nature of these growth mechanisms is that while material addition in both cases occurs by an essentially layer-by-layer mode, in neither case is the formation of critical nuclei (i.e. clusters stable against break up either to the vapour phase or with the atoms staying at the surface) of the growth rate controlling species (the group III) necessary for growth of the GaAs solid phase to occur. Indeed, the incorporation rate of the group III from the vapour phase in these simulations is near unity as seen in fig. 2, in conformity with the experimental situation for GaAs homoepitaxy under the usually employed growth conditions. As such, these growth mechanisms are fundamentally distinct from the conventional notion of growth via the nucleation and growth mechanism[4-6]. No nucleation barrier to growth is involved here. This simple point has unfortunately not been appreciated by the MBE growth community and reference to the mechanism of MBE growth as the classic "nucleation and growth" mechanism continue to appear. Indeed, the behavior of the reflection-high-energy-electron-diffraction (RHEED) intensity dynamics has been argued by Joyce and co-workers to provide evidence for MBE growth being the classic nucleation and growth mechanism[8]. We have found[4-6] such claims to be unfounded.

The nature of the effective surface migration process during MBE growth is shown in fig. 3 for an illustrative set of growth conditions corresponding to the RLI growth mechanism. Note the periodic behavior of the effective

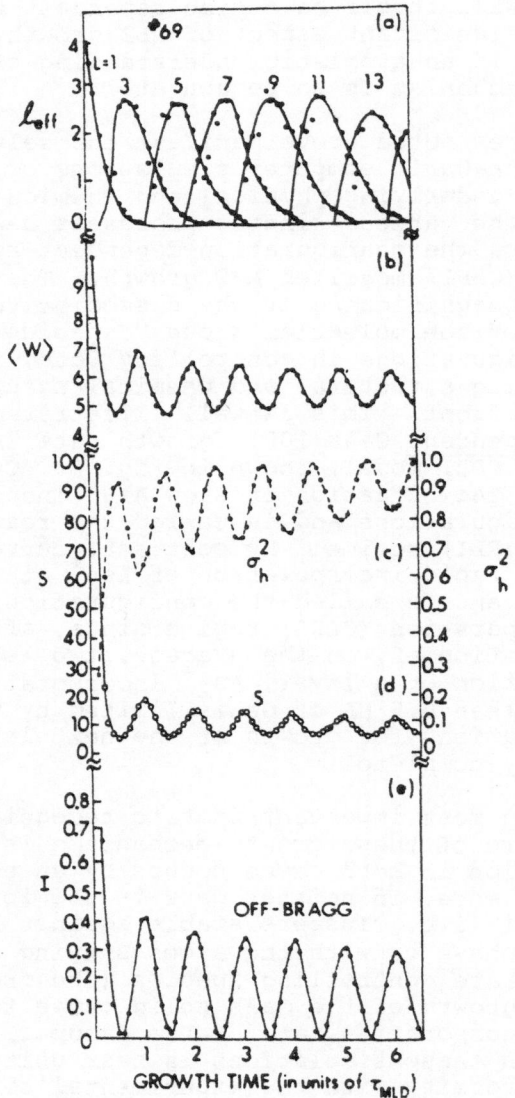

Fig. 3. Some illustrative results of the type of
information on growth and surface morphology
obtained from computer simulations. The particular
results correspond to growth conditions in the RLI
growth mechanism regime. (a) the effective surface
migration length of Ga in each layer during growth.
(b) the average terrace width at the growth front
in one of the principal surface unit cell
directions. (c) the mean square fluctuation in the
growth front step density distribution. (d) the
smoothness parameter as defined in the text. (e)
the behavior of the RHEED specular beam intensity
calculated in the kinematical approximation for the
condition $q \cdot d_{As-As} = \pi$ as explained in the text.
Growth time is measured in units of the time for
growth of a monolayer of Ga (τ_{ML}).

42

migration length during the course of growth of each layer
and the fact that the period coincides with the incorporation
of a monolayer's worth of the group III atoms. The damping
seen is caused by the growth front profile statistically
becoming progressively rougher as growth continues, as seen
in the time dependent behavior of the average terrace width
($<W>$) and the mean square fluctuation (σ_h^2) in the step
height distribution, shown in fig. 3. Note once again the
damped periodic nature of both these quantities and their
expected inverse relation. The period once again is seen to
coincide with the period of incorporation of a single layer
of the group three atoms. Ghaisas and Madhukar have combined
these measures of the lateral and vertical degree of "surface
smoothness" (i.e. $<W>$ and σ_h^2) into an overall measure
defined as[4,6]

$$S = \frac{<W_I>\ <W_J>}{h^2}$$

in which $<W_I>$ and $<W_J>$ are the average terrace widths along
the directions of the surface unit cell axes. Fig. 3 also
shows the corresponding behavior of S. Note that the maxima
in S coincide with incorporation of a monolayer. This is a
consequence of the growth being via the RLI mechanism. At
high arsenic pressures corresponding to the CLRI growth
mechanism regime, the maxima in the smoothness S and the Ga
monolayer incorporation do not necessarily coincide even
though the oscillation period is the same[4,5]. This is a
matter of considerable significance for pragmatic usage of
RHEED as we shall see later. Although space considerations
do not permit an exhaustive display of a variety of results,
these illustrative examples of the type of direct information
provided by the computer simulations should be sufficient to
also recognize that the growth mechanisms and the resulting
atomistic nature of the growth front are, through the
attendant kinetics of growth, intimately dependent upon the
user controlled growth parameters, T_S, P_{As_4} and group III
flux.

The growth condition dependence of the time dependent
structural nature of the growth front has consequences for
the dynamics of the intensity of the various diffraction
beams in (RHEED). Through a comparison of the predicted
dependence on the growth conditions and corresponding
measurements much insight can be gained into the nature of
the growth process. We take up this aspect in the next
section, finishing this section with a few illustrative
examples of the formation of GaAs/Al_xGa_{1-x}As(100) interfaces
and the structural and chemical nature of the interfaces
revealed by the simulations.

II.B. Formation of Interfaces

Let us consider the behavior of GaAs interfaces formed
with an alloy $Al_{x_0}Ga_{1-x_0}$As. The most commonly employed value
of x_0 is about 0.3. Consequently we provide some
illustrative results of the computer simulations[8] for 4 ML
wide GaAs(100) well layers alternating against 4 ML wide
$Al_{0.33}Ga_{0.67}$As(100) layers. In fig. 4 are shown the Ga and
Al concentrations in each layer of the structure at the end
of growth. The sequence of growth shown is for growth of the

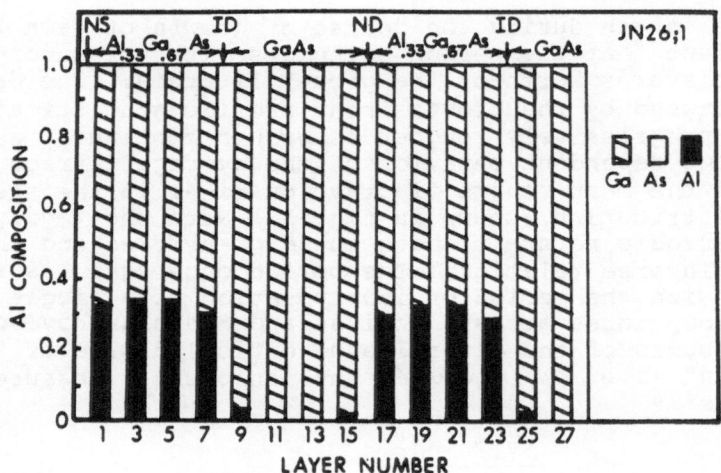

Fig. 4. Shows the Al (black), Ga (cross hatched) and As
(white) layer concentrations for simulated growth
of $Al_{0.33}Ga_{0.67}As/GaAs/Al_{0.33}Ga_{0.67}As(100)$
structure with each layer of width 4 ML. Symbols N
and I stand for normal and inverted, and S and D
for static and dynamic.

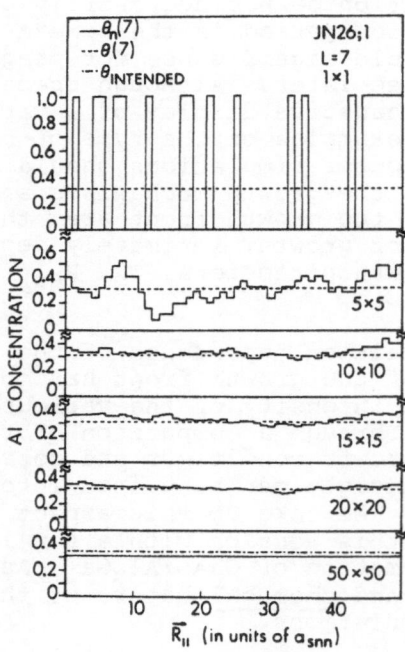

Fig. 5. Local average Al concentration in layer 7 of fig.
4, calculated on various lateral length scales
indicated by (m x m) in units of the surface
nearest neighbor (a_{SNN}). The size of the simulated
substrate matrix is (50x50) and the dash-dot line
indicates the intended overall layer composition of
0.33 whereas the actual overall composition (broken
lines) is near 0.27.

first $Al_{0.33}Ga_{0.67}As(100)$ layer on the static surface of the GaAs "substrate". Note the mixing of the Ga and Al atoms between the interfacial layers 7 and 9, and 15 and 17, defining the inverted (dynamic) and normal (dynamic) interfaces. In addition, one also sees a certain degree of fluctuation away from 0.33 in the overall average Al concentration in layers 1 through 5 as well. This is a conseuence of the difference in the Ga and Al surface migration behavior. Naturally, if the growth of the intended $Al_{0.33}Ga_{0.67}As$ layer were allowed to continue to large thickness, a situation not practical in computer simulations due to the computational needs, the Al concentration would settle at 0.33 once a steady state between the various kinetic processes is reached. Nevertheless, these results reveal another feature of formation of $GaAs/Al_x Ga_{1-x} As(100)$ interfaces - namely, even in regions where mixing of Al with a Ga interfacial layer may not occur, such as growth in the regions of large terraces of a well prepared static GaAs(100) surface, fluctuations in the <u>overall average</u> Al concentration of the alloy barrier atomic planes <u>in the growth direction</u> nevertheless occur at and near the interfacial region. It was pleasing to see that some evidence in support of the pre-dictions was presented in the talk by Drs. Gower and Fewster on the TEM and X-ray diffraction work of the Phillips group.

A far more significant aspect of the Al concentration behavior relates in fact to the spatial distribution of the Al atoms <u>within</u> a given atomic plane. This <u>lateral</u> (as opposed to "vertical" meaning in the growth direction) distribution of Al atoms needs to be examined on various length scales defining a <u>local average</u> Al concentration. The length scales of relevance are determined by the nature and size of the probe particles involved in the particular experiments employed to probe the nature of the interfaces. This point has been emphasized[9-12] and demonstrated[13] in our work on the interpretation of the photoluminescence behavior of $GaAs/Al_x Ga_{1-x} As$ quantum well structures for which the size of the exciton is shown to be the relevant length scale. We shall take up this, not particularly well appreciated, point in sec. (IV). Here, in fig. 5, we show the fluctuations in the local Al concentration, calculated on various length scales, as a function of the lateral position for layer 7 forming the first inverted interface of fig. 4. One notes that significant fluctuations in the local Al concentration can occur. Naturally, the larger the spatial scale on which the local average concentration is of relevance, the smaller the degree of fluctuation in the local average concentration percieved by the probe particle. The spatial scale in fig. 5 is denoted as an area (mxm) where the distance is measured in units of the (100) surface nearest neighbor (a_{SNN}). In fig. 6 is shown the behavior of the local Al concentration on a (10x10) scale for the various layers of the quantum well structure. Note that fluctuations occur for layers away from the immediate interface layers and will be seen to be of utmost significance in controlling the optical and electrical characteristics of narrow quantum wells involving significant peneration of the confined electron, hole, exciton, etc. wave functions into the alloy barrier regions[13]. Finally, in fig. 7 is shown the behavior of the local Al concentration on a (10x10) scale for each

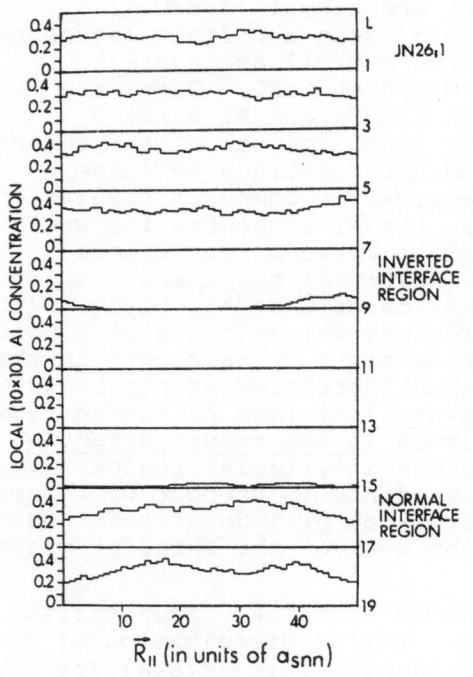

Fig. 6.

Shows the local average Al concentration behavior in the lateral direction on a (10x10) a_{SNN} scale for different layers constituting the quantum well of fig. 4.

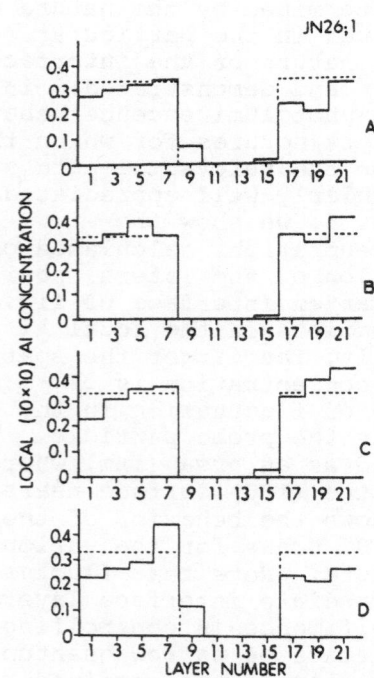

Fig. 7.

Shows the local Al concentration on a (10x10) a_{SNN} scale for each atomic layer of the quantum well of fig. 4 at four different lateral spatial locations. One clearly recognizes that fluctuations in the band edge discontinuity must follow.

layer at four different lateral positions in the sample. Note the fluctuations away from the intended ideal quantum indicated by the broken lines. It thus follows that fluctuations in the band edge discontinuity are likely to be a dominant feature, intrinsic to the kinetics of growth. This fact has never been appreciated in the literature and in sec. IV we provide experimental evidence to support our contention[9-13].

III. RHEED: COMPUTER SIMULATION PREDICTIONS AND EXPERIMENTS

The knowledge of the growth condition dependent dynamics of the atomistic structural and chemical nature of the surfaces and interfaces gained from computer simulations can be fruitfully converted into measured properties - both, real time (i.e. during growth) as well as those obtained from ex-situ examination of the grown structure via structural (x-ray, electron microscope, etc.) optical (absorption, luminescence, Raman, etc.) and electrical methods. The primary in-situ, real-time, probe, available in most MBE systems due to its compatibility with the growth chamber geometry, is reflection electron diffraction, generally carried out with the primary electron beam energy of order 10 KeV. Hence the name reflection high energy electron diffraction (RHEED). In this section we focus on a few aspects of the RHEED intensity measurements[7,14-31] and the types of atomistic features of RHEED that can be obtained from the computer simulations of the growth process if they were coupled with a reliable theory of RHEED. Such a comparison can potentially make RHEED intensity dynamics a powerful, easy to use, real time tool for examining and establishing correlations between growth kinetics and the resulting nature of interfaces, quite apart from its pragmatic value in determining reproducible and optimized growth conditions.

III.A. GaAs(100) Homo-epitaxy

The surge of interest in RHEED intensity behavior began with reports that damped oscillations in the intensity of any given feature of the RHEED pattern are found upon commencement of growth, the oscillation period coinciding with the growth of a monolayer[14-17]. A few illustrative examples of the specular beam intensity oscillation and recovery[28] behavior during GaAs(100) homo-epitaxy under As-stabilized (2x4) reconstruction conditions recorded in our laboratory[17,30] are shown in fig. 8. The data were taken with the electron beam azimuth along the 2-fold reconstruction direction at various angles of incidence denoted by θ and at different substrate temperatures for a fixed As_4 pressure and Ga flux (i.e. the growth rate). The details of the specular beam intensity behavior clearly illustrate that two fundamental effects are intertwined in the observations;

(a) Growth condition effects,
(b) Diffraction condition effects.

Progress in understanding the information content of the RHEED intensity dynamics thus requires development of

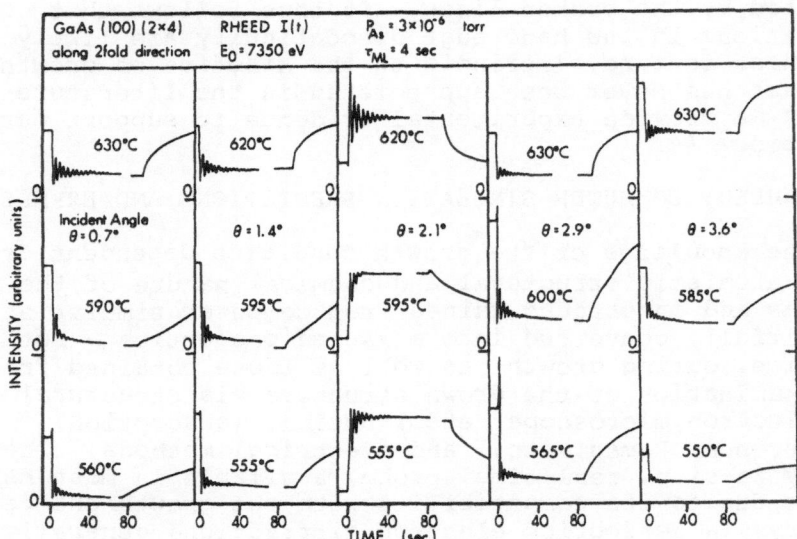

Fig. 8. Some illustrative measurements of the specular beam
intensity during GaAs(100) homoepitaxy. Note the
dependence of starting intensity, dynamics, and
recovery upon growth termination, on the angle of
incidence for a fixed growth condition indicating
the presence of multiple scattering, dynamical
scattering, thermal diffuse scattering, refraction
and beam penetration effects. Note also the change
in intensity dynamics for a fixed diffraction
condition with varying growth condition. The GaAs
monolayer growth time is 4 sec.

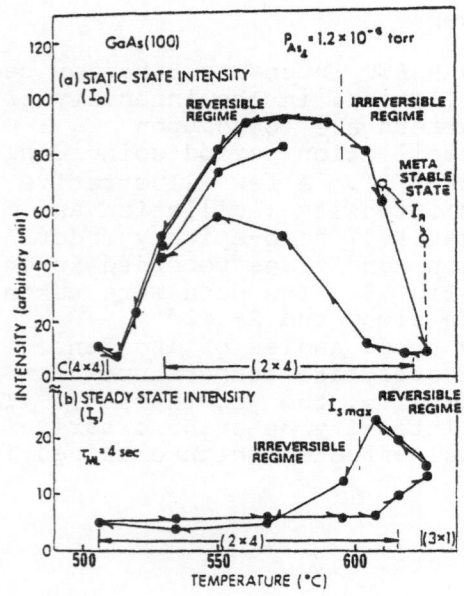

Fig. 9.

Panel (a) - behavior
of the specular beam
for GaAs(100) static
surface for the 2-
fold pattern of the
As-stabilized (2x4)
surface. Panel (b) -
behavior of the
steady state intens-
ity (I_S) at the same
As_4 pressure as in
panel (a) for a
monolayer deposition
time (τ_{ML}) of 4 sec.

appropriate theory of electron diffraction as well as a sufficiently large and sytematic data base. Considerable data base has been generated in our laboratory over the past three years and, in conjunction with the data base of other phenomena which, in this author's view, would take the next few years to sort out. It is thus no wonder that at the present time no definite and clear answers to most issues that may be raised are available. Indeed, as might be expected at this early stage, attempts to interpret limited sets of data appearing in sequential publications, have generally caused more confusion than clarity. It is unfortunate that a large part of this confusion arises from imprecise use of well established concepts and terminology, to some degree or another, on the part of all workers. Some of this is perhaps inherent to the beginnings of a new field. It is, nevertheless, now imperative that greater thought be given and care exercised to avoid putting in the literature imprecise and vague, though often intuitively appealing, terminology, expressions and explanations.

It is not the intent of this section to provide either a short review of the status of RHEED or even to compile a comprehensive list of the various issues and controversies already created, whether real, imagined or unnecessarily exaggerated. Rather, we wish to indicate here the potential of the subject in bringing about a connection between growth conditions and the attendant growth kinetics controlled surface structural and chemical nature. Thus in the following we restrict ourselves to presentation of data which, on one hand, bear upon usage of RHEED for pragmatic purposes, such as reproducibly establishing growth conditions[23,24,29], and on the other, also show the challenging problems of surface kinetics that need to be tackled. At the same time, we will present consequences of the computer simulation generated surface structure dynamics for the RHEED intensity behavior calculated within the kinematical theory of diffraction. We cannot emphasize enough that in so doing there is no implication that the kinematical theory is adequate for examining most of the challenging issues, particularly the need to sort out the growth condition and diffraction condition related effects. It is well known that the nature of the electron-solid interaction is sufficiently strong so as to demand a fully dynamical theory of electron diffraction and to harp on this aspect is merely an exercise in the art of making the self evident appear profound. Our use of the kinematical theory is thus motivated by, (i) the unavailability of a reliable dynamical theory, in spite of claims to the contrary[32], and (ii) with a view of generating a basic skeleton from which a more substantial, and quantitatively reliable analysis of the phenomena can be subsequently developed. The kinematical theory will fail in certain qualitative aspects as well so that its usefulness for many of the qualitative features generated rests in the hands of those who do not lose sight of its limitations.

While much has been published on the dynamics of the RHEED intensity behavior in the past three years, not sufficient attention has been paid to the behavior of the static surface intensity itself as a function of the

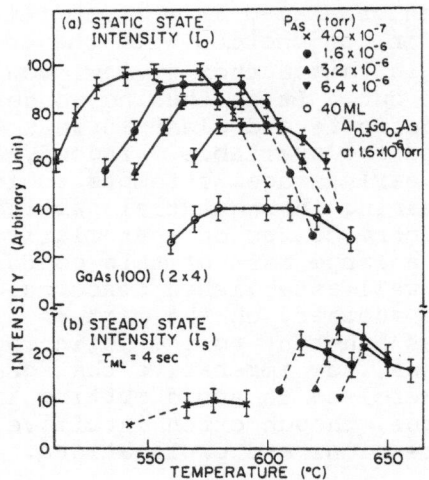

Fig. 10. Shows the variation, with substrate temperature, of
the specular beam intensity for static GaAs(100)
surface and in the steady state for growth at 0.25
ML/sec., at different As_4 pressures. Measurements
are for the 2-fold pattern of the As-stabilized
(2x4) surfaces and at the shallowest angle of
incidence (O=0.7°) of fig. 8. The irreversible
regions are not shown for clarity of figure. Open
circles correspond to the surface of a 40 ML thick
$Al_{0.3}Ga_{0.7}As(100)$ layer grown on GaAs(100). Note
the presence of a maximum in I_s(max.) at $T_s \sim 635°C$
and $P_{As} \sim 3x10^{-6}$ Torr for a growth rate of 0.25
ML/sec.

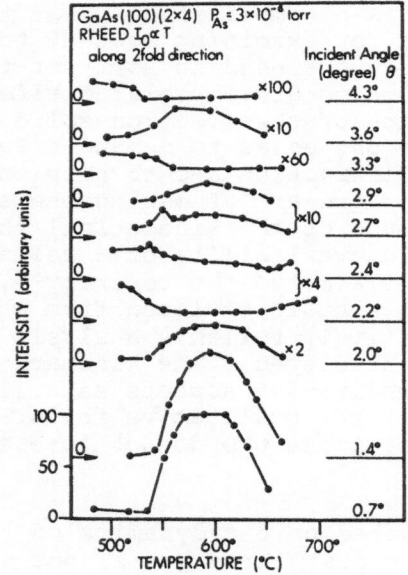

Fig. 11.

Shows the behavior of
the specular beam
intensity as a func-
tion of substrate
temperature at
different angles of
incidence. The
zeroes on the left
hand side set the
scale origin for the
corresponding angle
of incidence noted on
the right. The scale
is set by the first
curve at θ - 0.7°.

substrate temperature (T_S) and group V pressure. We therefore begin by presenting a few results of fairly exhaustive measurements[20-25,30,31] of the static surface intensity as a function of substrate temperature and As_4 pressure at different, but fixed, diffraction conditions. In Fig. 9 is shown the behavior of the specular beam intensity (I_O) from the 2-fold pattern of the GaAs(100) surface in the As(2x4) reconstruction regime as a function of T_S at the fixed As_4 pressure (P_{As}) shown. Note that in the overall "cap like" behavior, regions of reversible and irreversible behavior are found with respect to change in T_S[22,23]. To the extent that the variation in the specular beam intensity at fixed diffraction conditions may be expected to be inversely related to change in surface smoothness caused by, in this case, a change in T_S, the behavior seen in fig. 9 is of practical value in preparing the smoothest GaAs(100) surfaces prior to growth of the quantum well structures[9-12]. Even if one were unwilling to put faith in such a relationship, the variation of I_O with T_S at different fixed P_{As} values, as shown in fig. 10, is of immense value in, at least, arriving at reproducible growth conditions from day to day in a given machine, without having to solely rely upon notoriously fickle measuring instruments such as ion gauges, thermocouples and pyrometers. Indeed, this has been a routine practice in our group for the past two years and the growth condition for every quantum well structure grown is checked against such intrinsic behavior of the GaAs surface. Such behavior, when further coupled with the behavior of I_O as a function of the angle of incidence (Θ), such as shown in fig. 11, and the phenomenon of arsenic controlled oscillations[29], further eliminate the uncertainties in the reproducibility of the desired (T_S, P_{As}) combination. While, as should be expected, considerable variation with diffraction condition is seen, it is nevertheless our view that RHEED, at these shallowest angle of incidence at which it is most sensitive to the surface atomic and bilayer steps, can be useful in guiding choice of growth conditions.

We now turn to the RHEED intensity behavior during growth. As stated earlier, while the complicated dynamics requires a sophisticated and implementable theory of electron diffraction, it may be useful to at least acquire a rudimentary feel for why periodic variations in the intensity might be expected and under what conditions[4,6]. Consider, for a moment, the growth process. If the growth is begun under growth conditions such that the effective migration length, l_{eff}, of the growth rate controlling specie is smaller than the average terrace width, $<W_O>$, of the starting surface then growth will occur through formation of clusters or islands on the terraces. As a result the step density distribution will begin to change. In consequence, the measured RHEED intensity, at whatever chosen diffraction condition for whichever chosen diffraction pattern feature (the specular spot, integral order streaks, or fractional order streaks), will also change. If growth is essentially via a layer-by-layer mode of material addition (as is the case with MBE growth under the most commonly employed growth conditions) then the step density distribution will undergo a cyclic behavior with addition of each layer's worth of material, the only departure in the identical cyclic nature of the step density arising from the statistical roughening of the growth front structural profile. We have already seen this behavior in the plot of $\langle W_I \rangle$, σ_h^2 and the smoothness parameter S given by the computer simulation results of fig. 3. It thus follows that the changes in the RHEED intensity must also show a cyclic behavior accompanied by damping of the amplitude. What does not necessarily follow from these simple considerations is whether the initial change in intensity upon growth initiation may be positive or negative and how it may depend upon the chosen diffraction conditions (compare column three of fig. 8 with the others). In fig. 3 we show the specular beam intensity calculated for the computer simulation results of the same figure under the condition $q_\perp d_{As-As} = \pi$ and for scattering only from the topmost exposed surface layers within the kinematical approximation. Here $q_\perp$ is the momentum transfer perpendicular to the surface and d_{As-As} is the distance between adjacent As(100) planes exposed at the surface. One notes that in fig. 3 the specular beam intensity maxima coincide with that of S, it oscillates with the same periodicity as the properties characterizing the surface morphology (i.e. $\langle W_I \rangle$, σ_h^2 and S), and exhibits a damping similar to that for the smoothness parameter S. The coincidence of the maxima in both with time of incorporation of a monolayer of Ga (τ_{ML}) is a consequence of the RLI regime of growth mechanism. At high arsenic pressure, i.e. in the CLRI regime, while the smoothness S and specular beam intensity calculated within the approximations noted above are in phase, their maxima do not necessarily coincide with the monolayer delivery time. Thus caution needs to be exercised even in attributing the RHEED oscillation maxima to completion of a monolayer and delivery of an integral number of monolayers.

Returning to measurements of RHEED dynamics, a feature of practical significance for growth of heterojunctions and thick quantum wells is the behavior of the value of the specular beam intensity in the steady state (I_S), i.e. when the oscillation amplitude on the detection scale has died out. In fig. 9, panel (b) is shown the behavior of I_S as a function of T_S during homoepitaxy of GaAs(100) at a monolayer deposition time (τ_{ML}) of 4 sec. All other conditions, including the diffraction measurement conditions, are the same as for the results for I_O shown in panel (a) of the same figure. Note that a maximum in I_S, (I_S (max.)), exists and, in contrast to the behavior of I_O, an irreversibility is found on the lower temperature side of the temperature at which I_S(max) occurs. In fig. 10, panel (b) is shown the behavior of I_S for different but fixed As_4 pressures corresponding to the same values as for the I_O behavior shown in panel (a) of the same figure.

III.B. $Al_xGa_{1-x}As(100)$ and Interface Formation

We now turn to a brief discussion of the RHEED intensity behavior of $Al_xGa_{1-x}As(100)$, $x \neq o$, and during formation of interfaces with GaAs. Fairly comprehensive studies[18-22,24,31] of these have been carried out and it is neither possible nor perhaps essential that even an illustrative example of each type of study be included here. We restrict ourselves to

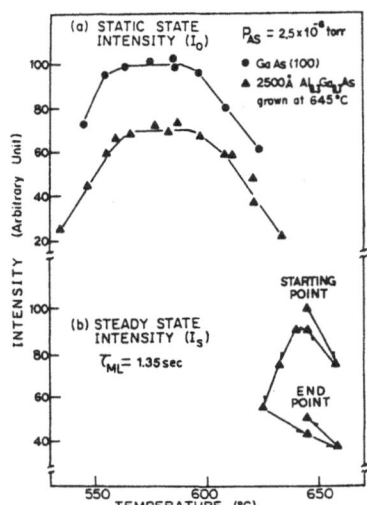

Fig. 12. Shows the behavior of the specular beam intensity
as a function of the substrate temperature at fixed
As_4 pressure of $2x10^{-6}$ Torr from a static
$Al_{0.3}Ga_{0.7}As(100)$ surface (panel a) and during
steady state growth (panel b) at a monolayer growth
time of 1.35 sec. The $Al_{0.3}Ga_{0.7}As$ layer was first
grown to a thickness of $2500°A$ at $645°C$ under the
same conditions. The measurements are for the 2-
fold pattern of the As-stabilized (2x4) surfaces
and at the shallowest angle of incidence, $\Theta=0.7°$.

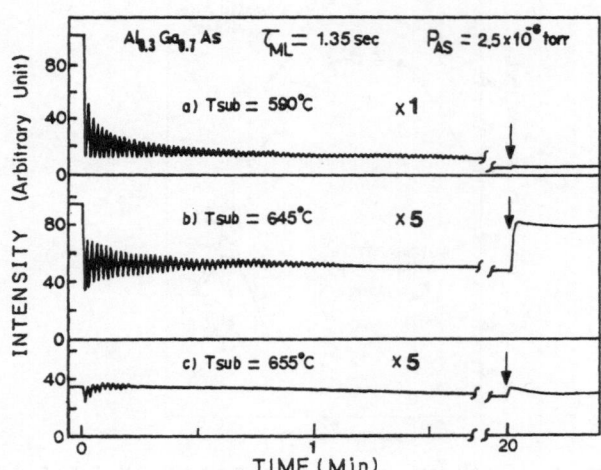

Fig. 13. Shows the dynamics of the specular beam intensity
during growth and after interruption, for
$Al_{0.3}Ga_{0.7}As(100)$ at a fixed As_4 pressure of 2.5 x
10^{-6} Torr and monolayer deposition time of 1.35
sec., for three different substrate temperatures.
The measurement conditions are the same as for fig.
12.

presentation of selected examples which help motivate some of the considerations for growth of quantum wells to be discussed in the next section.

Studies of the behavior of I_O, I_S and I_R of $Al_xGa_{1-x}As$ (100) analogous to those illustrated for GaAs(100) in the preceding section have been carried out at USC. In fig. 12 is shown the I_O and I_S behavior of $Al_{0.3}Ga_{0.7}As$(100) surface in the As-stabilized (2x4) region for the 2-fold reconstruction direction and at angle of incidence $\theta=0.7^O$ which is near the condition for maximum sensitivity to As-As steps. The I_O measurements were carried out[31] on a 2500°A thick $Al_{0.3}Ga_{0.7}As$(100) layer grown on a GaAs epilayer at 645°C and $_{ML}$ = 1.35 sec. For ready comparison, the I_O behavior of GaAs(100) under the same conditions is reproduced. A "cap like" behavior, similar to GaAs is observed although the absolute intensity is lower. The steady state intensity shows irreversibility on the low temperature side of a maximum, similar to GaAs. The I_S(max.), however, occurs at about 15°C higher substrate temperature than for GaAs under identical growth conditions. In fig. 13 is shown the specular beam dynamics for deposition of $Al_{0.3}Ga_{0.7}As$(100) at three different substrate temperatures for the same growth rate and As_4 pressure. Note the qualitative difference in the behavior at 655°C which is the very high temperature side of the I_O plateau of the starting GaAs surface where the intensity is very low due to the degradation of the GaAs surface.

The generally lower overall intensity of $Al_xGa_{1-x}As$(100) static and dynamic surfaces at growth conditions which give high GaAs intensities is a consequence of the inherent slower

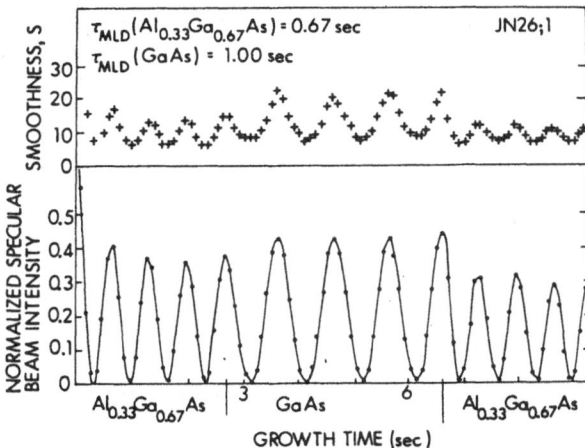

Fig. 14. Shows the dynamics of the smoothness parameter (S) and specular beam intensity, calculated within the kinematical approximation for an angle of incidence corresponding to the first out-of-phase condition for As to As adjacent surface planes, obtained from the computer simulations results for the $Al_{0.3}Ga_{0.7}As$/GaAs/ $Al_{0.3}Ga_{0.7}As$(100) quantum well structure of fig. 4.

migration kinetics and faster reaction kinetics of Al as compared to the Ga atoms. This causes roughening of the surfaces. This is borne out in fig. 14 which shows the behavior of the smoothness parameter and the specular beam corresponding to the computer simulation results of fig. 4 for growth of GaAs/$Al_{0.3}Ga_{0.7}As$(100) system. A difference between GaAs and $Al_{0.3}Ga_{0.7}As$(100) growth, consistent with the higher effective migration length of Ga, is clearly seen.

The preceding illustrative examples point to the significance of the kinetics of growth in controlling both, the structural and chemical nature of the interfaces. While a variety of potential means of doing so by external means (e.g. light sources, etc.) can be concieved, work is only beginning in these areas and one may look forward to some interesting results in the future. In the meantime, a simple device, suggested by the computer simulations of the growth front surface morphology dynamics, to improve the structural quality of the surface on which an interface is to be formed, is to merely interrupt the growth and let the instantaneous growth front relax towards a smoother surface profile i.e. one with larger average terrace widths and lower step density. To the extent that one may put faith in the qualitative consistency between the behavior of the smoothness parameter and the corresponding specular beam intensity behavior calculated of necessity in the simplest of approximation at the present time, the same may be inferred from measurements of the specular beam intensity behavior.

Indeed, the beneficial influence of growth interruption is already manifest in the examples of GaAs/AlAs multiple quantum well (MQW) growth shown in fig. 15. One notices that growth interruption, particularly after growth of the GaAs layer, tends to sustain the structural quality of both interfaces for growth of much larger number of ultra-thin quantum wells. One also notes the difference in the behavior of $(GaAs)_2/(AlAs)_2$ and $(GaAs)_6/(AlAs)_6$ MQW in that, the slow migration kinetics of Al makes the growth front of a 6ML AlAs layer so much worse than for a 2ML AlAs layer that recovery may take an unacceptably long time, if indeed it can occur at all given the gettering of background impurities by the highly reactive Al. This also indicates that for intrinsic kinetic reasons, as well as extrinsic reasons, growth interruption is no panacea for all the ills and its usefulness lies in its judicious application based upon a clear understanding of the relationship between growth condition and growth kinetics of the system and structure under consideration.

IV. GROWTH AND EXAMINATION OF QUANTUM WELLS

In this section we present some results on the growth, photoluminescence (PL) and PL excitation (PLE) spectra of $Al_{0.3}Ga_{0.7}As$(100) single quantum wells grown on the basis of the GaAs I_O and I_S behavior. The objective is (1) to examine which of the expected consequences for the nature of interfaces implied by the computer simulation generated surface smoothness behavior and its qualitative connection with the RHEED intensity may be consistent with that extracted from PL and PLE, and (2) to examine the PL line width dependence on the GaAs quantum well width for SQW

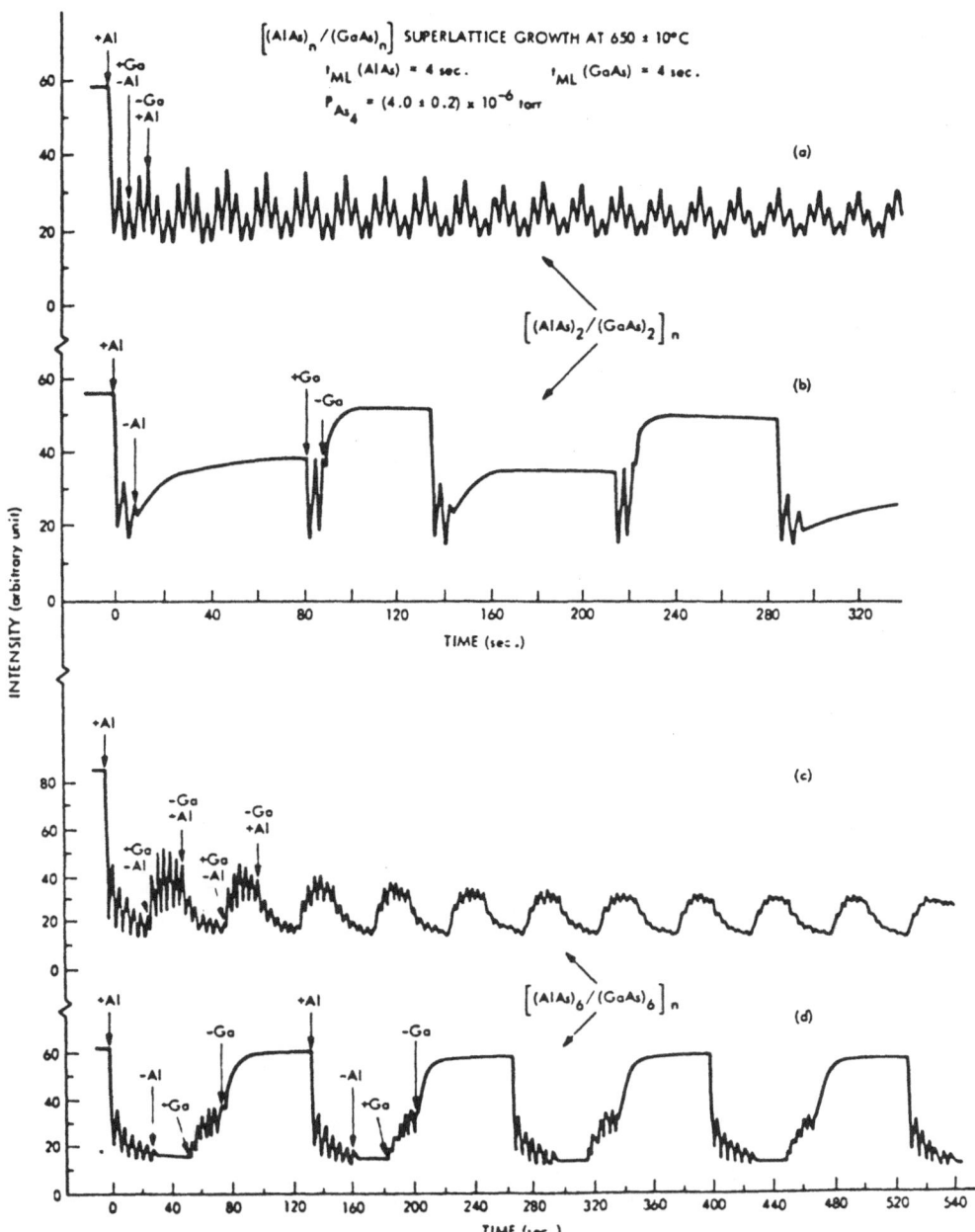

Fig. 15. Shows the specular beam intensity dynamics during
formation of (GaAs)$_n$/(AlAs)$_n$ multiple quantum well
structures following the conventional practice of
no growth interruption (curves a and c) and with
growth interruption (curves b and d) for n=2 and 6.
The growth conditions employed correspond to the
unique combination of T_S=650°C, P_{As_4} = 4x10^{-6} Torr
and GaAs and AlAs growth rates of 0.25 ML/sec. for
which the static GaAs, and the steady state GaAs
and AlAs surface simultaneous reach the maximum
intensity at the first out-of-phase diffraction
condition (see ref. 21).

structures grown under _identical_ growth conditions specified via the observed RHEED static and dynamic intensity behavior. The latter set of systematic growths at least ensure that the growth kinetics of the quantum wells are the same so that the behavior of the optical properties may be related, with much higher probability, to only the change in the well thickness.

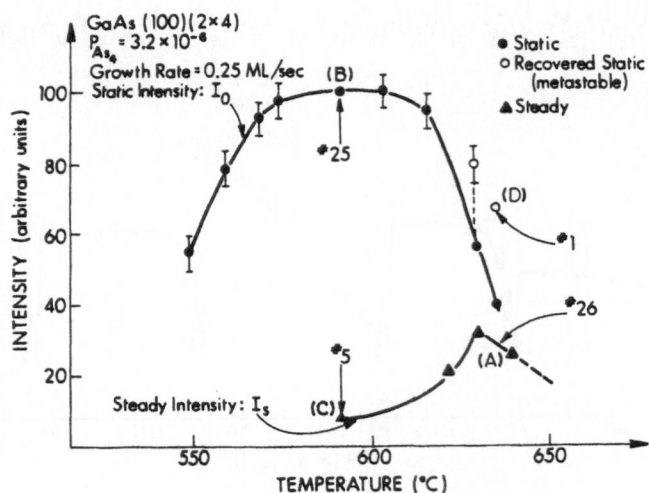

Fig. 16. Behavior of the specular beam intensity for static and steady state (at a growth rate of 0.25 ML/sec.) As-stabilized (2x4) GaAs(100) surfaces as a function of substrate temperature at a fixed As_4 pressure of 3.2×10^{-6} Torr. Points (A) through (D) indicate the conditions of growth of the $GaAs/Al_{0.3}Ga_{0.7}As(100)$ single quantum wells discussed in the text. Note that point (D) corresponds to the smoother, but metastable, surface achieved upon growth termination for GaAs growth at (A). Also note that the particular combination of growth conditions corresponds to the maximum of $I_S(max)$ as a function of As_4 pressure.

In fig. 16 is shown the I_O and I_S behavior of GaAs as a function of substrate temperature at a pressure of 3.2×10^{-6} Torr in our Φ-400 MBE growth chamber. The growth rate during I_S measurement was 0.25 ML/sec. We first consider the comparative PL behavior of four SQW with identical structures grown at the four points marked (A) through (D) in fig. 16. Points (A) and (C) are relevant for growth of samples following the conventional practice of no growth interruption. Their comparison should provide some insight into whether the growth front smoothness expected on the basis of the RHEED specular beam intensity measured at the shallowest angle of incidence consistent with maximum sensitivity to As to As steps is indeed realized. Points (B) and (D) in fig. 16 are relevant for growth employing the technique of growth interruption introduced some three years ago in the context of growth of GaAs/InAs(100) system[33,34], and since employed for growth of the $GaAs/Al_xGa_{1-x}As(100)$ system[35-38]. The expectation, based upon the structural

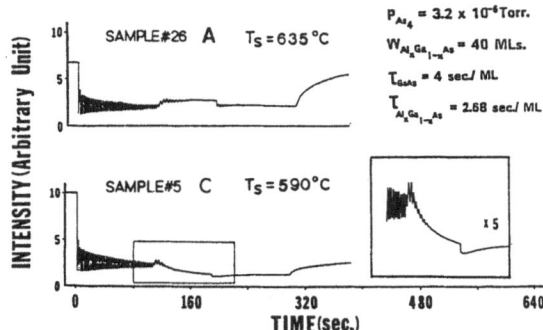

SQW $Al_xGa_{1-x}As/GaAs$

SAMPLE#26 A $T_S = 635\,°C$

SAMPLE#5 C $T_S = 590\,°C$

$P_{Al_4} = 3.2 \times 10^{-6}$ Torr.

$W_{Al_xGa_{1-x}As} = 40$ MLs.

$\tau_{GaAs} = 4$ sec./ ML

$\tau_{Al_xGa_{1-x}As} = 2.68$ sec./ ML

INTENSITY (Arbitrary Unit)

TIME (sec.)

Fig. 17. Shows the specular beam dynamics during growth of $Al_{0.3}Ga_{0.7}As/GaAs/Al_{0.3}Ga_{0.7}As$ (100) single quantum well structures grown at points (A) and (C) of fig. 16. The GaAs well layer is 20 ML thick and the barrier layers are 40 ML thick.

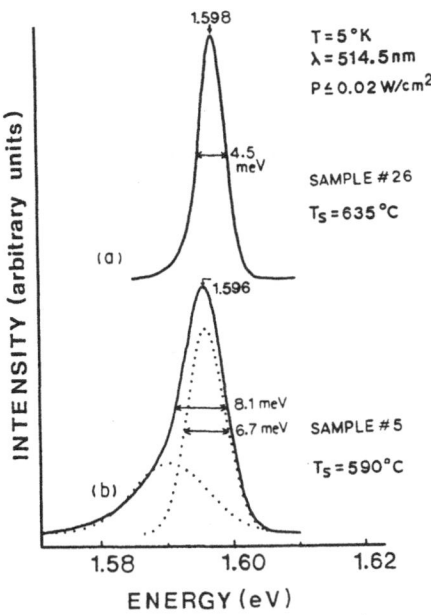

$T = 5\,°K$
$\lambda = 514.5$ nm
$P \le 0.02$ W/cm^2

SAMPLE # 26
$T_S = 635\,°C$

SAMPLE # 5
$T_S = 590\,°C$

INTENSITY (arbitrary units)

ENERGY (eV)

Fig. 18. Shows the 5°K photoluminescence behavior of single quantum well sample numbers 26 and 5 grown at points (A) and (C), respectively, of fig. 16.

smoothness notion expected of the specular beam intensity, would be that samples grown at point (B) should have better interfaces (at least the normal interface) than those grown at point (D).

Let us begin by comparing samples grown at points (A) and (C). Fig. 17 shows the specular beam behavior for 20 ML wide GaAs well sandwiched between 40 ML of $Al_{0.3}Ga_{0.7}As$ barrier layers. The monolayer growth times are 4 sec. for GaAs and 2.68 sec. for $Al_{0.3}Ga_{0.7}As$. Note the degradation of intensity during growth of the GaAs well layer at the lower temperature, thus also leading to a worse normal interface upon subsequent deposition of $Al_{0.3}Ga_{0.7}As$. At the As_4 pressure and $Al_{0.3}Ga_{0.7}As$ growth rate employed, no major difference in the bahavior of $Al_{0.3}Ga_{0.7}As$ intensity after 40 ML growth is observed so that, at face value, the inverted interfaces in both cases would be expected to be comparable in so far as their structural quality is concerned. However, as we have seen in the computer simulation results of sec. II (see figs. 4 through 7), the chemical quality determined by the spatial distribution behavior of Al would be expected to be worse for both, the inverted and the normal interfaces at the lower temperature due to the greater disparity between the Al and Ga surface kinetics. Thus for both structural and chemical reasons one might expect the PL behavior of the higher temperature growth near $I_s(max)$ to be better. Fig. 18 shows the 5°K luminescence behavior of these two SQW. One readily sees that the RHEED intensity based expectations are indeed realized[12].

Next, let us compare samples grown at point (B) and (D). Fig. 19 shows the specular beam intensity behavior during growth of these samples, both consisting of 21 ML intended GaAs well sandwiched between 40 ML barriers of $Al_{0.3}Ga_{0.7}As$, as before. The growth rates are the same as before as well. Consistent with points (B) and (D) of fig. 16, one sees that the GaAs starting and recovered intensities are higher for the lower temperature growth. On the basis of the structural smoothness alone implied, at best, by the specular beam intensity, one would expect that sample # 6 grown at point (B) should show higher quality PL behavior. The PL behavior, shown in fig. 20 however, cannot be readily interpreted to conform to this expectation due to the appearance of fine structure in the luminescence[10,11]. We comment on the origin of the fine structure below, but first simply note that even the individual line widths of the fine structure lines is narrower for the higher temperature growth in spite of the lower structural smoothness implied by the expected inverse correlation between the specular beam intensity and structural smoothness. One may thus tend to conclude that the aforementioned inverse correlation does not exist and that believing otherwise was misleading. This, however, is not necessarily the case. The point to recognize is that RHEED is sensitive only to the structural quality whereas the PL behavior is sensitive also to the chemical quality of the interfaces[10-12]. At the lower temperature (point B of fig. 16), the greater disparity of Al and Ga surface kinetics coupled with the slower absolute migration kinetics of Al implies that greater compositional fluctuation of Al at the interfaces is the result, as seen in the computer

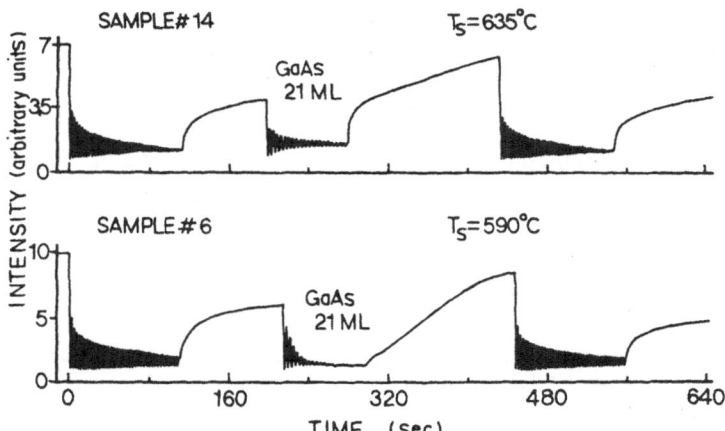

Fig. 19. Shows the specular beam dynamics during growth of
$Al_{0.3}Ga_{0.7}As/$ $GaAs/Al_{0.3}Ga_{0.7}As(100)$ single quantum
well structures grown at conditions denoted by
points (B) and (D) of fig. 16. The GaAs well layer
is 21 ML thick and the barrier layers are 40 ML
thick.

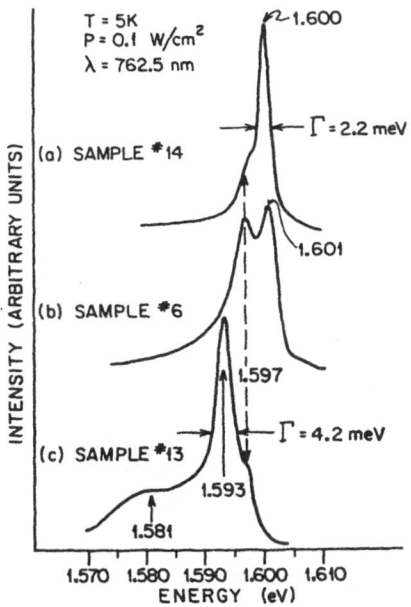

Fig. 20. Shows the 5°K photoluminescence behavior of single
quantum well sample numbers 14 and 6 grown at
points (D) and (B), respectively, of fig. 16.
Shown also is the behavior of sample # 13 grown
with identical structure at the same conditions as
for sample # 14 (i.e. point D of fig. 16), except
that growth was interrupted only at the normal
interface. The fine structure corresponds to
regions of well fluctuating by ± 1ML in the sample,
as discussed in the text.

simulations. Thus whatever advantages might have been expected to accrue from smoother GaAs profiles are negated. Fluctuations in the band edge discontinuity and short ranged alloy disorder type scattering is thus higher in the lower temperature interrupted growth sample[10,11].

The fine structure separation seen in fig. 20 coincides well with the theoretically expected separation between the n=1 electron to HH exciton recombination luminescence from wells differing in width by 1 ML. The emissions at 1.600 eV and 1.597 eV correspond to wells of thickness 20 ML and 21 ML, respectively. The sample thus consists of regions of at least these two well thicknesses. This is consistent with the expected consequence of growth interruption which allows the instantaneous terrace widths to become larger as the recovery process proceeds. Thus, if the recovered average terrace width, <W>, were significantly larger than the exciton size, R_{exc}, then the exciton locally experiences a well defined well width, even though on a larger length scale it may fluctuate by +1 ML in the sample. By contrast, in the absence of growth interruption, the instantaneous small terraces at the growth front get frozen in upon formation of the interfaces. Consequently, with <W> < R_{exc}, the exciton experiences an averaged out "width" (i.e. interfaces graded over at least one atomic layer), the recombination luminescence thus exhibiting only a single broader line with no fine structure (fig. 18). Shown in fig. 20, curve (c) is also the behavior of a SQW grown under identical growth conditions and with identical structure as the higher temperature interrupted growth (curve a), but with growth interruption only at the normal interface. Comparison of the two clearly indicates that under the growth conditions employed, it is the interruption at the inverted interface which is primarily responsible for the improvement of the luminescence behavior[11]. This is also consistent with the shift of the dominant emission peak in curve (C) to 1.593 eV which is close to the emission from 22 ML wide well regions. The broad peak at 1.581 eV is impurity related.

The PL excitation behavior[10] of samples 14 and 6 is shown by broken lines in fig. 21. One clearly observes the n=1 electron to light hole exciton recombination lines associated with the two heavy hole exciton lines (i.e. the two different well thickness regions in the sample.). In addition, the step like structure between the heavy and light hole related emission manifests the step like nature of the density of states of quasi-two dimensionally confined particles. It is also the region where the 2s state of the heavy hole exciton would give emission and it is possible that such emission is riding on top[10].

From the above it follows that neither for the conventional practice of growth without interruption, nor for the newer approach of growth interruption, is the commonly employed model of the PL line widths arising from fluctuations in the well width applicable. Indeed, within this model, the line width (Γ) is expected to vary nearly as ($\Delta d_w/d_w^3$) where d_w and Δd_w are the well width and fluctuations in well width, respectively. Consequently, the soundness of this prevalent notion and model can be further

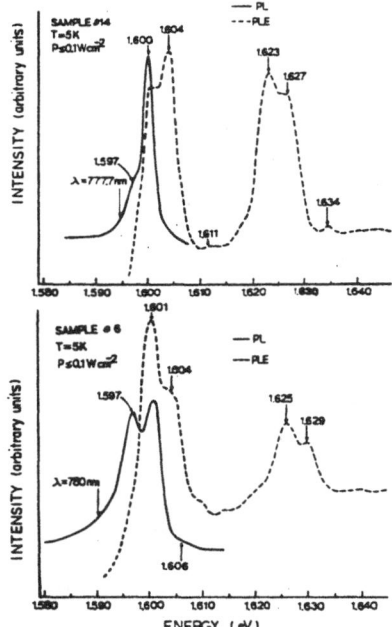

Fig. 21. Shows the 50K photoluminescence excitation spectra
(broken lines) of sample numbers 14 and 6 with
detection set as indicated by the wavelength value
set against the solid line (PL) curves, included
for comparison.

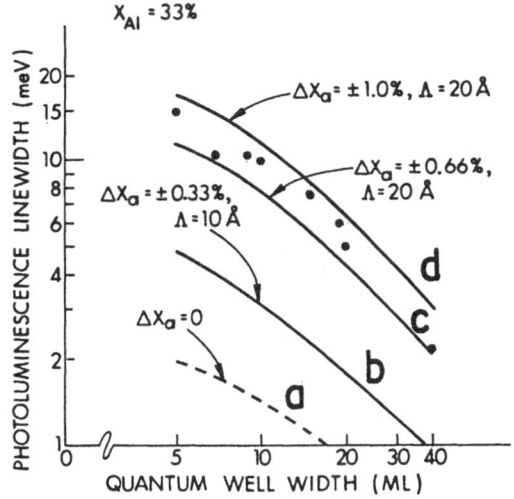

Fig. 22. Dependence of the n=1 electron – heavyhole luminescence line width on the quantum well width (d_w)
of single quantum wells grown, without interruption
at point (A) of fig. 16. The solid lines are the
theoretical values obtained on the basis of
fluctuations in the band edge discontinuity and
short ranged alloy disorder. ΔX_a and Λ represent
the amplitude and correlation length of local Al
concentration fluctuations. Broken line shows the
theoretical values for short ranged disorder only.

experimentally checked by determining the line width behavior of samples grown without interruption under identical growth conditions but with varying intended well thicknesses only. A systematic study of precisely the aforementioned has been carried out in our group and the results are shown in fig. 22. All these samples were grown at the conditions corresponding to point (A) of fig. 16. Note that on the log-log scale, the line width behaves closer to d_w^{-1} rather than d_w^{-3}. It is thus clear that the notion of well width fluctuations for systems involving an alloy as the barrier layer is simply not applicable. The solid lines shown in fig. 22 are results of the line width calculated on the basis of the physical model that it is the fluctuation in the band edge discontinuity arising from local average Al composition fluctuations, and the shorter ranged alloy disorder type of scattering arising from the mean square fluctuations about the local average, which control the line width. While we proposed this model some time ago, the theory of the line width has been developed only recently. The symbol Δx_a denotes the amplitude of <u>local average</u> Al composition fluctuation on a correlation length scale denoted by Λ in a Gaussian distribution. The broken line corresponds to the calculated line width in the absence of any fluctuations in the local average Al concentration i.e. no band edge discontinuity fluctuations, but due only to the alloy disorder scattering. One notes that the calculated dependence of the line width on well thickness agrees rather well with the experiments. Additionally, the needed magnitude of local Al concentration fluctuation of order one percent is quite consistent with the known uncertainties inherent in both, the technology and intrinsic kinetics of MBE growth. The correlation length Λ also is quite reasonable and consistent with the results for alloys.

V. SUMMARIZING REMARKS

It is hoped that this chapter has provided some indication of both, the types of fundamental issues as well as approaches that are under investigation to gain a better understanding of the nature of the MBE growth process and its implications for realization of high quality multi-interface structures of thin films. Although much of the discussion was confined to the GaAs/Al$_x$Ga$_{1-x}$As(100) system, it is to be recognized and appreciated that each and every one of the basic considerations of surface kinetics involved here and their consequences are equally relevant to the growth and understanding of <u>any</u> of the other systems discussed at this workshop. Naturally, each class of systems introduces its own special aspects in addition, such as the role of strain in lattice mismatched systems - semiconductor/semiconductor or metal/metal. Although results on GaAs/Al$_x$Ga$_{1-x}$As(100), used as a test system for precise formulation of concepts, issues and answers, have also found some correspondence already in some of the work on structural aspects of this system represented at this workshop in the talks on X-ray diffraction and electron microscope studies, in the longer run it is the cross-fertilization of the basic framework emerging from the combination of the computer simulation and fundamental experimental studies with the other material systems which would have made this inter-disciplinary research workshop a worthwhile endeavour for the participants and organizers alike.

ACKNOWLEDGEMENTS

As with any research effort, particularly following the multipronged approach undertaken at USC, one is indebted to many present and past colleagues. I should like to particularly acknowledge the contributions of Drs. M. Thomsen and S.V. Ghaisas to the Monte-Carlo computer simulations, of Drs. P. Chen, T.C.Lee, M.Y.Yen and Mr. N.M.Cho to the RHEED studies, Dr. F. Voillot and Messrs. J.Y.Kim and W.C.Tang to the optical studies, and Dr. S.B. Ogale to the theory of PL linewidths. The assistance of Mr. P.G. Newman in MBE growth is gratefully acknowledged. It is the collective work of these students and associates which has made this article possible. Finally, the able typing of the manuscript by Mrs. Arsho Apardian is thankfully acknowledged. Financial support for the work reported here has been provided by The Office of Naval Research, The Air Force Office of Scientific Research, and the Joint Services Electronics Project, and is gratefully acknowledged.

REFERENCES

1. For MBE of III-V Semiconductors, see for example, A.C. Gossard, Thin Films: Preparation and Properties, Eds. K.N. Tu and Ro. Rosenberg (Ac. Press, New York, 1982).

2. For an overview of modulated structures, see "Synthetic Modulated Structures", Eds. L.L. Chang and B.C. Giessen (Ac. Press, New York, 1985).

3. See also, Proceedings of the First International Conference on Metastable and Modulated Semiconductor Structures, Dec. 1982, Pasadena, Ca., Eds. F.J. Grunthaner and A. Madhukar, Jour. Vac. Sci. Tech. B1, p.217 (1983).

4. A. Madhukar and S.V. Ghaisas, CRC Critical Reviews in Solid State and Materials Sciences (To appear, 1987).

5. S.V. Ghaisas and A. Madhukar (To be published).

6. S.V. Ghaisas and A. Madhukar, Phys. Rev. Lett. 56, 1066 (1986).

7. J.H. Neave, B.A. Joyce, P.J. Dobson and N. Norton, App. Phys. A31, 1, (1983); ibid A34, 1 (1984).

8. M. Thomsen and A. Madhukar, Jour. Cryst. Growth (To appear).

9. F. Voillot, A. Madhukar, J.Y. Kim, P. Chen, N.M. Cho, W.C. Tang and P.G. Newman, App. Phys. Lett. 48, 1009 (1986).

10. F. Voillot, A. Madhukar, W.C. Tang, M. Thomsen, J.Y. Kim, and P. Chen, App. Phys. Lett. (Jan. 26, 1987 issue, In Press).

11. A. Madhukar, P. Chen, F. Voillot, M. Thomsen, J.Y. Kim, W.C. Tang and S. V. Ghaisas, Jour. Cryst. Growth (In Press).

12. F. Voillot, J.Y. Kim, W.C. Tang A. Madhukar and P. Chen, Paper presented at the 2nd International Conference on Superlattices, Microstructures and Microdevices, Aug. 1986, (Goteborg, Sweden). Superlattices and Microstructures (In Press).

13. A. Madhukar, S.B. Ogale, W.C. Tang, and F. Voillot, Phys. Rev. Lett. (Submitted).
14. J.J. Harris, B.A. Joyce and P.J. Dobson, Surf. Sc. Letts., 103 L90 (1981); ibid C.E.C. Wood, 108, L441 (1981).
15. J.M. Van Hove, C.S. Lent, P.R. Pukite and P. Cohen, Jour. Vac. Sc. Tech. B1, 741 (1983); ibid B3, 546 (1984).
16. T. Sakamoto, H. Funabashi, K. Ohta, T. Nakagowa, N.J. Kawai and T. Kojima, Japanese Jour. App. Phys. 23, L657 (1984).
17. B.F. Lewis, T.C. Lee, F.J. Grunthaner, A. Madhukar, R. Fernandez and J. Maserjian, Jour. Vac. Sc. Tech. B2, 419 (1984).
18. A. Madhukar, S.V. Ghaisas, T.C. Lee, M.Y. Yen, P.Chen, J.Y. Kim, and P.G. Newman, Proceedings of the SPIE Conference, Los Angeles, (Jan. 20-25, 1985), Vol. 524, 78 (1985).
19. A. Madhukar, T.C. Lee, M.Y. Yen, P.Chen, J.Y. Kim, and P.G. Newman, App. Phys. Lett. 46, 1148 (1985).
20. T.C. Lee, M.Y. Yen, P. Chen and A. Madhukar, Jour. Vac. Sc. Tech. A4, 884 (1986).
21. M.Y. Yen, T.C. Lee, P. Chen and A. Madhukar, Jour. Vac. Sc. Tech. B4, 590 (1986).
22. T.C. Lee, M.Y. Yen, P. Chen and A. Madhukar, Surf. Sc. 174, 55 (1986).
23. P. Chen, A. Madhukar, J.Y. Kim and T.C. Lee, App. Phys. Lett. 48, 650 (1986).
24. P. Chen, J.Y. Kim, A. Madhukar and N.M Cho, Jour. Vac. Sc. Tech. B4, 890 (1986).
25. P. Chen, A. Madhukar, J.Y. Kim and N.M. Cho, Paper presented at the 18th International Conference on Physics of Semiconductors, Aug. 1986 (Stockholm, Sweden). To appear in the proceedings.
26. B.A. Joyce, P.J. Dobson, J.H. Neave, K. Woodbridge, J. Zhang, P.K. Larsen and B. Bolger, Surf. Sc. 168, 423 (1986).
27. P.K. Larsen, P.J. Dobson, J.H. Neave, B.A. Joyce, B. Bolger and J. Zhang, Surf. Sc. 169 176 (1986).
28. B.F. Lewis, F.J. Grunthaner, A. Madhukar, T.C.Lee and R. Fernandez, Jour. Vac. Sc. Tech. B3, 1317 (1985)
29. B.F. Lewis, R. Fernandez, A. Madhukar and F.J. Grunthaner, Jour. Vac. Sc. Tech. B4, 560 (1986)
30. P. Chen, N.M. Cho and A. Madhukar (Unpublished).
31. N.M. Cho, A. Madhukar and P. Chen, (To be published).
32. The Philips group (Drs. B.A. Joyce, P.J. Dobson, J.H. Neave, P.K. Larsen and co-workers), referring to the theory of RHEED given by P.A. Maksym and J.L. Beeby (Surf. Sc. 110, 423 (1981)), has repeatedly asserted that the theoretical treatment exists for analysis of the data, though it has never been used by them. The authors themselves have not made such claims. Work on the theory of RHEED in our group at USC has been underway for the past two years and revealed some time ago that the results of the above noted theory are non-convergent. Similar conclusions have independently been reached by Dr. S.Y. Tong of University of Wisconsin, Milwaukee (private communication). Discussions with Drs. Beeby and Maksym at this workshop indicate that their most recent work, presented at this workshop, may

not suffer from earlier difficulties. However, this, as
well as the theory developed at USC (M. Thomsen and A.
Madhukar, unpublished) which predicts oscillations, need
to be adequately tested before they may be considered
suitable. There is thus the possibility that reliable
and implementable theories of RHEED may become available
in the near future. In the meantime, it would be well

for prospective users to refrain from originating and
perpetuating claims far beyond those that the authors
themselves are willing to make.

33. F.J. Grunthaner, B.F. Lewis, A. Madhukar, T.C.Lee, M.Y.
    Yen and R. Fernandez, Invited paper presented at the 1st
    International Conference on Superlattices,
    Microstructures and Microdevices, Aug. 1984 (Urbana,
    Illinois).
34. F.J. Grunthaner, M.Y.Yen, R. Fernandez, T.C. Lee, A.
    Madhukar and B.F. Lewis, App. Phys. Lett. $\underline{46}$, 983
    (1985).
35. N. Sano, H. Kato, M. Nakayama, S. Chika and H. Terauchi,
    Japanese Jour. App. Phys. $\underline{23}$, L640 (1984).
36. H. Sakaki, M. Tanaka and J. Yoshino, Japanese Jour. App.
    Phys. $\underline{24}$, L417 (1985).
37. T. Fukunaga, K.L. Kobayashi and H. Nakashima, Japanese
    Jour. App. Phys. $\underline{24}$, L510 (1985).
38. T.Hayakawa, T. Suzima, K. Takahashi, M. Kondo, S.
    Yamamoto, S. Yano and T. Hijikata, App. Phys. Lett. $\underline{47}$,
    52 (1985).

DIFFRACTION STUDIES OF EPITAXY:

ELASTIC, INELASTIC AND DYNAMIC CONTRIBUTIONS TO RHEED

P.I. Cohen, P.R. Pukite, and S. Batra

Department of Electrical Engineering
University of Minnesota
Minneapolis, MN 55455

ABSTRACT

Reflection high-energy electron diffraction (RHEED) is widely used for studies of epitaxy. It is an exceedingly surface sensitive technique because at low glancing angles, high energy electrons are able to interact strongly with the last few atomic layers of a solid. Yet for the same reason it is complicated to interpret, and dynamic (multiple) scattering theory must be used to analyze the measured diffracted intensities. Unfortunately these methods require large machine calculations, insist upon perfect surface periodicity, include only elastic scattering, and, in any event, are still in their infancy. Based on experience with low-energy electron diffraction, we expected that the shape of the diffracted beams would be amenable to kinematic analysis. Further RHEED has the important advantage that the scattering geometry can be chosen to allow only a few beams to contribute to the diffraction. Diffraction data were measured during epitaxial growth on vicinal and singular GaAs surfaces. Kinematic analysis is then used to understand the shape of the diffracted beams. We emphasize that no attempt is made to interpret the absolute intensities of the pattern. The role of inelastic and dynamic processes is described. The results indicate that kinematic theory describes the main features of the RHEED intensity oscillations that are observed during epitaxial growth as well as the angular profile of the integral order diffracted beams when growth is stopped. The method is to choose scattering geometries so that the kinematic angular dependence is obtained.

INTRODUCTION

Reflection high-energy electron diffraction (RHEED) is one of the most powerful tools for the study of epitaxial growth. It is an _in situ_ technique that provides real time control over growth parameters. RHEED is routinely used to set substrate temperature, constituent fluxes, and to determine the conditions under which epitaxy is obtained. It is exceedingly sensitive to surface morphology, allowing measurement of surface diffusion, sublimation rates, surface misorientation, and terrace length distributions. It is sensitive to surface reconstructions even at the high temperatures required for epitaxy. But the diffraction patterns are quite complicated, with a quantitative understanding only recently being developed.

This paper is written largely in response to discussions at the Fourth International Conference on Molecular Beam Epitaxy held in York and at this NATO Workshop. At issue is whether the structure observed along the (00) streak is explicable in terms of dynamic (multiple scattering) theory or in terms of inelastic effects or in terms of kinematic theory. Our view is that kinematic theory can be used to understand the _shape_ of a diffracted beam if one is careful to make use of scattering geometry to distinguish kinematic from dynamic effects.

The next section describes the predictions of kinematic theory. An important point, that is often not realized, is that this in fact includes multiple scattering within a terrace and only neglects multiple scattering between terraces. To observe this ideal behavior special care must be taken in sample preparation. The necessary experimental procedures are described in the third section. In the fourth and main section we show measurements from GaAs(001) surfaces and compare the results to the predictions of the kinematic theory. We show that kinematic analyses, with few exceptions, explain the main features of the data. The main theme is that epitaxial growth invariably takes place on a vicinal surface and that RHEED, if properly used, can characterize the underlying step structure. With this basis one can begin to understand the particular growth mechanisms of epitaxy.

KINEMATIC ANALYSIS

Ewald's Construction

Ewald's construction describes the combination of energy and momentum conservation that gives the geometry of the diffraction. In this construction a sphere of radius $k = 2\pi/\lambda$ is superposed on the reciprocal lattice of the surface. For a semi-infinite periodic, ideal surface, the reciprocal lattice is a family of parallel lines. The intersection of the Ewald sphere with this ideal reciprocal lattice consists of points arranged along circles. The diffraction pattern that results is shown in Fig. 1a.

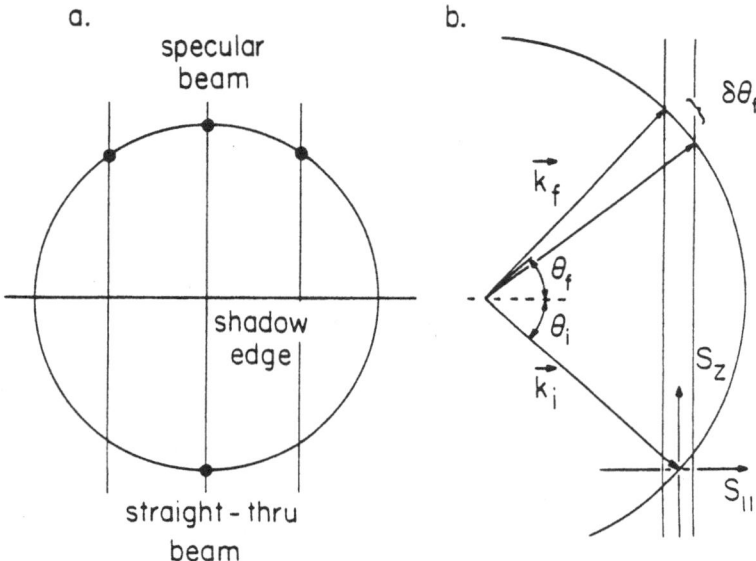

Fig. 1a. Diffraction pattern that would appear on the phosphor screen for a perfect low-index surface and a perfect diffractometer. Above the shadow edge the intersection of the Ewald sphere with the sharp reciprocal lattice of the surface is indicated. The position of the portion of the beam that misses the sample is indicated below the shadow. This pattern is very nearly obtained if an MBE grown GaAs(100) surface is annealed in an As flux.

Fig. 1b. Ewald construction that gives a streaked specular beam for the case of a reciprocal lattice rod broadened by disorder. The scattering geometry and notation of the analysis are given. This is a side view of the geometry in Fig. 1a. Note that a uniformly broadened rod is not what is obtained for the case of random surface steps. But this gives the main idea for the formation of a streak in real systems.

Three important electron trajectories are illustrated here: the specular beam, two non-specular beams, the straight-thru beam and the shadow edge. The specular beam obeys the condition that the glancing angle of incidence and reflection are equal, i.e. $\theta_i = \theta_f$. The position of this beam must be modified for the case of a vicinal surface, as will be described shortly. The straight-thru beam is that part of the incident beam which misses the sample and strikes the phosphor screen directly. It is essential for measurement of the scattering geometry. The shadow edge corresponds to the lowest take-off angle from which inelastics can leave the surface. Though not well defined the shadow edge can be used to estimate the diffraction angles. This simple construction is modified by disorder on a real surface, but to a large extent the diffraction pattern still corresponds to the intersection of a reciprocal lattice with the Ewald sphere. Our goal here is to demonstrate how well one can map out the modified reciprocal lattice and then to determine the disorder on a GaAs surface during molecular beam epitaxy.

The formation of streaks when there is disorder is shown in Fig. 1b. If the perodicity is not perfect, then parallel momentum conservation is relaxed and the reciprocal lattice lines are broadened into, for example, rods. The intersection of the sphere with the rod indicates the allowed variation in $\underline{S}_{\parallel}$, where the momentum transfer $\underline{S} = \underline{k}_f - \underline{k}_i$. Because the glancing angle of incidence, θ_i, is small, the result is that the intersection with the Ewald sphere is a family of parallel streaks. The diffraction is thus very sensitive to disorder in the same direction as the incident beam. By contrast it is much less sensitive to disorder perpendicular to the beam direction. Figure 1b shows the Ewald construction for a surface which is a mosaic of domains with random, small angle grain boundaries. This gives a reciprocal lattice of rods whose thickness in momentum space is $2\pi/L$ where L is the mean domain diameter. Note that the thickness of the rod is constant for all momentum transfers, S_z, perpendicular to the surface. The angular length of the corresponding streak would decrease smoothly as $1/(\sin \theta_i)$. This is seldom observed since, as we will show, the dominant disordering mechanism on GaAs surfaces is the presence of random monatomic steps.

Vicinal Surfaces

For most epitaxial films the substrate is prepared by sawing a wafer from a boule and then polishing the surfaces to be as smooth as possible and oriented in a desired direction. Typical commercial GaAs(100) wafers are oriented to within 6 mrad from the (100) plane and seldom less than 1 mrad. If after polishing and heat cleaning in a vacuum, these surfaces exhibited a surface that was a staircase with roughly equal terrace lengths, then kinematic theory would give split diffraction beams (Henzler, 1979). Qualitatively, there are specific angles of incidence, termed out-of-phase conditions, in which waves scattered from adjacent terraces are π out of phase. At these angles the specular beam should be split into two nearly equal components. At other angles the intensity is unevenly distributed between the two components, with one predominating at a condition where waves scattered from adjacent terraces are in phase (a Bragg angle). Thus, any real surface is expected to exhibit structure in the distribution of intensity along an integral order RHEED streak because of the underlying staircase that results from unintentional or intentional misorientation from a low-index plane. Before giving a detailed description of the expected intensity variation, it should be noted that the staircase can be remarkably disordered and still give observable splitting. The precise condition is that the probability of a terrace of length L be peaked at a value greater than zero and less than the distance resolveable by the instrument. In general, though, the larger the mean square fluctuation in the distribution of terrace lengths, the weaker the splitting will be (Pukite et al., 1985).

The Ewald construction for a vicinal surface, shown in Fig. 2, illustrates these ideas in more detail. Following Henzler (1979), the surface can be thought of as the convolution of a finite terrace with an infinite grating of the step edges, inclined at an angle θ_c to the low index plane. The reciprocal lattice is the Fourier transform of this convolution and equals the product of the Fourier transform of the finite terrace and the transform of the grating of step edges. This product reciprocal lattice is graphically shown in Fig. 2, where the dashed lines indicate the broadened rod due to a finite sized terrace and the inclined slashes correspond to the reciprocal lattice of the longer grating of step edges. Since the product of the two are taken, one does not expect that slashes extending past the dashed envelope will contribute very much intensity to the diffraction. If the surface is not a perfectly regular staircase the slashes will not be as sharp. In the figure, the Ewald sphere is superposed on the product reciprocal lattice, and one can see that there are two strong intersections with the staircase reciprocal lattice.

To determine the kinematic scattering angles from a staircase, place

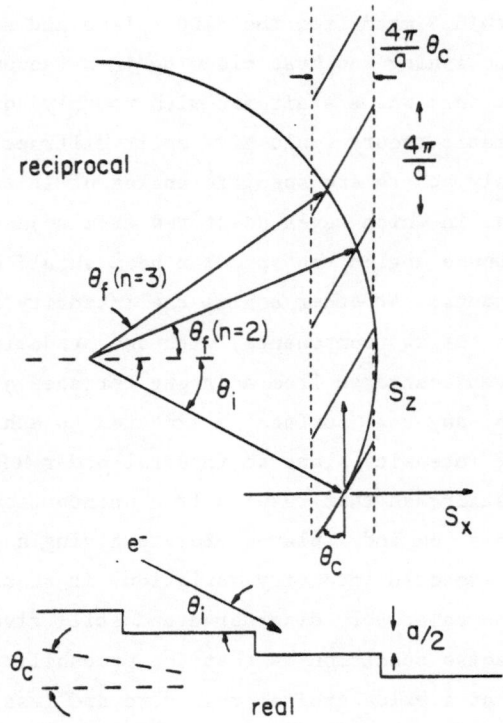

Fig. 2. Ewald construction showing the
formation of split diffraction
beams due to a regular staircase.
The dashed envelope is roughly 2π
divided by the terrace length.
The short lines are the reciprocal
lattice due to the step edges.
The diffracted beams corresponding
to the n = 2 and n = 3 order beams
(of the staircase) are indicated.

the center of the Ewald sphere at $(-k \cos \theta_i, 0, k \sin \theta_i)$ and choose an
arbitrary point on the Ewald sphere to be $(S_x, 0, S_z)$. For the incident
beam direction chosen to be down the staircase, then the reciprocal lattice
of the staircase edges is given by

$$S_x = \theta_c (S_z - 4\pi n/a) \qquad (1)$$

where a is the lattice constant and n = 0,1,2,3,.... . Similarly one can
write the momentum transfers as

$$S_x = k \cos\theta_f - k \cos\theta_i \qquad (2)$$

$$S_z = k \sin\theta_f + k \sin\theta_i . \qquad (3)$$

Combining Eqns. 1,2 and 3 and simplifying to the case of small angles, one finds that for a staircase misoriented by θ_c and for a glancing angle of incidence of θ_i directed down the staircase (all angles measured with respect to the low index plane) that there should be a peak in diffracted intensity at θ_f given by:

$$\theta_f^2 + 2\theta_c\theta_f = \theta_i^2 - 2\theta_c\theta_i + \frac{8\pi n}{ka}\theta_c \quad , \qquad (4)$$

where n = 0, 1, 2, 3,... . A similar result could be obtained for any arbitrary incident azimuth with respect to the surface staircase, but this is given for simplicity of exposition. In the discussion section we will compare the measured values of peak positions to this calculation. One should note that this result relies on the diffracted peaks being well separated. If peaks at different values of θ_f are broad and overlap then it will be difficult to compare to Eq. 4. In that case the data must be fit to models of the staircase disorder as described by Pukite et al. (1985).

There are several variations of Eq. 4 that are useful. First, to separate this kinematic effect from bulk processes, one should examine the change in θ_f when θ_i is changed. If inelastic processes are involved, and in particular bulk Kikuchi-like processes are involved, then the angle of the final diffracted beam should be independent of incident angle. After an inelastic process (a related argument can be offered for an elastic surface resonance), the phase relation of the scattered electron with respect to the incident beam is lost. If a range of scattered electrons is then considered, the strongest peaks will appear at bulk Bragg scattering conditions. These features will be fixed relative to the crystal geometry and therefore the shadow edge. Instead, if a peak in diffracted intensity along a streak is described by the kinematic result of Eq. 4, then by changing the incident angle from $\theta_{i,1}$ to $\theta_{i,2}$, the position of the final peak intensity should change from $\theta_{f,1}$ to $\theta_{f,2}$. Letting the variations be given by $\Delta\theta_i$ and $\Delta\theta_f$ and mean values $\langle\theta_i\rangle$ and $\langle\theta_f\rangle$, respectively, the kinematic expectation is

$$\Delta\theta_f = \frac{\langle\theta_i\rangle - \theta_c}{\langle\theta_f\rangle + \theta_c}\Delta\theta_i \quad . \qquad (5)$$

From this it is apparent that for a crystal misoriented by an amount that is close to 2°, that there will be maxima that do not appear to change when the incident angle is changed, mimicking bulk-like processes. In general

though, the final peak angles are strongly dependent on the incident
angles. Both cases will be exhibited later.

A clear way to show how quantitatively eq. 4 agrees with measurements
is to use eq. 4 to find the difference in the glancing angle, θ, between
the split specular beam that should be observed for a surface staircase.
In the language of the Ewald construction of Fig. 2, this corresponds to
the separation in angle, $\delta\theta_f$, determined by the intersection of the Ewald
sphere with two adjacent order slashes of the reciprocal lattice of the
staircase. This is illustrated in Fig. 2. Then in eq. 4, keeping θ_i fixed
and finding the final angles corresponding to n and n+1, one obtains that

$$\delta\theta_f = \frac{4\pi}{ka} \frac{\theta_c \cos\phi}{\langle\theta_f\rangle + \theta_c \cos\phi} \quad , \qquad (6)$$

where $\theta_c \cos\phi$ is the projected misorientation when the incident beam is
directed at an arbitrary azimuthal angle and zero chosen to be down the
staricase. Alternatively this can be inverted to give:

$$\frac{1}{\delta\theta_f} = \frac{ka}{4\pi} + \frac{ka}{4\pi} \frac{\langle\theta_f\rangle}{\theta_c \cos\phi} \qquad (7)$$

which is a convenient form for comparison to the data. To show that the
structure that is observed along an integral order streak can be explained
in terms of kinematic theory, we require quantitative agreement with eqns.
4 - 7.

There are two important considerations that are implicit in this
formulation of the problem. First, the angles to be used in these
calculations are external angles, not the refracted internal angles.
Second, multiple scattering within a terrace is included, though multiple
scattering between terraces is neglected. To see these points let $\underline{r}_i$ be a
vector from the origin to the leftmost edge of the ith terrace. Then
neglecting multiple scattering <u>between</u> terraces, the diffracted amplitude
is given by

$$A(\underline{S}) = \sum_{\underline{r}_i} e^{i \underline{S} \cdot \underline{r}_i} A_i(\underline{S}) \qquad (8)$$

where $A_i(\underline{S})$ is the dynamically scattered amplitude from the ith terrace.
The important point is that $A_i(\underline{S})$ is still allowed to contain all multiple

scattering processes within the ith terrace and the underlying atoms. The
fundamental point is that the inelastic mean free path for RHEED is of the
order of the mean terrace size so that this assumption is valid. To see
that only external angles should be used in computing the positions of the
kinematic interference maxima, note that the extra path lengths contained
in the phase factor i$\underline{Sr}_i$ are external to the crystal. In other words the
phase change at each terrace scattering is identical and gives no net
contribution. (The refracted angles need to be used when using a measured
value of the diffracted intensity to account for changes in the momentum
transfer across a diffraction angular profile.) Both of these assumptions
have been used successfully before by Henzler (1979) and Lagally (1984).

Singular Surfaces

We define surfaces for which the average terrace length of the
macroscopic misorientation is less than - 0.5 mrad to be singular. This
corresponds to the instrumental resolution of our diffractometer. On these
surfaces the picture of the diffraction is qualitatively different than the
diffraction from a vicinal surface. In the present case the surface can be
thought of as composed of a random arrangement of up and down steps. There
is not usually the long range periodicity of the staircase. Fig. 3 shows a

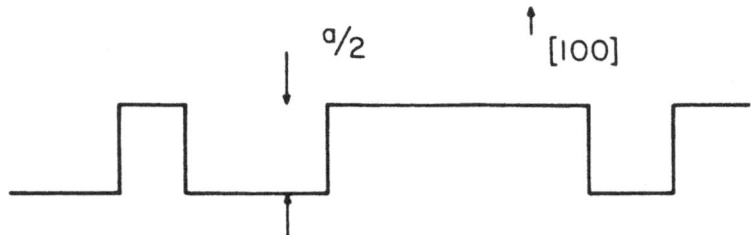

Fig. 3. Schematic of a random arrangement of up and down
steps distributed among two levels of a (100)
surface. The step height is from an As layer to an
As layer. Larger numbers of levels can be treated,
but without affecting the qualitative description.

schematic of a singular (100) surface in which the step height between two
arsenic layers is a/2 and in which only two surface levels are involved.
This would be the case, for example, if GaAs were deposited as two-
dimensional clusters on top of an otherwise featureless plane. The basic
diffraction mechanism is still the same. But now at out-of-phase
conditions, because the step distribution is less periodic than a
staircase, the reciprocal lattice is just broadened rather than split. At

a Bragg angle, where scattering from different surface levels is in phase, the diffracted beam remains sharp.

To calculate the diffraction from a singular surface, one needs to work out the correlation functions that describe the distribution of surface steps. For certain random distributions this has been worked out by Lent and Cohen (1984) and by Pukite et al. (1985) and by Lu and Lagally (1982) and by Pimbley and Lu (1985). The pair correlation function, $C(\underline{u})$, is the probability that there is a scatterer at the origin and also a scatterer $\underline{u}$ away. The diffracted intensity in the kinematic approximation is the Fourier transform of $C(\underline{u})$. The correlation function of a surface consisting of two levels is shown in Fig. 4. At $\underline{u} = 0$ the correlation

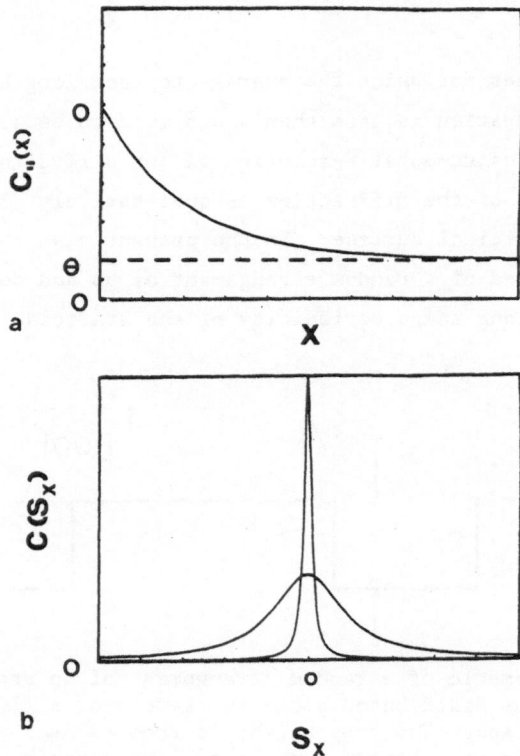

Fig. 4. The top panel shows the probability of finding a scatterer at the origin and one x away. The bottom panel is the Fourier transform of this function, showing a separation into a central spike (a delta function broadened by the finite instrument response of the diffractometer) and a broad part due to the disorder.

function is θ, the fraction of surface sites occupied or coverage. This decreases to the asymptotic value, θ^2, since the probability that two well separated sites are occupied is uncorrelated. This correlation function can be written as a decreasing part due to the disorder and a constant that corresponds to the coverage of the periodic lattice (Lent and Cohen, 1984). The diffracted intensity, the Fourier transform, is then a delta function plus a broad part. In practice the delta function is broadened slightly by the instrumental response. Including the coverage and angle of incidence dependencies, Lent and Cohen (1984) obtained:

$$I(\underline{S}_\parallel) = G_o(\underline{S}_\parallel)\{\theta^2 + (1-\theta)^2 + 2\theta(1-\theta)\cos(S_z d)\} +$$
$$+ G_1(\underline{S}_\parallel) \ 2\theta(1-\theta) \ [1-\cos(S_z d)]. \tag{9}$$

where d is the step height and G_0 and G_1 are functions that depend upon the explicit form of the disorder. This formulation is quite general and makes no assumption about the disordering processes.

Equation 9 depends importantly on scattering geometry and surface coverage. This means that during crystal growth as the fraction of surface scatterers changes, one should see changes in the distribution of intensity in the shape of a diffracted beam. Further, by changing the scattering geometry one should be able to check to determine whether changes in the dynamic diffraction processes are affecting a measurement.

The diffraction from a two-level surface, according to eq. 9, is illustrated in Fig. 5. Here the intensity vs S_x is plotted at several S_z values. Equations 2 and 3 can be used to relate momentum transfers to scattering angles. Note that $S_z = 2k\theta_i$ exactly at the position of the specular beam. These curves are calculated for the case of half coverage. At this value of θ, the central spike vanishes at the out-of-phase condition. Had we plotted one of these angular profiles as a function of θ at fixed scattering angles, one would observe that the broad component and the central spike would change in relative proportion and that the broad part would change in width. This ideal behavior of Fig. 5 is very difficult to observe in practice, due largely to the difficulty in obtaining a two level system. It will be compared to a close approximation in the next section.

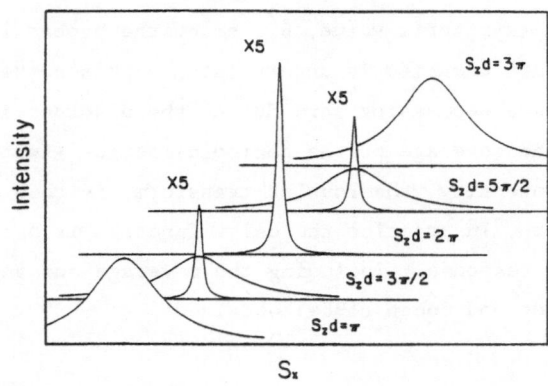

Fig. 5. Scans of the intensity of the specular beam vs S_x.
Here $S_z = 2k\theta_i$ and $S_x = k(\cos\theta_f - \cos\theta_i)$. After Lent and
Cohen, 1984.

EXPERIMENTAL

Epitaxial films were grown on GaAs(100) wafers in a Physical
Electronics MBE 400 system. The samples were etched in a 5:1:1 solution of
H_2SO_4, H_2O_2, and H_2O for 5 min. A 2mm × 2mm section was cleaved from the
wafer and soldered to a molybdenum block with In. The small sample size
reduces the flux variation over the surface, reduces the range of angles
subtended by the detector, and reduces the effect of thermal stress.
Before diffraction measurements were made, a buffer of at least 0.15 μm
GaAs was grown using a growth rate of 0.3 μm/h with an As:Ga flux ratio of
about 3:1.

The diffractometer consisted of an electron gun that provided a 10 keV
beam with an angular divergence of 0.5 mrad. The diffracted beams were
focused at the phosphor screen to minimize the contributions to the
instrument response of the parallel diameter of the beam and of the range
of angles subtended by the detector (Park et al., 1971 and Van Hove et al.,
1983). This is the simplest way of achieving a high instrument response.
A slightly better way would be to maximize the resolution of the split
diffracted beam from a small staircase. The glancing incident (polar)
angle could be varied by a combination of sample motion and beam
deflection. The incident azimuth could be varied by rotating the sample
about the normal. The light from the phosphor screen was imaged onto a slit

aligned perpendicular to the long direction of RHEED streaks so that an effective one-dimensional correlation function was measured (Marienhoff and Henzler, 1984). The light was detected by a photomultiplier. The scattering angles were determined by measuring the azimuthal position of the sample relative to symmetry directions and from the separation between the specular beam, straight-through beam, and shadow edge. The sample to screen distance was 32 cm. Finally, all measurements were made with azimuthal angles 11° from symmetry azimuths.

The angular distribution of intensity in the diffraction pattern can be measured by two different methods. First, the pattern can be left fixed and the detector moved manually across the phosphor screen. For profiles in which the intensity distribution along the entire length of the streak is desired, this method is used, as in this paper. The second method is to leave the detector fixed and to move the diffraction pattern with a pair of electromagnetic coils. This method has the advantages of being less sensitive to the granularity and nonuniformity of the phosphor screen. Using the second method our instrument can resolve splitting of the diffracted beam from a surface misoriented from the (100) by as little as 0.5 mrad. Assuming at least two terraces are involved in the diffraction, this means that the instrument is sensitive to coherent diffraction over distances as large as 1 μm. One should recognize that an important component of the instrument response to this large distance is the flatness of the sample. Bowing of the lattice planes or random up and down steps among many levels can mask the intrinsic response of the diffractometer, even if the sample is well oriented.

RESULTS AND DISCUSSION

Vicinal Surface

The character of steps on the substrate surface controls molecular beam epitaxy. On AlGaAs the misorientation of (100) substrates structure determines the surface morphology (Tsui et al., 1985 and Pukite et al., 1986). The electrical properties quantum well structures are asymmetric, also depending upon the substrate steps (Radulescu et al., 1986). We suspect that the incorporation of dopants should also be affected. The growth of high quality GaAs on Ge or Si relies on substrate steps to eliminate antiphase disorder (Pukite and Cohen, 1987). Control of substrate steps reduces the propagation of dislocations in strained layer epitaxy (Otsuka et al., 1986). In short, to understand the nucleation and growth of compound semiconductors by MBE one needs to first characterize the static GaAs surface and then follow the morphology and order during the growth itself. In this section we will present the diffraction patterns that are measured from vicinal GaAs surfaces. We will show that explanations relying on bulk diffraction mechanisms are inconsistent with the data. We will show that kinematic theory can explain the main features of the results.

To see the role of bulk-like processes, one needs to determine whether the structure along the specular streak occurs at fixed final scattering angle θ_f while changing the incident angle, θ_i. If this were the case then the cause would be a Kikuchi-like process. According to Eqs. 4 and 5, data from both 6 mrad and 2° misorientations must to be examined. The results of our measurements are shown in Figs. 6 and 7 for GaAs(100)

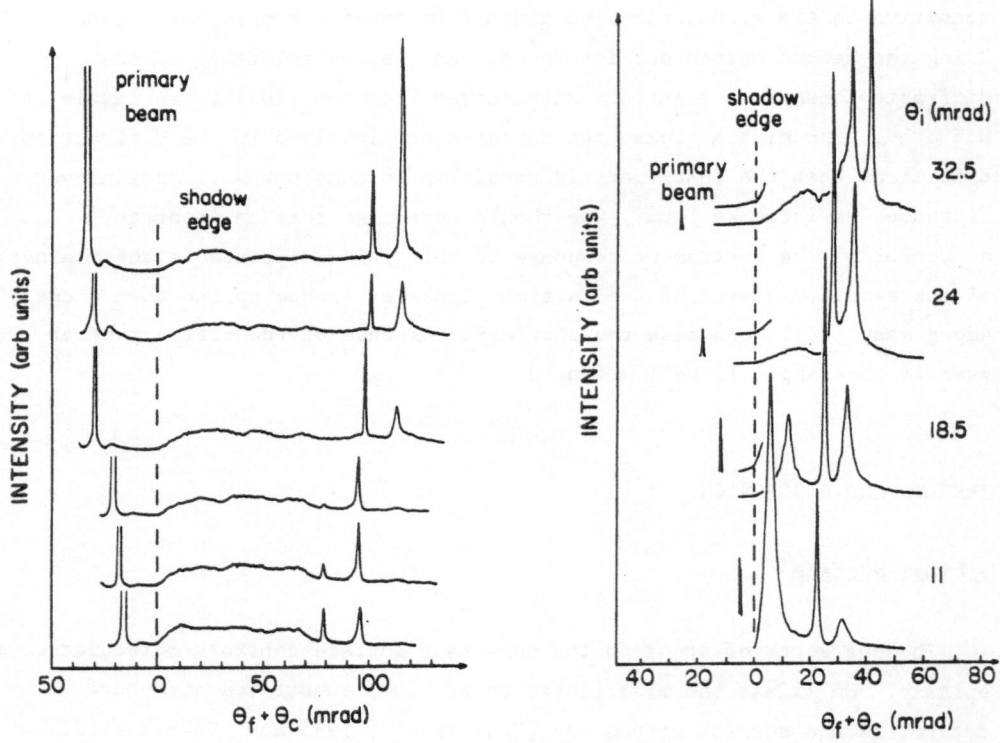

Fig. 6. Scans of the intensity along the (00) streak for different angles of incidence. For these measurements the detector is moved mechanically along the length of the streak. Note that the shadow edge occurs at $\theta_f = -\theta_c$. For this 2° misorientation the intensity appears to be fixed relative to the crystal.

Fig. 7. Scans similar to Fig. 6, but for a 6 mrad misorientation. For this well oriented surface, the apparent trading of intensity between peaks fixed relative to the bulk crystal seen above is not present.

surfaces in a (1×1) reconstruction misoriented toward the <100> direction by 6 mrad and misoriented toward the (111)B by 2°, respectively. Both figures show profiles of the intensity along the length of the specular streak for several incident angles. The data is plotted so that the shadow edges, indicated by the short line at the point of inflection, are all aligned. The shadow edges each correspond to $\theta_f = -\theta_c$. First, one should note that in both cases the surface is sufficiently well ordered and the diffractometer has sufficient resolution that the peaks in the data are well separated. This makes it simpler to compare to kinematic calculation. Second, one can see that the peaks in Fig. 6, the 2° misorientation, appear fixed relative to the shadow edge. As the angle of incidence is changed, it appears that the distribution of intensity just moves from one peak to another. By contrast, the peak positions in the data of Fig. 7 clearly move as the incident angle is changed. In these latter curves it is also observed that the distribution of diffracted intensity seems to move from peak to peak. Since the peak positions are not fixed with respect to the shadow edge, the data of Fig. 7 is inconsistent with bulk-like processses. Moreover, both sets of data are consistent with Eqs. 4 and 5.

For a better comparison to Eq. 4, in Fig. 8 we plot θ_f vs. θ_i from the data of Fig. 7, the 6 mrad misorientation. For these data the scattering angles, θ_i and θ_f, were measured by determining the distance on the screen from the position of the shadow edge. Most of the scatter in the points arises because the inflection point is not so well defined for the small samples used in this measurement. A better way to measure θ_i and θ_f is to assume that Eq. 4 is valid (which of course could not be done here) and then determine the angles from the position of the peaks and the straight-thru beam (Pukite et al., 1984a). For Fig. 8, the solid curves are a calculation from Eq. 4 and the open circles are the data. Each value of n corresponds to a different intersection of Ewald sphere with one of the slashes due to the reciprocal lattice of the grating of step edges as described in Fig. 2. The isolated points far from the n = 0 and n = 1 curves deviate from the kinematic calculation because these peaks are broad and weak. Overall, the agreement with kinematic calculation is excellent. Other comparisons with this data can be made. For example Pukite et al. (1984b) showed that by measuring the angular separation between the split peaks as a function of θ_i, one could determine the local surface misorientation to within 0.1 mrad. The kinematic theory is quantitatively correct.

Not only does the kinematic theory give the correct θ_i and θ_f

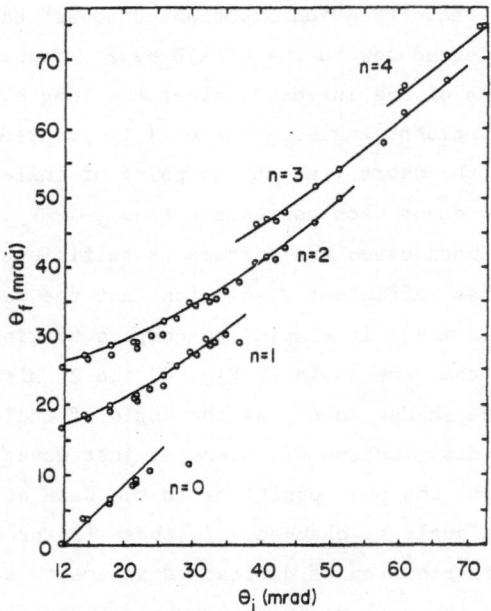

Fig. 8. Plot, corresponding to Eq. 4, of the angular
position of the structure in the (00) streak vs
the incident angle. All angles are measured
with respect to the low index plane and not the
macroscopic surface. For this 6 mrad surface,
the position of the intensity peaks changes
with the incident angle, indicating that these
are not related to bulk Kikuchi processes. The
indices n correspond to the order of the
diffraction process from the terraces of the
staircase. The solid lines are calculated from
Eq. 4 assuming only that $\theta_c = 6$ mrad.

dependence, but it correctly predicts $\delta\theta_f$ for the vicinal surface when the
azimuthal direction of the incident electron beam is varied. To see this
the data from samples misoriented by 5 mrad, 2°, and 6° is plotted vs the
azimuthal incident angle ϕ. The reference is chosen so that when the
incident electron beam is pointing down the staircase, i.e. $\phi = 0$. This
comparison of Eq. 6 with experiment is shown in Fig. 9. In this figure, we
have plotted the difference between the final angles, $\delta\theta_f = \theta_{f,1} - \theta_{f,2}$,
where $\theta_{f,1}$ and $\theta_{f,2}$ are the intersections of the Ewald sphere with the

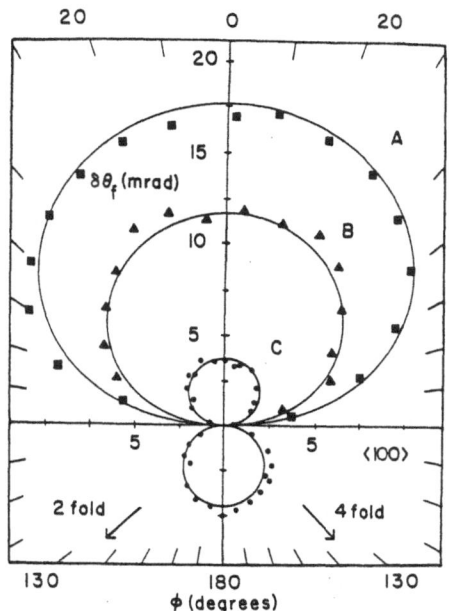

Fig. 9. Separation in the glancing takeoff
angle of the diffracted beams vs
azimuthal angle of incidence for three
misorientations. The closed symbols
are measurements and the curves
calculation of the kinematic theory.
Samples A, B, and C are misoriented
from the (100) by 2.1°, 1.1°, and
5 mrad respectively. The incident
angle is set at 51 mrad (after Pukite
et al. 1984).

inclined slashes of the reciprocal lattice (of the step edges) that are
separated by n = 1. Note that when the incident azimuth ϕ is not zero,
these two Ewald intersections are also separated azimuthally, but that
difference is not plotted. In the figure the closed symbols are data and
the solid curves are calculations from Eq. 6, with both measurements and
calculation made at the out-of-phase angle of 51 mrad. Once again, even
though the dynamic (multiple scattering) conditions are changing rapidly as
the incident azimuthal angle is varied, the agreement with kinematic
calculation is striking.

<u>Singular</u> <u>Surface</u>

From an analytical point of view we distinguish between singular and
vicinal surfaces (Lent and Cohen, 1984 and Pukite et al., 1985). A vicinal
surface is constructed by forming a staircase with an infinite number of
low-index terraces. Though very general types of disorder can be included,
it is difficult to analyze situations in which different surface levels are
not treated alike. By contrast, a singular orientation is one in which a
perfect surface coincides with a low-index plane. When this surface is
rough, diffraction from a small number of surface levels will broaden the
beams in a fashion that is qualitatively different than the split beams
observed from a staircase. For this type of surface it is much easier to
calculate the diffraction when the levels should be considered different.
For example, in an adsorption experiment, one needs to be able to treat the
case of an adsorbate island distribution that is different than the island
of the underlying substrate.

In the limit of small misorientations, these two types of surfaces are
very similar. When the range of angles in the incident beam is larger than
the peak separation due to the surface misorientation, a vicinal surface
cannot be distinguished from a perfect low-index plane. For our
diffractometer, surfaces misoriented by less than about 0.5 mrad cannot be
resolved. In fact with current GaAs technology, this is also the limit to
which a surface can be polished flat over a macroscopic area. We define
these surfaces to be effectively singular, and in this section their
diffraction is described. We will show that Eq. 9 describes diffraction
from these surfaces both during growth and when growth is interrupted.

The clearest comparison with the model can be made when the surface
consists of scatterers distributed among two levels. Lent and Cohen (1986)
prepared a close approximation to a two-level surface by despositing
submonolayer quantities of GaAs onto a GaAs(100) surface at moderately low
temperatures. These temperatures were sufficiently high that the adsorbed
molecules could dissociate and order locally, yet too low for migrations
over the distances required for the adatoms to attach to the widely
separated steps of the underlying substrate. The intensity along the
length of the (00) streak was then measured vs θ_f for a variety of incident
angles θ_i. Reiterating our concerns over the influence of multiple
scattering: these data were taken at an azimuthal angle of incidence away
from a symmetry azimuth, and we require that data at many incident angles,

over which the multiple scattering conditions change, give the same
results. The measurements are shown in Fig. 10 and Fig. 11.

In Fig. 10 GaAs was deposited on GaAs(100) to give $\theta = 0.4$ monolayers

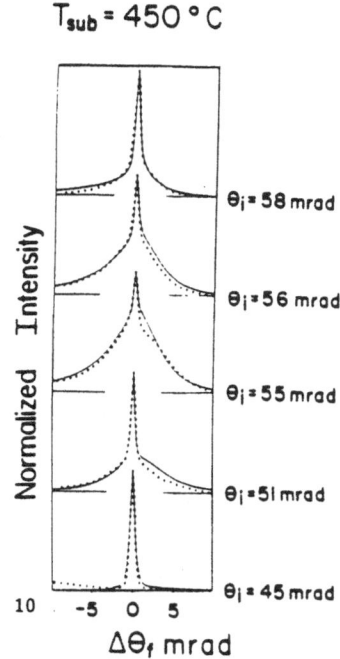

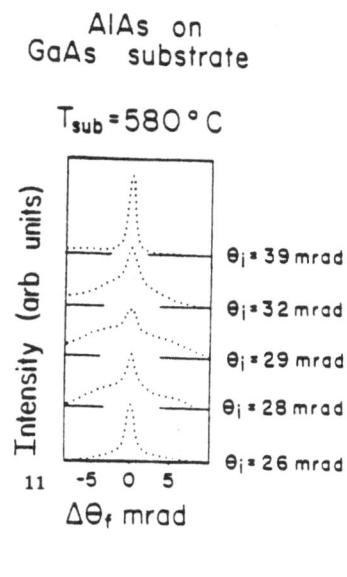

Fig. 10. Scans along the length of the (00) streak for a submonolayer
          amount of GaAs deposited on nearly singular GaAs(100). The dots
          are a portion of the data; the solid curve is a fit. The results
          should be compared to Figs. 4 and 5.
Fig. 11. Scans along the length of the (00) streak for a submonolayer of
          AlAs deposited on a nearly singular GaAs(100) substrate.

in the top level of an approximately two-level surface. In Fig. 11, a
submonolayer amount of AlAs was deposited on GaAs. Both sets of data show
that near a Bragg angle the peak is sharp and away from the Bragg, towards
the out-of-phase, the peaks are broad, giving rise to longer streaks. Fig.
11 perhaps shows the distinction between the two components of the
diffraction most clearly. There is a central spike that is the delta
function due to the long range order convolved with the instrument
response. And there is a broad part due to the step disorder. To

demonstrate that the angular dependence of Eq. 9 accurately describes the
data, the curves of Fig. 10 were fit to Voight profiles (Puerta, 1981).
Using fixed functional forms for the central spike and for the broad part,
it was determined whether the angular dependence of Eq. 9 could be
satisfied. The results are the solid curves in Fig. 10. One can see
excellent agreement was obtained over a range of angles. The results
indicate that in this case the average terrace length was 630 Å. More
details are given by Lent and Cohen (1986).

An example from Fuchs et al. (1985) for a completely different set of
diffraction conditions and on a different low-index GaAs surface, is shown
in Fig. 12. For this data, a cut and polished GaAs(110) was etched and

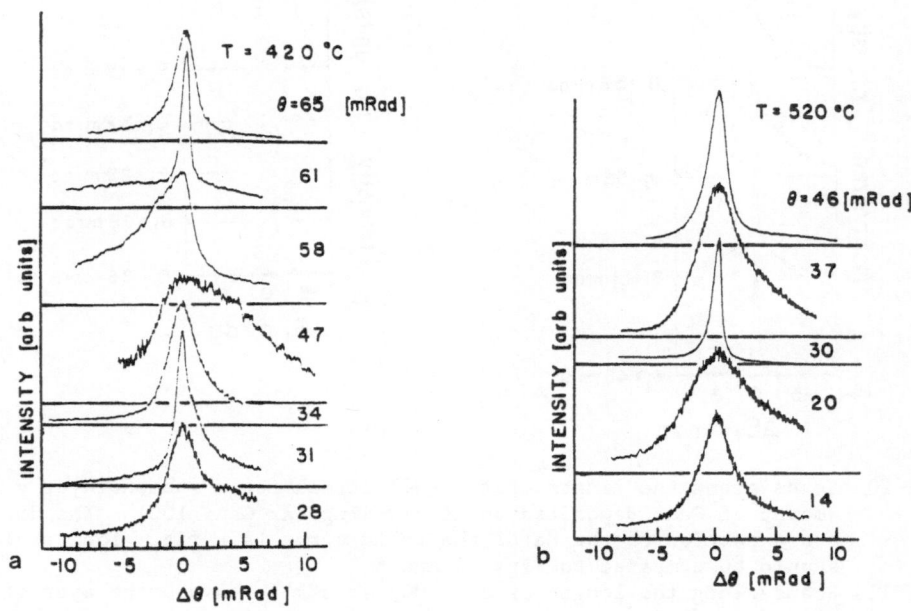

Fig. 12. a.)The left panel shows scans along the specular beam for a
         GaAs(110) surface on which GaAs had been deposited at low
         temperature. The position of the out-of-phase corresponds to
         single layer steps. Upon annealing the curve near 47 mrad
         sharpens, but as seen in the right panel (b), new out-of-phase
         conditions emerge, indicating double layer steps.

heat cleaned in an As flux following standard (100) procedures. The
results, shown in Fig. 12, indicate that after an initial deposition of 20
monolayers of GaAs onto a surface at 420 °C, the surface consists of single
layer GaAs steps. But following an anneal to 500 °C, the surface is
transformed into one with double layer GaAs steps. The temperature at
which the data are measured is not an issue. The crucial result is that by
heat treatment alone the surface step distribution is changed from single

to double layer and that the diffraction follows the expected form. Similar conclusions have been reached on vicinal GaAs(110 surfaces by Ranke (1983) from adsorption studies and by Wowchak (1985) from diffraction measurements of beam splitting. Fig. 12a shows scans along specular (00) beam after the initial adsorption at 420 °C. This surface plus adsorbed layers does not approximate the ideal, two-level surface as well as the data of Figs. 10 and 11, and so does not exhibit the clear separation between central spike and broad component. Yet one can see that near the out-of-phase condition (46 mrad) the peak is broad and that away from this angle becomes sharper, as expected. The data of Fig. 12b shows the same scans after the surface is annealed. Now the peak is sharp at the double layer Bragg angle of 46 mrad and becomes broad only at the double-layer out-of-phase conditions. Because of the double step height the additional Bragg conditions are introduced as reflected in the data. It is difficult to imagine that some other mechanism could also explain these dependencies of the diffraction on scattering geometry so simply.

Intensity Oscillations

When growth is initiated onto a static GaAs(100) surface, cyclic variations in the specular intensity are observed with period equal to the time to deposit a monolayer of material (Wood, 1981; Harris and Joyce, 1981; Van Hove et al., 1983; Sakamoto et al., 1984; Lewis et al., 1986). These oscillations reflect an alternation between smooth and rough surface morphologies during the layer-by-layer growth of GaAs. A variety of measurements indicate that quantitative interpretation following Eq. 9 is valid. For example, Van Hove and Cohen (1987) showed that the migration length of Ga at approximately half coverage could be determined by measuring the width of the diffracted beams. Their result is in good agreement with the oscillation damping method of Neave et al., (1985). One could also compare the amplitude of intensity oscillations vs θ_i with what one would calculate based on Eq. 9. There are several difficulties with this procedure. First, since all surfaces are misoriented it is not possible to reach a condition of exactly half coverge. Second, during growth it is not likely that only two levels on the surface will be occupied. Third, multiple scattering should affect the initial change in diffracted intensity as other beams will be strong while the surface is relatively smooth, even if non-symmetric incident azimuths are used. Finally, one does not know that the island distribution will give the same functional form during growth, though that appears to be the case.

Nonetheless one can check the fundamental soundness of the method by
determining the amplitude of intensity oscillations as a function of θ_i
over a range of scattering angles where resonance effects and other dynamic
scattering processes have minimal impact. One wants to check if the main
dependencies given by Eq. 9 hold.

The comparison is shown in Figs. 13 and 14. In Fig. 13 measurements of

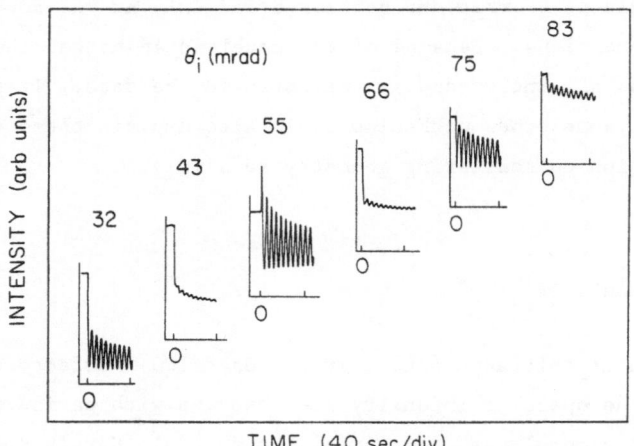

Fig. 13. Intensity of the specular beam vs time after starting GaAs growth.
At t = 0 the Ga shutter was opened. The angle of incidence is
indicated for each plot. Note the striking variation of the
amplitude with angle.

the specular intensity are plotted as a function of time for several
glancing angles of incidence θ_i. The absolute zero of each curve and θ_i
for each curve are also indicated. Though the initial transient and the
phase of the oscillations is not yet understood, we can compare the
amplitude of the oscillations to the calculation of Eq. 9. This is shown
in Fig. 14, where
we plot the ratio of the first minimum to the first maximum. When the
oscillations are weak, this ratio will approach unity; when the
oscillations are strong, this ratio is significantly less than unity. For
comparison to Eq. 9, we calculate the ratio of $I(\underline{S})$ evaluated at $\theta = 1/2$ to
$I(\underline{S})$ evaluated at $\theta = 1$, where for each we set $S_\parallel = 0$. The result is a

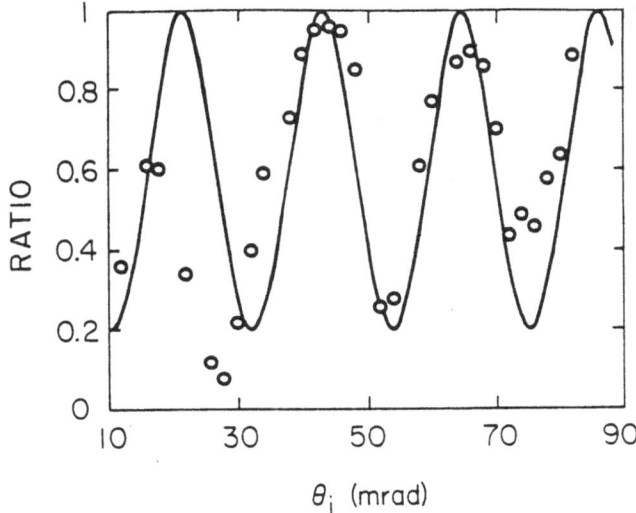

Fig. 14. From the data shown in Fig. 13, the ratio of the first minimum to first maximum are plotted vs θ_i. When the intensity oscillations are weak, this ratio is near unity. The curve is a calculation using Eq. 10, a one parameter fit.

ratio given by:

$$\frac{I(\theta = 1/2)}{I(\theta = 1)} = \frac{1}{2}(1 + \cos S_z d) + \frac{G_1}{G_o}\frac{1}{2}(1 - \cos S_z d) \qquad (10)$$

This is plotted in Fig. 14 with the one adjustable parameter G_1/G_o chosen to fit the data. The agreement of the main θ_i dependence of the calculated ratio with the data is quite striking. Note that there is a shift at low angles. For similar data taken from intensity oscillations measured during the growth of Ge on cleaved GaAs(110) there is no shift. We think that there is now no longer any question that the main features of the intensity oscillations are due to single-scattering interference that arises from the extra path lengths between scatterers at different surface levels.

There are still other complications that need to be understood. For example, there is typically a rapid decrease in the specular beam at the onset of the growth, though this depends on angle as seen in Fig. 13. In LEED Horn and Henzler (1986) can explain a similar drop by the inclusion of a third level in the interference calculation. Second the phase of the intensity oscillations of Fig. 13 depends upon θ_i. The problem is even worse when one considers that the growth process invariably takes place on a vicinal surface. As described earlier, a vicinal surface gives split diffraction beams at all azimuthal angles, though not always resolved. If one measures the intensity oscillations of these two beams, one obtains different phases. Both the amplitude and phase depend on scattering angle. Under conditions when the two components are not resolved the different interference conditions can contribute. For example if in the growth of GaAs there are, as suggested by Ghaisas and Madhukar (1985), regions of uncovered Ga, then the path lengths and scattering factors giving rise to the split beams will be different. But this by itself does not explain the data. At this point all we can say about intensity oscillations from vicinal surfaces (i.e. all GaAs(100) surfaces) is that the n = 1 order beam oscillates more strongly than the n = 2 order beam over the range $25 < \theta_i < 35$ mrad and independent of azimuth. At other incident angles, the split diffraction beams oscillate with nearly the same amplitude.

One needs to consider nonspecular beams when data is taken with the incident beam aligned along a symmetry condition. First one expects that multiple scattering be more important since the scattered intensity is shared by more strong propagating beams. Second the Ewald sphere intersects the nonspecular diffraction rods at different kinematic conditions. For example, the specular beam might be at a Bragg angle while a non specular beam would not, and at worse be near an out-of-phase angle. Since electrons are conserved, even kinematic analysis would predict the specular beam to oscillate in time. For similar reasons, the intensity variation of different beams should not be in phase -- the kinematic interference conditions would place the intersection of the Ewald sphere and the nonspecular rods at different diffraction conditions. At this point, we see no advantage to measuring the nonspecular beam variation.

CONCLUSION

Electron diffraction in a reflection mode was thought to be best suited to studies of epitaxy primarily because of convenience. But now it is also apparent that because of its intrinsic sensitivity to surface morphology and long range order, RHEED can give quantitative information on the growth kinetics. The complication is that the diffraction patterns contain kinematic, dynamic, and inelastic features. We have chosen scattering geometries to minimize dynamic processes such as the enhancements that are commonly observed (Larsen, 1987). Then by studying the systematic variation in diffraction as a function of scattering angles, we have separated out the kinematic contribution to the diffraction. We show that the diffraction is not fixed relative to the bulk crystal as would be the case for Kikuchi-like processes. Further, a major factor in the diffraction from GaAs(100) is that all surfaces are to some extent vicinal and impact the growth and the qualitative appearance of the diffraction. The major feature giving rise to the oscillations in the specular intensity is the path length interference that results from the presence of surface steps.

ACKNOWLEDGEMENTS

This work was supported by the National Science Foundation grant number DMR-8319821 and by the MN Center for Microelectronics and Information Sciences. We are grateful to A.M. Wowchak, N. Zhou, and G.J. Whaley for their assistance.

REFERENCES

Fuchs, J., Van Hove, J.M., Pukite, P.R., Whaley, G.J., and Cohen, P.I., 1985, in "Layered Structures, Epitaxy, and Interfaces," J.M. Gibson and L.R.Dawson, eds., 37:437, Materials Research Society Proceedings.

Ghaisas, S.V. and Madhukar, A., 1985, J. Vacuum Sci. and Technol., B3:540.

Harris, J.J., and Joyce, B.A., 1981, Surf. Sci., 103:L90;

Henzler, M., 1979, in "Electron Spectroscopy for Surface Analysis," ed. H. Ibach (Springer, Berlin).

Henzler, M., and Marienhoff, P., 1984, J. Vac. Sci. Technol., B2:346.

Horn, M. and Henzler, M., 1986, private communication.

Lagally, M.G., 1984, in "Methods of Experimental Physics: Surfaces", eds. R.L. Park and M.G. Lagally (Academic Press, N.Y.)

Larsen, P.K., Meyer-Ehmsen, G., Bolger, B., and Hoeven, A.-J., 1987, J. Vacuum Science and Technol., in press.

Lent, C.S. and Cohen, P.I., 1984, Surf.Sci., 139:121.

Lent, C.S. and Cohen, P.I., 1986, Phys. Rev., B33:8329.

Lewis, B.F., Lee, T.C., Grunthaner, F.J., Madhuker, A., Fernandez, R., and Maserjian, J., 1986, J. Vacuum Science Technol., B4:560.

Lu, T.-M. and Lagally, M.G., 1982, Surf. Sci., 120:47.

Neave, J.H., Dobson, P.J., Joyce, B.A., and Zhang, J., 1985, Appl. Phys. Lett., 47:100.

Otsuka, N., Choi, C., Nakamura, Y., Nagakura, S., Fischer, R., Peng, C.K., and Morkoc, H., 1986, in "Heteroepitaxy on Silicon," eds., J.C.C. Fan and J.M. Poate, (Materials Research Society, Pittsburgh) 67:85.

Park, R.L., Houston, J.E., and Schreiner, D.G., 1971, <u>Rev. Sci. Instrum.</u>, 42:60.

Pimbley, J.M. and Lu, T.-M., 1985, <u>J. Appl. Phys.</u>, 58:2184.

Puerta, J., and Martin, P., 1981, <u>Appl. Optics</u>, 20:3923.

Pukite, P.R., Van Hove, J.M., and Cohen, P.I., 1984a, <u>J. Vac. Sci. Technol.</u>, B2:243.

Pukite, P.R., Van Hove, J.M., and Cohen, P.I., 1984b, <u>Appl. Phys. Lett.</u>, 15:456.

Pukite, P.R., Lent, C.S., and Cohen, P.I., 1985, <u>Surf. Sci.</u>, 161:39.

Pukite, P.R. and Cohen, P.I., 1987, <u>J. Cryst. Growth</u>, in press.

Radulescu, D.C., Wicks, G.W. Wicks, Schaff, W.J., Calawa, A.R., and Eastman, L.F., 1986, presented at the Fourth Int. Conf. on Molecular Beam Epitaxy (York).

Ranke, W., 1983, <u>Physcia</u> <u>Scripta</u>, T4:100.

Sakamoto, T., Funabashi, H., Ohta, K., Nakagawa, T., Kawai, N., and Kojima, T., 1984, <u>Jpn. J. Appl. Phys.</u>, L23:657.

Saluja, D., Pukite, P.R., Batra, S., and Cohen, P.I., 1987, <u>J. Vac. Sci. Technol.</u>, submitted.

Tsui, R.K., Curless, J.A., Kramer, G.D., Peffley, M.S., and Rode, D.L., 1985, <u>J. Appl. Phys.</u>, 58:2570.

Van Hove, J.M., Pukite, P.R., Cohen, P.I., and Lent, C.S., 1983, <u>J. Vac. Sci. Technol.</u>, A1:609.

Van Hove, J.M. and Cohen, P.I., 1987, J. Cryst. Growth, in press.

Wood, C.E.C., 1981, <u>Surf. Sci.</u>, 108:L441

Wowchak, A.M. and Cohen, P.I., presented at the National Symposium of the American vacuum society, Houston, 1985.

SOME ASPECTS OF RHEED THEORY

P.A. Maksym

Department of Physics
University of Leicester
Leicester LE1 7RH, U.K.

INTRODUCTION

Reflection high energy electron diffraction (RHEED) is becoming increasingly important in the study of surface growth and surface morphology. For example, RHEED is the principal tool used for in situ monitoring of molecular beam epitaxial (MBE) growth. Indeed, the discovery that RHEED intensity oscillations can be observed during MBE growth [1] has led to a dramatic upsurge of interest in the technique. Another important advance is the development of RHEED imaging methods [2,3] which offer a direct probe of surface morphology. These advances in experimental technique have been paralleled by developments in the theory of RHEED [4,5] with the result that meaningful theory/experiment comparisons are beginning to be done [6]. The purpose of this contribution is to review the current status of the theory and to assess its future prospects.

Many early RHEED intensity calculations are based on Bethe's n beam dynamical diffraction theory [see, for example, 7,8]. The idea behind this is to expand the electron wave function and the scattering potential as 3D Fourier series, thus reducing the calculation to a matrix eigenvector problem. While this approach enables transmission electron diffraction problems to be solved quite efficiently it is less efficient when applied to the reflection case. It turns out that reflection problems are best handled by retaining the Fourier expansions only in the two dimensions parallel to the surface [4]. This reduces the Schroedinger equation to a system of ordinary differential equations which can be solved quite efficiently. The resulting theory is similar in some respects to the theory of LEED [9] and in fact some of the methods of LEED theory can be applied to RHEED [5]. All recently published RHEED intensity calculations have been done by the 2D Fourier expansion method. The method itself can be implemented in various ways: Maksym and Beeby [4,5], Ichimiya [10] and Marten and Meyer-Ehmsen [11] have independently developed programs based on this principle.

In the following the approach of Maksym and Beeby is considered in more detail. After a brief overview of their calculational procedure the main steps in the computation of RHEED intensities are fully explained. Some of the practical details associated with the computations are also described, mainly to illustrate the nature of the computational tasks involved. The description of the theory is followed by an account of some applications.

First, results for a single layer of arsenic atoms are used to illustrate
the need for a dynamical diffraction theory. Next, some results of theory/
experiment comparisons are discussed, the two surfaces considered being the
MgO(001) surface and the Si(001):2H surface. The last applications mentioned are to more complicated structures, namely stepped surfaces and the
2 x 4 reconstructed surface of GaAs(001). Finally, the current theoretical
results are critically assessed and some future objectives are discussed.

THEORY

Overview of RHEED Intensity Calculations

The present generation of RHEED theories is concerned with eleastic
diffraction by a surface which is periodic in the two directions parallel
to it (x and y). In this case the electron wave function and the potential
can be Fourier expanded in the form

$$\psi(\underline{r}) = \exp(i\,\underline{k}_{\shortparallel}\cdot\underline{\rho}) \sum_{\underline{\kappa}} \psi_{\underline{\kappa}}(z)\, \exp(i\,\underline{\kappa}\cdot\underline{\rho})$$

$$V(\underline{r}) = \sum_{\underline{\kappa}} V_{\underline{\kappa}}(z)\, \exp(i\,\underline{\kappa}\cdot\underline{\rho})$$

(1)

where $\underline{\rho}$ is a position vector parallel to the surface, z is the perpendicular
direction, $\underline{\kappa}$ is a 2D reciprocal mesh vector and $\underline{k}_{\shortparallel}$ is the parallel component
of the incident electron wave vector. Substituting equation (1) into the
Schroedinger equation leads to a coupled system of ordinary differential
equations for the $\psi_{\underline{\kappa}}$:

$$\frac{d^2}{dz^2}\,\psi_{\underline{\kappa}} + k_{\underline{\kappa}}^2\,\psi_{\underline{\kappa}} = \frac{2m_0}{\hbar^2} \sum_{\underline{\kappa}^1} V_{\underline{\kappa}-\underline{\kappa}^1}\,\psi_{\underline{\kappa}^1}$$

(2)

The $k_{\underline{\kappa}}$ are the perpendicular components of the diffracted electron wave
vectors:

$$k_{\underline{\kappa}}^2 = k^2 - \left|\,\underline{k}_{\shortparallel} + \underline{\kappa}\,\right|^2$$

(3)

For high energy electrons there are some small, but significant, corrections
to the Schroedinger equation which are caused by relativistic effects [12].
Thus the magnitude, k, of the incident electron wave vector is determined
from the electron kinetic energy, E, as

$$k^2 = \frac{2m_0 E}{\hbar^2}\left\{1 + \frac{E}{2m_0 c^2}\right\},$$

where m_0 is the rest mass of the electron. In addition the potential is
increased by a factor of $(1 + E/m_0 c^2)$ so that

$$V_{\underline{\kappa}} = \left(1 + \frac{E}{m_0 c^2}\right)\overline{V}_{\underline{\kappa}}$$

where the $\overline{V}_{\underline{\kappa}}$ are the actual Fourier components of the scattering potential.

To calculate the RHEED intensities it is necessary to solve equation 2.
Although this equation could be solved as it stands it is convenient to make
the further substitution

$$\psi_{\underline{K}}(z) = Q_{\underline{K}}^{+}(z) \, \exp(i\,k_{\underline{K}}\,z) + Q_{\underline{K}}^{-}(z) \, \exp(-i\,k_{\underline{K}}\,z) \tag{4}$$

Roughly speaking, the $Q_{\underline{K}}^{+}$ and $Q_{\underline{K}}^{-}$ correspond to upward and downward propagating waves (in the vacuum the correspondence is exact). One advantage of using the $Q_{\underline{K}}^{\pm}$ is that this allows an economical solution of equation 2 in the form of an integral equation [4]. Another advantage is that it enables the scattering properties of various 'layers' of the structure to be described by a simple matrix formalism. This enables the results obtained by solving equation 2 for the individual layers to be combined into the solution for the entire structure. Equation 2 then only needs to be solved for the distinct layers of the structure with consequent savings in computer time [5]. The word 'layer' is used here to mean any 2D periodic slab of the structure, for example a single layer of atoms, a single layer of unit cells, a stack of unit cell layers or indeed the entire structure itself. There are at least three different matrices which can be used to describe scattering by a layer. The transfer matrix M and its inverse N relate the amplitudes of the waves at the top and bottom of the layer. In defining these matrices it is convenient to absorb the phase factors which occur in the definition of the $Q_{\underline{K}}^{\pm}$. Thus $R_{\underline{K}}$ and $I_{\underline{K}}$ are used to represent the values of $Q_{\underline{K}}^{+} \exp(i\,k_{\underline{K}}\,z)$ and $Q_{\underline{K}}^{-} \exp(-i\,k_{\underline{K}}\,z)$ respectively at the top of the layer and $X_{\underline{K}}$ and $T_{\underline{K}}$ represent the corresponding values at the bottom of the layer (see figure 1). The transfer matrices are then defined by the equations

$$\begin{bmatrix} \underline{R} \\ \underline{I} \end{bmatrix} = M \begin{bmatrix} \underline{X} \\ \underline{T} \end{bmatrix} \qquad\qquad \begin{bmatrix} \underline{X} \\ \underline{T} \end{bmatrix} = N \begin{bmatrix} \underline{R} \\ \underline{I} \end{bmatrix} \tag{5}$$

where $\underline{R}$ is a vector composed of the $R_{\underline{K}}$ and similarly for $\underline{I}$, $\underline{X}$ and $\underline{T}$. In practice the transfer matrices are computed from the solutions of equation 2. The third matrix used to describe the scattering properties of layer is the scattering matrix S. This relates the amplitudes of the waves entering and leaving the layer. It is defined by the equation

$$\begin{bmatrix} \underline{R} \\ \underline{T} \end{bmatrix} = S \begin{bmatrix} \underline{X} \\ \underline{I} \end{bmatrix} \tag{6}$$

and it can be computed from the sub-matrices of M and N. Once the scattering matrices of the individual layers have been computed they can be combined to form the scattering matrix of the full structure including the surface and substrate. This gives the amplitudes of the RHEED beams from which the intensities may be found.

In summary, then, the computation of RHEED intensities involves the following steps:

1. computation of the potential

2. computation of the transfer matrices

3. computation of the scattering matrices

4. combination of the individual scattering matrices.

It is essential to use the scattering matrices for steps 3 and 4. The alternative of multiplying transfer matrices is prone to numerical error.

Computation of the Potential

The potentials of the individual atoms are obtained by Fourier transforming electron scattering factors which are available in the literature

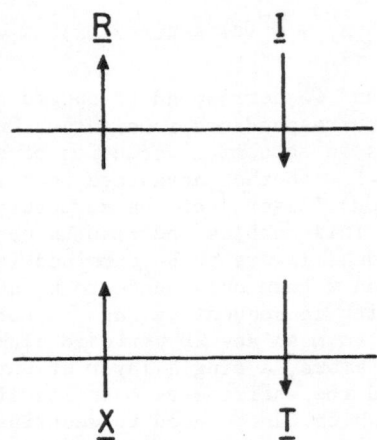

Fig. 1. Schematic illustration of the definition of the beam ampli-
tudes $\underline{R}$, $\underline{I}$, $\underline{X}$, $\underline{T}$. The upward arrows represent amplitudes
derived from the $Q_{\underline{K}}^{+}$ and the downward arrows represent
amplitudes derived from the $Q_{\underline{K}}^{-}$. The boundaries of the
layer are represented by the horizontal lines.

[16]. Once the atomic potentials have been found the scattering potential
of the structure is constructed by summing the potentials of the individual
atoms. The potential obtained in this way needs to be corrected to allow
for the effects of absorption and lattice vibrations. In addition the
volume average potential usually needs to be adjusted to bring the experimental and theoretical peak positions into correspondence. Maksym [6] has
discussed these points in detail for the case of the MgO(001) surface.

Computation of Transfer Matrices

The transfer matrices are computed by using the integral equation
approach of Maksym and Beeby [4]. This enables the $Q_{\underline{K}}^{+}$ and $Q_{\underline{K}}^{-}$ to be found
from a coupled system of integral equations. Repeatedly starting the inte-
gration with the appropriate initial conditions gives the columns of the
transfer matrices directly.

Computation of Scattering Matrices

When n Fourier components of the wave function (beams) are retained the
size of the matrices M, N and S is $2n$ x $2n$. Each matrix consists of four
n x n sub-matrices:

$$M = \begin{pmatrix} M_1 & M_2 \\ M_3 & M_4 \end{pmatrix} \qquad N = \begin{pmatrix} N_1 & N_2 \\ N_3 & N_4 \end{pmatrix} \qquad S = \begin{pmatrix} S_1 & S_2 \\ S_3 & S_4 \end{pmatrix}$$

As is easily shown from its definition, the sub-matrices of S can be found
from the relations

98

$$S_1 = N_1^{-1} \qquad\qquad S_2 = M_2 \, M_4^{-1}$$

$$S_3 = N_3 \, N_1^{-1} \qquad\qquad S_4 = M_4^{-1} \tag{7}$$

In principle the sub-matrices of S could also be computed entirely from those of M or N, however the results of numerical tests indicate that use of equation 7 is less prone to numerical error.

Combination of Scattering Matrices

The layer doubling algorithm of LEED theory [9] is used to combine the scattering matrices. This gives the sub-matrices of the S matrix of a double layer in terms of the S sub-matrices of the individual layers. Suppose that the two individual layers are labelled α and β (figure 2) and that the amplitudes of the beams at the boundaries of the composite layer are R, I and X, T while those of the beams of the junction of the layers are U, W. Then, by definition, the beam amplitudes satisfy the equations

$$\begin{pmatrix} R \\ W \end{pmatrix} = S^\beta \begin{pmatrix} U \\ I \end{pmatrix} \qquad \begin{pmatrix} U \\ T \end{pmatrix} = S^\alpha \begin{pmatrix} X \\ W \end{pmatrix} \qquad \begin{pmatrix} R \\ T \end{pmatrix} = S^{\alpha\beta} \begin{pmatrix} X \\ I \end{pmatrix}$$

where S^α and S^β are the S matrices of the individual layers and $S^{\alpha\beta}$ is the S matrix of the composite layer. Eliminating U and W from the first two equations leads to the relations

$$S_1^{\alpha\beta} = S_1^\beta (1 - S_2^\alpha S_3^\beta)^{-1} S_1^\alpha$$

$$S_2^{\alpha\beta} = S_2^\beta + S_1^\beta (1 - S_2^\alpha S_3^\beta)^{-1} S_2^\alpha S_4^\beta$$

$$S_3^{\alpha\beta} = S_3^\alpha + S_4^\alpha (1 - S_3^\beta S_2^\alpha)^{-1} S_3^\beta S_1^\alpha \tag{8}$$

$$S_4^{\alpha\beta} = S_4^\alpha (1 - S_3^\beta S_2^\alpha)^{-1} S_4^\beta$$

These relations constitute the layer doubling algorithm.

The algorithm is actually used at two different stages in the calculation of RHEED intensities. First, the α and β layers are taken to be identical, and initially S^α and S^β are set to the S matrices of a single layer of unit cells. Then one pass of the layer doubling algorithm gives the S matrix of a stack of 2 unit cell layers, a subsequent pass gives S for a stack of 4 unit cell layers and so on. Thus the S matrix of the entire substrate is computed quite rapidly. Then the layer doubling algorithm is used again for the final stage of the calculation where the S matrices of the substrate and surface are combined to give the S matrix of the full structure.

Practical details

Although the calculational procedure just described works quite well it can be made more efficient by exploiting the symmetries of the surface substrate. There are three commonly occurring symmetries which are easily taken into account. The first two are mirror symmetries. When the RHEED pattern itself has left-right symmetry certain columns of the transfer matrices are identical so the number of distinct matrix elements is reduced by approximately one half. A second possibility for exploiting mirror

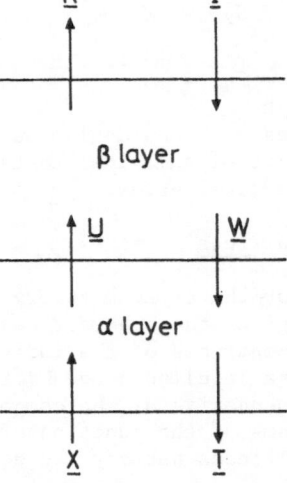

Fig. 2. Schematic illustration of the beam amplitudes at the bound-
aries of a composite layer. The vectors $\underline{R}$, $\underline{I}$, $\underline{X}$ and $\underline{T}$ are
defined as in figure 1. The intermediate amplitudes $\underline{U}$ and
$\underline{W}$ are derived from the $Q_{\underline{k}}^{+}$ and $Q_{\underline{k}}^{-}$ respectively.

symmetries arises when any layer has a mirror plane parallel to the surface.
Then only two of the S sub-matrices are distinct. The final symmetry which
can be used occurs when the individual layers of a unit cell differ by no
more than a translation. In this case it is not necessary to integrate the
Schroedinger equation through the full width of the unit cell layer [5].

Another feature which affects computer time requirements is the number
of beams used to represent the electron wave function and the number of
Fourier components used to represent the potential. Obviously if too many
beams are used then accurate results will be obtained but at great computational cost. In contrast, if too few beams are used then the results
obtained will be inaccurate. Since the computer time needed scales as n^3
the choice of beam set requires some care.

Generally the beam set has to be selected by trial and error. The
number of beams required depends on the atomic species involved and on the
conditions of incidence. For structures composed of light atoms remarkably
few beams are needed provided that the diffracted beams are well separated
in $\underline{k}$ space. This is illustrated in figure 3 which shows some computed RHEED
intensities for 10 keV electrons incident on the MgO(001) surface in the
[010] azimuth. All the curves are rocking curves, that is plots of the
intensities of a particular beam against the glancing angle of the incident
beam. Each curve has been computed with a different number of beams or
potential coefficients. The full curves were computed using all the potential coefficients which could couple the beams but only 7 beams and 9
potential coefficients were used to compute the broken curve. Nevertheless
the broken curve is clearly similar to the 19 beam curve. In fact the
corresponding peak heights typically differ by $\sim 3\%$, except for the peaks
at $5.55°$ and $7.1°$ which differ by $\sim 7\%$ and $\sim 10\%$ respectively. Computations
for structures containing heavier atoms or for incident beam azimuths other
than [010] can require more beams. For example, at least 17 are needed to
compute rocking curves for electrons incident on GaAs(001) in the [110]

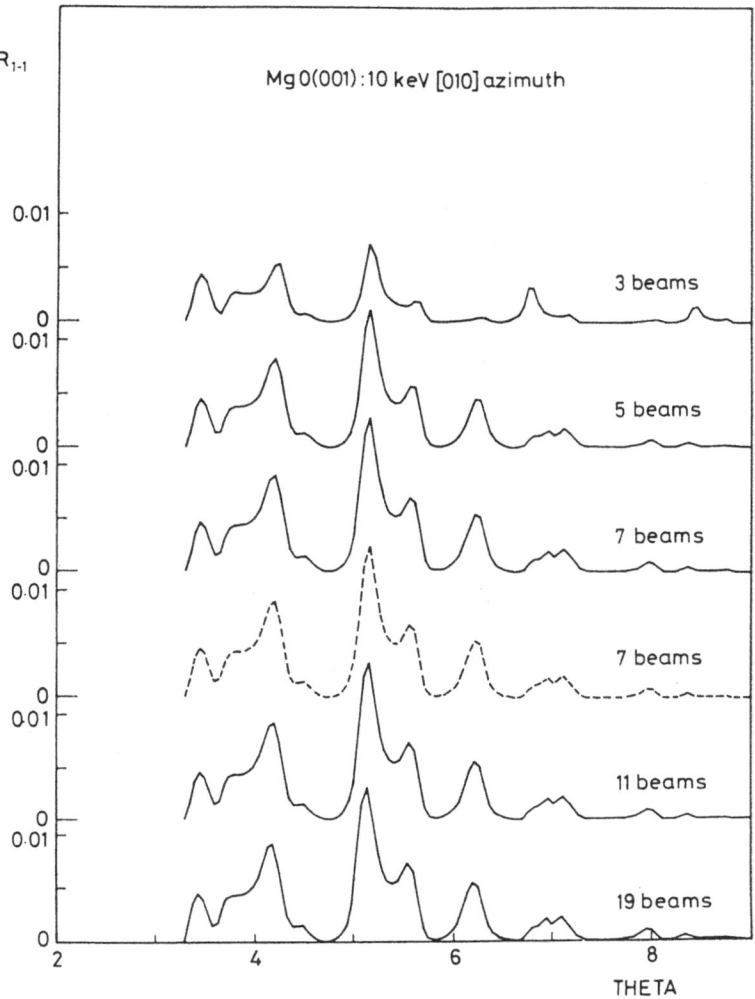

Fig. 3. Computed rocking curves for 10 keV electrons incident on the MgO(001) surface in the [010] azimuth. The curves show the reflectivity of the 1-1 beam as a function of the glancing angle of the incident beam. Each curve was computed with a different number of beams or potential coefficients. The full curves correspond to a fully coupled beam set but only 9 potential coefficients were used to compute the broken curve.

azimuth, the exact number depending on the accuracy required.

A final practical point concerns numerical accuracy. Evanescent beams, which contain exponentially growing or decaying components can only be represented over a limited range which is related to machine precision. This sets an upper limit on the integration range which can be used to compute the transfer matrix. Often the limit is larger than the width of any of the layers, however if very strongly evanescent beams need to be included this is not the case. The layers are then split into segments each of which is within the limit and the layer S matrices are computed by combining segment S matrices with the layer doubling algorithm. This was done in the calculations leading to figure 4.

APPLICATIONS

Need for a Dynamical Theory

Electrons interact strongly with atoms so dynamical effects are very important in electron diffraction. Even for a single layer of atoms the results obtained from a dynamical theory can be very different from those obtained with a single scattering (kinematical) theory. This is illustrated in figure 4 which shows dynamic and kinematic specular beam rocking curves for a single layer of arsenic atoms. The electron energy is 10 keV and the electrons are incident in the <110> azimuth. The kinematic curve is clearly featureless and the kinematic reflectivity is greater than unity when the glancing angle is less than about 1.5°. In contrast the dynamic curve has a series of peaks which are associated with the emergence of various beams in the zeroth Laue zone. (Only zeroth zone beams were included in the calculations leading to figure 4). Arrows indicate the beam emergence angles. The peaks correspond to resonances caused by multiple scattering of electrons while they are in the atomic layer. Thus in this case, at least, a fully dynamical theory is needed to compute the absolute reflectivities.

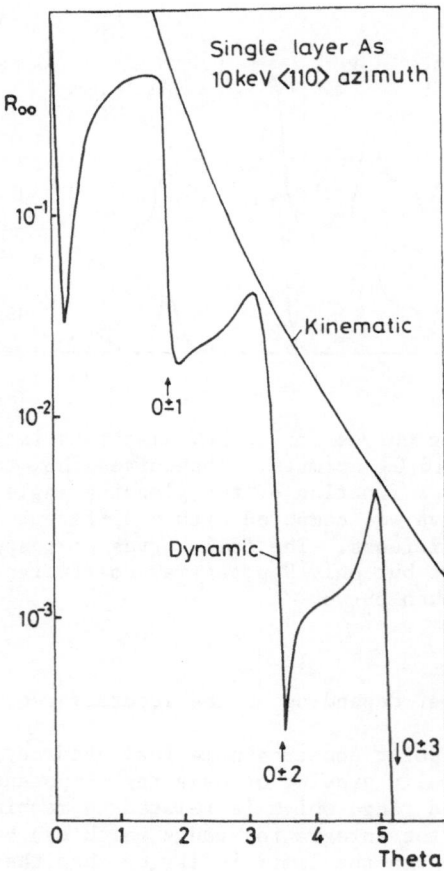

Fig. 4. Computed dynamic and kinematic specular beam rocking curves
         for 10 keV electrons incident on a single layer of As atoms
         in the <110> azimuth. The arrows indicate the emergence
         angles of various beams in the zeroth Laue zone.

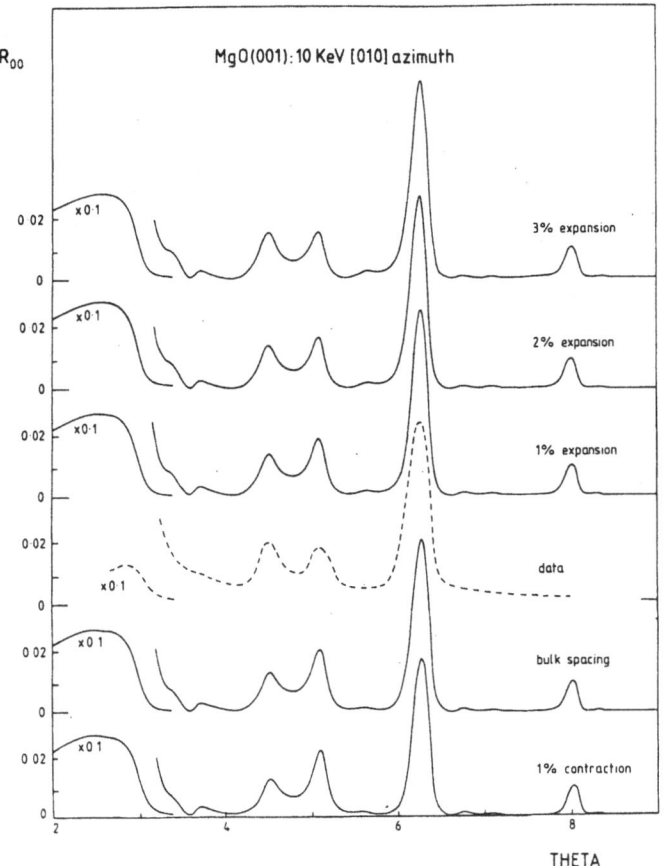

Fig. 5. Calculated and experimental rocking curves for 10 keV
electrons incident on the MgO(001) surface in the [010]
azimuth. Each curve shows how the specular reflection
coefficient (R_{00}) depends on the glancing angle of the
incident beam. The full curves show absolute reflection
coefficients calculated for various values of the relaxation. The broken curve shows the experimental relative
reflection coefficient.

The MgO(001) Surface

The MgO(001) surface has been thoroughly studied by LEED [15, 16, 17,
18] and it is accepted that its structure can be described by at most two
parameters. One is the relaxation parameter which describes the spacing of
the top two atomic layers. The other is the rumpling parameter which describes the differential relaxation of the surface, that is the possibility
of the oxygen sub-layer moving upwards out of the surface plane together
with the magnesium sub-layer moving in towards the bulk. The relaxation is
known to be an expansion which is less than 3% of the bulk layer spacing
but the value of the rumpling is less certain and has been the subject of
some controversy [18, 19].

RHEED intensity calculations for the MgO(001) surface indicate that

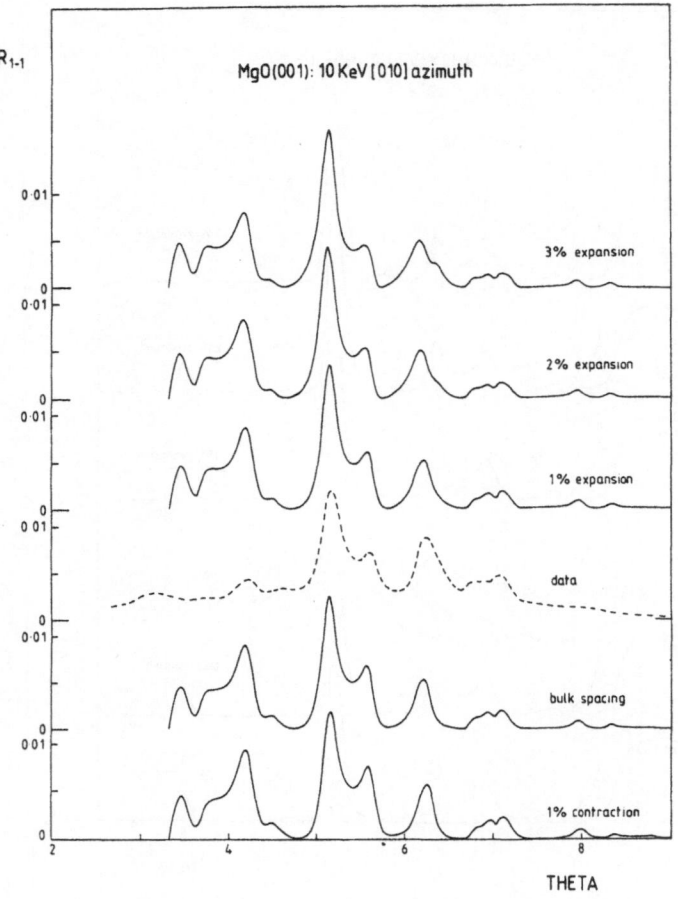

Fig. 6. Calculated and experimental rocking curves for 10 keV
electron incident on the MgO(001) surface in the [010]
azimuth. Each curve shows how the side beam reflection
coefficient R_{1-1} depends on the glancing angle of the
incident beam. The full curves show absolute reflection
coefficients calculated for various values of the relaxation. The broken curve shows the experimental relative
reflection coefficient.

the relaxation and rumpling can be determined independently from data collected with the incident beam along appropriate azimuths [6]. This is
because RHEED is very insensitive to rumpling when the incident beam
azimuth is along one of the [n10] directions with n even. The precise
reason for this will be discussed shortly but first the theory/experiment
comparison for the [010] incident beam azimuth will be presented. The
results are shown in figures 5 and 6. Each set of curves is for 10 keV
electrons. The full curves are calculated for various values of the relaxation and the dashed curve shows data collected by Ichimiya and Takeuchi
[20]. Both the experimental and theoretical curves have a series of peaks
which are mostly due to Bragg diffraction effects [4]. The agreement
between theory and experiment is clearly good. Detailed comparison of the
curves enables the relaxation to be determined and an upper limit of a 3%
expansion is found. This is consistent with the LEED result.

Fig. 7. Two perspective views of the rumpled MgO(001) surface
(schematic). The top view represents what would be seen
at grazing incidence in the [010] azimuth. The bottom
view is similar but corresponds to the [110] azimuth.
Note that the rumpling is most clearly visible when the
incident beam azimuth is [110].

Now the selective sensitivity to rumpling will be explained. It is a
direct consequence of the grazing incidence geometry of RHEED which causes
the Laue zones to be well separated in k space. As discussed by Maksym and
Beeby [4] this has the further consequence that the coupling between the
beams within a zone is much stronger than the inter-zone coupling. Hence
the RHEED intensities are most sensitive to the intra-zone potential coefficients. In terms of real space this means that RHEED is much more
sensitive to the structure in the direction perpendicular to the incident
beam azimuth than to structure in the parallel direction. Physically, the
electrons will 'see' almost a projection of the structure onto the plane
perpendicular to both the surface and the incident beam azimuth. In the
case of MgO(001) the azimuths most sensitive to rumpling can be found by
examining the inter-Laue zone potential coefficients [6], however it is
simpler to look at the structure in real space. This is illustrated in
figure 7 which shows two perspective views of the MgO(001) surface as seen
from grazing incidence. The upward displaced circles represent oxygen atoms
and the downward displaced circles represent magnesium atoms. The top view
corresponds to incidence in the [010] direction. In this case the rows of
atoms parallel to the incident beam azimuth consist of alternate oxygen
atoms and magnesium atoms. However the rumpling is not easy to see because
in projection the alternate displacements obscure each other. The bottom
view corresponds to incidence in the [110] direction. In this case the
atomic rows in the direction parallel to the incident beam azimuth are composed either of oxygen atoms or magnesium atoms. In projection the rumpling
can clearly be seen as an alternate upward and downward displacement of each
row. Some computed rocking curves for rumpled and non-rumpled MgO surfaces
are shown in figure 8 ([010] azimuth) and figure 9 ([110] azimuth). It is
evident that only the curves for the [110] azimuth are significantly sensitive to rumpling. The selective sensitivity is a potential advantage of
RHEED data analysis because fitting parameters one at a time is more
efficient than fitting them all together.

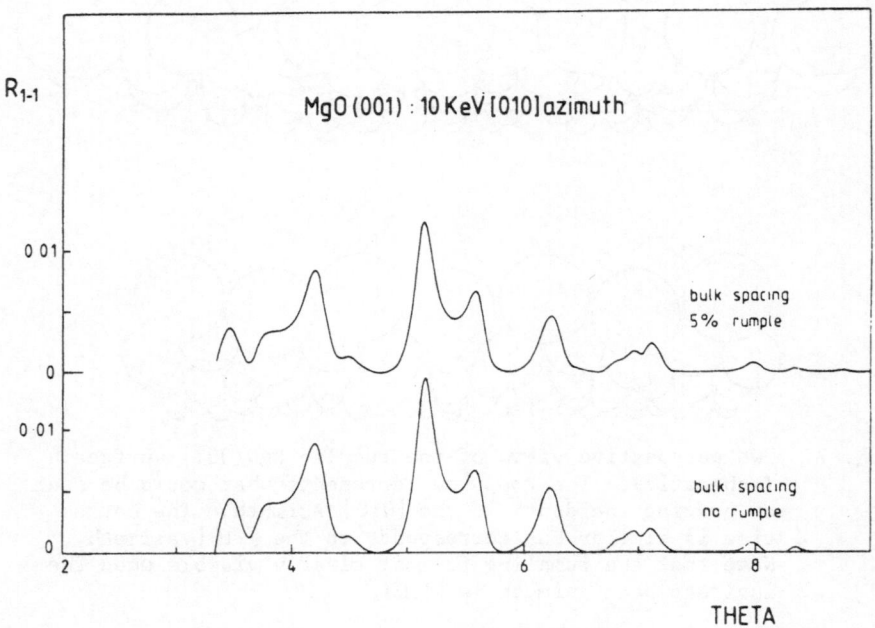

Fig. 8. Side beam rocking curves for rumpled and non-rumpled MgO(001) surfaces. The similarity of the curves indicates that the RHEED intensities are insensitive to rumpling when the incident beam azimuth is [010].

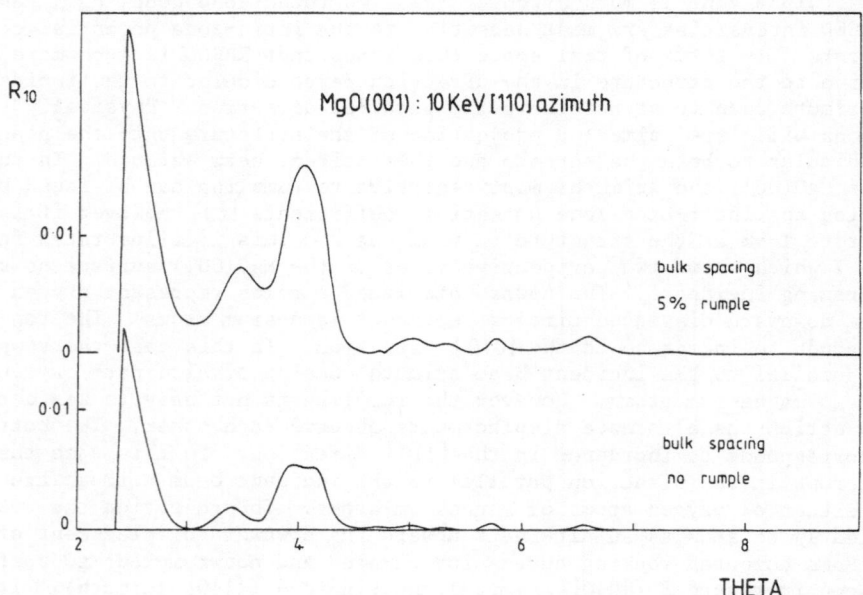

Fig. 9. Side beam rocking curves for rumpled and non-rumpled MgO(001) surfaces. The two curves are noticeably different showing that the RHEED intensities are sensitive to rumpling when the incident beam azimuth is [110].

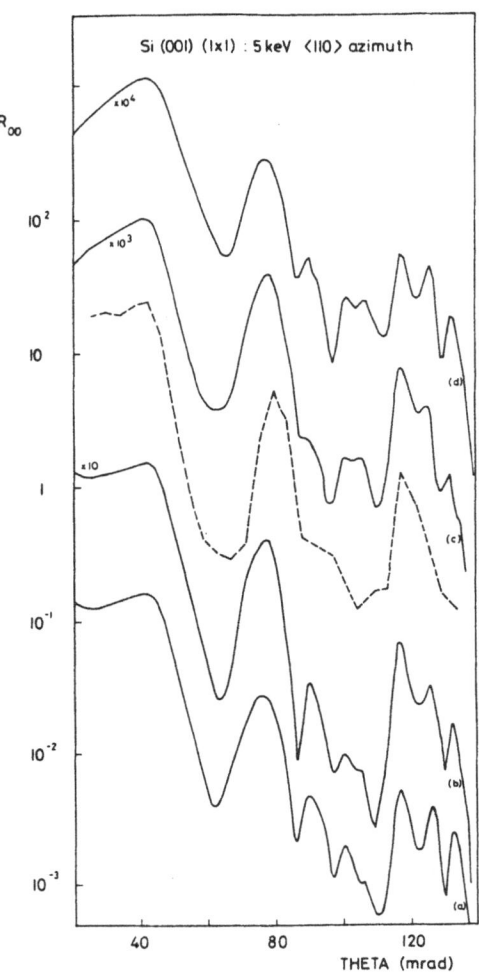

Fig. 10. Calculated and experimental specular reflectivities for electrons incident on the Si(001)-(1x1) surface in the <110> azimuth. The full curves show calculated absolute reflectivities. Curves (a) and (b) are for the hydrogenated surface, assuming 50% and 80%, respectively, contributions from the perpendicular domains, while curves (c) and (d) are for the clean surface assuming 80% and 50%, respectively, contributions from the perpendicular domains. The broken curve shows the experimental specular reflectivity in arbitrary units.

The Si(001) : 2H Surface

A bulk silicon unit cell has four atomic layers parallel to its (001) surface. Two of these are inequivalent so even the ideally terminated Si(001) surface would consist of two domain types which, in equilibrium, would be present in roughly equal proportions. This introduces a slight complication into the RHEED data analysis as it is necessary to average the computed intensities over the two domain types. A further complicating factor is that the Si(001) surface undergoes a (2x1) reconstruction and the exact geometry of this is still unknown. However the reconstruction can be removed by exposing the surface to hydrogen [21]. It is this hydrogenated (1x1) structure that is discussed here.

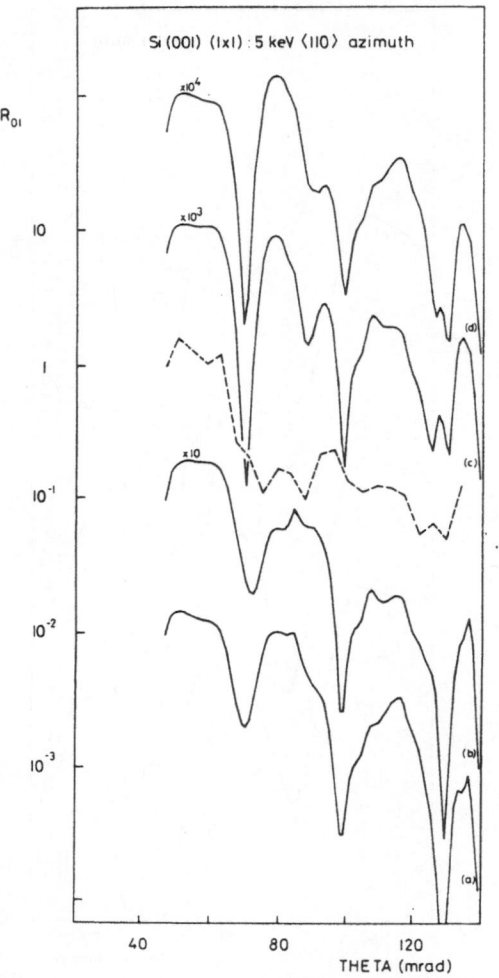

Fig. 11. Calculated and experimental side beam reflectivities for
electrons incident on the Si(001)-(1x1) surface in the
<110> azimuth. The full curves show calculated absolute
reflectivities. Curves (a) and (b) are for the hydrogenated surface, assuming 50% and 80%, respectively,
contributions from the perpendicular domains, while curves
(c) and (d) are for the clean surface assuming 80% and
50%, respectively, contributions from the perpendicular
domains. The broken curve shows the experimental side beam
reflectivity in arbitrary units.

This surface has recently been the subject of a RHEED study by Ashby
et al [22]. These authors have collected and analysed RHEED intensity data
for 5 keV electrons incident in the <110> and <010> azimuths. Figures 10
and 11 show their results for the <110> azimuth. The full curves are calculated and the dashed curves are experimental. It is clear that the
agreement between theory and experiment is quite good for the specular beam
(figure 10) but only fair for the first side beam (figure 11). Only the
specular beam will be considered in detail here. In each figure there are
two curves for the hydrogenated surface (a, b) and two for the ideally terminated surface (c, d). Within each pair, curves a and d correspond to a
50/50 average of the intensities from the two domain types and curves b, c
correspond to a weighted average with 80% of the intensity coming from the

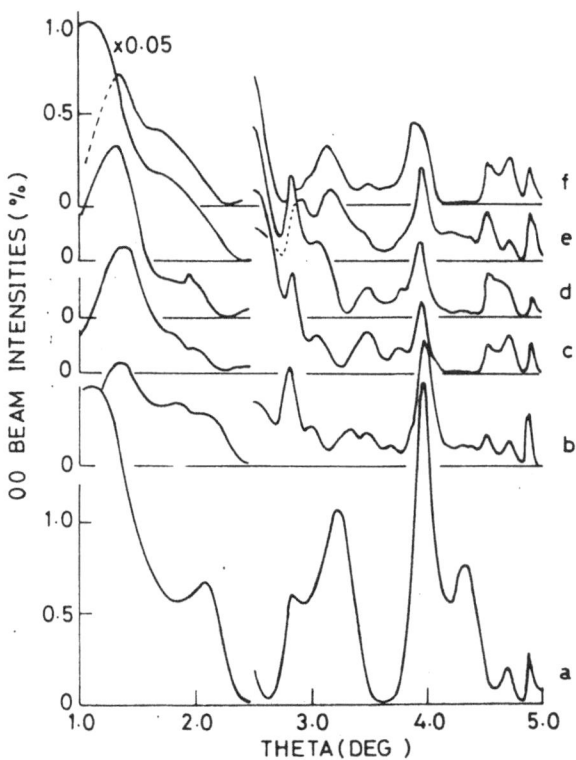

Fig. 12. Specular RHEED intensity versus glancing angle, θ, of the
incident beam when the azimuth is parallel to the step
edges. Curve (a) is for a flat Al(001) surface and curves
(b, e) are for stepped surfaces with N_p = 5. The number of
topmost layer atoms ranges from 1 to 4 for curves (b, e)
respectively. Curve (f) is for a stepped surface with
N_p = 7 and n_o = 3.

domains with their surface chains of bonds perpendicular to the incident
beam azimuth. There are obvious differences between all four theoretical
curves showing that the RHEED intensities are sensitive to both the hydrogen
and the averaging procedure. Upon inspection of figure 10 it can be found
that the best agreement between theory and experiment is obtained when the
surface is taken to be hydrogenated and a weighted average is used. Both of
these results are surprising. The former indicates that the RHEED intensities are sensitive to the hydrogen even though it scatters weakly. The
latter seems to indicate that the surface was composed mostly of one domain
type, as discussed by Sakamoto in these proceedings. However Ashby et al
felt that this was unlikely for their particular surface. Instead they
suggest that the need for weighted averaging is a consequence of surface
disorder. The surface silicon atoms could move more easily in the direction
parallel to the surface chains of bonds than in the direction perpendicular
to them. As a result the surface disorder could be anisotropic. Combined

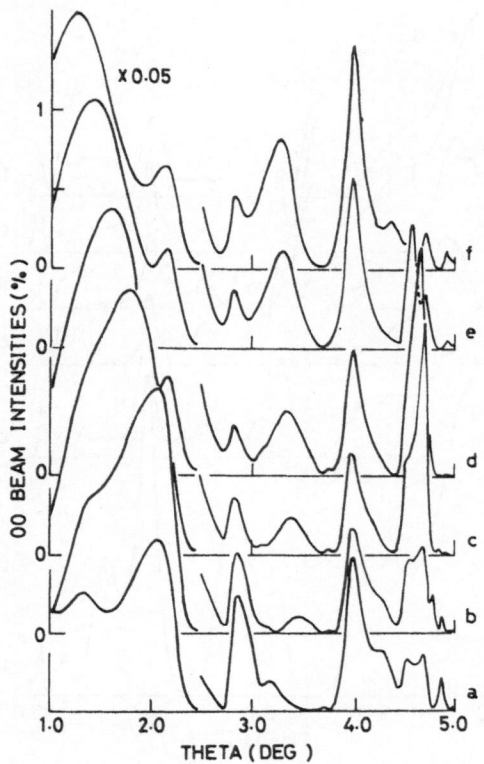

Fig. 13. Specular intensity versus θ for stepped Al(001) surfaces
with the incident beam azimuth perpendicular to the step
edges. All the curves correspond to N_p = 7 but n_o ranges
from 1 to 6 for curves (a) to (f) respectively.

with the enhanced sensitivity of RHEED to the structure in the direction
perpendicular to the incident beam azimuth, this could explain the need for
weighted averaging. Details are given in reference [22].

More Complicated Structures

The major difficulty in calculating electron diffraction intensities
for complicated structures is that many beams are needed to represent the
electron wave function. Since the computer time needed scales as n^3 this
severely limits the range of structures which can be considered. In RHEED,
however, the coupling between the Laue zones is weak so to a first approximation only the zeroth Laue zones need be considered. This considerably
reduces the size of the beam set and enables more complicated structures to
be investigated, at least to the level of understanding the physics.
Furthermore in favourable cases, such as MgO, the results so obtained are
adequate to analyse real data.

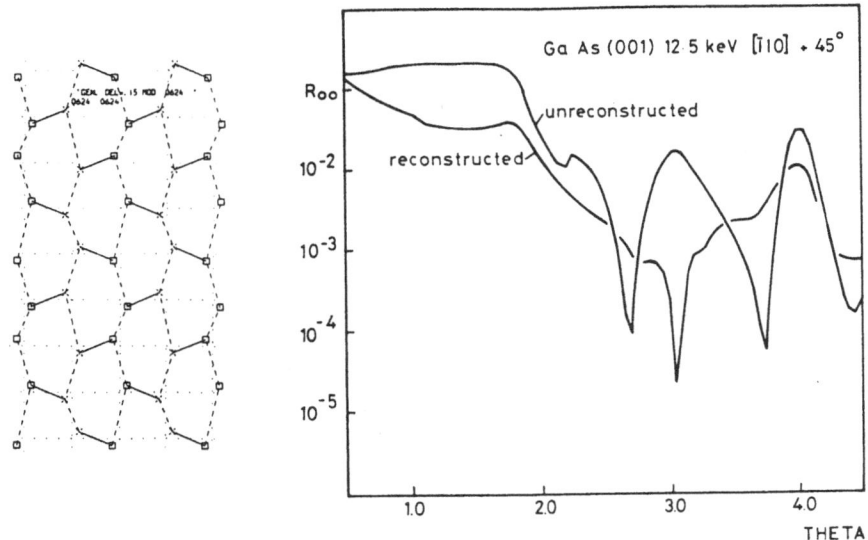

Fig. 14. Specular rocking curves for reconstructed and unreconstructed GaAs(001) surfaces. The diagram on the left
shows the particular 2 x 4 reconstruction corresponding
to the curve on the right. Note the difference between
the two curves.

Stepped Surfaces. Dynamical RHEED intensity calculations for stepped
Al(001) surfaces have been done by Kawamura and Maksym [23]. They consider
model surfaces which have periodic arrays of steps. Each period is N_p atoms
long and there are n_O atoms in the topmost atomic layer of the steps. The
electron energy is taken to be 10 keV. One of the main findings is that the
RHEED intensities are most affected by the steps when the incident beam
azimuth is parallel to the step edges. This is, of course, another mani-
festation of the selective structural sensitivity of RHEED. Some rocking
curves for the parallel case are shown in figure 12. It is clear that the
peak heights for the stepped surfaces are much smaller than those for the
flat surface (curve (a)). When the incident beam azimuth is perpendicular
to the step edges the rocking curves are not so drastically affected (figure
13). Kawamura and Maksym have examined the peak height variations in detail
and they report that some peak heights are oscillatory functions of n_O.
This is reminiscent of the oscillations observed during MBE growth.

The GaAs(001)-(2 x 4) surface. The arsenic stabilized GaAs(001)
surface is commonly observed to have a 2 x 4 reconstruction. Since RHEED is
used to monitor MBE growth it is natural to enquire about the effect of the
reconstruction on the RHEED intensities. The main difficulty in answering
this question is that the detailed geometry of the 2 x 4 reconstruction is
not understood. Current opinion is that the reconstruction is caused by
dimerization of the surface arsenic atoms and in addition the dimers are
expected to the tilted and twisted. However even within these constraints
many different geometries are possible, corresponding to the many arrangements of the dimers in a 2 x 4 cell. Figure 14 shows one possible arrangement. The full lines represent the tilted twisted dimers, the crosses and
squares marking the positions of the up and down atoms, respectively, of each
dimer. The displacement of the atoms from their ideal positions are 0.6 Å
in each of the x, y and z directions; the [110] direction is up the page.
Figure 14 also shows a specular beam rocking curve calculated for the

indicated structure, together with one calculated for the unreconstructed As terminated surface of GaAs(001). The curves are taken from unpublished work by M G Knibb. There are obvious differences between them indicating that the RHEED intensities are sensitive to the reconstruction. In principal, then, it would be possible to analyse RHEED data to investigate the atomic geometry of the reconstruction. Work is in progress but a definitive geometry has not yet emerged.

DISCUSSION

The recent progress in the theory of RHEED is most encouraging but much work remains to be done. At present only the elastic component of the RHEED intensities can be calculated realistically, however it is well known that there are other components as well. This is evident from the appearance of experimental RHEED patterns which tend to consist of streaks superimposed on a background of Kikuchi lines. These features should provide additional information about the surface so it is important to understand them.

The underlying reason for the streaking is the combination of high electron energies and grazing incidence which is peculiar to RHEED. Under these conditions the angles between the lowest order reciprocal lattice rods and the Ewald sphere tend to be small. As a result fairly small changes in either the energy or the parallel momentum of the electrons lead to relatively large changes in the perpendicular momentum. Thus, in principle, any diffuse or inelastic scattering mechanism can contribute to the streaks. The problem is to discover which of the many possible mechanisms are the most important.

Another problem associated with inelastic effects concerns the interpretation of experimental rocking curves. The data shown in the last section for the MgO(001) and Si(001) surfaces almost certainly contains an inelastic component as well as the elastic component. Yet it can be well fitted using a theory in which only elastic diffraction is taken into account. One possible explanation of this is that the rocking curves for elastic and inelastic diffraction are very similar [6, 22]. This would be very useful if it was true in general as there would then never be any doubts about possible inelastic components in RHEED data. Clearly, the problem merits further investigation.

The quantitative understanding of diffuse and inelastic scattering effects should be the next objective in the development of RHEED theory. Of the two, diffuse scattering is probably the more important as understanding this would be of great help in studies of surface growth and surface morphology. The necessary calculations could be done in principle but the sheer amount of computer time required is an obstacle. However, although detailed computer modelling may not be possible in the near future it should certainly be possible to investigate the relevant physics, either by solving simplified models or by developing approximate theories.

ACKNOWLEDGEMENTS

I would like to thank Professor J L Beeby and M G Knibb for many useful discussions. The rocking curves for GaAs were computed by M G Knibb. This work was supported by the UK Science and Engineering Research Council.

REFERENCES

[1] J. H. Neave, B. A. Joyce, P. J. Dobson and N. Norton, Appl.Phys. A31:1 (1983).

[2] N. Osakabe, Y. Tanishiro, K. Yagi and G. Honjo, Surface Sci. 109:353 (1981).

[3] M. Ichikawa and K. Hayakawa, Japan J. Appl. Phys. 21:154 (1982).

[4] P. A. Maksym and J. L. Beeby, Surface Sci. 110:423 (1981).

[5] P. A. Maksym and J. L. Beeby, Surface Sci. 140:77 (1984).

[6] P. A. Maksym, Surface Sci. 149:157 (1985).

[7] R. Colella and J. F. Menadue, Acta. Cryst. A28:16 (1972).

[8] K. Britze and G. Meyer-Ehmsen, Surface Sci. 77:131 (1978).

[9] J. B. Pendry, "Low Energy Electron Diffraction", Academic, New York (1974).

[10] A. Ichimiya, Japan J. Appl. Phys. 22:176 (1983).

[11] H. Marten and G. Meyer-Ehmsen, Surface Sci. 151:570 (1985).

[12] C. J. Humphreys, Repts. Progr. Phys. 42:1825 (1979).

[13] P. A. Doyle and P. S. Turner, Acta. Cryst. A24:390 (1968).

[14] P. A. Maksym and J. L. Beeby, Applications of Surface Sci. 11/12:663 (1982).

[15] C. G. Kinniburgh, J. Phys. C. (Solid State Phys.) 8:2382 (1975).

[16] C. G. Kinniburgh, J. Phys. C. (Solid State Phys.) 9:2695 (1976).

[17] M. R. Welton-Cook and W. Berndt, J. Phys. C. (Solid State Phys.) 15:5691 (1982).

[18] T. Urano, T. Kanaji and M. Kaburagi, Surface Sci. 134:109 (1983).

[19] T. Gotoh, S. Murakami, K. Kinosita and Y. Murata, J. Phys. Soc. Japan 50:2063 (1983).

[20] A. Ichimiya and Y. Takeuchi, Surface Sci. 128:343 (1983).

[21] S. J. White and D. P. Woodruff, J. Phys. C. (Solid State Phys.) 9:L451 (1976).

[22] J. V. Ashby, N. Norton and P. A. Maksym, Surface Sci. 175:604 (1986).

[23] T. Kawamara and P. A. Maksym, Surface Sci. 161:12 (1985).

SUPERLATTICES AND SUPERSTRUCTURES GROWN BY MOCVD

Naozo Watanabe, Yoshifumi Mori and Hiroji Kawai

Sony Corporation Research Center
174, Fujitsuka-cho, Hodogaya-ku
Yokohama, Japan

ABSTRACT

MOCVD technollogy has, in recent several years, been remarkably advanced to meet wide range of requirement of the most active forefront of III-V compound semiconductor technology. It has recently been successfully used in growing various kinds of abrupt structures such as superlattices, quantum wells, modulation doped structures and HBT's and so on. Abrupt structures have also been grown with materials other than GaAs/AlGaAs. There are fundamental differences in the growth temperature dependence of the structure of the quantum wells between MOCVD and MBE, which reflect the difference in the growth mechanisms between the two methods. In addition to the sophisticated structures, an abrupt structure device such as HIFET (hetero-interface FET: so-called HEMT) has been produced as a practical commercial product.

INTRODUCTION

MOCVD technology has for some years been suspected whether it can be qualified as a method to grow abrupt structures, because clusters or defects are usually included in the barrier wall of the superlattice. In the photoluminescence spectra of the multi-quantum wells grown by MOCVD in the pioneering work by N.Holonyak et al.[1], there were observed spurious peaks, which were afterwards ascribed to the defects in the barrier. Several years later, Frijlink and Maluenda demonstrated that they could grow two quantum wells of GaAs/$Al_{0.5}Ga_{0.5}As$ with an intervening thin barrier, which did not show any appreciable spurious photoluminescence peaks[2]. Their result was encouraging and stimulated a new tide of research in the field of the quantum structure of III-V semiconductors with MOCVD.

Various reactors and growth systems have been devised to realize abrupt structures, and quantum structures of GaAs/AlGaAs, mostly similar to that grown by MBE, have been grown to test the capability of MOCVD. In couple of recent years, MOCVD has catched up MBE in growing basic superstructures and has started to be applied to new structures. However, there seems to be a fundamental difference between the two methods in the elementary reaction process, which can be inferred from the difference in the temperature dependence of the grown structures. III-V materials other than AlGaAs series have been grown and their superstructures have also been grown by MOCVD. It has been adopted for production of HIFET's (hetero- interface FET: so-called

HEMT), a typical device incorporating an abrupt hetero-interface, as well as for semiconductor lasers.

MOCVD AND GROWTH OF SUPERSTRUCTURE

MOCVD is an acronym of metalorganic chemical vapor deposition. It is a chemical vapor deposition process in which at least one species of the source materials is an organometallic compound and the reaction is an organic reaction process. Because the process is usually used for epitaxial growth of semiconductor materials, it is also called MOVPE (metalorganic vapor phase epitaxy), more restrictively. Some people favor OM (organometallic) instead of MO (metalorganic) because organomtallic compounds are used in the process. There are then four namings of possible four combinations of MO or OM with CVD or VPE. A typical example of the reaction is as follows:

$$(1-x) \; Ga(CH_3)_3 +$$
$$AsH_3 \rightarrow Al_xGa_{1-x}As \downarrow + 3 \; CH_4 \uparrow \qquad (1)$$
$$x \;\; Al(CH_3)_3 +$$

Ethyl compounds are also widely used instead of methyl compounds. Changing the metals to In, or As to P or Sb, various combinations of compounds have been grown. Alkyl compounds are usually liquid and should have considerable vapor pressure appropriate for keeping a necessary partial pressure over the substrate. The mixed gas is carried over the heated substrate and the reaction (1) takes place on the surface. How is the detailed elementary process is not clear at present, which will be discussed later in this paper.

MOCVD for growing Abrupt Structure

In an early stage of the development of MOCVD, no special precaution has been taken to grow abrupt structures. In a conventional vertical reactor as seen in Fig.1[3], for example, a relatively large chamber was used with a large volume above the susceptor. Convectional circulation and gas phase reaction take place in this space, which degrade the abruptness of the interface and produce small particles, which are eventually included in

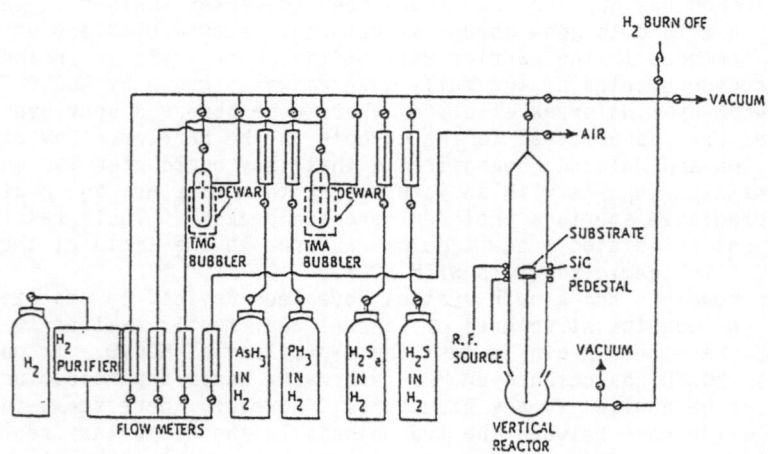

Fig. 1 Schematic of a typical MOCVD apparatus with a vertical reactor constructed by Manasevit and Simpson[3].

116

the barrier wall. Even if the convection and the reaction are suppressed by some means the large volume has a possibility of stagnating and unfavorably mixing incoming gas species and degrading the abruptness.

There have been various new attempts to overcome these effects. The most novel attempt may be the "Chimney Reactor" by Leys et al.[4], which is shown in Fig.2. In this reactor, the gas flows upward from the inlet located below the susceptor which is attached to the inside wall of the reactor. Because the hotter susceptor zone is located above the cooler inlet zone, there is no convectional circulation. With this "Chimney Reactor" superlattices and superstructures are grown and quantum wells as thin as 7 A were clearly observed in a dark field image taken by electron microscopy. This fact demonstrated the importance to suppress convection circulation for growing superlattices.

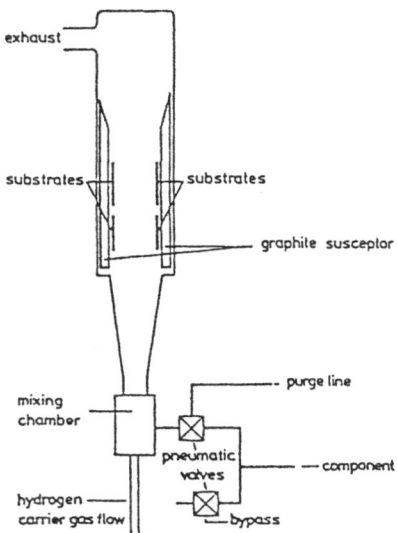

Fig. 2 "Chimney type reactor" designed by Leys et al. to grow heterostructures with abrupt interface[4].

In the down flow reactor, the space upstream the susceptor must be minimized to reduce convection circulation, which is also effective for quick change of the gas composition above the substrate. An example of such a reactor is one used by Kobayashi and Fukui[5], which is shown in Fig.3. Three wafers can be set in one run and the dead space is reduced. They grew an $(InAs)_n(GaAs)_n$ superlattice on an InP substrate. The superlattice with n down to 1 was grown. While the superlattice with n of 1 has to be viewed with some doubt in its perfectness, their result shows the vertical downflow reactor can be used to grow such an ultimate superlattice, when necessary precautions are taken for its basic structure and its operating procedure.

A guide line for the growth of superstructures in MOCVD would be summarized as follows:

1) Space upstream the susceptor should be as small as possible.
2) Gas flow speed should be high.
3) Flow pattern should be laminar and turbulence should be avoided.
4) Low pressure may be effective.

All the items above are intended to make quick change of the gas composition

above the substrate and/or to avoid disturbance of the flow pattern, during
the growth of the interface. Any other means to help to achieve such a
purpose would be effective for the growth of the abrupt interface. For the
growth of the quasi-monolayer interface, the whole gas composition should
favorably be changed within a fraction of the time necessary for the growth
of one monolayer. Whether the actual reactor is vertical or horizontal would
not have much concern so long as the necessary conditions are fulfilled.

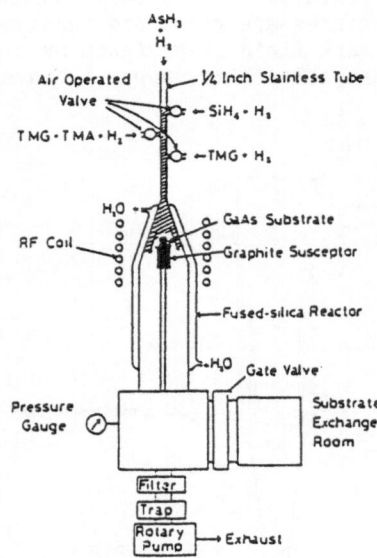

Fig. 3 A typical down flow reactor, where the upstream space was
minimized for the growth of abrupt structure[5].

Thickness Control in Thin Layer Growth

In-situ monitoring of the thickness and/or the composition of the
epitaxial layer is desirable for the precise control of the thickness and
composition of the growing layer. In MOCVD there seem to be no other means
than optical. Electron diffraction (RHEED) is effective , at least in
laboratory, to monitor the layer-by-layer growth process in MBE. This
technique is not applicable in MOCVD.

Ellipsometry was investigated as a means to monitor the growh process in
MOCVD by Hottier et al.[6]. They used the technique to determine the
transition layer thickness at the GaAs/AlGaAs(AlAs) interface. Their result
indicated the interface transition layer thickness less than 25A would be
difficult to resolve by this method. Kawai et al. investigated simple
reflectometry for in-situ monitoring of the thickness and the refractive
index of the grown layer[7]. It may be further refined for in-situ monitoring
of the grown layer in the range of 10 nm or some nm.

These methods are based on the interference effect due to the difference
between the refractive indices of the substrate and the thin grown layer and
due to the thickness of the layer. The resolution limit is determined by the
wavelength in the layer. The refractive index is a parameter of the bulk
material and it is doubtful that the optical means can be used in evaluating
the transition layer thickness less than 10 atomic layers.

Superlattices with unit layer thickness of few monolayers have been
grown in a growth system without any in-situ monitoring. The thickness is
controlled by timing the switching of the mass flow controller, with a
sequene controller. A one monolayer superlattice (InAs/GaAs)[8] and a two

monolayer superlattice (GaAs/AlAs)[9] have been grown simply by controlling the switching time.

Novel Approaches for Monolayer Growth

There have been several reports for layer-by-layer growth in MBE. In MOCVD there have recently been reported two new trials for realizing layer-by-layer epitaxy. One was called flow modulation epitaxy (FME) and reported by Kobayashi et al.[10]. TMG and arsine were switched alternately and carried to the substrate and one monolayer was grown in one cycle. It was essential to keep a low pressure of arsine in the TMG cycle. It is an interesting trial, though it is not certain whether one monolayer epitaxy is really controlled or controllable. Another approach is one reported by Doi et al.[11]. It was called switched laser MOVPE (SL-MOVPE). TMG and arsine were alternately carried to the substrate and irradiating the substrate in TMG cycle with an Ar laser stepwise monolayer epitaxy (SME) was made possible under a wide allowable range of conditions: substrate temperature and so on. Photochemical reaction is essential in this process, which improves the controllability of the monolayer epitaxy.

These approaches are encouraging to realize monolayer epitaxy in MOCVD. The photochemical docomposition helps the surface reaction in the latter and the epitaxial growth is possible at a temperature as low as 400C, and an allowable range of the monolayer growth condition is relaxed by the photochemical effect. These approaches combined with the high controllability of the MOCVD described in other parts of this report can be expected to realize a realistic monolayer epitaxy process.

BASIC SUPERSTRUCTURES GROWN BY MOCVD

Superlattices observed by Transmission Electron Microscopy

Sputtering Auger Electron Spectroscopy (SAES) is a most reliable means to explore an in-depth compositional profile of the layered structures. It has been widely used for investigating the layered structure 5nm or more thick. Layered structures as thin as 2.5nm was also analyzed by SAES, but the observed profile was severely deformed. Its resolution is limited to about 1.5nm by escape depth of auger electrons and by Ar sputtering process. Ultrathin layered structures recently grown by MOCVD is far beyond the resolution limit of SAES.

Transmission Electron Microscopy (TEM) is a sole means to be applicable to such a ultrathin layer, though it cannot provide a compositional profile. The wafer containing a superlattice on a (001) plane of GaAs is cleaved in a (110) plane and thinned further to some 10nm thick by ion etching for TEM observation. Fig.4 is a dark field image and a lattice image of a superlattice $(GaAs)_{17}(AlAs)_{11}$ [12]. From the image the interface abruptness is seen to be of about one monolayer and the roughness is also of the same order. We must, however, be careful because the TEM image is a kind of an average over the whole thickness of the wafer, several 10nm. Fig.5 is a dark field image over an area of 200nm of a superlattice $(GaAs)_2(AlAs)_2$. Though the image is also an average over the whole thickness of the wafer, the figure confirms us that the interface abruptness of about one monolayer are grown by MOCVD. In this figure it should be noticed that there is no disordering over the whole area of 200nm square: disordering means merging of two layers or branching out to two layers. Fig.6 shows a lattice image and an electron diffraction pattern of the superlattice[9]. The electron diffraction pattern also assures the superlattice has a period of 4 monolayers.

Fig. 4 Dark field and lattice images observed by transmission
electron microscopy (TEM) of an MOCVD grown superlattice
$(GaAs)_{17}(AlAs)_{11}$.

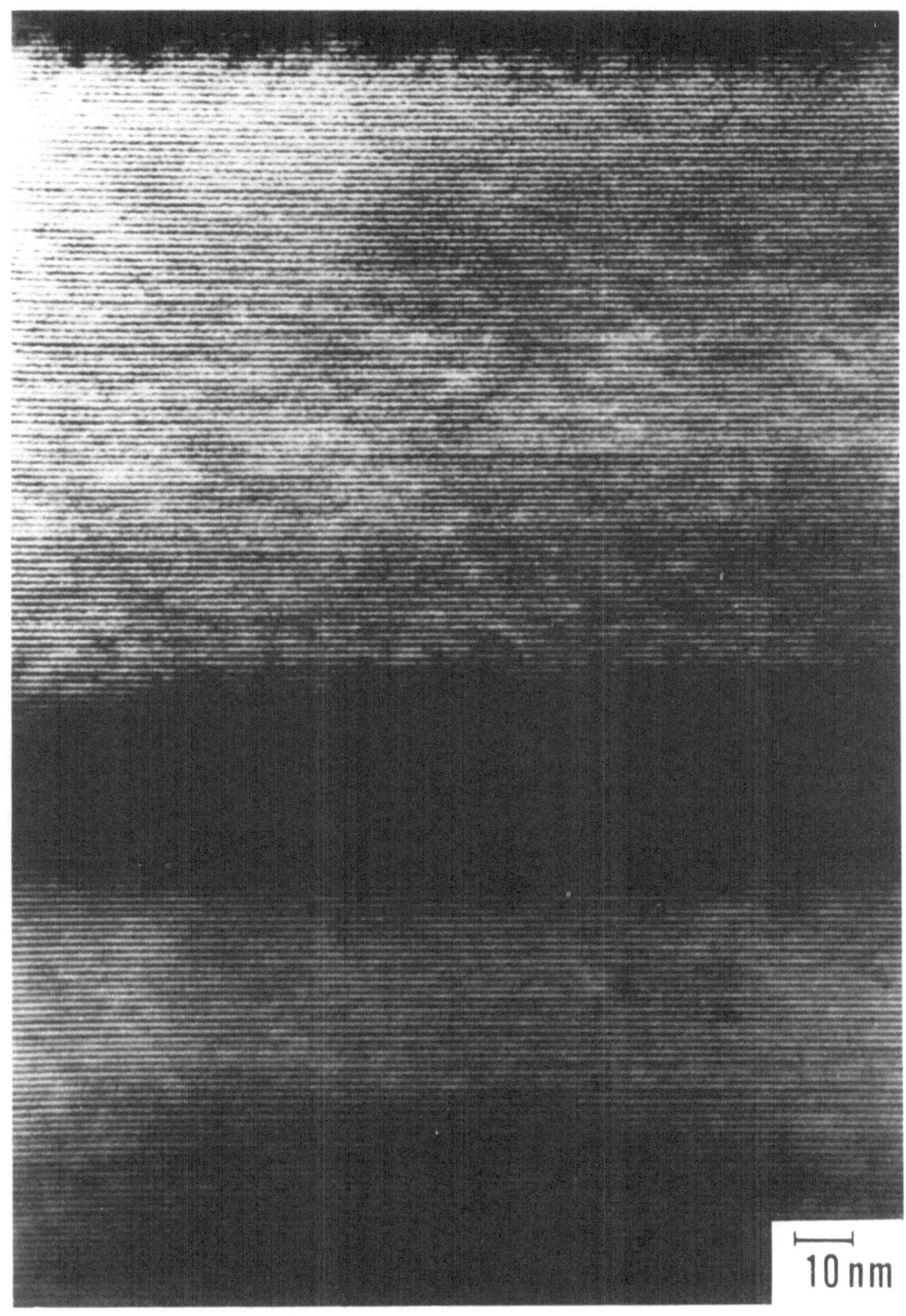

10 nm

Fig. 5 TEM dark field image of a $(GaAs)_2(AlAs)_2$ ultrathin
superlattice over an area of 200 nm square.

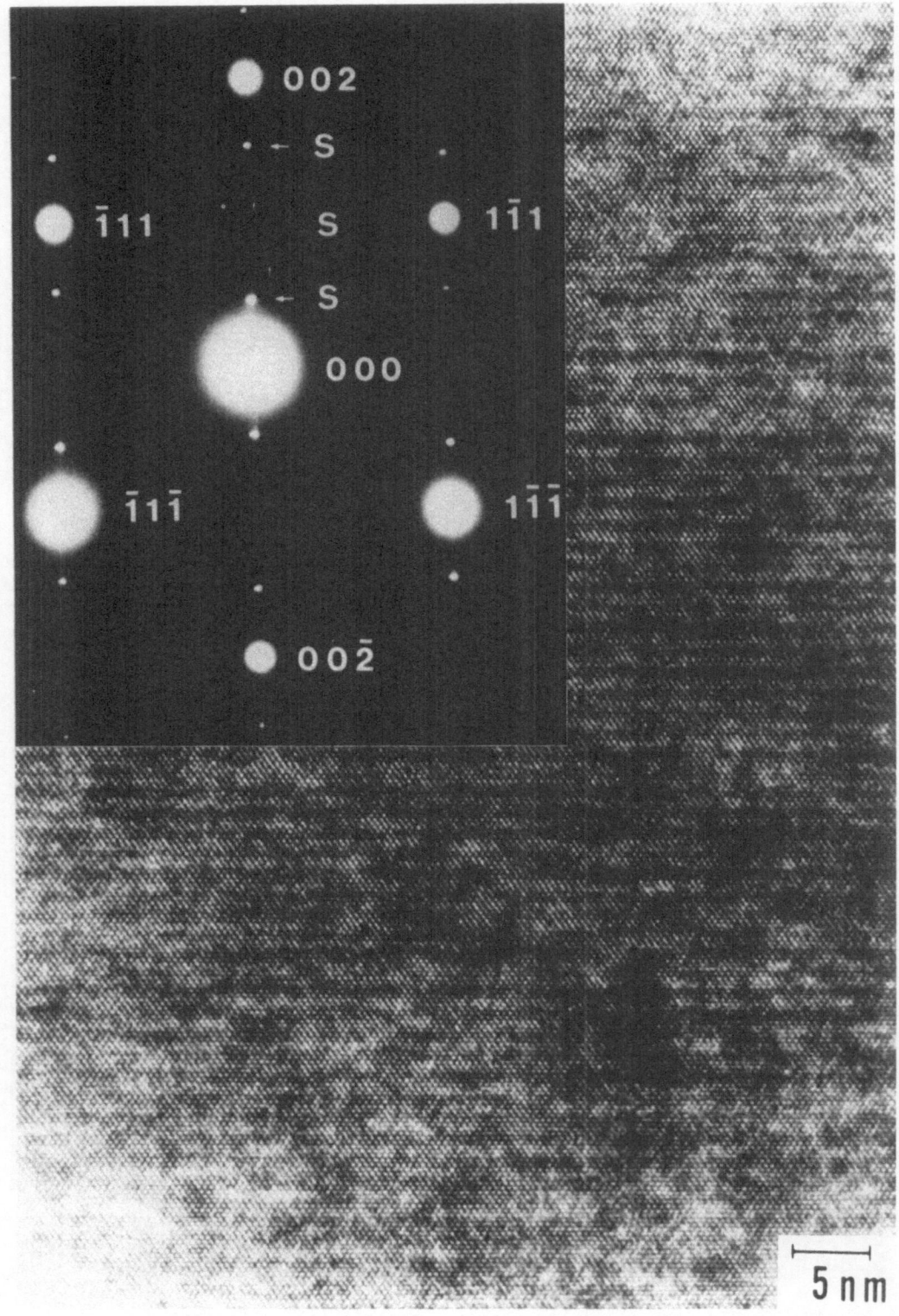

Fig. 6 A lattice image and an electron diffraction pattern of
the same superlattice as in Fig. 5.

Single and coupled Double Quantum Well: Photoluminescence Study

Photoluminescence (PL) is a simple but powerful means to investigate ultrathin quantum wells. The ground state of the quantum well, GaAs/AlGaAs, is determined by the interface band discontinuity and by the size of the quantum well. The wavelength of the PL line of quantum wells, GaAs/$Al_{0.54}Ga_{0.46}$, is plotted against the well width in Fig.7[13]. The experimental data are plotted together with a theoretically calculated curve. The agreeement of the data with the calculated curve is excellent. Possible interface degradation shifts the PL wavelength in the same way as the variation of the thickness of the well. The result indicates the controllability of the thickness as well as the interface degradation is within about one monolayer.

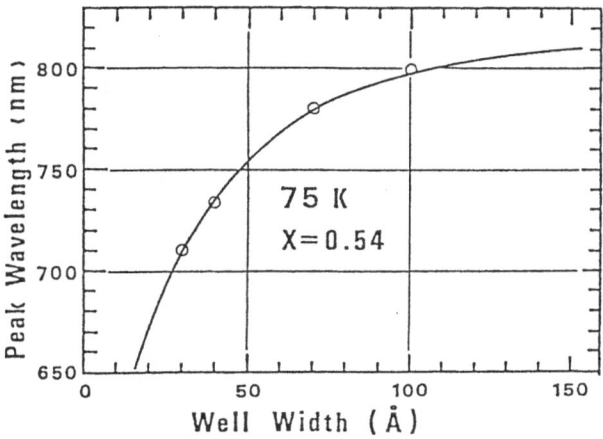

Fig. 7 Photoluminescence peak wavelength versus well width of the quantum well, GaAs sandwiched by $Al_{0.54}Ga_{0.46}As$; Circles are measured data and the curve is calculated[13].

A coupled double well is an artificial one dimensional two atom molecule with a variable interatomic distance. It is not only a test vehicle of the controllability of MOCVD but also that for checking the band discontinuity parameter at the heterointerface and related band parameters[14),15)]. Two quantum wells, GaAs/$Al_{0.5}Ga_{0.5}As$, 3.0nm thick are grown with an intervening barrier of thickness from 1.2 to 4.0nm[14)]. The ground states of the separate quantum wells are coupled via the barrier and the state is split to form a doublet: one symmetric and one antisymmetric state. Of the eight combinations of the two conduction band states with the four valence band states, two of the heavy hole band and two of the light hole band, optical transitions are allowed between four combinations. One main peak and three satellite shoulders are observed in the PL spectra as seen in Fig.8. The three shoulders are quenched at low temperature, 77K for example, due to thermalization of electrons and holes to the ground states. The shoulders are observed at higher temperature, room temperature for example.

The spectra systematically vary with the thickness of the intervening barrier. Their transition energies are analyzed and compared with the result theoreticlly calculated using appropriate band parameters and band discontinuity parameters. The data can be used to check the band parameters and the band discontinuity parameters. MOCVD can grow such fine structures with sufficient controllability.

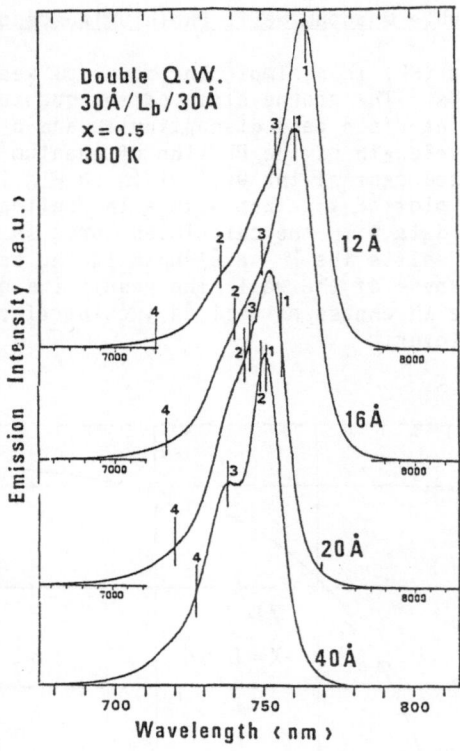

Fig. 8 Photoluminescence spectra of the $GaAs/Al_{0.5}Ga_{0.5}As$ coupled double quantum well. The intervening barrier thickness is changed from 1.2 to 4.0 nm. Marks 1~4 indicate four allowed transitions[14].

Single Quantum Wells: Other than GaAs/AlGaAs

GaInAsP materials have been studied for long wavelength laser materials. Razeghi et al. grew quantum wells with this material, GaInAs/InP[16]. The quantum well as thin as 8A was grown and the its PL line was clearly observed.

GaInAs/AlInAs (grown on InP) is expeted to be a promising material for high frequency and/or high speed devices because of its high electron mobility. Its recent result in devices' aspect will be decribed in a later section of this report. The quantum wells, GaInAs/AlInAs, as thin as 12A were grown and they showed narrow PL lines, Fig.9[17].

GaInP/AlInP (grown on GaAs) is promising for short wavelength lasers. The shortest wavelength lasers of all III-V materials are expected to be grown with this material. The quantum wells, GaInP/(AlGa)InP, as thin as 5A were grown and their PL lines were clearly observable, Fig.10[18].

The application field of MOCVD has been and is expanding to wider area. The quantum wells have been grown with various material systems as well as epitaxial layers for electronic and optical devices. These materials expand the usable wavelength range of the lasers and the frequency range of the electronic devices.

124

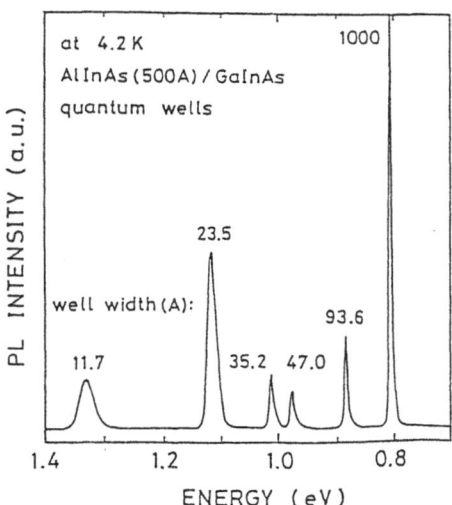

Fig. 9 Photoluminescence spectra of the GaInAs/AlInAs quantum wells.
Numerals in the figure are width of the quantum wells[17].

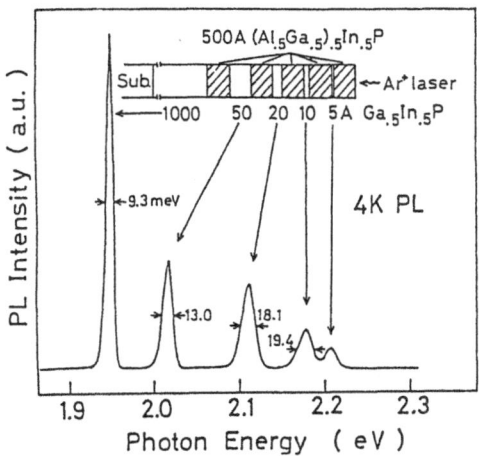

Fig. 10 Photoluminescence spectra of GaInP/(AlGa)InP quantum wells.
The peak from the narrowest well of 5 A is clearly seen[18],[19].

GROWTH MECHANISM IN MOCVD

In the preceeding sections the recent MOCVD processes in application to the growth of the thin layered structure were described and the recent progress in the growth of the basic thin layered structure was reviewed. In this section we will discuss on the growth mechanism in MOCVD based on the study of the growth temperature change of the quantum well grown by MOCVD.

Single Quantum Well Structure and Growth Temperature

Superlattices and quantum wells are grown in MOCVD, usually at temperatures higher than in MBE. The ultrathin superlattice $(GaAs)_2(AlAs)_2$ described in a preceeding section was actually grown at 780C. It is in marked contrast with the case of MBE, where the thin superlattices are usually difficult or unconceivable to grow at such a high temperature. A series of quantum wells of different sizes were grown at temperatures from 670C to 820C to investigate the growth temperature dependence of the well structure[9].

The PL spectra taken on samples containing quantum wells of varied sizes are shown in Fig.11. The most remarkable point to be noticed is that the linewidth of the PL peak becomes sharper as the growth temperature increases up to 820C. It is contrary to the case of MBE, where the superlattices are more degraded with the increasing growth temperature and the growth of superlattices are difficult or impossible at temperatures higher than 650C. The interface smoothness was apparently improved with the growth temperature. Even in the quantum well grown at the lowest temperature, 670C, where the interface is roughest, the PL linewidth can be interpreted assuming that the interface has only hills of one monolayer height. The apparent PL linewidth of the quantum well grown at higher temperature is therefore a fraction of one monolayer, and it is necessary to introduce some detailed structure to explain the apparent interface roughness of less than one monolayer.

There are two possibilities to explain the apparent interface roughness less than one monolayer. The first possibility is to assume the interface is actually flat with only occasional small hills. It must, then, be assumed that once a nucleus of growth is formed on the growing plane a whole plane should be covered instantly with one crystal layer. The number of isolated nuclei should decrease with the growth temperature in this model. It is not a natural assumption. The second possibility is to assume that there are many hills or valleys of one monolayer height over the growing plane as seen in Fig.12, but their lateral sizes become smaller as the growth temperature rises. The exciton in the well has a size of about 25nm, and the interface would effectively flat for the exciton if the size of the hills are much smaller than the size of the exciton[9],[20].

The dependence of the PL spectra on the growth temperature was analyzed on the following assumption: 1) The interface has many hills of one monolayer step. 2) The hills consist of smallest unit hills, whose size is a function of the growth temperature. 3) The fraction of the exciton size covered by the hill level area fluctuates statstically around its average of 50%, which results in a fluctuation of the effective well thickness seen by the exciton. The dependence of the PL spectra on the growth temperature is explained as follows: As the size of the smallest unit hill decreases with the increasing growth temperature, the fluctuation of the coverage ratio decreases and the fluctuation of the effective well thickness decreases.[9],[20]

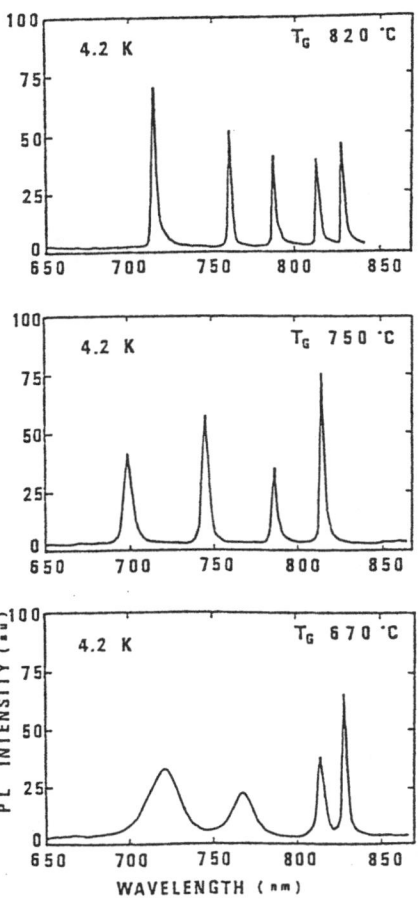

Fig. 11 Photoluminescence spectra of GaAs/AlGaAs quantum wells of
sizes 3.0 to 15.0 nm, grown at temperatures from 670 to
820 C. The apparent linewidth of the peak becomes narrower
as the growth temperature increases[9].

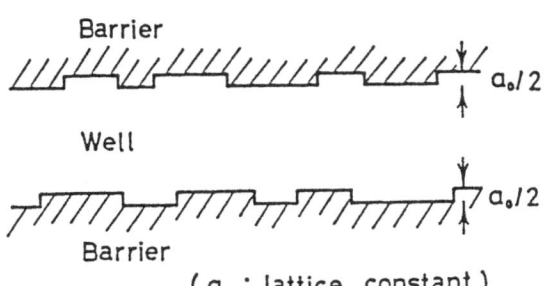

Fig. 12 A model of the heterointerface structure.

Growth Mechanism in MOCVD: MOCVD VS MBE

The heterointerface structure GaAs/GaAlAs grown by MOCVD was discussed in the last section based on the data of the PL spectra of the quantum wells grown at various temperatures from 670 to 820C. Following Singh et al.'s formulation[20] we calculated the width of the PL line expected from the quantum wells with the various hill mosaics on the interface. The PL linewidth was plotted against the well width with the smallest unit size of the hill as a parameter in Fig.13[9]. The experimental linewidth data were also plotted in the figure using different symbols for the data obtained at

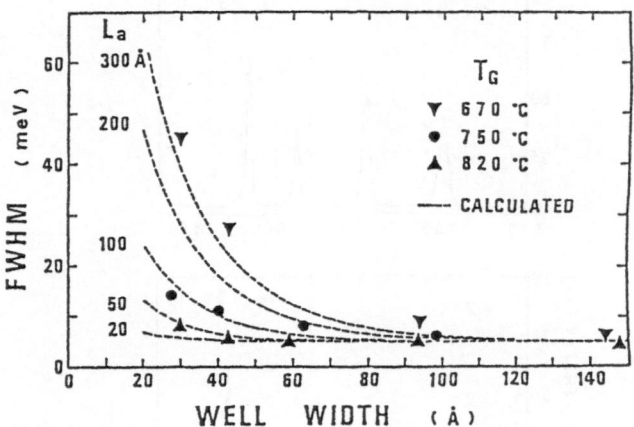

Fig. 13 Line width of the PL peak from quantum wells, GaAs/Al$_x$Ga$_{1-x}$As (x=0.5 for ▼ and ▲, 0.6 for ●). Dotted curves represent the calculated result for the well with the interfaces with hills and valleys of monolayer height as shown in Fig. 12. L$_a$ denotes the smallest unit size of the hills[9].

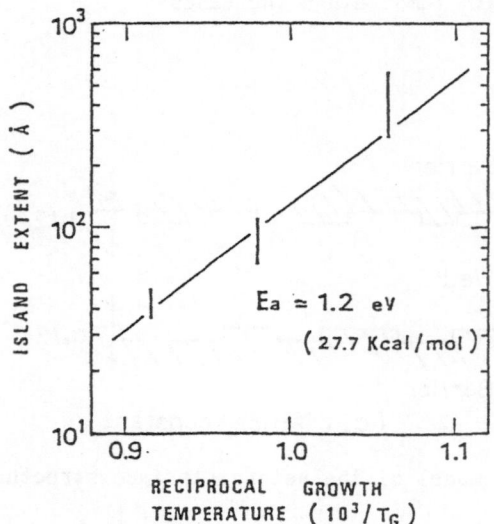

Fig. 14 Linear hill size estimated from Fig. 13 is plotted against the reciprocal growth temperature.

the three growth temperatures. The data corresponding to one particular growth temperature lie near one curve corresponding to a particular hill size. The unit hill size was related to the growth temperature and plotted in Fig.14[9]. It is a relation of the linear dimension of the hill, that is reiprocal of the linear step density, against the reciprocal temperature. The activation energy is 1.2eV, and the activation energy of the two dimensional step formation is 2.4eV twice the one dimensional value.

The smaller hill formation on the interface plane at the higher growth tmeperature is a natural model, but the question is why the overall superlattice structure is retained, while many small hills are formed on the heterointerface. In MBE the superlattice can never been grown at temperatures as high as 820C. To explain this situation we proposed a model of the elementary process of crystal growth in MOCVD[9].

GaMe$_3$ (Me=CH$_3$) molecules are carried to the growth surface by a carrier gas hydrogen in MOCVD process. They will be cracked and lose Me bases partly or completely. Our model for this process is that GaMe$_3$ will lose two Me bases and reach the surface as GaMe* [9]. This ad-molecule will wonder over the surface, and finally it will react on the surface with As or AsH* to form one fixed unit of GaAs. The whole reaction process is pictorially drawn in Fig.15. When the reaction takes place on the flat plane, it will make a nucleus originating a new hill. The hill formation rate is limited by this reaction. The overall growth rate, however, is determined by the feed rate of the reactants that reach the surface by diffusion in the stagnant layer. The last Me base is thought to have high bonding energy of 3.4eV[21] and it is thought to be a reason why the activation energy of hill formation is high of 2.4eV.

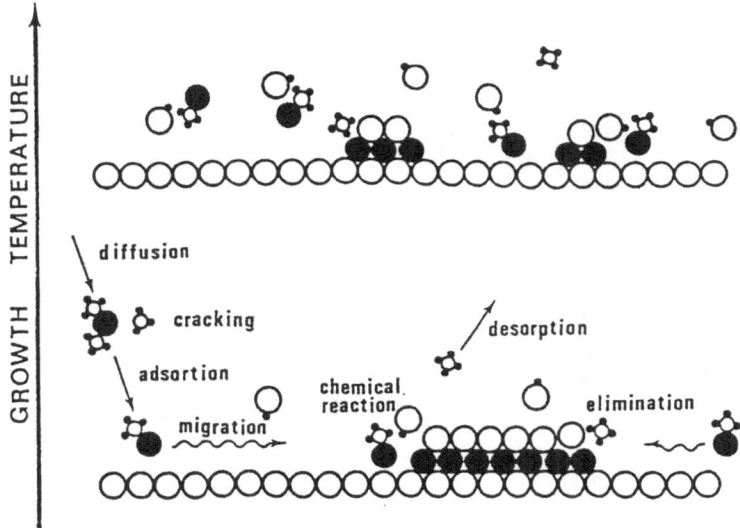

Fig. 15 A model of the elementary process in MOCVD. Larger open and closed circles, smaller open and closed circles represent arsenic, gallium, carbon and hydrogen atoms, respectively.

The model described in the preceeding paragraph is important to explain the difference of the elementary growth process and the resultant difference in the growth temperature dependence of the interface structure between MOCVD and MBE. In MBE the metal ad-species would be a Ga atom and the atom size is small compared to the case of MOCVD where the ad-species was proposed to be a a GaMe* molecule. The Ga ad-atom at the surface is very active and may exchange position with a deeper lying metal atom, Al for example, which can be thought to be a reason why the superstructures cannot be conserved in MBE at higher growth temperature.

There appeared some results relating to our proposed growth model. S.P.Denbaars et al. measured decomposition products from TMG flowing into the hot wall furnace. The main product was methane and it was twice the original TMG up to 450C and increased to three times above that temperature[22]. Their result was obtained in quasi-equilibrium in the hot walled furnace and may not be directly applicable to the superlattice growth, where the gas flow rate is very high and the reactor has a cold wall. The fact that the methane product was twice the TMG up to 450C in quasi-equilibrium may be considered to be a partial support of our model. Tirtowidjojo and Pollard made a thermodynamic calculation of equilibrium gas phase species in a TMG-AsH$_3$-H$_2$ system[23]. Their result was that the most abandant Ga containing gas phase species between 750K and 1100K is Ga(CH$_3$). These results seem to be indirectly supporting our model.

SUPERSTRUCTURES AND DEVICES GROWN BY MOCVD

Various kinds of structures and devices have been grown by MOCVD and one of them, HIFET (hetero-interface FET) has already been actual commercial products. We will review various kinds of superstructures and related devices made by MOCVD.

Ultrathin Superlattice

Superlattices, (GaAs)$_n$(AlAs)$_n$, with n from 25 down to 1 have been grown and their physical properties have been investigated. It may be of some doubt whether the superlattice with n of unity was actually grown. The Raman scattering experiment used to measure the phonon spectra of the superlattice, however, revealed the phonon energy of the superlattice intended to be (GaAs)$_1$(AlAs)$_1$ is clearly different from that of the random alloy Al$_{0.5}$Ga$_{0.5}$As[24]. The superlattice (GaAs)$_2$(AlAs)$_2$ shows a clear TEM image as described.

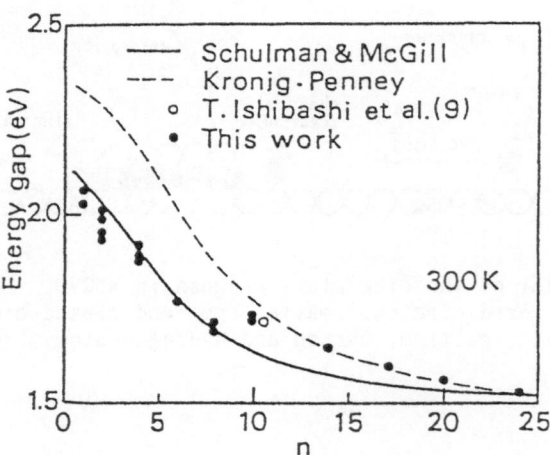

Fig. 16 Direct energy gap (Γ band) of the (GaAs)$_n$(AlAs)$_n$ superlattice versus the number of monolayers (n) per slab[24].

The simple Kronig-Penney model analysis of the band structure of the quantum well predicts that the band edge of the superlattice $(GaAs)_n(AlAs)_n$ becomes indirect as n decreases and crosses 10, because the direct band edge of GaAs has a smaller mass and the ground level in the quantum well rises faster than the other indirect band edges. The band edge of the electronic level in the ultrathin layer superlattice (UTSL) seems to be direct for n down to 2, because the PL efficiency is as high as that of the superlattice with higher n value at liq.He temperature[24]. It is not in agreement with the prediction based on the Kronig-Penney model and the tight binding calculations gave the better agreement with the experimental result in the UTSL with n less than about 10 as seen in Fig.16. The result also predicts that the direct band gap material, with a gap of 2.0 eV, can be realized with GaAs-AlAs series when the UTSL is formed. Phonons in the UTSL also behaves differently from the bulk material. They are pinned by the UTSL and the electron-phonon interaction is also expected to be different from the bulk material[25].

Multi-quantum Well Laser and GRIN-SCH Laser

Multi-quantum well laser was a start of MOCVD for growing superlattices and superstructures. MOCVD was once suspected to have problems for growing thin barrier layers[1]. It has remarkably been advanced to wipe out such views. The multi-quantum well laser was investigated to make lasers at wavelength shorter than attainable with the mixed alloy compound. The multi-quantum well laser with a well layer containing Al was also investigated to shorten the lasing wavelength[26]. The ultrathin superlattice described in the last section may be used for this purpose.

Another laser of interest is a GRIN-SCH (graded index separate confinement) laser. The threshold current was remarkably reduced incorporating a graded index layer in combination with a narrow quantum well. The very low threshold current laser of this kind was also grown by MOCVD[27]. This concept is effective to reduce threshold current and may find wide applications especially when the laser is incorporated in the integrated circuit (OEIC).

Modulation doping and HIFET (hetero-interface FET): GaAs/AlGaAs

Most of the experiments on modulation doping have been done by MBE. Andre et al reported enhanced electron mobility in the modulation doped structure grown by MOCVD[28]. Still better results have been reported by Fukui and Kobayashi[29], and by Mori et al.[30]. Fig.17 is a plot of electron mobility against temperature for the modulation doped structure with a spacer layer of 10 nm and a sheet carrier concentration of 5.2×10^{11} cm^{-2}[30]. The mobility is 7500, 148,000 and 400,000 cm^2/v·s at room, liquid nitrogen and liquid helium temperature, respectively. They are lower than the best record of the structure grown by MBE, but still high enough for actual device applications. Fig.18 compares the mobility at 77K of the modulation doped interface carriers with the data published by Hiyamizu et al.[31].

HIFET (hetero-interface FET) is a MISFET using an AlGaAs layer as an insulator layer (HEMT=n channel HIFET). It is so named because the essential difference of this device from the MOSFET is in the use of the crystalline insulator and the crystalline hetero-interface at the channel. The channel conduction type and V_{th} can be controlled by doping the insulator (AlGaAs) with positive ions (donor) for n-type or with negative ions (acceptor) for p-type. The HIFET device is fabricated from the wafer grown by MOCVD. Fig.19 is a plot of the noise figure data at 12 GHz for four n-HIFET devices with different gate lengths[32]. The HIFET is already produced as a commercial device by MOCVD.

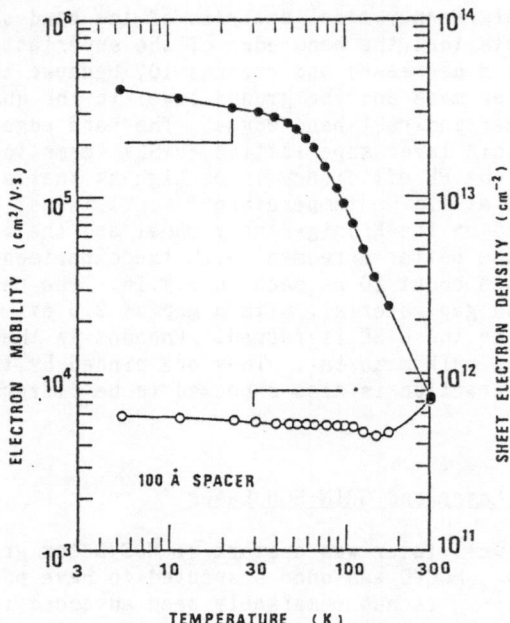

Fig. 17 Two-dimensional electron mobility against temperature in
the modulation doped GaAs/n-AlGaAs structure with a spacer
of 10 nm[30].

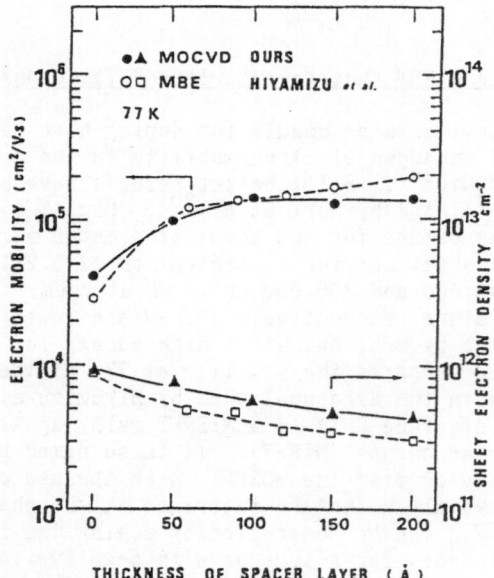

Fig. 18 Comparison of the electron mobility in the modulation doped
structure grown by MOCVD[30] and MBE[31].

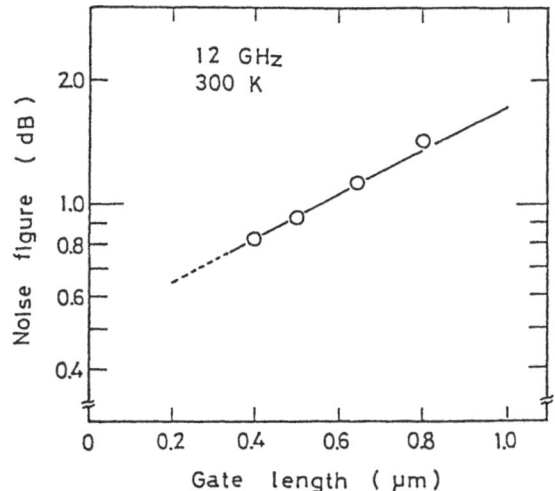

Fig. 19 Noise figure versus the gate length in the n-HIFET (GaAs/
AlGaAs) measured at 12 GHz[32].

Modulation doping and HIFET: InGaAs/AlInAs

InGaAs grown lattice matched to InP has high electron mobility and is
considered to be a hopeful material for high frequency and/or high speed

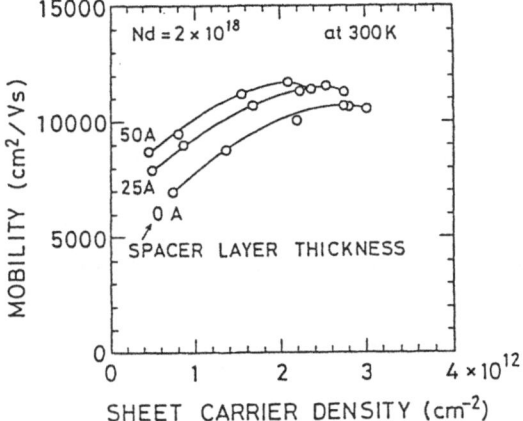

Fig. 20 Room temperature electron mobility in the GaInAs/n-AlInAs
modulation doped structure with various spacer layer
plotted against the sheet electron density[17].

devices. Its advantage in application to HIFET is not only in its high
electron mobility but also in its high energy band discontinuity at the
interface with AlInAs. The sheet carrier concentration can be increased
owing to its high band gap discontinuity. The high carrier concentration
being combined with the high electron mobility, a much higher quality factor
can be realized in this material combination compared with GaAs/AlGaAs.

Figure 20 is a recently obtained mobility-sheet carrier concentration
relation[28], which is a first data obtained by MOCVD of this kind. It should
be noticed that the mobility continues to increase after the sheet carrier
concentration has passed 10^{12} cm^{-2}, and the highest mobility of 11,700 cm^2/v·s
was obtained at a sheet carrier concentration of 2.6×10^{12} cm^{-2}. The mobility
at 77K was 57,200 cm^2/v·s, which was not so high, as it is well known, limited
by alloy scattering. The HIFET was fabricated from the wafer with the gate
length of 1 μm and the transconductance was 404 mS/mm.

Hetero-bipolar Transistor

The hetero-bipolar transistor (HBT) was proposed many years ago by
H.Kroemer. Using a large gap semiconductor material as an emitter, the
injection efficiency of the emitter is expected to be improved and the base
resistance can be lowered because the base can be highly doped without
significantly increasing the counter injection from the base to the emitter.
The HBT structure can be formed by GaAs/AlGaAs, because this material system
can make a high quality heterojunction using AlGaAs as a large gap emitter.
As with other heterojunction electronic devices, the HBT has also been grown
mostly by MBE. Dubon et al. reported on a HBT AlGaAs(n)/GaAs(p)/AlGaAs(n)
grown by MOCVD that its characteristics depended on the doping profiles of Zn
in the base and the graded profile of Al in the emitter-base junction[33].
They obtained a high transconductance of 5500 in their HBT.

Taira et al. made a systematic study of the effect of the emitter
grading on the transistor characteristics: how the emitter spike is reduced
and how the hole confinement changes with the grading in the emitter
barrier[34]. MOCVD can make detailed control of the Al profile and/or the Zn
profile to numerically analyze their effect on the electrical
characteristics. They also used the HBT structure to estimate the band
discontinuity parameter from the dependence of the collector current on the
emitter-base bias voltage and on the temperature. They estimated the band
offset is $\Delta E_c = 0.6 \ \Delta E_g$ for x < 0.4[35].

MOCVD will possibly be a main stream technology, when the HBT is to be
produced as commercial products.

Hot Electron Transistor

The hot electron transistor (HET) is a very sophisticated device and
is expected to have advantage in that it is a unipolar device and the charge
carrier of the emitter current and the charge in the base are both electrons
and the base resistance can be reduced owing to the electrons' high mobility.
Scattering of the injected electrons in the base layer deprives the electrons
of their energy and they readily lose energy and become the base current.
The probability of the scattering is very high and it is very difficult to
obtain a high current gain. High current gain of 2 at liq.He temperature,
high as HET, was attained by the HET grown by MOCVD[36]. It is a high value,
a bit better than the one attained by MBE.

SUMMARY AND CONCLUSION

Recent Progress of MOCVD technology was reviewed, placing main stress
on the growth of superlattices and superstructures. Most of the structures
grown by MBE and some structures not grown by MBE have been grown by MOCVD.

There is some fundamental difference in the growth mechanism and in its elementary process. It is reflected in the difference in the growth temperature dependence of the interface structure. Ultrathin superlattices and devices based on abrupt interface structure grown by MOCVD were reviewed.

The difference in the growth mechanism between MBE and MOCVD were discussed on a tentatively proposed model. Its microstructural difference should be clarified for better understanding of the growth mechanism. Whatever the difference, MOCVD will more and more widely be used as an actual means for device production.

REFERENCES

1. N. Holonyak, Jr., W. D. Laidig, B. a. Bojak, K. Hess, J. J. Coleman, P. D. Dapkus, and J. Bardeen, Alloy Clustering in $Al_xGa_{1-x}As$-GaAs Quantum-Well Heterostructures, Phys. Rev. Lett., 24:1703 (1980).
2. P. M. Frijlink, and J. Maluenda, MOVPE Growth of $Ga_{1-x}Al_xAs$-GaAs Quantum Well Heterostructureas, Japan. J. Appl. Phys., 21:L574 (1982).
3. H. M. Manasevit, and W. I. Simpson, The Use of Metal-Organics in the Preparation of Semiconductor Materials, J. Electrochem. Soc., 116:1725 (1969).
4. M. R. Leys, C. Van Opdorp, M. P. a. Viegers, and H. J. Talen-van der Mheen, Growth of Multiple Thin Layer Structures in the GaAs-AlAs System Using a Novel VPE Reactor, J. Cryst. Growth, 68:431 (1984).
5. N. Kobayashi, and T. Fukui, Selectively Doped GaAs/n-AlGaAs Heterostructures Grown by MOCVD, Extended Abstracts of the 16th (1984 International) Conference on Solid State Devices and Materials, P671 (1984).
6. F. Hottier, J. Hallais, and F. Simondet, In situ monitoring by ellipsometry of metalorganic epitaxy of GaAlAs-GaAs superlattice, J. Appl. Phys., 51:1599 (1980).
7. H. Kawai, S. Imanaga, K. Kaneko, and N. Watanabe, Complex refractive indices of AlGaAs at high temperature measured by in situ reflectometry during growth by metalorganic chemical vapor depositon, to be published in J. Appl. Phys., 61 on January 1 (1987).
8. T. Fukui, and H. Saito, $(InAs)_1(GaAs)_1$ Layered Crystal Grown by MOCVD, Japan. J. Appl. Phys., 23:L521 (1984).
9. N. Watanabe, and Y. Mori, Ultrathin Layers GaAs/GaAlAs grown by MOCVD and their Structural Characterization, Surface Science, 174:10 (1986)
10. N. Kobayashi, T. Makimoto, and Y. Horikoshi, Flow-Rate Modulation Epitaxy of GaAs, Japan. J. Appl. Phys., 24:L962 (1985).
11. A. Doi, Y. Aoyagi, S. Iwai, and S. Namba, Stepwise Monolayer Growth of GaAs by Switched Laser Metal Organic Vapor Phase Epitaxy, Extended Abstracts of the 18th (1986 International) Conference on Solid State Devices and Materials, P739 (1986).
12. K. Kajiwara, H. Kawai, K. Kaneko, and N. Watanabe, Structure of MOCVD grown AlAs/GaAs heterointerfaces observed by transmiossion electron microscopy, J. Appl. Phys., 24:L85 (1985).
13. H. Kawai, K. Kaneko, and N. Watanabe, Photoluminescence of AlGaAs/GaAs quantum well grown by metallorganic chemical vapor deposition, J. Appl. Phys., 56:463 (1984).
14. H. Kawai, K. Kaneko, and N. Watanabe, Doublet state of resonantly coupled AllGaAs/GaAs quantum wells grown by metalorganic chemical vapor deposition, J. Appl. Phys., 58:1263 (1985).
15. N. Watanabe, and H. Kawai, Single and coupled double-well GaAs/AlGaAs and energy dependent light hole mass, to be published in J. Appl. Phys., vol.60 on 15 November (1986).
16. M. Razeghi, J. Nagle, and C. Weisbuch, Optical studies of GaInAs/InP quantum wells, Proc. 11th Int. Symp. GaAs and Related Compounds, Biarritz (1984), Inst. Phys. Conf. Ser., 74:379 (1985).
17. M. Kamada, H. Ishikawa, M. Ikeda, Y. Mori, and C. Kojima, Selectively

doped AlInAs/GaInAs heterostructures grown by MOCVD and their application to HIFETs (heterointerface FETs), 13th International Symposium on Gallium Arsenide and Related Compounds (Las Vegas) 1986.

18. M. Ikeda, K. Nakano, Y. Mori, K. Kaneko, and N. Watanabe, MOCVD Growth of AlGaInP at Atmospheic Pressure Using Triethylmetals and Phosphine, J. Cryst. Growth, 77:380 (1986).

19. M. Ikeda, pivate communication.

20. J. Singh, K. K. Bajaj, and S. Chaudhuri, Theory of photoluminescence line shape due to interfacial quality in quantum well structures, Appl. Phys. Lett. 44:805 (1984).

21. M. G. Jacko, and S. J. W. Price, The Pyrolysis of Trimethyl Gallium, Canadian Journal of Chemistry, 41:1560 (1963).

22. S. P. DenBaars. B. Y. Maa, P. D. Dapkus, A. D. Danner and H. C. Lee, Homogeneous and heterogeneous thermal decomposition rates of trimethylgallium and arsine and their relevance to the growth of GaAs by MOCVD, J. of Cryst. Growth, 77:188 (1986).

23. M. Tirtowidjojo, and R. Pollard, Equilibrium gas phase species for MOCVD of $Al_xGa_{1-x}As$, J. of Cryst. Growth, 77:200 (1986).

24. A. Ishibashi, Y. Mori, M. Itabashi, and N. Watanabe, Optical properties of $(Alas)_n(GaAs)_n$ superlattices grown by metalorganic chemical vapor deposition, J. Appl. Phys., 58:2691 (1985).

25. A. Ishibashi, Y. Mori, M. Itabashi, and N. Watanabe, A fundamentally new aspect of electron-phonon interaction in $(AlAs)_m(GaAs)_n$ ultrathin-layer superlattices, 18th International Conference on the Physics of Semiconductors (Stockholm), Tu 3-1, 33 (1983).

26. H. Kawai, O. Matsuda, and K. Kaneko, High Al-content visible (AlGa)As multiple quantum well heterostructure lasers grown by metalorganic chemical vapor depositon, Japan. J. Appl. Phys. Lett., 22:L727 (1983).

27. S. D. Hersee, M. Karakowski, R. Blondeau, M. Baldy, B. de Cremoux, and J. P. Duchemin, Abrupt OMVPE Grown GaAs/GaAlAs Heterojunctions, J. of Cryst. Growth, 68:383 (1984).

28. J. P. Andre, A. Briere, M. Rocchi, and M. Riet, Growth of (Al,Ga)As/GaAs Heterostructures for HEMT Devices, J. of Cryst. Growth, 68:445 (1984).

29. N. Kobayashi, and T. Fukui, Improved 2deg Mobility in Selectively Doped GaAs/N-AlGaAs Grown by MOCVD Using Triethyl Organimetallic Compounds, Electron. Lett., 20:887 (1984).

30. Y. Mori, F. Nakamura, and N. Watanabe, High electron mobility in the selectively doped heterostructures grown by normal pressure metalorganic chemical vapor depositon, J. Appl. Phys., 60:334 (1986).

31. S. Hiyamizu, J. Saito, K. Nanbu, and T. Ishikawa, Improved Electron Movility Higher than 10^6 cm^2/Vs in Selectively doped GaAs/N-AlGaAs Heterostructures Grown by MBE, Japan. J. Appl. Phys., 22:L669 (1983).

32. K. Tanaka, H. Takakuwa, F. Nakamura, Y. Mori, and Y. Kato, Low-noise Microwave HIFET Fabricated using Photolithography and MOCVD, Electron. Lett., 22:487 (1986).

33. C. Dubon, R. Azoulay, P. Desrousseaux, J. Dangla, A. M. Duchenois, M. Hountondji, and D. Ankri, Effects of MOCVD growth conditions on DC characteristics of GaAlAs-GaAs double heterojunction bipolar transistors, Proc. 11th Int. Symp. GaAs and Related Compounds, Biaritz (1984), Inst. Phys. Conf. Ser., 74:175 (1985).

34. K. Taira, C. Takano, H. Kawai, and M. Arai, Emitter grading in AlGaAs/GaAs Heterojunction bipolar Transistors grown by metalorganic chemical vapor depositon, to be published in Appl. Phys. Lett., vol. 60 on November 10 (1986).

35. K. Taira, C. Takano, H. kawai, and M. Arai, Band offsets deduced from AlGaAs/GaAs heterojunction bipolar transistors, (unpublished).

36. I. Hase, H. Kawai, S. Imanaga, K. Kaneko, and N. Watanabe, MOCVD-grown AlGaAs/GaAs hot electron transistors with a base width δf 30 nm, Electron. Lett., 21:757 (1985).

GROWTH OF INDIUM PHOSPHIDE/INDIUM GALLIUM ARSENIDE STRUCTURES BY

MOCVD USING AN ATMOSPHERIC PRESSURE REACTOR

S.J. Bass, S.J. Barnett, G.T. Brown, N.G. Chew, A.G. Cullis,
M.S. Skolnick, and L.L. Taylor

Royal Signals and Radar Establishment
St. Andrews Road, Malvern, Worcs. WR14 3PS, U.K.

ABSTRACT

An atmospheric pressure reactor has been used to grow indium phosphide/indium gallium arsenide epitaxial layers. Growth pauses were used to sharpen the heterojunction interfaces. In order to study heterojunctions, two dimensional gas and quantum effects, remotely doped alloy layers, quantum wells (QWs) and multiple quantum wells (MQWs) were grown. The structures were studied by Hall measurements, magneto-resistance effects, photoluminesence (PL), optical absorption, double crystal x-ray diffraction and transmission electron microscopy (TEM).

The best indium phosphide grown had a mobility of 135000 cm^2 $sec^{-1}v^{-1}$ at 77K with n = 3 x $10^{14}cm^{-3}$. Indium gallium arsenide alloy compositions were reproducible with a mobility of 82000 cm^2 $s^{-1}v^{-1}$ at 77K and n = 6 x $10^{14}cm^{-3}$. The optimum growth temperature was 660-680°C. The x-ray diffraction rocking curve width (FWHM) was 21 arc secs and the PL linewidth at 2K was 1.85 meV.

The remotely doped structures showed good two dimensional electron gas (2 DEG) effects at the heterojunction with well resolved Shubnikov de Haas oscillations and a 2 DEG mobility of 85000 $cm^2v^{-1}s^{-1}$ at 6.5 x 10^{11} cm^{-2} and 4.2K. Well defined quantum wells could be grown even with growth times as short as 3 seconds. Quantum wells had a PL linewidth of 5.3 meV at 150Å width and 2K and 5.5meV at 100Å at zero carrier density. The quantum wells showed 2 DEG mobilities of up 95000 cm^2 $sec^{-1}v^{-1}$ at carrier densities of 9 x $10^{11}cm^{-2}$. Multiple 100Å quantum wells have been grown with up to 50 repeats and InP spacers of 50-500Å. The PL linewidths of MQWs were similar to those of single QWs. Well resolved heavy and light hole peaks were observed in optical absorption with line widths of 10 meV. X-ray diffraction showed that the periodicity of the MQWs was well-controlled but there was a periodic strain in a nominally lattice matched structure. TEM images showed that the microscopic uniformity of the MQWs was good. The alloy/InP interface was never as sharp as the InP/alloy interface.

INTRODUCTION

There is now considerable research interest in low dimensional solid (LDS) structures in III-V compounds to study heterojunction, two dimensional electron gas (2 DEG) and quantum effects. Examples of these structures are high electron mobility transistors (HEMTs), , quantum wells (QWs), multiple quantum wells (MQWs) and superlattices. III-V compound devices are of great commercial importance and LDS structures offer the prospect of a number of novel devices as well as a significant improvement of existing devices. LDS material can be grown by vapour phase epitaxy (VPE), liquid phase epitaxy (LPE), metalorganic chemical vapour deposition (MOCVD) or molecular beam epitaxy (MBE). Of these MOCVD and MBE are by far the most widely used. In the Ga-Al-As system recent results from MBE have been impressive with atomic layer control of growth and excellent electrical and optical properties. To achieve these results a very advanced technology is required with equipment engineered to a very high standard. MOCVD has also given good results and could well be a better production technique.

For the In-Ga-As-P III-V compounds MOCVD is the most widely used. There has been relatively little work on MBE of indium and phosphorus compounds. Using a volatile element like phosphorus in a high vacuum system presents some problems. Recently there have been good results using solid source MBE (1). There have also been good results using chemical beam or metal-organic MBE (2). In this technique organometallics or hydrides are used in place of the elements.

Using MOCVD there are many variations possible in starting materials (precursors), reactor design, control systems, growth procedures and operating pressure. A number of workers (3,4) used the low pressure MOCVD reactor usually operating at 0.1 atmospheres. The advantages claimed are less unwanted side reactions and more uniform growth. The atmospheric pressure reactor (5-9) is simpler to operate and can also give very good results. A comprehensive review of MOCVD has been given by Ludowise (10). In this paper we describe what has been achieved using an atmospheric pressure reactor to grow indium phosphide and indium gallium arsenide thick epitaxial layers and LDS structures: the choice of precursors, equipment, design and growth procedures are also considered. Electrical measurements, photoluminescence (PL), X-ray diffraction, optical absorption and transmission electron microscopy (TEM) were used to characterise the material and structures grown.

MATERIALS

The basic reaction for MOCVD is thought to be:

$$M(CH_3)_3 + XH_3 \xrightarrow[\text{SUBSTRATE}]{\substack{\text{HEAT} \\ H_2\,\text{CARRIER} \\ \text{GAS}}} MX + 3CH_4$$

METAL ALKYL   HYDRIDE   III-V COMPOUND

It is an irreversible cracking reaction which takes place on the heated substrate and substrate support but nowhere else in the reactor. A large excess of the group V hydride is generally used for the best results. Some of the excess is decomposed and deposits on the reactor. Little is actually known about the reaction mechanism. Experiments have been done (11) to analyse the gas composition after reaction but there has been no satisfactory method of studying the growing surface. The approach to growth is largely empirical. Alloys can readily be grown by using mixtures of alkyls and hydrides. Gaseous dopants such as hydrogen sulphide or dimethylzinc can also be added to the gas stream. For all this work the methyl alkyls, trimethylgallium (TMG) and trimethylindium (TMI) were used. The ethyl alkyls, triethylgallium (TEG) and triethylindium (TEI), have also been widely used. However, unlike the methyl alkyls the latter can decompose readily by the elimination of alkenes:

$$M\begin{array}{c}\diagup \text{Alk}\\\diagdown \text{Alk}\\ \diagdown \text{Alk}\end{array}\xrightarrow{\text{HEAT}} M\begin{array}{c}\diagup \text{H}\\\diagdown \text{Alk}\\ \diagdown \text{Alk}\end{array} + \text{Alkene} \xrightarrow{\text{HEAT}} \underset{\text{METAL}}{M} + 1.5\,H_2 + 2\,\text{Alkene}$$

These reactions can take place at temperatures as low as 100°C and can produce an undesirable premature decomposition of the alkyl. It is claimed that higher alkyls can be used at lower growth temperatures and give less carbon contamination. There is little evidence that carbon contamination is a problem in gallium and indium III-Vs. The ethyl alkyls are probably better suited to low pressure reactors.

Another unwanted side reaction with indium compounds is the formation of polymers (12) by the reaction:-

$$\text{In}\begin{array}{c}\diagup \text{CH}_3\\\diagdown \text{CH}_3\\ \diagdown \text{CH}_3\end{array} + P\begin{array}{c}\diagup \text{H}\\\diagdown \text{H}\\ \diagdown \text{H}\end{array}\longrightarrow \left[\begin{array}{cc}\text{CH}_3 & \text{H}\\ | & |\\ -\text{In} & - P-\end{array}\right] + 2\,CH_4$$

In early work a white deposit was observed in the low temperature regions of the reactor where the phosphine is mixed with the alkyls. Polymer formation could be avoided by cracking the phosphine to elemental phosphorus (13) or using adduct compounds of the indium alkyls (14,15). The problem is much worse with the higher alkyls of indium than with TMI (16). Polymer formation is connected with the purity of the materials and has completely disappeared with the latest precursors and equipment. Group III metal alkyls are Lewis acids and form adduct compounds with Lewis bases such as phosphorus alkyls or amines (17). Although polymer formation is no longer a problem adducts could still be useful precursors if they are safer to handle, easier to purify or a precursor of lower vapour pressure is required. Hydrides are the usual source of group V elements though there has recently been work using group V alkyls (18).

EQUIPMENT

Several different designs of MOCVD growth systems are now commercially available. The growth system used here was developed and built at RSRE. It was designed for a wide range of III-V compounds and dopants. The system was computer controlled to grow complex structures and multilayers.

A horizontal reactor (figure 1) was used with a water cooled outer envelope to reduce parasitic side reactions. For the growth of gallium arsenide good results could be obtained without the cooling water. The seed pedestal was graphite and coated with silicon carbide. The large cross section pedestal should give uniform heating and a reproducible measurement of temperature with a thermocouple. A number of reactor designs use a removable liner for cleaning. In this design the whole tube could easily be removed, cleaned and reloaded.

The reactor design was based on the stagnant layer model of Eversteyn

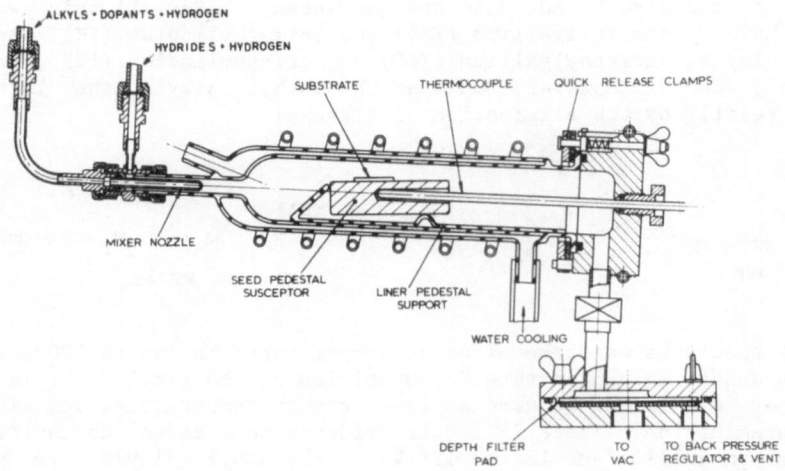

Figure 1: MOCVD Reactor

et al (19). The wedge shaped insert in the tube is meant to give a better gas flow over the substrate. Theoretical and experimental studies (20) of gas flow in reactors show that the gas flow is streamlined. It is very important that the gas flow is uniform to achieve uniform growth. Flow studies of this reactor using smoke show that there is a vortex in front of the growth zone. This will limit the rate at which gas composition can be changed to produce a sharp interface. The problem could be partly overcome by using pauses in the growth procedure which allow one component to be flushed from the reactor before the next component was switched in. A number of streamline reactors have been designed but care must be taken to see that the vapour components are well mixed if non uniform growth is to be avoided.

To control the alloy composition, precision control of the vapour concentrations is necessary. The performance available with electronic flow controllers is of the order of $\pm$ 0.1%. Composition control of the solid should be of this order with lattice mismatch ($\Delta a/a$) of $\pm$ 0.00007. The alkyls are generally supplied in stainless steel bottles with a bubbler tube. These are heated in a thermostatic bath controlled to $\pm$.02°C (Figure

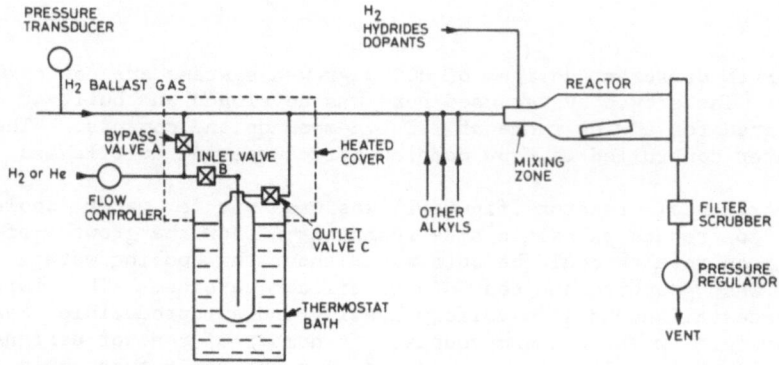

Figure 2: Schematic diagram of flow system for evaporating, metering and switching alkyls to the reactor

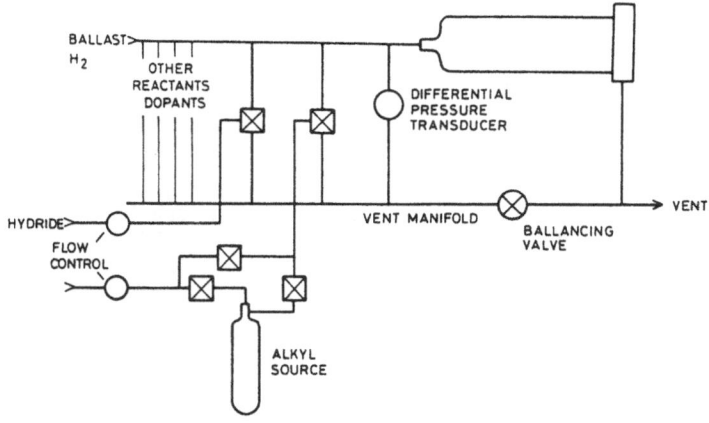

Figure 3: Schematic diagram of vent/run system for fast switching of alkyls
to reactor

2). To initiate growth the valves are sequenced and operated
electronically. Bypass valve A is closed then bottle valves B and C are
opened. To grow thin layers it is necessary to do this without surges or
pauses. The system pressure must be carefully stabilised and the
evaporators preflushed to stabilise the pressure and equilibrate the vapour
pressure. An alternative way to switch gas composition is the vent/run (21)
system a version of which is shown in Figure 3. The alkyl and hydride flows
are run to vent and switched into the reactor when growth is required. In
this system careful stabilisation and balancing of the pressure are also
required. To grow thin layers growth times are only a few seconds. In
small scale systems the gas flows are low and pulsing effects from the valve
operation are a problem (22). The problem could be decreased by increasing
the flow rates and decreasing the vapour pressures. This is difficult with
TMG because of its high volatility. The problem could be overcome by using
a different gallium compound or adduct or using a diffusion cell (8). Extra
ballast flow can be added to the flow from the bubbler to increase the
volume of gas switched and reduce the effects of dead spaces and valve
pulses.

GROWTH OF INDIUM PHOSPHIDE

The reactor was normally operated with total flow of 7.5 1 min^{-1} giving
a gas velocity of 100 mm s^{-1}. After flushing the system to stabilise
pressures, the cleaned and loaded reactor tube was then connected to the
system. After further flushing with hydrogen the substrate was heated to
750°C under a hydrogen phosphine/atmosphere. With indium phosphide it is
essential at all times to maintain an adequate over-pressure of phosphorus
or phosphine to prevent phosphorus loss and surface damage. This bakeout
stage was performed to outgas and dry the tube and pedestal and also to
reduce or evaporate any oxide films on the substrate. There was no facility
in this system to remove any surface films by gaseous etching with hydrogen
chloride.

Growth conditions for good indium phosphide are shown in table 1.
Optimum growth temperature for good electrical properties is around 600°C.
The carrier concentration rise with growth temperature. At a temperature of

TABLE 1

Indium Phosphide
Optimum Growth of 4um layer

Growth temperature	605°C
TMI conc.	1.0×10^{-4} mole fractions
Phosphine conc.	3×10^{-3} mole fractions
Growth rate	2.9um hr^{-1} or 8Å s^{-1}
Electron conc. 298K	$2.7 \times 10^{14} cm^{-3}$
Electron conc. 77K	$2.7 \times 10^{14} cm^{-3}$
Mobility 298K	$5000 cm^2 v^{-1} s^{-1}$
Mobility 77K	$135000 cm^2 v^{-1} s^{-1}$

680°C the carrier concentration is 3-6 x $10^{15} cm^{-3}$. The TMI vapour concentration is calculated from the vapour pressure but is probably high by a factor of 2. The layer thickness decreases by 10% over a growth length of 40 mm. The growth rate is directly proportional to the alykl vapour pressure and almost independent of other factors for a given total flow and reactor geometry. With available precursors, 77K mobilities of >100000 $cm^2 sec^{-1} v^{-1}$ were readily obtainable. For the best electrical properties InP should be grown at a low growth rate, at 600°C, with as high a concentration of phosphine as is possible.

ALLOY GROWTH

To grow indium gallium arsenide layers a buffer layer of indium phosphide was first grown as has been described. The alloy growth was initiated by similtaneously turning off the phosphine and turning on the trimethylgallium and arsine. TEM studies (Figure 4) showed that structures grown above 700°C had defects at the interface. Interfaces grown in the lower temperature ranges were defect free. The defects at higher temperatures probably resulted from a loss of phosphorus due to the high vapour pressure of InP. Optimum growth conditions and best material properties are shown in table 2. The Ga/In ratio in the solid is directly proportional to the Ga/In ratio in the vapour. The composition could easily be adjusted to the lattice matched x = 0.467 gallium. To grow very thin layers the grown rate could be reduced by reducing the alkyl concentration: 8Å sec^{-1} was about the practical limit with the present equipment. The electrical properties were largely independent of temperature over the range 620°C to 700°C in growth temperature (23). The composition of the layers was accurately determined by double crystal x-ray diffraction. The layers

TABLE 2

Gallium Indium Arsenide	
Optimum of Lattice Matched Alloy	4um Layer
Growth temperature	680°C
TMI Conc.	1×10^{-4} mole fractions
TMG Conc.	4.8×10^{-5} mole fractions
Arsine conc.	1.2×10^{-3} mole fractions
Growth rate	5.8um hr^{-1} or 16Å s^{-1}
Electron conc. 298K	$7 \times 10 \ cm^{-3}$
Electron conc. 77K	$6 \times 10 \ cm^{-3}$
Mobility 298K	$11300 cm \ v^{-1} s^{-1}$
Mobility 77K	$82000 cm \ v^{-1} s^{-1}$
X-ray rocking curve width	21 arc secs
PL line width 2K	1.85meV

142

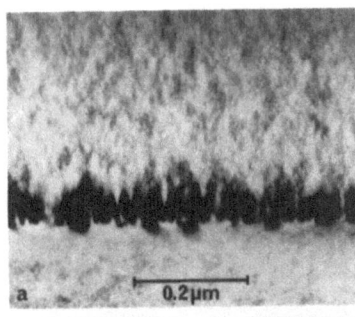

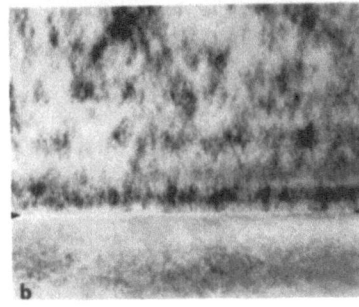

Fig.4: Cross-section TEM
micrographs of InP/
InGaAs interface for
alloys grown at (a)
717C°,(b) 596°C

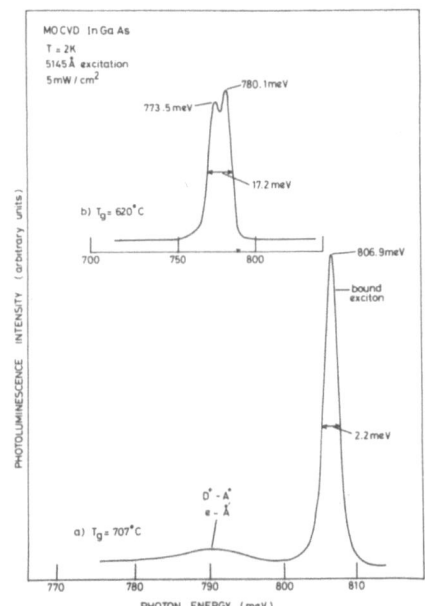

Figure 5: PL of alloys grown
at high and low
temperatures

showed a variation in composition of 0.2% over a growth length of 40 mm. For the best material the x-ray rocking curve width was 21 arc seconds. This was about 5 arc seconds wider than a theoretical simulation.

Photoluminesence at 2K (Figure 5) was strong with a best peak width of 1.85meV at low pump power (5mW cm^{-2}) decreasing to 1.5 meV at 25mW cm^{-2}. The donor-acceptor pair peaks were always weak. There was a marked deterioration of the photoluminescence spectra at growth temperatures below 640°C. The intensity dropped, the line width increased and there was a shift to lower energies. A possible explanation of this effect is spinodal decomposition. This is a partial separation into two phases which can take place at lower growth temperatures due to a miscibility gap in the phase diagram (24). Figure 6 shows TEM micrographs of material grown at high and low temperatures. The layers grown at low temperature showed very strong long range contrast features which are thought to be due to phase separation. This phenomena has been reported by other workers (25, 26). There are also reports (27) of material of good luminescence properties grown at the much lower temperature of 550°C. The significantly lower temperature than in our experiments may be due to a number of factors including the use of a low pressure reactor, a lower growth rate, use of ethyl alkyls or a higher V/III ratio in the vapour.

Both InP and the alloys can be controllably doped n and p type with sulphur or zinc added as hydrogen sulphide or diethylzinc.

MODULATION DOPED STRUCTURES FOR 2 DEG STUDIES

Both low pressure MOCVD (28, 29) and atmospheric pressure MOCVD (30-33)

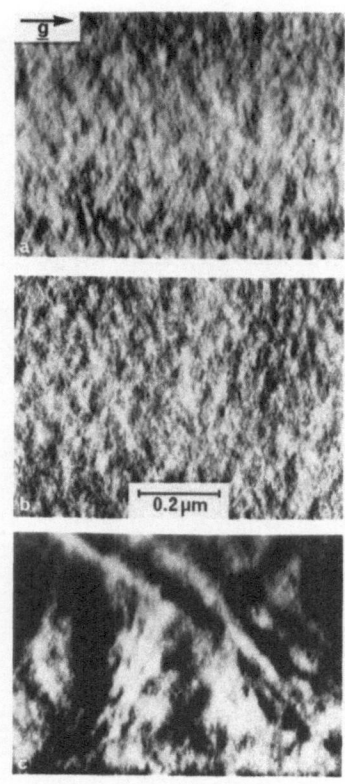

Figure 6: Cross-sectional TEM images of alloy layers grown at (a) 707°C, (b) 717°C, (c) 596°C

1 x 10^{17}	InP	1000Å
U.D. InP SPACER		100Å
U.D. In$_{1-x}$Ga$_x$A$_s$		1000 Å
U.D. InP BUFFER		2000Å

S.I. SUBSTRATE

Figure 7: Modulation doped structure

have been used to grow heterostructures QWs and MQWs. For this work the structure shown in Figure 7 was grown, using the optimum growth conditions listed in the previous section, and was used to study heterojunctions and 2 DEG properties. The active layer was a 1000Å of alloy. A doped layer of InP provided the electrons for the 2 DEG at the interface. The doped layer could be on top, underneath or on both sides of the alloy. There was a spacer layer between the alloy and doped InP so as to separate the charges of the ionised impurities from the 2 DEG region. To produce a sharper interface there was a pause in the growth between the alloy and the InP. During the pause all alkyls and hydrides were turned off for 5-10 seconds. These structures showed well defined Shubnikov De Haas oscillations (34). Measurements were made of carrier density and mobility against temperature (Figure 8). At low temperatures the mobility was constant as a function of temperature which indicated a 2 DEG. For experiments where the doping level was too high there was some parallel conduction in the InP and this was shown by an initial decrease of n with temperature. There was also a persistent photoconductivity effect. This could be used to generate extra electrons so that a whole series of measurements at different carrier concentrations could be made on the same sample (Figure 9) both for normal and inverted HEMT structures. The mobility rose with electron concentration

144

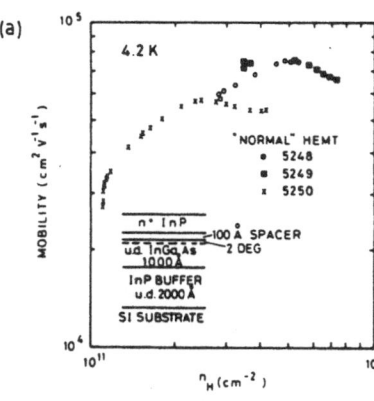

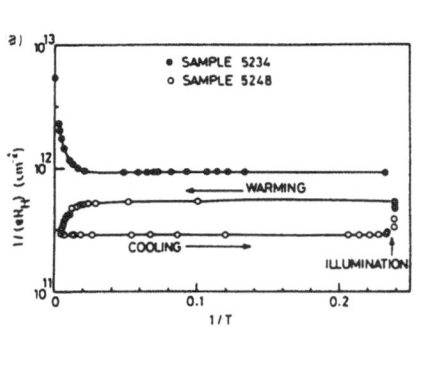

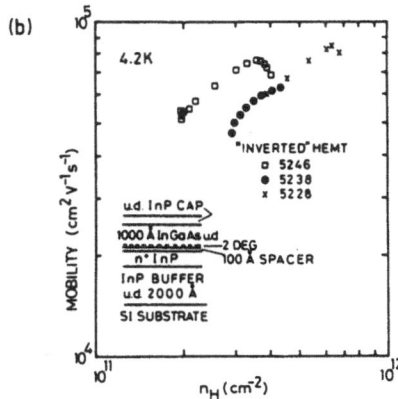

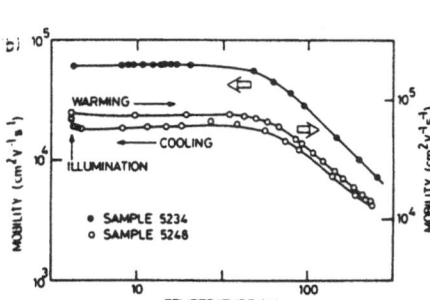

Figure 8: Variation of carrier density (a) and mobility (b) with temperature for modulation-doped heterostructures

Figure 9: Mobility versus carrier density at 4.2K using persistent photoconductivity effect

until a second sub-band filled and reduced the mobility. The best 2 DEG mobility obtained was $85000\,cm^2sec^{-1}v^{-1}$ at carrier concentration of $6.5 \times 10^{11}cm^{-2}$. In the InP/InGaAs system, unlike the GaAs/GaAlAs system, there was essentially no difference between the normal and inverted structures. Both interfaces of the alloy were electrically the same.

QUANTUM WELLS

Undoped and modulation doped quantum wells 50-200Å thick were grown (35) by the same technique as used for the thicker structures. Figure 10 shows a TEM micrograph of a layer estimated at 50Å from the thick layer growth but measured 75Å. The layer was well defined and microscopically uniform although the growth time was only 3 seconds. Cathodoluminescence studies showed that the uniformity over the growth area was good. The narrowest well so far grown had a width of 60Å. Indeed, this appeared to be the limit of the existing system due to gas switching transients. The quantum wells show strong photoluminescence. The properties of some of the wells are shown in table 3. There was some variation in peak position for wells of the same width which could be due to a variation in alloy composition of up to 2%. Even in layers grown without doping the lines were broadened by free carriers generated by the persistent photoconductivity effect. Using the depletion layer of a Schottky diode (36) these carriers could be removed to narrow the linewidth by several meV. The linewidth can

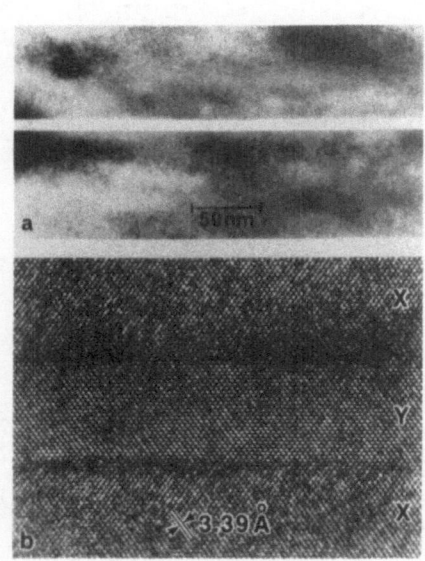

Figure 10: Cross-sectional TEM images of 75Å quantum well structure (a) diffraction contrast image (g=200), (b) high resolution lattice image in (110) projection

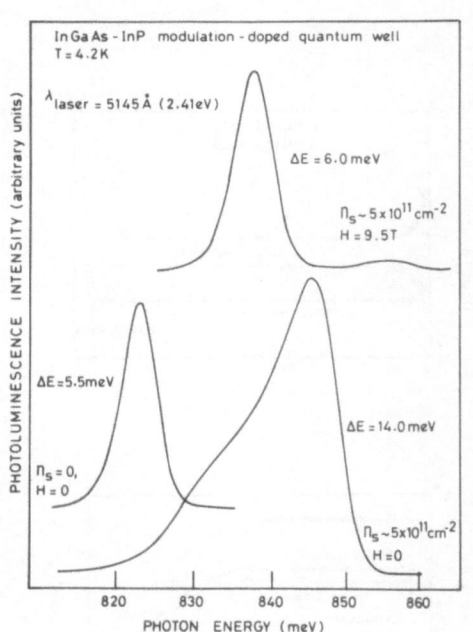

Figure 11: Demonstration of interface quality in 100Å QW using both a magnetic field and Schottky barrier depletion to remove free carrier broadening

Table 3

Photoluminesence properties of quantum wells
Nominally lattice matched

Run	Well width Å (measured)	Linewidth meV	Peak energy meV
5371	100	7.9 (no carriers)	882.9
5282	100 (110)	4.6 (H=5.2T) (no carrier broadening)	885.4
5383 (mod. doped)	100	5.5 (no carriers)	864.5
5385	150	5.3	845.3
5253	50 (70)	14	942.6
5597	25 (60)	16.1	977.3

146

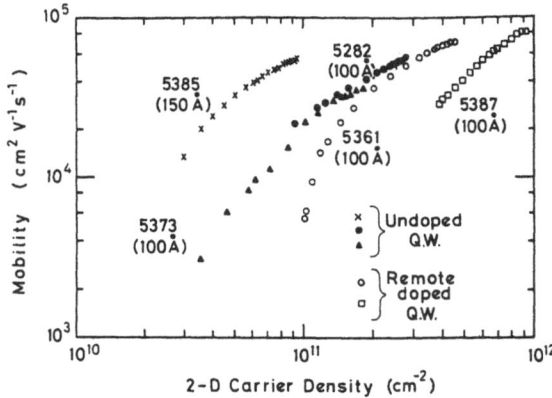

Figure 12: Mobility versus carrier concentration at 4.2K in undoped and modulation doped quantum wells using persistent photo conductivity effect

also be narrowed by using a magnetic field (Figure 11).

The 2 DEG mobility in the quantum wells could be measured by using the modulation doped structures and the persistent photoconductivity effect (37) (Figure 13). In this case the mobility did not saturate because of the wider sub-band separation. The highest mobility obtained was 95000 $cm^2V^{-1}s^{-1}$ at a carrier concentration of $9 \times 10^{11}cm^{-2}$. This high mobility shows that the heterojunctions must be of very good quality. Very good PL linewidths (5.5 meV) were obtained for the modulation doped quantum wells in depletion under reverse bias showing that there was no significant disruption of the interfaces by the remote doping.

MULTIPLE QUANTUM WELLS

Multiple quantum wells (MQWs) were grown using the same growth conditions as used for single wells. The computer was programmed to repeat the structure as many times as required. Figure 13 shows cross section TEM micrographs of 50 x 100Å indium gallium arsenide layers separated by 200Å of InP. Microscopic uniformity of thickness was very good. Close examination of this and the TEM micrographs of other MQWs and single QWs showed that the alloy/InP interface is not as smooth or clearly defined as the InP/alloy interface. Other workers (38) have examined interfaces using ellipsometry and reported that there appears to be an intermediate alloy layer at the alloy/InP interface. This was either not present or very much reduced at the inverse junction. A further structure was grown in which the alloy layers were separated by only 50Å of InP. Here the layers were not as uniform in thickness. The initial growth of InP on alloy was probably uneven but flattened out as the growth continued.

Double crystal diffractometry of these multiple wells showed well defined superlattice peaks. Figure 14 shows the rocking curve produced by 8 x 100Å wells separated by 500Å of InP. Thicknesses were estimated from the bulk growth rate. Computer simulations of this structure based on the dynamical theory were carried out. The peaks were of near theoretical width indicating good uniformity in terms of the periodicity. The measured periodicity was 660Å, which was in good agreement with the TEM measurement of 550Å + 110Å. However the fact that there were strong superlattice reflections indicated that there was a periodic strain in the layer structure. Theoretical simulations based on a mismatched alloy layer did not correspond to the experimental curve. A more likely explanation is that

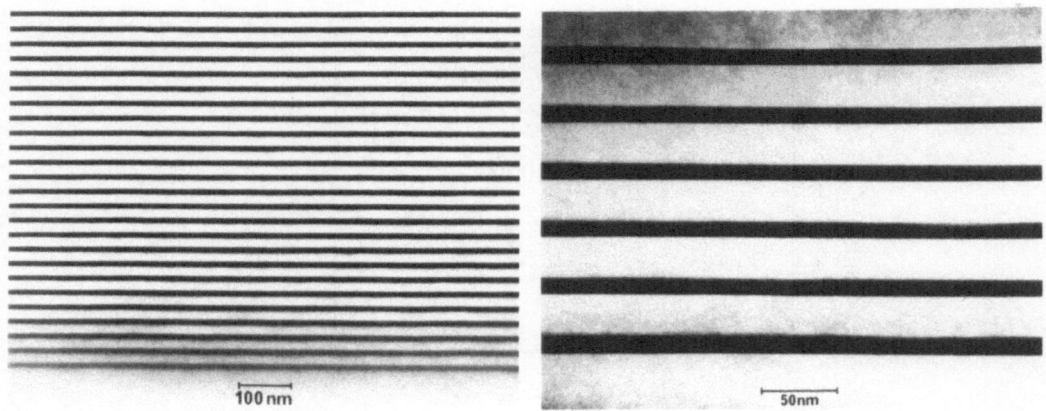

Figure 13: Cross-sectional TEM images of 50x100Å alloy layers with 500Å
InP spacers at intermediate and high magnification (200 dark field)

a small amount of arsenic has been incorporated in the InP layers and a
simulation with 6% arsenic gave a reasonable fit to experimental data. In
the MOCVD growth of indium gallium arsenide, the arsenic is incorporated 50-
100 times more ·efficiently than the phosphorus. Even a small trace of
residual arsenic in the reactor would give an appreciable concentration in
the InP. This could also account for the observation of an intermediate
alloy between the alloy and InP as the effect would be strong near the
interface.

Single quantum wells were too thin to give measureable optical
absorption but data on optical absorption from the well could be obtained
from photoconductivity (PC) (39) and photoluminescence excitation
spectroscopy (PLE). Multiple wells with more than 8 wells gave well defined
absorption spectra. Figure (15) shows the optical absorption spectra for 50
x 100Å MQW at four different temperatures from 10 to 100K. The n = 1 heavy
and light hole exciton peaks are clearly resolved. The weak peak between

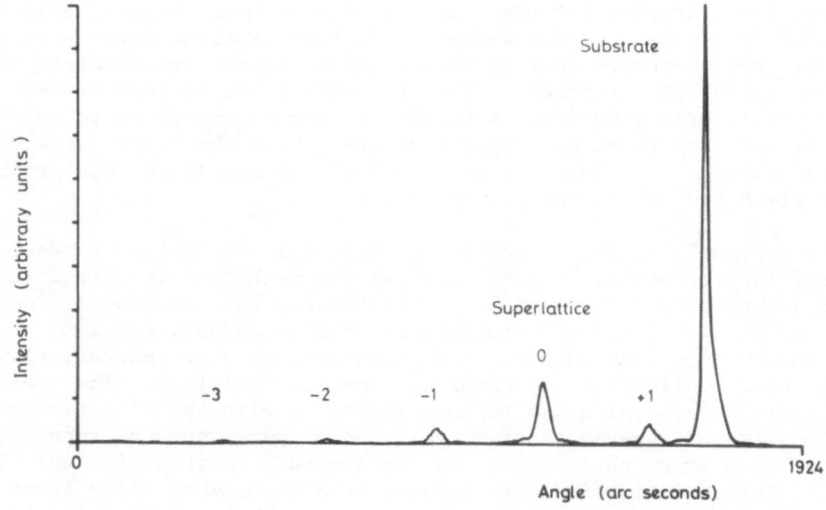

Figure 14: Double crystal X-ray diffraction curve of 8x100Å layers with
500Å InP spacers. 400 reflection Cu K α1 radiation

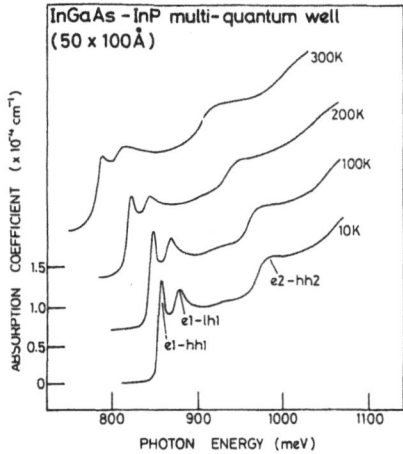

Figure 15: Series of optical absorption spectra for 50x100Å quantum wells at temperatures from 10-300K

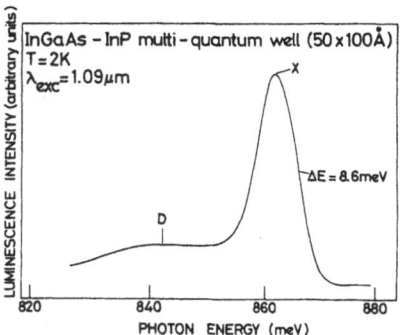

Figure 16: Photoluminescence of MQWs at 2K

n = 1 and n = 2 regions is probably the weakly allowed n = 1 electron to n = 3 heavy hole peak.   The low temperature linewidth of the n = 1 heavy hole peak (el-hhl) of 8 meV at 10K is indicative of the high quality of the structure particularly bearing in mind that it was averaged over fifty wells.   Comparison with the absorption spectra of single quantum wells measured by PC and PLE where linewidths of 15 meV were found shows that there was no deterioration in quality. In fact some improvement is found in MQWs, compared to single QWs.   Any interwell fluctuations must have been small.   In single wells narrower linewidths were found in PL than in PC or PLE.   This was attributed to the observation of bound excitons in PL with a Stokes shift of 10 meV relative to the free excitons observed in PC or PLE In the MQW sample on the other hand the PL linewidth of 8.6 meV (Figure 16) was close to that found for absorption.   Taken together with the small Stokes shift of 5 meV, we conclude that relatively weakly localised excitons are observed in the PL of MQWs.   The minor peak to the lower energy in figure labelled D probably arose from free to bound recombination at a defect or impurity.

CONCLUSION

The atmospheric MOCVD system can grow InP, indium gallium arsenide, single and multiple quantum wells of very high quality.   The heterojunction quality, as shown by TEM, electrical and photoluminescence properties is very good. So far this equipment has not been able to grow layers less than 50Å wide.   This does not appear to be intrinsic to the process but an instrumental problem.   There is still room for optimisation in the growth rate and the gas switching procedure.

REFERENCES

1.   P.A.Claxton,   J.S.Roberts,   J.P.R.David,   C.M.Sotomayor-Torres, M.S.Skolnick, P.R.Tapster and K.J.Nash, MBE Conference, York, 1986. To be published in J.Cryst.Growth
2. W.T.Tsang, J.Electr.Mater. 15 (1986) 235
3. M.Razeghi, M.A.Poisson, J.P.Larivain and J.P.Duchemin, J.Electr.Mater.

$\underline{12}$ (1983) 371

4. A.P.Roth, M.A.Sacilotti, R.A.Masut, A.Machado and P.J.Darcy,
J.Appl.Phys. $\underline{60}$ (1986) 2003

5. S.J.Bass and M.L.Young, J.Cryst.Growth, $\underline{68}$ (1984) 311

6. K.W.Carey, Appl.Phys.Letts. $\underline{46}$ (1985) 89

7. K.T.Chan, L.D.Zhu and J.M.Ballantyne, Appl.Phys.Letts. $\underline{47}$ (1985) 44

8. A.Mircea, R.Mellet, B.Rose, D.Robein, H.Thibierge, G.Laroux, P.Daste,
S.Godefroy, P.Ossart and A.M.Pougnet, J.Electron.Mater. $\underline{15}$ (1986) 205

9. C.C.Hsu, R.M.Cohen and G.B.Stringfellow, J.Cryst.Growth $\underline{63}$ (1983) 8

10. M.J.Ludowise, J.Appl.Phys. $\underline{58}$ (1985) R31

11. C.A.Larsen and G.B.Stringfellow, J.Cryst.Growth $\underline{75}$ (1986) 247

12. R.Didchenko, J.E.Alix and R.H.Toenikoetter, J.Inorg.Nucl.Chem. $\underline{14}$ (1960)
35

13. J.P.Duchemin, M.Bonnet, G.Beuchet and F.Koelsch, Inst.Phys.Con.Ser. 45,
Chapter 1 (1978) P10

14. R.H.Moss and J.S.Evans, J.Cryst.Growth $\underline{55}$ (1981) 129

15. S.J.Bass, C.Pickering and M.L.Young, J.Cryst.Growth, $\underline{64}$ (1983) 68

16. G.B.Stringfellow, J.Cryst.Growth $\underline{68}$ (1984) 111

17. S.J.Bass, M.S.Skolnick, H.Chudzynska and L.Smith, J.Cryst.Growth $\underline{75}$
(1986) 221

18. C.A.Larsen, C.H.Chen, M.Kitamura, G.B.Stringfellow, D.W.Brown and
A.J.Robertson, Appl.Phys.Lett. $\underline{48}$ (1986) 1531

19. F.Eversteyn, P.J.Severin, C.H.J. v.d. Brekel and H.L.Peek,
J.Electrochem.Soc. $\underline{117}$ (1970) 925

20. L.J.Giling, J.Electrochem.Soc. $\underline{129}$ (1986) 634

21. R.J.M.Griffiths, N.G.Chew, A.G.Cullis and G.C.Joyce, Electr.Letts. $\underline{19}$
(1983) 988

22. E.J.Thrush, J.E.A.Whiteaway, G.Wale-Evans, D.R.White and A.G.Cullis,
J.Cryst.Growth $\underline{68}$ (1984) 412

23. S.J.Bass, S.J.Barnett, G.T.Brown, N.G.Chew, A.G.Cullis, A.D.Pitt and
M.S.Skolnick, Intnl.Conf. on Cryst.Growth 8, York 1986 - to be published in
J.Cryst.Growth

24. B. de Cremoux, P.Hirtz and J.Ricardi, Inst.Phys.Conf.Ser. $\underline{56}$ (1981) 115

25. K.L.Fry, C.P.Kuo, R.M.Cohen and G.B.Stringfellow, Appl.Phys.Letts. $\underline{46}$
(1985) 955

26. K.W.Carey, Appl.Phys.Letts. $\underline{46}$ (1985) 89

27. K.H.Goetz, D.Bimberg, H.Jurgensen, J.Selders, A.V.Solomonov, G.F.Glinski
and M.Razeghi, J.Appl.Phys. $\underline{54}$ (1983) 4543

28. M.Razeghi and J.P.Duchemin, J.Vac.Sci.Technol. $\underline{B1}$(2) (1983) 262

29. M.Razeghi, J.P.Duchemin, J.C.Portal, L.Dmowski, G.Remeni, R.J.Nicholas
and A.Briggs, Appl.Phys.Letts. $\underline{48}$ (1986) 712

30. A.W.Nelson, R.H.Moss, J.G. Regnault, P.C.Spurdens and S.Wong,
Electr.Letts. $\underline{21}$ (1985) 331

31. C.P.Kuo, K.L.Fry and G.B.Stringfellow, Appl.Phys.Letts. $\underline{47}$ (1985) 855

32. J.J.Coleman, G.Costrini, S.J.Jeng and C.M.Wayman, J.Appl.Phys. $\underline{59}$ (1986)
428

33. F.Scholz, P.Wiedemann, K.W.Benz, G. Trankle, E.Lach, A.Forchel, G.Laube
and J.Weidlen, Appl.Phys.Letts. $\underline{48}$ (1986) 811

34. M.J.Kane, D.A.Anderson, L.L.Taylor and S.J.Bass, J.Appl.Phys. $\underline{60}$ (1986)
657

35. M.S.Skolnick, P.R.Tapster, S.J.Bass, N.Apsley, A.D.Pitt, N.G.Chew,
A.G.Cullis, S.P.Aldred and C.A.Warwick, Appl.Phys.Lett. $\underline{48}$ (1986) 1455

36. M.S.Skolnick, K.J.Nash, P.R.Tapster, S.J.Bass and A.D.Pitt - to be
published

37. M.J.Kane, D.A.Anderson, L.L.Taylor and S.J.Bass - to be published in
Appl.Phys.Letts. 17 Nov. 1986

38. M.Erman, J.P.Andre and J.LeBris, J.Appl.Phys. $\underline{59}$ (1986) 2019

39. M.S.Skolnick, P.R.Tapster, S.J.Bass, A.D.Pitt, N.Apsley and S.P.Aldred,
Semicond.Sci.Technol. $\underline{1}$ (1986) 29

Crown Copyright © HMSO 1986

MOCVD GROWTH OF NARROW GAP

LOW DIMENSIONAL STRUCTURES

M. Razeghi

Thomson-CSF-LCR
B.P. 10
91410 Orsay France

ABSTRACT

Very thin epitaxial layers of III-V semiconductor compounds of accurately controlled thickness and composition can be prepared by low pressure metal organic chemical vapor deposition growth technique. The interfaces between these layers significantly influence their electronic and optical properties and, ultimately, device performance.

INTRODUCTION

During the past ten years extensive theoretical and experimental studies of low dimensional structures in two distinct contexts have been performed:

- the low-dimensional meaning less than three, and the properties of the dimensional electron or hole gas in different III-V semiconductor heterojunctions is now a mature, even aging subject.
- Low as come to small, as in the physics of semiconductors, where the feature size of the sample is small compared with intrinsic quantum length scale associated with carriers in semiconductors. This is given by de-Broglie wavelength defined as $\lambda = h/p$ where h is Planck's constant and p the carrier momentum, typically given by $p^2/2m^* = KT$.

Low-dimensional structures (LDS) have become essential elements in modern device technology.

There is now need for a wide range of semiconducting materials, especially silicon, III-V and II-VI compounds, to be grown as thin single crystal films. Thickness has to be controlled, increasingly more accurately, and uniformity of thickness can be vital. Epitaxial layers as thin as 10 Å or less are now needed. Excellent homogeneity and purity is required. It is desirable to have no misfit dislocations present, and to have a very sharp boundary between the substrate and epitaxial layers. A major goal of solid state physics and solid state technology is the perfection of materials and substrates in which charge carriers have long time, low scattering, high mobilities and controlled densities, such materials have allowed the elucidation of the electronic structure of solids and the development of semiconductor electronics and photonics devices.

Metal organic chemical vapor deposition (MOCVD) has established itself as a unique and important epitaxial crystal growth technique yielding high quality LDS for fundamental semiconductor physics research and useful semiconductor devices, both electronic and photonic. The growth of semiconductor III-V compounds results by introducing metered amounts of the group III alkyls and the group V hydrides into a quartz tube which contains a substrate placed on an RF-heated carbon susceptor. The hot susceptor has a catalytic effect on the decomposition of the gaseous products and the growth therefore primarily takes place at this hot surface. MOCVD is attractive because of its relative simplicity compared to other growth methods. It can be produce heterostructures, multiquantum wells (MQW) and superlattices (SL) with very abrupt switch on and switchoff transitions in composition as well as in doping profiles in continuous growth by rapid changes of the gas composition in the reaction chamber. The technique is attractive in the ability to grow uniform layers, low-background of doping density, sharp interfaces and the potential for commercial application. MOCVD can prepare multilayer structures with thickness as thin as a few atomic layers. This allows the study and device applications of two-dimensional electron gas (TDEG) [1], two-dimensional hole gas (TDHG) [2] transport and quantum size effect (QSE) [3] in variety of III-V compound semiconductors, heterojunctions and multilayers [4]. It also makes possible to "engineer the band gap" by growing a predetermined alloy composition and doping profile. As a result on entirely new class of electronic and photonic devices are realized. Another recent advance is the ability to grow strained layers, superlattices in which the crystal lattices of the two materials are not very closely matched. In that case there is a built in strain in each layer [5]. This paper is restricted to the growth, characterization, quality, defects and interfaces of narrow gap LDS grown by low pressure metalorganic chemical vapor deposition (LP-MOCVD) growth technique, for optoelectronic and microwave devices.

EXPERIMENTAL DETAILS

The narrow gap III-V compounds arbitrarily defined to have an energy gap E_g < 2.0 eV discussed in this review are listed together with properties relevant to their crystal growth and some of their physical parameters are listed in table I.

They all melt at high temperatures and have a strong tendency to decompose well below their melting points. In consequence, controlled melt growth of these compounds is difficult and most efforts have been directed towards vapor growth as a means for producing both heterojunctions and LDS.

Elemental vapor phase epitaxy usually requires growth temperatures within the range 700-900°C. It has been recognized increasingly that growth at high temperature thermodynamically favors the formation of a multitude of impurity- defect complexes which act as electron traps and pin the Fermi level far from both the conduction and valence band edges of these compounds and hence they often exhibit compensating properties. The lower growth temperature (500-630°C) for LP-MOCVD of narrow band gap III-V compounds, a feature shared with molecular beam epitaxy (MBE), has stimulated substantial interest in this preparative route as a potential way of either eliminating or controlling detrimental impurities and defects. This control is particularly important where the material is the active medium in optoelectronic devices.

Table I - Properties of III-V compounds

Compound	E_g (eV)	a_o (Å)	m_e^*	$\bar{n}$	χ (eV)
InAs	0.36	6.057	0.15 m_0	3.178	4.9
InP	1.35	5.869	0.077 m_0	3.45	4.35
GaAs	1.42	5.653	0.067 m_0	3.655	4.07
GaP	2.2	5.451	0.82 m_0	3.452	4.3
$Ga_{0.47}In_{0.53}As$	0.75	5.869	0.41 m_0	3.56	–
$Ga_{0.49}In_{0.51}P$	1.9	5.653	–	–	–
$Ga_xIn_{1-x}As_yP_{1-y}$	(0.75-1.35)	5.869	–	–	–
$Ga_{x'}In_{1-x'}As_{y'}P_{1-y'}$	(1.42-1.9)	5.653	–	–	–

A review by Ludowise [6] presented practical information about the apparatus and technique as well as more fundamental details concerning the interrelation of the several parameters which control the deposition and material quality.

LP-MOCVD was first applied to the growth of the InP using the metal alkyls triethylindium $((C_2H_5)_3In$, TEI) and phosphine PH_3 as the source of phosphorous (P) by Duchemin [6]. A review by Razeghi [7] described advances in the field prior to 1984. The developments since then for the growth of LDS of narrow gap III-V compounds, form the basis of present review.

Starting Materials

The compounds used as sources of the group III and group V elements for this study are listed in Table II. Their vapor pressure details are given in Annex 1 [9].

Triethylindium (TEI) and triethylgalium (TEG) have been used as group III sources. Hydrides pure Arsine (AsH_3) and pure phosphine (PH_3) have been used as group V sources. Diethylzinc (DEZ_n) is used for P-type doping and sulphides H_2S or silane H_4Si have been used for n-type doping [16]. Pure hydrogen H_2 and pure nitrogen N_2 have been used as carrier gas. The presence of N_2 is necessary to avoid the parasitic reaction between TEI and AsH_3 or PH_3. The presence of H_2 is necessary to avoid the deposition of carbon. TEI and TEG are contained in stainless steel bubbler, which are held in controlled-temperature baths at 31 and 0°C respectively. An accurately metered flow of nitrogen (N_2) for TEI and purified H_2 for TEG is passed through the appropriate bubbler. To ensure that the source material remains in vapor form, the saturated vapor that emerges from the bottle is immediately diluted by a flow of H_2. The mole fraction, and thus the partial pressure of the source species is lower in the mixture and is prevented from condensing in the stainless steel pipe work. The flow rates

Table II - Starting materials

Group III sources	Group V sources	P-Type dopant sources	n-type dopant sources
$Ga(C_2H_5)_3$	AsH_3	$(C_2H_5)_2Zn$	SH_2
$Ga(CH_3)_3$	$(CH_3)As$ $(C_2H_5)_3As$	$(CH_3)_2Zn$	SiH_4
$In(C_2H_5)_3$	PH_3	$(C_5H_5)Mg$	$(C_2H_5)_2Te$
$In(CH_3)_3$	$(CH_3)P$ $(C_2H_5)_3P$	$(CH_3)_2Cd$	$(C_2H_5)_2Se$

of the hydrides, H_2 and N_2 were controlled by mass-flow-controllers within 0.2%. The metal alkyl or hydride flow can be either injected into the reactor or into the waste line by using the three-way valves. In each case, the source flow is first switched into the waste line to establish the flow-rate and then switched into the reactor.

Reactor Design

The most important part of the growth apparatus is the deposition chamber. There are two basic reactor types, a vertical design originated by Manasevit and Simpson [10] in which the gas flow is perpendicular to the substrate surface, and the horizontal version developed by Bass and Oliver [11] in which the gas flow is parallel to the substrate surface. Both types have been used successfully at low and atmospheric pressures using several different substrate heating methods involving radio-frequency induction, resistance, radiant, and laser heating.

The advantages of a low pressure MOCVD process are:

i)   elimination of parasitic nucleations in the gas phase,
ii) reduction of out-diffusion (i.e. the solid state diffusion of impurities from the substrate through active layers or from one active layer to another).
iii) reduction of autodoping (i.e. the doping of an epitaxial layer by volatile impurities that originate from the substrate).
iv) improvement of the interface sharpness and impurity profiles.
v)   lower growth temperature.
vi) thickness uniformity and compositional homogeneity.
vii) elimination of memory effect.

The presence of vortices (behind the susceptor) and the dead volumes (sharp corners in reactor inlet), will act as sources of unwanted materials which cannot be removed easily. So during the growth of a sharp heterojunction or a steep doping profile, by switching a flow with another chemical composition, the original composition is still present in the vortices and trapped gases. The slow out diffusion from these parts will smear out the doping or heterojunction profiles. This is called the memory effect. We can eliminate such trapped gases by rapid evacuation, i.e. working at low pressure.

In relation to reactor design, the main advantages of the horizontal reactor is considered to be the uniform gas flow achieved over a slightly angled susceptor (7-15°) which leads to uniform deposition as the reactants are depleted from the gas flow.

The most important feature in the growth of LDS is the arrangement for mixing the gases in order to inhibit the prereaction between the constituent gas flow. The reactor design has concentrated on introducing the gases to the reactor separately. A good mixing between the metal alkyls and hydrides through a delivery tube near to the heated substrate is vital. So the use of N_2 which eliminates the prereaction enables good mixing to take place and allows the efficient gas mixer designed for quaterneries growth to be utilised.

Growth Procedure

The LP-MOCVD growth of III-V compounds can be explained by the rapid transport of reactive species from the pipeline to the deposition zone by forced flow (in this case, the control of flow dynamics, flow mixing and the adjustment of the flow to geometrical effects and temperature changes is essential). Then the transport of reactive species in the deposition zone to the hot substrates by diffusion (the knowledge about development of concentration profiles, the effect of thermal gradients on diffusion and the effect of annihilation or creation of molecular species on diffusion is crucial).

Since the growth rate in MOCVD is limited by mass transport of the group III growth component, flow dynamics coupled with diffusion govern the deposition rates and there- with, the gas phase depletion. A profound knowledge of flow patterns and concentration profiles in the reactor are essential for optimisation of the reactor construction. In this respect two important factors are homogeneous growth on large surface areas and minimization of gas memory effects, which is essential in the growth of LDS and sharp interfaces. For that, flow region must be laminar not turbulence in order to achieve control over growth process, and must develop its pattern in a controlled way when it enters the reactor. Also when the gas is heated to the process temperature, no instabilities due to natural convection (buoyancy) in the boundary layer may occur. The term boundary layer is used here to define the regions of rapidly increasing compositional, thermal or momentum gradients perpendicular to the substrate.

Flow Patterns

One of the important factors in the MOCVD reactor is the flow pattern. We can specify the macroscopic gas movements in the MOCVD reactor as follows:

1) Laminar flow contrary to the turbulence,

2) Diffusion contrary to the free convection (buoyancy).

In order to specify the circumstances in which the different types of flow occur, we need to introduce the concept of the Reynolds number for the first case, and Rayleigh number for the second cases.

The Reynolds number (R_e) (dimensionless) of reactor flow is [12]:

$$R_e \simeq \frac{\rho \overline{V} d}{\eta}$$

where d is the diameter of the tube (m), V is the average flowrate (m/sec), ρ the density (kg/m^3) and η the dynamic viscosity (kg/m.sec) of the gas. Laminar flows typically have low Reynolds number. At higher Reynolds number a transition takes place from laminar to turbulent.

When R_e is small (less than 100) the flow regime is laminar. In laminar flow, the velocity at a fixed position is always the same. Each element of reactive species travel smoothly along a simple well-defined path [13]. Each element starting at the same place follows the same path.

When R_e is high, the flow becomes turbulent, none of these features is retained. The flow develops a highly random character with rapid irregular fluctuations of velocity in both space and time. In this case, an element of gas flow follows a highly irregular distorted path. Different elements starting at the same place follow different paths, since the pattern of irregularities is changing all the time. [14].

The Rayleigh number (R_a) dimensionless of reactor flow is:

$$R_a \simeq \frac{g_\alpha \; Cp \; p^2 h^3 \; \Delta T}{\eta K}$$

where α is the coefficient of thermal expansion in (1/T), g is gravity constant (9.81 m.sec^{-2}), Cp is specific heat (J.kg^{-1}.K^{-1}), ρ is density (Kg.m^{-3}), h is free hight above susceptor (m), ΔT = T(susceptor) − T(reactor wall), η is dynamic viscosity (Kg.m^{-1}.sec^{-1}) and K is thermal conductivity (J.m^{-1}.sec^{-1}.K^{-1}). When $R_a < 1700$, the gas is stable, for $R_a > 1700$ free convection occurs.

With high Rayleigh number the free convection occurs, which affects the mass transfer, growth rate and homogeneity. Usually convection occurs between hot substrate and reactor cold wall. The cause of convection is the action of gravitational field on the density field on the density variations associated with temperature variations. The heavy cold gas is situated above light hot gas. If the former moves downwards and the latter upwards, there is a release of potential energy which can provide kinetic energy for the motion. There is thus a possiblity that the equilibrium will be unstable.

From the kinetic gas theory, it follows that both the dynamic viscosity η and the thermal conductivity K are independent of reactor pressure. But the density ρ is pressure dependence, which means that $R_e \propto \rho$ and $R_a \propto \rho^2$. So, the consequence is that lower pressures stabilize the convection behaviour of the gas flows, and laminar flows are easily obtained.

In conclusion, for the growth of high quality III-V semiconductor LDS with sharp interfaces, the following remarks on reactor design is crucial:

1. Laminar flows free of convection should exist:

   a) using horizontal reactor,
   b) working at low pressure,
   c) decreasing the reactor diameter,

2. no temperature gradient should be present across the susceptor,

3. Eliminate the memory effect:

    a) the geometry of the reactor is such that no vortices can develop,
    b) no dead volumes are present inside the reactor,
    c) elimination of sharp corners in reactor inlet, where the laminar flow can go by without having a strong interaction, also behind the susceptor.

EXPERIMENTAL RESULTS

InP

    High quality Inp layers have been grown by LP-MOCVD [15]. The carrier concentration as low as 10^{14} cm^{-3} with electron hall mobility as high as 5,600 cm^2.v^{-1}.s^{-1} at 300 K and 100,000 cm^2.v^{-1}.s^{-1} at 77°K have been currently obtained. The purity of starting material, the quality of substrate and preparation of substrate before growth are responsible for quality of epitaxial layers. The growth temperature is between 500 up to 650°C. The InP and related compounds can be doped P-type by using diethylzinc (DEZ) or dimethylzinc (DMZ). When the flowrate of DEZ is kept constant, the free carrier concentration varies exponentially with 1/T, where T is growth temperature [16].

    InP and related compounds epilayers grown by LP-MOCVD can be doped n-Type using H_2S or SiH_4. When the flowrate of H_2S is kept constant, the free-carrier concentration varies exponentially with 1/T like in the case of DEZ. The free carrier concentration in the epilayer decreases when the growth temperature increases.

    In the case of SiH_4, when the flowrate is kept constant, the free carrier concentration increases with increasing growth temperature. At higher temperature, the decomposition of the SiH_4 is more efficient. We found that for high doping level $\simeq 10^{18}$ cm^{-3} (such as laser) it is better to use H_2S. Using H_2S the epitaxial layers are less compensated. Silicon in III-V compounds is amphoter, incorporation of Si in InP depends on the ratio of III/V elements. But the diffusion coefficient of Si is less than S, so for modulation doping it is better to use Si.

    Application of InP. The increasing interest in radar systems operating at 94 GHz has created a demand for microwave sources with high power, high efficiency and low noise. A promising candidate for these requirements is the InP Gunn diode which according to theory [17] can oscillate at frequencies in excess of 94 GHz in the fundamental mode and deliver powers greater than 50 mW. It has been known, both from theory and experiments, that Gunn diode made from InP are superior to those in GaAs with respect to output power and efficiency. This superiority is principally due to the fact that the velocity field curve for InP exhibits a larger peak to valley velocity ratio than does the one for GaAs (V_p/V_v = 4 for InP and V_p/V_v = 2.4 for GaAs [17]). The energy separation between the two valleys is ΔE, which is about 0.31 eV for GaAs and 0.53 eV for InP. The threshold field ε_T defining the onset of negative differential resistivity is 3.2 kV/cm for GaAs and 10.5 kV/cm for InP. The peak velocity V_p is about 2.2×10^7 cm/s for high purity GaAs and 2.5×10^7 cm/s for high purity InP. The cutoff frequency of InP is approximately 200 GHz at 3 dB compared with 100 GHz for GaAs. High power and high efficiency InP Gunn diodes were fabricated from a 3 layer (N$^+$/N/N$^+$) structure grown by LP-MOCVD. An integral heat sink technology was developed to produce several well controlled mesa diameters in the 30 to 60 μm range. In order to reduce the parasitic inductance of connections, diode chips were packaged using different banding wire geometries.

CW output power up to 100 mW with 2.5% efficiency at 94 GHz have been obtained [19].

$Ga_{0.47}In_{0.53}As-InP$

GaInAs lattice matched to InP is a potentially useful material for microwave and optoelectronic devices. The energy gap of $Ga_{0.47}In_{0.53}As$ is 0.75 eV (λ = 1.67 μ), the effective electron mass of 0.041 m_0, the velocity of electron which is much higher than that in GaAs at a low field, and a large difference between Γ and L band ($\Delta E=\Gamma_8 - L_6 = 0.55eV$). High quality GaInAs lattice matched to InP substrate has been grown by LP-MOCVD over 10 cm^2 of InP substrate [20]. The layer with 3 μm thick exhibits photoluminescence linewidth of less than 2 meV at 2°K. The electron mobility such as 12,000 $cm^2.v^{-1}.s^{-1}$ at 300°K, 100,000 $cm^2.v^{-1}.s^{-1}$ at 77°K and more than 260,000 $cm^2.v^{-1}.s^{-1}$ at 2°K for a carrier concentration of 2 x 10^{15} cm^{-3} have been measured.

The two-dimensional hole gas at the interface of GaInAs-InP with mobility of 10,500 $cm^2.v^{-1}.s^{-1}$ at 2°K have been measured for the first time [2].

Very high quality $Ga_{0.47}In_{0.53}As-InP$ heterojunctions, quantum wells and superlattices have been grown by LP-MOCVD. Excitation spectroscopy shows evidence of strong and well resolved excitons peaks in the luminescence and excitation spectra of GaInAs-InP quantum wells. Optical absorption show room-temperature exciton in GaInAs-InP superlattices [25]. The quantum well as thin as 8 Å with a photoluminescence linewidth of 9 meV has been grown.

Negative differential resistance at room temperature from resonant tunnelling in GaInAs-InP double barrier heterostructures grown by LP-MOCVD has been observed for the first time [26]. Fig. 1 shows the "edge-on" transmission electron microscopy (TEM) characterization on a resonant tunnelling structure of GaInAs-InP of a 28 Å thickness grown under the same conditions. The result shows that the thickness of the triple layer is generally regular [28].

Application. The use of the $Ga_{0.43}In_{0.57}As$ lattice-matched to InP for long-wavelength detectors is now well established. Among the known structures, the simple and reliable P|N has become one of the more attractive enabling the fabrication of high-sensitivity PIN-FET optical receivers as well as that of broad-area general-purpose photodiodes [21].

The InP/InGaAs/InP double heterostructures are grown by LP-MOCVD [22]. The structure consists of three nominally undoped layers deposited on a n-Type InP substrate as follows:

- 1.5 μm thick undoped InP buffer layer,
- 3 μm thick undoped GaInAs absorption layer,
- 1.5 μm thick undoped InP window layer,

in which the standard thickness of 1.5 μm InP window layer has been decreased in some experiments up to 0.5 μm in order to enhance the responsivity in the near-infra-red spectrum beyond 0.9 μm. The background impurity concentration deduced from C-V (capacitance-voltage) measurements is routinely obtained in the range of 10^{15} cm^{-3}, suitable to achieve the low capacitance required for PIN-FET implementation. The P-n junction is

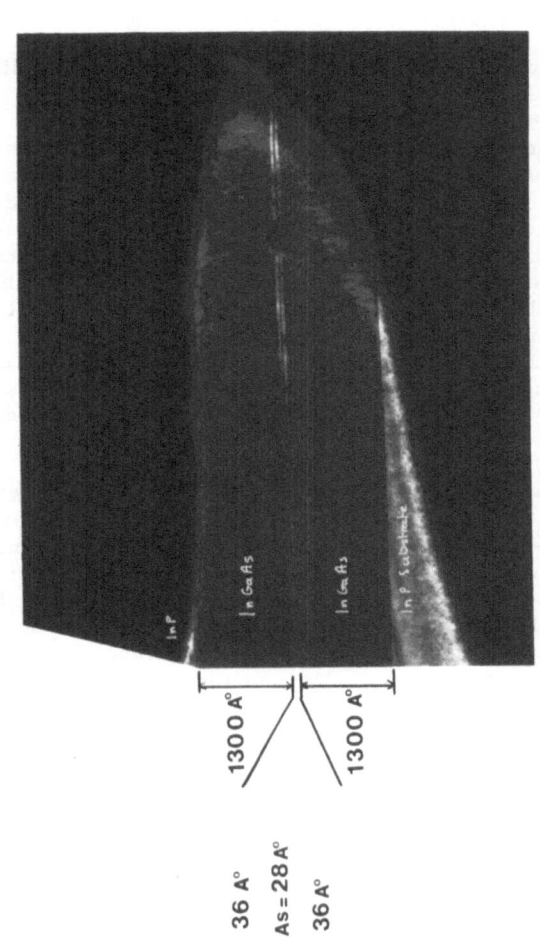

InP = 36 A°

InGa As = 28 A°

InP = 36 A°

Fig. 1 : "Edge-on" Transmission electron microscopy characterization on a resonant tunnelling structure of GaInAs-InP of a 28 Å GaInAs layer sandwiched between two InP layers of 36 Å thickness.

next formed by Zn diffusion using the semiclosed box technique [23] and is located through the InP window in the GaInAs layer typically 0.5 to 1 μm away from the upper heterointerface. The ohmic contacts are made by sputtered Au-Zn alloyed at 425°C for 30 s on the P side, and by sputtered Au on the n side. Mesas are finally defined by a chemical etch in a Br_2 : HBr : H_2O solution. Fig. 2 gives a histogram of the dark current for one wafer grown by LP-MOCVD over 10 cm^2 of InP substrate, comparing with a histogram of the dark current for one wafer grown by LPE over 1.0 cm^2 of InP substrate. The value of dark current density less than 10^{-5} A/cm^2 has been achieved for uncoated, unpassivated diode with mesa technology. The quantum efficiency measured as a function of the wavelength using lock-in techniques was 70% in the 1.0 to 1.6 μm range, which corresponds to a nearly 100% internal efficiency for devices without an antireflection coat. The bandwidth has also been accurately determined with units mounted in a microwave package and we measured a 3 GHz-3 dB cutoff frequency at the bias joint.

The etch-pit observation technique has been used for the study of the crystalline perfection of PIN heterostructure. Fig. 3 shows the transition region after chemical etching of an undoped GaInAs-InP double heterostructure for PIN diode. Before the selective revelation, we form a chemical level with a very low angle [24]. By using this technique, one can see whether misfit dislocations are generated at the interfaces between GaInAs and InP layers and also whether the substrate-epilayer interface is defect free. The density of dark current of PIN diode depends directly to the quality of the epitaxial layers and interfaces. This figure shows that etch-pit density (EPD) was the same in the substrate as in the InP buffer and window layers, and the interfaces are defect free.

GaInAs-InP JFETs

$Ga_{0.47}In_{0.53}As$ has a favorable band structure for the development of millimeter wave FETs. A high electron velocity can be achieved even at high doping levels, which will give high transconductance and high cutoff frequency. Since GaInAs has too small bandgap to form a sufficiently high and leakage free Schottky barrier, different solutions have been proposed to make the gate: a high bandgap material such as InAlAs or InP (lattice-matched) or GaInP, GaAs and GaP (lattice-mismatched) system [5] can be grown on top of the channel layer. One can also use a PN junction obtained either by diffusion, by ion implantation or by etching a P-type epitaxial layer.

The JFETs GaInAs-InP were fabricated using a chemical etching technique. The structure consists of:

- 0.5 μm thick undoped InP buffer layer,
- 0.2 μm thick GaInAs, n-doped, $N_D - N_A \simeq 10^{17}$ cm^{-3}, channel layer,
- 0.5 μm thick InP, ZN-doped, $N_A - N_D \simeq 10^{18}$ cm^{-3} layer,
- 0.5 μm thick GaInAs, Zn-doped, $N_A - N_D \simeq 10^{19}$ cm^{-3} contact layer.

These layers are successively grown on a semi-insulating InP substrate at 550°C. JFETs with 0.5 μm gate lengths, 150 μm gate widths and 1.5 μm source drain spacing have been fabricated. The transconductance as high as 260 ms/mm has been achieved [27]. In spite of high diffusion coefficient of Zinc, no problem of Zinc in the channel has been detected and an electron velocity as high as 3.6×10^7 cm/s was deduced from the transconductance.

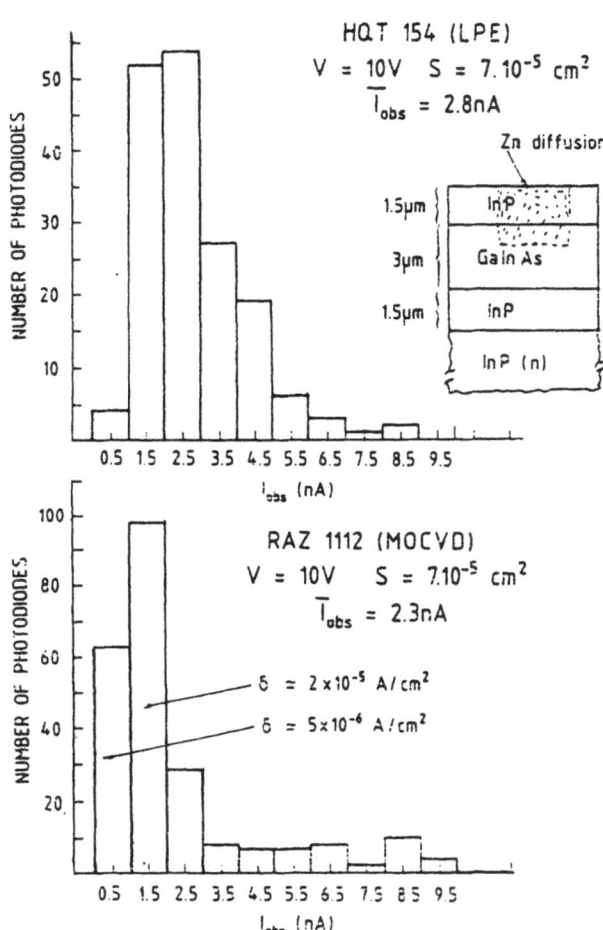

Fig. 2 : Histogram of dark current at −10 V bias for one wafer grown by LP-MOCVD over 10 cm^2 of InP substrate comparing with a histogram of the dark current for one wafer grown by LPE over 1.5 cm^2 of InP substrate.

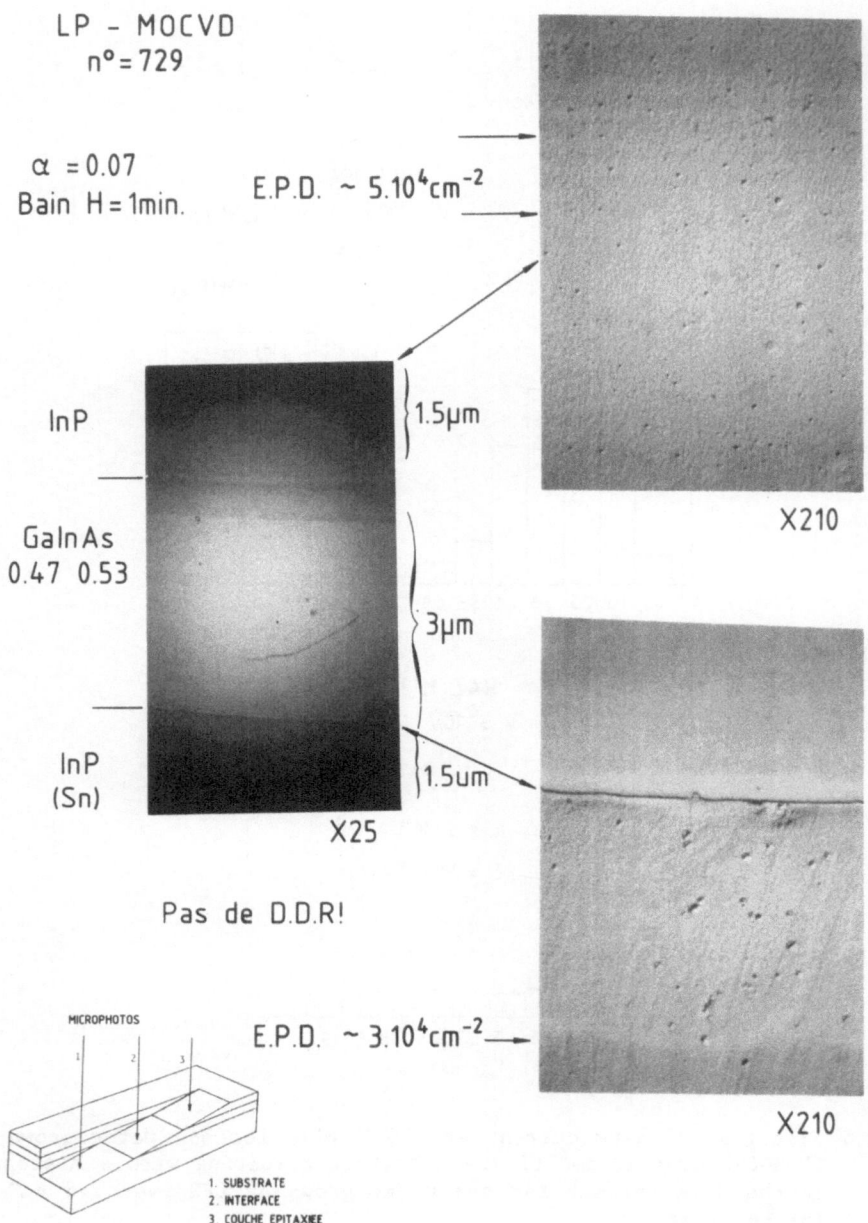

LP - MOCVD
n° = 729

α = 0.07
Bain H = 1min.

E.P.D. ~ 5.10^4cm^{-2}

X210

InP

{1.5µm

GaInAs
0.47 0.53

}3µm

InP
(Sn)

}1.5um

X25

Pas de D.D.R!

MICROPHOTOS

1. SUBSTRATE
2. INTERFACE
3. COUCHE EPITAXIEE

E.P.D. ~ 3.10^4cm^{-2}

X210

Fig. 3 : The depth of a chemical level, realized on an epitaxial
GaInAs-InP DH (for PIN diode) in a bromine solution.

GaInP-GaInAs-InP MESFETs

A $Ga_{0.49}InP_{0.51}/Ga_{0.47}In_{0.53}As/InP$ MESFETs has been fabricated from material grown by LP-MOCVD, materials structures consisting of n-Type GaInAs of 1500 Å thick, doped to 3×10^{17} cm^{-3} with sulphur and undoped GaInP of 800 Å thick with a carrier concentration of 10^{16} cm^{-3} grown at 550°C onto (100) oriented Fe doped semi-insulating InP substrates. Large geometry FETs with 2 μm gate lengths, 150 μm gate widths and 5 μm source drain spacing have been fabricated. The source and drain contacts on the GaInP layer consist of evaporated Au-Ge-Ni. Pt/Ti/Pt/Au was used for gate contact. The gate pads were finally isolated from the active area by under etching at the same time as the component isolation by mesa etching down to the semi-isolating InP substrate. The transconductance of 50 ms/mm has been obtained.

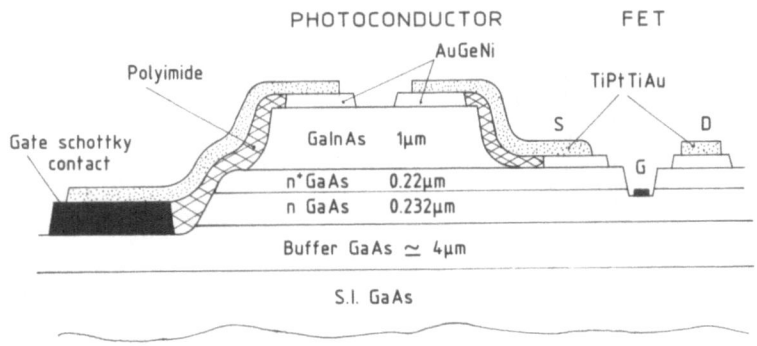

Fig. 4 : Schematic cross section of integrated photoreceiver.

Integrated photoreceiver and FET

A monolithic integrated circuit consisting of a $Ga_{0.47}In_{0.53}As$ planar photoconductive detector (suitable for 1.3-1.55 μm wavelength optical communication systems) associated with a GaAs field effect transistor, have been fabricated using strained heteroepitaxies [19]. The strained heteroepitaxy is constituted of an undoped GaInAs layer grown on a classical GaAs FET epitaxy (Fig. 4). The strained heteroepitaxies did not degrade the noise performance of the photoconductor in the gigahertz frequency range.

Vapor Pressures

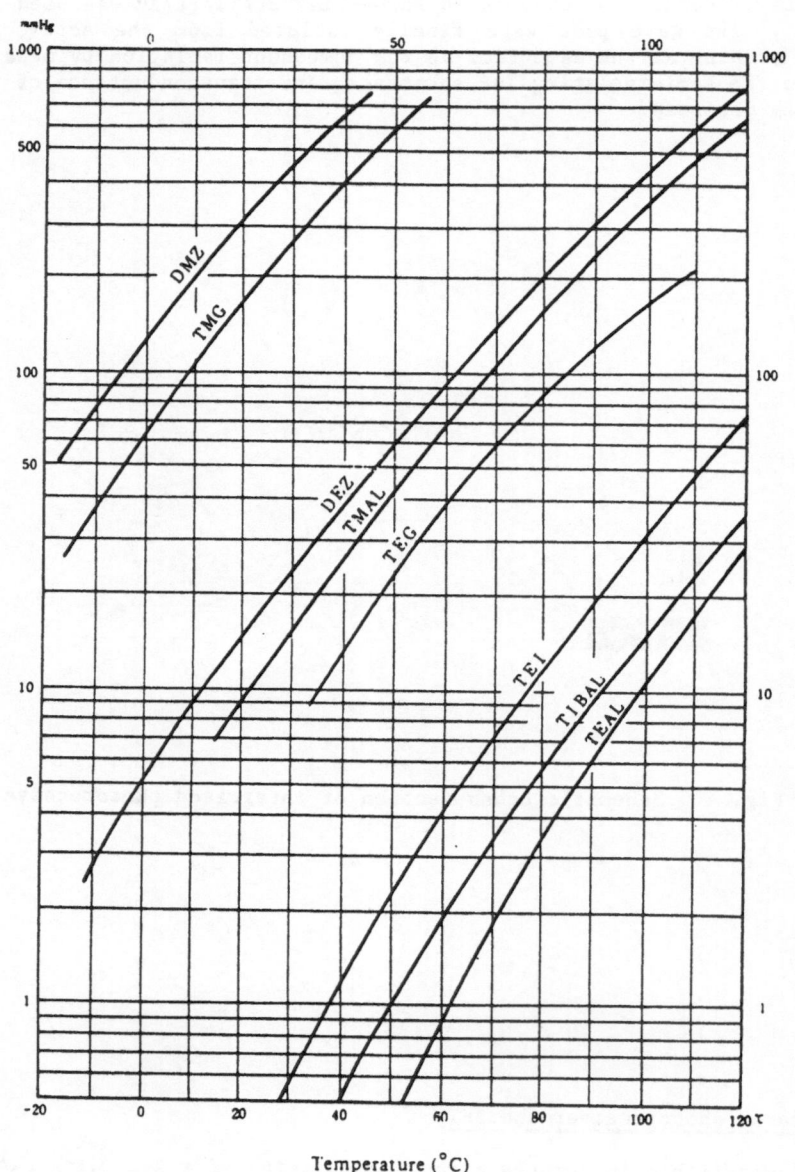

Temperature (°C)

DIETHYLZINC

ACRONYM . . . . DEZ

FORMULA . . . . $(C_2H_5)_2Zn$

FORMULA WEIGHT 123.49

METALLIC PURITY . . . . . 99.9999 wt % (min) zinc

APPEARANCE . . . . Clear, colorless liquid

DENSITY . . . . 1.198 g/mL @ $30^{\circ}C$

MELTING POINT . . . . . . $-30^{\circ}C$

VAPOR PRESSURE . . . . . . 3.6 mm Hg @ $0.0^{\circ}C$
16 mm Hg @ $25.0^{\circ}C$
760 mm Hg @ $117.6^{\circ}C$

BEHAVIOR TOWARD ORGANIC
SOLVENTS . . . . Completely miscible, without reaction, with aromatic
and saturated aliphatic and alicyclic hydrocarbons.
Forms relatively unstable complexes with simple ethers,
thioethers, phosphines and arsines, but more stable
complexes with tertiary amines and cyclic ethers.

STABILITY TO AIR . . . . . Ignites on exposure (pyrophoric).

STABILITY TO WATER . . . . Reacts violently, evolving gaseous hydrocarbons,
carbon dioxide and water.

STORAGE STABILITY. . . . . Stable indefinitely at ambient temperatures when
stored in an inert atmosphere.

DIMETHYLZINC

ACRONYM . . . . DMZ

FORMULA . . . . $(CH_3)_2Zn$

FORMULA WEIGHT 95.44

METALLIC PURITY . . . . . 99.9999 wt % (min) zinc

APPEARANCE . . . . Clear, colorless liquid

DENSITY . . . . 1.386 g/ml @ $10.5^{\circ}C$

MELTING POINT . . . . . . $-29.0^{\circ}C$

VAPOR PRESSURE . . . . . . 123mm Hg @ $0.0^{\circ}C$
372mm Hg @ $25.0^{\circ}C$
760mm Hg @ $44.0^{\circ}C$

BEHAVIOR TOWARD ORGANIC
SOLVENTS . . . . Completely miscible, without reaction, with aromatic
and saturated aliphatic and alicyclic hydrocarbons.
Forms relatively unstable complexes with simple ethers,
thioethers, phosphines and arsines, but more stable
complexes with tertiary amines and cyclic ethers.

STABILITY TO AIR . . . . . Ignites on exposure (pyrophoric).

STABILITY TO WATER . . . . Reacts violently, producing gaseous hydrocarbons,
carbon dioxide, water and oxygenated zinc compounds.

STORAGE STABILITY. . . . . Stable indefinitely at ambient temperatures when stored
in an inert atmosphere.

TRIETHYLINDIUM

ACRONYM . . . . TEI

FORMULA . . . . $(C_2H_5)_3In$

FORMULA WEIGHT 202.01

METALLIC PURITY . . . . . . 99.9999 wt% (min) Indium

APPEARANCE . . . . Clear, colorless liquid

DENSITY . . . . 1.260 g/mL @ 20^oC

MELTING POINT . . . . -32^oC

VAPOR PRESSURE . . . . . . 1.18 mm Hg @ 40^oC
4.05 mm Hg @ 60^oC
12.0 mm Hg @ 80^oC

BEHAVIOR TOWARD ORGANIC
  SOLVENTS . . . . Completely miscible, without reaction , with aromatic
and saturated aliphatic and alicyclic hydrocarbons.
Forms complexes with ethers, thioethers, tertiary
amines, -phosphines, -arsines and other Lewis bases.

STABILITY TO AIR . . . . . Ignites on exposure (pyrophoric).

STABILITY TO WATER . . . . Partially hydrolyzed; loses one ethyl group with
cold water.

STORAGE STABILITY. . . . . Stable indefinitely at ambient temperatures when
stored in an inert atmosphere.

TRIMETHYLINDIUM

ACRONYM . . . . TMI

FORMULA . . . . $(CH_3)_3In$

FORMULA WEIGHT 159.85

METALLIC PURITY . . . . . 99.999 wt % (min) indium

APPEARANCE . . . . White, crystalline solid

DENSITY . . . . 1.568 g/ml at 19^oC

MELTING POINT . . . . . . 89^oC (192.2^oF)

BOILING POINT . . . . . . 135.8^oC (276.4^oF)/760 mm Hg
67^oC (152.6^oF)/12 mm Hg

VAPOR PRESSURE. . . . . . 15 mm Hg @ 41.7^oC (107^oF)

STABILITY TO AIR. . . . . Pyrophoric, ignites spontaneously in air.

SOLUBILITY. . . . . Completely miscible with most common organic solvents.

STORAGE STABILITY . . . . Stable indefinitely when stored in an inert atmosphere.

TRIMETHYLGALLIUM

ACRONYM . . . . TMG

FORMULA . . . . $(CH_3)_3Ga$

FORMULA WEIGHT 114.82

METALLIC PURITY . . . . . . 99.9999 wt % (min) gallium

APPEARANCE . . . . Clear, colorless liquid

DENSITY . . . . 1.151 g/mL @ 15°C

MELTING POINT . . . . . . -15.8°C

VAPOR PRESSURE . . . . . . 64.5 mm Hg @ 0.0°C
226.5 mm Hg @ 25.0°C
760 mm Hg @ 55.8°C

BEHAVIOR TOWARD ORGANIC
SOLVENTS . . . . Completely miscible, without reaction, with aromatic and saturated aliphatic and alicyclic hydrocarbons. Forms complexes with ethers, thioethers, tertiary amines, tertiary phosphines, tertiary arsines and other Lewis bases.

STABILITY TO AIR . . . . . Ignites on exposure (pyrophoric)

STABILITY TO WATER . . . . Reacts vigorously, forming methane and Me_2GaOH or $[(Me_2Ga)_2O]x$.

STORAGE STABILITY. . . . . Stable indefinitely at ambient temperatures when stored in an inert atmosphere.

TRIMETHYLGALLIUM

ACRONYM . . . . TEG

FORMULA . . . . $(C_2H_5)_3Ga$

FORMULA WEIGHT 156.91

METALLIC PURITY . . . . . . 99.9999 wt % (min) gallium

APPEARANCE . . . . Clear, colorless liquid

DENSITY . . . . 1.0586 g/mL (at 30°C)

MELTING POINT . . . . . . -82.3°C

VAPOR PRESSURE . . . . . . 16 mm Hg @ 43°C
62 mm Hg @ 72°C
760 mm Hg @ 143°C

BEHAVIOR TOWARD ORGANIC
SOLVENTS . . . . Completely miscible, without reaction, with aromatic and saturated aliphatic and alicyclic hydrocarbons. Forms complexes with ethers, thioethers, tertiary amines, tertiary phosphines, tertiary arsines and other Lewis bases.

STABILITY TO AIR . . . . . Ignites on exposure (pyrophoric)

STABILITY TO WATER . . . . Reacts vigorously, forming ethane and Et_2GaOH or $[(Et_2Ga)_2O]_x$.

STORAGE STABILITY. . . . . Stable indefinitely at room temperature in inert atmosphere.

TRIETHYLINDIUM

ACRONYM	TEAL
FORMULA	$(C_2H_5)_3Al$
FORMULA WEIGHT	114.16
METALLIC PURITY	99.999 wt % (min) aluminum
APPEARANCE	Clear, colorless liquid
DENSITY	0.835 g/ml @ 25°C
MELTING POINT	-52.5°C
VAPOR PRESSURE	0.5 mm Hg @ 55°C 1.0 mm Hg @ 62°C 760 mm Hg @ 186°C
BEHAVIOR TOWARD ORGANIC SOLVENTS	Completely miscible, without reaction, with aromatic and saturated aliphatic and alicyclic hydrocarbons. Forms complexes with ethers, thioethers, tertiary amines, tertiary phosphines, tertiary arsines and other Lewis bases.
STABILITY TO AIR	Ignites on exposure (pyrophoric).
STABILITY TO WATER	Reacts violently, evolving gaseous hydrocarbons, carbon dioxide and water.
STORAGE STABILITY	Stable indefinitely at ambient temperatures when stored in an inert atmosphere.

TRIMETHYLALUMINUM

ACRONYM	TMAL
FORMULA	$(CH_3)_3Al$
FORMULA WEIGHT	72.08
METALLIC PURITY	99.9999 wt % (min) aluminum
APPEARANCE	Clear, colorless liquid
DENSITY	0.752 g/mL @ 20°C
MELTING POINT	15°C
VAPOR PRESSURE	9 mm Hg @ 20°C 69 mm Hg @ 60°C 760 mm Hg @ 127°C
BEHAVIOR TOWARD ORGANIC SOLVENTS	Completely miscible, without reaction, with aromatic and saturated aliphatic and alicyclic hydrocarbons. Forms complexes with ethers, thioethers, tertiary amines, tertiary phosphines, tertiary arsines and other Lewis bases.
STABILITY TO AIR	Ignites on exposure (pyrophoric).
STABILITY TO WATER	Reacts violently.
STORAGE STABILITY	Stable at ambient temperatures when stored in an inert atmosphere.

REFERENCES

1. M. Razeghi, J.P. Duchemin and J.C. Portal, Appl. Phys. Letters, 46:46 (1985).
2. M. Razeghi, P. Maurel, A. Tardella, L. Donomski, D. Gauthier, J.C. Portal, J. Appl. Phys. (schedule for September 1986).
3. M. Razeghi, J.P. Hirtz, U.O. Ziemelis, C. Delalande, B. Etienne and M. Voos, Appl. Phys. Letters, 43:585 (1983).
4. M. Razeghi, P. Maurel, F. Omnes, L. Donomski, J.C. Portal, Appl. Phys. Letters, 48:1267 (1986).
5. M. Razeghi, P. Maurel, F. Omnes, E. Thörngren, NATO Workshop on low-dimensional structure. St-Andrew, U.K. 29 July-2 August (1986).
6. M.J. Ludowise, J. Appl. Phys., 58:R31 (1985).
7. J.P. Duchemin, M. Bonnet, F. Koelsch, and D. Huighe, J. Electrochem. Soc., 126:1134 (1979).
8. M. Razeghi, Chapter 12 of "Technology for chemicals and materials for electronics", ed. Howells, London (1984).
9. Alphagaz, TEXAS ALKYLS Production.
10. H.M. Manasevit and W.I. Simpson, J. Electrochem., 118:C291 (1971).
11. S.J. Bass and P. Oliver, Inst. Phys. Conf. Ser. 33b:1 (1977).
12. D.J. Tritton, "Physical Fluid Dynamics", Van Nostrand Reinhold (UK) Co.Ltd (1982), University Press, Cambridge.
13. L. Prandtl and O.C. Tietgens, "Applied Hydro-and Aeromechanics", Mc-Grew-Hill/Doner (1934).
14. A.H. Shapiro, Shape and Flow (Heinemann) (1961).
15. M. Razeghi and J.P. Duchemin, J. Cryst. Growth, 64:76 (1983).
    M.A. Poisson, C. Brylinski, and J.P. Duchemin, Appl. Phys. Lett., 46:476 (1985).
16. M. Razeghi, "semiconductors and semimetals", vol. 22, ed. W.T. Tsang (1985).
17. S.M. Sze, "Physics of semiconductor devices", John Wiley & Sons (1981).
18. Ridely, Appl. Phys. 48:754 (1977).
19. M.A. Poisson, C. Brylinski, G. Coloner, D. Osselin, S. Hersee, F. Azon, D. Lechevallier, J. Lacombe, J. Electron Lett., 20:n°25/26, 1061 (1984).
20. M. Razeghi and J.P. Duchemin, J. Vac. Sci. Tech. B, Vol. 1, n°2 (1983).
21. C.A. Burrus, A.G. Dental and T.P. Lee, Opt. Commun, 38:124 (1981).
22. P. Poulain, M. Razeghi, K. Kazmierski, R. Blondeau, P. Philippe, Electronics Letters, Vol. 21, n°10, 441 (1985).
23. K. Kazmierski, A. Huber, G. Morillot and B. Decremoux, Upn, J. Appl. Phys., 23:628 (1984).
24. R.A. Huler, M. Razeghi, G. Morillot, GaAs and related compounds, FRANCE, 26-28 September (1984), Inst. Phys. Conf. Ser. N°74, 41.
25. M. Razeghi, J. Nagle, P. Maurel, F. Omnes and J.P. Pocholle, Appl. Phys. Letters (to be published).
26. M. Razeghi, A. Tardella, R.A. Danies, A.P. Long, M.J. Kelly, E. Britton, C. Boothroyd and W.M. Stehbs, Electronic Letters (to be published).
27. J.Y. Raulin, E. Thorngren, M.A. Poisson, M. Razeghi, G. Colomer, Appl. Phys. Letters (to be published).
28. M.J. Kelly, C.B. Boothroyd and W.M. Stobhs (private communication). These results will be published.
29. M. Razeghi, J. Ramdanis, D. Vessiele, M. Decoster, M. Constant and J. Vanbremeersch, Appl. Phys. Lett., 49:215 (1986).

THE PREPARATION OF MODULATED SEMICONDUCTOR STRUCTURES

BY LIQUID PHASE EPITAXY

E. Bauser

Max-Planck-Institut fuer Festkoerperforschung

Heisenbergstr. 1, D-7000 Stuttgart 80, FRG

INTRODUCTION

The paper discusses the particular advantages which liquid phase epitaxy (LPE) offers for the production of low-dimensional structures. It deals with the progress which growth techniques and growth apparature have made. It describes present and future capabilities of the LPE technique in the preparation of semiconductor superlattices, quantum well structures and heterostructures. The possibilities are described for controlling the morphologies of surfaces and interfaces by means of different crystal growth mechanisms.

Usually the sizes of modulated semiconductor structures and low-dimensional structures must be precise on an atomic scale. Liquid phase epitaxy (LPE) offers specific advantages for the preparation of layers of high perfection. Experimental results, which will be shown, demonstrate the advantages of LPE in the preparation of modulated structures. Possibilities overcoming difficulties in the application of LPE will be discussed.

Advantages of LPE

The main advantage of LPE is that solution growth occurs when conditions are close to equilibrium.[1] The supersaturations required for growth are extremely low. It is, therefore, possible to control precisely the morphology of the growth interface. LPE layers with atomically entirely flat surface areas have been obtained, for example, and layers have been grown whose surfaces show trains of monomolecular and equidistant steps which cover regions of up to many square millimeters in size.[2] Step interdistance and step height at these surfaces may be in the ratio of more than 10^4 to 1. Although such growth interfaces are almost atomically flat they permit uniform, non-intermittant crystal growth. Growth faces of this kind have an exceedingly high morphological stability and are, therefore, favorable to multilayer growth.[3] The material, which results owing to the uniform motion of monostep trains, is extremely homogeneous both in composition and in the distribution of dopants and impurities. In addition, this homogeneity of the crystal establishes high structural perfection of the material.

LPE allows high purity layers to be produced particularly in those cases in which the solvent species are constituents of the semiconductor-compound or -alloy. Moreover, it is possible to grow by LPE very thin layers with thicknesses in the nm range.[4,5] On the other hand, growth rates are sufficiently high and permit the growth of layers whose thicknesses may exceed 100 μm. Multilayers can therefore be prepared which contain in one stack thin <u>and</u> thick layers.[5] Also, selective-area growth on partially masked substrates is easily achieved. No deposition on the masked areas will occur.[2] The growth of single- and multilayers on patterned substrates constitutes a further possibility; and one of the most attractive possibilities seems to be the seeded, single crystalline lateral growth over insulating layers.[6] Preliminary experiments with silicon growth over thermally oxidized silicon produced a film length-to-thickness ratio > 45:1.

Another advantage of LPE which may balance still existing difficulties consists in being able to make consumption of original material very low. Solutions can be resaturated for many applications and recycled. Compared to MOCVD the LPE technique is inherently safer because the raw materials and waste products are less toxic and not pyrophoric.[1]

Difficulties in LPE

One of the difficulties with LPE lies in achieving a precise orientation of the growth interface and in maintaining its reproducibility. Due to the low supersaturation required, 2-D nucleation does not occur during LPE growth. Hence, the crystallographic orientation of the interface determines the total density of available growth steps, provided the substrates are dislocation-free. The range of favorable misorientations of the substrate growth face as a whole in relation to a low indices face extends only from zero to about 0.1° in LPE.[1,2] For substrate misorientations by more than 0.1°, the probability increases that the growth steps will develop bunches during their lateral motion. The interfaces may then show the well-known growth terraces, an undesirable result of the LPE-typical low supersaturation. When substrate misorientation is zero, misorientation steps are absent. When, in addition, the substrate is dislocation-free, perpendicular macroscopic growth (PMG) does not occur in common LPE experiments. If, however, the substrate contains dislocations, these usually act as sources of equidistant monosteps in LPE. And, at considerable distances from their sources, the step edges straighten. In cases of low substrate dislocation density, trains of equidistant and straight monosteps may develop and cover at the surfaces of the LPE grown layers, regions of up to several square millimeters .[7] Surfaces of this kind display an appropriate interface morphology for the growth of modulated semiconductor structures. It is precisely this interface morphology that must at least locally be maintained during the growth of multilayer- and superlattice structures. It is, therefore, necessary to accommodate the design of LPE crucibles to this requirement.

Another difficulty and critical part of the LPE process is related to the growth apparatus. It is caused by the movement of the solution or the substrate at the beginning and at the end of the required growth period.[1] The introduction of multilayer growth by LPE in horizontal graphite boats with movable sliders has been a successful step forward in the mechanical technology of LPE.[8] The majority of LPE growth systems in use today apply graphite slider boats of various forms, many of which are automated. Most of the LPE boats used today for growth of III-V compounds can be regarded as descendants of the original idea published by Panish et al.[9] and Alferov et al.[10] The difficulties which still remain with the slider boats consist in the relative movement of parts of the crucible containing the substrates or solutions. A boat project that avoids relative movements of parts of the crucible close to the growth interface and which is at the

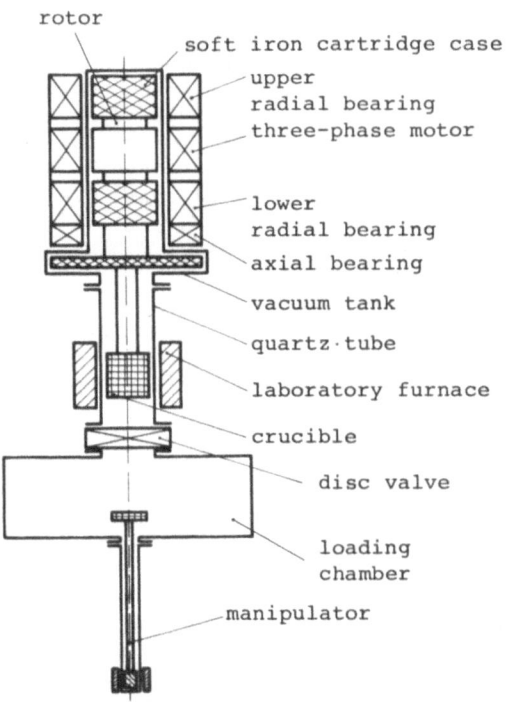

rotor
soft iron cartridge case
upper
radial bearing
three-phase motor
lower
radial bearing
axial bearing
vacuum tank
quartz tube
laboratory furnace
crucible
disc valve
loading
chamber
manipulator

Fig. 1. LPE centrifuge with contactless electromagnetic suspension of the rotor, schematic
longitudinal section, after 13.

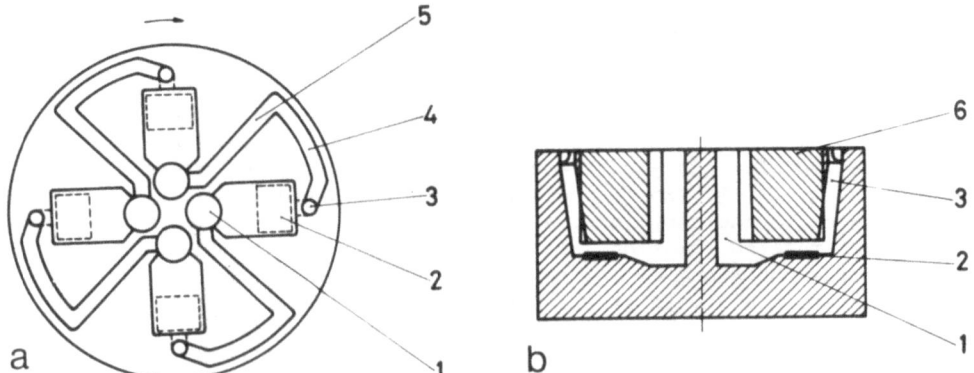

Fig. 2. Crucible of the LPE centrifuge (a) cross section, (b) longitudinal
section. The crucible itself contains no moving parts. The solutions
are transported by centrifugal forces. (1) containers for solution,
(2) substrates, (3), (4) and (5) solution-flow channels, (6) substrate holder blocks.

same time appropriate for multilayer growth has been outlined by Scheel.[11] The boat revolves about an axis inclined at ~10° to the horizontal. It transports the solutions in an Archimedes screw and applies gravity to establish a closed cycle for each of the solutions. Another difficulty which arises particularly in growth on large area substrates consists in the incomplete removal of solutions. In order to overcome all such difficulties, further work must be devoted to the technical development of growth systems. An experiment directed towards overcoming the difficulties is described in the following.

CENTRIFUGAL LPE TECHNIQUE

The centrifugal technique represents an experiment to adapt the LPE technique to the requirements of
(I)   multilayer growth with precisely defined and unperturbed interface morphology,
(II) growth on large area substrates,
(III) high capacity of the growth system and high throughput for production purposes.
This technique utilizes centrifugal forces in order to move the solutions.[12] No moving parts are necessary for the crucible for the sake of solution transport. Consequently, there is no abrasion from the crucible that might cause defects during layer growth, and nor does scratching occur that might injure the growth interface. Rapid rotation of the crucible about a vertical axis ensures fast transport of the solutions by centrifugal forces and produces brief contact between the solution and the substrate. It is therefore possible to grow extremely thin layers. The transport of the solutions is complete and no residue remains. A mixing of solutions of different composition is unlikely.

After each layer growth the solutions are resaturated and completely recycled inside the crucible. Growth conditions are therefore identical for each individual layer of, for example, a superlattice. Automatic operation of the entire system is easy. The system is suitable for upscaling in wafer dimension and capacity. Fig. 1 shows a longitudinal section of the epitaxy centrifuge. The crucible is mounted on the lower end of a vertical rotor. The rotor has at its upper end a contactless, electromagnetic suspension inside a thin-walled vacuum tank.[13] The bearing magnets and the rotor drive are outside the vacuum tank. Both the operation of the epitaxy centrifuge[14] and the epitaxial growth process[2] are computer controlled. The process is reliable. Therefore, after charging the crucible and choosing a growth programme the automated epitaxy apparatus produces single- or multilayers and needs no supervision.

A graphite boat of simple design is shown schematically in Fig. 2. This crucible has four containers for the solutions indicated by 1. The growth procedure of a npnp-multilayer, for example, may be described by referring to Figs. 2-3. Two of the containers which are opposite to each other and close to the substrates A and C in Fig. 3, are filled with n- and p-doped saturated solutions, respectively. Two substrate wafers are placed in positions A and C, and two undoped wafers for resaturation of the solution are positioned at B and D. When the crucible starts to rotate the solutions slip from the container position 1 to the growth position in which they cover the substrates 2. The solutions then fill the gaplike channel above the substrates A and C (see Fig. 3). In this particular example the boat then rotates at a frequency of 230 min^{-1}. Reducing the temperature of the revolving crucible results in the growth of epitaxial layers on the substrates A and C. An increase in the rotational frequency of the crucible causes the solutions to withdraw from the substrates and to move upwards through flow channel 3 into storage channel 4. The solutions remain there as long as the crucible rotates at higher rates,

174

e.g. in this example: 600 min^{-1}. When the rotational frequency is reduced, the solutions fall through the channels 5 (Fig. 2a) into the inner containers close to B and D. When the frequency of the boat is again increased up to 230 min^{-1}, the solutions move into the gaps above the wafers B and D. These wafers are of undoped material, and as the temperature of the crucible is increased, the solutions become resaturated at the expense of wafers B and D. During following increases, decreases and again increases of the rotational frequency of the boat, the solutions move to the substrates C and A. There the next epitaxial layers grow during times of temperature reduction. The growth cycle for pn- and np-structures can be repeated many times during one run.

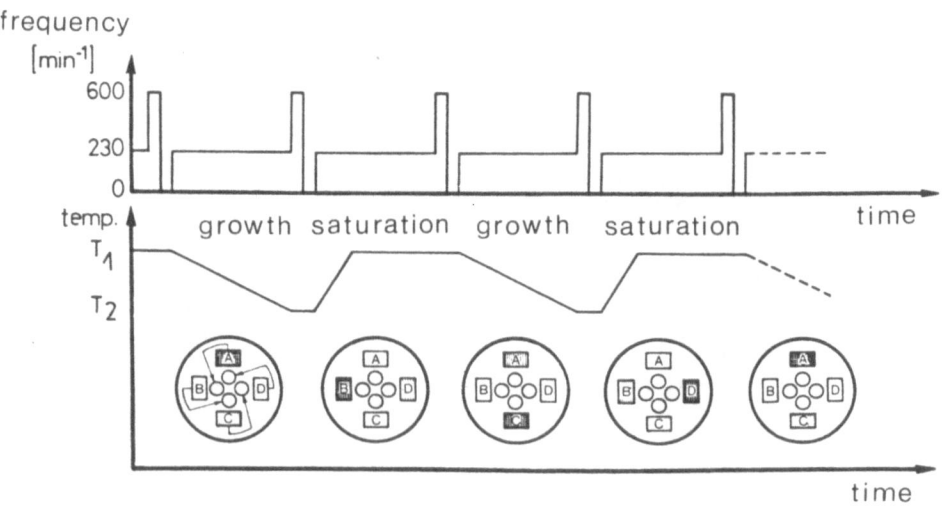

Fig. 3. Frequency- and temperature-time diagrams for the preparation of n-p-n-p multilayers in the centrifuge. While the n- and p-doped solutions cover substrates A and C, respectively, the temperature is reduced from T_1 to T_2, and n- and p-type epitaxial layers grow on substrates A and C. After the layer growth on substrates A and C, the solutions are moved through the channels 3, 4 and 5 (see Fig. 2) and via the containers 1 on to the wafers at B and D (see Fig. 3). While the solutions cover wafers B and D, the temperature is raised again from T_2 to T_1 and the solutions get resaturated at the expense of the undoped wafers B and D. The resaturated n- and p-doped solutions are then moved to substrates C and A, respectively. While the solutions cover substrate C and A, the temperature is again reduced from T_1 to T_2, and n- and p-doped epitaxial layers grow on substrates C and A, and so on.

Multilayers, patterned substrates

Silicon nipi-structures of high quality[15] have been obtained by using the centrifugal LPE technique. The result of one of the first successful growth experiments in the LPE centrifuge is shown in Figs. 4 and 5. The precision in the preparation of doping multilayers has meanwhile been considerably improved.[16] Subsequently, silicon LPE growth on patterned

Fig. 4.
Silicon multilayer
grown from In- and
In:As-solutions to
produce an alterna-
ting p-n structure.
Scanning electron
micrograph (SEM) of
a photoetched clea-
vage face.

substrates, on partially masked substrates, as well as the growth of high
quality silicon on insulators has been studied. In addition to single
layers, also multilayers can be grown on these substrates.[17] Figure 6a-c
shows the result of an attempt in which, before the epitaxial growth, a
stripe pattern was routinely fabricated in a thermally oxidized wafer. The
patterned sample was then etched in Ga at 500°C. There occurs thin underetching along the stripe edges. After removing the thermal oxide from the
ridges, the centrifugal system was applied to grow a thick buffer layer,
and on top of that an n-p-n multilayer of 24 p-type and 24 n-type layers.
The surfaces of the epitaxially grown stripes are smooth and planar. Equidistant monosteps, only slightly bent, extend across the width of the
stripes, as shown in the optical micrograph of Fig. 7.

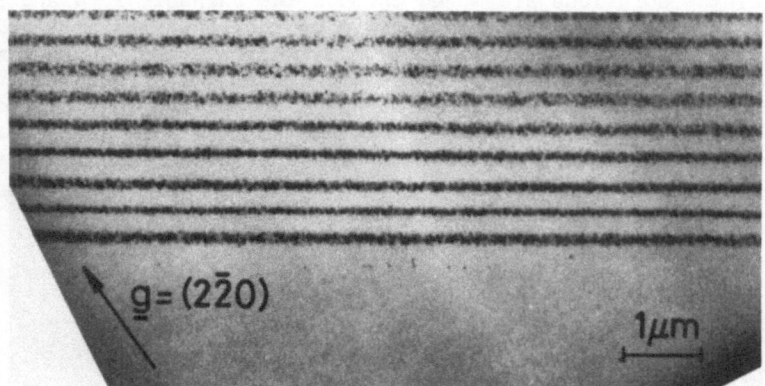

Fig. 5. Transmission electron micrograph of the multilayer shown in
Fig. 4. Cross-section specimen, electron beam parallel to the
interface planes. Dark areas correspond to As-doped layers, g
diffraction vector.

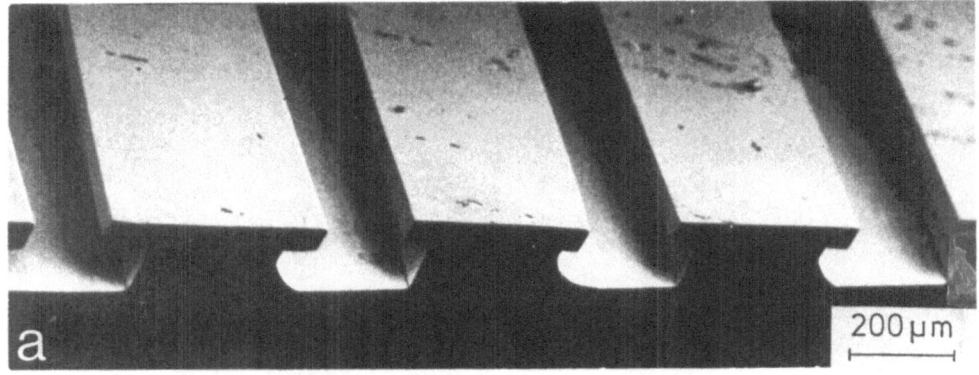

Fig. 6. Si-multilayers grown on profiled substrate. SEM-micrographs.
a) cleavage face and line profile, oblique view (after 17).

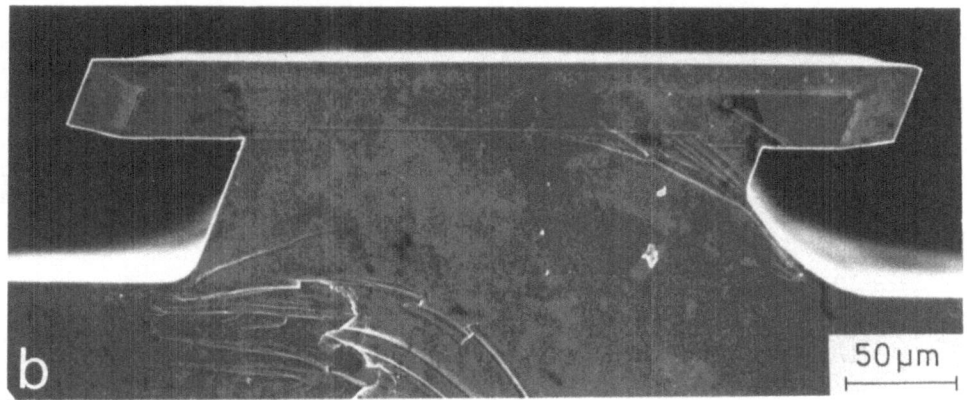

b) magnified section of a). The (111) cleavage face was photoetched to obtain a visible contrast at the interface substrate/
bufferlayer and at the interfaces of the p-n multilayers.

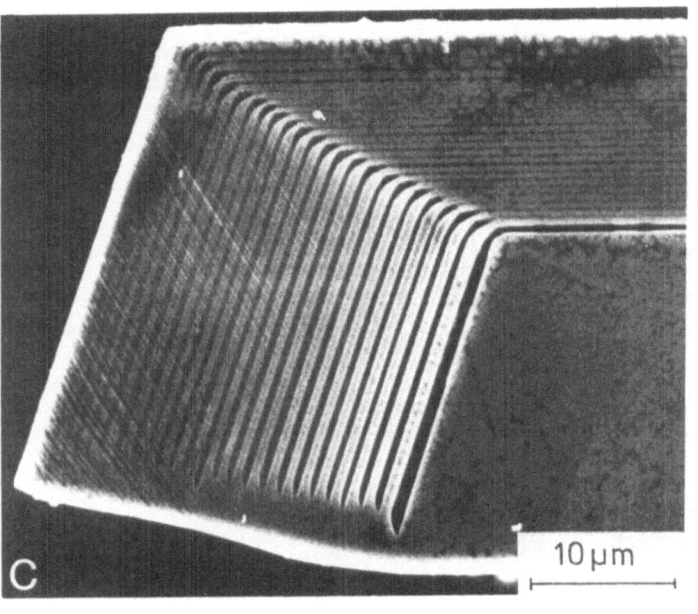

c) magnified section
   of b), showing
   the multilayer
   around the left
   side of the
   "table-like" pro-
   file.

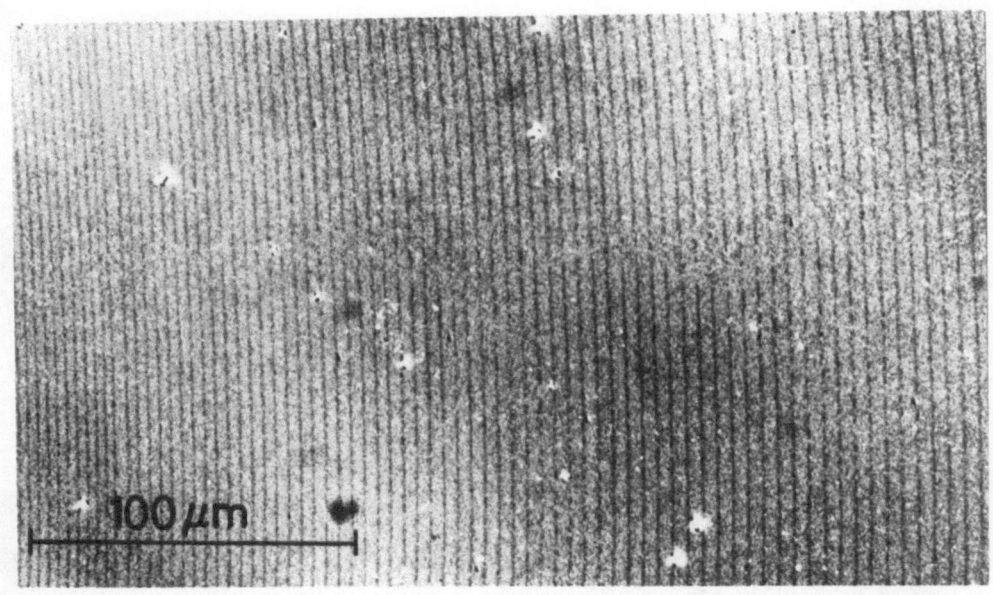

Fig. 7. Train of equidistant monomolecular growth steps at the surface of a multilayer which was grown on a stripe-patterned substrate. The monosteps originate from a step source at the left. N-DIC optical micrograph.

Selective growth

Even if very thick epitaxial islands are deposited, selective growth on partially SiO_2-masked substrates occurs without any competing nucleation or undesired growth on the masked areas. Examples of selective silicon LPE growth are shown in Figs. 8 and 9. The samples show Si islands, with the masked areas in-between, free of nuclei. The selective growth in the windows occurred uniformly and resulted in smooth, in some cases even planar, surfaces of the selectively grown epitaxial islands. Similar as in the example shown in Fig. 6, the epitaxial islands can be used as buffer layer region for the deposition of multilayers.

Fig. 8. a) High and distinctly faceted silicon islands grown in rectangular shaped windows of the SiO_2 mask. The silicon substrate is (100) oriented. The masked areas in-between the epitaxial islands are free of nuclei. SEM micrograph.

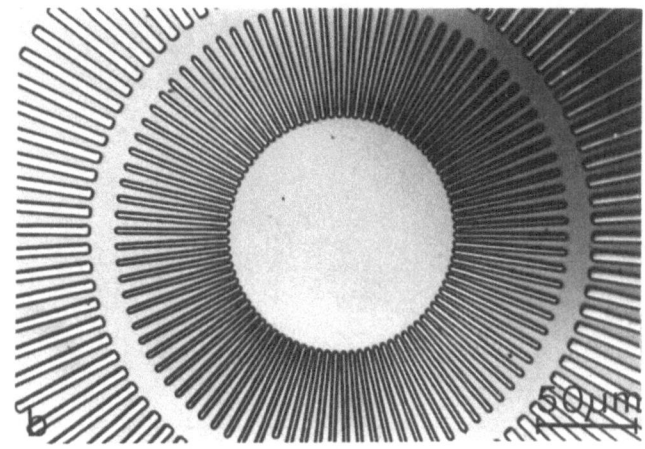

Fig. 8. b) LPE silicon, grown from In-solution in
windows of different shape and different
orientation. The substrate is (100)-oriented.
N-DIC micrograph.

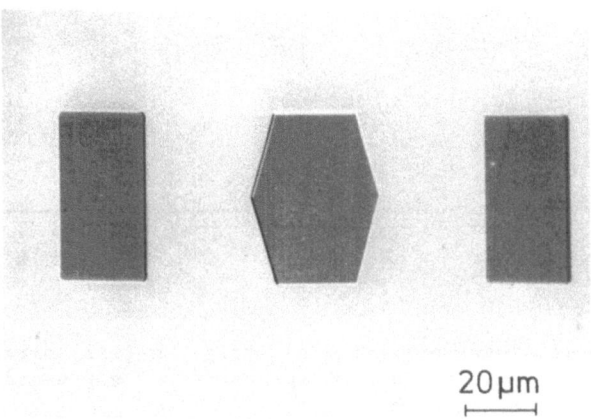

20 µm

Fig. 9. Uniform selective growth in rectangular windows.
Lateral growth occured over the long vertical
edges of the middle window. Substrate orientation
is (111). SEM micrograph.

Single crystalline seeded lateral overgrowth

Lateral overgrowth over masked areas occurs readily in LPE. It is
partly due to a strong anisotropy of growth rates at low temperatures. The
absence of nuclei on the masked areas between the epitaxial islands is a
necessary condition for making possible unperturbed lateral growth over
the masking layer. In our experiments, this layer consisted of SiO_2.
Figure 10 shows an example of selective LPE with lateral growth over a
SiO_2-stripe patterned sample. The edges of the epitaxially grown stripes
are straight if the stripes are oriented parallel with a (110) direction.
For other stripe orientations, the edges of the stripes may at low growth
temperatures become pronged. Very thick or very thin single crystal films
of Si on SiO_2 can be grown, as required. A cross section of a very thick
single crystal stripe is shown in the micrograph of Fig. 11. This micro-

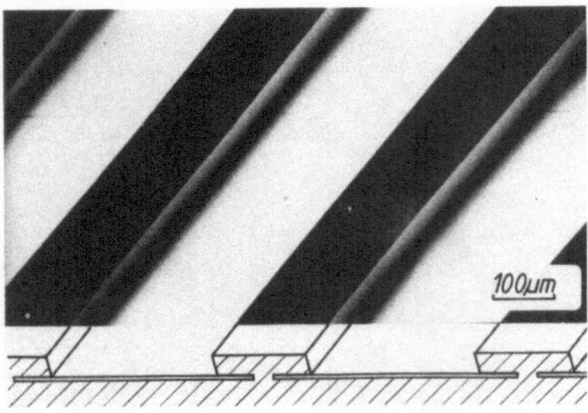

Fig.10. Stripes, laterally grown over thermal oxide. Stripe orientation 110 in the (111) plane. SEM micrograph.

Fig.11. Thick epitaxial Si-stripe on (111) oriented oxidized and stripe-patterned substrate. SEM micrograph.

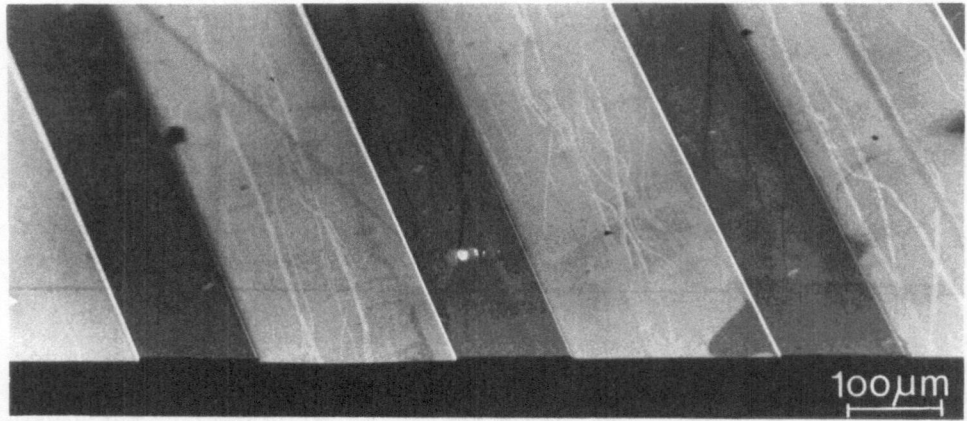

Fig.12. Thin monocrystalline and defect-free epitaxial Si-stripes on (111) oriented substrate. SEM micrograph.

Fig.13. Cross section of a sample with thin defect-free overgrowth lamella.
Growth from In-solution, T = 950°C-850°C. The single crystalline
lamella extends over the SiO$_2$ by more than 100 μm. SEM micrograph.

Fig.14. Plan view of the thin and defect-free silicon film
which grew laterally over the SiO$_2$. Optical micrograph, N-DIC. The silicon lamella which grew over
the oxide is only about 3 μm thick and therefore
optically transparent. It shows visible interferences
of thin layers. The seeding stripe is the uniformly
grey area which is marked by dashed lines. The overgrown layer is over 75 μm wide on one side.

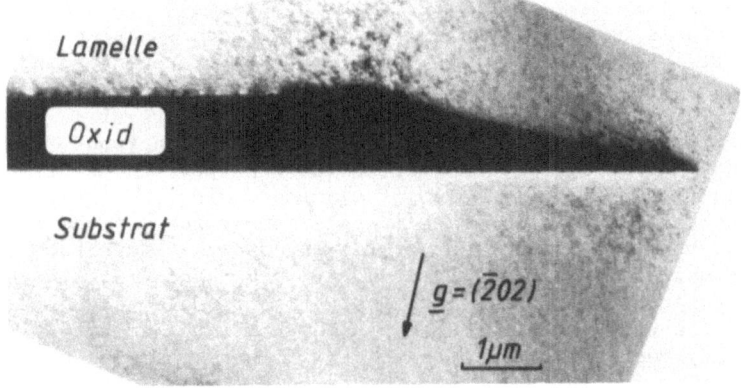

Fig.15. TEM cross section of the seeding site (at right hand
side) and the overgrowth area. The black bar is the
SiO$_2$. No defects are visible in the Si overgrowth
showing the superior crystal quality. Only tiny areas
indicating stress induced contrast are visible at the
upper oxide edge. The tapered edge of the oxide results
from underetching of the photoresist.

graph reveals the seeding site with thick films grown on both sides over the oxide. Figure 12 shows the result of a growth experiment with thin Si films on SiO_2. The stripes in this sample are oriented parallel to a (110) direction. A cross section taken from a similar sample and shown in Fig. 13 will give details. On the left hand side of Fig. 13, the overgrowth lamella extends over the SiO_2 by more than 100 µm.[2] The asymmetry in overgrowth width on the two sides of the stripe is, among other parameters, due to substrate misorientation. The thickness uniformity of the LPE-grown Si films on oxide layers is apparent from the uniformly distributed growth steps at their surfaces and is confirmed by the optical micrograph of Fig. 14 which shows interference fringes of the optically transparent Si lamella grown over the SiO_2. The area of the seeding stripe below the epitaxially grown material is marked by dashed lines.[2]

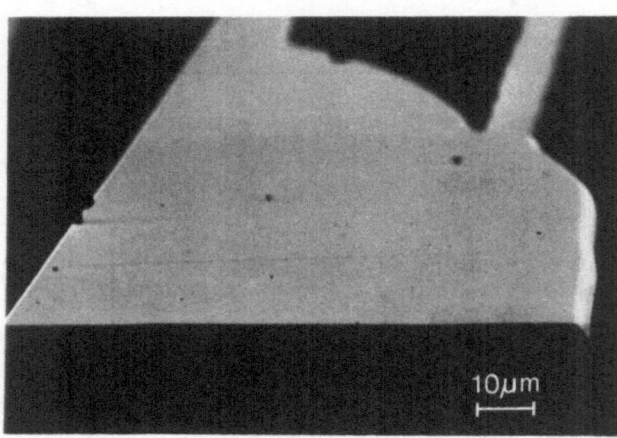

Fig. 16.
Growth of monocrystalline Si, seeded at two stripes, over SiO_2. The oxide is buried below a single crystalline lamella. The surface of the lamella is planar. There is no corrugation at the surface which might indicate, where the lamella from the two seeding areas have met and joined. SEM micrograph of cleaved sample.[2]

10µm

A detailed examination of the overgrown areas in a high voltage transmission electron microscope (HVTEM) confirmed the absence of defects on both sides of the buried SiO_2 and in the area near and at the via hole in the SiO_2. A TEM micrograph which demonstrates the typical structural perfection of these areas in LPE samples is shown in Fig. 15. The masking oxide between two stripes may be entirely buried below the laterally growing monocrystalline lamella from neighboring seeding sites. As shown in Fig. 16, there is no corrugated area in the surface of the overgrowth that might indicate where the two lamellae from the different seeding areas have met and joined. The seeded lateral overgrowth by LPE has interesting aspects with respect to 3-D integration. It does not require lateral temperature gradients locally to drive the solidification front from the seeding areas and across the sample. In the LPE approach the temperature distribution is uniform at each point of time during the growth process. Growth is achieved by reducing the overall temperature of the sample as uniformly as possible. Advantages of the low growth temperature and of temperature uniformity during the selective growth consists in the fact that the structures obtained are entirely free of defects and that the surfaces of the films obtained by lateral overgrowth are very flat and smooth. It is, therefore, possible to grow in one epitaxial process the semiconductor on the insulator and then a multilayer on the laterally grown semiconductor. It has been shown previously that AlGaAs/GaAs quantum well structures of high quality can be grown by LPE in a standard slider boat system.[4] This result raises hopes that further improvements are achievable in the quality of 2D semiconductors by carefully designing and controlling the structure of the growth interfaces. A few studies to illustrate this intention are reported in the following.

MICROSCOPIC GROWTH MECHANISMS IN LPE

Up to now in-situ observations of growing interfaces in semiconductor LPE are not possible. But in an optical microscope it is possible to photograph precisely growth step patterns of surfaces of LPE layers at different stages of growth. A limitation of this observation technique is, however, given by the lateral resolution of optical microscopes of ≈ 0.2 μm. The vertical resolution of good Normarski-Differential Interference Contrast (N-DIC) microscopes which is approximately at 2 Å, is on the other hand sufficiently high to observe monosteps on semiconductor LPE-grown surfaces.

Many observations of as-grown surfaces of LPE layers confirm that their growth proceeds by a lateral microscopic growth (LμG) process. On a microscopic scale, new ad-atoms are attached only at the edges of steps, which exist already at the growth interface, or at defects, e.g. dislocations, which emerge at the growth interface. Genuine 2-D nucleation at an atomically flat and defect-free interface does not occur during properly controlled LPE.

Continuous addition of molecules at the edges of steps causes these steps to move laterally, approximately parallel to the growth plane. Macroscopic growth occurs in a perpendicular direction to the growth interface as a result of a superposition of lateral microscopic growths. When enough steps are available, growth proceeds easily: all steps move laterally and thus contribute to the perpendicular macroscopic growth (PMG). This general process comprises several possible growth mechanisms. Which of these growth mechanisms will dominate, depends on the geometrical configuration of the steps initially present at the growth interface.

The orientation of the growth face at a given instant with respect to the crystal lattice is, therefore, of significant influence in LPE growth. The chosen orientation is nearly always close to a low indices plane, e.g. the (100) plane or the (111) plane. The relation between growth mechanism and orientation of growth face becomes apparent when studying LPE layers grown on spherically shaped substrates. In Fig. 17, the influence of the substrate orientation on the formation of growth mechanisms is summarized. The figure shows schematically a cross section of an LPE layer grown on a spherical substrate.

FACET GROWTH ON DISLOCATION-FREE SUBSTRATES

The use of an exact low indices surface of a dislocation-free crystal can impede LPE growth on that surface completely. Misorientation steps are absent, and 2-D nucleation does not occur under usual LPE conditions. It is difficult to obtain zero-substrate misorientation over large areas, but it is easy to obtain it locally. For LPE studies, a silicon substrate may be chosen whose surface is shaped spherically and is directed convexly towards the fluid, with the (111) pole in its center. During LPE on a substrate like this an atomically flat 111 plane develops around the (111) pole. It increases in size as lateral growth proceeds in the slightly misoriented areas around the 111 plane. Atomically flat silicon 111 faces of several mm^2 size have been obtained in preliminary experiments, as shown in Fig. 18. Faces of this origin are naturally smoother and of better planarity than polished surfaces even of the finest quality. They may have a variety of technical applications. Care must be taken that no defects are created in the substrate crystal during the shaping and polishing of the substrates and the formation of 2-D nuclei on the atomically flat plane must be prevented during LPE growth. Systematic studies in LPE including 2-D nucleation on otherwise atomically flat planes of dislocation-free substrates have not yet been reported.

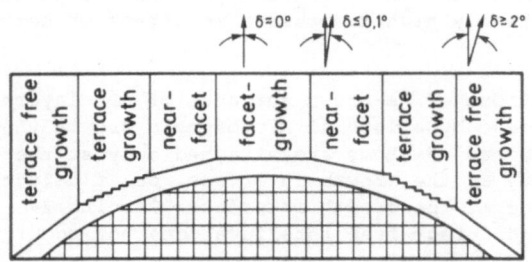

Fig. 17. Schematic representation of diverse growth mechanisms due to
different orientations of the growth face on a spherically shaped
substrate with LPE layer. Examples of the surface morphology of
GaAs LPE layers pertaining to the marked values of misorientation
are shown in the figures indicated.

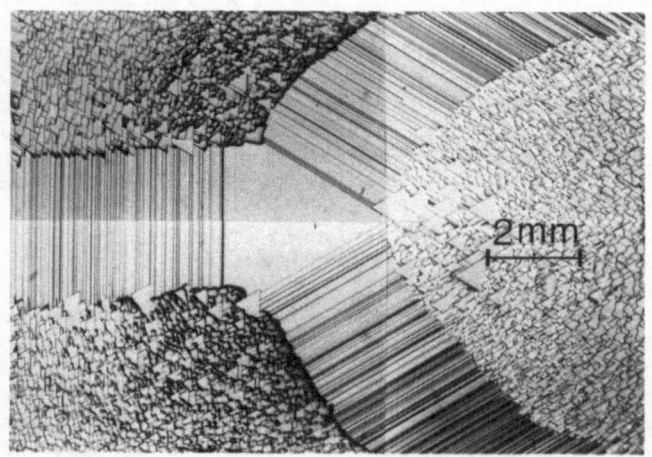

Fig. 18. Surface of a Si-LPE layer grown from Bi-solution on a (111)-
oriented dislocation-free Si substrate, whose surface was spherically shaped. The triangular shaped central facet is atomically
flat. N-DIC micrograph.[18]

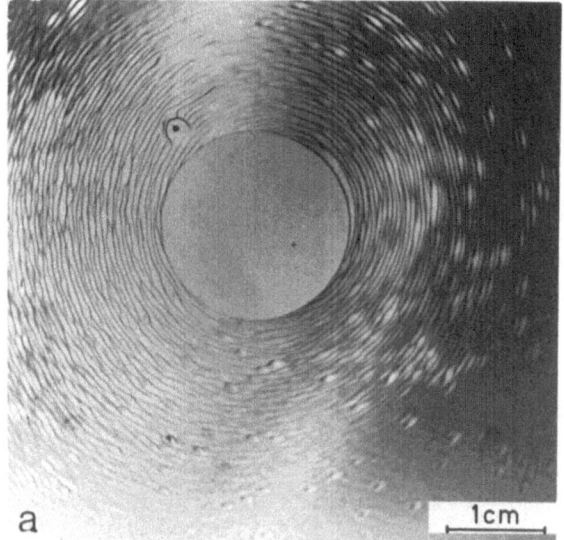

Fig. 19.
a) Surface of GaAs layer grown on a spherical substrate with the (100) pole in its center. The substrate has dislocations.

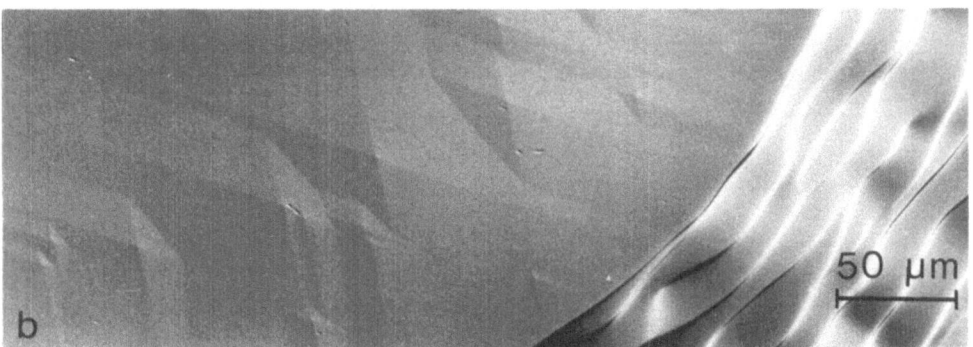

Fig. 19. b) Flat pyramids on (100) facet of GaAs LPE layer. The apex of each pyramid marks a dislocation step source. The flatness of large parts of the facet results from the strictly regular pattern of growth steps, compare Figs. 20-23.

DISLOCATION-CONTROLLED FACET GROWTH

Dislocation-controlled facet growth develops in growth faces where misorientation steps are absent and where dislocations emerge at the growth interface. This condition exists, for example, around a (111) or a (100) pole of a spherical substrate which contains dislocations. (See Fig. 19a). The substrate surface may be assumed to be atomically flat close to the pole. The steps required for growth in these areas are then provided by dislocations which terminate in the growth face. Steps originating from dislocations arrange themselves around the emergence points of the dislocations. They appear in either spirals or closed loops. Every dislocation which acts as a step source therefore gives rise to a flat cone or pyramid at the growth interface, as shown in Fig. 19b. The exactly regular patterns of monomolecular steps, which characterize dislocation-controlled facet growth and are created by the dislocation step sources, may be seen by optical microscopy using the N-DIC technique. Figures 20 and 21 are micrographs of monosteps on a 100 and 111 facet, as-grown, of a GaAs or respectively, a Si LPE layer. Monosteps may be even more distinctly revealed by decoration, enabling the monostep patterns to be easily studied over large areas, as shown in Figs. 22 and 23. The flatness in atomic

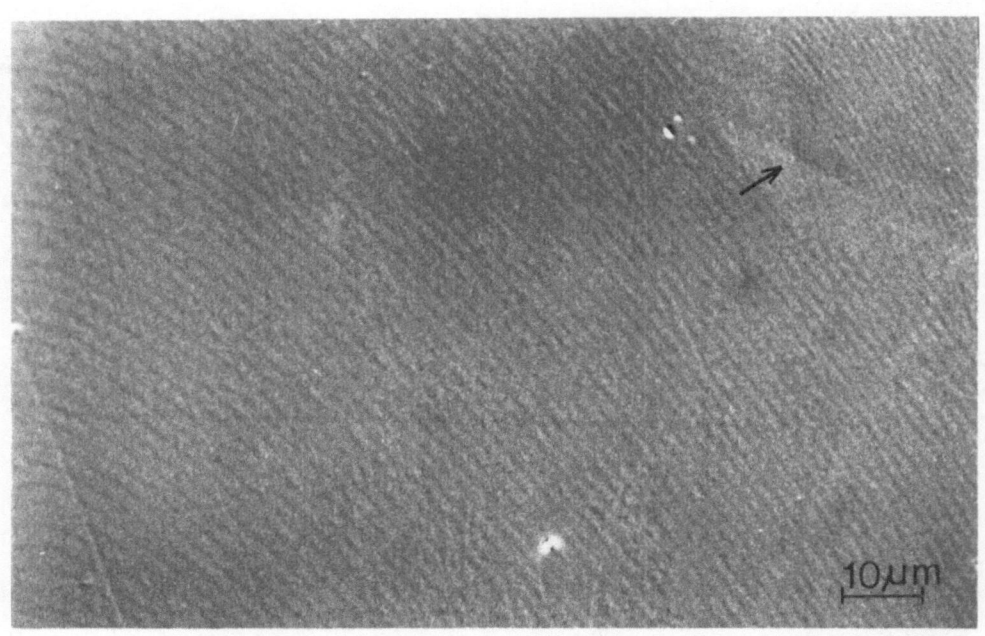

Fig.20. Monosteps on a (100) facet of GaAs LPE layer, as-grown, N-DIC micrograph. Emergence point of dislocation marked by arrow.

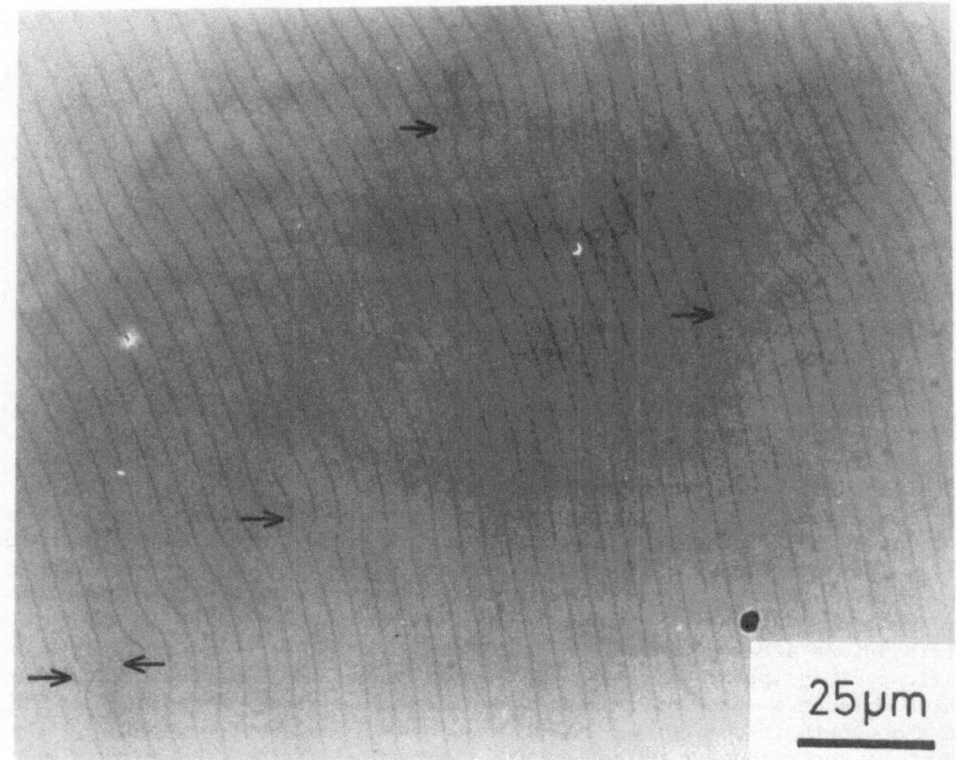

Fig.21. Monosteps on (111) facet of Si grown from Bi-solution. The steps moved from left to right. Emergence points of dislocations marked by arrows. N-DIC micrograph.[18]

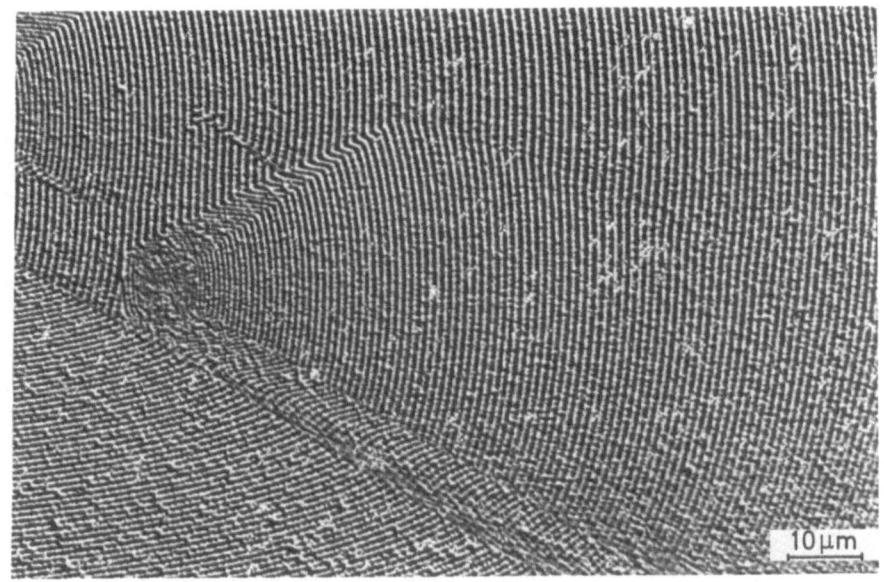

Fig.22. Monomolecular steps, decorated, on (100) facet of GaAs LPE layer.
N-DIC micrograph.[3]

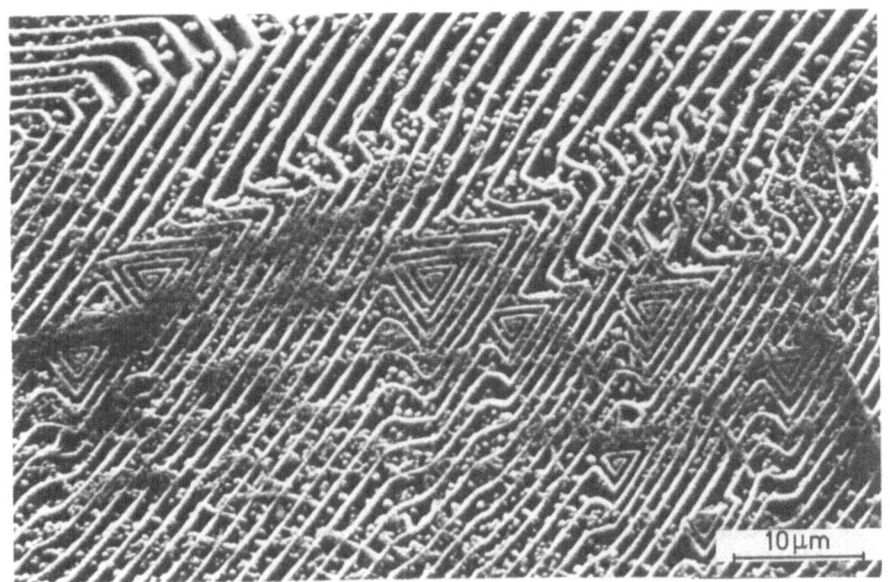

Fig.23. Monomolecular steps, decorated on (111) facet of GaAs LPE layer,
SEM micrograph.[3]

187

dimensions and the well-ordered structure of such surfaces are evident from Figs. 19 through 23. These Figures show that the average interstep distances of monomolecular steps are between about 1000 and 5000 nm; occasionally even larger interstep distances in trains of monosteps have been observed. It is essential for growth interfaces of this kind to be almost atomically flat. But they permit uniform, non-intermittent crystal growth, and in addition, they have an outstandingly high morphological stability. This means that if epitaxial growth with the regular growth step patterns as described continues on a crystal surface, the step patterns will be maintained. They remain essentially unchanged even if the subsequent epitaxial layer grows in thickness to many microns. The highly regular atom arrangement at the surface of the growth face combined with the excellent morphological stability of the growth interface makes dislocation-controlled facet growth an optimal mechanism for the design of semiconductor heterointerfaces, i.e. superlattices and multi quantum wells. Dislocations, which are the persistent sources for the desired regular growth step trains, may in future experiments perhaps be suitably positioned in dislocation-free substrates or even replaced by other step sources. An understanding of the nucleation mechanism in detail is necessary for improved control of the interface morphologies.

Investigations of dislocation step sources show that every dislocation can act as a step source provided that it emerges in an area which is free of competing steps. The analysis of growth step sources by electron microscopy reveals that dislocations are active as sources of growth steps even if they lack a Burgers vector component perpendicular to the growth interface.[7] This observation supports what Bethge and Keller have suggested, namely that during evaporation concentric surface step structures on alkali halides nucleate at edge dislocations.

The step generation process at "screw" dislocations is comprehensively described in the classical Frank model. The nucleation mechanism at dislocations which do not have a Burgers vector component perpendicular to the growth interface is still being discussed. Explanations have been proposed based on electrostatic forces, impurity effects, surface relaxation effects, and dissociation of the dislocation into partial dislocations.[19] The dissociation model is briefly described here, since electron microscopic observations of silicon - see Fig. 24 - have given it particular support. In the dissociation model proposed by Strunk,[7] a step of less than elementary height extends between the two partial dislocations, as illustrated in Fig. 25a-c. This step is formed because the partial Burgers vectors of type a/6 <112> have a component perpendicular to the growth face while the total Burgers vector does not.

When atoms become attached to this submolecular step (Fig.25b and c), an elementary step is formed at 1 and at the same time another submolecular step occurs at 2. This may immediately produce another laterally growing step. The dissociation thus provides for the possibility of steady nucleation of new growth steps and may explain the high efficiency of such step sources. It should be noted that this nucleation process is essentially different from the intermittent operation of a screw dislocation dipole although topologically it appears to be a similar source.

NEAR-FACET GROWTH

The growth steps which promote near-facet growth are not provided by dislocations. They are rather "misorientation steps" and therefore lack the outstanding regularity which is typical of step trains provided by dislocation step sources. The difference is illustrated in Fig. 26. The height of the growth steps is at the most that of a few lattice distances.

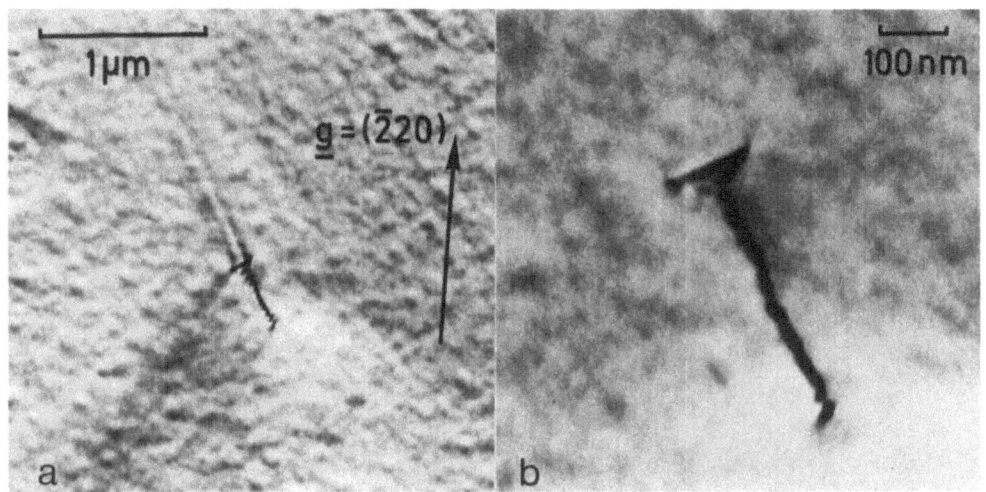

Fig.24. Transmission electron micrograph of a growth step source in a silicon epitaxial layer grown from Ga solution. Foil normal to (111). (a) image of the dislocation in the center of its three-sided growth pyramid. (b) Magnified view of the dislocation. The dissociation into partials is especially wide near the emergence of the dislocation at the growth face.

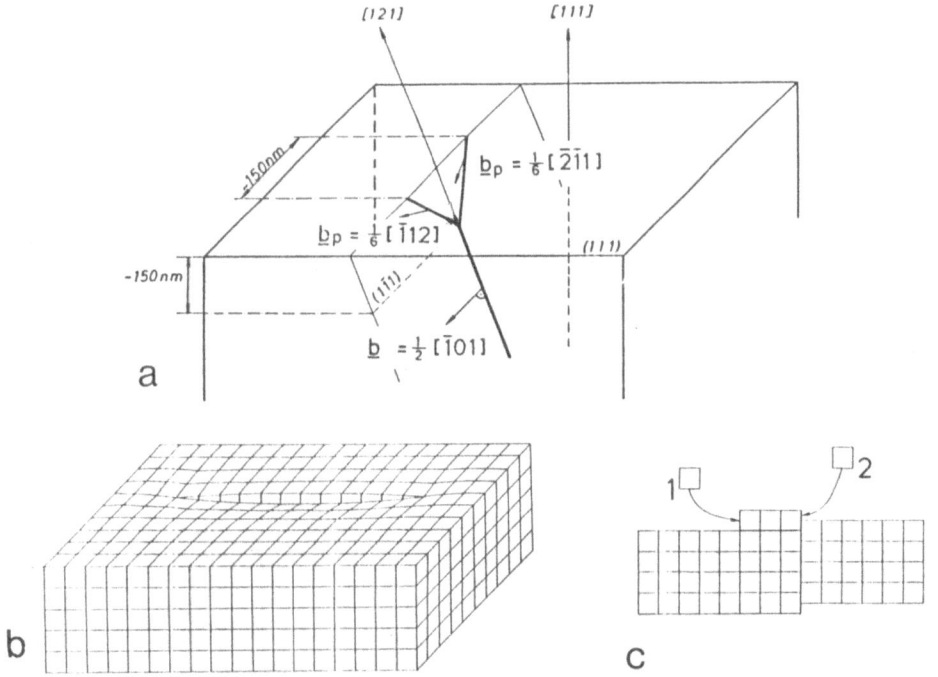

Fig.25. (a) Possible arrangement and Burgers' vectors of a dislocation such as shown in Fig. 24. (b) Model of a step source which is formed by a dislocation without a Burgers vector component perpendicular to the growth face. A submolecular step extends between the emergence points of the two partial dislocations of the dissociated perfect dislocation. (c) Cross section through (b), illustrating the permanent action of the subatomic step as a nucleation center.[7]

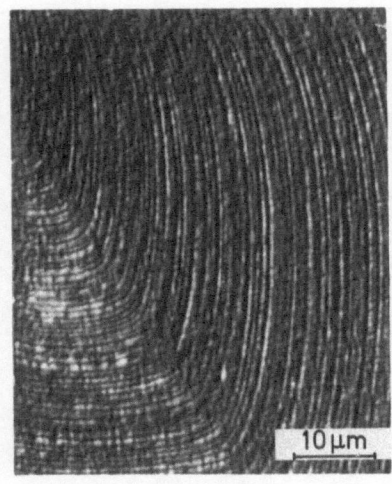 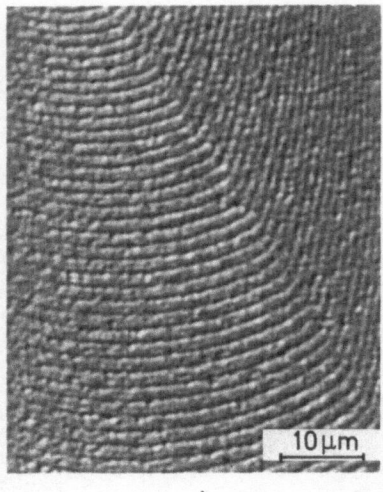

a
b

Fig. 26. (a) Near-facet growth, (b) dislocation-controlled facet growth.
In near-facet growth, the step pattern is not as strictly regular
as in dislocation-controlled facet growth. The growth steps in
(a) and (b) are decorated.

The interstep distances are roughly between 10^2 and 10^3 nm for substrate
misorientations by angles δ between 0.1° and 0.015° with regard to a low
index face. The surfaces of the layers which represent the typical interface morphology are, therefore, still very smooth from a macroscopic point
of view. Material obtained by near-facet growth is therefore suitable for
many technical applications. The disorder at the surface (compare Fig.
26), and hence at the growth interface, is, however, obviously greater in
near-facet growth than in dislocation-controlled facet growth. The higher
surface and interface disorder in near-facet growth, combined with a decreased morphological stability makes this growth mode appear less favorable when used specifically for the preparation of modulated semiconductor structures.

TERRACE GROWTH

Substrate misorientations of more than 0.1° and less than 2° usually
bring terrace growth into operation during LPE. Terraces are bunched steps
with heights from a few lattice distances upwards. Often very distinct
terraces develop, when a low degree of supersaturation is used in order to
obtain very thin layers. Terrace growth is unfavorable not only due to the
terraced growth face but also for the reason that local fluctuations in
dopant concentration and in the alloy composition are very likely to develop in the bulk of the material when terraces persist at the growing
interface. These morphological and chemical nonuniformities impair the
properties of modulated structures. It is therefore crucial in practical
LPE work to systematically prevent terrace formation in those areas of a
substrate which are provided for the preparation of multilayer structures.
Studies are presently being undertaken to lay out possible experimental
routes.

190

According to a theory formulated by Rode, a critical orientation of the substrate exists, preventing the formation of terraces during epitaxial growth.[20] The gradual change from a terraced surface to a terrace-free surface with increasing misorientation can indeed be observed, for instance, on the surfaces of LPE layers grown on spherically shaped substrates (Figs. 19a, right hand side and 27). A procedure for evaluating the critical angle is described, and angles for the critical misorientation are expected to cover the range of up to several degrees and to increase with greater layer thickness. According to Rode's theory, the surface of thick layers, grown on critically oriented substrates, exhibits a long wavelength sinusoidal structure. This long wavelength sinusoidal structure may be understood as a symptom of strong morphological instability which can make the growth of modulated structures more difficult. Experiments to evaluate the possible advantages of terrace-free growth are under way.

Fig. 27. GaAs LPE layer grown on a spherical substrate. Surface area near wafer edge, showing the transition from a terraced surface to a terrace-free surface. N-DIC micrograph.

INFLUENCE OF DIFFERENT CRYSTAL GROWTH MECHANISMS ON THE PRECISION OF MODULATED STRUCTURES

The continuous flow of equidistant monosteps, with step-to-step distances very large compared to the step height ($>10^3$:1), may be considered an appropriate mechanism for the preparation of modulated structures. A growth mechanism of this kind exists, for example, in dislocation- controlled facet growth. It may be achieved in LPE without any competing 2-D nucleation on the atomically flat treads between the step edges, even if the step edges are some ten microns apart. Such an interface can be regarded as atomically flat. During growth its step pattern, although it moves, remains basicly unchanged. Growth interrupts, if they were applied to this LPE growth mode, would be incapable of changing the step distribution. At any moment, therefore, whenever growth is stopped, equidistant monosteps are present at the interface. It may also be expected in dislocation-controlled facet growth that the step patterns in the growth sequence GaAs/AlGaAs and AlGaAs/GaAs are basically the same and that, hence, no difference in morphology is to be expected for the two interfaces. It should be mentioned, though, that this expectation is allowable only in dislocation-controlled facet growth and in areas sufficiently distant from the step sources. Different morphologies were detected, for example, by HRTEM in some MBE grown heterostructures. In LPE near facet growth, differences in the GaAs/AlGaAs and AlGaAs/GaAs interfaces are also to be expected.

A growth mode designated "step flow" for MBE growth on vicinal surfaces has been reported recently by J.H. Neave et al.[21] and B.A. Joyce et al.[22] This MBE growth mode manifests itself by a constant response in RHEED specular beam intensity. Step flow occurs in MBE when the mobility of surface atoms is high enough for them to reach the step edges. Growth temperatures are then slightly elevated compared to the growth mode involving 2-D nucleation which causes RHEED oscillations.

RHEED intensity oscillations permit an accurate control of layer thickness in MBE, and the uncertainty in quantum well width may be restricted to one monolayer or even less. To reduce the step density and thus the roughness of the interfaces, growth interrupts have been applied. Growth interrupts, however, appear to be effective only in cases when 2-D nucleation is involved in the growth mode. The island size has then been shown to increase.[23] This increase in island size, which is coupled with an improvement of the optical properties of quantum wells has been directly demonstrated in several experiments by Christen et al.[24] When the MBE growth occurs exclusively via the "step flow" mode, growth interrupts cause no considerable alteration of the interface morphology. In the MBE step-flow mode the interface is likely to consist of more or less regular step trains without island characteristics. In this case, therefore, the properties of MBE grown quantum wells do not depend on growth interruptions.

SUMMARY

Optical microscopy of LPE grown surfaces permits the direct mapping of surface step patterns down to the monolayer height of steps, provided that the step-to-step distances exceed ~0.2 μm. Monostep trains with step-to-step distances of more than 0.2 μm can be achieved at the surfaces of LPE layers. Since surfaces of this kind are morphologically extremely stable, observations of layer surfaces may give realistic images of interface morphologies.

It is evident that among the LPE-typical growth mechanisms near-facet growth, terrace growth and terrace-free growth are unfavourable for the growth of modulated structures. An optimal growth potential, however, exists in dislocation-controlled facet growth, provided the dislocation step sources are sufficiently distant. An examination of the growth step patterns shows that the interfaces are atomically almost flat. The strictly regular monostep trains have an exceptional morphological stability. Interfaces with these attributes, therefore, are most appropriate for the growth of modulated structures.

The maintenance of interfaces of highest perfection in slider boats during multilayer growth has proved to be difficult. Therefore, a centrifugal LPE system was developed and tested. The results of numerous LPE growth experiments with silicon indicate that the centrifugal technique offers new potentialities for the preparation of high quality modulated structures.

ACKNOWLEDGEMENTS

Parts of this work result from a joint project in which H.P. Strunk and
his group participated and which was supported by the German Federal Ministry
of Research and Technology. Valuable discussions with H.J. Queisser and
the pleasant cooperation with H.P. Strunk, D. Käß, G. Schweitzer, R. Linnebach,
M. Warth, W.H. Appel, P. Koroknay, B. Kunath, A. Müller, K.S. Löchner,
E. Kisela, I. Wührl-Petry and M. Punschke-Smyrek are gratefully acknowledged.
The author wishes to thank B. Krämer for typing the manuscript.

REFERENCES

1 P.D. Greene, Liquid-Phase Epitaxy, in: ISSCG, 6th International Summer
   School on Crystal Growth, Lecture Notes Vol. 1, p. 187, Edinburg, July
   1986.
2 E. Bauser and H.P. Strunk, Silizium-Epitaxieschichten mittels Lösungs-
   transport durch Fliehkraft und deren strukturelle und elektrische Charak-
   terisierung überwiegend mit elektronenmikroskopischen Methoden, For-
   schungsbericht BMFT FB T 86 - 142, in press.
3 E. Bauser and H.P. Strunk, Microscopic Growth Mechanisms of Semiconductors: Experiments and Models, J. of Crystal Growth $\underline{69}$: 561 (1984).
4 K. Kelting, K. Koehler, and P. Zwicknagl, Luminescence of $Ga_{1-x}Al_xAs/$
   GaAs single quantum wells grown by liquid phase epitaxy, Appl. Phys.
   Lett. $\underline{48}$: 157 (1986).
5 H. Hillmer, G. Mayer, A. Forchel, K.S. Löchner and E. Bauser, Optical
   time-of-flight investigations of ambipolar carrier transport in GaAlAs
   using GaAs/GaAlAs double quantum well structures, Appl. Phys. Lett. $\underline{49}$:
   948 (1986).
6 E. Bauser, D. Käss, M. Warth and H.P. Strunk, Silicon layers grown on
   Patterned Substrates by Liquid Phase Epitaxy, in: Mat. Res. Soc. Symp.
   Proc. Vol. 54: 267, 1986 Materials Research Society.
7 E. Bauser and H.P. Strunk, Dislocations as Growth Step Sources in Solution
   Growth and their Influence on Interface Structures, Thin Solid Films
   $\underline{93}$: 185 (1982).
8 H. Kressel and H. Nelson, Properties and Applications of III-V Compound
   Films Deposited by Liquid Phase Epitaxy, in: Physics of Thin Films,
   Vol. 7, Academic Press, New York (1973).
9 M.B. Panish, I. Hayashi and S. Sumski, Double-Heterostructure Injection
   Lasers with Room-Temperature Thresholds as low as 2300 A/cm^2, Appl. Phys.
   Letts. $\underline{16}$: 326 (1970).
10 Zh.I. Alferov, M.V. Andreev, E. Korolkov, E.L. Portnoi and D.N. Tretyakov,
   Coherent Radiation of Epitaxial Heterojunction Structures in the AlGa-
   GaAs System, Soviet Phys. Semiconductors $\underline{2}$: 1289 (1969).
11 H.J. Scheel, A New Technique for Multilayer LPE, J. of Crystal Growth
   $\underline{42}$: 301 (1977).
12 E. Bauser, L. Schmidt, K.S. Löchner and E. Raabe, Liquid Phase Epitaxy
   Apparatus for Multiple Layers utilizing Centrifugal Forces, Japan. J.
   Appl. Phys. $\underline{16}$, Suppl.16-1: 457 (1977).
13 G. Schweitzer, A. Traxler, H. Bleuler, E. Bauser and P. Koroknay, Magnetische
   Lagerung einer Epitaxiezentrifuge bei Hochvakuumbedingungen, Vakuum-
   Technik $\underline{32}$: 70 (1983).
14 G. Schweitzer, Regelungstechnik in der Mechanik: Anwendung auf Magnetlager,
   SIA-Zeitschrift $\underline{9}$: 275 (1983).
15 D. Käss, M. Warth, W. Appel, H.P. Strunk and E. Bauser, Silicon Multilayers
   grown by Liquid Phase Epitaxy, in: "Silicon Molecular Beam Epitaxy",
   Proc. Vol. 85-7, J.C. Bean, ed., The Electrochem. Soc., Pennington NJ,
   USA (1985), p. 250.
16 H.P. Trah et al., to be published.

17 E. Bauser, D. Käss, M. Warth and H.P. Strunk, Silicon layers grown on patterned substrates by Liquid Phase Epitaxy, Mat. Res. Soc. Symp. Proc. 54: 267 (1986).

18 W.H. Appel, Flüssigphasenepitaxie von Silizium: Wachstumskinetik und Eigenschaften der Schichten, Thesis, Universität Stuttgart (1985).

19 E. Bauser and H. Strunk, Analysis of Dislocations Creating Monomolecular Growth Steps, J. of Crystal Growth 51: 362 (1981).

20 D.L. Rode, Surface Dislocation Theory of Reconstructed Crystals, Phys. Status Solidi (a) 32: 425 (1975).

21 J.H. Neave, P.J. Dobson and B.A. Joyce, Reflection high-energy electron diffraction oscillations from vicinal surfaces - a new approach to surface diffusion measurements, Appl. Phys. Lett. 47: 100 (1985).

22 B.A. Joyce, P.J. Dobson, J.H. Neave and K. Woodbridge, Rheed Studies of Heterojunction and Quantum Well Formation during MBE Growth - From Multiple Scattering to Band Offsets, Surface Science 168: 423 (1986).

23 J. Christen, D. Bimberg, T. Fukanaka and H. Nakashima, Direct Imaging of Monolayer Islands at GaAs/GaAlAs Interfaces, in: Sol. State Devices 1986, to be published.

GROWTH AND STRUCTURE OF COMPOSITIONALLY MODULATED
AMORPHOUS SEMICONDUCTOR SUPERLATTICES AND HETEROJUNCTIONS

L. Yang and B. Abeles

Exxon Research and Engineering Co.
Annandale, N.J. 08801

INTRODUCTION

Over the past fifteen years research on single-crystal semiconductor superlattices has grown to be a major subfield of semiconductor physics. Thus, it is remarkable that it is only in the past three years that the first published reports of superlattices made from the prototypical amorphous semiconductor, amorphous hydrogenated silicon (a-Si:H), have appeared.[1-4] The main reason for the late start of the amorphous semiconductor superlattice field may have been the widely held preconception that epitaxial growth of single crystals is required for the synthesis of high quality multilayer structures, with uniform atomically abrupt layers. Recent transmission electron microscopy work[5-7] on the structure of amorphous superlattices has amply demonstrated that highly regular superlattices can be made from the a-Si:H family of materials, synthesized by low temperature ($\leq 300°C$) plasma-assisted chemical vapor deposition (PCVD).

Research on the new amorphous superlattice materials is now going on in many laboratories throughout the world.[8] The ability to make interfaces which are nearly atomically abrupt has made it possible to observe new electronic phenomena in amorphous semiconductors which include quantum size effects,[1,9-12] enhanced photoluminescence[13] and photoconductivity,[14] transfer doping[15] and electroabsorption.[16,17] These experiments are providing new insights into fundamental properties of amorphous materials and are expected to result in new technological applications in solar cells,[18,19] electroluminescent diodes[20] and others.

The ability to make amorphous superlattices with a large number of uniformly repeating interfaces makes it possible to deduce structural information from measurements of average bulk properties of superlattice films. Interface effects are clearly brought out when the repeat distance of the superlattice becomes comparable to the widths of the interface region. Using this technique, the composition and bonding at the interfaces and density of interface defects has been determined from RBS[21], N[15] nuclear reaction[21,22], IR[21,23], Raman scattering[24] optical absorption,[10,15] electroluminescence,[26] and electroabsorption measurements.[16,17]

Another technique to study the electronic structure of the interfaces is by photoemission spectroscopy of single heterojunctions.[27,28] Spectroscopy of the Si-2p core levels provides information on the nearest neighbor coordination of the Si atoms. From measurements of core level spectra on heterojunctions in which the thickness of the overlayer film is varied, it is possible to deduce the width of the interface region. Another important property of heterojunctions which is determined by photoemission is the offset of the valence bands. This parameter is critical in determining the transport and optical properties of the superlattices.

In this paper we focus on amorphous multilayers based on a-Si:H made by PCVD (Sec. 2). To monitor the growth of the multilayers we have utilized in-situ optical reflectance from the surface of the growing film[29] (Sec. 3). Modeling of the measured reflectance provides information on the structure of the interfaces and the optical constants and thicknesses of the sublayers. In Sec. 4 we describe the use of IR spectroscopy for determining the distribution and bonding of hydrogen at the interfaces of a-Si:H/a-SiN$_x$:H and a-Si:H/a-SiO$_x$:H superlattices. In Sec. 5 we present new results of photoemission measurements on a-Si:H/a-SiO$_x$:H heterojunctions[23]. This system is particularily interesting because the large contrast in the electronic properties of a-Si:H and a-SiO$_x$:H makes it possible to resolve readily the electronic structure of the interfaces. Moreover, because a-Si:H forms an atomically abrupt interface with a-SiO$_x$:H we were able to study quantum confinement effects in a-Si:H down to monolayer dimensions.

GROWTH AND COMPOSITION

The superlattices consisting of layers of a-Si:H alternating with layers of a-SiN$_x$:H, a-SiO$_x$:H and a-Ge:H are made by a plasma assisted CVD process in which the composition of the reactive gas in the plasma

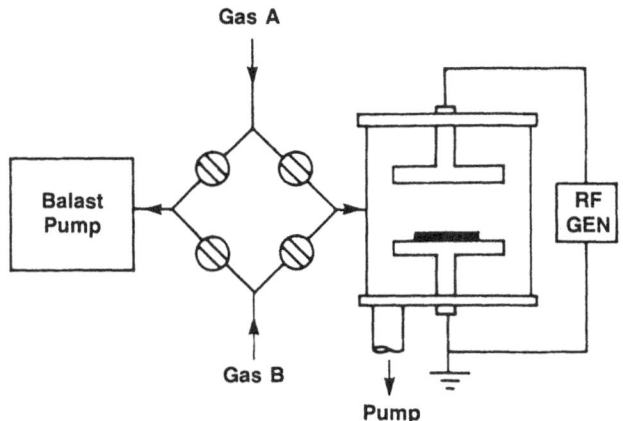

Gas A

Balast
Pump

RF
GEN

Gas B

Pump

Figure 1   Plasma reactor for growth of multilayers. Gases A and B
               are exchanged in the reactor by synchronously opening and
               closing two pairs of valves. The balast pump makes it
               possible to exchange the gases without interrupting the
               flow.

reactor is changed periodically.[1,30] The substrates are held at ~ 240°C
and mounted on the anode of a 13.6 MHz capacitive reactor as shown in
Fig. 1. The gases, at a typical flow rate of 85 scc/min and pressure of
30 mTorr, are exchanged in the reactor with a time constant of ~ 1 sec
which is short compared to the time (~ 3sec) needed to grow a monolayer. Films can be made either with interrupting the plasma during the
gas exchange, or without plasma interruption.

Table 1 lists the reactive gas mixtures used for making the different hydrogenated amorphous materials, as well as their chemical
compositions measured by RBS and N^{15} nuclear resonant reaction techniques. In the case of the multilayers average compositions were measured because the depth resolution of these techniques was insufficient
to resolve compositional variations on the scale of the sublayer thicknesses. While the average composition (Si, N, O) in the multilayers

Table 1

Gas Mixture	Composition
SiH_4	$SiH_{0.1}$
$GeH_4 + 9H_2$	$GeH_{0.6}$
$SiH_4 + 5\ NH_3$	$SiN_{.95}H_{.60}$
$SiH_4 + 50\ NO_2$	$SiO_{1.9}H_{.08}$

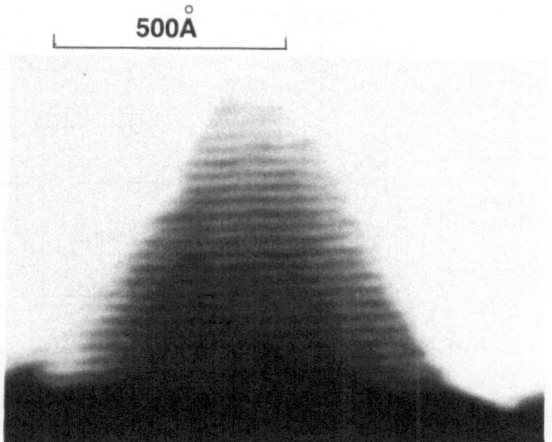

500Å

Figure 2   Electron micrograph of conical section at top surface of a 440 layer-thick a-Si:H/a-Ge:H superlattice[5] with repeat distance of 27Å. Dark bands in image are the a-Ge:H layers.

agreed with those calculated from compositions measured on thick films, the multilayers had ~ 10^{15} cm^{-2} extra hydrogen per bilayer concentrated at the interfaces.

Transmission electron microscopy[5-7] and X-ray diffraction[30] have provided convincing evidences that these films form layered structures with sublayer thicknesses as small as ~ 10Å. An example of a TEM micrograph of an a-Si:H/a-Ge:H superlattice[5] is shown in Fig. 2. The conical TEM section was prepared from the upper most section of a 440 layer sample having a repeat distance of 27Å. The reason why the layers are so uniform is not understood at present. One could speculate that dissociated species in the plasma, atomic hydrogen for example, scour the growth surface selectively etching away protrusions.

IN-SITU OPTICAL REFLECTANCE

In the optical set up used (shown in Fig. 3) a linearly polarized He-Ne laser beam (6328Å, 1 mw power) incident at 67° from the normal was reflected from the surface of the growing film and separated into s-polarized (E field normal to plane of incidence) and p-polarized (E field parallel to plane of incidence) components by means of a beam splitter and detected by Si photodiodes. In the following we illustrate the use of the optical technique to monitor the growth of a-Si:H/a-SiN$_x$:H, a-Si:H/a-SiO$_x$:H and a-Si:H/a-Ge:H superlattices and we discuss in detail opitcal modeling of the nitride superlattices.

In Fig. 3 is shown the time dependence of the p-reflectance for a

198

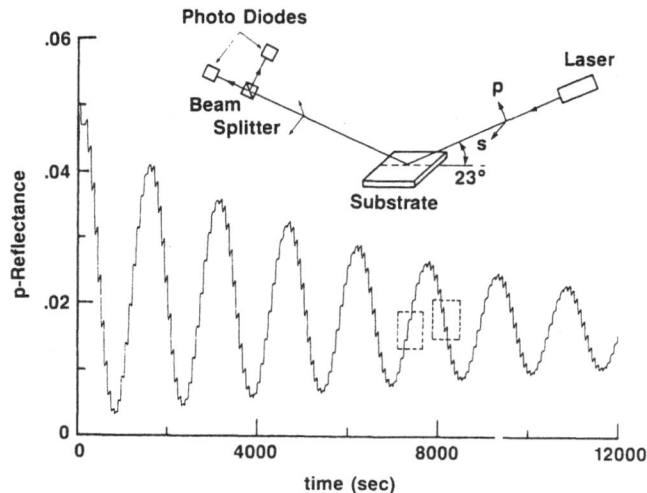

Figure 3   The p-polarized reflectance as a function of time from
growing a-Si:H/a-SiN$_x$:H superlattice film. The regions
indicated by the rectangles are shown on an expanded scale
in Fig. 4a and 4b. Insert shows schematic of optical set
up.

growing a-Si:H/a-SiN$_x$:H superlattice[29] on a c-Si substrate. The superlattice, with a repeat distance of 55Å, was made by alternating a 22.5
sec plasma discharge of SiH$_4$ with that of 37 sec of the SiH$_4$-NH$_3$ mixture
(Table 1) and interrupting the plasma for 10 sec after switching the
gases. The interference fringes in Fig. 3, given by the attenuated
sinusoidal wave, differ from those of a homogeneous film in that they
exhibit fine structure which is due to the finite thickness of the
layers. The fine structure corresponding to the rising and falling
portions of the interference fringes, (dashed rectangles in Fig. 3) is
shown on an expanded scale in Fig. 4a and 4b, respectively. The interference fringes for the s- reflectance had a slightly shorter period,
and the fine structure was not as pronounced as in the p- reflectance.
The observed difference in the periods of the s and p interference
fringes was consistent with the expected uniaxial properties of the
superlattice.

To model the optical reflectance of the superlattice we used a
recursion relation which allowed us to calculate the reflectance of an
arbitrary number of different layers from the reflectance of a single
layer on top of a substrate.[29]. The simplest description of the superlattice is given by an ideal two-layer model consisting of plane parallel alternating layers of a-Si:H and a-SiN$_x$:H forming abrupt inter-

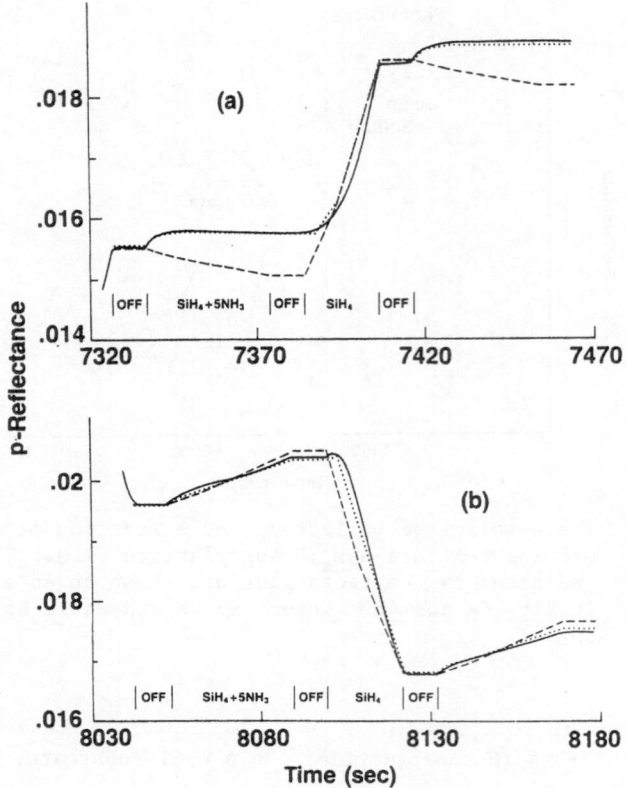

Figure 4   Expanded sections of p-polarized reflectance indicated by the rectangles in Fig. 3 corresponding to rising (a) and falling (b) portions of the interference fringes. The experimental data is given by the full curve, the ideal two-layer model by the dashed curves and the model corresponding to Fig. 5 by the dotted curve. The time sequences of switching of the gases and the interruption of the plasma are indicated by the vertical lines.

faces. The dashed lines in Fig. 4a and 4b are the calculated reflectance for such a model in which for the indices of the a-Si:H layers and a-SiN$_x$:H layers we assumed the bulk values of 4.3 and 2.05 respectively, and the layer thicknesses and absorption coefficient were used as adjustable parameters to fit the period and amplitude of the the interference fringes. We note that there are major discrepancies between the ideal two-layer model and the observed reflectance fine structure.

Examination of TEM crossection of a-Si:H/a-SiN$_x$:H superlattices[6,7] suggests that a more realistic model must take into account roughness of the interfaces which has a scale of ~ 10Å normal to the plane of the

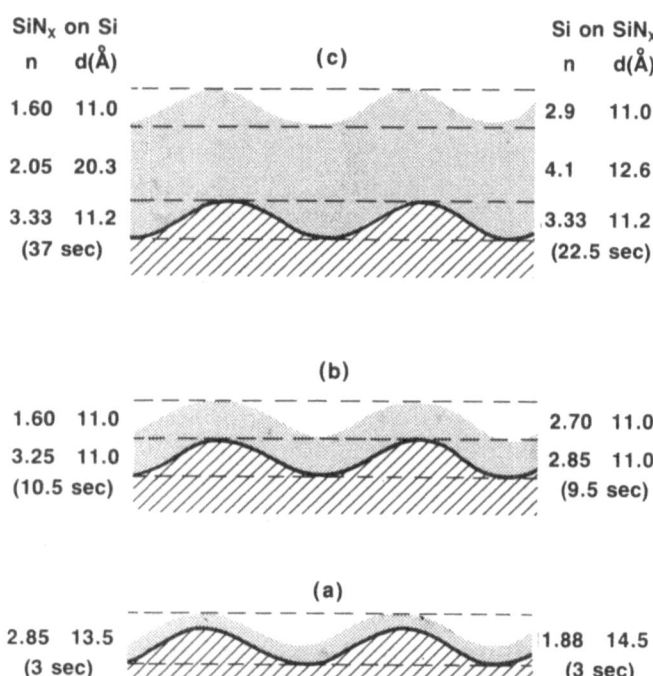

SiN$_x$ on Si   (c)   Si on SiN$_x$
  n   d(Å)   n   d(Å)

1.60   11.0   2.9   11.0

2.05   20.3   4.1   12.6

3.33   11.2   3.33   11.2
(37 sec)   (22.5 sec)

                                 (b)

1.60   11.0   2.70   11.0
3.25   11.0   2.85   11.0
(10.5 sec)   (9.5 sec)

                                 (a)

2.85   13.5   1.88   14.5
(3 sec)   (3 sec)

Figure 5   Schematic of three stages in the growth of an interface
            (a) top layer is thinner than the vertical roughness
            length scale; (b) top layer is equal to the roughness
            length scale; (c) top layer is thicker than the roughness
            length scale. The interface layers and bulk layer are
            indicated by the dashed lines and the corresponding opti-
            cal constants are listed for the interfaces Si on SiN$_x$ and
            SiN$_x$ on Si.

layers and a lateral scale of ≥ 100Å. A schematic representation of the
interface roughness and the successive stages in the formation of an
interface are shown in Fig. 5. As long as the thickness of the growing
top layer is less than the roughness length scale normal to the layers
we model the interface between vacuum, top layer and bottom layer by a
single composite layer as indicated by the two parallel dashed lines
(a). When the thickness of the top layer becomes equal or greater than
the roughness length scale we divide the layer into two (b) or three (c)
regions, consisting of a composite layer of the top layer and vacuum, a
bulk top layer, and a composite layer of the top layer and the bottom
layer. Using the above model of interface formation we obtained a
considerably improved fit to the experimental data as shown by the
dotted curves in Fig. 4. Equally good fit as in Fig. 4 was obtained for
all the 175 periods of the superlattice indicating that the deposition
process is remarkably reproducable from period to period.

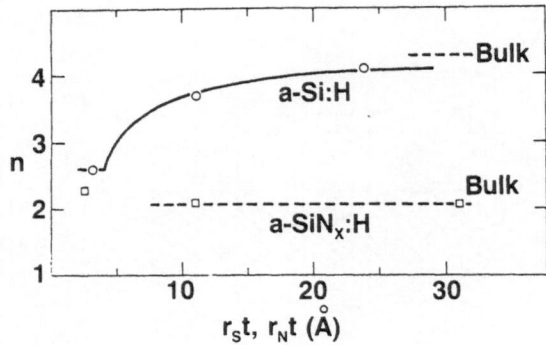

Figure 6   The indices n_S and n_N of growing a–Si:H and a–SiN$_x$:H top
layers as a function of layer thickness. The solid curve
is a fit of Eq. (2).

We next show that the values of the indices n of the composite
layers are consistent with the optical indices n_S and n_N of the a–Si:H
and a–SiN$_x$:H layers. To this end we used the effective medium relation

$$n^2 = x_S\, n_S^{\ 2} + x_N\, n_N^{\ 2} + x_V \tag{1}$$

where x_S, x_N and x_V are the volume fractions of a–Si:H, a–SiN$_x$:H and
vacuum in the composite layers. Equation (1) is valid for the case of a
composite of flat discs with the E–field in the plane of the discs (zero
depolarization effects) which we expect to be a good approximation for
the interfaces. The values of x were chosen so as to be consistent with
a symmetric interface topography. Thus, for instance in the case of a
two component interface $x = \frac{1}{2}$. In Fig. 6 we have plotted the calculated values of the indices n_S and n_N of the growing top layers as a
function of their thicknesses. We note that in the first 1–2 monolayers
the value of n_S is appreciably lower, while n_N is slightly higher than
the respective bulk values. Both the decrease in n_S, as well as the
increase in n_N, are consistent with modified bonding of the interface
monolayers. There is evidence that the interfaces Si on SiN$_x$ and Si–
vacuum[31] each have a monolayer of bonded hydrogen. The extra H is
expected to result in a substantial decrease in the index. Accordingly,
we modeled the a–Si:H layer by two Si interface layers of total thickness Δ with index $n_{S,i}$ and a bulk region with index $n_{S,b}$. The index n_S
of the a–Si:H layer is then given by the effective medium relation,

$$n_S^{\ 2} = (\Delta\, n^2_{\ S,i} + (r_S\, t - \Delta)\, n^2_{\ S,b})/r_S t \tag{2}$$

where r_S is the deposition rate, and t the growth time. The solid line
in Fig. 6 is calculated from Eq. 2 with $n_{S,i}$ = 2.6, $n_{S,b}$ = 4.3 and Δ =
4Å. The underlaying assumption in Eq. (2) that the dielectric properties of the two Si interface layers depend only on the nearest neighbor
bonds, is based on the dielectric theory of Phillips which has been
applied to Si based alloys by Aspnes and Theeten.[32]

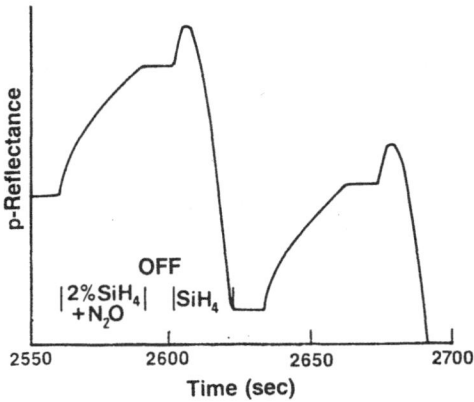

Figure 7   P-polarized reflectance from a growing a-Si:H/a-SiO$_x$:H
                  (20 Å/30Å) superlattice film. Gas flow periods and
                  interruption of the plasma are indicated by vertical
                  lines.

Figure 7 shows the reflectance fine structure for a-Si:H/a-SiO$_x$:H
superlattice corresponding to the falling side of the interference
fringe. We note that the peak immediately after turning on the SiH$_4$
plasma is much more pronounced than the corresponding peak in the a-
Si:H/a-SiN$_x$:H superlattice (Fig. 4). This peak, which we attribute to
interface roughness, is enhanced in the oxide system because of the
large contrast in refractive indices between a-Si:H (n=4.3) and a-SiO$_x$:H
(n = 1.48). The steep rise in reflectivity at the beginning of the a-
SiO$_x$:H deposition is mainly due to the growth of ~ 10Å oxide layer by
plasma oxidation of the underlaying silicon layer and the subsequent
slower rise is due to growth of the oxide by CVD. In the nitride superlattice growth of the nitride by nitridation was not a significant
factor.

Figure 8 shows the reflectance from a-Si:H/a-Ge:H superlattices in

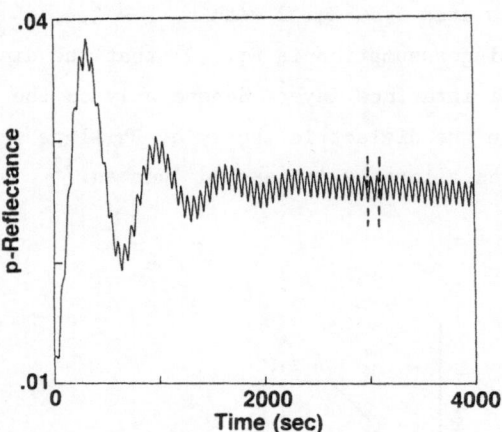

Figure 8   P-polarized reflectance as a function of time from a
            growing a-Si:H/a-Ge:H superlattice film. Section indica-
            ted by dashed lines is shown on an expanded scale in Fig.
            9.

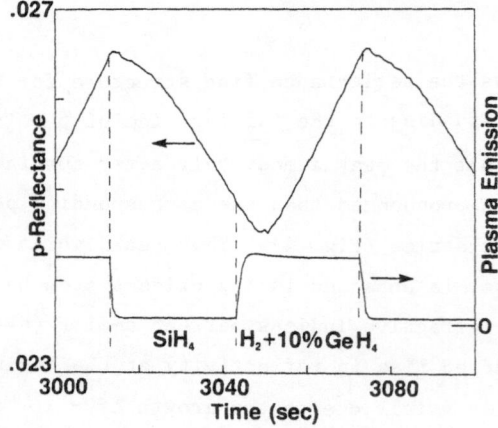

Figure 9   Expanded section of p-polarized reflectance from a-Si:H/
            a-Ge:H superlattice indicated by dashed lines in Fig. 8.
            Also shown in the figure is the emission intensity from
            the discharge. The time sequence of switching gases and
            the interruption of the plasma are indicated by the verti-
            cal lines.

which 30 sec of SiH_4 is alternated with 30 sec of GeH_4 - H_2 mixture, without interrupting the plasma. We note that the interference fringes are rapidly damped because of the large absorption by the a-Ge:H layers. The amplitude of the fine structure is small in this system because of the small difference in the indices of a-Si:H (4.3) and a-Ge:H (4.75). Fig. 9 shows an expanded section of the reflectance as well as the plasma emission intensity measured by a photodiode. The dashed vertical lines indicate the switching times. The reflectance curve exhibits a transition region $\sim$ 10 Å at the interfaces which is attributed to the interface roughness effects as discussed earlier. In addition, because of the finite gas exchange time of $\sim$ 1 sec in the reactor, which shows up as a transition region of $\sim$ 3 sec in the plasm emission intensity, there is a compositionally graded region $\sim$ 3Å at each interface.

BONDING OF HYDROGEN AT INTERFACES

Difference in the concentration and chemical bonding environment of hydrogen in bulk amorphous material and at the vicinity of interfaces can be clearly brought out by comparing the infrared transmission spectra measured on superlattice samples in which sublayer thicknesses are varied in a systematic manner. In Figs. 10 and 11 are shown the IR transmission spectra of the Si-H stretch bands for a series of a-Si:H/a-SiN_x:H[21] and a-Si:H/a-SiO_x:H[23] superlattices. In the a-Si:H/a-SiN_x:H series, the repeat distance of the superlattice was varied while keeping the ratio of the two sublayer thicknesses constant while in the a-Si:H/a-SiO_x:H series only the a-Si:H layer thickness was varied while the a-SiO_x:H thickness was kept roughly constant at $\sim$ 13Å. In bulk a-Si:H and in the superlattices with large a-Si:H layer thickness, d_S, the 2000 cm^{-1} mode dominates indicating that H is bonded primarily as monohydride. With decreasing d_S the interface Si-H bands become more prominent. The 2080 cm^{-1} band is associated with dihydride or H bonded on the surfaces of microvoids[33] in a $\sim$ 20Å region at the interface which is formed when a-Si:H is deposited on a-SiO_x:H or on SiN_x:H. The enhancement of vibrational band 2155-2175 cm^{-1} in the Si/SiN_x superlattices (Fig. 10) and the 2100-2300 cm^{-1} band in Si/SiO_x (Fig. 11) is due to Si-H vibrations with Si back bonded to nitrogen and oxygen atoms respectively at the interfaces. The 2100-2300 cm^{-1} band is associated only with the interfaces because we observed no Si-H vibrations in bulk a-SiO_x:H films. From quantitative analysis of the IR and N^{15} nuclear resonant reaction measurements it was concluded that there is $\sim$ 10^{15} cm^{-2} extra hydrogen

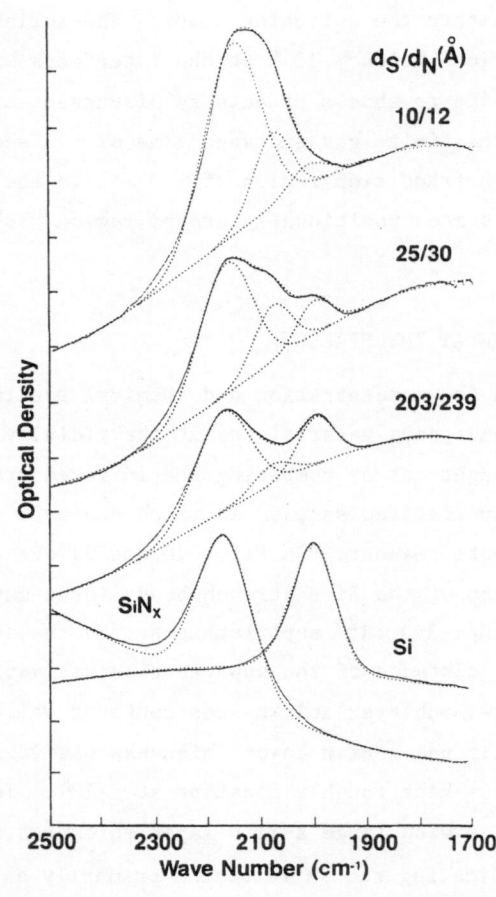

Figure 10   S-H stretching bands in a-Si:H/a-SiN$_x$:H superlattices and bulk a-Si:H and a-SiN$_x$:H films. The Si and SiN$_x$ layer thicknesses d_S and d_N are indicated. The gausian curves and their sum (dotted curves) are fitted to the experimental data (full curves).

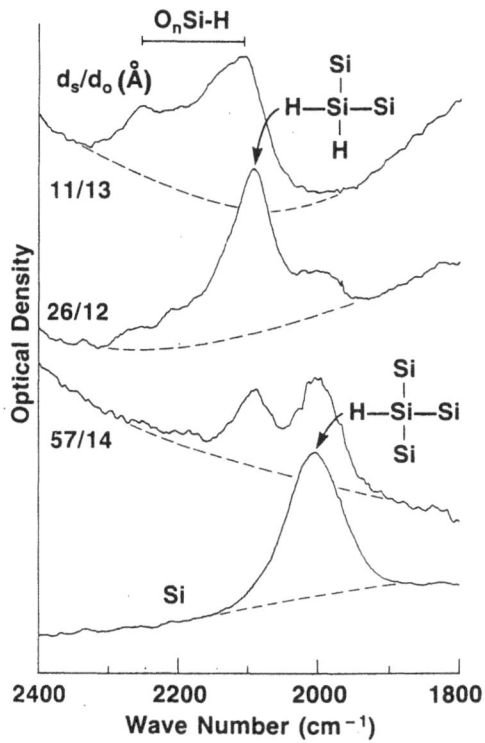

Figure 11   Si-H stretching bands in a-Si:H/a-SiO$_x$:H superlattices and bulk a-Si:H film. The Si and SiO$_x$ sublayer thicknesses d$_S$ and d$_o$ are indicated.

bonded at the interfaces per each bilayer.[17,21] The effect of the interface hydrogen on the core levels is discussed in the next section.

PHOTOEMISSION SPECTROSCOPY OF a-Si:H/a-SiO$_x$:H INTERFACES

Because photoemission is a surface sensitive probe measurements of interfaces are made on single heterojunctions in which the overlayer film is thin enough ($\lesssim$ 20Å) for photoelectrons to be emitted from both members of the heterojunction.[34] The heterojunctions were deposited on a 400Å a-Si:H buffer layer on top of polished stainless steel substrates. The Si on SiO$_x$ heterojunctions were made by depositing first a 60Å a-SiO$_x$:H layer on top of the a-Si:H buffer layer followed by the a-Si:H overlayer. The SiO$_x$ on Si heterojunctions were made by depositing the a-SiO$_x$:H overlayer directly on top of the a-Si:H buffer layer. The plasma was turned off for several seconds before depositing the overlayer to allow the gases to be completely exchanged. To achieve precise timing in the ignition of the plasma we used a Tesla coil. After the

207

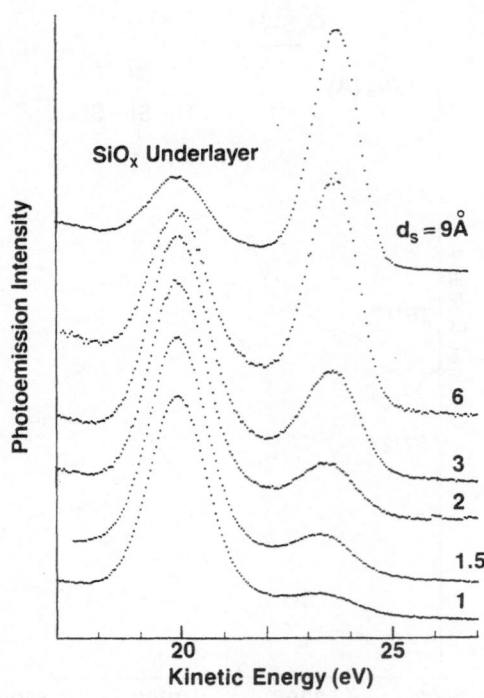

Figure 12   Si-2p core level spectra for Si on SiO_x measured at 125 eV photon energy.   Thickness d_S of a-Si:H overlayers are indicated.

deposition the reactor was pumped down to its base pressure of 5×10^{-9} torr and the sample was transferred under UHV to the measurement chamber.   The photoemission was measured with light from the UV ring at NSLS, Brookhaven National Laboratory.   The combined resolution of the monochromator and CMA was 0.3 eV.

<u>Core Levels Si-2p</u>

The Si-2p core level spectra for the Si on SiO_x and SiO_x on Si heterojunctions with different overlayer thicknesses are shown in Figs. 12 and 13 respectively.   The thicknesses of the overlayers were taken to be equal to the product of deposition time and deposition rate where the deposition rate was determined from in situ optical reflectivity measurements made separately on thick layers.   The growth of the a-Si:H overlayer in Fig. 12 manifests itself by an abrupt (within the first monolayer) chemical shift of the overlayer photoemission peak to a kinetic energy 3.7 eV above that of the a-SiO_x:H underlayer.   This chemical shift is due to the different Si coordination in the two materials, four nearest Si neighbors in the a-Si:H overlayer, compared to

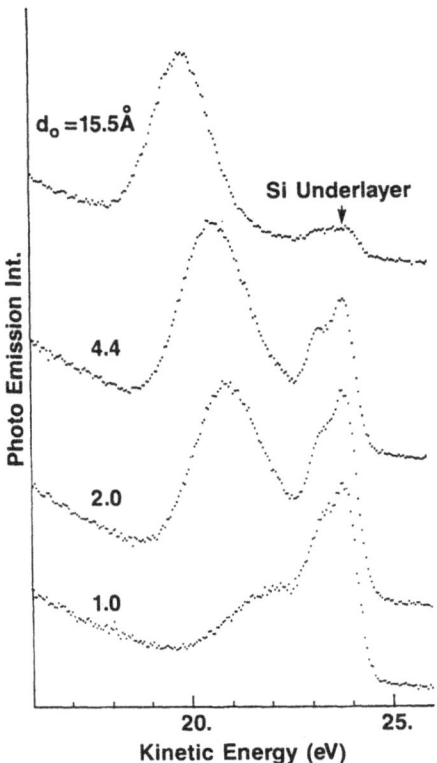

Figure 13   Si-2p core level spectra for SiO_x on Si measured at 125 eV
photon energy. Thickness d_0 of SiO_x overlayers are indicated.

four nearest O neighbors in the $a-SiO_x$:H underlayer (0.9 eV chemical
shift per oxygen bond). H bonded to Si gives rise to a much smaller
chemical shift, i.e. ~ 0.3eV per Si:H bond.[31] The near total absence of
any sub-oxide derived Si-2p structure in Fig. 7 indicates that the
overlayer Si atoms bond almost exclusively to Si in the underlayer, that
oxygen takes almost no part in the bonding to the overlayer and that the
interface is atomically abrupt. The large mismatch between the Si and
SiO_x networks is relaxed by ~ 1×10^{15} cm^{-2} extra hydrogen at the interface. On the other hand, the growth of the $a-SiO_x$:H overlayer in Fig.
13 manifests itself by a considerably more gradual downward shift in the
overlayer photoemission peak indicating that this interface has a graded
composition.

To extract quantitative information on the thickness of the overlayers from the photoemission data, systematic variations in photoemission intensity arising from misalignment of sample, change in the beam

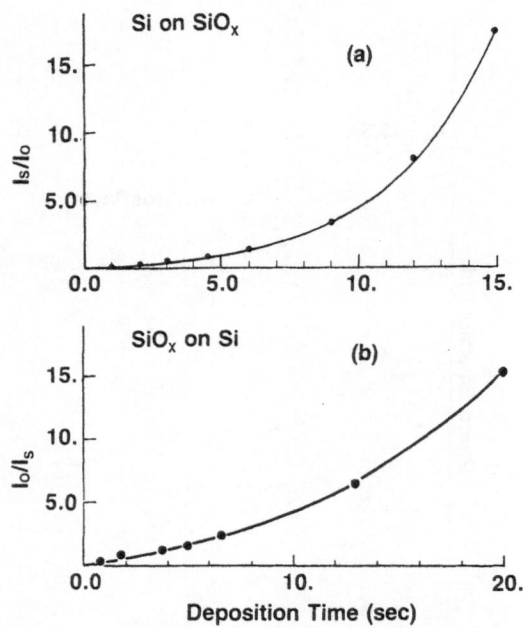

Figure 14   Ratio of overlayer to underlayer integrated intensities as
a function of deposition time. (a) Si on SiO$_x$ interface,
(b) SiO$_x$ on Si interface. Full curves were calculated
from Eqs. 3 and 4.

position or CMA were eliminated by dividing the integrated overlayer
intensity by the integrated underlayer intensity. This ratio is plotted
vs. deposition time in Fig. 14a and b for the two types of heterojunctions. From simple considerations of the photoelectron detection geometry shown in Fig. 15, the intensities of the overlayer, I_2, and underlayer I_1, are given by:

$$I_2 = M_2 N_2 \lambda_2 \int \cos \Theta \ [1-\exp \ (-d_2/\lambda_2 \cos \Theta)] \ d\omega \qquad (3)$$

$$I_1 = M_1 N_1 \lambda_1 \int \cos \Theta \ \exp \ [-d_2/\lambda_2 \cos \Theta] \ d\omega \qquad (4)$$

where M is the optical matrix element, N is the density of Si atoms, λ
is the electron escape depth, d_2 the overlayer thickness, Θ the angle
between the direction of electron escape path and the normal to the film
and the integration is over the acceptance cone of the CMA. The solid
curves in Fig. 14 were calculated using for the overlayer thicknesses
values determined from in-situ optical reflectance measurements and
using the escape depths as adjustable parameters. The best fit values
for the escape depths were λ_S = 4.5Å for Si and λ_0 = 16Å for SiO$_x$. The

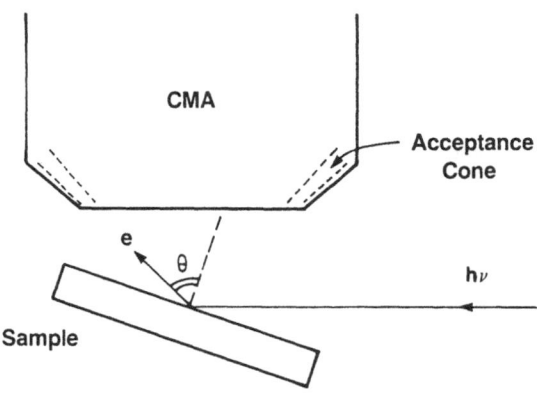

Figure 15   Schematic diagram of photoelectron detection geometry.

escape depth of 4.5Å at a kinetic energy of 25 eV is in agreement with
published values[31] for a-Si:H and the value of 16Å for SiO_x is consistent with the large band gap (9ev) of the oxide. The good agreement
between experimental and calculated values of I_2/I_1 is a check on the
validity of our overlayer thickness determination. Moreover, it confirms our assumption that for thicknesses larger than a monolayer the
overlayer forms a continuous film with uniform thickness.

To determine the compositional variation of the SiO_x on Si interface we assumed each overlayer in Fig. 13 can be derived from the pro-
ceeding one by adding an incremental new layer on top without changing
the underlaying material. Based on this model we find that the width of
the interface, defined by the distance over which the composition
changes from $SiO_{0.2}$ to $SiO_{1.8}$, is ~ 5 Å. The fact that this interface
is relatively wide is not surprising, since the first few monolayers of
a-SiO_x:H grow by oxidation of the underlaying a-Si:H (Sec. 3), a process
that involves diffusion.

Another difference between the Si on SiO_x and SiO_x on Si interfaces
is that the spin orbit splitting (0.6 eV) in the a-Si:H overlayers is
smeared out (Fig. 12), while in the a-Si:H underlayer it becomes progressively more resolved with increasing overlayer thickness (Fig.
13). A simple explanation for this is that the top few monolayers of
the growing a-Si:H layer contribute a broad core line, as a result of
being strongly disordered and hydrogenated.[31] Plasma oxidation consumes
the defective Si layer and the hydrogen in it, leaving behind a less
disordered a-Si:H layer with a correspondingly narrower line shape.
Evidence that this oxidation process results in extra hydrogen in the
oxide is provided by the IR spectra in Sec. 4.

Valence Band

Further evidence that Si on SiO_x forms an atomically abrupt interface is provided by the valence band measurements shown in Fig. 16. The
peak at 114 eV kinetic energy corresponds to the O-2p non-bonding
states[35] and the shoulder at the high kinetic energies corresponds to
the Si-3p states[34] of the Si overlayers. The curves in Fig. 16 were
shifted horizontally to bring into coincidence the O-2p peaks. In this
way the Si valence band is measured relative to the $a-SiO_x$ valence band
and small variations due to Fermi level shifts and variations due to
irreproducibility in monochromator setting were eliminated. The a-Si:H
valence band is brought out more clearly in Fig. 17 by subtracting from
the heterostructure signal the SiO_x underlayer signal. We note that the
shape of the Si-3p peak for different overlayer thicknesses is unchanged
from that in bulk a-Si:H (also shown in Fig. 17). In Fig. 18 we have
plotted the offset energy ΔE_v between the a-Si:H and the $a-SiO_x$:H valence band maxima as a function of Si overlayer thickness. The valence
band maximum is defined by the intersection of the dashed lines as
illustrated in Fig. 17 for one of the heterojunctions. Except for the
thinnest a-Si:H overlayer, ΔE_v = 3.95 $\pm$ 0.08 eV. We also measured the
offset energy in the SiO_x on Si heterojunctions and found ΔE_v = 4.10 $\pm$
.08. Thus our photoemission measurements on single heterojunctions
indicate that the offsets in the valence bands for the two interfaces
are, within our experimental error, the same. We note that the interpretation of electroabsorption measurements on a-Si:H/$a-SiO_x$:H superlattices suggested that the Si on SiO_x offset energy is 170 meV larger than
for the opposite interface[17], which is outside the experimental error of
our photoemission measurements. A likely cause for this discrepancy is
that the multilayers differ slightly structurally from single heterojunctions.

Because there is a 4 eV offset in the valence bands, the a-Si:H
overlayer forms a deep potential well for both electrons and holes. The
absence of any variation in the valence band offset with overlayer
thickness, except for the thinnest overlayer, in Fig. 18 leads us to
conclude that the quantum shift associated with the valence band is
smaller than our experimental uncertainty of 80mV. Thus the large blue
shift observed in the optical absorption of ultra-thin a-Si:H
layers[1,10,11] must be attributed to quantum shifts of the conduction
band alone. For instance in a a-Si:H/$a-SiO_x$:H superlattice with 10Å
thick a-Si:H layers we found the optical bandgap to be 2.12eV compared
to 1.73 eV for bulk a-Si:H. Assuming that the quantum shift is given by

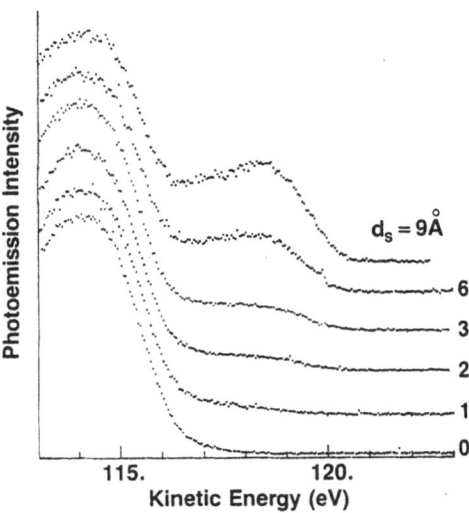

Figure 16   Photoemission intensity vs kinetic energy from top of valence bands of Si on SiO_x heterojunctions for different a-Si:H overlayer thicknesses d_S measured at 125 eV photon energy. Also shown valence band of $a-SiO_x$:H ($d_S=0$).

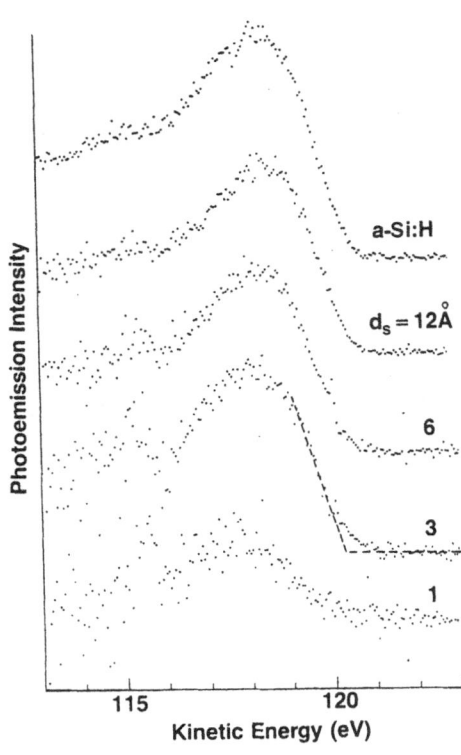

Figure 17   Photoemission intensity from a-Si:H overlayer vs kinetic energy. Photoemission from $a-SiO_x$:H underlayer was subtracted.

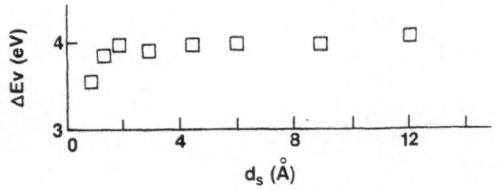

Figure 18   Offset between a–Si:H and a–SiO$_x$:H valence bands, ΔE_v, as a function of a–Si:H overlayer thickness d_s.

the simple relation, $h^2/8m_e^* d_s^2$, the blue shift of 0.39 eV in the optical gap can be accounted for by an electrons effective mass $m_e^* = 1.0$ m_o, where m_o is the free electron effective mass. The fact that this value of the effective electron mass is considerably larger than 0.3 m_O deduced from transport measurements in a–Ge:H/a–Si:H superlattices[12] is likely due to a disorder related decrease in the quantum shift in nitride and oxide superlattices. A much smaller quantum shift in the valence band compared to that of the conduction band implies that the effective mass of holes must be at least an order of magnitude larger than that of electrons. This interpretation is consistent with the fact that the free electron mobility in a–Si:H (13.6 cm^2/v sec) is more than an order of magnitude larger than that of holes[36] (0.6 cm^2/v sec). Moreover, a large ratio of hole to electron masses is also consistent with calculations by Zdetsis et al[37] who find that the density of states at the top of the valence band is about an order of magnitude higher than the density of states at the bottom of the conduction band.

L$_{23}$ Absorption Edge

The L$_{23}$ edge is due to transitions from Si–2p core levels to unoccupied conduction band states. In a–Si:H (as well as in c–Si) the conduction band is modified by strong core hole exciton interactions giving rise to a steep absorption edge.[34]. The L$_{23}$ spectra of the Si on SiO$_x$ heterojunctions obtained by partial yield measurements are shown in Fig. 19. This edge, shown for the Si on SiO$_x$ heterojunctions in Fig. 15 is well resolved from the underlayer a–SiO$_x$:H exciton peak[38] (105 eV) and conduction band edge (107 eV), once again demonstrating the abruptness of the interfaces. We define the absorption threshold energy E_T as the energy at which the absorption edge assumes half its value. E_T is plotted versus d_s in Fig. 20. Since the core–hole exciton in a–Si:H is not resolved from the conduction band, the core hole exciton binding energy E_{ex} in a–Si:H is defined as the difference between the optical band gap E_G determined from measurements in the visible, and the photoemission band gap E_p. The photoemission gap is defined as $E_G = E_T$

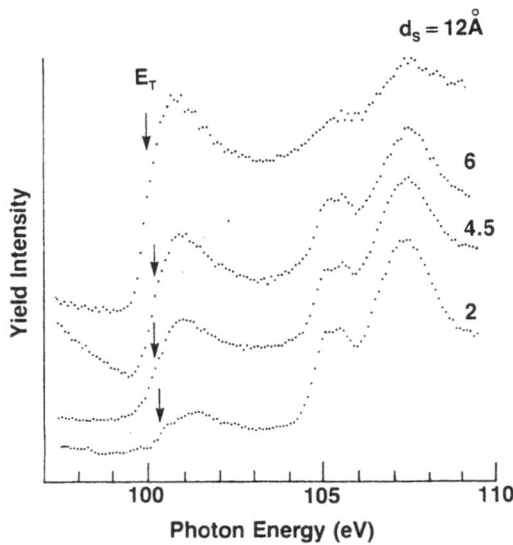

Figure 19 L_{23} absorption edge as a function of photon energy for Si on SiO_x heterojunctions with different a-Si:H overlayer thicknesses. The arrows indicate the a-Si:H L_{23} edge energy E_T.

– E_B where E_B is the Si-2p binding energy referred to the valence band edge. Our value of E_{ex} = 0.33 ± 0.15 eV for large overlayer thicknesses (see also ref. 27) is in agreement with values reported by Wesner et al[39] and Ley et al,[32] but is much smaller than the value 0.95 eV reported by Evangelisti et al.[40] The heterojunction energy level diagram for thick overlayers is shown in Fig. 21.

The increase in the threshold energy E_T in Fig. 20 for d_S<6Å is attributed to a transition from three dimensional to two dimensional behavior of the core hole exciton interaction. Evidence that the variation in E_T in the range d_S>2Å is not due to Si-H or Si-O bonding comes from the fact that the Si-2p core level is centered at the same binding energy as it is in the thick a-Si:H overlayers (Fig. 12). For d_S≲2Å the overlayer core level peak shifts to higher binding energies presumably due to a large prevelance of Si-H bonds which in turn may strongly affect E_T. We use the following geometric argument to deduce the range of the exciton interaction energy. As long as the potential well is wider than the exciton diameter $2r_{ex}$, we expect E_T to be independent of well width d_S, while at the same time we expect the conduction band edge E_c, and consequently the exciton binding energy $E_{ex} = E_c - E_T$, to increase as $1/d_S^2$. When $d_S \leq 2r_{ex}$ the exciton becomes two-dimensional, the value of E_{ex} the exciton becomes two-dimensional, the value of E_{ex}

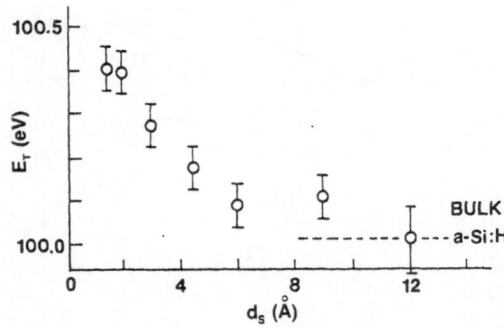

Figure 20 L_{23} absorption energy E_T in a-Si:H as a function of a-Si:H overlayer thickness d_s.

saturates and E_T will increase at the same rate as E_c. Based on this argument we equate the well width at which E_T begins to rise sharply, d_s = 6 Å in Fig. 20, with the exciton diameter.

The large value of the core hole exciton energy (few tenths of electron volts compared to the few milielectron volts for shallow valence excitons and donor impurity levels) in both crystalline and amorphous semiconductors has been the subject of theoretical controversy for many years.[40,41] In the absence of a complete theory of core hole excitons we use a simple effective mass model in which the exciton binding energy E_{ex} and radius r_{ex} are given by:[42]

$$E_{ex} = E_R \, m_e*/m_o \, \varepsilon_{eff}^2 \qquad (4)$$

$$r_{ex} = r_B \, m_o \, \varepsilon_{eff}/m_e* \qquad (5)$$

where E_R = 13.6 eV is the Rydberg energy, r_b = 0.52 Å is the Bohr radius and we have introduced an effective dielectric constant ε_{eff} which takes into account incomplete screening of the core hole. Setting in Eq. 4 for E_{ex} = 0.33 eV and m_e*/m_o = 1 we obtain ε = 6.4 and from Eq. 5 we obtain r_{ex} = 3Å in agreement with the core hole exciton diameter of 6Å determined from Fig. 17.

CONCLUSION

A wide range of amorphous semiconductor superlattices with nearly atomically abrupt interfaces can be prepared by plasma assisted CVD. In situ optical reflectivity provides a simple tool for monitoring growth of the multilayers and yields information on sublayer thicknesses, their optical constants and macroscopic structure of the interfaces. IR

216

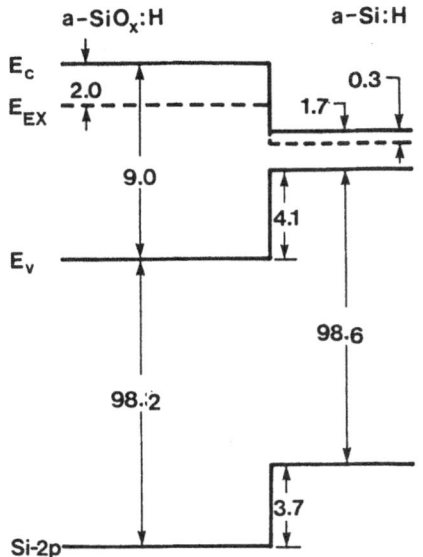

Figure 21   Energy diagram of a-Si:H/a-SiO$_x$:H heterojunction showing
the relative positions of Si-2p core levels, valence band
edges, E$_v$, conduction band edges E$_c$ and core hole exciton
levels E$_{ex}$.

spectroscopy and N^{15} nuclear reaction profiling of a-Si:H/a-SiN$_x$:H and
a-Si:H/a-SiO$_x$:H superlattices show ~ 10^{15} cm^{-2} extra hydrogen bonded
primarily at the Si on SiO$_x$ and SiN$_x$ interfaces. Hydrogen relieves the
large lattice mismatch between the layers and pacifies defects. Si-2p
core level spectroscopy show that the Si on SiO$_x$ interface is atomically
abrupt while the SiO$_x$ on Si interface is about 5Å wide. The offset
between the a-Si:H and a-SiO$_x$:H valance bands is 4.0 eV for both types
of interfaces. From the fact that the offset energy is independent of
a-Si:H overlayer thickness down to monolayer dimensions we conclude that
quantum shifts of the valence band edge are an order of magnitude smaller than those for the conduction band indicating a corresponding in-
equality between the electron and hole effective masses. From the
dependence of the L$_{23}$ absorption edge energy on a-Si:H layer thickness,
we deduce a radius of 3Å for the core hole exciton in a-Si:H.

ACKNOWLEDGEMENTS

    We thank P. D. Persans for his contributions to the in situ optics,
W. Eberhardt and D. Sondericker for contributions to the photoemission
work and H. Stasiewski for technical assistance in all aspects of the
work.

REFERENCES

1. B. Abeles and T. Tiedje, Phys. Rev. Lett. 51:2003 (1983).
2. M. Hirose and S. Miyazaki, J. Non-Cryst. Solids 66:327 (1984).
3. M. Hundhausen, L. Ley and R. Carius, Phys. Rev. Lett. 53:1598 (1984).
4. J. Kakalios and H. Fritzsche, Phys. Rev. Lett. 53:1602 (1984).
5. H. Deckman, J. H. Dunsmuir, and B. Abeles, Appl. Phys. Lett. 46:171 (1985).
6. R. Cheng, S. Wen, J. Feng, and H. Fritzsche, Appl. Phys. Lett. 46:592 (1985).
7. C. C. Tsai, Materials Research Society Symposia Proceedings 70, 1986 (in print).
8. Proceedings 18th International Conference on Physics of Semiconductors Stockholm, 1986; Proc. 11th International Conference on Amorphous and Liquid Semiconductors, Editors F. Evangelisti and J. Stuke North Holland, Amsterdam 1985.
9. T. Tiedje, B. Abeles, P. D. Persans, B. G. Brooks, and G. D. Cody, J. of Non-Cryst. Solids 66:345 (1984).
10. J. Kakalios, H. Fritzsche, N. Ibaraki, and S. R. Ovshinsky, J. Non-Cryst. Solids 66:339 (1984).
11. Z. M. Chen, J. N. Wang, X. Y. Mei, and G. L. Kong, Solid State Commun. 58:379 (1986).
12. P. D. Persans, B. Abeles, J. Scanlon, and H. Stasiewski, in Proceedings of the Seventeenth International Conference on the Physics of Semiconductors, San Francisco, 1984, edited by D. J. Chadi and W. A. Harrison (Springer, New York, 1985), p. 499. C. R. Wronski, P. D. Persans and B. Abeles, Appl. Phys. Lett. (in print).
13. T. Tiedje, B. Abeles, and B. Brooks, Phys. Rev. Lett. 54:254 (1985).
14. C. R. Wronski, T. Tiedje, P. D. Persans, B. Abeles, and M. Hicks Appl. Phys. Lett. (in print).
15. T. Tiedje and B. Abeles, Appl. Phys. Lett. 45:179 (1984).
16. C. B. Roxlo, B. Abeles, and T. Tiedje, Phys. Rev. Lett 52:1994 (1984).
17. C. B. Roxlo and B. Abeles, Phys. Rev. B 34:2522 (1986).
18. S. Tsuda et al, Conference Record of 18th Photovoltaic Specialists Conference, IEEE, New York 1985, p. 1295.
19. R. A. Arya, A. Catalano, J. O'Dwood, J. Morris, and G. Wood ibid ref. 18 p. 1710.
20. D. Kruangam, M. Deguchi, T. Endo, W. Guang, P. H. Okamoto, and Y. Hamakawa 18th Int. Conf. Solid State Devices and Materials Tokyo 1986.
21. B. Abeles, L. Yang, P. D. Persans, H. Stasiewski, and W. Lanford, Appl. Phys. Lett. 48:168 (1986).

22. W. A. Lanford and B. Abeles, Nuclear Instr. and Methods in Physics Research B15, (1986).

23. B. Abeles, L. Yang, E. Eberhardt, and C. B. Roxlo, Proceedings 18th Int. Conf. on Physics of Semiconductors, Stockholm (1986).
24. P. D. Persans, A. F. Ruppert, B. Abeles, and T. Tiedje, Phys. Rev. B32:5558 (1985).
25. C. B. Roxlo, B. Abeles, and P. D. Persans, J. Vac. Sci. Tech. (in print).
26. T. Tiedje, B. G. Brooks, and B. Abeles, AIP Conf. Proc. 120:416 (1984).
27. B. Abeles, I. Wagner, W. Eberhardt, J. Stohr, H. Stasiewski, and F. Sette, AIP Conf. Proc. 120:394 (1984).
28. F. Evangelisti, P. Fiorini, C. Giovannella, F. Patella, P. Pefetti, C. Quaresima, and M. Capizzi, Appl. Phys. Lett. 44:764 (1984).

29. L. Yang, B. Abeles, and P. D. Persans, Appl. Phys. Lett. 49:631 (1986).
30. B. Abeles, T. Tiedje, T. Liang, H. W. Deckman, H. W. Stasiewski, J. C. Scanlon, and P. M. Eisenberger, J. Non-Cryst. Solids 66:351 (1984).
31. L. Ley, J. Reichardt, and R. L. Johnson, in Proceedings of the 17th Int. Conf. Phys. of Semiconductors, San Francisco, 1984, Edited by D. J. Chadi and W. A. Harrison (Springer NY, 1985).
32. J. C. Phillips, Phys. Rev. Lett. 20:550 (1968), D. E. Aspnes and J. B. Theeten, J. Appl. Phys. 50:4928 (1979).
33. H. Wagner and W. Beyer, Solid State Commun. 48:585 (1983).
34. L. Ley in the Physics of Hydrogenated Amorphous Silicon Vol. II J. D. Joannopoulos and G. Lucovsky, Editors (Springer, Berlin 1984) p. 61.
35. G. Hollinger and F. S. Himpsel, J. Vac. Sci. Technol. A1, 640 (1982).
36. T. Tiedje, J. M. Cebulka, D. L. Morel. and B. Abeles, Phys. Rev. Lett. 46:1425 (1981).
37. D. Zdetsis, E. N. Economu, D. A. Papaconstantopoulus, and N. Flytzanis, Phys. Rev. B31:2410 (1985).
38. M. L. Knotek, R. H. Stulen, G. M. Loubriel, V. Rehn, R. A. Rosenberg, and C. C. Parks, Surface Science 291 (1983).
39. D. Wesner and W. Eberhardt, Phys. Rev. B28:7087 (1983).
40. F. Evangelisti, F. Patella, R. A. Riedel, G. Margaritondo, P. Fiorini, P. Perfetti, and C. Quaresima Phys. Rev. Lett 53:2504 (1984).
41. A. Quattropani, F. Bassani, G. Margaritondo, and G. Tinivella, Noovocimento 51B:N.2 355 (1979).
42. N. W. Ashcroft and N. D. Meermin, Solid State Physics, Saunders College, Philadelphia (1976).

ATOMIC LAYER EPITAXY OF COMPOUND SEMICONDUCTORS

Markus Pessa

Department of Physics, Tampere University of Technology
P.O. Box 527, SF-33101 Tampere, Finland

Atomic layer epitxy (ALE) is a relatively new method of making low-dimensional overlayer structures of semiconductor and insulator compounds[1]. ALE makes use of the difference between chemical and physical adsorption of molecular species brought onto the substrate surface as alternate molecular beam or vapor pulses. The growth proceeds stepwise in a layer-by-layer fashion resulting in "digital epitaxy" where film thickness is precisely controlled. It also provides advantages for coating several large-area substrates simultaneously in an industrial environment[2].

The characteristic feature of the simplified layer-by-layer model of ALE[3] is that the number of layers deposited is determined solely by the number of operational cycles performed. Since all the beams are turned off for a short time between two successive pulses, a thermodynamical equilibrium is approached at the end of each reaction step. This is not normally true in other vapor phase deposition methods, where the deviation from equilibrium is actually the driving force for the film growth.

Although ALE has the attractive feature of producing one complete monolayer of a compound film per operational cycle in a self-limiting manner, achieving such a growth strictly in practice requires optimal growth conditions[4-6]. Growth is affected by the stability of the surface layer after different pulses, surface defects, non-unity sticking coefficients and details of the reaction mechanism. All these factors evidently depend upon the material to be grown, its crystal structure and the reactants used, making the choice of growth parameters an important, non-trivial procedure.

Epitaxy of II-VI compounds, including $CdTe$[3,4], $CdMnTe$[7,8], $ZnTe$[9], $ZnSe$[10] and $ZnMnSe$[11], has been achieved using the principle of ALE by evaporating elements from effusion cells in ultra-high vacuum, as in the molecular beam epitaxy (MBE) method. III-V compounds have been grown by applying metallo-organic chemical vapor deposition (MOCVD) ALE and chloride ALE. Using these low-pressure ALE techniques high-quality epilayers of $GaAs^{12-18}$, $AlAs^{13}$, $InAs^{17,18}$, $GaInAs^{18}$, InP^{19}, GaP^{19} and $GaInP^{19}$ have been prepared.

With regard to polycrystalline and amorphous materials, the ac thin--film electroluminescent displays made by Lohja Ltd have demonstrated unequivocally that ALE can be used to grow uniform deposits of good structural integrity both mechanically and dielectrically speaking. These qualities are of importance in many other fields, too.

ALE has begun to attract considerable attention as a method for producing thin films of high quality. Fabrication of ternary compounds and low--dimensional (magnetic) structures is of particular interest. ALE may also prove a vechile for investigating fundamental aspects of compound semiconductor growth. However, it still remains to be demonstrated that the perfection of structure and control of dopants required of materials for electronic devices can in fact be achieved by ALE.

REFERENCES

1. C.H.L Goodman and M. Pessa, J. Appl. Phys. **60**, R65 (1986)
2. T. Suntola and J. Hyvärinen, Ann. Rev. Mater. Sci. **15**, 177 (1985)
3. M. Pessa, O. Jylhä and M. A. Herman, J. Crystal Growth **67**, 255 (1984)
4. M. A. Herman, M. Vulli and M. Pessa, J. Crystal Growth **73**, 403 (1985)
5. M. A. Herman, O. Jylhä and M. Pessa, Cryst. Res. Technol. **21**, 841 (1986) and Cryst. Res. Technol. **21**, 969 (1986)
6. T. Pakkanen, M. Lindblad and V. Nevalainen, *in The Extended Abstracts of The First Symposium on Atomic Layer Epitaxy*, Espoo, Finland, 1984, p. 14
7. M. A. Herman, O. Jylhä and M. Pessa, J. Crystal Growth **66**, 480 (1984)
8. M. Pessa and O. Jylhä, Appl. Phys. Lett. **45**, 646 (1984)
9. T. Yao and T. Takeda, Appl. Phys. Lett. **48**, 160 (1986)
10. T. Yao, T. Takeda and R. Watanuki, App.l Phys. Lett. **48**, 1615 (1986)
11. M. Pessa, *in Abstracts of The MRS 1986 Fall Meeting*, Boston, USA, p. 632
12. J. Nishizawa, H. Abe and T. Kurabayashi, J. Electrochem. Soc., **132**, 1197 (1985); J. Nishizawa, H. Abe, T. Kurabayashi and N. Sakuri, J. Vac. Sci. Technol. **A4**, 706 (1986); J. Nishizawa and T. Kurabayashi, J. Cryst. Soc. Japan **28**, 133 (1986)
13. S.M. Bedair, M.A. Tischler, T. Katsuyama and N.A. El-Masry, Appl. Phys. Lett. **47**, 51 (1985)
14. M.A. Tischler and S.M. Bedair, Appl. Phys. Lett **48**, 1681 (1986)

15. A. Doi, Y. Aoyagi and S. Namba, Appl. Phys. Lett. **48**, 1787 (1986)

16. A. Usui and S. Sunakawa, Jap. J. App.l Phys. **25**, L212 (1986)

17. M.A. Tischler, N.G. Anderson and S.M. Bedair, Appl. Phys. Lett. **49**, 1199 (1986)

18. M.A. Tischler and S.M. Bedair, J. Crystal Growth **77**, 89 (1986)

19. A. Usui and H. Sunakawa, *in Proceedings of The 13th Int. Symp. on GaAs and Related Compounds,* Las Vegas, 1986

REFLECTION HIGH-ENERGY ELECTRON DIFFRACTION INTENSITY OSCILLATION - AN EFFECTIVE TOOL OF Si AND Ge_xSi_{1-x} MOLECULAR BEAM EPITAXY

Tsunenori Sakamoto, Kunihiro Sakamoto, Satoru Nagao[*], Gen Hashiguchi[**], Katsuya Kuniyoshi[***] and Yoshio Bando[†]

Electrotechnical Laboratory, Tsukuba, Ibaraki, 305 Japan
  On leave from [*]Mitsubishi Chemical Industries
  Research Center, [**]Chuo University, [***]Meiji
  University
[†]National Institute for Research in Inorganic Materials, Tsukuba, Ibaraki, 305 Japan

INTRODUCTION

Molecular beam epitaxy (MBE) is becoming an important technique for growing epitaxial Si based films. The first advantage of Si MBE is a low growth temperature, usually in the range of 400 to 800°C, which is much lower than that required for conventional techniques. The lower growth temperature reduces diffusion of dopants. The second advantage is an excellent control of doping distribution which is essential for high speed VLSI. The third advantage is an ability to fabricate heterojunction and superlattice structures. Examples are Ge_xSi_{1-x} strained-layer super-lattices[1-5], metal silicides[6,7], Si on insulators[8,9] and Si hetero-junctions with III-V compound semiconductors[10,11]. Among them, hetero-epitaxy of Ge_xSi_{1-x}/Si attracts much attention because it can add to conventional Si integrated circuits exciting possibilities of hetero-junction devices. Modulation-doped Ge_xSi_{1-x}/Si strained-layer hetero-structures showed two-dimensional carrier gas properties and enhanced mobilities for electron[2] and hole[3]. Recently, n-channel[4] and p-channel[5] modulation-doped field effect transistors were successfully fabricated.

In order to bring these abilities of Si MBE to the full, it is necessary to monitor and control the crystal growth at an atomic-order precision because electric and optical properties of microstructures are very sensitive to layer thickness, alloy composition, impurity doping and interface roughness.

Reflection high-energy electron diffraction (RHEED) is a powerful in situ probe for a surface structure and has been used in studies of films prepared by MBE. Since the first observation of intensity oscillation of RHEED specular-beam during the MBE growth of GaAs[12,13], many researchers have investigated its properties and found that it was caused by the two-dimensional layer-by-layer growth and a period of the oscillation corresponded to a monolayer growth[14,15]. This has led to an in situ observation of the monolayer growth and a precise control technique of layer thickness and interface roughness (phase-locked epitaxy technique)[16,17]. The RHEED intensity oscillation has also been observed during Si epitaxy[18-20] and Ge_xSi_{1-x}/Si strained-layer heteroepitaxy[21].

In this paper some topics on MBE growth of Si and Ge_xSi_{1-x} are described from a viewpoint of the RHEED intensity observation. We lay stress on the RHEED intensity oscillation during the growth on Si(001) 2x1 single-domain structure. Properties of the Ge_xSi_{1-x}/Si strained-layer heterostructure and strained-layer superlattice are also described.

EXPERIMENTAL

Experiments were made in an ion-pumped MBE system. The base pressure and the pressure during growths in the growth chamber were 1×10^{-7} and 1×10^{-6} Pa, respectively. Si and Ge beams were evaporated from high purity single-crystalline Si by a 2kW electron gun and from high purity single-crystalline Ge by a PBN Knudsen cell heated up to $1250^\circ C$, respectively. A substrate was heated by passing a current directly through the wafer. While most experiments used well-oriented Si(001) substrates of $50 \times 10 \times 0.5 mm^3$ in size, limited works have also been done with Si(111) substrates. Prior to loading, wafers were degreased and finally boiled in an $HCl:H_2O_2:H_2O$ (1:1:4) solution to form a protective thin oxide film[22]. Conditions of the RHEED observation are as follows. The acceleration voltage of an incident electron-beam was 40keV and the full width at half maximum (FWHM) of the beam on a RHEED phosphor screen was about 200μm. The incidence angle on the surface was usually 0.9° (off-Bragg condition). Intensity of a particular RHEED pattern spot was measured via an optical fiber by a photo-multiplier. A scanning magnetic field was applied on diffracted electron beams to measure the intensity profiles of RHEED spots.

RHEED INTENSITY OSCILLATION DURING Si MBE GROWTH

RHEED intensity oscillation during growth on Si(001) substrate

To begin with, we show the longest RHEED intensity oscillation which has been obtained during the growth on Si(001) substrate. As shown in Fig. 1, the intensity oscillation was damped very slowly and more than

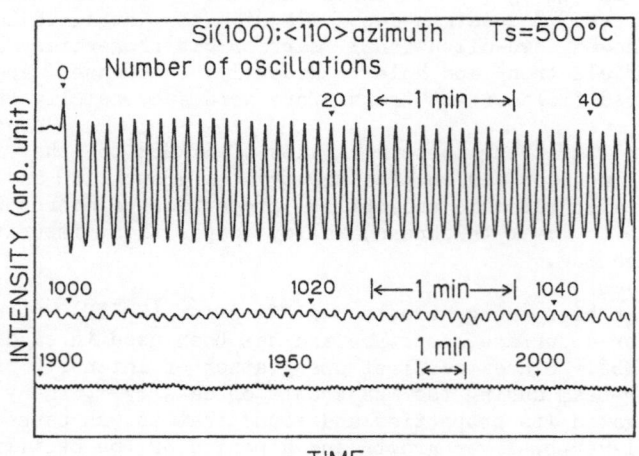

Fig.1. RHEED intensity oscillations of the specular-beam during the Si MBE growth on Si(001) at $500^\circ C$ taken from <110> azimuth.

2200 periods of oscillation was observed for over three hours. Because one period of the oscillation during Si growth observed from <110> azimuth corresponded to a biatomic-layer ($a_0/2$=0.27nm) growth as discussed later, we could count more than 4400 atomic layers, about 600nm in thickness, by monitoring RHEED intensity oscillation. Thus we could control the thickness of grown layer up to 600nm at the accuracy of an atomic layer. This precise layer-thickness control is one of the great advantages offered to Si MBE by the measurement of RHEED intensity oscillations.

Surface cleaning and RHEED intensity oscillation

We next discuss the surface preparation conditions to obtain such stable RHEED intensity oscillations. The first step in a growth chamber is a surface cleaning process. The protective thin oxide film was decomposed at 800°C with the help of a low flux Si beam (about 3×10^{13}atom/cm^2·s) incident on the surface[23] Figure 2 shows the RHEED intensity variation of the specular-beam and RHEED patterns during the cleaning of Si(001) surface. The RHEED pattern changed from an initial weak 1x1 streaky one via a 2x1 one with well-developed half-order streaks to a final 2x1 one with relatively strong spots on the zeroth Laue ring. The RHEED intensity also changed along with the change in the RHEED pattern and finally saturated at a level stronger than that of before cleaning. We consider that the saturation of intensity is a sign of the completion of surface cleaning.

After the surface cleaning, a substrate was cooled down to 500°C and a Si buffer layer was deposited. Figure 3 shows the RHEED intensity variation during the buffer layer growth. Although the intensity showed oscillations, their amplitude were very small and diminished to near zero after about only 40 periods of oscillations, which was much poorer than that shown in Fig. 1. The increase of average intensity shown in the figure indicated that the surface was smoothened during the buffer layer growth. It was concluded from these observations that an as-cleaned Si surface was not atomically flat, and that the cleaning process described above was insufficient to obtain the strong and lasting RHEED intensity oscillations.

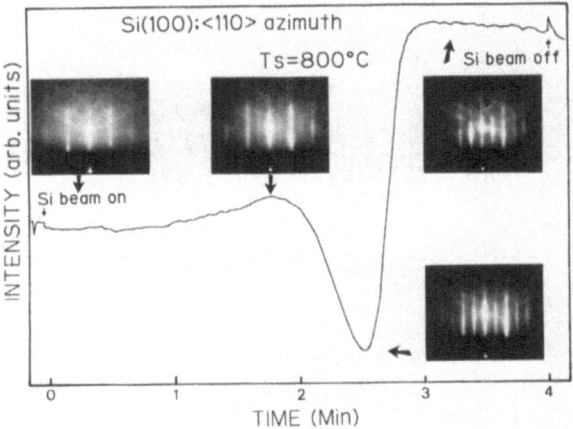

Fig.2. RHEED intensity variation of the specular-beam and corresponding RHEED patterns during the cleaning of Si(001) surface.

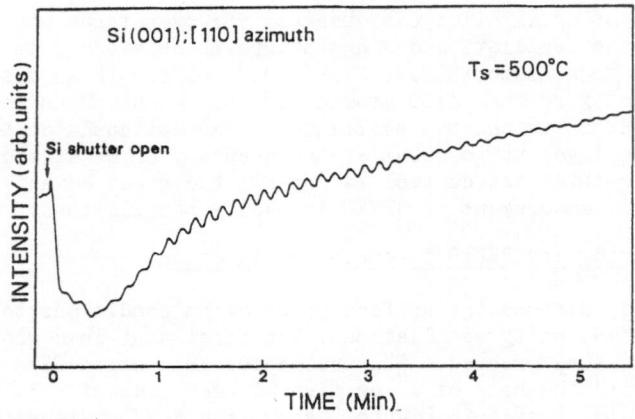

Fig.3. RHEED intensity variation of the specular-beam during the Si MBE growth at 500°C on an as-cleaned Si(001) surface.

High temperature annealing and RHEED intensity oscillation

It was found that Si substrates must be annealed prior to growth at the temperature much higher than that of during growth in order to get stable RHEED intensity oscillations[18]. After the growth of undoped buffer layers of 100-200nm thick, the substrates were annealed at various conditions. Figure 4 shows the RHEED intensity oscillations during the growth at 400°C after the annealing. With a weak annealing condition of 900°C for 15sec, a small asymmetric oscillation was observed. With the increase of annealing temperature and time, the amplitude of oscillations increased rapidly and their period doubled. We measured the intensity and the FWHM

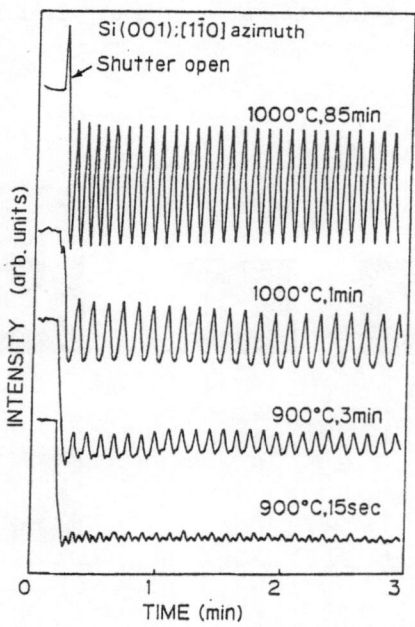

Fig.4. RHEED intensity oscillations of the specular-beam during the Si(001) MBE growth at 400°C after the annealing of various conditions.

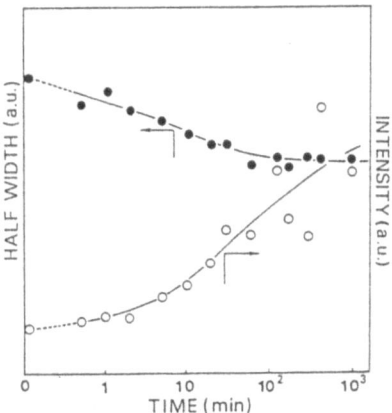

Fig.5. Intensity and FWHM of the RHEED specular-beam versus
annealing time at 1000°C. RHEED-observation temperature is
300°C.

of specular-beam to investigate quantitatively the effect of annealing at
1000°C. Figure 5 shows the results. The intensity increased and the FWHM
decreased with the increase of annealing time, indicating that the
Si(001) surface became atomically flat and the average domain size of a
terrace became bigger during the annealing.

Formation of the single-domain Si(001) 2x1 structure

We checked the RHEED patterns before and after the annealing to
examine the surface reconstruction[19]. Figure 6 shows RHEED patterns of an
as-grown Si buffer layer surface, (a) from the [$\bar{1}$10] azimuth and (b) from
the [010] azimuth. Since the diffraction pattern on the half-order Laue

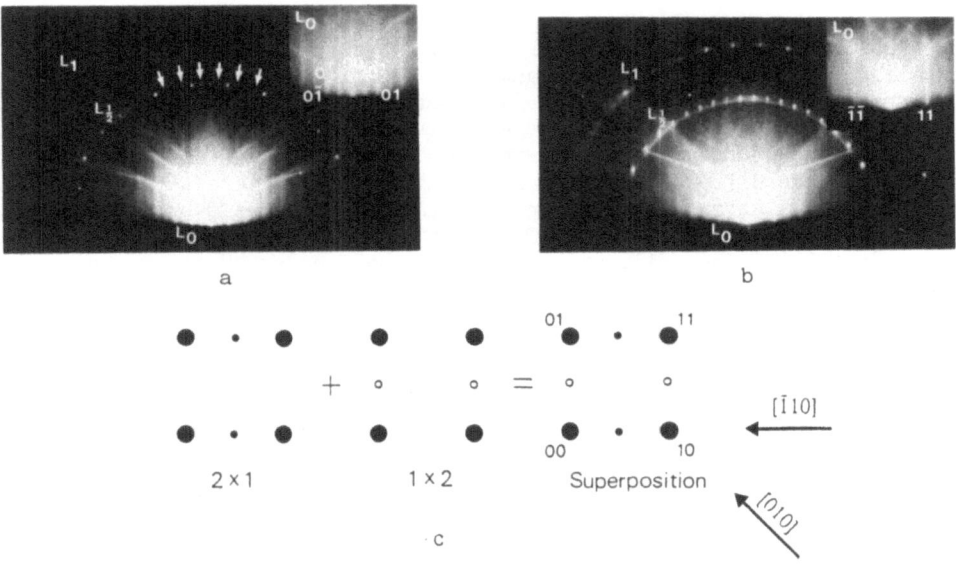

Fig.6. Typical RHEED patterns after a buffer layer growth on
Si(001) surface obtained from (a) [$\bar{1}$10] and (b) [010]
azimuths. Upper inset figures show magnified views of the
zeroth Laue zone. (c) Reciprocal lattices of Si(001) surface structure, 2x1, 1x2 and superposition of 2x1 and 1x2.

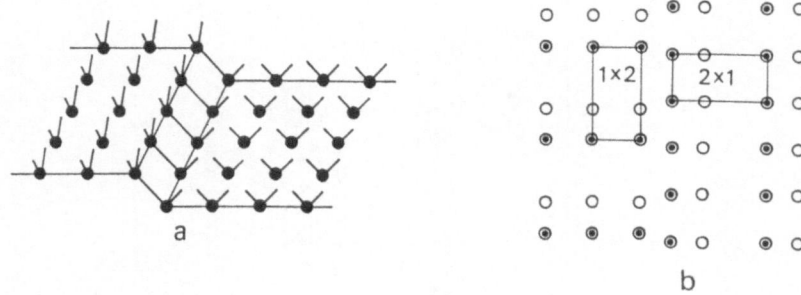

Fig.7. Schematic illustration of a two-domain Si(001) 2x1 reconstructed-surface with a monatomic-height step. (a) dangling-
bonds and (b) surface reconstructions.

ring ($L_{1/2}$) showed two-fold periodic spots with equal intensities, the
reconstruction was assigned to the well-known two-domain Si(001) 2x1
structure. As shown in the reciprocal lattices of Fig. 6(c), observed
RHEED patterns were consistent with the superposition of two orthogonal
superlattices reconstructed in 2x1 and in 1x2. It was explained in the
real lattice space, as illustrated in Fig. 7, that there were many steps
of a monatomic height in the as-grown surface and that the mixture of
2x1 and 1x2 reconstructions were observed as dangling-bonds on both sides
of a monolayer step were at the right angles to each other.

After the annealing of the sample at 1000°C for 20min, the RHEED
pattern changed substantially. Figure 8 shows RHEED patterns after the
annealing. First the pattern changed from a streaky one to a spotty one
with bright diffraction spots aligned on the zeroth Laue ring (L_0)
indicating that the surface became atomically flat. More essential change
was that the RHEED pattern observed from the [$\bar{1}$10] and from the [110]
azimuth became different, which had not been seen before the annealing.
From the [$\bar{1}$10] azimuth, as shown in Fig. 8(a), the half-order Laue ring
($L_{1/2}$) disappeared. From the [110] azimuth, as shown in Fig. 8(b), the
half-order Laue ring still existed but the half-order spots on the zeroth
Laue ring ((1/2 0) and ($\bar{1}$/2 0)) weakened on the contrary. Figure 8(c)
shows the RHEED pattern from the [010] azimuth. Compared with Fig. 6(b),
it is obvious that one series of half-order spots on the half-order Laue
ring (indicated by arrows) disappeared. Because these patterns were consistent with a reciprocal lattice of Fig. 8(d), it was found that the
surface was mostly covered with 2x1 reconstructed-domains. This surface
structure is called a single-domain Si(001) 2x1 structure. In the real
lattice space, as shown in Fig. 9, small domains separated by monatomic-
layer-height steps were developed to larger domains separated by biatomic-
layer-height steps and dangling-bonds of surface atoms were aligned in a
single direction by the high temperature annealing. Although a single-
domain Si(001) 2x1 structure had been observed in a vicinal (001)
surface[24,25], this was the first observation of a stable single-domain
structure on a well-oriented (001) surface. The formation of single-domain
structure was essential for the observation of a stable RHEED intensity
oscillation as discussed later.

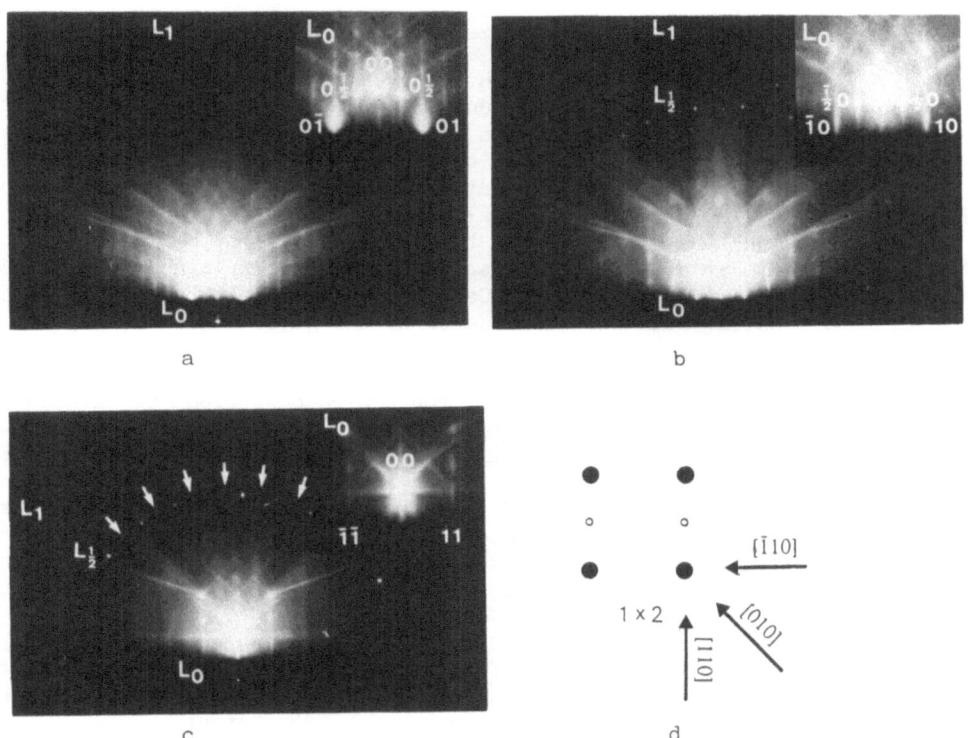

Fig.8. RHEED patterns after the annealing at 1000°C for 20 min, taken from (a) [$\bar{1}$10], (b) [110] and (c) [010] azimuths. Upper inset figures show magnified views of the zeroth Laue zone. (d) Reciprocal lattice of Si(001) 1x2 surface structure.

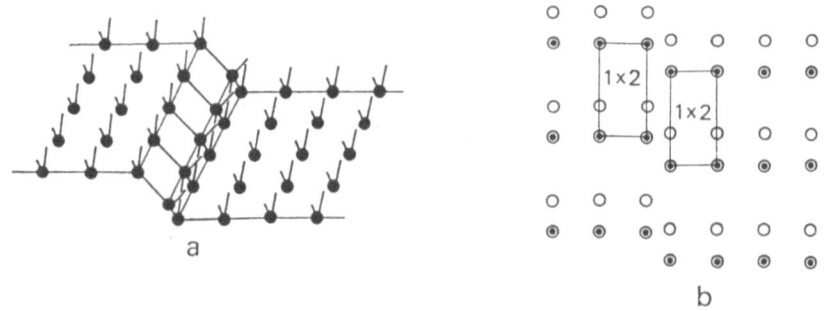

Fig.9. Schematic illustration of a single-domain Si(001) 2x1 reconstructed-surface with a biatomic-height step. (a) dangling-bonds and (b) surface reconstructions.

231

Mechanism of RHEED intensity oscillation during MBE on Si(001)

The RHEED intensity oscillation observed on Si(001) showed peculiar dependence on electron beam incident azimuth. Looking from the <100> azimuth, one period of oscillation corresponded to a monatomic-layer growth, from the <110> azimuth, on the other hand, the period was doubled and it corresponded to a biatomic-layer growth. The mechanism of RHEED intensity oscillation is usually explained by the surface "rough-smooth transition" that the electron beam reflectivity of a smooth surface is larger than that of a rough surface. This model explained the RHEED intensity oscillation during the GaAs growth very well but it can not explain the peculiar azimuthal dependence of the Si(001) growth.

In order to investigate the mechanism of RHEED intensity oscillations during the Si(001) growth, three RHEED spots were monitored simultaneously using three sets of optical-fiber monitoring system[20]. Figure 10 shows the oscillations taken from three different diffraction spots in the RHEED patterns for the [010] azimuth. The upper inset figures show the reciprocal lattice of Si(001) 2x1+1x2 and the corresponding diffraction pattern schematically :(a) large solid circles, (b) small solid circles, and (c) small open circles represent the bulk diffraction-spot, the 2x1 and the 1x2 reconstruction-related spots, respectively. The phase difference between the two reconstruction-related traces (b) and (c) was almost 180°, indicating that the 2x1 and the 1x2 reconstruction appeared alternately during the growth. Because the surface reconstruction changes from 2x1 to 1x2 by a monatomic-layer growth, one period of oscillation in both traces (b) and (c) corresponded to a biatomic-layer height growth. Oscillation maximum of the specular-beam shown in the trace (a) coincided with the each maximum of both traces (b) and (c) which corresponded to the flat surface after the completion of a monolayer growth. This indicated that the monolayer-mode oscillation of specular-beam observed from the [010] azimuth reflected the surface "rough-smooth transition".

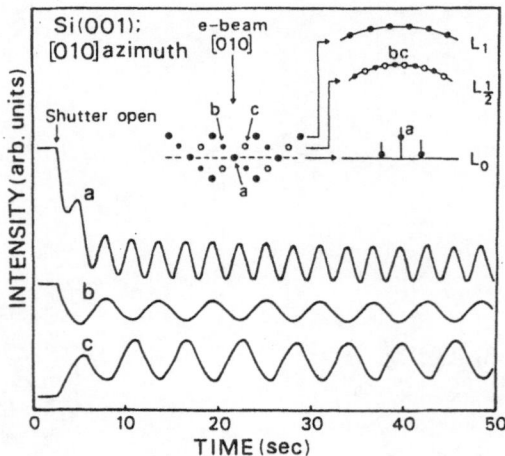

Fig.10. RHEED intensity oscillations of the three different diffraction spots taken in the [010] azimuth pattern simultaneously: (a)specular-beam (b)2x1 reconstruction spot, and (c)1x2 reconstruction spot. Inset figures show a reciprocal lattice of (100)2x1+1x2 and a corresponding diffraction pattern schematically: (a) large solid circles, (b) small solid circles, (c) small open circles represent the bulk diffraction, 2x1 and 1x2 reconstruction-related spots, respectively.

A similar experiment was done for the RHEED intensity oscillation observed from the [1$\bar{1}$0] azimuth as shown in Fig. 11. The anti-phase oscillations of the two reconstruction-related traces (b) and (c) confirmed the alternate growth of the 2x1 and the 1x2 reconstructed surfaces. One period of oscillation of the specular-beam shown in the trace (a) coincided with that of (b) and (c), which corresponded to a biatomic-layer-height growth. This indicated that the bilayer-mode RHEED intensity oscillation observed from the [1$\bar{1}$0] azimuth was related to the change in surface reconstructions.

This peculiar biatomic-layer mode oscillation observed from the <110> azimuth can be explained by the difference in reflectivity between the 2x1 and the 1x2 reconstructed-surfaces. Figure 12 shows the intensity of specular-beams for the variation of electron-beam incident angle observed from the [1$\bar{1}$0] and the [110] azimuths which showed the 1x2 and the 2x1 patterns, respectively. It should be noted that the reflection intensity of both 2x1 and 1x2 reconstruction showed a sharp maximum at the incident angle of about 16mrad, where we usually observed the RHEED intensity oscillation, and that the intensity of the 1x2 reconstruction was twice as large as that of the 2x1 reconstruction. Thus a bright 1x2 surface and a dark 2x1 surface appeared alternately during the growth on a single-domain Si(001) surface. We considered that the alternating growth of the 2x1 and 1x2 reconstruction was the major reason for the bilayer-mode oscillation assuming that the reflectivity difference caused by the "2x1-1x2 transition" was larger than that caused by the "rough-smooth transition". This was the first observation of RHEED intensity oscillation caused by the change in surface reconstructions.

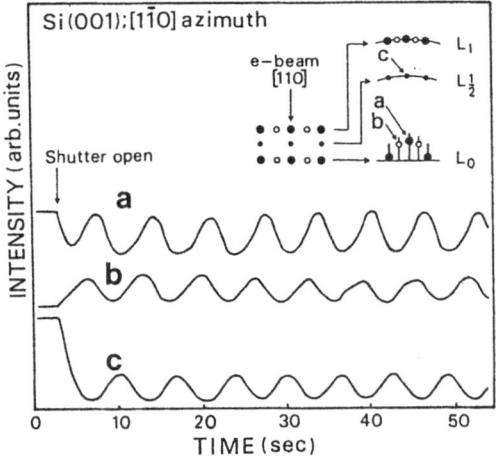

Fig.11. RHEED intensity oscillations of the three different diffraction spots taken in the [1$\bar{1}$0] azimuth pattern simultaneously: (a)specular-beam (b)2x1 reconstruction spot, and (c)1x2 reconstruction spot. Inset figures show a reciprocal lattice of (100)2x1+1x2 and a corresponding diffraction pattern schematically: (a) large solid circles, (b) small open circles, (c) small solid circles represent the bulk diffraction, 2x1 and 1x2 reconstruction-related spots, respectively.

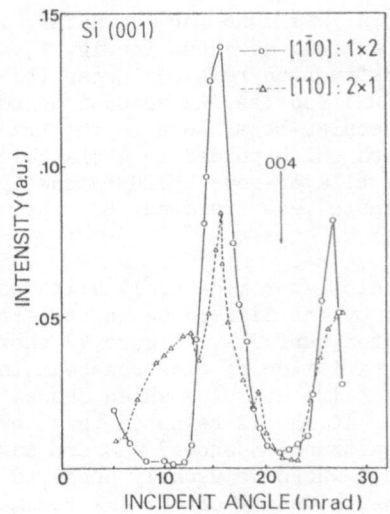

Fig.12. RHEED specular-beam intensity after the annealing versus electron-beam incident angle observed from the [1$\bar{1}$0] azimuth (1x2 reconstruction) and from the [110] azimuth (2x1 reconstruction).

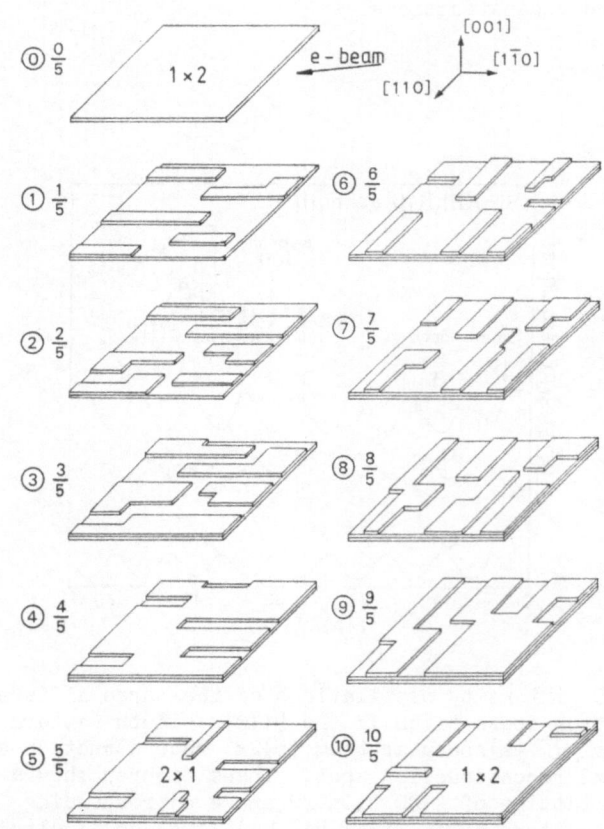

Fig.13. Schematic picture of a Si(001) surface during the layer-by-layer growth on the 2x1 single-domain structure. 1x2 and 2x1 reconstructed-surfaces appear alternately.

A schematic surface morphology during the layer-by-layer growth on the single-domain Si(001) 2x1 structure is illustrated in Fig. 13. On the "bright" surface reconstructed in 1x2 ⓪ , elongated islands of monatomic height with the longest dimension parallel to the [1͞10] direction appeared ① - ④ , because the growth rate of surface steps was assumed to be faster for the direction of dangling-bonds of surface atoms. Finally the surface was almost covered with the "dark" 2x1 reconstruction ④ - ⑤ . Then elongated–islands parallel to the [110] direction appeared on ∙the 2x1 reconstruction ⑤ - ⑨ until the surface was mostly covered with the "bright" 1x2 reconstruction again ⑩ . Thus the alternating reconstructions of 1x2 and 2x1 appeared during the growth on the single-domain Si(001), which caused the bilayer-mode RHEED intensity oscillation observed from the [1͞10] azimuth. From these discussions it was concluded that the formation of single-domain structure was essential for observing the stable RHEED intensity oscillation.

According to the mechanism of RHEED intensity oscillation described above, the intensity maximum of the specular-beam shown in the trace (a) of Fig. 11 should coincide in phase with that of the 1x2 reconstruction spot shown in the trace (c) but in our experiment the phase of trace (a) advanced to that of trace (c). Calculations considering multiple scattering[26-28] showed that the electron-beam reflectivity of stepped surfaces strongly depended on the beam incident azimuth with respect to the step direction. The intensity of specular-beam whose incident azimuth was parallel to the step edges was lower than that of perpendicular to the step edges. As this theory neglected the effect of surface reconstructions, we applied it to modify our model. Because steps of 2x1 islands grown on a 1x2 structure were mostly parallel to the beam-incident azimuth, as shown in Fig. 13, the intensity of a specular-beam decreased rapidly by the effects of both reconstructions and steps. Thus the phase-advance of the specular-beam to the 1x2 reconstruction spot was explained by the effect of alternating step directions.

Application of the single-domain Si(001) 2x1 structure to the hetero-epitaxy of GaAs

The potential benefits of GaAs growth on Si substrates has precipitated recent surge of activity in this area[10,11]. There are two main difficulties associated with the growth of GaAs on Si. The first is that of lattice mismatch, GaAs has a 4% larger lattice constant than that of Si. The second difficulty is the problem of antiphase-disorder. The zincblende lattice is composed of two interpenetrating FCC sublattices, one

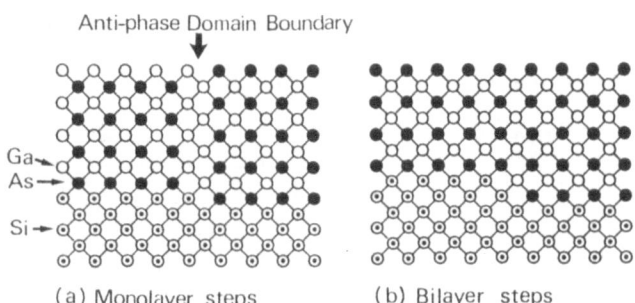

(a) Monolayer steps   (b) Bilayer steps

Fig.14. Schematic cross-sections of GaAs/Si(001) structures. (a) an antiphase-domain-boundary exists at a monolayer-height step of Si surface, (b) while it does not appear at a bilayer-height step.

for the cation and one for the anion for polar semiconductors. Since Si is a nonpolar semiconductor, uniform ordering of the polar semiconductor sublattices is not ensured. Upon crossing an antiphase-boundary, the sublattices are interchanged. An antiphase-disorder appears at a monolayer step of Si surface as shown schematically in Fig. 14(a), while it does not appear in principle at a bilayer step as shown in Fig. 14(b). Thus a single-domain Si(001) 2x1 structure which is mostly covered with biatomic-height terraces avoids antiphase-disorder. Many works have showed that antiphase-disorder-free GaAs was grown only on Si(001) tilted toward the <110> azimuth but unsuccessful on well-oriented Si(001)[29,30]. The high temperature annealing which forms a stable single-domain structure on well-oriented Si(001) surface will make it possible to obtain antiphase-disorder-free GaAs grown on well-oriented Si(001).

RHEED INTENSITY OSCILLATION DURING THE Ge_xSi_{1-x} STRAINED-LAYER EPITAXY ON Si SUBSTRATES

Growth of Ge_xSi_{1-x} on Si(001) substrates

Ge_xSi_{1-x} is a promising material for the fabrication of heterojunction devices of Si-based structure. Although the lattice mismatch between Si and Ge is relatively large ($\sim 4\%$), recent progress in Si MBE made it possible to achieve the pseudomorphic growth of Ge_xSi_{1-x}/Si on Si substrates up to 1 μm in thickness[1]. However, to the authors' knowledge, no reports had appeared on the RHEED intensity oscillation during Ge_xSi_{1-x} strained-layer heteroepitaxy prior to our recent work[21].

The Ge_xSi_{1-x} films were grown at 450°C on the Si (001) 2x1 single-domain surface. Films with various Ge mole fractions were grown by controlling the evaporation rate of a Si source and that of a Ge source. Typical RHEED intensity oscillations of the specular-beam observed from the [1$\bar{1}$0] azimuth are shown in Fig. 15. The growth of Ge_xSi_{1-x} was preceded by the Si growth for several oscillations to determine the alloy composition. Because the sticking coefficients of Si and Ge under this growth condition were almost unity, the mole fraction x of Ge in Ge_xSi_{1-x} layer was easily obtained from oscillation frequencies as follows,

$$x = [f(GeSi)-f(Si)]/f(GeSi) \qquad (1)$$

where f(Si) and f(GeSi) denote the RHEED intensity oscillation frequencies during Si layer and Ge_xSi_{1-x} layer growth keeping the Si evaporation rate constant, respectively. While the oscillation amplitude and the average intensity were almost unchanged during the Si growth, they began to decrease after the Ge_xSi_{1-x} growth started. The decrease took place more rapidly with the increase of Ge mole fraction in the film. In the growth of Ge on Si at 450°C, i.e. x=1.0, only two periods of oscillation was observed as shown in the trace (a) of Fig. 16.

These damping behavior of oscillations were explained by the clustering during the Ge_xSi_{1-x} growth as follows. When the Ge mole fraction was small, the RHEED pattern during the Ge_xSi_{1-x} growth was almost the same as that of during the Si growth. However, with the increase of the Ge mole fraction, the RHEED pattern changed rapidly from the initial streaky one to the final spotty one as shown in Fig. 16. It was concluded from these observations of RHEED patterns that the decrease of the oscillation amplitude during the Ge_xSi_{1-x} growth was a result of a change in growth mode from a two-dimensional layer-by-layer one to a three-dimensional one which was caused by the clustering of Ge_xSi_{1-x}. The reason is not clear why the three-dimensional growth was promoted by the increase of Ge mole fraction. Considering the fact, however, that more than tens periods of

236

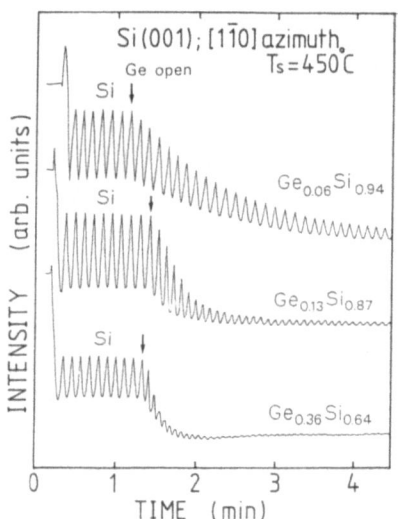

Fig.15. RHEED intensity oscillations of the specular-beam during the Ge_xSi_{1-x} growth on Si(001) at $450^\circ C$ taken from the [1Ī0] azimuth for various Ge mole fractions. The Si growth preceded the Ge_xSi_{1-x} growth to determine alloy compositions.

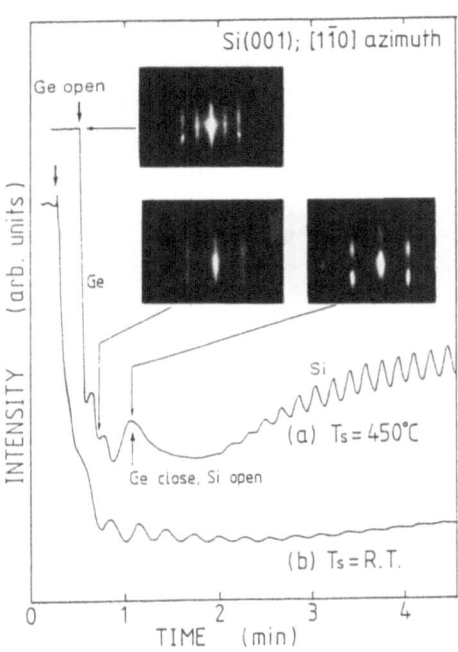

Fig.16. RHEED intensity oscillations during the Ge growth on Si(001) (a) at $450^\circ C$ and (b) at the room temperature. RHEED patterns during the growth at $450^\circ C$ are also shown.

oscillation was observed during the homoepitaxial growth of Ge at $350^{\circ}C$[31], in contrast with the poor oscillatory property of the Ge/Si strained-layer growth, it could not say that Ge itself had strong tendency to cluster. These results suggested that the motive force to the clustering had some relation to the lattice mismatch between Si substrates and Ge_xSi_{1-x} epilayers.

In the growth of Ge on Si at the room temperature, however, more than 10 periods of oscillation was observed clearly as shown in the trace (b) of Fig. 16. This indicated that a lower substrate temperature which suppressed the surface migration was preferable for the layer-by-layer growth of Ge, because Ge atoms on a Si surface had relatively large cohesive energy which would cause a three-dimensional growth. Asai et al.[32] examined the growth of Ge on Si(001) using Auger electron spectroscopy and showed that Ge grew in layer-by-layer fashion up to three layers at $350^{\circ}C$ and more than six layers at the room temperature. Their data were almost equal to our results obtained by the observation of RHEED intensity oscillations.

Figure 17 summarizes experimental data on the critical Ge_xSi_{1-x} layer thickness for various Ge mole fractions at which the RHEED intensity oscillations almost faded away. Works on the critical layer thickness of strained-layer growth obtained from both calculations[33-35] and experiments[1] are also shown in the figure. Although what we measured was the

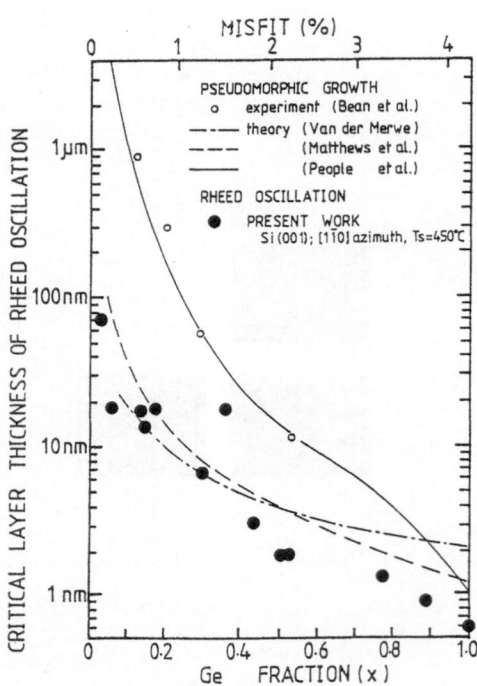

Fig. 17. Solid circles are summary of the experimental data on the thickness of Ge_xSi_{1-x} layer grown at $450^{\circ}C$ where RHEED intensity oscillation almost fades away. Data on the critical layer thickness for the pseudomorphic growth are also shown: open circles - Bean et al. (experiment)[1]; solid line - People et al. (theory)[35]; dashed line - Matthews et al. (theory)[34]; dot-dashed line - Van der Merwe (theory)[33].

"critical layer thickness" of RHEED intensity oscillations, our results were roughly equal to works on the "critical layer thickness" of pseudo-morphic growth. The relationship between the two kinds of critical layer thicknesses is not clear yet. However, considering that the three-dimensional growth took over the two-dimensional layer-by-layer growth at the critical layer thickness of RHEED intensity oscillations, the coincidence of two kinds of critical layer thicknesses suggested that the three-dimensional growth originated in the defect formation in strained-layer.

Oscillation recovery during the Si overgrowth on Ge_xSi_{1-x} layer

Another interesting feature of the growth of Ge_xSi_{1-x}/Si hetero-structures was the smoothening of surface during Si overgrowth. In the trace (a) of Fig. 16 we stopped the Ge growth and opened the Si source shutter after the RHEED pattern changed to spotty one. Although no oscillations were observed at the beginning of the Si overgrowth, a weak oscillation appeared and gained in its amplitude gradually with the increase of Si-overgrown layer thickness. We can consider from this result that the atomically rough surface through the three-dimensional growth of Ge was smoothened by Si atoms during the initial stage of Si overgrowth and then the RHEED intensity oscillation was observed on the smoothened Si-over-grown surface. This interpretation is illustrated schematically in Fig. 18. With the help of the oscillation recovery, we could apply the "phase-locking technique" to the growth of Ge_xSi_{1-x}/Si strained-layer superlattices as described later.

Growth of Ge_xSi_{1-x} on Si(111) substrates

Figure 19 shows the RHEED intensity oscillations during the strained-layer growth of Ge_xSi_{1-x} on Si(111) substrates. The oscillation amplitude increased at the first stage of the Ge_xSi_{1-x} growth before it began to decrease rapidly. The mechanism of oscillation enhancement is not clear yet. However, it could be thought that incorporated Ge atoms acted as nuclei of the two-dimensional growth on a Si(111) surface where poorer oscillations were observed during the Si homoepitaxy compared with that of on Si(001).

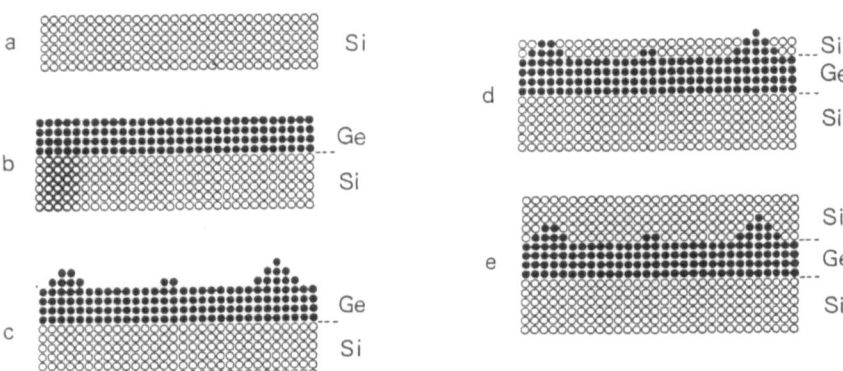

Fig.18. Schematic illustration of the Ge growth on Si(001) at 450°C derived from the RHEED intensity oscillation shown in Fig. 16(a).

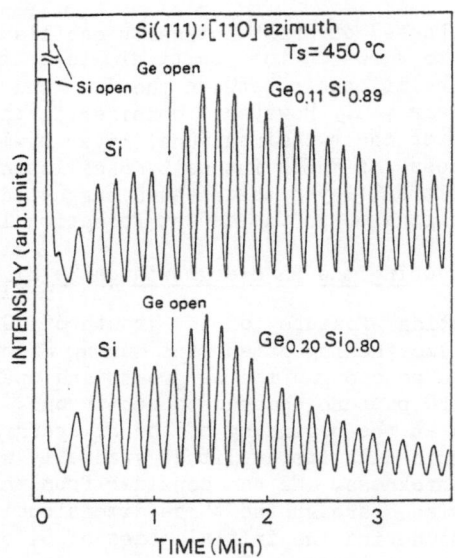

Fig.19. RHEED intensity oscillations of the specular-beam during Ge_xSi_{1-x} growth on Si(111) at $450^\circ C$ taken from the [110] azimuth for various Ge mole fractions. The Si growth preceded the Ge_xSi_{1-x} growth.

Growth of Ge_xSi_{1-x}/Si strained-layer superlattice on Si(001) substrates

We fabricated Ge_xSi_{1-x}/Si strained-layer superlattices on Si(001) substrates monitoring the RHEED intensity oscillations. Figure 20 shows the RHEED intensity oscillations during the growth of $Ge_{0.25}Si_{0.75}$/Si strained-layer superlattices at $450^\circ C$. The Ge source shutter was opened and closed alternately at every fifth peak of oscillation, i.e., the strained-layer superlattice was composed of 10 atomic layers of Si and 10 atomic layers of $Ge_{0.25}Si_{0.75}$ since a period of oscillation corresponded to a bilayer-growth.

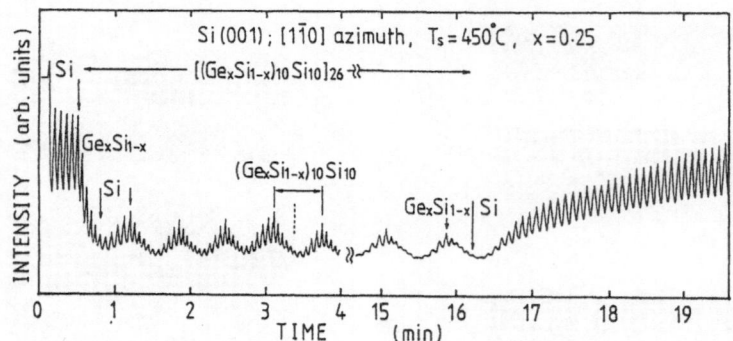

Fig.20. RHEED intensity oscillation during the growth of $(Ge_{0.25}Si_{0.75})_{10}$/$Si_{10}$ strained-layer superlattice structure on Si(001) at $450^\circ C$. Using "phase-locked epitaxy" technique, 26 periods of strained-layer superlattice was grown without any growth interruption.

Though it is needless to say that the precise controllability of the layer thickness is one of advantages of the oscillation-monitoring-method, more important point is that we can control the interface roughness using the "phase-locked epitaxy" technique. Since the flat surface after the completion of a layer growth showed peaks in the oscillation, the Ge_xSi_{1-x}/Si interfaces in the strained-layer superlattice shown in Fig. 20 were controlled to be smooth.

As described previously the oscillation amplitude during the Ge_xSi_{1-x} growth decreased rapidly but recovered to some extent during the Si growth. This surface smoothening by the Si-overgrown layers helped the long observation of the RHEED intensity oscillation during the growth of strained-layer superlattices. We could obtain 26 periods of $(Ge_{0.25}Si_{0.75})_{10}/Si_{10}$ strained-layer superlattice, i.e. 520 atomic layers (70nm), monitoring the RHEED intensity oscillation as shown in Fig. 20. The total thickness of the $Ge_{0.25}Si_{0.75}$ layers (35nm) was much larger than the critical thickness of RHEED intensity oscillation (~8nm). Figure 21 shows the Ge mole fraction dependence of RHEED intensity oscillation during the growth of Ge_xSi_{1-x}/Si strained-layer superlattices. With the increase of the Ge mole fraction, the oscillation amplitude damped rapidly during the Ge_xSi_{1-x} growth but recovered slowly during the Si growth.

Observation of Ge_xSi_{1-x}/Si strained-layer interface

The Ge_xSi_{1-x}/Si interfaces in a strained-layer superlattice were examined by a JEM-4000FX high-resolution transmission electron microscope (HRTEM) operating at 400keV. The specimen was cleaved and thinned parallel to (110) plane. The RHEED intensity oscillation monitored during the growth is illustrated in Fig. 22. The 20 monolayers of $Ge_{0.26}Si_{0.74}$

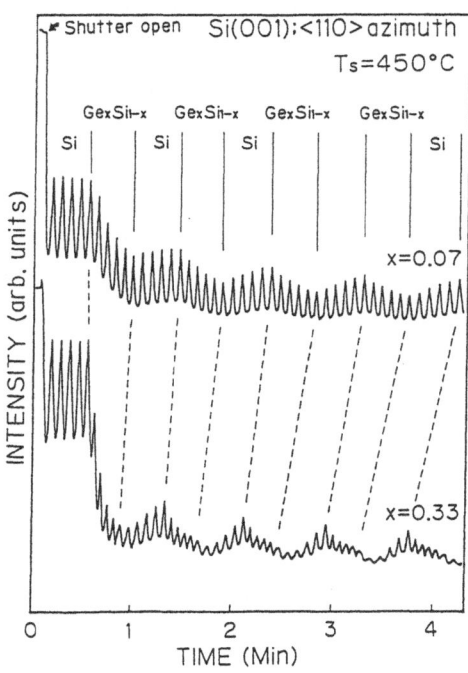

Fig.21. RHEED intensity oscillations during the growth of Ge_xSi_{1-x}/Si strained-layer superlattices on Si(001) at $450°C$ for various Ge mole fractions.

(2.7nm) and the 40 monolayers of Si (5.4nm) were deposited alternately. In this case because the surface of Si substrate had not been prepared to form a perfect single-domain structure, the oscillation recovery during the Si layer growth was insufficient and we could not use the "phase-locking technique" more than four periods of strained-layer superlattice layers. The oscillation indicated that the interface of the Ge_xSi_{1-x} on Si was smoother than that of the Si on Ge_xSi_{1-x} and that the both interfaces got rougher with the progress in growth.[1] This tendency is also shown in

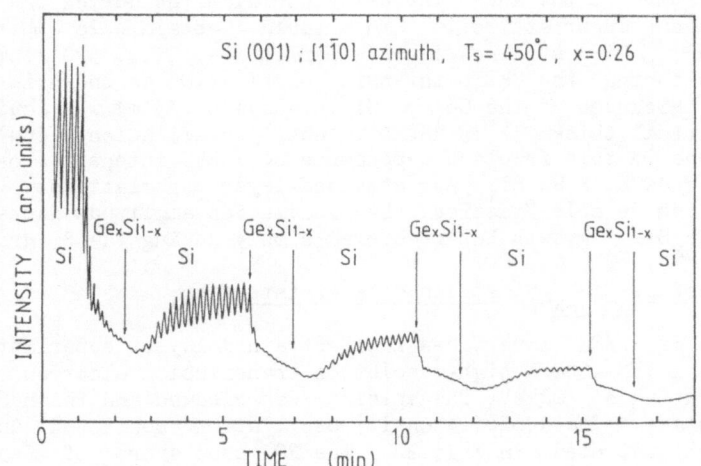

Fig.22. RHEED intensity oscillation during the growth of $(Ge_{0.26}Si_{0.74})_{20}/Si_{40}$ strained-layer superlattice structure on Si(001) at 450°C for a HRTEM observation.

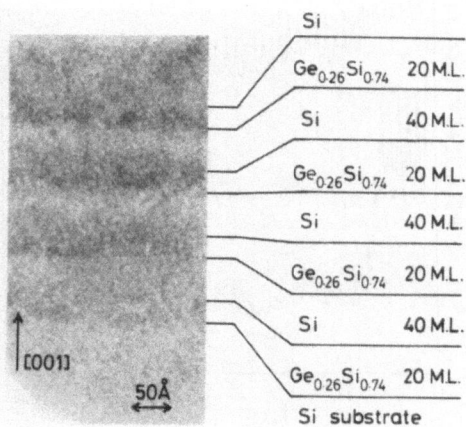

Fig.23. A HRTEM image of the $[(Ge_{0.26}Si_{0.74})_{20}/Si_{40}]_4$ strained-layer superlattice structure whose RHEED intensity oscillation during the growth is shown in Fig. 22.

the cross-sectional HRTEM image of Fig. 23 where a defect-free four period strained-layer superlattice is shown. Although the Ge_xSi_{1-x} and Si layers presented a weak contrast to each other, the interface of the Ge_xSi_{1-x} on Si was clearer than that of the Si on Ge_xSi_{1-x}. On careful examination of the four Ge_xSi_{1-x} on Si interfaces, it was found that the second interface was clearly smoother than the third and fourth interfaces. These results were in accordance with that obtained from the RHEED intensity oscillation. By counting the layer thickness from the HRTEM image it was proved directly that the one period of oscillation shown in Fig. 22 corresponded to the growth of biatomic-layer.

The RHEED intensity oscillation and the HRTEM image showed the difference in roughness between the Ge_xSi_{1-x} on Si interface and the Si on Ge_xSi_{1-x} interface, which suggests that it is necessary to use the Ge_xSi_{1-x} on Si interface to improve electrical properties of a modulation-doped structure. Similar results were obtained on the strained-layer heterostructure of Ge_xSi_{1-x}/Si(111)[36]. A growth interruption during the growth of GaAs/AlAs superlattices made the heterointerface smooth and the RHEED intensity oscillation recover. In the case of Ge_xSi_{1-x}/Si strained-layer heterostructure, however, the RHEED intensity did not recover after the growth interruption at $450^\circ C$ indicating that the surface was not made smooth. The search for a growth procedure which will make the Ge_xSi_{1-x} surface smooth and at the same time prevent the degradation of crystal quality and the diffusion of Ge atoms at the Ge_xSi_{1-x}/Si interface is left for the future tasks.

CONCLUSION

RHEED and its intensity oscillations were applied to the in situ monitoring of Si MBE. A high-temperature annealing was found to change a surface into the single-domain Si(001) 2x1 structure, in other words monatomic-layer steps on a Si(001) surface changed into biatomic-layer steps, which was essentially necessary for the observation of a stable RHEED intensity oscillation. The alternating surface reconstructions of 2x1 and 1x2 were observed during the MBE growth. The formation of a single-domain Si(001) 2x1 surface is also important to avoid the antiphase-domains in GaAs layers overgrown on Si substrates. However, the annealing condition as high as $1000^\circ C$ will spoil an important advantage in Si MBE of the low temperature growth. Investigations of annealing conditions, such as a laser annealing, to obtain the single-domain structure without affecting the built-in structures are left for the future tasks. The RHEED intensity oscillations were also observed during the strained-layer heteroepitaxy of Ge_xSi_{1-x} on Si(001). The duration of oscillation had a similar dependence on the Ge mole fraction to the critical thickness of the pseudomorphic growth. The roughened Ge_xSi_{1-x} surface was found to be smoothened during the Si overlayer growth. Ge_xSi_{1-x}/Si strained-layer superlattice was grown monitoring the RHEED intensity oscillation without any growth interruption. We confirmed that the RHEED intensity oscillation opened the door to the atomic-order control of the Si and Ge_xSi_{1-x}/Si MBE.

ACKNOWLEDGMENT

The authors would like to thank Drs. N.Hashizume, K.Ohta, T.Nakagawa and T.Kojima of their institute and Prof. T.Kawamura and prof. K.Tomizawa for their helpful discussions. Thanks are due to Dr. T.Tsurushima and Prof. S.Gonda for their continuous encouragements.

REFERENCES

1. J.C.Bean, L.C.Feldman, A.T.Fiory, S.Nakahara and I.K.Robinson, Ge_x-Si_{1-x}/Si strained-layer superlattice grown by molecular beam epitaxy, J. Vac. Sci. & Technol. A2: 436 (1984).

2. H.Jorke and H.J.Herzog, Mobility enhancement in modulation-doped $Si_{1-x}Ge_x$ superlattice grown by molecular beam epitaxy, J. Electrochem. Soc. 133: 998 (1986).

3. R.People, J.C.Bean and V.D.Lang, Modulation doping in Ge(x)Si(1-x)/Si strained layer heterostructures: effect of alloy layer thickness, doping set back, and cladding layer dopant concentration, J. Vac. Sci. & Technol. A3: 846 (1985).

4. H.Daembkes, H.J.Herzog, H.Jorke, H.Kibbel and E.Kaspar, The n-channel SiGe/Si modulation-doped field-effect transistor, IEEE trans. Electron Devices ED-33: 633 (1986).

5. T.P.Pearsall and J.C.Bean, Enhancement- and depletion-mode p-channel Ge_xSi_{1-x} modulation-doped FET's, IEEE Electron Device Lett. EDL-7: 308 (1986).

6. S.Saito, H.Ishiwara and S.Furukawa, Double heteroepitaxy in the $Si(111)/CoSi_2$/Si structure, Appl. Phys. Lett. 37: 203 (1980).

7. R.T.Tung, J.M.Gibson, and A.F.J.Levi, Growth of strained-layer semiconductor-metal-semiconductor heterostructures, Appl. Phys. Lett. 48: 1264 (1986).

8. H.Ishiwara and T.Asano, Silicon/insulator heteroepitaxial structures formed by vacuum deposition of CaF_2 and Si, Appl. Phys. Lett. 40: 66 (1982).

9. A.Munoz-Yague and C.Fontaine, Molecular beam epitaxy of insulating fluoride-semiconductor heterostructures, Surf. Sci. 168: 626 (1986).

10. S.Nishi, H.Inomata, M.Akiyama and K.Kaminishi, Growth of single domain GaAs on 2-inch Si(100) substrate by molecular beam epitaxy, Jpn. J. Appl. Phys. 24: L391 (1985).

11. R.Fischer and H.Morkoc, III-V semiconductors on Si substrates: new direction for heterojunction electronics, Solid State Electron. 29: 269 (1986).

12. J.J.Harris, B.A.Joyce and P.J.Dobson, Oscillations in the surface structure of Sn-doped GaAs during growth by MBE, Surf. Sci. 103: L90 (1981).

13. C.E.C.Wood, RED intensity oscillations during MBE of GaAs, Surf. Sci. 108: L441 (1981).

14. J.H.Neave, B.A.Joyce, P.J.Dobson and N.Norton, Dynamics of film growth of GaAs by MBE from Rheed observation, Appl. Phys. A31: 1 (1983).

15. J.M.Van Hove, C.S.Lent, P.R.Pukite and P.I.Cohen, Damped oscillation in reflection high-energy electron diffraction during GaAs MBE, J. Vac. Sci. & Technol. B1: 741 (1983).

16. T.Sakamoto, H.Funabashi, K.Ohta, T.Nakagawa, N.J.Kawai, T.Kojima and Y.Bando, Well defined superlattice structures made by phase-locked epitaxy using RHEED intensity oscillation, Superlattices and Microstructures 1: 347 (1985).

17. T.Sakamoto, H.Funabashi, K.Ohta, T.Nakagawa, N.J.Kawai and T.Kojima, Phase-locked epitaxy using RHEED intensity oscillation, Jpn. J. Appl. Phys. 23: L657 (1984).

18. T.Sakamoto, N.J.Kawai, T.Nakagawa, K.Ohta and T.Kojima, Intensity oscillations of reflection high-energy electron diffraction during silicon molecular beam epitaxial growth, Appl. Phys. Lett. 47: 617 (1985).

19. T.Sakamoto and G.Hashiguchi, Si(001)-2x1 single-domain structure obtained by high temperature annealing, Jpn. J. Appl. Phys. 25: L78 (1986).

20. T.Sakamoto, T.Kawamura and G.Hashiguchi, Observation of alternating reconstructions of silicon(001) 2x1 and 1x2 using reflection high-energy electron diffraction during molecular beam epitaxy, Appl. Phys. Lett. 48: 1612 (1986).

21. K.Sakamoto, T.Sakamoto, S.Nagao, G.Hashiguchi, K.Kuniyoshi and Y.Bando, Reflection high-energy electron diffraction intensity oscillations during Ge_xSi_{1-x} MBE growth on Si(001) substrates, submitted to Jpn. J. Appl. Phys.

22. A.Ishizuka and Y.Shiraki, Low temperature surface cleaning of silicon and its application to silicon MBE, J. Electrochem. Soc. 133: 666 (1986).

23. K.Kugimiya, Y.Shirafuji and N.Matsuo, Si-beam radiation cleaning in molecular-beam epitaxy, Jpn. J. Appl. Phys. 24: 564 (1985).

24. R.Kaplan, LEED study of the stepped surface of vicinal Si(001), Surf. Sci. 93: 145 (1980).

25. N.Aizaki and T.Tatsumi, In situ RHEED observation of selective diminution at Si(001)-2x1 superlattice spots during MBE, Surf. Sci. 174: 658 (1986).

26. T.Kawamura, P.A.Maksym and Iijima, Calculation of RHEED intensities from stepped surfaces, Surf. Sci. 148: L671 (1984).

27. T.Kawamura and P.A.Maksym, RHEED from stepped surfaces and its relation to RHEED intensity oscillations observed during MBE, Surf. Sci. 161: 12 (1985).

28. T.Kawamura, T.Sakamoto and K.Ohta, Origin of azimuthal effect of RHEED intensity oscillations observed during MBE, Surf. Sci. 171: L409 (1986).

29. H.Kroemer, K.J.Polasko and S.C.Wright, On the (110) orientation as the preferred orientation for the molecular beam epitaxial growth of GaAs on Ge, GaP on Si and similar zincblende-on-diamond systems, Appl. Phys. Lett. 36: 763 (1980).

30. N.Otsuka, C.Choi, L.A.Kolodziejski, R.L.Gunshor, R.Fischer, C.K.Peng, H.Morkoc, Y.Nakamura and S.Nagakura, Study of heteroepitaxial interfaces by atomic resolution electron microscopy, J. Vac. Sci. & Technol. 134: 896 (1986).

31. T.Kojima, K.Ohta, T.Sakamoto and T.Nakagawa: Preprints of the 33rd Spring Meeting of Japan Society of Applied Physics and of the Related Societies, Chiba, April, 1986, 4p-V-12.

32. M.Asai, H.Ueba and C.Tatsuyama, Heteroepitaxial growth of Ge films on the Si(100)-2x1 surface, J. Appl. Phys 58: 2577(1985).

33. J.H.Van der Merwe, Crystal interfaces. part II. finite overgrowth, J. Appl. Phys 34: 123 (1962).

34. J.W.Matthews and A.E.Blakeslee, Defects in epitaxial multilayers I. misfit dislocations, J. Cryst. Growth 27: 118 (1974).

35. R.People and J.C.Bean, Calculation of critical layer thickness versus lattice mismatch for Ge_xSi_{1-x}/Si strained layer heterostructures, Appl. Phys. Lett. 47: 322 (1985), [Erratum; 49: 229 (1986)].

36. T.Tatsumi and N.Aizaki: Preprints of the 33rd Spring Meeting of Japan Society of Applied Physics and of the Related Societies, Chiba, April, 1986, 4p-V-14.

RHEED INTENSITY OSCILLATIONS AND THE EPITAXIAL GROWTH OF QUASI-2D MAGNETIC

SEMICONDUCTORS

L. A. Kolodziejski and R. L. Gunshor

School of Electrical Engineering
Purdue University
West Lafayette, Indiana 47907

A. V. Nurmikko

Division of Engineering
Brown University
Providence, Rhode Island 02912

N. Otsuka

Materials Engineering
Purdue University
West Lafayette, Indiana 47907

ABSTRACT

Reflection high energy electron diffraction (RHEED) intensity oscillations have been observed in the II-VI compound ZnSe and in the magnetic semiconductor MnSe. These RHEED intensity oscillations provided monolayer resolution in the growth of superlattice structures designed to explore the effects of reduced dimensionality on the magnetic ordering of MnSe. Using molecular beam epitaxy, the heretofore hypothetical zincblende MnSe has been grown. Although "thick" layers of MnSe are antiferromagnetic, thin layers of one, three, and four monolayers were found to exhibit paramagnetic behavior. The growth, use of the RHEED intensity oscillations, and magneto-optical characterization of these unique, highly lattice-mismatched structures is discussed. In addition nucleation characteristics, obtained via RHEED intensity oscillations, are described at a II-VI compound/III-V compound interface for the ZnSe/GaAs heterojunction.

INTRODUCTION

Utilizing the non-equilibrium growth technique of molecular beam epitaxy (MBE), new opportunities exist in the materials development of the family of II-VI compounds and associated diluted magnetic semiconductors. Although interest in the II-VI compounds for potential opto-electronic device applications has existed for some time, realization of the potential for these materials has been hindered by a propensity for defect generation combined with problems associated with dopant incorporation. Most of the previous experience with II-VI materials has been based on equilibrium growth techniques. As an example of the possibilities afforded by non-equilibrium growth methods, heretofore "hypothetical" zincblende MnSe has been grown by MBE. This magnetic semiconductor has been incorporated in strained-layer superlattice structures and grown as strain-relieved epitaxial layers. Of particular emphasis in this paper is the use of RHEED intensity oscillations for the fabrication of a series of superlattice structures configured to investigate the effect of reduced dimensionality on the magnetic ordering of MnSe. It was found that a superlattice structure containing ultrathin layers of MnSe (on the order of one unit cell) exhibited substantial paramagnetic behavior even though this material was expected to be antiferromagnetic. Initial experiments involving the controlled variation of the individual MnSe layer thickness indicated that the MnSe/ZnSe material system, combined with the flexibility of the MBE growth technique, provided a novel opportunity to study magnetic effects in systems of reduced dimensionality.

The RHEED intensity oscillations were originally employed as a necessary tool to achieve a controlled approach to quasi-2D magnetic layers. In the course of those experiments it became apparent that the RHEED intensity variations, in their variety and repeatability, provided a potentially rich plethora of information relative to the kinetics of the MBE growth process. The growth parameters of the II-VI materials differed substantially from that of the better understood GaAs. For the II-VI compounds the use of relatively low substrate temperatures and nearly identical constituent fluxes suggested that the migration of both cation and anion species are important to the growth dynamics. The RHEED intensity oscillations described herein are sufficiently different from those observed in other material systems that further work, combined with modified Monte Carlo simulations[1] and associated RHEED diffraction calculations, are necessary to fully interpret the experimental data.

DILUTED MAGNETIC SEMICONDUCTORS

Diluted magnetic semiconductors (DMS) are II-VI compounds in which some fraction of the host cation sites are occupied by one of the transition metal elements such as Mn. Interesting and potentially useful magnetic properties result from the exchange interaction between the magnetic moments of the 3-d electrons of Mn and the extended states of the II-VI semiconductor. The entire family of II-VI compounds is available for incorporation of the magnetic ion, thus providing access to wavelengths extending from far infrared to the near ultraviolet. These various ternary compounds span a wide range of lattice parameters, which is an important factor as one considers opportunities for new heteroepitaxial structures. Recent work[2,3] has demonstrated the feasibility for the epitaxial growth of diluted magnetic semiconductors by the technique of molecular beam epitaxy. Of more significance was the demonstrated growth of multiple quantum well and superlattice structures in the $(Cd,Mn)Te$[4-10] and the $(Zn,Mn)Se$[3,9,10] material systems.

METASTABLE ZINCBLENDE MnSe

The molecular beam epitaxial growth of the pseudo-binary material system ZnSe-MnSe has resulted in the investigation of the metastable zincblende crystal structure over a large range of alloy fractions which are unavailable by conventional bulk equilibrium growth techniques. Thick ($\sim$1-3 μm) epilayers of zincblende $Zn_{1-x}Mn_xSe$ have been grown by MBE over the entire ($0 < x \leq 0.66$) composition range, whereas bulk crystals exist with pure zincblende crystal structure only up to $x < 0.10$[11]. Extrapolation of the lattice parameter and bandgap data as a function of the Mn content provides a prediction of these material properties for the hypothetical zincblende MnSe (see Fig. 1 and 2). From these data obtained on zincblende epilayers of $(Zn,Mn)Se$, zincblende MnSe is

248

predicted to have a lattice constant of 5.93Å and a bandgap value of 3.4 eV. Using the non-equilibrium growth technique of MBE, we have grown the heretofore "hypothetical" zincblende semiconductor MnSe.

Three separate effusion cells containing elemental Zn, Mn, and Se were used to grow all of the structures to be described. A selenium-to-zinc flux ratio of approximately unity was used for the growth of ZnSe, whereas a selenium-to-manganese flux ratio of approximately three was used for the growth of MnSe. The Se flux was measured to be 3.4×10^{14} molecules/cm^2·sec with a quartz crystal monitor placed beside the substrate position. The substrate temperature of 400°C was calibrated through the use of the observance of two eutectic phase changes for Au on Ge (356°C) and Al on Si (577°C). The above combination of fluxes and substrate temperatures provided a growth rate of approximately 1 to 1.4 Å/sec for the two materials. In all cases the superlattice structures were separated from the GaAs (100) substrate by a ZnSe buffer layer approximately 1.5 μm in thickness.

Reflection high energy electron diffraction (RHEED) provided the first observation of the zincblende crystal structure for MnSe. As the Zn shutter was closed and the Mn shutter was opened to begin the growth of MnSe on ZnSe, a

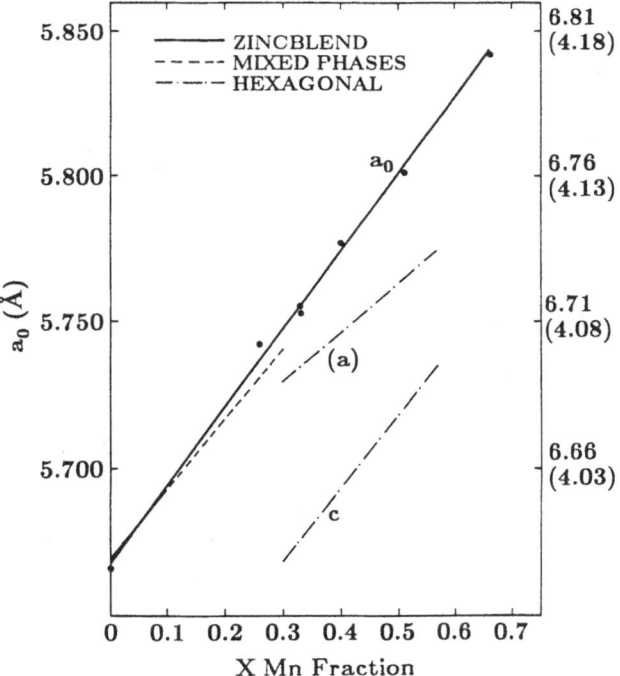

Fig. 1. Lattice parameter (a_0) as a function of the Mn mole fraction (x) for zincblende $Zn_{1-x}Mn_xSe$ epilayers. The broken lines represent data obtained on bulk crystals[10] which contain mixed phases of zincblende and hexagonal crystal structure.

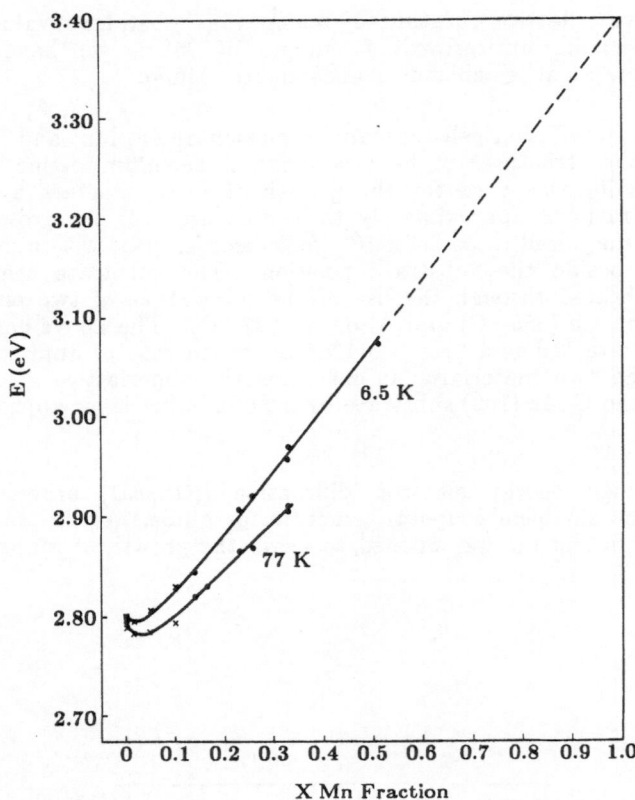

Fig. 2. Near-bandgap energy (E_g) variation versus the Mn content (x) for $Zn_{1-x}Mn_xSe$ epilayers. Data are shown at 77 K and 6.5 K with the low temperature data extrapolated to yield a predicted bandgap value of 3.4 eV for zincblende MnSe.

virtually instantaneous increase in the intensity of the two-fold reconstruction streaks was observed[12]. The RHEED pattern shows a high degree of ordering as demonstrated by the presence of Kikuchi lines and, with the exception of somewhat more intense reconstruction lines, is quite similar to the pattern observed from ZnSe. In all cases the zincblende crystal structure was confirmed by examination of cross-sectional specimens by transmission electron microscopy (TEM). Electron diffraction patterns corresponded only to zincblende phases with no indication of the wurtzite or rock-salt crystal structure. The metastable zincblende MnSe has been· grown in strained-layer superlattice structures with thicknesses ranging from 3 to 32 Å. When grown as a "thick" epilayer (400 Å) the zincblende crystal structure remains in the presence of strain-relieving interfacial misfit dislocations.

RHEED INTENSITY OSCILLATIONS

The zincblende MnSe served as the barrier material when combined with ZnSe in superlattice structures. The ability to tailor the thickness of, and spacing between layers of the magnetic semiconductor provided an opportunity to study magnetic ordering as the dimension varies from 3D to quasi-2D. It was crucial in a study of this type that the MnSe layer thickness be controlled to one monolayer. Although RHEED intensity oscillations have not been previously reported in II-VI compounds, they were employed during the superlattice fabrication to obtain the requisite one monolayer resolution. The behavior of RHEED oscillations in various III-V compounds, group IV semiconductors, and metals are discussed elsewhere in this volume.

The RHEED intensity oscillations to be described were all obtained through monitoring the intensity variation of the specular spot. The 10 kV electron beam, operating at an emission current of 1 mA, was incident on the sample with $\theta = 1^\circ$ in the off-Bragg condition. Both [100] and [110] azimuths have been studied. The intensity of the specular spot was detected by an S-1 photocathode via an optical fiber containing a light collecting lens with a 0.5 mm aperture.

The RHEED intensity oscillations were observed during the homoepitaxial growth of ZnSe on ZnSe and during the lattice mismatched heteroepitaxial growth of ZnSe on MnSe (4.7% mismatch) and ZnSe on GaAs (0.25 % mismatch). One period of oscillation was found to be equivalent to one monolayer of growth as confirmed by transmission electron microscopy. It was interesting to compare the homoepitaxy of ZnSe with the growth of ZnSe on MnSe under identical growth conditions. In the case of homoepitaxy (Fig. 3), six or seven oscillation periods are observed after interruption and re-initiation of growth. Whether the growth of ZnSe was interrupted in the presence of either a Zn or Se flux, the specular intensity was seen to decrease. Alternatively, closing both Zn and Se shutters resulted in an increase in the specular spot intensity, suggesting recovery had occurred. In contrast to homoepitaxy, the nucleation of ZnSe on MnSe (Fig. 4) exhibited up to 30 oscillation periods of enhanced amplitude. In this case the growth of MnSe was terminated by closing the Mn shutter prior to the nucleation of ZnSe.

Nucleation characteristics of ZnSe on GaAs bulk substrates and MBE-grown GaAs *epilayers* has also been studied utilizing RHEED diffraction patterns and intensity oscillations. A two-dimensional nucleation of ZnSe occurred on GaAs epilayers in striking contrast to the three-dimensional nucleation on GaAs substrates. This difference in nucleation character was evidenced in the evolution of the RHEED diffraction patterns[13], as well as intensity oscillations observed during the early stages of nucleation. Although As and Ga was not present in the II-VI MBE system, the as-grown GaAs epilayer surface (grown in a separate MBE system in our laboratory) was preserved using arsenic-passivation techniques[13]. After the growth of the GaAs epilayer, the Ga shutter was closed and the substrate heater was turned off. As the substrate temperature approached room temperature, an amorphous arsenic layer was deposited, as evidenced by a lack of gallium peaks in Auger electron spectroscopy (AES) spectra; a diffuse RHEED pattern confirmed the amorphous nature of the arsenic layer. With the arsenic layer serving as protection, the sample was then transferred in air to the II-VI MBE system. Following thermal desorption, the RHEED pattern of the GaAs epilayer showed similar reconstruction features as those typically exhibited by a GaAs bulk substrate subsequent to oxide desorption. When ZnSe is nucleated on a GaAs substrate, we[3,13], as well as other groups[14,15,16] have reported observing a spotty RHEED pattern indicative of three-dimensional nucleation; eventually the spots elongated into streaks. Nucleation of ZnSe on a GaAs epilayer resulted in a strongly streaked RHEED pattern after only 9 seconds (The growth rate was 1.3 Å/sec.). Reconstruction lines appeared after 9 seconds when using an epilayer, whereas minutes elapsed before distinct reconstruction lines appeared in the case of a substrate. The two-dimensional nucleation mechanism on the epilayer was corroborated by observations of strong RHEED intensity oscillations (Fig. 5). In contrast the intensity variation of the specular spot when nucleating on a GaAs substrate was similar to measurements reported for three-dimensional nucleation

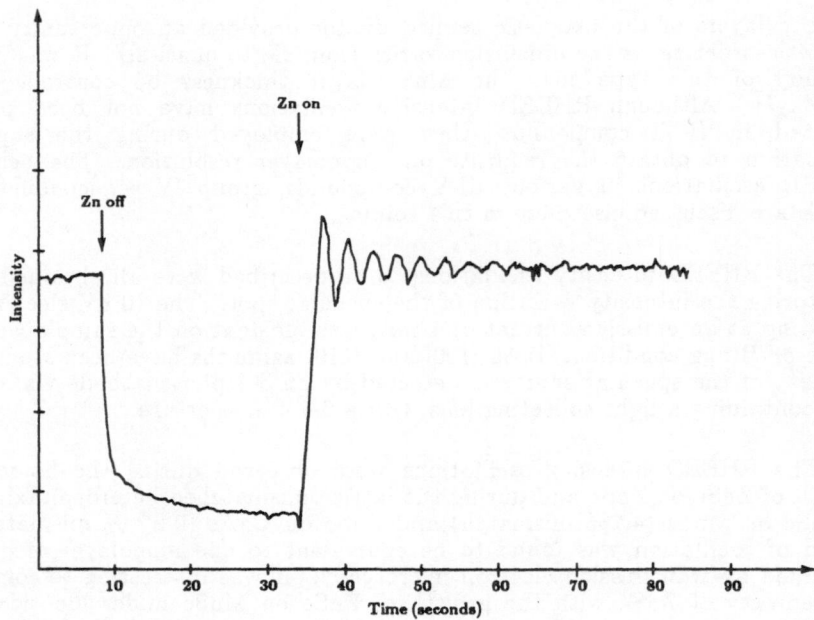

Fig 3.   RHEED intensity oscillations observed when the Zn
          shutter is closed to interrupt ZnSe growth and when
          the Zn shutter is re-opened to resume growth ([100]
          azimuth at 400 °C substrate temperature).

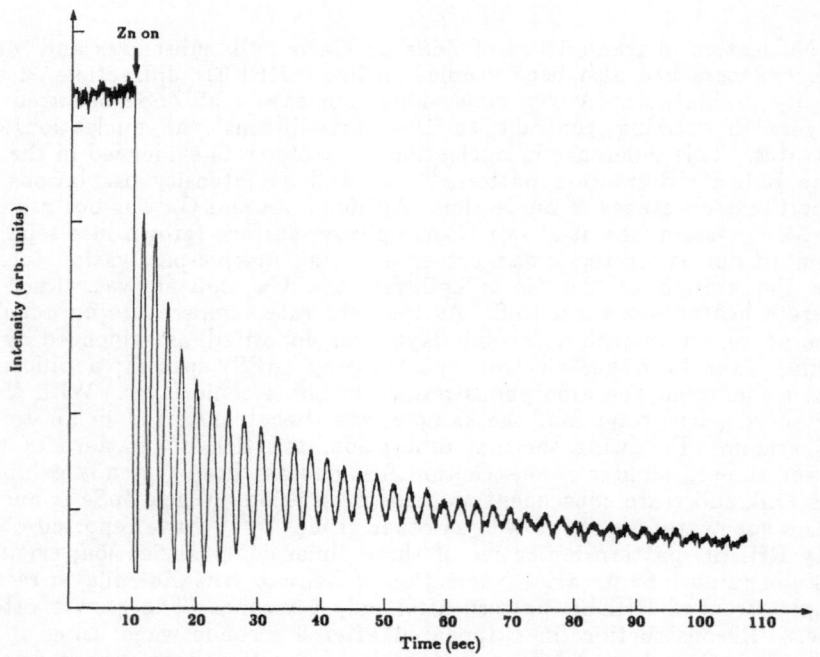

Fig. 4.   RHEED intensity oscillations during nucleation of
          ZnSe on a MnSe layer which has been recovered in the
          presence of a Se flux ([100] azimuth at a 400 °C
          substrate temperature).

252

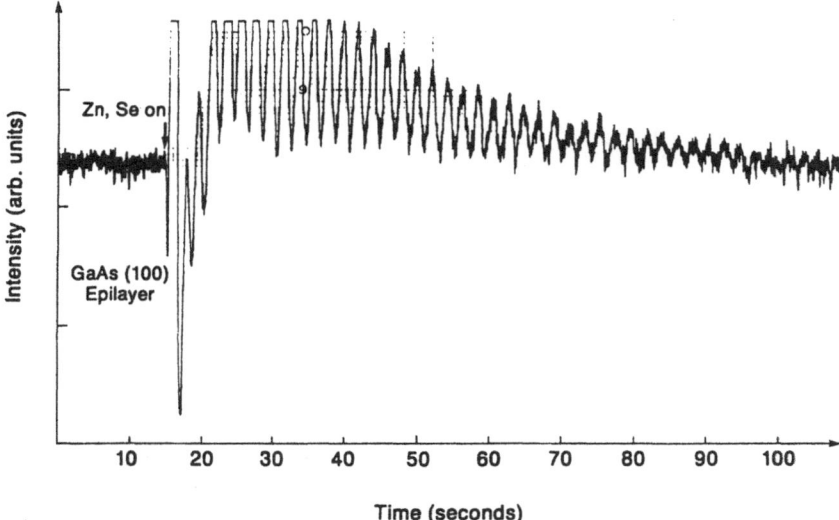

Fig. 5. RHEED intensity oscillations obtained during nucleation of ZnSe on a GaAs epilayer at a substrate temperature of 320°C and viewed in the [110] azimuth. (Some positive peaks were cut off due to recorder bias limitations.)

of InGaAs on GaAs epilayers[17]. As seen in Fig. 6, RHEED intensity oscillations were absent as growth was initiated on a GaAs substrate.

To explore possible origins contributing to the differences in nucleation behavior, the microstructure of the interfaces for ZnSe grown on both GaAs epilayers and substrates were examined using cross-sectional high resolution electron microscopy (HREM). The ZnSe film grown on a GaAs epilayer exhibited a featureless image of the interface which appears as an atomistically flat boundary over wide areas with a perfectly coherent contact between the two layers (Fig. 7). On the other hand, for growth on a GaAs substrate the interface showed wavy, step-like boundary images which indicated the presence of small pits and steps. From the HREM images it is suggested that a major factor contributing to the occurrence of either two- or three-dimensional nucleation is the relative density of nucleation sites such as surface dislocations and step edges. An MBE-grown epilayer is expected to possess a lower density of surface defects than found on a chemically etched substrate.

In Fig. 8 is seen the effect of alternating the cation species without growth interruption during the fabrication of a binary superlattice consisting of MnSe/ZnSe. In this particular binary superlattice each period consisted of four monolayers of ZnSe separated by three monolayers of MnSe. The layer thicknesses were determined by counting the number of oscillation periods and confirmed by high resolution transmission electron microscope imaging of cross-sectional specimens. Throughout the entire growth of the five period superlattice, intensity oscillations were seen in both the ZnSe and MnSe layers. Initiation of the growth of the first MnSe layer resulted in an increase of the specular spot intensity. Although difficult to see in this figure, it was often possible to detect

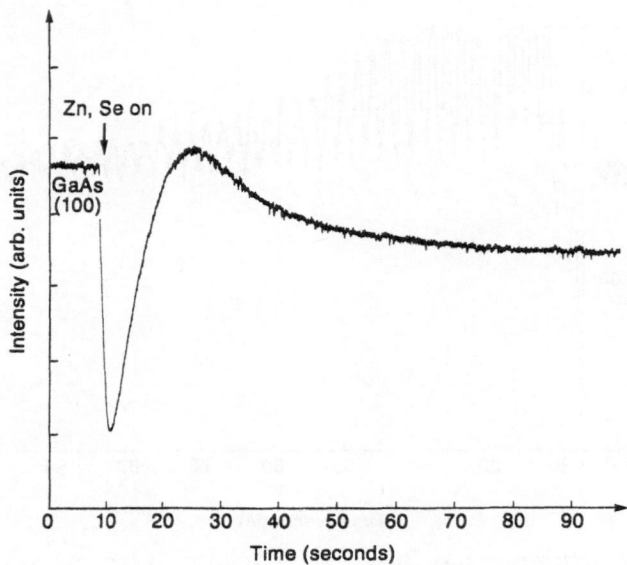

Fig. 6. Intensity variation of the specular spot reflection in the [110] observed during nucleation of ZnSe on a GaAs bulk substrate (400° C substrate temperature). The higher substrate temperature would be expected to favor two-dimensional nucleation.

the presence of periodic oscillations superimposed on the rising intensity. As the Zn shutter was closed and the Mn shutter was opened at an intensity oscillation maxima for ZnSe, the layer-by-layer growth of MnSe resulted in the observed intensity oscillations. It is interesting to note the enhanced persistence of the oscillations resulting from alternating the two binary materials. The observation of RHEED intensity oscillations during the growth of the magnetic semiconductor MnSe enabled one to control the MnSe layer thickness with one monolayer resolution.

MAGNETIC SEMICONDUCTOR SUPERLATTICES

A series of comb-like superlattice structures consisting of 30 to 100 periods were grown with MnSe layer thicknesses of one, three, and four monolayers; the MnSe layers were separated by approximately 45 Å of ZnSe. A fourth superlattice structure to be discussed consisted of 30 periods containing ten monolayers of MnSe alternated with 24 Å of ZnSe.

Figure 9 shows the RHEED intensity oscillations recorded during fabrication of a comb structure containing barrier layers consisting of four monolayers of MnSe. This sequence of intensity variation was obtained during the 25th period; essentially identical features were observed during the first and last period. The first detail to note is the reduction of the specular spot intensity as the Zn shutter was closed to interrupt the ZnSe growth. Upon re-opening the Zn shutter, the specular beam intensity was found to oscillate as the ZnSe growth resumed. At a

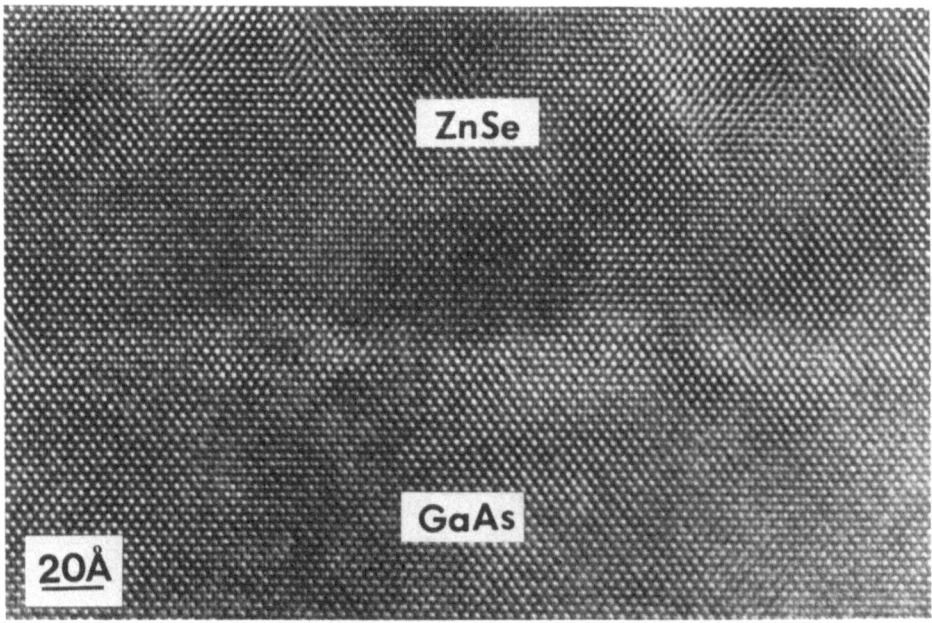

Fig. 7. HREM image of the ZnSe-GaAs epilayer interface.
Measurements were performed using the 1 MeV TEM
at the Tokyo Institute of Technology.

maxima of the oscillation, the Zn shutter was closed and the Mn shutter opened.
As described above, four oscillation periods were observed to be superimposed on
the overall intensity increase. The growth of the MnSe layer ended as the Mn
shutter was closed and the Zn shutter was opened. Oscillations of enhanced
amplitude (as compared to those seen when ZnSe is nucleated on ZnSe) persisted
for 16 oscillation periods, constituting the completion of one superlattice period.
The shuttering sequence, wherein every layer was nucleated at a maximum of the
intensity oscillation, was intended to optimize the interface smoothness during
continuous growth. The persistence of oscillations extending throughout the
entire superlattice growth implied minimization of the number of layers involved
in the growth front. To explore possible effects of growth interruption on the
magnetic behavior of these superlattices, two of the structures (the comb
superlattices containing one and three monolayers of MnSe) were fabricated by
two different growth sequences. One method was as described in the discussion of
Fig. 5; the second sequence involved interruption of growth at each interface. For
the one monolayer comb structure, each MnSe layer was interrupted by closing
the Mn shutter for 60 seconds while each ZnSe layer was interrupted by closing
both shutters for 10 seconds. The three monolayer comb structure had each layer
interrupted for 12 seconds (with the same shutter arrangement during
interruption as described above). In all cases of interrupted growth the shutter
arrangement was chosen such that an increase in the specular spot intensity
occurred during interruption suggesting that each layer was recovered prior to
nucleation of the next layer.

The magnetic behavior of the comb superlattices was determined from
photoluminescence (PL) measurements in the presence of an external magnetic
field. The magnetization was deduced from the decrease in energy of the ground
state transition in PL resulting from an externally applied magnetic field (Zeeman
shift). The magneto-optical shifts of the PL spectra originate from an exchange
interaction between the extended states of the superlattice and the net magnetic

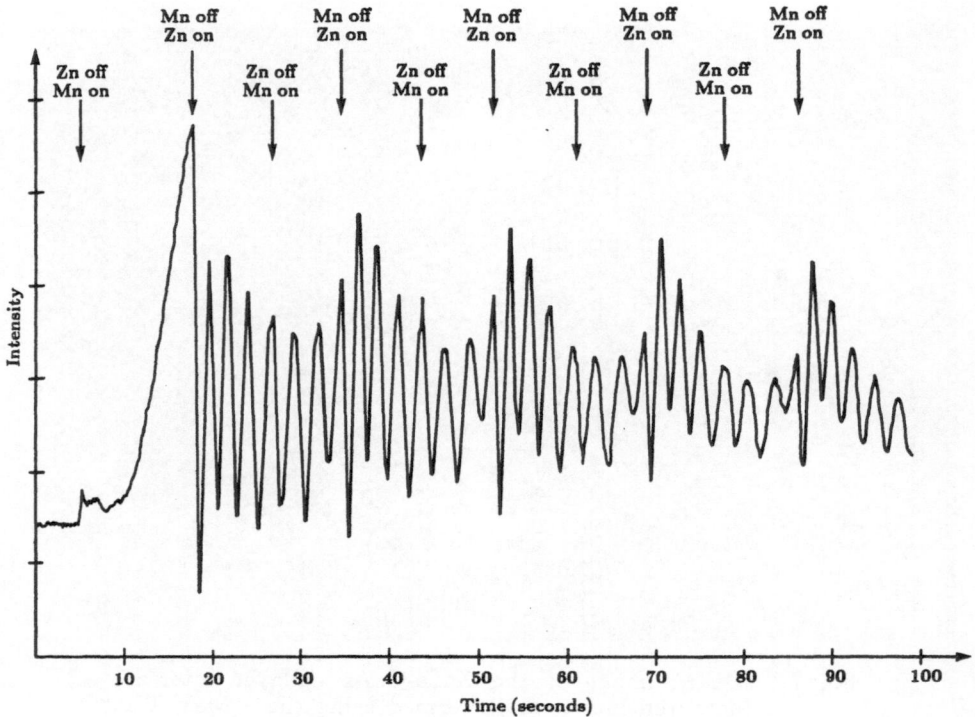

Fig. 8. RHEED intensity oscillations observed during growth of a MnSe/ZnSe binary superlattice structure. The Se source shutter remained open throughout. The superlattice is grown on a thick ZnSe buffer layer ([100] azimuth and a 400 °C substrate temperature).

moments of the Mn ions in the thin MnSe layers. Calculations based on a Kronig-Penney model employed a bandgap energy value for zincblende MnSe of 3.4 eV, a value determined by extrapolation of optical data obtained on zincblende (Zn,Mn)Se epilayers (Fig. 1). (The bandgap published for rock-salt MnSe is 2.5 eV[18].) The calculations provided zero magnetic field ground state transition energies to within 10 meV of the measured energies.

The rock-salt crystal structure of MnSe is known to exhibit antiferromagnetic ordering at low lattice temperatures[19]. Recent susceptibility measurements[20] performed on Bridgman-grown bulk crystals of $Zn_{1-x}Mn_xSe$ existing in the wurtzite phase, indicate an increase in antiferromagnetic ordering (at low temperature) as the Mn mole fraction is increased (x<0.5). Magneto-optic measurements on zincblende MBE-grown epilayers show a reduction of Zeeman shift with increasing Mn mole fraction; the decreasing red shift is attributed to the tendency for antiferromagnetic ordering. The similarity in magnetic behavior between zincblende and wurtzite (Zn,Mn)Se crystals is not surprising as nearest neighbor distances are the same for the two crystal structures. Anticipating no spectral red-shift for a superlattice structure containing "antiferromagnetic" MnSe, Fig. 10 shows a surprising degree of paramagnetic behavior. The sample characterized in the figure is a comb superlattice consisting of three monolayers of

MnSe separated by 45Å of ZnSe. The PL data shown corresponds to emission from the ground state transition with B=0 Tesla and 5 Tesla at 1.8 K (excitation at 0.325 μm). Fig. 11 provides a summary of measured Zeeman shifts for six

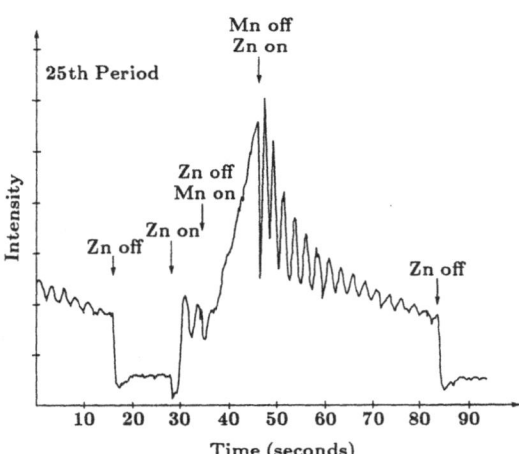

Fig. 9.   RHEED intensity oscillations recorded during the 25th period of a comb superlattice having four monolayers of MnSe per period (total of 30 periods). The specular spot is viewed in the [100] azimuth at a substrate temperature of 400 °C.

MnSe/ZnSe superlattice structures, including data for the one and three monolayer comb superlattices fabricated using the aforementioned interrupted growth technique. The difference in the total number of Mn ions per period in the various comb structures is normalized by plotting the exciton shift per monolayer, $\Delta E/N$ versus the number of MnSe monolayers (N) per superlattice period. The first data point to note is that obtained on a superlattice containing ten contiguous monolayers of MnSe per period. We interpret these MnSe layers as behaving three-dimensionally, as they exhibit the expected lack of paramagnetic behavior. As the MnSe layer thickness is reduced to the quasi-2D limit (one monolayer), the increased red shift is interpreted as a tendency for weakened antiferromagnetic ordering. One speculation as to the origin of the frustration of antiferromagnetic ordering is based on the presence of interface disorder. Such interfacial disorder (of approximately one monolayer) is inherent to the layer-by-layer growth process associated with MBE. In an attempt to improve the interface smoothness, interrupted growth techniques were used during fabrication for two samples. The magneto-optical data obtained on these two samples is denoted by the squares in Fig. 11. The essentially unchanged paramagnetic behavior in both cases suggests that interface disorder is not a prime factor. However, as yet there have not been definitive experiments completed to correlate interface smoothness with growth interruption in the II-VI materials. Further work is required to unambiguously relate dimensionality to magnetic ordering.

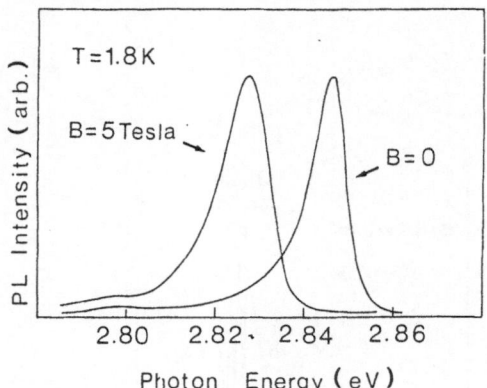

Fig. 10. Photoluminescence data showing emission of the ground state transition obtained in the presence of both zero magnetic field and a 5 Tesla magnetic field at 1.8 K, Faraday geometry. The comb superlattice contained three monolayers of MnSe separated by 45Å of ZnSe.

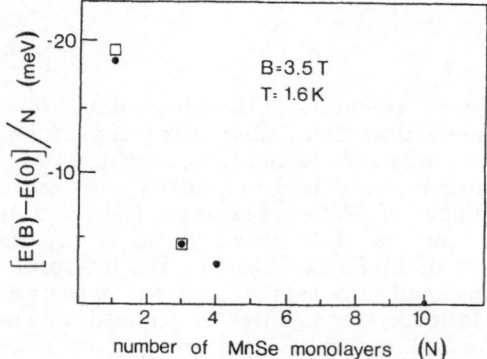

Fig. 11. Zeeman shift per monolayer of MnSe versus the number of MnSe monolayers per superlattice period (B_z = 3.5 Tesla, T = 1.6 K). The data identified by the "□" was measured on the two superlattices fabricated using growth interruption techniques at each interface.

ACKNOWLEDGEMENT

The authors gratefully acknowledge the following people for their contributions to this work: S. Datta, M. R. Melloch, Y. Hefetz, D. Lee, R. Venkatasubramanian, C. Choi, J. Qiu, G. Studtmann, M. Vaziri, and D. Lubelski. The work at Purdue was supported by Office of Naval Research Contract N00014-82-K0S63, and Air Force Office of Scientific Research Grant 85-0185. The work at Brown was supported by Office of Naval Research Contract N00014-83-K0638.

REFERENCES

1. R. Venkatasubramanian, N. Otsuka, S. Datta, L. A. Kolodziejski, and R. L. Gunshor, "Monte Carlo Simulation of Growth of II-VI Semiconductors by MBE," Materials Research Society Symposium, Dec. 1-5, 1987, Boston.

2. L. A. Kolodziejski, T. Sakamoto, R. L. Gunshor, and S. Datta, "Molecular Beam Epitaxy of $Cd_{1-x}Mn_xTe$," Appl. Phys. Lett. 44:799 (1984).

3. L. A. Kolodziejski, R. L. Gunshor, T. C. Bonsett, R. Venkatasubramanian, S. Datta, R. B. Bylsma, W. M. Becker, and N. Otsuka, "Wide Gap II-VI Superlattices of ZnSe-ZnMnSe," Appl. Phys. Lett. 47:169 (1985).

4. L. A. Kolodziejski, T. C. Bonsett, R. L. Gunshor, S. Datta, R. B. Bylsma, W. M. Becker, and N. Otsuka, "Molecular Beam Epitaxy of Diluted Magnetic Semiconductor (CdMnTe) Superlattices," Appl. Phys. Lett. 45:440 (1984).

5. L. A. Kolodziejski, R. L. Gunshor, S. Datta, T. C. Bonsett, M. Yamanishi, R. Frohne, T. Sakamoto, R. B. Bylsma, W. M. Becker, and N. Otsuka, "MBE Growth of Films and Superlattices of Diluted Magnetic Semiconductors," J. Vac. Sci. Technol. B3:714 (1985).

6. R. N. Bicknell, R. Yanka, N. C. Giles-Taylor, D. K. Blanks, E. L. Buckland, and J. F. Schetzina, "CdMnTe-CdTe Multilayers Grown by Molecular Beam Epitaxy," Appl. Phys. Lett. 45:92 (1984).

7. R. N. Bicknell, R. W. Yanka, N. C. Giles-Taylor, D. K. Blanks, E. L. Buckland, and J. F. Schetzina, "Properties of CdMnTe-CdTe Superlattices Grown by Molecular Beam Epitaxy," J. Vac. Sci. Technol. B3:709 (1985).

8. L. A. Kolodziejski, R. L. Gunshor, N. Otsuka, X. C. Zhang, S. K. Chang, and A. V. Nurmikko, (100)-Oriented Superlattices of $Cd_{0.76}Mn_{0.24}Te$ on (100) GaAs," Appl. Phys. Lett. 47:882 (1985).

9. R. L. Gunshor, L. A. Kolodziejski, N. Otsuka, and S. Datta, "ZnSe-ZnMnSe and CdTe-CdMnTe Superlattices," Surf. Sci. 174:522 (1986).

10. L. A. Kolodziejski, R. L. Gunshor, N. Otsuka, S. Datta, W. M. Becker, and A. V. Nurmikko, "Wide Gap II-VI Superlattices," IEEE on Quantum Electronics QE-22:1666 (1986).

11. D. R. Yoder-Short, U. Debska, and J. K. Furdyna, "Lattice Parameters of $Zn_{1-x}Mn_xSe$ and Tetrahedral Bond Lengths in $A^{II}_{1-x}Mn_xB^{VI}$ Alloys," J. Appl. Phys. 58:4056 (1985).

12. L. A. Kolodziejski, R. L. Gunshor, N. Otsuka, B. P. Gu, Y. Hefetz, and A. V. Nurmikko, "Two-dimensional Metastable' Magnetic Semiconductor Structures," Appl. Phys. Lett. 48:1482 (1986).

13. R. L. Gunshor, L. A. Kolodziejski, M. R. Melloch, M. Vaziri, C. Choi, and N. Otsuka, "Nucleation and Characterization of Pseudomorphic ZnSe Grown on Molecular Beam Epitaxially-Grown GaAs Epilayers," Appl. Phys. Lett. 50:xxx (1987).

14. T. Yao, S. Amano, Y. Makita, and S. Naekawa, "Molecular Beam Epitaxy of ZnTe Single Crystal Thin Films," Jpn. J. Appl. Phys. 15:1001 (1976).

15. R. M. Park, N. M. Salansky, "Surface Structures and Properties of ZnSe Grown on (100) GaAs by Molecular Beam Epitaxy," Appl. Phys. Lett. 44:249 (1984).

16. T. Yao and T. Takeda, "Growth Process in Atomic Layer Epitaxy of Zn Chalcogenide Single Crystalline Films on (100) GaAs," Appl. Phys. Lett. 48:160 (1986).

17. B. F. Lewis, T. C. Lee, F. J. Grunthaner, A. Madhukar, R. Fernandez, and J. Maserjian, "RHEED Oscillation Studies of MBE Growth Kinetics and Lattice Mismatch Strain-Induced Effects During InGaAs Growth on GaAs (100)," J. Vac. Sci. Technol. B2:419(1984).

18. D. L. Decker and R. L. Wild, "Optical Properties of α-MnSe," Phys. Rev. B 4:3425 (1971).

19. J. W. Allen, G. Lucovsky, J. C. Mikkelsen, "Optical Properties and Electronic Structure of Crossroads Material MnTe," Solid State Commun. 24:367(1977).

20. J. K. Furdyna, R. B. Frankel, and U. Debska, unpublished.

MAGNETIC INTERFACE PREPARATION AND ANALYSIS

Ulrich Gradmann and
Marek Przybylski*

Physikalisches Institut
Technische Universitaet Clausthal
D-3392 Clausthal-Zellerfeld

1. INTRODUCTION

Recent interest in artificial metallic superlattices[1-4] has two faces: The first aim is to study, for the case of metals, the unusual physical properties resulting from the periodic modulation, which became such a fascinating field for the case of semiconductor lattices. Secondly, metallic superlattices seem to be convenient systems for the study of interface phenomena, of size effects in thin films and of twodimensional systems, which can easily be observed in superlattices because of their strong multiplication. However, this analysis of interface phenomena can be done more reliably using single films, for which growth modes and structures of interfaces can be controlled much better than in superlattices. Physical properties, however must then be detected with monolayer or submonolayer sensitivity. For the case of interface magnetism this analysis using single ultrathin films has been done for many years using high sensitivity Torsion Oscillation Magnetometry (TOM) both in air[5,6] (ATOM) and in UHV[7,9] (UTOM), and recently using Conversion Electron Moessbauer Spectroscopy[10] (CEMS). The present paper reports on this type of experimental interface magnetism using thin films, with special emphasis on the preparation of flat epitaxial film structures, bounded by well-defined, diffusion-free, atomically sharp plane interfaces. These should be used for any straightforward analysis of basic phenomena, because they form the subject of theoretical models, which now are available for a broad variety of magnetic thin film and interface phenomena as finite temperature size effects[5-9,11,12], the local structure of low temperature magnetization near surfaces and interfaces[13-15], magnetic hyperfine field oscillations[16] and enhanced magnetic moments in monolayers and interfaces[17]. Magnetic surface and interface anisotropies[18,19], which are of special technical interest, also should be analyzed starting from ideal interfaces.

In the present paper, we first discuss experimental methods and theoretical models for ideal metallic film and interface preparation and analysis. Based on this analysis of preparation problems, some recent examples of experimental magnetic interface analysis are reported. It will be shown that additional structural information results from the magnetic analysis.

---

*Permanent address: Solid State Physics Department,
Academy of Mining and Metallurgy, Krakow, Poland.

2. EPITAXIAL GROWTH

The driving force for epitaxial growth is the dependence of interfacial energy on the crystalline orientation of a film with respect to its single crystal substrate. As this dependence is a universal phenomenon, epitaxial growth, in principle, is a universal phenomenon, too, the only real conditions being that at the temperatures needed for sufficient mobility of the growing material ("epitaxial temperatures") no interdiffusion or even melting takes place. These conditions being given, we further ask for the conditions for the desired layer-by-layer growth. We start with a short discussion of the methods to test for it.

2.1 Experimental Growth Mode Analysis

The standard test for layer-by-layer growth is performed using Auger Electron Spectroscopy (AES). For the case of layer-by-layer growth, the Auger amplitude I of the film material increases as a function of mean film thickness in linear sections, with breaks at the discrete values d_n where the n-th monolayer has been completed . Auger amplitudes I_n at the breaks are given by

$$I_n = I_o[1 - \exp(-d_n/\lambda \cdot \cos\alpha)], \tag{1}$$

where λ is the mean free path of the Auger electrons for inelastic scattering (for numerical data on λ compare [21]) and α the mean angle of acceptance of the Auger analyzer. The Auger amplitude of the substrate decreases in a similar way. Any deviation from the ideal mode results in a decrease of the film amplitude in comparison with eqn.(1) and increase of the substrate amplitude.

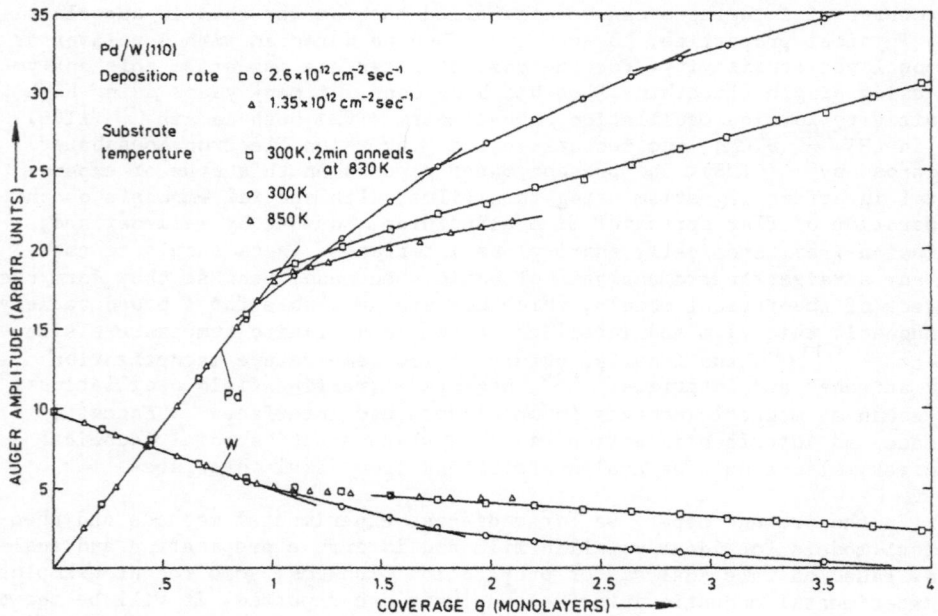

Fig. 1: Epitaxial growth of Pd on W(110)[22] . Pd and W
Auger amplitudes versus Pd coverage on a W(110)
surface.

This is shown in fig. 1 for the growth of Pd on W(110)[22], which grows layer by layer at 300 K at least up to n = 3 layers, whereas at 800 K threedimensional crystals are formed on top of one flat monolayer. Diffusion to the substrate as a cause for the reduction of the film signal must be ruled out by other methods (in the example by flash desorption). This combination of layer-by-layer growth at low temperatures with islanding at higher temperatures has been observed for several metallic systems, for a review compare[23]. The Auger method works only for the first few monolayers.

A good qualitative criterion is given by reflexion high energy electron diffraction (RHEED), where diffraction streaks indicate atomically flat surfaces, whereas diffraction spots indicate islands. For a quantitative evaluation compare[24]. It is difficult to distinguish between layer-by-layer growth and film structures with flat islands, however.

The most promising method to test for layer-by-layer growth seems to be the observation of oscillations in electron diffraction features, as discussed for the case of RHEED in several articles in this book[2]. The method has been applied mainly to semiconductors, its application to metals seems promising. Further, the same method can be applied for Low Energy Electron Diffraction (LEED)[25]. The most promising approach seems to be a Spot Profile Analysis (SPA)[24,25] as a function of film thickness. Layer-by-layer growth then should be characterized by periodic changes between sharp spots indicating the plane surface of completed monolayers and broadened spots between. Details of film growth then should be taken from a quantitative analysis of these broadened spots. As a special merit of this method, the data can be analyzed quantitatively by kinematic theory[24,25]. Transfer widths of 1000 nm have been reported for RHEED[24], of 200 nm for LEED[26].

Additional information on growth mechanisms from electron microscopy is out of the scope of this review; compare other articles in this book[2].

2.2 Growth Mode Models

Most metal films are prepared by evaporation and condensation in UHV. This condensation proceeds far from thermodynamic equilibrium. Any models to describe it must therefore be taken as rough guides to understand the processes and to find appropriate conditions to grow films layer by layer. A final test on the structures should always be done using the methods mentioned in the last section.

Equilibrium considerations. A first attempt to understand growth modes has been given by Bauer[27] in terms of classical nucleation theory, using the tacit assumption that both the critical nucleus and the overcritical crystallites form equilibrium structures, that means their shape is determined by minimization of the surface enthalpy. If γ_s and γ are the free surface enthalpies of bulk substrate and film materials, resp., and if the total surface/interface enthalpy γ_n of n parallel layers on the substrate is conveniently split in γ and an interface enthalpy $\gamma_{i,n}$ according

$$\gamma_n = \gamma_{i,n} + \gamma \qquad (2),$$

then the first monolayer wets the substrate (nucleation-free growth of this monolayer) for

$$\Delta\gamma_1 = \gamma_{i,1} + \gamma - \gamma_s < 0$$

Conversely for $\Delta\gamma_1 > 0$ the material of the first layer (and the following one) forms threedimensional crystallites on the otherwise free surface (Volmer-Weber(VM)-mode, island growth).

Layer-by-layer (Frank-van-der-Merwe(FM)) growth takes place if each layer wets the preceding one, that is if $\gamma_{i,n+1} < \gamma_{i,n}$ for all n. Strain energies are included in $\gamma_{i,n}$ which as a rule increase monotoneously with n, and certainly do so in pseudomorphic growth systems. FM growth therefore appears as a rare exception. Layer growth in pseudomorphic systems is impossible. What must be expected as a rule is $\gamma_{i,n+1} < \gamma_{i,n}$ for $n < N$ in combination with $\gamma_{i,N+1} > \gamma_{i,N}$. In this case threedimensional nucleation occurs on top of N monolayers (Stranski-Krastanov(SK)-mode). Experimental examples of SK-growth are discussed in[23]. Note that SK-growth with large N cannot be distinguished from FM-growth using AES as discussed in the previous section (compare fig.1).

The role of supersaturation. A general rule in film condensation is that continuous films of given thickness can be obtained by condensation on low temperature substrates with high growth rates. This applies for epitaxial growth, too. A transition from SK-to FM-growth can be induced in certain epitaxial systems by lowering the substrate temperature or increasing the growth rate, eg. Cu or W(110)[22], compare fig.1. For further examples compare[23]. This was explained in an extension of Bauers approach by Markov and Kaishew[28,29]. They started from the fact that the size of the critical nucleus is determined by the supersaturation $\Delta\mu$ of the condensing beam. Asking for a condition that the height of the critical nucleus equals that of one monolayer, they resulted in

$$\Delta\gamma < \Delta\mu/c \cdot \delta^2, \tag{4}$$

where $C = 2$ and $\sqrt{3}$ for fcc(100) and (111) films, respectively, and δ is the nearest neighbour distance. For supersaturation above a critical value $\Delta\mu_{cr} = c \cdot \delta^2 \cdot \Delta\gamma$ the film nucleates by monolayer nuclei, resulting in what the authors call "layer growth" and can be regarded as a reasonable approximation to layer-by-layer growth.

The key role of $\Delta\mu$ for the growth mode has been checked quantitatively for γ-Fe(111)-films on Cu(111) using AES[30,31]. The growth mode, as characterized from Auger intensities versus film thickness, showed to be dependent on $\Delta\mu$ only, independently of whether $\Delta\mu$ was changed by the growth rate or by the temperature of the substrate. From the critical supersaturation $\Delta\mu_{cr}$ for the transition to layer growth, $\Delta\mu_{cr}/C \delta^2 = 3460$ erg/cm could be determined, being 2.5 times larger than the estimated value of $\Delta\gamma = 1380$ erg/cm . Despite of this quantitative discrepancy, the key role of $\Delta\mu$ was clearly confirmed.

Growth kinetics after nucleation. If in the non-wetting case the critical nucleus is of monolayer height, it remains as a problem of growth kinetics whether post-nucleation growth remains twodimensional or goes to the third dimension, as indicated by equilibrium considerations. A comprehensive discussion of these kinetic conditions of the 2D-3D-transition in epitaxial thin films has been given by Stoyanov and Markov[23] in a pair bonding model for fcc(100) and (111) films on homosymmetric substrates. The main physical ideas can be explained from the following simple model; compare fig.2:

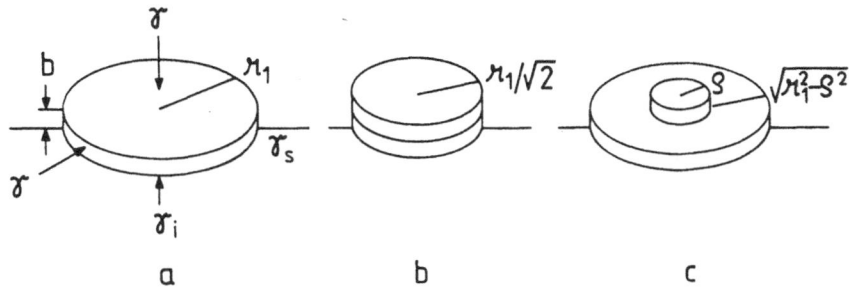

Fig. 2: Comparison of monolayer and double layer
islands of equal volume.

Compare the total surface energy Γ for three film crystallites of
equal volume, (a) a circular monolayer disc of height b and of radius r_1,
(b) a double layer of radius $r_1/\sqrt{2}$, and (c) a monolayer disc of radius
$\sqrt{r_1^2 - \rho^2}$ with a second layer nucleus of radius ρ on top of it. If the edge
energies are approximated conveniently by surface energies of circular
walls, the total surface energies of the three nuclei are given by

$$\Gamma_a = \pi r_1^2 \Delta\gamma + 2\pi b r_1 \gamma;$$

$$\Gamma_b = \frac{1}{2}\pi r_1^2 \Delta\gamma + 2\sqrt{2}\pi b r_1 \gamma \quad \text{and}$$

$$\Gamma_c = \pi(r_1^2 - \rho^2)\Delta\gamma + 2\pi b\left[\sqrt{r_1^2 + \rho^2} + \rho\right]\gamma,$$

respectively. It can easily be shown that $\Gamma_a = \Gamma_b$ for
$r_{1,c} = 4(\sqrt{2} - 1)\cdot b \cdot(\gamma/\Delta\gamma)$. The monolayer disc is stable for $r_1 < r_{1,c}$,
it becomes metastable with respect to the double layer for $r_1 > r_{1,c}$.
However, a transition from (a) to (b) must go via the intermediate situation (c). For $\rho \ll r$, the increment $\Delta\Gamma = \Gamma_c - \Gamma_a$ is given by

$$\Delta\Gamma = -\pi\rho^2\Delta\gamma + 2\pi\rho\gamma \tag{5}$$

$\Delta\Gamma$ is shown as a function of ρ in fig.3. It shows a maximum

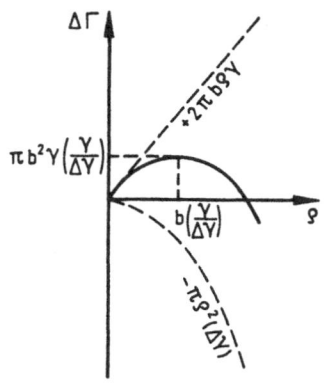

Fig. 3: Incremental energy
$\Delta\Gamma = \Gamma_c - \Gamma_a$ of
the second-layer nu-
cleus, fig.2 c.

$$\Delta \Gamma_{max} = \pi b^2 \gamma (\gamma/\Delta\gamma), \tag{6}$$

which forms an energy barrier for second monolayer nucleation. For typical values of $\gamma = 1.0$ Jm^{-2}, b = 2 Å, $(\gamma/\Delta\gamma) = 5$, we get $r_{1c} = 16.6$ Å and $\Delta\Gamma_{max} = 4$ eV. This barrier may well prevent a second layer nucleation before coalescence of monolayer islands; as a consequence, a complete metastable monolayer can be formed. As for subsequent layers the $\Delta\gamma$ is smaller than for the first one, one expects layer growth in the following according eqn.6 and metastable layer-by-layer growth as a whole. If however the substrate temperature is sufficiently high or the evaporation rate is small enough, the islands can transform before the monolayer patches coalesce and island growth is observed. The role of supersaturation for the layer-island transition becames quite clear. For the details compare[23]; the gross experimental observations are in fair agreement with the model.

In semiconductor superlattices similar components are frequently combined with $(\Delta\gamma/\gamma) \ll 1$, resulting in large values of $\Delta\Gamma_{max}$ according eqn(6) and metastable layer-by-layer growth of both components on one another. In metal films and superlattices, where both the chemical and the strain contributions to $\Delta\gamma$ are larger, the considerations given above give a hint for achieving layer-by-layer growth for $\Delta\gamma > 0$, too: Metastable layer-by-layer growth can be stabilized by high growth rates and low substrate tempertures. It is a matter of experimental experience whether these temperatures suffice to establish single crystalline order in the growing film or not.

2.3 Misfit Dislocations

The misfit f between substrate and growing film lattices is either accommodated entirely by elastic strain (pseudomorphism, coherent growth) or is shared between misfit dislocations and strain ε[32-39]. This can be shown by calculating U_ε, the energy associated with misfit accommodation by elastic strain, and U_d, the energy associated with misfit dislocation lines. The elastic strain ε^* which minimizes the sum of these two energies is the optimum misfit strain. The condition for pseudomorphism is obtained by equating ε^* to the misfit f. The elastic energy per unit area of film is

$$U_\varepsilon = E\varepsilon^2 h (1 - \nu). \tag{7}$$

The energy associated with a square network of dislocations which are in edge orientation and have Burgers vectors in the film plane is

$$U_d = \frac{2(f - \varepsilon)}{b} \frac{Gb^2}{4\pi(1 - \nu)} (\ln\frac{h}{b} + 1), \tag{8}$$

where b is the magnitude of the Burgers vector, and G the shear modulus. In elastically isotropic materials $E = 2 G(1 + \nu)$. The optimum elastic strain predicted from Eqs.(7) and (8) is therefore

$$\varepsilon^* = \frac{b}{8\pi h(1 + \nu)} (\ln\frac{h}{b} + 1). \tag{9}$$

The film thickness h_c at which it becomes energetically favorable for misfit dislocations to be made is

$$h_c = \frac{b}{8\pi f(1 + \nu)} (\ln\frac{h}{b} + 1)$$ (10)

This equation has been used to construct the lowest line C in fig. 4.

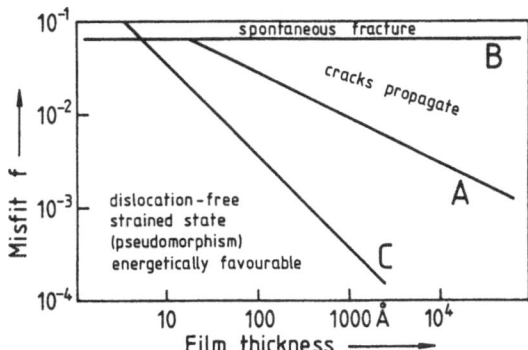

Fig. 4: Predictions of misfit dislocations and cracking in epitaxial
films. Below C, given by eq.(10)
using b = 2 Å, pseudo-
morphism is stable. Above A,
cracks propagate; above B, the
film spontaneously fractures
(from [38], modified)

Below it, pseudomorphic films are stable, above it, strain may be released if misfit dislocations can be generated. Experimental analysis of large misfit metallic systems (Ag(111) on Cu(111), f_{AgCu} = + 13.3 %[40]; Fe(110) on W(110), f_{FeW} = -9.4 %[41]; Ni(111) on R (0001), f_{NiRe} = -9.8 %[42]) show the generation of periodic misfit dislocation networks in the first atomic layers, which accommodate the misfit completely. Strong periodic lattice distortions fade out at $h \approx a/f$, resulting in unstrained surfaces for h > 50 Å . For medium misfit, e.q. Ni on Cu(f_{NiCu} = -2.5 %[36,37]) , nonpe-riodic networks of misfit dislocations are formed for larger h(h $\approx$ 10 Å in Ni on Cu) , with an indication of residual strain of the order 10^{-3} for thicker films[36,37]. The generation of misfit dislocations seems to be hin-dered by work hardening of the films[37]. For low misfit systems, pseudomor-phic strain persists in equilibrium up to large values of h_c , or even over h_c as the driving force for misfit generation becomes smaller and generation may be prevented. Further lines are included in fig. 4[38] above which cracks propagate or the film fractures spontaneously, resp. , if no misfit dislocations are formed to release the pseudomorphic strain. This can occur in semiconductor systems[38,44], where the Peierls-Nabarro stress for misfit movement is high[38,44]. In metals, this has not been observed, as expected from their good plasticity. As a whole, for low misfits we expect dislocation-free structures with far-reaching pseudomorphic strain, whereas large misfits are connected with strong periodic distortions, con-fined to a thin interface sheet, the volume of thicker films being free from strain.

3. MAGNETIC INTERFACE ANALYSIS

Experimental analysis of magnetic interface and surface phenomena can now be done by a couple of methods like thin film magnetometry [3-7,9], XPS [46], FMR [47,48], magnetooptical methods [49], spin-polarized neutron reflexion [50] and Moessbauer spectroscopy [51,52]. In addition, surfaces can be analysed most powerful by spin polarized electron spectroscopies [45,53]; these are out of the scope of this article, which concentrates on internal interfaces. Most of these experimental methods are discussed in other contributions [3,4,45-50] to this book .

The present contribution reports on two experimental methods of magnetic interface analysis which have been used in our laboratory. The first one is Torsi Oscillation Magnetometry (TOM) of these ferromagnetic films, which has been used for a long time at atmospheric pressures, with film preparation in a separate UHV-system [54-59,61-63], for reviews compare [6,54,60] and recently with preparation and magnetometry in situ in one single UHV-system [42,64-66], for reviews compare [7,9,19]. At present, this is the only method which gives direct access to magnetic moment and magnetization with submonolayer sensitivity [42,56]; the instrument detects changes in moment below 1 % of one monolayer Ni, in UHV, during film growth [42]. A special advantage of the in situ-TOM is the possibility to measure the magnetic background of the epitaxial substrate (Re(0001), W(110)) directly before magnetic film preparation, in order to take finally the film properties from the difference before and after film growth and so to come free from the magnetic impurities in the substrate, which form a serious problem for monolayer experiments. A second unique feature of the method is that both magnetic moment and magnetic anisotropies of the films result independently from the analysis. They can be analyzed in terms of surface magnetization and magnetic surface anisotropies, as shown below. Note that FMR [47,48] measures anisotropies only, which can be analyzed only indirectly in terms of magnetization if crystalline anisotropies including surface anisotropies can be neglected in comparison with shape anisotropy (spectacular statements on enhanced magnetization in CuNi-multilayers [70], which were in contradiction with former magnetometry in Ni-films on Cu [36], resulted from a misinterpretation of FMR-measurements neglecting surface anisotropies).

Local analysis cannot be done using magnetometry. For the case of Fe, unique possibilities of local analysis are given by Moessbauer spectroscopy, if two conditions can be fulfilled: First, the sensitivity should be high enough to take a Moessbauer spectrum of one monolayer Fe57 in one day or less; secondly, one should be able to deposit the Moessbauer isotope Fe57 in one single atomic layer (probe layer) only. It has been shown recently [43,44,46] that this can be realized by combining Conversion Electron Moessbauer Spectroscopy (CEMS) with sophisticated preparation methods in one single UHV-system (monolyer-probe in situ CEMS). The intimate connexion between problems of preparation and film structure and of magnetic analysis, which forms a guiding idea of this paper, becomes extremely important. In turn , CEMS provides unique informations on film structures, too, which cannot be taken from other experiments [71].

In the following sections, experiments with TOM and with CEMS are reported, performed with layer-by-layer grown single crystal epitaxial films, consisting of few parallel atomic layers, starting from the monolayer. They are called "thin films" in the following.

3.1 Torsion Oscillation Magnetometry of Thin Films

The <u>working principles</u> of TOM[36,42,56,72] are shown in fig.5:
The sample, suspended in a homogeneus magnetic field on a thin torsion
filament, performes torsion oscillations on the vertical axis, the period T
of which is measured using a laser. Damping by air in the atmospheric
pressure TOM (ATOM), by eddy current in single crystal metal substrates
like Re(0001) in the UHV-TOM (UTOM), must be compensated, in the case of

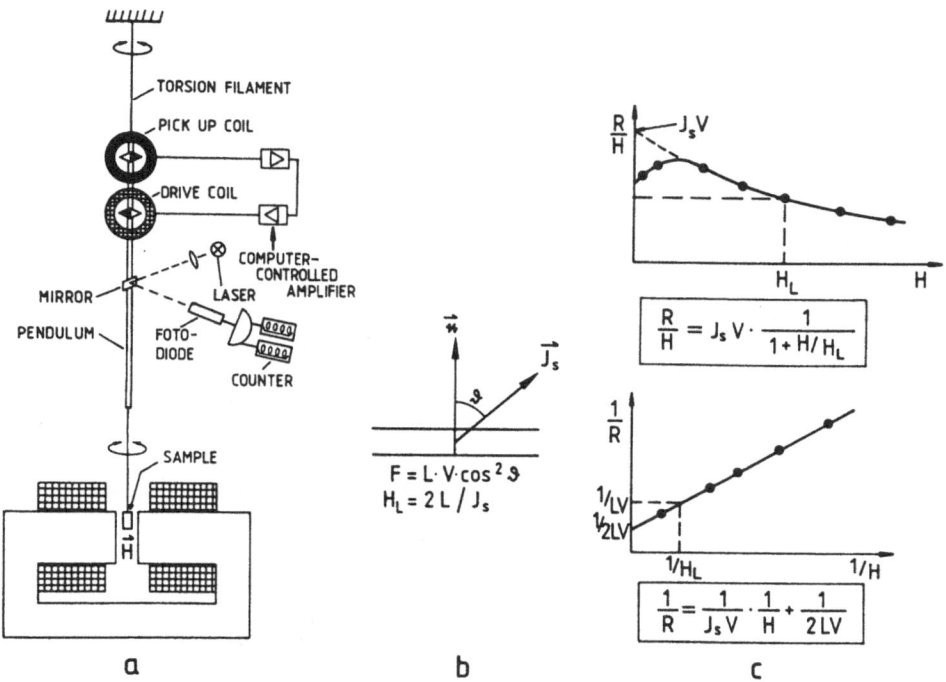

Fig. 5: Working principles of TOM.
    (a) Schematic description of the instrument.
    (b) Definition of film anisotropies.
    (c) Evaluation of measurements.

UTOM by a combination of pick-up and drive coil with a computer-controlled
feedback system. For a magnetic probe, T is determined by the coupling of
the magnetic moment $m = J_s V$ to the field, $J_s V \cdot H$, and by the coupling of
m to the lattice, given by crystalline anisotropies, including surface
anisotropies [18,19] as leading contributions for the case of thin films. In
a first approximation, the anisotropic part of the free energy is given by
$F = L \cdot V \cdot \cos^2 \vartheta$ (fig.5b); anisotropies are described by an anisotropy constant L or an anisotropy field $H_L = 2L/J_s$. The magnetic change of T is
transformed to a magnetic contribution to the directional moment, R, which
is connected with $m = J_s V$ and L by the equations given in fig.5c. Magnetic moment and anisotropy result independently from a plot of R/H versus
H or of 1/R versus 1/H. For both UTOM and ATOM, the detection limit is
below 10^{-1} Wb·m (10^{-2} ML Ni(111) on samples used).

    <u>Magnetic Size Effects.</u> A first application of TOM is the measurement
of spontaneous magnetization and Curie-temperature in thin films as a
function of the number D of atomic layers. This has been performed by ATOM
for 48Ni/52Fe(111)[5,6] and Co[6,55] in Cu(111), and by UTOM for Ni(111) on

Re(0001)[42]. For both systems it could be shown that in the ground state the magnetic moment in the boundary layers deviates only slightly from the bulk one. The results shown in fig.6a and b therefore represent the bare "size effect" caused by finite temperature thermal excitations. For $J_s(T,D)$, good agreement with spin-wave theory calculations could be established (fig.6a). In accordance with predictions of Allan[73], $T_c(D)$ follows a power law $T_c(\infty)-T_c(D) \sim D^{-\lambda}$, with $\lambda = 1.27 \pm 0.2$[42].

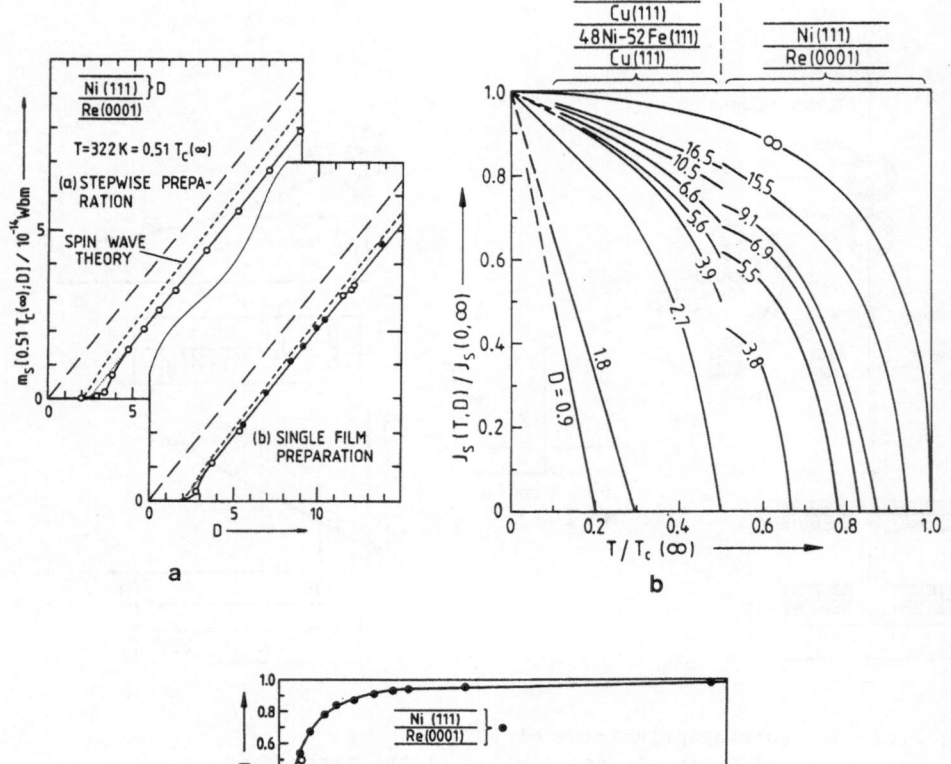

a

b

c

Fig. 6: Magnetic size effects.
    (a) The magnetic moment m_s of Ni(111)-films on Re(0001), followed in UTOM during film growth at 322 K($0.51\ T_c(\infty)$) follows spin-wave theory to a good approximation; slight differences only occur for different types of preparation[42].
    (b) The dependence of $J_s(T,D)$ on temperature T and number of atomic layers D follows the same line for 48Ni/52Fe(111) in Cu^5 and Ni(111) on Re(0001)[42] (compilation from[7]).
    (c) The Curie-temperature $T_c(D)$ becomes a function of thickness D[7,42].

Interface magnetization. The change of magnetic moment of a ferromagnetic film by contact with a nonmagnetic or antiferromagnetic coating could be measured in ATOM by comparison of film pairs with different coatings[54,57-59]. This is shown in fig.7 for a Mn-coating on 60Ni/40Fe(111); the decrease of moment by the equivalent of 1.5 "Dead Layers" results from an antiferromagnetic coupling of the first Mn-layer to the NiFe-film. Similar results were obtained for a C-coating.

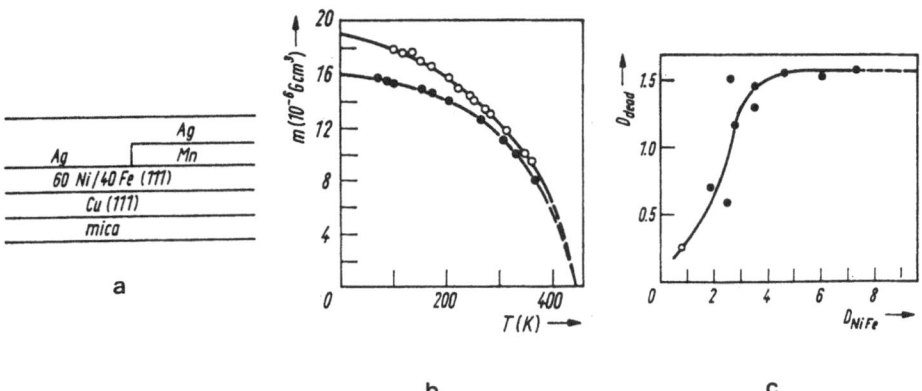

a

b

c

Fig. 7: Interface magnetization by ATOM[59].
(a) One part of a 60Ni/40Fe(111) film is coated by Ag, the other by Mn.
(b) For a film consisting of D = 2.5 layers, the magnetic moment of the Mn-coated part (●) is lowered with respect to the Ag-coated part (o). The Curie-Temperature remains unchanged (for comparison, $T_C(\infty)$ = 870 K for bulk material).
(c) The decrease of magnetic moment by Mn, expressed in terms of number of "dead" layers on NiFe, D_{dead}, becomes independent on D_{NiFe} for D_{NiFe} > 4; the saturation value D_{dead} = 1.5 characterizes the bulk interface.

In UTOM, the measurement can easily be done by following the change of magnetic moment during coating processes. This is shown in fig.8a for the coating of a 10.5 layers Ni(111)-film on Re(0001) by Cu[66]. The moment decreases within the first 4 layers Cu by 0.8 equivalents of 1 ML Ni(111) (1 ML Ni(111) ⇌ 5.8 x 10^{-15}Wbm), the main effect taking place in the first monolayer. Caused by size effects, this decrease, measured at RT, depends on D_{Ni} (fig.8b), saturating at 0.6 units for "thick" films; this is in excellent agreement with band calculations of Tersoff and Falicov[74]. For the case of Pd-coating, the moment increases, compare fig.8c. This results from a magnetic polarization of the nearly ferromagnetic Pd-coating; a quantitative version of this explanation has been given by Mathon[75,76]. The effect of a Pd-substrate is 3 times larger than of a Pd-coating. Apparently, the susceptibility of Pd(111) on Re(0001) is further enhanced by pseudomorphic strain by the misfit f_{RePd} = + 0.3%[64].

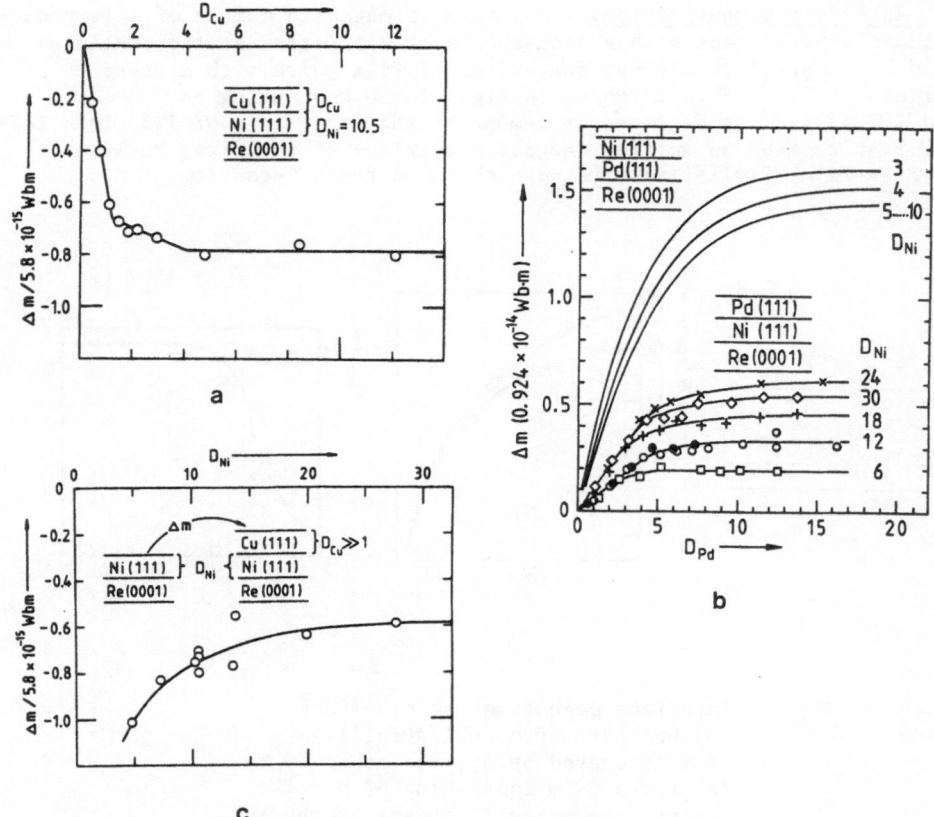

Fig. 8: Surface magnetization by UTOM[64,66].
(a) The magnetic moment of a Ni(111)-film
on Re(0001) decreases when coated by
Cu (IML Ni(111) $\hat{=}$ 5.8 10^{-15}Wbm)[66].
(b) The decrease from fig.(a) saturates
for thick Ni-films at 0.6 units[66].
(c) The magnetic moment of Ni(111)
is increased by contact with Pd, both
as coating and as substrate[64].

Magnetic exchange anisotropy. Magnetometry of thin films is an excellent tool to analyze unidirectional magnetic exchange anisotropies[77],
caused by contact of a ferromagnetic crystal with an antiferromagnetic
matrix (for a modern review compare[78]). Exchange anisotropy shows up as a
shifted hysteresis loop of the ferromagnetic component, as shown in fig.9a
for a D_{NiFe} = 2.5 layers 60Ni/40Fe(111)-film coated by Mn[59]. As expected,
the shift disappeared at the Néel temperature of the Mn-coating. A remarkable phenomenon was found by plotting the remanent magnetization m_r of
the ferromagnetic film (coated by Mn) as a function of temperature; as
shown in fig.9b, m_r shows a steep minimum at the Néel temperature (130 K)
of the Mn-coating (slightly enhanced above the bulk T_N = 96 K by spurious interdiffusion with NiFe). This was tentatively explained as an ac-
field like demagnetization of the NiFe-film by critical fluctuations in
the Mn-coating ("critical exchange coupling").

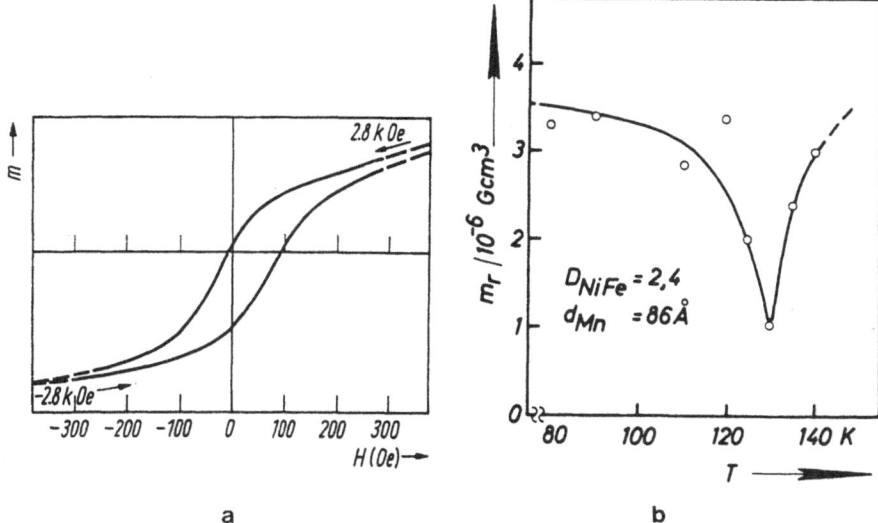

a   b

Fig. 9: Exchange anisotropy in NiFe/Mn-interfaces.
(a) Hysteresis of 2.5 layers 60Ni/40Fe(111)
on Cu(111) coated by Mn, shifted by
exchange interaction with the antiferromagnetic Mn-coating (measured at 80K)[59].
(b) The remanent moment m_r of the ferromagnetic NiFe-film shows a steep minimum
at the Néel temperature of the antiferromagnetic coating[57].

Néel-type magnetic interface anisotropies. It has been shown many
years ago by L. Néel[18] that the strong break of local lattice symmetry
given by a surface or interface of a magnetic crystal should result in
strong magnetic surface anisotropies. The most efficient way to determine
these anisotropies experimentally is magnetometry of thin films by TOM.
This has been shown many years ago by ATOM for NiFe(111)-films in
Cu[5,6,55,60]and recently by UTOM for Ni(111) on Re(0001)[64-66]. For a review
compare[19]. The mode of analysis is shown in fig.10a. Let the total anisotropy of the film be composed of bulk contributions and in addition a surface anisotropy, that is a magnetic contribution to surface energy

$$\sigma = K_s \cdot \cos^2 \vartheta \qquad (11),$$

characterized by an suface anisotropy constant K_s or, alternatively, an
anisotropy field $H_s = 2 K_s/(\delta \cdot J_s)$, where δ is the distance of atomic
layers. Then the total anisotropy field of the film, as a function of D,
is given by

$$\mu_o H_L = (J_s + \mu_o H_V) + \frac{1}{D} (\mu_o H_s^{(1)} + \mu_o H_s^{(2)}) \qquad (12)$$

where J_s and $\mu_o H_V$ are shape and bulk crystalline anisotropies, resp.,
and $\mu_o H_s^{(i)}$ the suface anisotropies of both surfaces (interfaces).

273

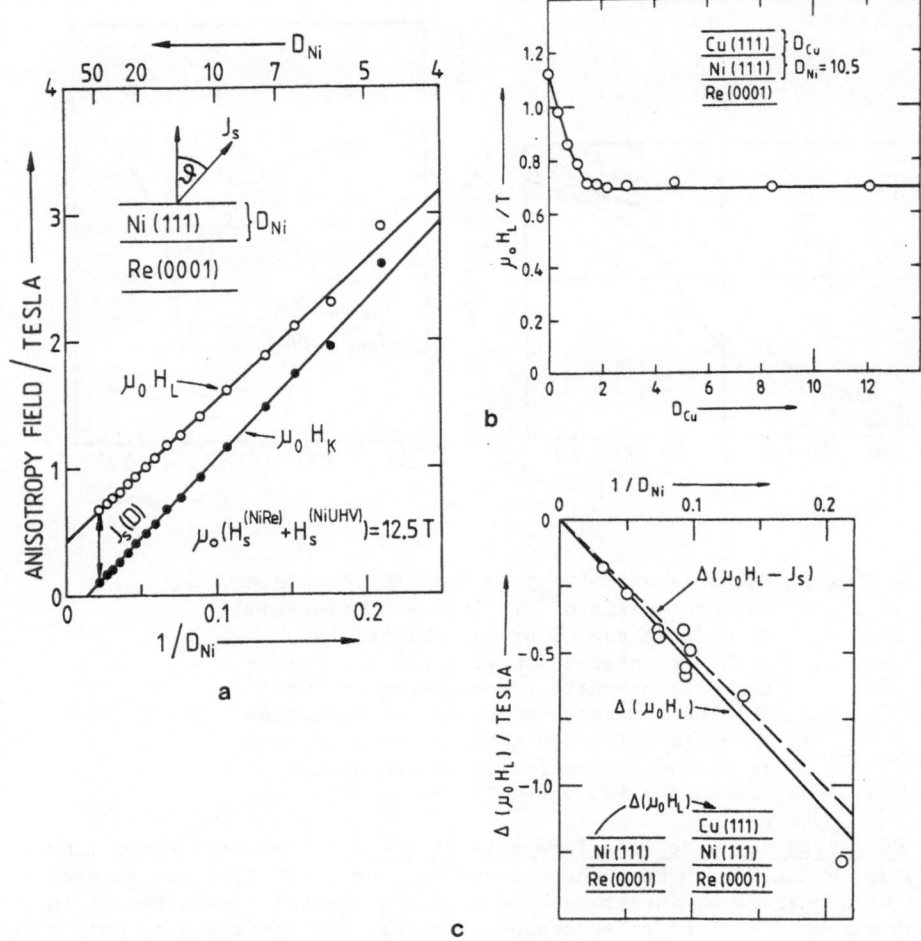

Fig. 10: Magnetic surface anisotropies[19,65,66].
(a) If the total anisotropy of films depends linearily on 1/D, magnetic surface anisotropies can be taken from the slope. (b) Changes of magnetic film anisotropies (anisotropy field H_L) by coating p.e. with Cu indicate changes of surface anisotropies. This interpretation can further be tested by plotting $\Delta(\mu_o H_L)$ versus 1/D; a linear dependence, as shown in fig.10c, shows that the interpretation is correct.

A linear dependence of $\mu_o H_L$ on 1/D, as shown in fig.10a for Ni(111) on Re(0001), can be taken as a confirmation of the model; at the same time, $\mu_o H_s^{(1)} + \mu_o H_s^{(2)}$ can be taken from the slope, which however should be taken for $\mu_o H_K = \mu_o H_L - J_s$ rather than $\mu_o H_L$ itself, because J_s depends on D, compare fig.6. Changes of $\mu_o H_L$ by coatings, which clearly come from changes in $\mu_o H_s$, can easily be followed in UTOM, as shown for the example of a Cu-coating in figs.10b and c. By using different combinations of substrates and coatings, single interface anisotropies could be determined[65,66,19]. Some results are given in table 1. For a review compare[19]. For additional in-plane anisotropies compare[67].

Table I: Magnetic surface anisotropies of Ni(111), given alternatively by anisotropy fields $\mu_0 H_s$, anisotropy constants K_s, or energies per surface atom $K_s/n^{(2)}$ where $n^{(2)} = (4/3)/a^2$ is the twodimensional density of atoms in a Ni(111) monolayer. The accuracy is of the order $\pm$ 1 Tesla for all experimental anisotropy fields.

Interface	$\mu_0 H_s$ Tesla	K_s mJ/m^2	$K_s/n^{(2)}$ μeV/atom
Experimental results, at 322 K [65]			
Ni(111)/UHV	9.9	0.48	160
Ni(111)/Cu(111)	4.6	0.22	74
Ni(111)/Pd(111)	4.6	0.22	74
Ni(111)/Re(0001)	4.0	0.19	65
Theoretical results			
Ni(111), R.T. Néel 1953[18]	−3.9	−0.19	−63
Ni(100), R.T. Néel 1953[18]	−2.1	−0.10	−33
Ni(100) monolayer, T=0 Fritsche et al. 1984[79]			+144

3.2 Conversion Electron Moessbauer Spectroscopy (CEMS)

The local character of Moessbauer spectroscopy makes it a unique probe for the local analysis of magnetic order. For the magnetic analysis of surfaces and ultrathin films, Conversion Electron Moessbauer Spectroscopy (CEMS) should be used. It has been used by Bayreuther et al[82] for the analysis of ultrathin Fe-films on Ag, which however grew by islands; straightforward analysis of the data was not possible. "Depth Selective" CEMS[83,84] uses energy losses of conversion electrons for local analysis and therefore has not the monolayer resolution which is needed for most basic problems of interface magnetism. The ultimate resolution comes out if monolayer probes of Fe57 can be used in films consisting otherwise of Fe56. The first approach to this type of analysis was done by Tyson et al[51]; however, they used transmission Moessbauer spectroscopy and therefore could not use monolayer probes by lack of sensitivity. The experiments to be discussed in the following have been done at Clausthal using a new Moessbauer spectrometer, capable of ultrasensitive CEMS in situ on layer by layer grown films in UHV[52,68], using monolayer probes of Fe57.

The Clausthal CEMS spectrometer is shown in fig.11a. Fe(110)-films including Fe57 monolayer probes can be prepared in UHV and analyzed in situ as shown by the spectra of fig.11b. Note the lateral shift of the inner lines in (B) and (C) in comparison with (A); this shift is proportional to quadrupole splitting ε, and therefore to the electrical field gradient V_{ZZ} at the Fe57 nucleus, too, which differs from zero (in the limits of accuracy) only when Fe57 is deposited in the first monolayer[52]. The disappearance of ε when Fe57 is deposited in the second layer shows first the rapid screening of electrical fields by conduction electrons. It secondly shows that no intermixing between the layers takes place, otherwise part of the atoms deposited in the 2nd layer should feel the field gradient present in the 1st one. Fe57-atoms stay in the layers where they are deposited. Local analysis of hyperfine fields can therefore be done in a straightforward way.

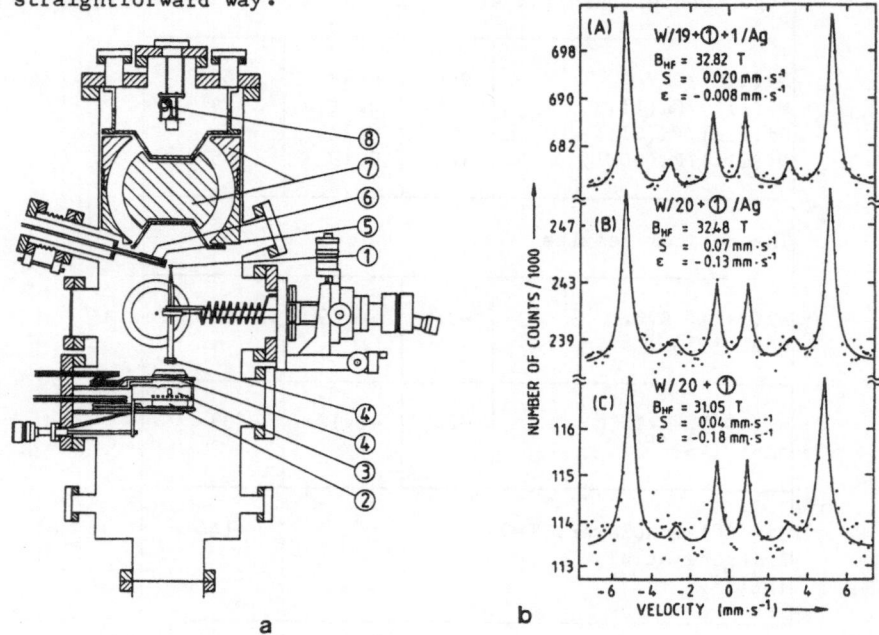

Fig. 11: (a) The Clausthal CEMS-Spectrometer[68]. Epitaxial film systems are grown on a W(110)-substrate ① (which is switched for preparation in position ④), from a MBE-system consisting of 6 BeO-crucibles ② with shutter ③ and 6 quartz crystal monitors ④. An additional monitor ④ in sample position is used for calibration. 6 metals including Fe56 and Fe57 can be grown below 10^{-10} Torr. After preparation, the sample is switched in position ①, where it is irradiated at grazing incidence from a moving Co57 Moessbauer-source ⑤, working in air, through a BeO-window ⑥. Conversion electrons are separated from background using a spherical mirror electron spectrometer ⑦, working as energy filter, and finally detected in a channeltron ⑧. (b) CEMS-spectra from Fe(110)-films on W(110), consisting of D = 21 monolayers, including one probe layer Fe57. For example, sample (A) consists of 19 layers Fe56, one probe layer Fe57 and one layer Fe56, coated by Ag; the second monolayer from a Ag-coated surface is probed. Correspondingly, the first monolayer is probed in (B) and (C) coated by Ag and uncoated, respectively. Spectra are taken in 50h, 20h and 8h, respectively. Basic Fe56 was prepared at 570K, the preparation temperature was lowered to 420K for Fe57. Spectra were taken at RT.

276

By following ε for Fe-atoms deposited in the first monolayer as a function of annealing temperatures, it could be shown that Fe-selfdiffusion starts only at temperatures above 470 K [52], in good accordance with bulk diffusion data.

The Fe(110)/Ag interface. Hyperfine fields B_{hf} were measured near the Ag-coated surface of 21-layers Fe(110)-films on W(110), using monolayer probes [69]. As shown in fig.12a, the temperature dependence of B_{hf} follows the bulk one for a central probe (W/9+③+9/Ag); for comparison, in the first monolayer an enhancement of B_{hf} at low temperatures by 1 Tesla is observed, in accordance with previous results of Tyson et al [51]. The thermally induced decrease, however, is enhanced roughly by a factor 2, as predicted many years ago by Rado [80]. As shown in fig.12b, the ground state B_{hf} increases monotoneously towards the Fe(110)-Ag-interface.

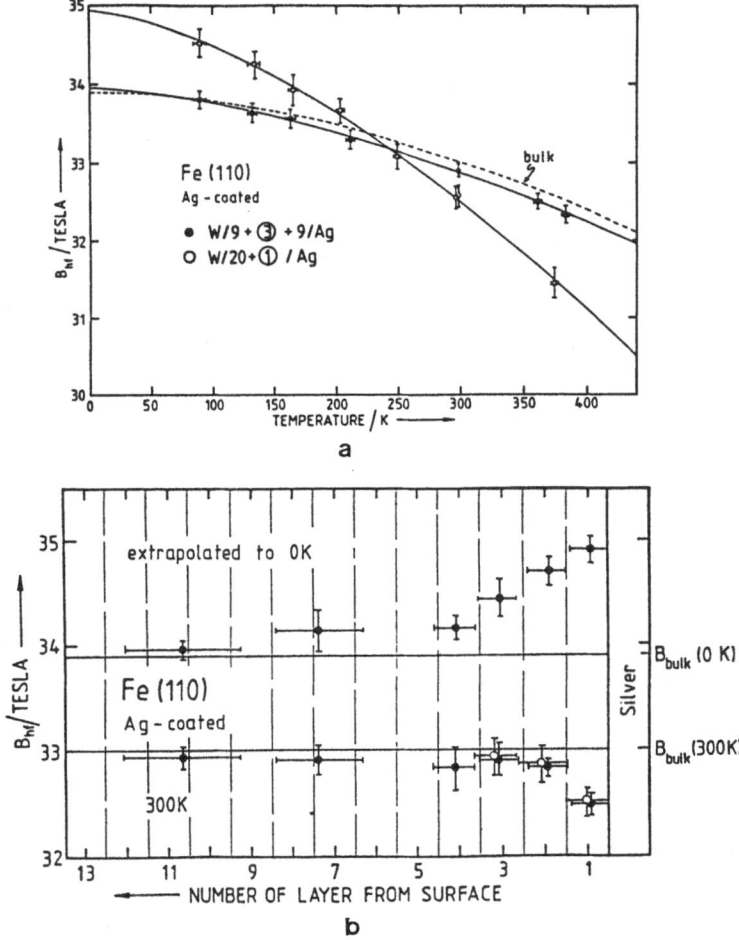

Fig. 12: Magnetic hyperfine field B_{hf} near the Fe(110)/Ag interface [69].
(a) Temperature dependence of B_{hf} in the first monolayer (o) and in the center of the film (•).
(b) B_{hf} at OK and at RT, respectivily. Preparation temperature started with 570K and was lowered to 420K starting from the probe layer (o) or from layer 11 (•), resp..

At 300 K, this ground state increase is slightly overcompensated by a thermally induced decrease. Two different methods of preparation were used for the 300 K measurements for the following reason[52]: The optimum crystalline order, as checked by the LEED pattern, was gained by preparation at 570K. All films were started with that temperature. However, to avoid $Fe57-Fe56$-interdiffusion, the temperature had to be lowered to 420 K at least from the probe layer on. In a first series of preparations, (o) in figs.12 and 13, the temperature was lowered starting from the probe layer only, to get the optimum quality of each single film. In a second series (•), the temperature was lowered for all films starting from layer 11, in order to achieve optimum comparability of the films. The results from both series, shown in figs. 12 and 13 for the Ag-covered and the free surface, resp., show excellent agreement.

The free Fe(110)-surface. As shown in Fig.13, the spatial variation of B_{hf} is much more expressed near the free Fe(110)-surface. Spatial oscillations of B_{hf} which have been predicted previously by Ohnishi et al[16] are seen again at RT and more clearly after extrapolation to T = 0. As discussed in [16], these oscillations come from the conduction electron contribution to B_{hf}.

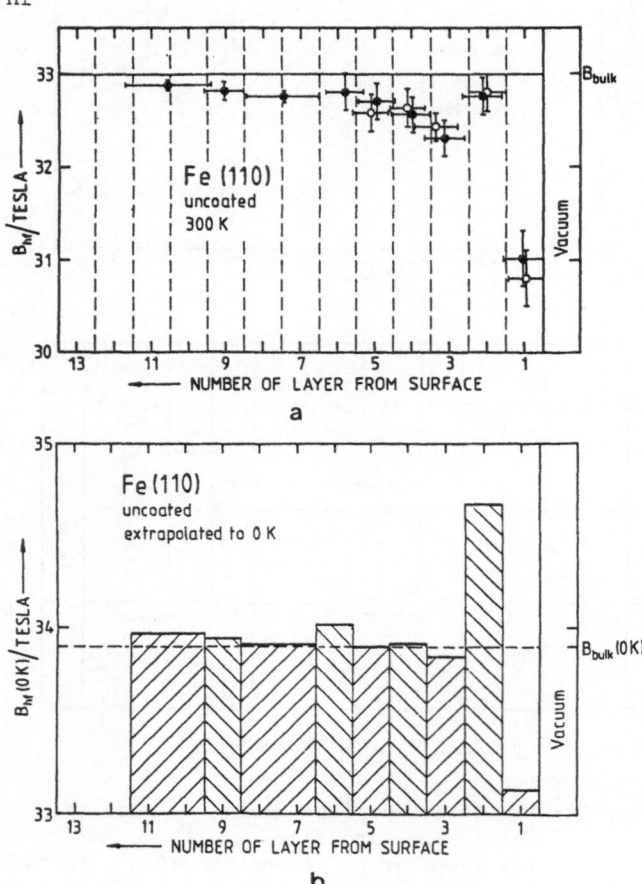

Fig. 13: Magnetic hyperfine field B_{hf} near the free Fe(110)-surface[69], (a) at RT (for two types of preparation compare fig.12), (b) extrapolated to 0 K. Spatial oscillations are clearly seen.

They result from the simultaneous cut off of the conduction electron gas in real space at the surface and in k-space at the Fermi-level. Correspondingly, they disappear when the conduction electron gas is restored in real space by a metallic coating, compare fig.12. For the ground state, FLAPW-claculations of Fu and Freeman[81] for the present surface Fe(110) give just the same type of oscillation, however with a threefold enhanced amplitude in comparison with the experiments.

The Fe(110)/W(110)-interface. As the films are prepared on W(110), they seemed to be appropriate for the monolayer-probe CEMS-analysis of the Fe(110)/W(110)-interface as well. However, basically new problems appeared in this interface, connected with interesting details of growth mode and interface structure, which in turn could be clarified further by the CEMS-analysis, giving a nice example of the tight interconnexion between structural and magnetic analysis.

The epitaxial growth of Fe on W(110) has been analyzed previously[41] using LEED and AES with the following results: (1) Frank van der Merwe growth was established by AES; (2) the two first monolayers are pseudomorphic, that means they are strained by f_{WFe} = + 9.5 % . Accordingly, these two first pseudomorphic layers are completed by the material of D = 1.64 bulk monolayers. (3) Above D $\approx$ 1.64, superstructure reflexions appear in the LEED pattern, indicating a periodic network of misfit dislocations, fading out above D $\approx$ 10. (4) AES shows no break at the completion of the first pseudomorphic monolayer; it therefore remains unclear whether the first layer is completed before the second starts or the first two layers grow simultaneously in one step. It remains unclear, too, whether the first two monolayers remain pseudomorphic in the thick film or if the cores of the misfit dislocations penetrate to the second or even the first Fe-layer. 500 K seemed to be an optimum preparation temperature. Surprisingly, the literal application of monolayer probe analysis to the W(110)/Fe(110) does not work, as can be seen from fig.14a[85]. Sample W/0.82 + 20/Ag, prepared at RT, neither showed the single component sextet structure to be expected when layer-by-layer growth would take place without subsequent intermixing, nor two components with equal amplitudes, which should indicate growth of the first two layers in one simultaneous step. Instead, we observe a two-component spectrum; the low-field component (B_{hf} = 20.9 T), which should be attributed to the first monolayer, contributes only 34 %, the high-field component (B_{hf} = 32.9 T), attributed to further "bulk" material, contributes 66 %. Apparently, interdiffusion or intermixing during film formation is enhanced by misfit dislocations. Accordingly, a reduction of preparation temperature to T_p = 90 K increases the low-field contribution to 64 %: Intermixing is hindered by low temperatures. (Note that at the same time the intensities change from $I_1 : I_2 : I_3 \approx$ 3:0:1 for T = 300 K[40] to 3:4:1 for T = 90 K. This indicates a switching of the magnetization[67]. As a whole, we observe a misfitdislocation enhanced intermixing (interdiffusion), which is shown even more clearly in fig.14c, where the relative contribution of the low-field component is given for nominal deposition of monolayer probes as indicated, for preparation at RT(●), at 500 K(o) and at 90 K(▨). Even when deposited in the 4th monolayer, Fe57 sample atoms appear in the 1st monolayer of the "thick" film with a probability of about 10 %.

The attribution of hyperfine fields to atomic layers in the film becomes even more conclusive from a CEMS-analysis of pure Fe57 films of variable thickness[71,86]. Moessbauer spectra, taken at 90 K are shown in fig.15 for some of these films. Four components can be seen, two of which, (a) and (c), are restricted to the monolayer region, the two others, (b) and (d), persist to higher values of D.

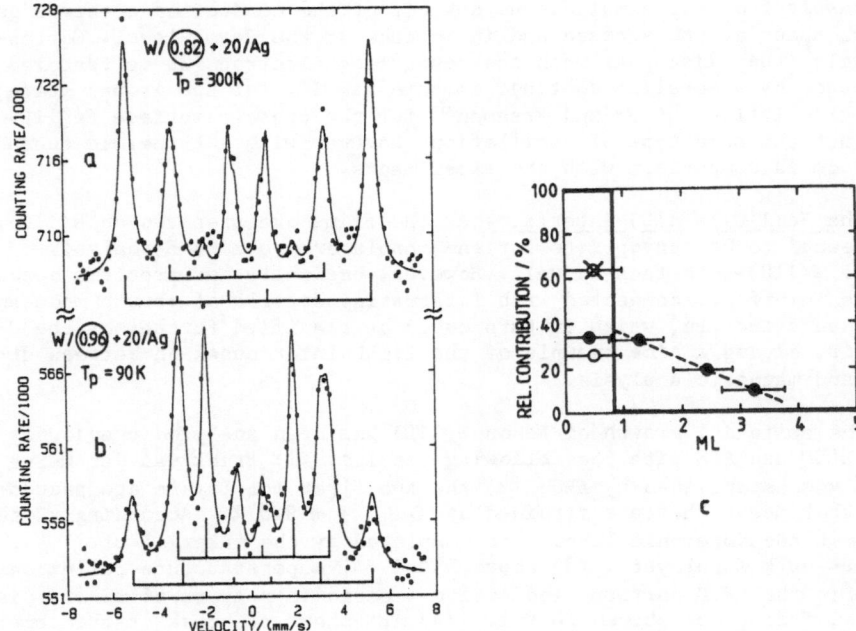

Fig. 14: Interface-dislocation enhanced interdiffusion in
Fe near the W(110)/Fe(110)-interface[85]. (a) In a
W/0.82 + 20/Ag film, prepared at 300 K, the relation of the high-field (32.9 T) to low field
(20.9 T) intensity is 0.66/0.34; (b) in a
W/0.96 + 20/Ag, film, prepared at 90 K, it is
0.36/0.64 (B_{hf} = 32.3 and 20.4 T, resp.). (c) The
relative contribution of the low field component
in W/x+1+y/Ag films (with x+1+y = 21) remains
finite, for T = 300 K (●), even for x = 2.6; results for preparation at 90 K (✘) and 500 K (○)
for comparison.

Fig.16 shows the evolution of these H components with increasing D. In
fig.16a, hyperfine fields B_{hf} are shown for films prepared at 500 K, measured both at 90 K and 300 K. The four components can be clearly followed
through all values of D. In comparison with the Fe(110)/Ag interface,
where B_{hf} deviates from the bulk values only by the order of 6 %, compare
fig.1, dramatic effects are observed when Fe(110) is in contact with the
transition metal W: In the first monolayer of thick films, B_{hf} is reduced
by ≈ 35 %, in the monolyer even by 64 %. It remains as a problem of future
research whether the magnetic moment per atom is reduced, too. In fig.16b
the relative contributions of all 4 components, taken from spectra at 90
K, are shown as a function of D, both for preparation at 500 K(○) and at
RT(●). They are compared with a simple model, where the film grows layer
by layer, starting with 2 pseudomorphic monolayers (completed at
D = 1,64). Components (a)-(d) are interpreted as follows (compare
fig.16b): (a) = first layer atoms, coated by Ag; (b) = first layer atoms,
coated by further Fe; (c) = second layer atoms, coated by Ag; (d) = second
layer atoms, coated by Fe, and further "bulklike" material. The fit of the
experimental data for 500 K-preparation is resonable, the fit for 300 K-
preparation is excellent.

280

The interpretation of the 4 components is conclusive. 300 K is an optimum temperature for an ideal growth of the first two monolayers. This does not contradict that for the growth of thicker films on optimum temperature of 500 K[41] or even 570 K[52] was found. Further, it becames quite clear that the first monolayer becomes finished before the second are starts, exactly at T = 300 K, approximately at 500 K.

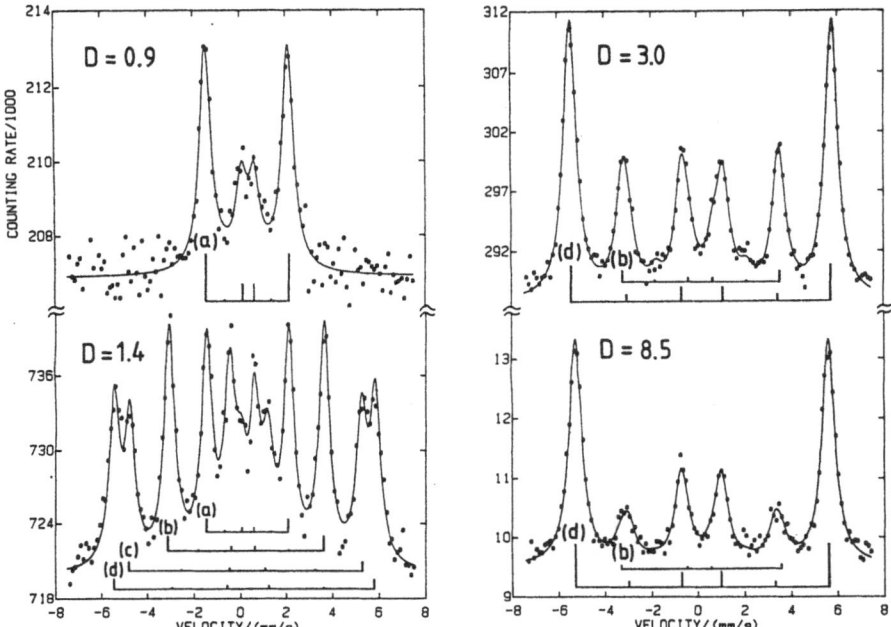

Fig. 15: Moessbauer spectra of pure Fe57-films[71,86]. Spectra of samples of type W/D/Ag, containing D layers of pur Fe57, were measured at 90 K. Films prepared at 500 K.

It allows to conclude that the first monolayer remains pseudomorphic in "thick" films, too ((a) + (b) = 0.82/D). Conversely, if the core of the misfit dislocations should penetrate to the first monolayer, (a) + (b) = 1.0/D should be expected, which is definitely not the case.

Note that the Fe(110) monolayer is included in the material of figs.15 and 16. It is ferromagnetic at 90 K, paramagnetic at RT; the Curie-Temperature is given by $T_C (1) = 280 + 10$ K. Details of monolayer magnetism will be published elsewhere.

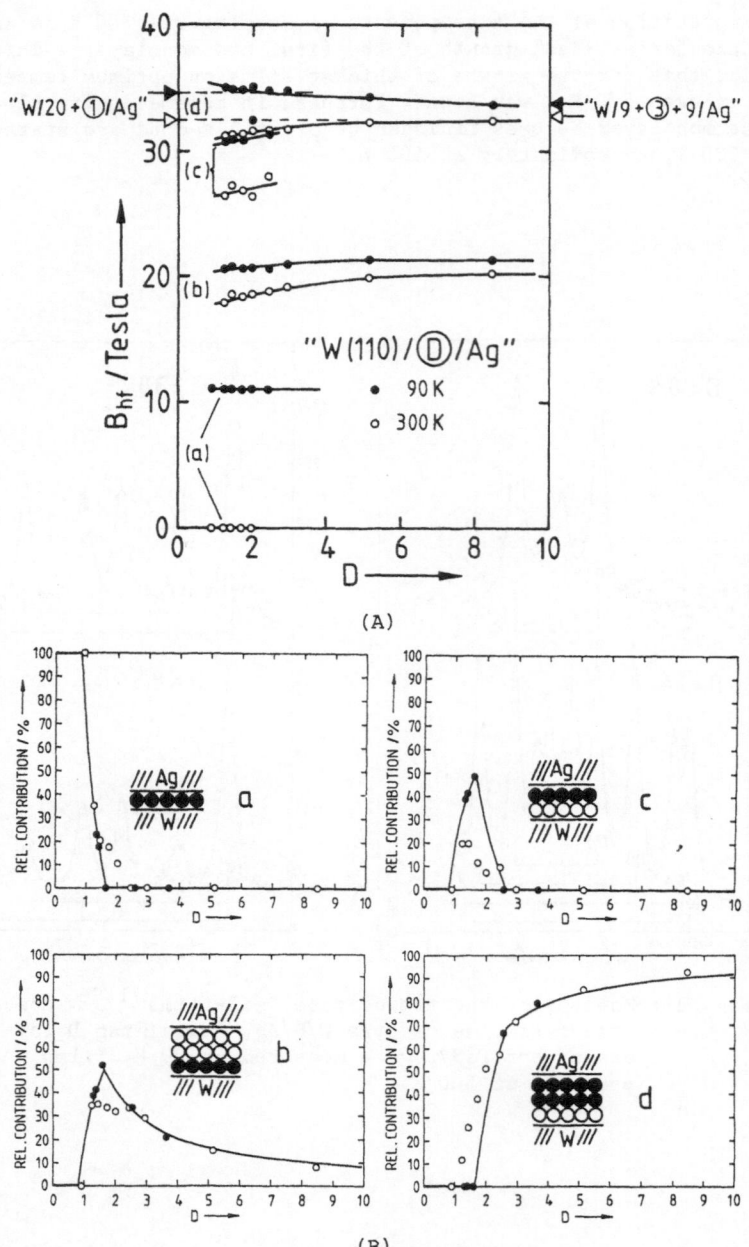

Fig. 16: Four components in the Moessbauer spectra
of pure Fe57-films on W(110), coated by
Ag(W(110)/D/Ag).
(A) B_{hf} of components (a)-(d), measured at
90 K (●) and 300 K (o), respectively, as a
function of D, the number of atomic Fe-lay-
ers. Films prepared at 500 K.
(B) Relative contributions of components
(a)-(d), for films prepared at 500 K (o) and
300 K (●), resp., in comparison with structural model, as indicated.

Finally, it is interesting to follow the contributions of first layer atoms, (a) + (b), as a function of D. This is shown in fig.17.

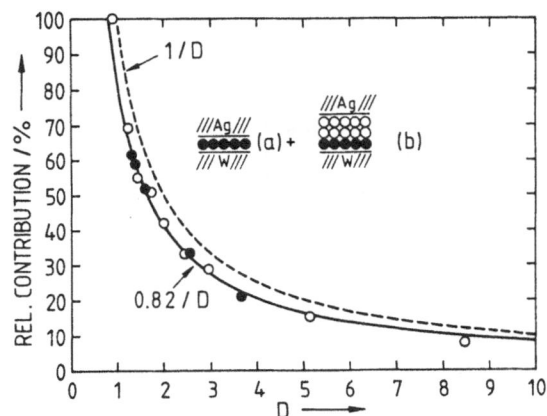

Fig. 17: The relative contributions of components
(a)+(b), compare fig. 16, are given by
0.82/D, not by 1/D. For Films prepared at
500 K (o) and 300 K (•), resp..

4. CONCLUSIONS

Experimental analysis of magnetic interface phenomena can be performed using carefully prepared and texted epitaxial single films. Experiments with an ultrasensitive Torsion Oscillation Magnetometer and with high resolution Conversion Electron Moessbauer Spectroscopy are complementary: TOM provides direct access to changes of magnetic moments and of interface anisotropies when the structure of the interface is changed. CEMS has the advantage of unique local resolution of hyperfine interaction parameters including the magnetic hyperfine field. Valuable structural informations are provided by the CEMS-analysis, too.

This work was supported by SFB 126 of the Deutsche Forschungsgemeinschaft and by the Stiftung Volkswagenwerk.

References

1. C. M. Falco and I. K. Schuller, in "Synthetic Modulated Structures/ VLSI", ed. by L. L. Chang and B. C. Giessen, Academic Press, Orlando (1985).
2. R. F. C. Farrow, S. S. P. Parkin, P. J. Dobson, J. H. Neave, A. S. Arrott, (ed), Thin-Film Growth Techniques for Low-Dimensional Structures, Plenum 1987.
3. J. Kwo in reference (2).
4. J. P. Renard in reference (2).
5. U. Gradmann and J. Mueller, Phys. Stat. Sol. $\underline{27}$, 313 (1986).
6. U. Gradmann, Appl. Phys. $\underline{3}$, 161 (1974).
7. U. Gradmann, R. Bergholz and E. Bergter, Thin Solid Films $\underline{126}$, 107 (1985).
8. L. M. Falicov and J. L. Moran-Lopez, (ed), "Magnetic Properties of Low-Dimensional Systems", Springer-Verlag (1986).
9. U. Gradmann in reference (8).
10. J. Korecki and U. Gradmann, Phys. Rev. Lett. $\underline{55}$, 2491 (1985).
11. For a theoretical review compare A. Corciover, G. Costache and D. Vaman, Solid State Physics (Academic Press (1972)) p. 237.
12. H. Hasegawa, Surface Science, to be published.
13. D. Wang, A. J. Freeman and H. Krakauer, Phys. Rev. $\underline{26}$, 1340 (1982).
14. J. Tersoff and L. M. Falicov, Phys. Rev. $\underline{B\ 26}$, 6186 (1982).
15. For a review of theoretical work on surface/interface magnetization compare articles of A. J. Freeman and R. H. Victora in reference (8).
16. S. Ohnishi, M. Weinert and A. J. Freeman, Phys. Rev. $\underline{B\ 30}$, 36 (1984).
17. C. L. Fu, A. J. Freeman, T. Oguchi, Phys. Rev. Lett. $\underline{54}$, 2700 (1985).
18. L. Néel, Compt. Rend. $\underline{237}$, 1468 (1953).
19. U. Gradmann, J. Magn. Magn. Mat. $\underline{54-57}$, 733 (1986).
20. T. E. Gallon, Surface Sci. $\underline{17}$, 486 (1969).
21. M. P. Seah and W. A. Dench, Surf. Interf. Analysis $\underline{1}$, 2 (1979).
22. W. Schlenk and E. Bauer, Surf. Sci. $\underline{93}$, 9 (1980).
23. S. Stoyanov and I. Markov, Surface Sci. $\underline{116}$, 313 (1982).
24. P. I. Cohen in reference (2); compare also C. S. Lent and P. I. Cohen, Phys. Rev. $\underline{B\ 33}$, 8329 (1986).
25. M. Henzler, Appl. Phys. $\underline{A\ 34}$, 205 (1984).
26. V. Scheithauer, G. Meyer, M. Henzler, Surface Sience, to be published.
27. E. Bauer, Z. Kristallogr. $\underline{110}$, 372 (1958).
28. I. Markov and R. Kaischew, Thin Solid Films $\underline{32}$, 163 (1976).
29. I. Markov and R. Kaischew, Kristall Tech. $\underline{11}$, 685 (1976).
30. U. Gradmann, W. Kuemmerle and P. Tillmanns, Thin Solid Films $\underline{34}$, 249 (1976).
31. U. Gradmann and P. Tillmanns, Phys. Stat. Sol. (a) $\underline{44}$, 539 (1977).
32. E. Bauer and J. H. van der Merwe, Phys. Rev. $\underline{B\ 33}$, 3657 (1986).
33. F. C. Frank and J. H. van der Merwe, Proc. Roy. Soc. $\underline{A\ 198}$, 216 (1949).
34. J. H. van der Merwe, J. Appl. Phys. $\underline{34}$, 117, 123 (1963).
35. J. W. Matthews, Phil. Mag. $\underline{6}$, 1347 (1961); $\underline{13}$, 1207 (1966).
36. U. Gradmann, Ann. Phys. (Leipzig) $\underline{17}$, 91 (1966).
37. J. W. Matthews and J. L. Crawford, Thin Solid Films $\underline{5}$, 187 (1970).
38. J. W. Matthews and E. Klokholm, Mat. Res. Bull. $\underline{7}$, 213 (1972).
39. For a review of coherent interfaces and misfit dislocations, compare J. W. Matthews in J. W. Matthews, (ed), Epitaxial Growth, part B, Academic Press (1975).
40. U. Gradmann, Phys. kondens. Materie $\underline{3}$, 91 (1964).
41. U. Gradmann and G. Waller, Surface Science $\underline{116}$, 539 (1982).
42. R. Bergholz and U. Gradmann, J. Magn. Magn. Mat. $\underline{45}$, 389 (1984).

43. P. J. Besser, J. E. Mee, P. E. Elkins and D. M. Heinz, Mat. Res. Bull. 6, 1111 (1971).
44. P. Haasen, Physical Metallurgy, Cambridge 1978.
45. H. C. Siegmann in reference (2).
46. A. S. Arrott, B. Heinrich, C. Liu and S. T. Purcell in reference (2).
47. G. A. Prinz in reference (2).
48. B. Heinrich, A. S. Arrott, J. F. Cochran, S. T. Purcell and N. Alberding in reference (2).
49. P. Gruenberg in reference (2).
50. R. Willis in reference (2).
51. J. Tyson, A. H. Owens, J. C. Walker and G. Bayreuther, J. Appl. Phys. 52, 2487 (1981).
52. J. Korecki und U. Gradmann, Phys. Rev. Letters 55, 2491 (1985).
53. R. Feder, (ed), Polarized Electrons in Surface Physics, World Scientific Publ., Singapore (1985).
54. U. Gradmann, J. Appl. Phys. 40, 1182 (1969).
55. U. Gradmann and J. Mueller, Czech. J. Physics B21, 553 (1971).
56. U. Gradmann, W. Kuemmerle and R. Tham, Appl. Physics 10, 219 (1976).
57. U. Gradmann and K. Salewski, Physica 86-88B, 1397 (1977).
58. U. Gradmann and K. Salewski, Physica 86-88B, 1399 (1977).
59. U. Gradmann and K. Salewski, Phys. Stat. Solidi (a) 39, 41 (1977).
60. U. Gradmann, J. Magn. Magn. Mat. 6, 173 (1977).
61. W. Kuemmerle and U. Gradmann, Solid State Communications 24, 33 (1977).
62. W. Kuemmerle and U. Gradmann, Phys. Stat. Sol. (a) 45, 171 (1978).
63. U. Gradmann and H. O. Isbert, J. Magn. Magn. Mat. 15-18, 1109 (1980).
64. U. Gradmann and R. Bergholz, Phys. Rev. Lett. 52, 771 (1984).
65. U. Gradmann, R. Bergholz and E. Bergter, IEEE Transactions Magnetics 20, 1840 (1984).
66. E. Bergter, U. Gradmann and R. Bergholz, Solid State Communications 53, 565 (1985).
67. U. Gradmann, J. Korecki and G. Waller, Appl. Phys. A 39, 1-8 (1986).
68. J. Korecki and U. Gradmann, Hyperfine Interactions, 28, 931 (1986).
69. J. Korecki and U. Gradmann, Europhys. Lett., 2 (8), 651 (1986).
70. J. B. Thaler, J. B. Ketterson and J. E. Hilliard, Phys. Rev. Lett. 41, 336 (1978).
71. M. Przybylski, U. Gradmann and J. Korecki, to be published.
72. R. Bergholz, Doctoral Thesis, Clausthal 1984.
73. C. A. T. Allan, Phys. Rev. B1, 352 (1970).
74. J. Tersoff and L. M. Falicov, Phys. Rev. B26, 6186 (1982).
75. J. Mathon, J. Phys. F 16, 669 (1986).
76. J. Mathon, J. Phys. F 16, L217 (1986).
77. W. H. Meiklejohn and C. P. Bean, Phys. Rev. 102, 1413 (1956) and 105, 905 (1957).
78. N. H. March, Ph. Lambin and F. Herman, J. Magn. Magn. Mat. 44, 1 (1984).
79. L. Fritsche, J. Noffke and H. Eckhardt, private communication.
80. G. T. Rado, Bull. Am. Phys. Soc. 2, 127 (1957).
81. C. L. Fu and A. J. Freeman, to be published; compare A. J. Freeman in reference (8).
82. G. Bayreuther and G. Lugert, J. Magn. Magn. Mat. 35, 50 (1983).
83. T. Shigematsu, H. D. Pfannes and W. Keune, Phys. Rev. Lett. 45, 1206 (1980).
84. W. Keune, Hyperfine Interactions 27, 111 (1986).
85. M. Przybylski, J. Korecki and U. Gradmann, to be published.
86. J. Korecki, M. Przybylski and U. Gradmann, to be published.

INCREASED MAGNETIC MOMENTS IN TRANSITION ELEMENTS THROUGH EPITAXY

A. S. Arrott, B. Heinrich, C. Liu[‡] and S. T. Purcell

Surface Science Laboratory, Department of Physics
Simon Fraser University
Burnaby, British Columbia, Canada

The elements of the first transition series, Ti, V, Cr, Mn, Fe, Co, Ni, and Cu, are of special importance for magnetism and metallurgy. The phase diagrams of these elements and their alloys with one another and elements such as C, Si and Al fill encylopedic volumes.[1-5] The correlation between atom size, by various measures, and magnetic moment has long been noted.[6] This correlation is readily seen by comparing the atomic concentrations of the first, second and third transition series elements as shown in Fig. 1, where the densities for the second and third series elements have been scaled for comparison with the first transition series. It seems that there is some missing density, excess volume, in Cr, Mn, Fe, Co, and Ni, all of which show ordering of magnetic moments. The main purpose of this paper is to argue that in the cases of Fe, Mn, Cr, and possibly V, artificially increasing the volume of these elements through controlled epitaxial growth may lead to higher magnetic moments and other technologically important magnetic properties. We like to call this atomic engineering, implying that we are building structures atom by atom.

We start with the case of Fe for the sake of introducing the subject of magnetism. Then we turn to Mn which is the main subject of our own work.

Iron

Fe continues to be most interesting both scientifically and technologically. Fe exists in three structural phases: body centered cubic, bcc, at high temperature from 1633 K to the 1753 K melting point and at low temperatures up to 1180 K; face centered cubic, fcc, from 1180 K to 1633 K; and hexagonal close packed for high pressures and moderate to low temperatures. Strictly speaking, the bcc phase becomes tetragonal (~1 ppm) below its ferromagnetic Curie temperature, T_c, at 1040 K. The magnetic moment of bcc Fe as measured by its saturation magnetization at low temperatures is 2.2 μ_B per atom. An extrapolation to Fe of the saturation moments of Ni (0.6 μ_B per atom) and Co (1.7 μ_B per atom) indicates the possibility of 2.8 μ_B per atom, which would decrease the cost of electrical equipment by more than this ratio. The first question is whether the application, somehow, of negative pressure could increase the lattice constant of Fe sufficiently to achieve this higher moment. An estimate of the volume expansion necessary is about 10 percent. The higher moment would be partially offset by the decreased atom density.

The addition of N to bcc Fe results in a tetragonal expansion of the lattice with an apparent increase in the average magnetic moment to 3 μ_B per atom in the structure $Fe_{16}N_2$.[7] N goes in and out of the bcc lattice in a cooperative manner expanding the lattice in one direction. At stoichiometry, 1/4 of the Fe

---

[*] This work is supported by the Natural Sciences and Engineering Research Council of Canada and in part presents the Ph. D. thesis of C. Liu, Simon Fraser Univerity, 1986.
[‡] Present address: Physics Department, Rice University, Houston, TX, U.S.A. 77251

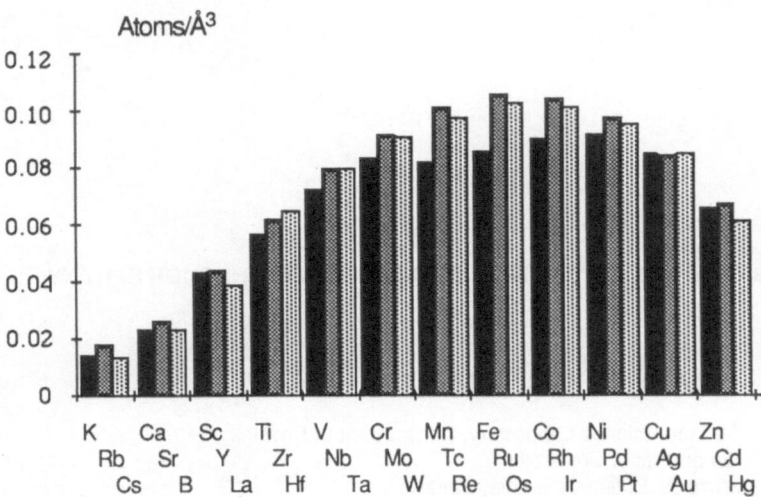

Fig. 1. Atomic concentration for the first three transition series elements. The second and third series have been multiplied by 1.40 and 1.44, respectively, for comparision with the first series to show the missing density in that series and therefore the correlation of increased volume with magnetism.

atoms are directly pushed upon by the N atoms. (Some might say that the N atoms donate electrons to the Fe atoms.) The rest, which get the benefit of a 14 percent increase in volume, appear to account for the increase in average magnetic moment.[8]

Addition of more N to Fe stabilizes an fcc lattice of Fe atoms in the structure Fe_4N. In this case 3/4 of the Fe atoms are pushed upon by the N atoms, see Fig. 2. The moments remain ferromagnetically aligned and the moment on the pushed upon Fe atoms remains ~2 μ_B per atom. But the other 1/4 of the Fe atoms benefit from an increase in volume and show a magnetic moment ~ 3 μ_B per atom.[9]

In cubic crystals, the ten states of 3d symmetry split into four states with E_g symmetry (lobes toward the six second neighhors of the bcc lattice), two with spin up and two with spin down, and six states of T_{2g} symmetry (lobes toward the eight first neighbors) three with spin up and three with spin down. If the eight electrons of metallic bcc Fe, outside its closed Ar core, were to occupy three of the T_{2g} bands of one spin and the four E_g bands with the eighth electron in the unpolarized 4s-4p band, one could obtain Fe with 3 μ_B per atom.

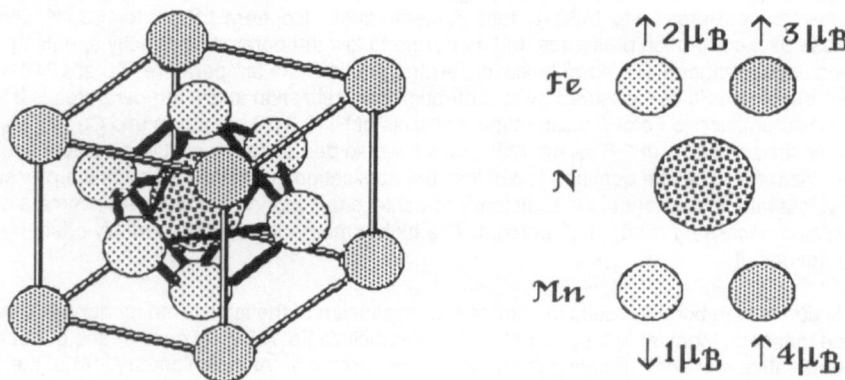

Fig. 2. The structure of Fe_4N and Mn_4N wherein an expanded fcc lattice of the transition metal is stablized by the N atom in the center of a regular octahedron composed of atoms of three of the four simple cubic sublattices of the fcc structure. The moment on the corner atoms increases to 3 μ_B for Fe and 4 μ_B for Mn.[9] Fe_4N is ferromagnetic. Mn_4N is ferrimagnetic.

In compounds with other elements one obtains up to 5 μ_B per Fe atom in the trivalent state. Such compounds have a low density of Fe atoms or the magnetic moments are arranged ferrimagnetically (such as in barium ferrite, the common household permanent magnet in which some moments point in opposing directions leading to a lower saturation moment) or antiferromagnetically in which case the moments cancel completely.

There are two immediate aims in engineering ferromagnetic materials with higher magnetizations. One is to produce a higher magnetic moment per atom without diluting the concentration of the magnetic atoms. The other is to obtain a ferromagnetic alignment of those moments. The magnetic properties of many thousands of intermetallic compounds containing the elements Cr, Mn and Fe stand as an indication that it may not be possible with conventional growth techniques. The structures that form as the result of equilibrium kinetics favor higher densities than are suitable for large moments and ferromagnetic interactions between nearest neighbors.

One might wonder why one is interested in higher moments for the first transition series elements when the rare earths have much larger moments, e.g. Gd (7 μ_B), Tb (9 μ_B), Dy (10 μ_B). The reasons are lower costs, higher Curie temperatures and higher atom densities. (Fe and Gd have almost the same weight densities, but Gd has almost three times the atomic weight.)

The effect of interstitial N in Fe is to expand the lattice by pushing from the inside. To the extent that atoms are less tightly bound at a surface, one could view a free surface as applying a negative pressure to the surface atoms. If the surface is stretched by epitaxial registry with a lattice of larger spacing, this too is a negative pressure of a sort. There is evidence both theoretical and experimental for obtaining metallic Fe with 3 μ_B per atom at surfaces or in thin layers grown epitaxially. This subject is of current interest.[10] The gains to be obtained from the increased moment of Fe are modest but of considerable economic potential. Fe(N) particles are considered for magnetic recording tapes.[11]

Manganese

Manganese as the element next to iron has one fewer electron. The low temperature α phase is stable up to 1000K. It is antiferromagnetic below 95K. The β phase forms from the α phase at 1000 K and is stable up to 1368 K as well as at low temperatures on quenching from above 1000 K. Both of these phases are quite complex. From 1368 K to 1406 K fcc γ-Mn is stable. If this phase is quenched, it undergoes a tetragonal distortion by ~4 per cent. From 1406 K to the 1450 K melting point bcc δ-Mn is stabilized by the vibrational entropy of this phase.

In the free atom Mn has five electrons in a half filled 3d shell and two electrons in the 4s shell. The moments of five 3d electrons are aligned parallel in the free atom in accordance with Hund's rule.[12] In certain metallic environments, Mn atoms continue to obey Hund's rule. In others the measurements of susceptibility and specific heat indicate that there is no thermodynamic spin degree of freedom, that is, if there are moments on individual Mn atoms at any instant of time, these are not retained long enough for any significant precession about local fields from some neighboring atoms. Magnetic moments can be deduced from susceptibility measurements by ploting $1/\chi T$ vs $1/T$ or from specific heat by calculating the magnetic entropy. If a material is magnetically ordered, the moment can be determined by neutron scattering. If a material is ferromagnetically ordered the moment can be measured directly by magnetic induction or by magnetic force. In principle a moment in the paramagnetic state can be measured by application of sufficiently large fields at sufficiently low temperatures. In fact this has been done for dilute alloys of Mn in a Ag lattice.[13] In this case the Mn atom does have 5 μ_B per atom as indicated also by the temperature dependence of the magnetic susceptibility and specific heat.

Note that Mn_4N has two types of sites as shown in Fig. 2. The presence of the N atom expands the fcc lattice of Mn atoms by pushing on its six nearest Mn neighbors. The Mn atoms at the corners of the cubic unit cell benefit from the expanded lattice without having to compete with the N atom for space. These Mn atoms have ~4 μ_B per atom while the others have ~1 μ_B per atom in the ferrimagnetic phase.[9] The volume per atom of the corner atoms is perhaps 10 percent larger than one would deduce, extrapolating to lower temperatures, from the lattice spacing of the high temperature γ phase. For a lattice in which all of the atoms are in equivalent sites, the volume per atom is unambiguous, but, for more complex structures or for amorphous materials, there always remains some freedom of choice as to what volume to assign to each atom. The history of this subject is covered by Pearson.[14]

The correlation of magnetic moment and volume for Mn in many environments has been reviewed by Mori and Mitsui.[15] As a measure of volume, Mori and Mitsui introduce the average Pauling[16] valence based on the spacing to each neighbor. Fig. 1 of their paper, showing a linear increase in moment with valence, could be augmented by many more points that would add further evidence for the correlation.

A measure of the volume per atom which is applicable even for non-crystalline solids, has been given by Gellatly and Finney.[17] This technique has been applied by Watson and Bennett[18] to the four sites in α-Mn. The assignment involves choosing the size of starting spheres. The assignment of equal radii minimizes the differences in calculated volumes. With this most conservative choice they find that sites I, II, III and IV have atomic volumes of 13.4, 13.1,12.5 and 11.6 Å^3, respectively.[19] The large sites I and II appear to have large magnetic moments, the small sites IV have none, and the moments of sites III are of intermediate size. There are major problems in deducing the actual moments in either the paramagetic state or the antiferromagnetically ordered state of such a complicated phase as α-Mn.[20]

The basic building block of α-Mn is the 12 atom hexatetrahedron shown in Fig. 3. These are the IV sites. If the hexatetrahedron is composed of 12 just touching spheres, the size of the large hole in the middle of the hexatetrahedron, Site I, is such that a sphere that just touches the 12 neighbors would have about twice the volume as the spheres in the hexatetrahedron. Such a sphere would have to have four large flats where it meets the four large Site II atoms at the hexagonal holes in the faces of the hexa-tetrahedron. In α-Mn the Site I atoms are placed on a bcc lattice. Each is surrounded by the 12 atoms of the hexatetrahedron and the four atoms of Site II. Each cluster of 17 atoms is surrounded by a sphere (almost) of 24 type III sites, the glue that holds together the 29 atoms of the bcc unit cell (58 atoms of the cubic unit cell). A colored picture of a ball and stick model of α-Mn is to be seen as the fronticepiece of Goldschmidt's "Interstitial Alloys".[21] That an element would form such a structure is quite amazing.

Mn can adapt to quite different environments. A large Mn site is likely to have a large moment, sometimes as large as 5 μ_B. If Mn were to be placed epitaxially on a substrate with an appropriately large lattice spacing, would Mn adapt by choosing the Hund's rule state with 5 μ_B per atom? This is the question that led us to obtain the molecular beam epitaxy (MBE) facility shown in Fig. 4 for the study of magnetic materials.

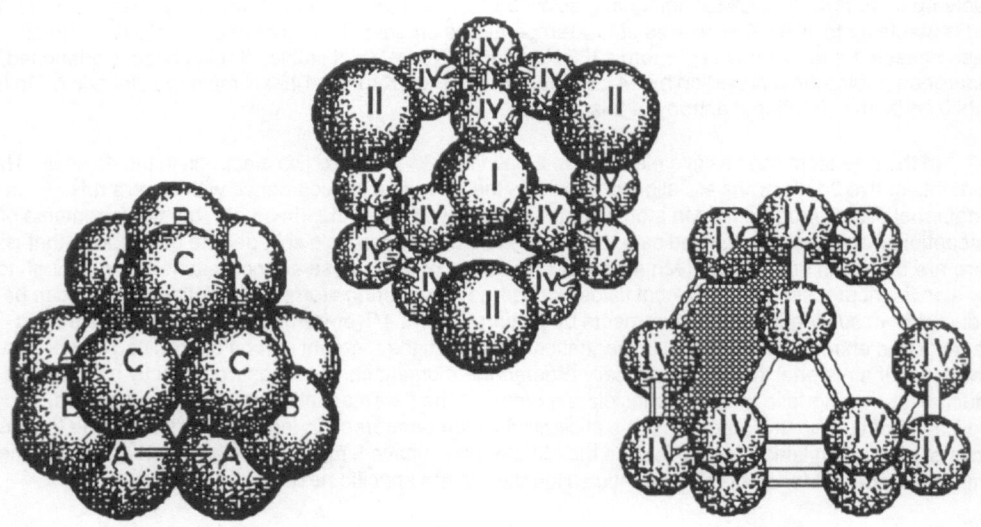

Fig. 3. The hexatetrahedron, the basic building block of α-Mn, is shown as a close packing of twelve balls on the left and as a ball and stick model on the right. If each of the four hexagonal faces were filled with an equal size ball, the struture would be close packed. In α-Mn these four atoms are removed and replaced by one in the center, with approximately twice the volume. (As the four atoms that are removed are half into the center, they count only as two.) The large atom in the center of the hexatetraderon joins to four other large atoms (of which only three are visible) through the hexagonal faces, as shown at the center.

We chose the close packed surface of Ru(0001) as the initial substrate because its spacing matches the nearest Mn neighbor distances in Mn_4N. Further, Ru(0001) is a well studied surface.[22] Our earliest work was with reflection high energy electron diffraction, RHEED, as our only in situ experimental tool. As we had not yet acquired the standard tools of surface analysis, we had to rely on what was known about the surface preparation of Ru.

Even if the first layer of Mn were to go down epitaxially on Ru, one might wonder, would expanding the lattice in two dimensions result in an expansion in the direction perpendicular to the surface. After all, Poisson's ratio comes close to conserving volume. But that is for linear elasticity with small variations from equilibrium. For large changes in volume, accompanied by a change in the magnetic state, linear elasticity no longer applies. The Hund's rule state for Mn would require using all five 3d wave functions of one spin. This state is necessarily spherical. Thus we expect that if Mn accommodates to the Ru lattice spacing, it will acquire a large moment as it does in Mn_4N and it will also propagate that spacing perpendicular to the surface. In this sense we view MBE as a technique for applying negative pressure to magnetic materials.

All of the above is the background for our study of the epitaxial growth of Mn. We report here the results of that study. This includes: the preparation of the substrate and analysis of its surface by Auger electron spectroscopy, AES, and by x-ray photoelectron spectroscopy, XPS; Mn growths at various temperatures; RHEED studies of the growth process and resulting structures; and AES and XPS studies of the ultra thin Mn films that we have produced. Of particular note is our use of the splitting of the 3s peak in XPS to deduce the magnetic moments of Mn atoms in various structures. Our initial work with Mn on Ru is supplemented by studies of alloys of Mn in Ag and in V, of α-Mn and of the bcc phase of Mn grown of the (001) surface of Fe. We have also grown Mn epitaxially on Ni(001), but we did not decipher the complex RHEED pattern.

We believe that the first two epitaxial layers of Mn on Ru, which are in an expanded lattice, have $5\mu_B$ per Mn atom. We have not yet carried out measurements that would determine what cooperative magnetic state results. Beyond the second layer a new phase of Mn forms that we believe is a Laves phase, closely related to α-Mn in that it is based also on the hexatetrahedron, but much simpler. This new phase, likely the Zn_2Mg structure, may be of lower energy than α-Mn.

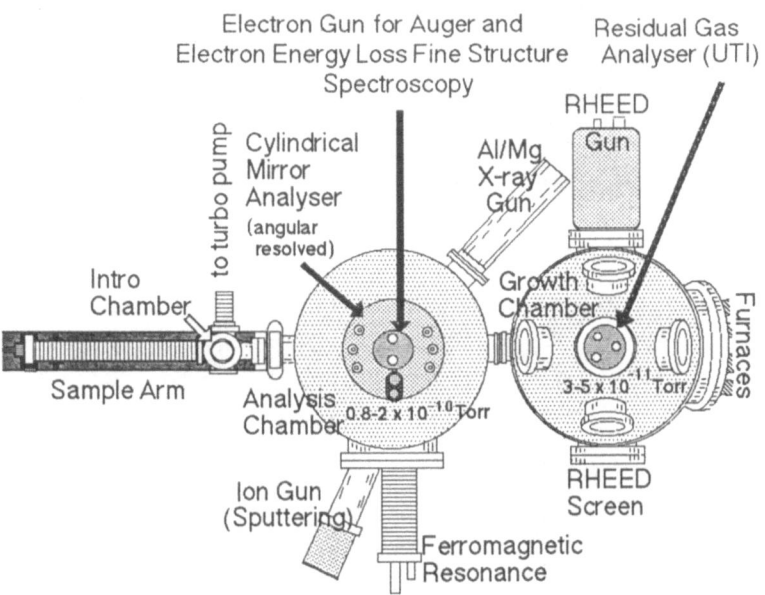

Fig. 4. Schematic layout of the PHI 400 MBE system with several tools of surface analysis. The sorption, turbo, ion, sublimation and cryo- pumps are not indicated. A unique feature of this system is the presence in the UHV of a 10 GHz cavity for ferromagnetic resonance.

Experimental Apparatus and Techniques

This research is carried out in our Physical Electronics (PHI) model 400 MBE facility equipped with RHEED, Residual Gas Analysis, RGA, and a double pass Cylindrical Mirror Analyser, CMA, for energy analysis of electrons emitted under electron beam excitation for AES or under Mg X ray excitation for XPS. The MBE machine allows evaporation from a series of sources on to a substrate held on the end of the sample arm which can be moved in and out of the several chambers, see Fig. 4.

RHEED allows characterization of the substrate surface, including surface roughness and the density of steps[23] as well as crystal structure . This is important in the preparation of the substrate which includes mechanically polishing outside the Ultra High Vacuum (UHV) system, followed by a sequence of Argon ion sputterings and recrystallization anneals in the UHV.

The growth of Mn on the (0001) surface of hexagonal Ru is monitored by RHEED at 10 keV. RHEED patterns for the Ru substrate and for a growth of 12 ML of Mn are shown in Fig. 5. The RHEED patterns of the substrate are those of a closepacked surface, having six fold symmetry. The interpretation of the position of the RHEED streaks can be carried out by simple kinematic diffraction theory applied to the top most layers. Our Ru crystal is a collection of six grains within an angular spread of 40 mrad. These are then necessarily vicinal surfaces, which means that the growth direction is at a small angle with respect to the surface normal. A vicinal surface provides steps for the nucleation of the epitaxial structure. The steps can be observed by sending the electron beam parallel to the step edges, which gives structure to the diffraction streaks (not visible in Fig. 5). The slight splitting of each streak is proportional to the vicinal angle, i.e. to the reciprocal of the mean step width. The importance of unidirectional steps in the growth of epitaxial layers is a subject of current investigations.[24] Clearly steps will influence cluster migration and rotation.

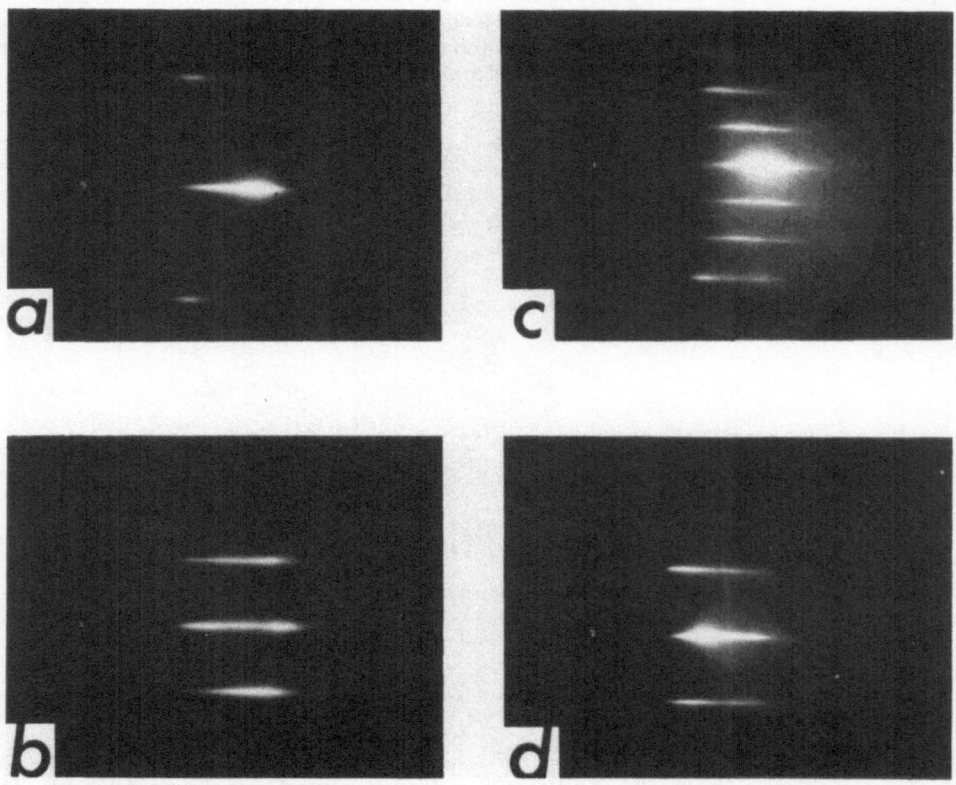

Fig. 5. RHEED patterns for the Ru(0001) substrate (a and b) and for a 12 ML (monolayer) deposit of Mn (c and d). The patterns are interpreted in Figs. 11 and 12.

We prepare the substrate by mechanical means, finishing with 1μm diamond. This is followed by sputtering with Ar$^+$ in the UHV system and annealing at 600° C which yields well defined RHEED patterns. These are never as sharp in photographs as they are to the eye on viewing the RHEED screen. The surface is then studied with AES and XPS to determine the residual impurities and to give a base line for the subsequent measurements of the thickness of the overlayers.

The furnaces use pyrolitic Boron Nitride crucibles for the evaporation of the elements Mn, Ag, Au and Fe. At temperatures over 1000° C the crucible gives off nitrogen, presumably through some mechanism that converts BN to some nitride with higher vapor pressure. Wire filaments of W or Mo are used to evaporate V,Cr, and Ni which are wrapped around the filaments. Alloys were made by coevaporation. The furnace temperatures are set by trials to find the right ratios of deposition as determined by XPS measurements after each trial. Mn is evaporated at ~610° C, (thermocouple reading). The deposition rate is ~1/2 ML (monolayers)/min. The RHEED patterns alter slightly in intensity and in sharpness during the growth of the first two monolayers of Mn, but there is little doubt that the growth is fully epitaxial. Beyond the first two monolayers, the RHEED pattern changes to one with three times as many streaks along the directions perpendicular to the close packed rows. This pattern corresponds to a close packed structure with each dimension increased by √3 and rotated by 30 or 90 degrees. It is called the (√3 x √3) R30° structure. We will argue below that this pattern arises from the formation of a Laves phase in the case of Mn.

The first two epitaxial layers of Mn have the same close packed structure as the Ru, but we do not know the stacking of those two layers with respect to one another or with respect to the Ru substrate. If Ru is an A layer and the first Mn layer is a B layer, is the second Mn layer a C layer or an A layer? How much difference would it make? Neither do we know the spacing in the growth direction. If one puts Mn down on a stepped Ru surface, it is the height of the Ru steps that would dominate the intensity distribution along the streaks in the RHEED pattern.

AES is mainly used to look for impurities on the substrate or in the evaporated overlayers. AES is used sparingly on the Mn deposits because it greatly increases the oxidation rate, presumably from the decompostion of CO_2. A typical AES spectrum of Mn on Ru is shown in Fig.6. Time dependences of the AES signals are shown in Fig. 7.

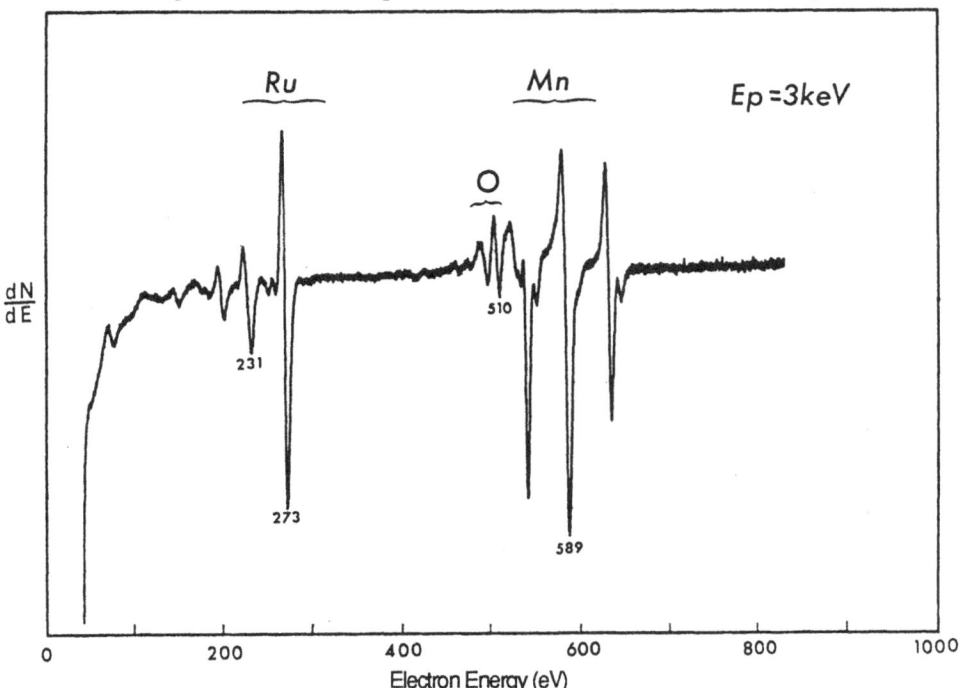

Fig. 6. An Auger Electron Spectroscopy survey for ~2 ML of Mn on Ru(0001). The primary electron beam energy was 3 keV. The modulation used was 6eV peak to peak.

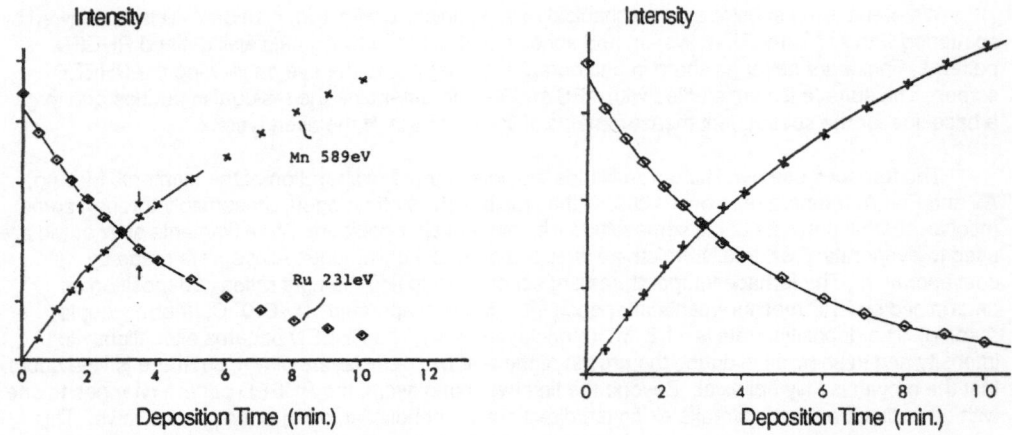

Fig. 7. The dependence of the AES signals from the Ru substrate and the Mn overlayers upon the Mn deposition time. The same data is fitted with straight line segments with breaks at the completion of the first and second Mn layers, marked by arrows, on the left and with exponentials on the right.

The major analytical tool in this research is XPS. The XPS spectrum for ~2 ML of Mn on Ru is shown in Fig. 8. The 3s peak of Mn, which is quite close to the 4s peak of Ru, is shown magnified by a factor of 10 at the right side of the spectrum. The Ru $M_{45}VV$ Auger electron peak appears at the left side of the spectrum at an apparent binding energy of 980 eV. The kinetic energy of this Auger electron is 273 eV as it leaves the surface. The kinetic energy of the electron coming from the Ru 3d peak at

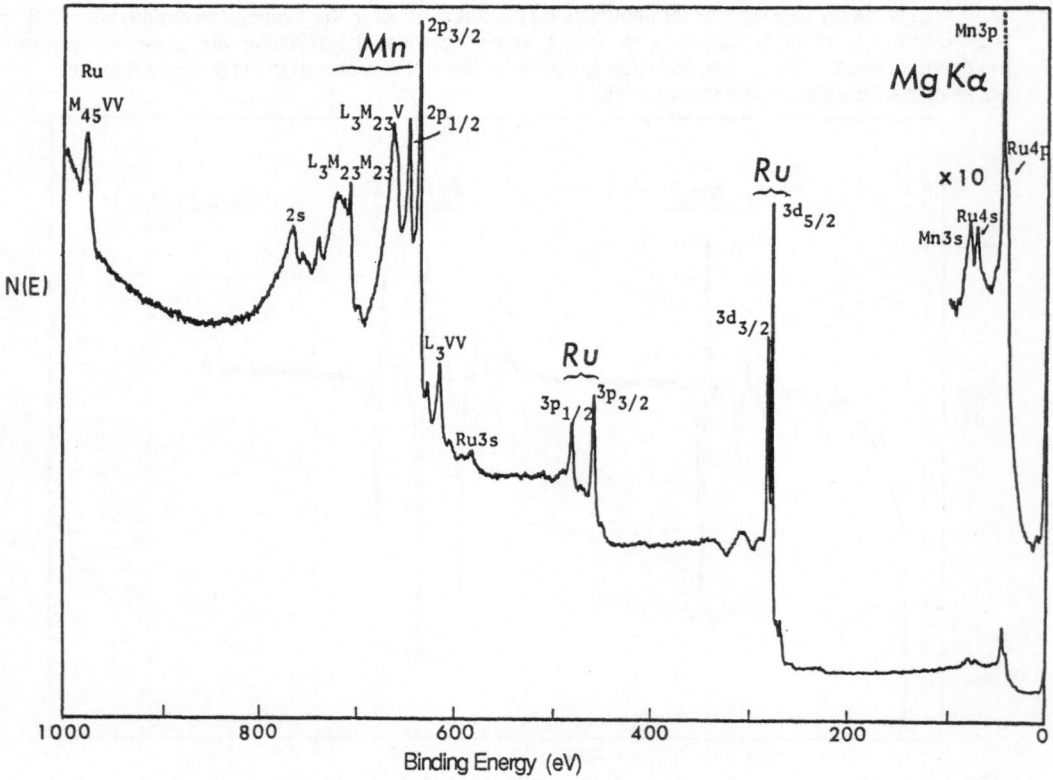

Fig. 8. XPS spectrum of a ~2 ML Mn thin film on Ru(0001) taken with Mg-K_α X ray radiation and a 50eV pass energy selected for the CMA. In the identification of the peaks, the symbol V stands for valence levels in the Auger excitations.

300 eV binding energy is 956 eV, as we use the Mg K$_\alpha$ line at 1256 eV. The intensity of the Ru M$_{45}$VV Auger electron peak decreases more rapidly than that of the Ru 3d peaks as the Mn coverage is increased, see Fig. 9. The exponential decreases in intensity can be used to measure the thickness of the Mn overlayer. As the apparent thicknesses are the same for the two different energy electrons, we have considerable confidence that we are growing uniform layers.

The rate of growth is controlled by setting the temperature of the Mn furnace. It is monitored by the Mn signal in the RGA. The absolute value is found from the XPS results, from the attenuation of the Ru signals, from the increase in the Mn signals and from the appearance of kinks in the time dependence of these signals at the completion of each of the first several layers. An example of AES intensities measured by periodically interrupting the growth process is shown in Fig. 7, where the same data is fit with straight line segments and to a smooth exponential. The Mn source was held at 615°C for this run, resulting in a growth rate of 0.57 ML/min. The break points are indicated at 1.75 and 3.5 min. As seen here the first Ru signal from the starting surface does not lie on the extrapolation back from the data taken after each Mn deposition. This inconsistency is a general rule, but the sign and magnitude of the deviation differs from run to run.

Because the material that results from the interruption of growth for AES measurement followed by resumption of deposition may not coincide with the results of continous deposition for the same total time, the experiment was carried out also in the following time consuming manner. The surface is sputter cleaned, annealed at 500°C, and cooled to growth temperature, 60°C. The deposition is carried out for 3.5 minutes and then the XPS and AES spectra were measured. The surface was again sputter cleaned, annealed and cooled. The next deposition was carried out for 7 min and XPS and AES again measured. And so on, for deposition times up to 18 minutes, with the results shown in Fig.10. While this sounds like doing it the hard way, we were motivated by the observation that the structure we obtain by the continous growth is quite different from that obtained by stopping and restarting the growth process. This difference is accompanied by increased oxidation in the latter process. As good as our vacuum is, it is not good enough to stop for the five or ten minutes necessary to carryout the spectral analyses, particularly when the sample is exposed to the electron beam for AES. Nevertheless, the XPS and AES intensities were consistent with layer by layer growth whether or not the growth was interrupted.

The vacuum is quite good. Between evaporations the ion gauge reads 3×10^{-11} torr, but this is close to its intrinsic limit. The cryopump and the ion pump are considerably aided by the gettering action of Mn which is sprayed around the system during evaporations and by thermal desorption during anneals. During evaporations the chamber vacuum rises to 10^{-9} torr due to H$_2$. The Mn source also gives off N$_2$, but this can be minimized by bringing the furnace to 1000°C for one hour before evaporation at ~610°C. Apparently the source accumulates nitrogen from the UHV system.

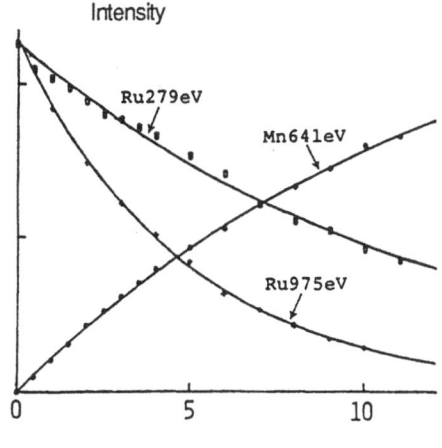

Fig. 9. XPS signals from Mn and Ru during the deposition fitted to exponentials.

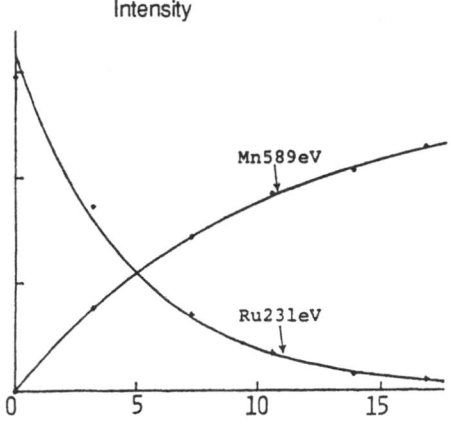

Fig. 10. AES signals where each deposition time started with a freshly prepared Ru substrate.

The reproducibility of the Mn evaporation rate from the controlled temperature of the furnance is most strikingly indicated by the change in the RHEED pattern that takes place during the growth of the third layer of Mn on Ru, see Fig 5. This change in pattern always appears 320 ± 10 seconds after the beginning of growth for our standard furnace temperature of 606°C. The growth rate is confirmed by deposition on the cleavage planes of the layered compound $NbSe_2$ which gives very sharp RHEED streaks. One monolayer of Mn is sufficient to completely remove any indication of the NbSe2 structure from the RHEED pattern. For the Mn furnace temperature of 606°C the time for disappearance of the $NbSe_2$ pattern in a series of trials was 135 ± 5 sec. This growth rate of slightly less than 1/2 monolayer per minute is typical of the rates we use in these studies.

The CMA can operate in an angular resolved mode for the detection of the electrons coming at various angles, e.g. perpendicular to or almost parallel to the surface. This is particularly useful in looking for variations in composition with depth below the surface, e.g., in distinguishing between oxygen on the surface and oxygen in the surface layers and in the alloy work that we describe below. In its regular mode the CMA accepts a conical shell of electrons with a half angle of 42 degrees. One of the disadvantages of our system is that the sample mount rotates about a point 3 cm. from the sample. In other words we do not have a proper goniometer which would be useful for fully exploiting the angular dependence of XPS.

RHEED patterns of Mn on Ru and Mn on Fe

The (0001) surface of Ru is a close packed layer with six fold symmetry. The primitive unit cell is shown as the grey diamond in Fig. 11a. The reciprocal lattice for electron diffraction from the surface layer is shown in Fig. 12a. The reciprocal lattice looks like the real lattice except for a reorientation of 30 degrees (or 90) and the replacement of the atoms by rods of intensity, corresponding to the fact that the real lattice appears very thin along the surface normal. The longest waves that are in phase with all of the atoms in the vertical and horizontal directions are indicated in Fig. 11a. The vertical periodicity is longer; correspondingly the reciprocal lattice rods in Fig. 12a are closer together in that direction. If every third atom is modified, the unit cell will increase in area by a factor of three and will rotate by 30 degrees, or 90 as shown by the grey diamond in Fig. 11b. This does not change the periodicity in the vertical direction. In the horizontal the periodicity is tripled and the reciprocal lattice becomes three times as dense as shown in Fig. 12b. The RHEED patterns of Fig. 5 correspond to Figs. 11 and 12. Patterns 5a and 5b go with Fig. 11a and Fig. 12a. Patterns 5c and 5d go with Fig. 11b and Fig. 12b. 5a and 5c correspond to the horizontal directions in Figs. 11 and 12. 5b and 5d correspond to the vertical directions.

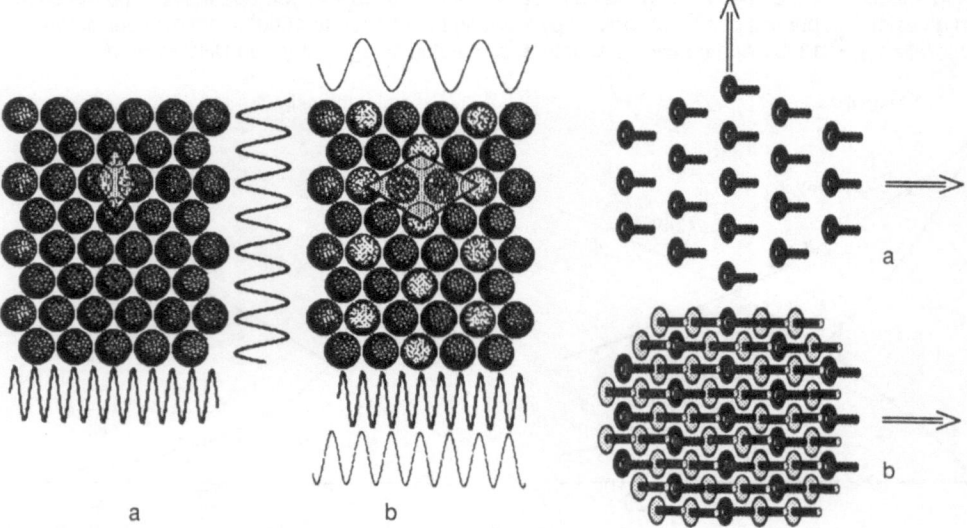

Fig. 11. The close packed lattice. (a) All atoms equivalent. (b) Every third atom is different. The unit cells are shown grey diamonds. The periodicities are indicated by the sine waves. The extra waves in (b) give the extra rods in the direction of the horizontal arrow in Fig. 12b.

Fig. 12. The reciprocal lattices (a) and (b) correspond to the real lattices (a) and (b) in Fig. 11. The first extra spot is the long wave in 11b, the second is the wave at the bottom

Note that the periodicity corresponding to the substrate is maintained in the growth of Mn. After the second layer one observes both of the extra waves indicated in Fig. 11b as well. Note also that the intensities are rather close for all three waves. If this were really the case for the Fourier components of the atomic density, one would be forced to the conclusion that only one atom in three was contributing to the intensity. Indeed this is just what one would expect for 1/3 coverage of the surface. But this pattern persists throughout the growth of Mn for as much as we have grown, which is 45 layers. We believe that one atom in each of the unit cells indicated by the grey diamond in Fig. 11b does actually stick up above the rest. We further believe that this is the fourth atom in a unit cell that has only three substrate atoms. We spectulate that Mn rearranges to increase its density, but does so in a way that accomodates the substrate.

An arrangement of three atoms and a hole in the space of three atoms in the sublayer is shown in Fig. 13. The three atoms sit on the ridges between pairs of atoms rather than nicely fitting into the close packed B sites that would normally follow an A layer in a hcp or fcc lattice. The hole is directly over an atom in the sublayer. An atom placed there would naturally stickup. A Mn would likely take a state with larger volume than that of the atoms that form the open hexagonal network. This arrangement would be consistent with the diffraction results. It is important to bear in mind that diffraction intensities for electrons are very complicated and deviate strongly from the results of simple kinematic theory that assumes weak interactions.

The next question is how could such an arrangement propagate. It seems most unlikely that all the atoms would sit on top of the ones under them. The packing of the next layer should differ from the first. The most likely places to put the next atoms are shown in Fig. 14 with and without the large B^2 atoms in place above the holes in the open hexagonal layer. The smaller B^1 atoms neatly surround the larger B^2 atoms. But as spheres, the B^1 atoms do not touch one another.

If the ratio of the diameters of the B^1 and B^2 atom were properly chosen, the centers of both the B^1 and B^2 atoms could lie in the same plane. If this were then a mirror plane, the next layer would repeat the first layer. This is the $CaCu_5$ structure. It is quite common for intermetallic systems where there is a very big difference in atomic sizes, as in the case of transition elements with rare earth elements. It does not occur for any two transition elements with one another, if we do not count Sc as a transition element. Part of the reason for this can be seen from Fig. 14. The B^1 atoms are much

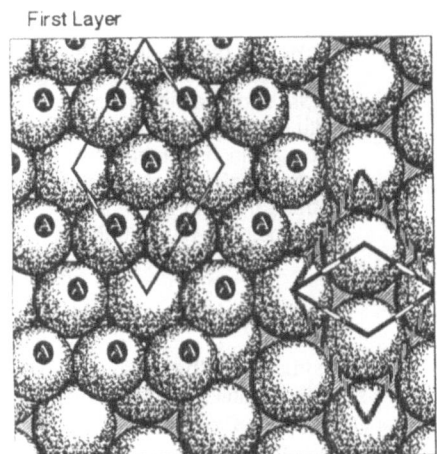

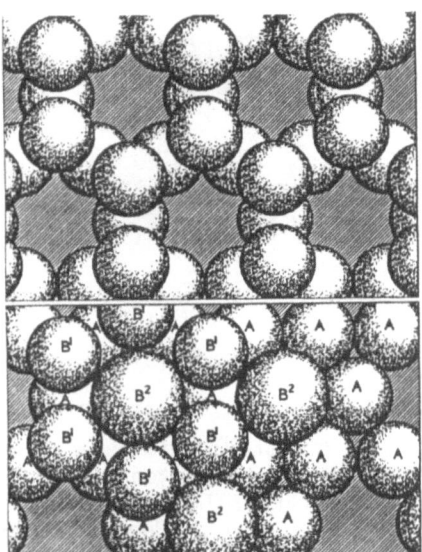

Fig. 13. The stacking of four closepacked sites in the overlayer in space of three closepacked sites in the sublayer. The three sites of the over layer, shown as occupied, are equivalent and form an open hexagonal network. The fourth site is shown as a hole in that network.

Fig. 14. If in stacking the second layer we let the center of the B^1 and B^2 atoms be in a common plane, which then becomes a mirror plane, we generate the $CaCu_5$ structure found mostly in transition metal-rare earth intermetallic compounds. This is unlikely for Mn.

too far apart for an atom like Mn which has the freedom to choose how big it should be for a given environment. It would pay to shrink the B^2 atoms a bit and to let every other B^1 atom become larger, rising above the plane of the small B^1 atoms by the same amount that the large B^2 atoms lie below the plane. This is shown in Fig. 15 where the B^2 atoms are now labled B^- and the larger of the B^1 atoms is labled B^+. The atoms in the next layer, labled C atoms, can then repeat the open hexagonal pattern of the A layer as indicated in Fig. 15. This structure can then be propagated.

There are as many ways to propagate this structure using the double layers (A, B^-,B,B^+) as there are ways to stack single close packed planes along the c axis, starting with hcp, fcc, dhcp, etc., including all possibilities of stacking faults. The hcp version of this stacking is called the Zn_2Mg structure. The fcc version is called the Cu_2Mg structure. Both are Laves phases, the most popular of all the intermetallic phases of transition metals. After all Mn has already demonstrated with the formation of α-Mn and β-Mn that it can form intermetallic phases with itself. The α-Mn structure is bcc with 58 atoms per cubic unit cell. That phase also occurs for $Al_{12}Mg_{17}$. Mg in its metallic hcp phase has an atomic volume that is 1.4 times that of Al in its metallic fcc phase.

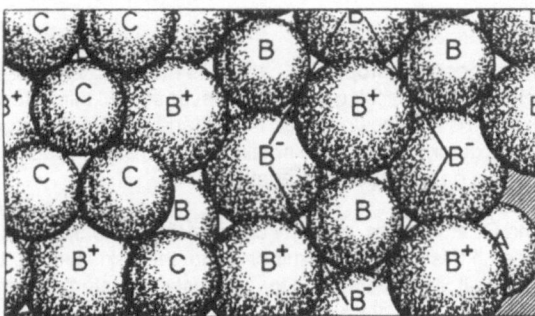

Third Layer   Second Layer

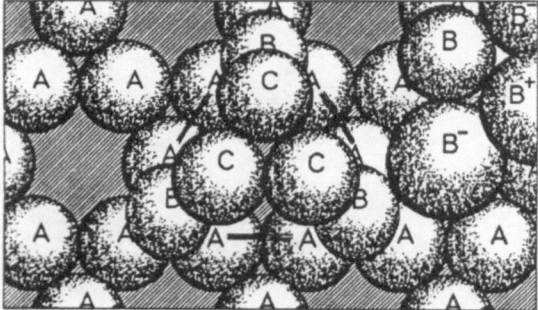

First Layer   Hexatetrahedron Second Layer

Fig. 15. The building up of the first three layers of a Laves phase. The first layer is an open hexagonal packing of small atoms, A. The next layer has the B^-, B and B^+ atoms at different heights. The B^- atom sits just above the hole in the A layer. The B and B^+ atoms sit in sites that are equivalent until the third layer forms another open hexagonal network with the hole over the B^+ atom which then is bigger than the B atom. Six atoms from the A layer, three B atoms and three C atoms form the hexatetrahedron that contains the B^- atom. Six atoms from the C layer, three B atoms and three A atoms form hexatetrahedra around the B^+ atoms. This hexatetrahedra is upside down with respect to the first one. These two orientations of hexatetrahedra share faces hexagonal faces with one another. The Laves phases are stackings of these hexatetrahedra. Each upward pointing hexatetrahedra is joined by four downward pointing hexatetrahedra.

It was most interesting to note in building this model that the structure of the Laves phase employs the same hexatetrahedron that occurs in α-Mn. This is seen in the building up of the structure in the center of Fig. 15. This similarity between α-Mn and the Laves phases was noted almost three decades ago by Kasper and Roberts[25] in their classic study of α-Mn and β-Mn. In comparing the two phases it seems that it would be difficult to argue for the more complicated α-Mn structure on the basis of packing alone. Arrott[26] put forth an argument for stabilizing that structure electronically through its most unusual reciprocal lattice, that generates 96 strong periodicities all of equal wave length. (The 48 {721}, the 24 {552} and the 24 {633} Fourier components are all strong.) But that argument, which relies completely on the cubic nature of the system, would fail in the limit of thin films. Perhaps in thin films, at least, the Laves phase has a lower energy than α-Mn, even without the Ru substrate.

All of the above is based upon the observation of the six fold symmetric RHEED pattern corresponding to Fig. 12b. The rest is purely conjectural, except that we have evidence from Reflection Electron Energy Loss Fine Structure that this is not α-Mn that we are growing. We have discussed REELFS in an accompanying paper.[27] Here we show the results, Fig. 16, for α-Mn and 15ML of the Mn Laves phase, which is how we refer to these films beyond 2 ML. The REELFS spectra show a shift in the position of the nearest neighbors to larger average spacing for the Mn Laves phase and a narrower line width consistent with lesser complexity in that phase than in α-Mn.

Magnitude Fourier Transform
(arb. units)

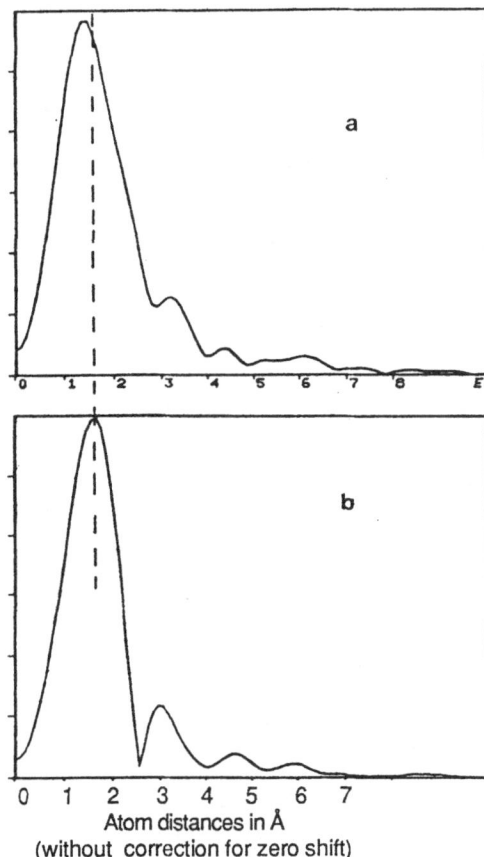

Atom distances in Å
(without correction for zero shift)

Fig. 16. Fourier inversion of the Reflection Electron Energy Loss Fine Structure spectra of (a) α-Mn and (b)15 ML of Mn on Ru(0001). The shift in the position of the first peak, which is a weighted measure of the nearest neighbor distance, between α-Mn and the Mn Laves phase may be as much as 0.15 Å. (The zero in the atom spacing should be corrected for the effect of phase shift in the electron reflection from neighboring atoms. This has not been done. It is assumed that this shift is the same for both Mn structures.) There are problems with oxidation of Mn during the protracted REELFS measurements, but nevertheless this result shows that the 15ML of Mn on Ru(0001) differs in more than its RHEED pattern. As the REELFS probes deeper than the surface layer, the difference is not just a surface effect. Note also the much sharper line for the Mn Laves phase, which is consistent with the lesser complexity of the Laves phase.

We would like to find a way to protect the Mn so that it could be removed from the vacuum system. We tried protecting the Mn with Ag, but this was a bad choice. Ag clearly creates havoc with the surface as the RHEED screen goes through a variety of patterns during the deposition of the first two layers of Ag on Mn.

We have not grown thick layers of Mn, primarily because of the broadening of the RHEED pattern that we observe for deposition times longer than the hour that we take to grow ~30 ML. When one now considers the problems encountered by Mn in growing as a Laves phase, it is not surprising that the patterns broaden. At each point on the crystal surface, there are three choices for the position of the holes in the open hexagonal network in the first layer of the Laves phase. In the second layer there is a choice between which of two sets of equivalent sites become the B and the B^+ atoms. This then determines the position of the open hexagonal C layer. But at every subsequent B layer there is another free choice. The difference between the fcc Cu_2Mg structure and the hcp Zn structure is in the choice of the B site in the third double layer from the original double layer of A and B sites shown in Fig.15. This gives a lot of freedom to make stacking faults. In addition these are vicinal surfaces with steps that will run the A half of the double layer into the B half at each boundary.

From RHEED studies, it appeared that the Laves phase of Mn was stable up to 450˚C. Mn on Ru(0001) showed definite signs of clumping above T/C=450˚C as read from the thermocouple (T/C) on the back of the substrate holder, which is an overestimate. (A free film illuminated uniformily from one side and radiating to the thermal void on the other with equal emissivity on both surfaces will be cooler than the source by $(1-2^{-1/4})$T, which in this case is 115˚C. While we have no independent measure of the gradient, we expect from optical pyrometry at higher temperatures that it is no more than half of this.) The clumped Mn begins to show definite diffraction spots at T/C=500˚C.

In our earliest work, before we had surface analysis, we started out growing Mn on Ru with the substrate held at temperatures typical for MBE work on GaAs. From the Mn lines of the RGA it became clear that we were desorbing Mn at a rate almost as fast as it was being deposited. Electron microscope pictures showed the formation of large islands of Mn for prolonged growth at T/C=550˚C. Yet it was the early depositions at T/C above 500˚C that first showed the development of the Laves phase diffraction pattern, on cooling below 450˚C. Under various conditions of time of growth, rate of deposition and temperature of the Ru substrate, the RHEED pattern matched that of Ru for T/C above 500˚C but on cooling several different RHEED patterns were observed. These were generally super-positions of the two extra streaks in the direction of nearest neighbors (see again Figs. 11 and 12) and one extra streak mid-way between them. Sometimes an extra streak appeared midway between the main streaks in the perpendicular direction in addition to the others. For growth below T/C=450˚C only the Laves phase pattern resulted. We had assumed that the growth was uniform below this temperature, but a more careful study using the ratio of the attenuation of the high and low energy Ru XPS lines showed that deviations from uniform, layer by layer, growth start already above T/C=240˚C. From all this it seems that the Laves phase of Mn is quite stable, particularly in that it persists well beyond the temperature range of uniform growth.

Recently we found for Ni that one can measure RHEED intensity oscillations in metallic growth at room temperature. On hind sight we can recall visually sensing variations of intensity during growth of Mn on Ru. We plan to return to Mn to study the time dependence of the RHEED intensities as we expect that, with the wide range of substrate temperatures available for the growth of the Laves phase, some variations in the growth mode should be detectable by this method. We have grown Mn on Ru(0001) at T/C= 265 K which is the current limit of our low temperature capability.

X ray Photoelectron Spectroscopy

XPS provides our most useful tool for investigating the magnetic nature of the Mn atoms in these new phases. We use the ideas presented by Veal and Paulikas[28] to interpret the 3s peak. These are discussed below. Here we note that it is important to measure also the behavior of the 3p and 2p peaks to distinguish between screening effects which affect all of these peaks and 3s-3d exchange effects that affect the 3s the most, the 3p peaks less and the 2p peaks almost not at all.

XPS is carried out using the 1253.6 eV Mg $K\alpha_{1,2}$ X ray line with a FWHM of 0.7 ev. The electron energy analyser is the PHI Model 15-255GAR with a hemispherical retarding grid system, a double-pass cylindrical mirror analyser and slotted aperture that can be inserted for angular resolution. The energy resolution of the system in the normal operating mode for XPS is 1.1 ev. Fig. 8 shows a scan for a ~2-layer Mn thin film on Ru over a range of emerging electron energies from ~250 eV up to ~1250 eV. As the highest energy electrons (on the right) are coming from the Fermi surface, the binding energy is measured backward toward the left. It is not particularly evident from Fig. 8 that the Mn 3s peak is split, as the energy resolution was relaxed to 1.5 eV for more rapid data collection on the full scan.

To gain better precision and improved statistics we use 1.1 eV resolution but take many passes (at least 8) through the portion of the spectrum of most interest. For this we employ a signal averager. Our statistics are limited by the stability of Mn with respect to oxidation, even in the UHV system. In Fig. 17 we show the change in the $2p_{3/2}$ peak of Mn from just after growth to a time 5 hours later. This peak is most sensitive to the change in the binding state of Mn due to oxidation. If the oxidation were to proceed, the peak near 639 eV would dissappear and the peak near 642 eV would fully develop. From following the time dependence of the oxidation we recognize the desireability of carrying out measurements in the first hour after growth. The amount of oxygen present can be determined by both XPS and AES, but the use of the electron gun in AES drastically increases the rate of oxidation of Mn.

For the Mn Laves phase grown epitaxially on Ru, for the bcc Mn structure grown epitaxially on Fe and for α-Mn which is prepared in bulk outside the vacuum system, there are no special problems in obtaining the XPS spectra. These are all homogenous systems. Any photoelectrons from the substrate that make it to the vacuum have suffered sufficient scattering that they contribute nothing to the XPS peaks, at most adding to the general background.

XPS Intensity

Fig. 17. The effect of oxidation on the Mn XPS $2p_{3/2}$ peak. The solid curve was taken just after growth. The dotted curve was taken 5 hours latter (including ~3 hours of XPS measurements). The analyser pass energy was set at 50 eV for both curves. The shoulder near 641.5 eV is due to oxide formation estimated from AES at 20 percent.

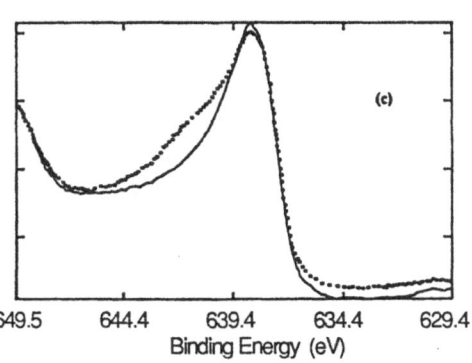

For the two layers of the expanded Mn structure on Ru, it is necessary to subtract the Ru peaks using scans of pure Ru suitably normalized to the reduced intensity of the Ru peaks from attenuation by the Mn overlayer. For alloys of Mn in Ag, Ag(Mn), and Mn in V,V(Mn), it is necessary to subtract the pure solvent spectra, suitably normalized. This is illustrated for Ag(Mn) in Fig. 18. Fortunately, as there are no oxidation problems with the pure solvents, these can be determined with great precision by collecting data over many hours.

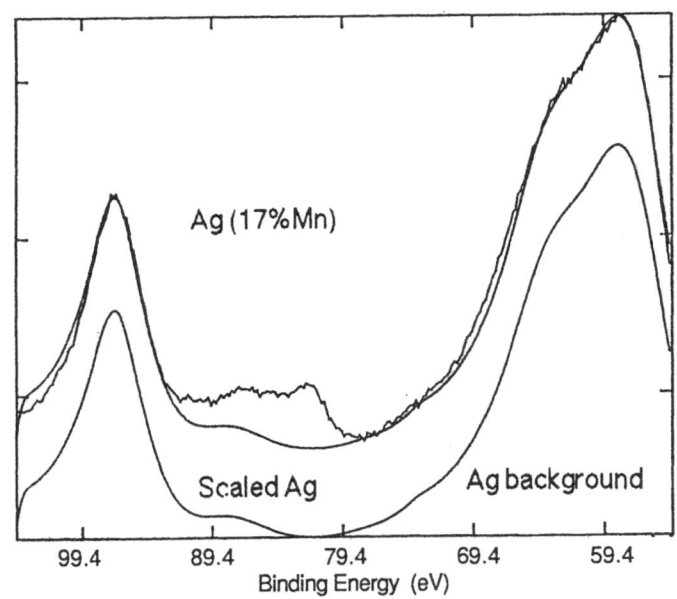

Fig. 18. XPS signals from Ag and Ag(17% Mn) illustrating the procedure for background subtraction used to retrieve the splitting of Mn 3s peak. The noisy line is the alloy data. The smooth lines are Ag data, as taken and normalized to the alloy data for subtraction.

Veal and Paulikas stress the importance of measuring the $2p_{3/2}$ peak in order to understand the role of screening effects . Ni provides an example. In Fig.19 the data for the $2p_{3/2}$ and the 3s peaks of Ni are fitted with Doniach-Sunjic lineshapes.[29] That the splitting in the Ni 3s peak is not an exchange splitting[30] is clear from the fact that the two splittings, according to the fits, are essentially the same, 6.15 ev for the $2p_{3/2}$ and 6.01 ev for the 3s.·

The Ni results are part of a resurvey of the transition metals 3p, 3s and $2p_{3/2}$ peaks which we carried out in order to see the results taken all on the same machine. The 3s peaks for Cu, Ni, Co, Fe, α-Mn, Cr, V and Ti are shown in Fig. 20. These curves are all broadened with respect to Cu. The difference between Cu and Ti is small. The difference between α-Mn and Fe is within the noise. Both show what looks like a split line even before fitting with line shapes. The splitting is about 4 ev. McFeely et al.[31] had previously shown that α-Mn has a splitting of ~4 eV. By comparison with 6 eV splittings seen for Mn compounds with 5 μ_B, McFeely et al. concluded that α-Mn has a magnetic moment of 3.5 μ_B. As both the magnetic moment of the 3d electrons and the overlap of the 3s and 3d wave functions would affect the magnetic contributions to the line broadenings and splittings, the data do not necessarily imply that these are the same in α-Mn and Fe.

There is broadening in Ni in addition to the screening peak split off by 6.15 eV shown in Fig. 19. This broadening for Ni, Co, Fe and α-Mn seems to correlate well with the magnetic moments of these metals as observed in cooperative magnetic states. The lines in Cr and V are definitely narrower than the more strongly magnetic elements. For the sake of comparision we have subtracted a normalized

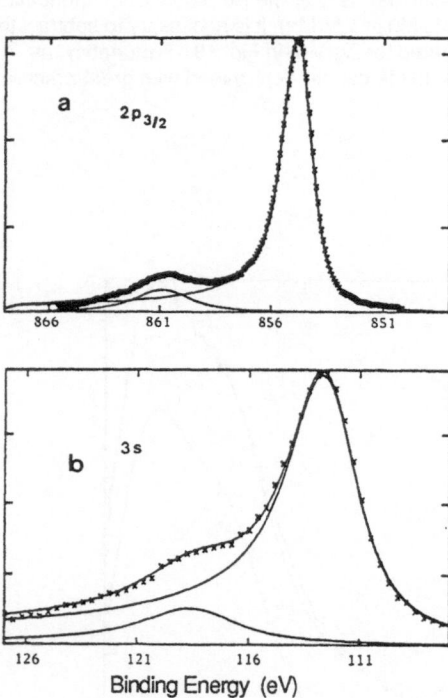

Fig. 19. (a) Ni $2p_{3/2}$ and (b) Ni 3s peaks fit with pairs of Doniach-Sunjic lineshapes. The data were taken with different resolutions, 25eV pass energy for the $2p_{3/2}$ and 50 eV for the 3s peaks. The splittings are (a) 6.15 eV and (b) 6.01 eV. The ratios of the peak intensities are (a) 6.6 and (b) 6.0. This illustrates the strong relation between the shapes of the $2p_{3/2}$ and the 3s peaks when screening is responsible for the line shapes and splitting.

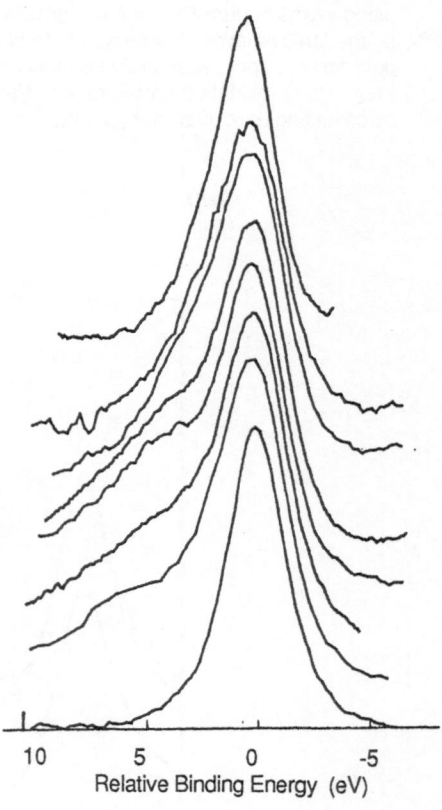

Fig. 20. The 3s peaks for the 3d transition elements Ti, V, Cr, α-Mn, Fe, Co, Ni and Cu in order from the top. The curves have been shifted and displaced upwards for the sake of comparision. The Cu curve is subtracted from each of the other peaks with the results shown in Fig. 21.

302

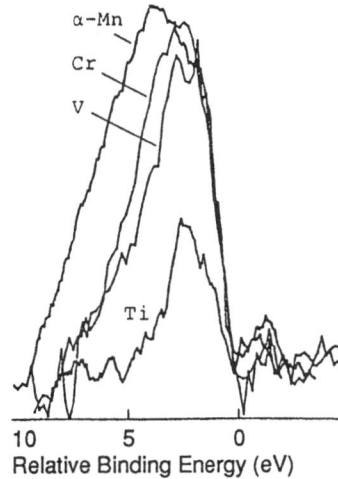
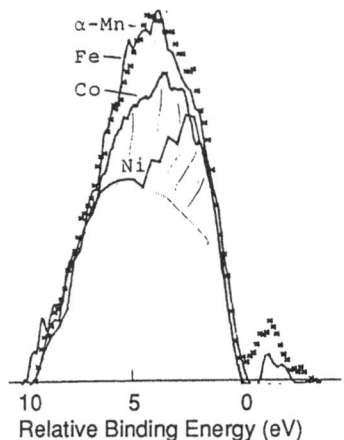

Relative Binding Energy (eV)   Relative Binding Energy (eV)

Fig. 21. The results of subtracting the Cu 3s peak from the XPS intensities of the transition elements using the data shown in Fig. 20. For comparision, the difference between Cu and α-Mn is shown twice. On the right it is indicated by the crosses which coincide with the data for Fe.

Cu 3s spectra from each of the 3s spectra of the other elements. This is not an altogether unambiguous process as there are background subtractions and the problem of how to align and normalize the intensities before subtraction. The results from aligning the low binding energy sides of the peaks are shown in Fig. 21.

In Fig. 22 we show the position of the principle 3p and 3s peaks for all the elements from Sc to Zn. Our own data is augmented by results from the literature.[32] Fig. 23 shows how well these two curves track one another. The point of this exercise is to show that the position of the principal 3s and 3p peaks are not sensitive to exchange interactions which should be three times as large for the 3s electrons. The comparision is made by passing straight lines through each set of data from Ti to Cu. For each element the deviation of the 3p energy from the one straight line is the same as the deviation of the 3s energy from the other line. The differences are quite small compared to the magnetic contributions to the line shapes shown in Figs. 20 and 21.

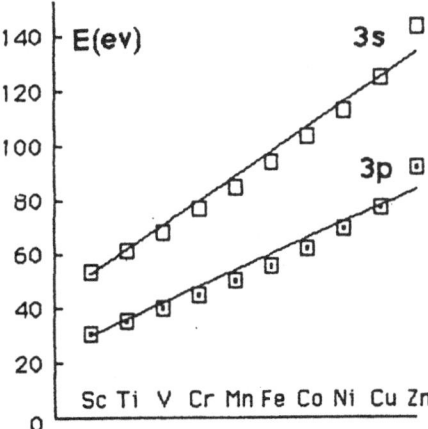
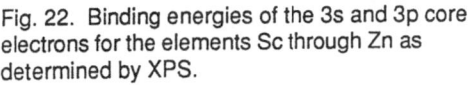
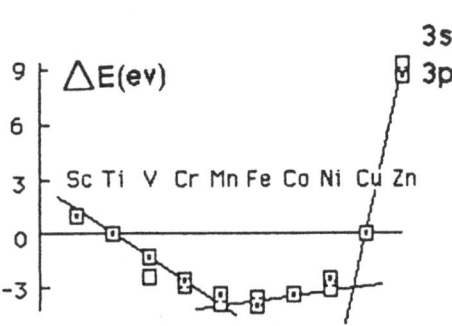

Fig. 22. Binding energies of the 3s and 3p core electrons for the elements Sc through Zn as determined by XPS.

Fig. 23. The correlation of the 3p and 3s peaks is shown by plotting the energy differences with respect to straight lines through Ti and Cu.

The line shapes for the 3s XPS peaks for α-Mn and the Mn Laves phase differ very little. This is not too surprising. What was puzzling was the result for the 3s XPS peaks for the expanded structure of the first two layers of Mn on Ru(0001). We were looking for an increase in the moment and hence an increase in the splitting of the 3s peak. The data for the Mn Laves phase and the first two layers of Mn are shown in Fig. 24. The 3s peaks of Mn are clearly split, but into how many peaks is in doubt. The $2p_{3/2}$ peaks of Mn do not show an obvious splittings, but an analysis using the Doniach-Sunjic line shape yields an apparent splitting of 1.3 eV with the smaller peak on the higher binding energy side of the main peak which has 2.2 times the area, see Fig. 25. We have analysed the 3s peaks guided by the shape of the $2p_{3/2}$ peak following the ideas of Veal and Paulikas. The results of this analysis are shown in Fig. 24 for the Mn Laves phase and for the expanded structure of the first two layers of Mn on Ru. While not precisely the same, these two curves show almost the same splitting, 4.30eV for the Mn Laves phase and 4.45 eV for the expanded structure.

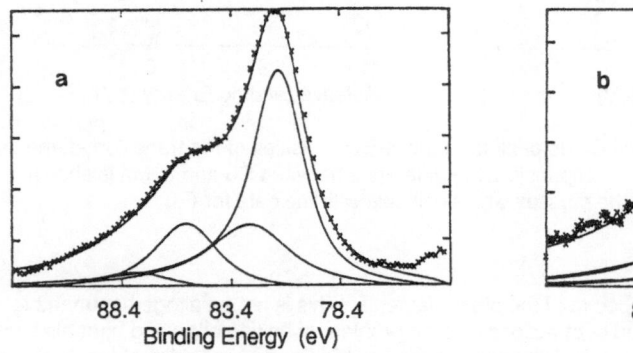
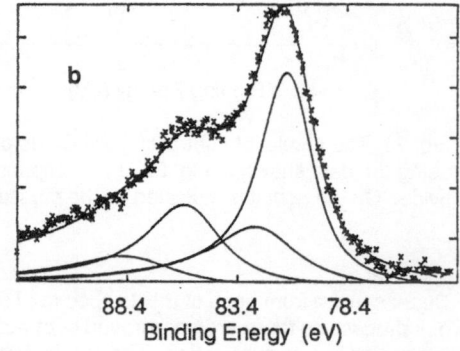

Fig 24. The 3s XPS peaks for (a) the Mn Laves phase and (b) 2 ML of Mn on Ru(0001) in an expanded structure. The data are fit with two peaks each composed of two parts taken from the fits to the $2p_{3/2}$ peaks as shown in Fig. 25.

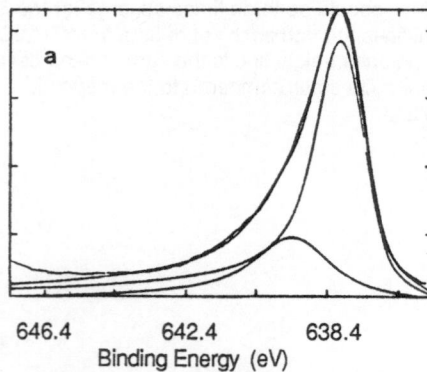
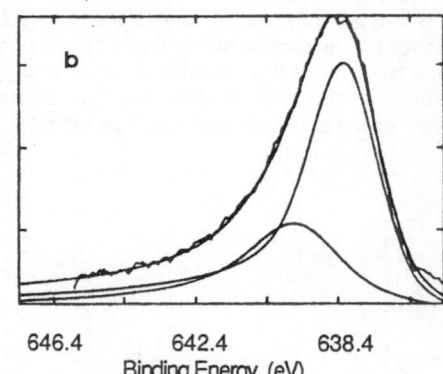

Fig. 25. The $2p_{3/2}$ XPS peaks for (a) the Mn Laves phase and (b) 2 ML of Mn on Ru(0001) in an expanded lattice. Two Doniach-Sunjic line shapes fit with splittings of 1.33 eV in (a) and 1.32 eV in (b) and ratios of 2.2 for the two peaks in (a) and 2.1 in (b). There is a slight shift in the positions of the main peaks of 0.16 eV.

We expected a larger splitting for the expanded structure for the reason that we anticipated a higher moment and a greater exchange interaction between the 3s and 3d electrons. Indeed we expected the difference in splitting to be proportional to the atomic moment. The extraction of values for the magnetic moment of α-Mn from thermodynamic data depends on one's model of the four inequivalent lattice sites. We know nothing about the thermodynamic properties of our new forms of Mn grown by epitaxy. But for Ag(Mn) there is little doubt in assigning a moment of $5\mu_B$ from the specfic heat and magnetic susceptibility.[33] It was this that led us to study the XPS of Ag(Mn).

Mn in Alloys

The Ag(Mn) alloys were grown by codeposition on Ru. Ag grows on Ru with an epitaxy of sorts. After doing some strange things in the growth of the first monolayer, Ag takes its own lattice parameter, but maintains the orientation of the Ru substrate. Ag grows with one of its [111] axes normal to the plane. The same growth behavior is found for codepostion of Ag(Mn).

The $2p_{3/2}$ peak of Mn in Ag is quite anomalous. Dilute amounts of Mn in Ag exhibit the largest recorded shift of a $2p_{3/2}$ peak of any element in any envirnment.[34] The data in Fig. 26 shows two $2p_{3/2}$ peaks for Ag(17% Mn), with the higher binding energy peak corresponding to the single shifted peak observed for more dilute alloys of Mn. The argument of Veal and Paulikas is that the shape of each part of the 3s peak split by exchange should have the same shape as the 2p peak but the chemical shifts need not be the same. In the case of Mn in Ag it is most likely that the shape we observe here for the $2p_{3/2}$ peak of Ag(17% Mn) peak is a superposition of the spectra for Mn atoms with different numbers of Ag neighbors. Each of the two peaks in the $2p_{3/2}$ spectrum should show the screening effects on the line shape which, following Veal and Paulikas, we would use to determine the line shape for each of the components of the 3s peak. We would not use the full $2p_{3/2}$ spectrum for that shape in the case of Ag(Mn).

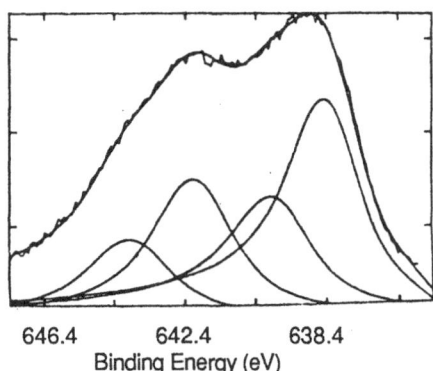

646.4   642.4   638.4
Binding Energy (eV)

Fig. 26. The $2p_{3/2}$ XPS peaks for Mn in Ag fit with two pairs of Doniach-Sunjic line shapes.

The 3s peak for Ag(Mn) grown on Ru is indistiguishable from that of the first two layers of Mn grown on Ru, see Fig 27. This result would fully satisfy the initial hypothesis of this work. If Mn in Ag is the same as the first two layers of Mn grown on Ru in an expanded structure, then in as much as Mn in Ag has a moment of $5\mu_B$, the same could be said of the expanded structure. The existence of the two peaks in the $2p_{3/2}$ spectra are not reflected in the 3s peak. For dilute alloys of Ag(Mn) only the higher binding peak is seen in the $2p_{3/2}$ spectra. It is the position of this peak that is not understood. We assume that this has nothing to do with the splitting of the 3s peak, which is 4 eV compared to the $2p_{3/2}$ shift of 6 eV.

Perhaps the Mn we measure with XPS is not in the Ag but is segregated to the upper surface of Ag. As a check of that possibility, we have grown the alloy by a codeposition in which for the last mono-layer only the Ag beam was activated. That this does not change the 3s spectra is some indication that Mn segregation is not the common cause of the similiar spectra for all these forms of Mn. Not being able to change the splitting by going towards $5\mu_B$, we next tried going the other way. In V(Mn) the Mn have an environment in which it has no moment according to the thermodynamic measurements.[35]

The codeposition of Mn and V does not produce epitaxy on the Ru substrate, but the deposited alloy goes down uniformly. The 3s peak of Mn in V is definitely less split than the other forms of Mn, but still there is some splitting left. This data is compared to the others in Fig. 27. In order to bring out the differences of the split-off peaks, we have subtracted a Cu 3s peak from each of the spectra after norm-alizing to match the amplitude and shifting to match the position of the main 3s peak in each of the Mn spectra. The results are shown in Fig. 28. We see evidence that it is the intensity of the split-off peak that is changing with our preconceived idea of how the magnetic moment would vary with atomic volume. We base much of our discussion on this observation. It should be noted that Mn grows epitax-ially in the bcc structure when deposited on Fe. Results for bcc Mn epitaxially grown on Fe are included. As the XPS results for the Mn Laves phase and α-Mn are indistinguishable, the latter are not shown .

Even without recourse to theoretical models, one would be tempted to read Fig. 28 as an indic-ation that the intensity of the split-off 3s peak in Mn is a measure of the magnetic moment. If that is the case, then it would follow that the expanded structure of the first two layers of Mn on Ru has the same moment as Ag(Mn). The peaks appear to decrease in order with atomic volumes with the Mn Laves phase and α-Mn next in size, then bcc Mn, and finally Mn in V with the smallest, but still finite moment.

There is a obvious interpretation of this observation. One could postulate that it is not the moment that is changing but only the fraction of the atoms that have that moment as Mn goes from one environment to another. This would still leave open the questions: What is the common moment that leads to a splitting of ~4 eV and what fraction of the atoms have that moment in the various materials?

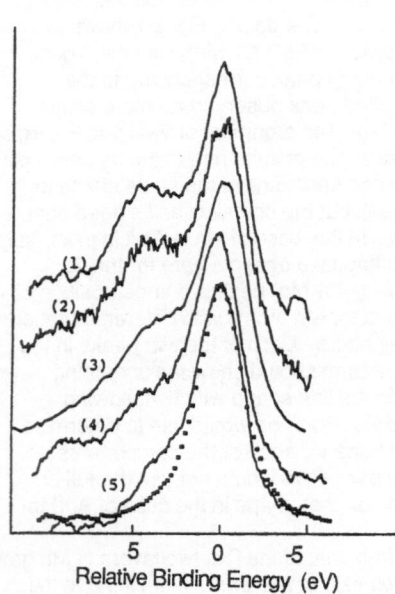

Fig. 27. The 3s XPS spectra for Mn in (1) Ag(Mn), (2) 2 ML of Mn on Ru(0001), (3) the Mn Laves phase and α-Mn, (4) bcc Mn on Fe(001) and (5) V(Mn). The 3s XPS spectra (shifted) for Cu (. . . . .) is included for comparision.

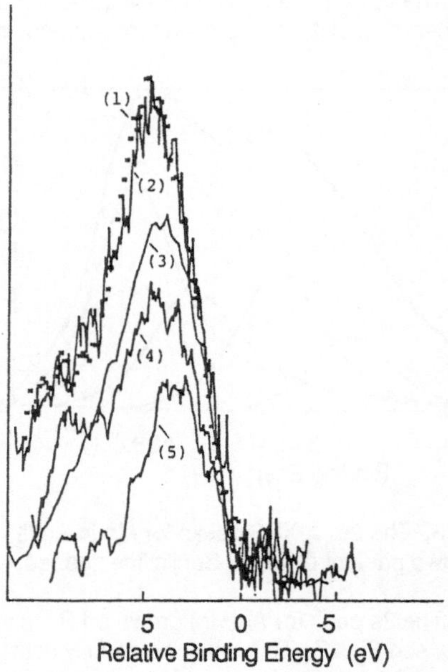

Fig. 28. The data of Fig. 27 with a normalized and shifted Cu 3s peak subtracted. While the peak positions remain almost constant, the magnitudes of the residual peaks decrease with decreasing magnetic moments from Ag(Mn) to V(Mn).

From studies of the splitting of the transition element diflourides and triflourides Veal and Paulikas deduced that the splitting of the 3s peak should go as $\Delta E_{3s}=(1.63\ eV)n$ for $n\leq 5$ and $\Delta E_{3s}=(-1.63\ eV)n + 16.3\ eV$ for $n\geq 5$, where n is the number of 3d electrons in the final state, assuming that Hund's rule applies in that final state. This is based upon the picture that the final state is screened, metastably, by an electron going into the d shell when an electron is ejected from a core hole. For MnF_2, the initial state is the Hund's rule state with $5\mu_B$ per atom. The final state is screened by an electron added to the 3d shell with its spin necessarily opposed to those from the ground state. The final state then has six electrons and a net moment of $4\mu_B$ per atom, see Fig. 29. The

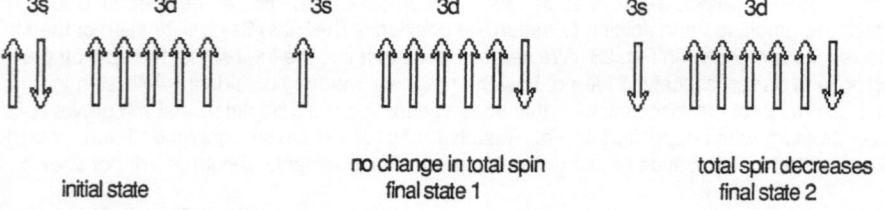

Fig. 29. The energy of final state 1 is lower than that of final state 2 by the exchange inteaction of a 3s electron with four aligned holes in the 3d shell. The lower energy means that the energy given to the exiting electron is larger, corresponding to a lower binding energy. In case 1 there is no change in the spin state. This is consistent with the correlation of the position of the lower binding energy peak with the position of the 3p peak.

306

exchange splitting of the 3s peak is due to the difference in energy depending on whether the remaining 3s electron has its spin parallel or antiparallel to the net spin of the six electrons in the 3d shell. For MnF_2 the splitting is 6.5 eV. By this argument a splitting of 4 eV would correspond to a net moment per atom of 2.5 μ_B in the final screened state and 3.5 μ_B in the initial state. This agrees with the conclusion of McFeeley et al. By this line of reasoning we are faced with the problem of how to explain the result for Ag(Mn), where we believe that Mn has a moment per atom of 5μ_B yet the splitting is only 4eV. Perhaps the calibration in terms of non-metallic compounds is not applicable to metals. But should we use Ag(Mn) to recalibrate the metals? We took this approach for Mn in our original discussion of these results.[36] We then explained Fig. 28 by saying that a fraction of the atoms have 5μ_B, either because of location in the lattice or else because of fluctuations in the magnitude of the moment on a time scale longer than the 10^{-15} sec corresponding to the line widths of the XPS peaks.

Yet it is a problem to explain that one finds a splitting of ~4 eV in Fe as well. One does not expect 5μ_B per atom in Fe unless it is effectively trivalent. If we were to use the Fe moment as a calibration then all of these Mn materials would have slightly more than 2μ_B per atom on the average. We have ignored this problem by saying that Fe is not Mn.

Another possibility is that we are wrong to use Ag(Mn) as a standard for two reasons. The $2p_{3/2}$ peak in Ag(Mn) is quite anomalous. If this is strange, it may be that the 3s peaks are strange as well. We do not have an answer to this objection. One might also ask whether the 5μ_B per atom in Ag(Mn) is all on the Mn atom or could there be a polarization of the surrounding Ag. "Giant" moment are found for transition elements in Pd, but that is for a nearly ferromagnetic host metal. It might be more likely to have a negative spin clothing of the Mn moments from the conduction electrons in Ag.

If we return to the picture of Mn with seven electrons outside the closed Ar core and five of these in d states obeying Hund's rule, then where are the other two electrons? For Mn as a solute, they could go into the conduction band of the solvent with a rearrangement of the conduction band charge density to keep the Mn atom electrically neutral. In all the cases where Mn is known to have 4 or 5μ_B per atom, there are other atoms around to take care of the rest of the electrons outside the Ar core. If one had pure Mn with 5μ_B per atom one would have to have a very large energy difference between the spin up and spin down electrons to drive two electrons from every Mn atom into the conduction band. Perhaps this would never happen for bulk Mn, but for 2 ML of Mn on the surface of Ru, the conduction band of the substrate itself might act as a reservoir to allow up two electrons per Mn atom to become itinerant.

In our search for higher moments in Mn through epitaxy we have made samples of 2 ML of Mn on Ru(0001). We have made bcc Mn stable at low temperatures. We also have a new form of Mn that we believe is a Laves phase. We believe that the 2 ML of Mn have large magnetic moments. Possibly they could be ferromagnetic. We do not yet know how to protect our 2 ML samples from oxidation and yet remain confident that the 2 ML are maintained as grown. It would be desireable to grow these samples in appropriate measuring equipment. It should be easier to determine the properties of the Laves phase and the bcc phase as it is likely that these can be grown thicker than we have so far tried. Then the problem of disturbing the structures on applying the protecting layers should not as serious as it seems to be for 2ML.

References

1. M. Hansen, "Constitution of Binary Alloys", 2nd ed. McGraw-Hill, New York (1958)
2. R. P. Elliott, "Constitution of Binary Alloys , First Supplement", McGraw-Hill, New York (1965)
3. F. A. Shunk, "Constitution of Binary Alloys, Second Supplement", McGraw-Hill, New York (1969)
4. W.B. Pearson, "A Handbook of Lattice Spacing and Structures of Metals and Alloys, Vol. 2", Pergamon Press, Oxford (1967) pp 55-75
5. P. Villars and L.D. Calvert, "Pearson's Handbook of Crystallographic Data for Intermetallic Phases" in 3 vols., American Society for Metals , Cleveland (1985)
6. W.B. Pearson, "A Handbook of Lattice Spacing and Structures of Metals and Alloys", Pergamon Press, Oxford (1958), pp 55-75; see also M. Shiga , Correlation Between Lattice Constant and Magnetic Moment in 3d Transition Metal Alloys in "1973- Magnetism and Magnetic Materials", American Institue of Physics, New York (1974), p 463-477
7. T.K. Kim and M. Takahashi, New Magnetic Material Having Ultrahigh Magnetic Moment, Appl. Phys. Lett. **20** , 492 (1972),

8. K. Mitsuoka, H. Miyajima, H. Ino and S. Chikazumi, Induced Magnetic Moment in Ferromagnetic Fe Alloys by Tetragonally Elongated Lattice Expansion, J. Phys. Soc. Jpn, 53:2381-2390 (1984)

9. B.C. Frazer, Magnetic Structure of Fe_4N, Phys. Rev. 112:75 (1958); also W.J. Takei, G. Shirane and B.C. Frazer, Magnetic Structure of Mn_4N, Phys. Rev. 119:122 (1961)

10. B.T. Jonker, K.-H Walker, E. Kisker, G.A. Prinz, and C. Carbone, Spin-polarized Photoemission Study of Epitaxial Fe(001) Films on Ag(001), Phys. Rev. Lett. 57:142 (1986); J. G. Gay and R. Richter, Phys. Rev. Lett. 56:2728 (1986); S.D. Bader, E.R. Moog and P. Grunberg, J. Magn. Magn. Mat. 53:L295 (1985); B.L. Gyorffy, A.J. Pindor, J. Staunton, G.M. Stocks and H. Winter, J. Phys F15:1337 (1985)

11 see for example: A. Tasaki, K. Tagawa, E. Kita, S. Harada and T. Kusunose, Recording Tapes Using Iron Nitride Fine Powder, IEEE Trans. Magnetics MAG-17:3026 (1981)

12. D.H. Martin, "Magnetism in Solids", M.I.T. Press, Cambridge, Mass (1967), p 122

13. G.E. Brodale, R.A.Fisher, N.E. Phillips and K. Matho, Approach to Magnetic Saturation in CuMn and AgMn, J. Magn. Magn. Mater, **54-57** 194 (1986)

14. W.B. Pearson, "The Crystal Chemistry and Physics of Metals and Alloys", Wiley-Interscience, New York (1972), pp 135-193

15. N. Mori and T. Mitsui, Localized Magnetic Moments and Pauling Valence in Manganese Metal, Some 3d-Transition Alloys and Intermetallic Compounds, J. Phys. Soc. Jpn, 25: 82 (1968); Mori and Mitsui give references to the early work on Mn intermetallic compounds.

16. L. Pauling in "Theory of Alloy Phases", The American Society for Metals, Cleveland, Ohio (1956), p 220

17. B. J. Gellatly and J. L. Finney, Characterization of Models of Multicomponent Amorphous Metals: The Radical Alternative to the Voronoi Polyhedron, J. Non-Crystall. Solids 50:313 (1982)

18. R.E. Watson and L.H. Bennett, "Alpha Manganese and the Frank Kasper Phases", Scripta Metall. 19:535-538 (1985)

19. Private communication R. E. Watson and L. H. Bennett

20. A. J. Bradley and J. Thewlis, Proc. Roy. Soc. A115:456 (1927); T. Yamada, Magnetism and Crystal Symmetry of α-Mn, J. Phys. Soc. Jpn, 28:596-609 (1970)

21. H.J. Goldschmidt, "Interstitial Alloys", Plenum Press, New York (1967)

22. see for example: T.E. Madey, R. Stockbauer, S.A. Flodström, J.F. van der Veen, F.J. Himpsel and D.E. Eastman, Photon-stimulated desorption from covalently bonded species: CO absorbed on Ru(001), Phys. Rev. B23:6847 (1980)

23. C. S. Lent and P. I. Cohen, Quantitative analysis of streaks in reflection high-energy electron diffraction, Phys. Rev. B33:8329 (1986); P. A. Maksym and J. L. Beeby, Surface Sci. 110, 423 (1981); T. Kawamura and P. A. Maksym, Surface Sci. 161, 12-24 (1985); J.B. Pendry, "Low Energy Electron Diffraction", Academic, London (1975), Chap. 4

24. E. Bauer and J. H. van der Merwe, Structure and growth of crystalline superlattices: From monolayers to superlattice, Phys. Rev. B33:3657 (1986)

25. J.S. Kasper and B.W. Roberts, Antiferromagnetic Structure of α-Manganese and a Magnetic Structure Study of β-Manganese, Phys. Rev. 101:537-544 (1956)

26. A. Arrott, Antiferromagnetism in Metals in "Magnetism, Vol II B", ed. G.T. Rado and H. Suhl, Academic Press, New York (1966) pp378-383

27. B. Heinrich, A.S. Arrott, J.F.Cochran, S.T. Purcell, K.B. Urquhart, N. Alberding and C. Liu, Epitaxial Growth and Surface Science Techniques Applied to the Case of Ni Overlayers on Single Crystal Fe(001), this volume.

28. B.W. Veal and A.P. Paulikas, X-Ray-Photoelectron Final-State Screening in Transition-Metal Compounds, Phys. Rev. Lett. 51:1995-1998 (1983); B.W. Veal and A.P. Paulikas, Final-state screening and chemical shifts in photoelectron spectroscopy, Phys. Rev B 31:5399-5416 (1985); B.W. Veal, D.E. Ellis and D.J. Lam, Molecular-cluster study of core-level x-ray photoelectron spectra: Application to $FeCl_2$, Phys. Rev. B 32:5391-5401 (1985)

29. S. Doniach and M. Sunjic, J. Phys. C3:285 (1970)

30. L.C. Davis and L.A. Feldkamp, Resonant photoemission involving super-Coster-Kronig transitions, Phys. Rev. B23:6239 (1981); see also later references in R. Clauberg, W. Gudat, W. Radlik, and W. Braun, Phys. Rev. B31:1754 (1985)

31. F.R. McFeely, S.P. Kowalszcyk, L.Ley and D.A. Shirley, Solid State Commun.15:1051 (1974)

32. C. D. Wagner, W. M. Riggs. L.E. Davis, J.F. Moulder and G.E. Mulenberg, "Handbook of X-Ray Photoeclectron Spectroscopy, (Perkin Elmer Corporation, Physical Electronics Division, Eden Pairie, Minnesota (1980); D. A. Shirley, R.L. Maartin, S.P. Kowalczyk, F.R. McFeeley and L. Ley, Phys. Rev. B15:544 (1977)

33. Fitting of the susceptibility and specific heat yeild a value of $4\mu_B$, see for instance F.W. Smith, Phys. Rev. B14:241 (1976), but it appears that the Mn atom has 5 μ_B and the conduction band electrons produce a negative spin clothing of $1\mu_B$ that is sufficiently well coupled to show up in the entropy and susceptibility. This subject is treated by D.C. Abbas, T.J. Aton and C.P. Slichter, Phys. Rev. B25:1474 (1982)

34. P. Steiner, F. Hüfner, N. Martinsson and B. Johansson, Core-Level Binding Energy Shifts in Dilute Alloys, Solid State Commun.37:73 (1981)

35. E. von Meerwall and D.S. Schreiber, Local Magnetic Fields in Vanadium-Manganese Alloy System, Phys. Rev. B3:1 (1971)

36. A.S. Arrott, B. Heinrich, C. Liu and S.T. Purcell, Deducing 3d Spin Polarization from 3s XPS in Ultrathin Metal Films grown by Molecular Beam Epitaxy, J. Magn. Magn. Mat. 54-57:1025 (1986); B. Heinrich, C.Liu and A.S. Arrott, Very Thin Films of Mn, Ag, and Ag(Mn) Epitaxially Deposited on Ru, J. Vac. Sci. Techol. B3:766 (1985)

GROWTH AND CHARACTERIZATION OF MAGNETIC TRANSITION

METAL OVERLAYERS ON GaAs SUBSTRATES*

G.A. Prinz

Naval Research Laboratory
Washington, DC

Metallization of semiconductors is an important aspect of thin film electronics. Epitaxial metal films, i.e. single-crystal metal films in crystallographic registry with the semiconductor substrate, are the ultimate refinement of metallization and offer significant opportunities for new and improved devices. Epitaxial Al films on GaAs were an early example, offering much lower noise in Schottky barrier diode applications than previous metallization techniques.[1] *Magnetic* metal films grown epitaxially on semiconductors open additional avenues for fundamental research as well as new electronic device development, by combining the properties of both magnetic materials and semiconductors in one hybrid structure.

GROWTH

The first system to be extensively studied is Fe grown on GaAs. The bcc α-Fe structure can be fitted to the GaAs zinc-blende structure as a c($\frac{1}{2}$x1) overlayer by compressing the α-Fe lattice 1.35%. This is illustrated in Fig. 1, which shows the locations of the atoms for an unreconstructed (110) surface of GaAs and α-Fe respectively. Although

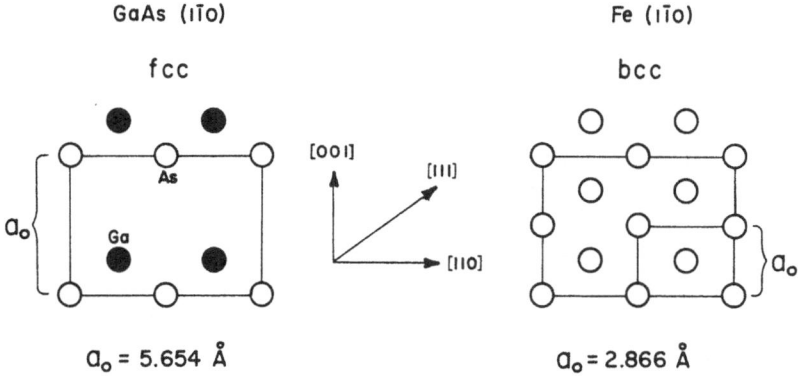

Fig. 1. Surface nets of (110) faces of GaAs and α-Fe.

* This work supported by the Office of Naval Research.

there is some reconstruction of an exposed surface of (110) GaAs[2] consisting primarily of a vertical displacement of the Ga atoms inwards (-0.45Å) and the As atoms outwards (+0.2Å) due to unsatisfied bonds, a detailed study of the interface structure during the initial stages of Fe growth has not yet been carried out. Model calculations[3] for Al on GaAs (110) suggest that the lowest energy locations are approximately midway between second-nearest neighbor Ga-As pairs in the "valleys" located between the zig-zag GaAs ridges which run parallel to [110] on the (110) face. Both calculations and experiments would be valuable for the Fe/GaAs (110) system since it will be seen that the magnetic properties of the films exhibit anisotropies which may derive from the initial growth patterns.

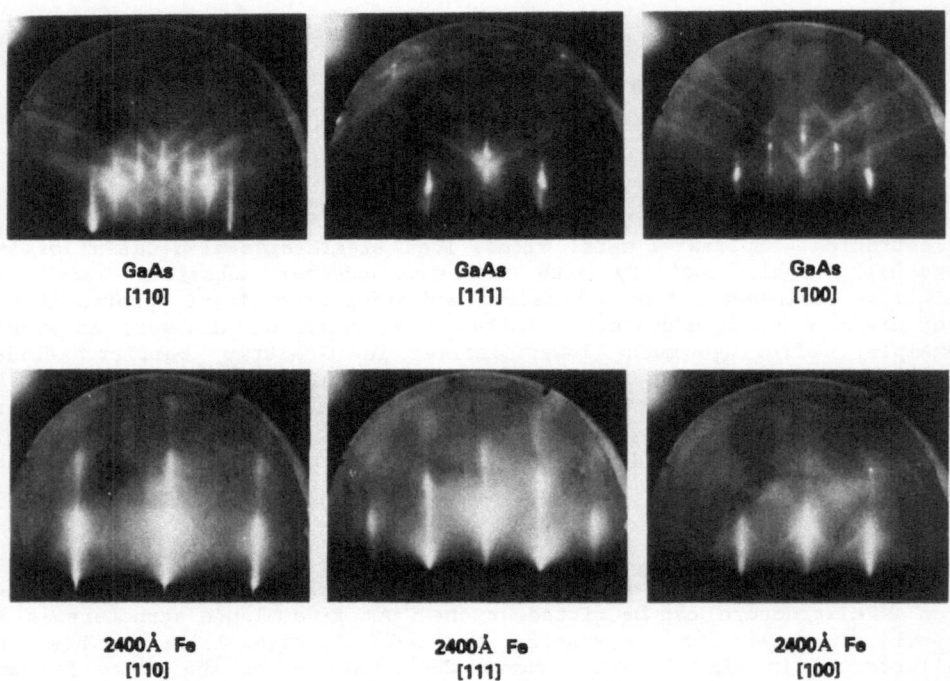

| GaAs | GaAs | GaAs |
| [110] | [111] | [100] |

| 2400Å Fe | 2400Å Fe | 2400Å Fe |
| [110] | [111] | [100] |

Fig. 2. RHEED patterns from (110) face of GaAs before growth and from (110) face of epitaxial Fe film with 10 kV electron beam directed along each of the indicated axes.

What is known about the growth can be seen in Fig. 2 which shows the Reflection High Energy Electron Diffraction (RHEED) patterns obtained from the surface before and after the Fe deposition. Deposition was carried out from a Knudsen cell using pyrolytic BN crucibles at flux rates of ≈ 10Å/min onto substrates held at 175°C. The substrates were prepared by chemico-mechanical polish using 0.5% Br:methanol with subsequent vacuum annealing at 585°C for 30 minutes. The RHEED patterns taken along the [110], [111] and [001] azimuths respectively show the 4-fold change in spacing one would expect along [110], the absence of change along [111] (since the repeat distances in both surface nets is the same perpendicular to this direction) and the 2-fold change along [001]. Measurement of the diffraction pattern spacing shows initial pseudomorphic growth, matching the GaAs template, and eventual transition to α-Fe lattice spacing at ≈200Å, the 1.3% mismatch presumably being taken up by the introduction of dislocations at the interface.

MAGNETIZATION

While the electron diffraction shows that the films being grown have the proper symmetry and spacing to correspond to a (110) face of bcc Fe, magnetic characterization showed several surprises. The first is seen in Fig. 3, which shows the magnetic moment/unit volume (magnetization-M) versus film thickness for a series of films grown under identical conditions. The most striking feature is the apparent decrease in M as the film thickness approaches zero. If one assumes that there is some interfacial region within which the magnetization has an exponential dependence of the form

$$M(Z) = M_o(1-e^{-Z/L_o}) \qquad (1)$$

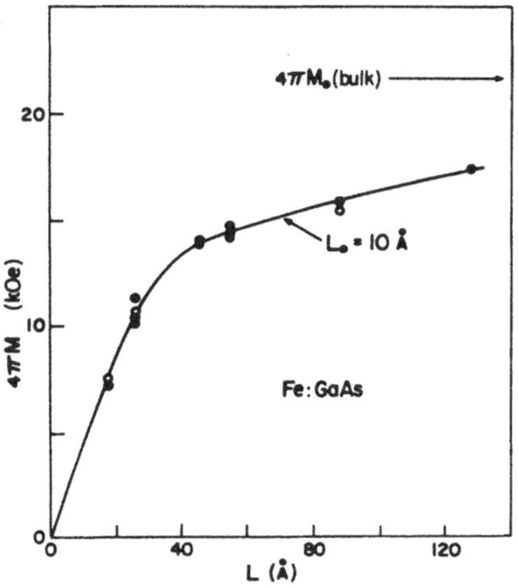

Fig. 3. Dependence of magnetization upon thickness for epitaxial Fe films on (110) GaAs, at 77°K (o) and 300°K (•).[4]

where Z is measured from the Fe/GaAs interface, and one integrates this over the thickness of a given film L, one obtains

$$M(L) = \frac{\int_0^L M(Z)\,dZ}{\int_0^L dZ} =$$

$$M_o \left\{1 - \frac{L_o}{L}\ (1-e^{-L/L_o})\right\} . \qquad (2)$$

This expression has been fitted to the data for a universal value of $L_o = 10\text{Å}$ for all the films measured. Since similar results were obtained for films regardless of final overcoating (Al, Ge or oxide) it was concluded that the decrease in magnetization arose from some mechanism at the Fe/GaAs interface which had an exponential decay depth of $\approx10\text{Å}$. This result had serious implications for the ferromagnetic resonance properties of the films.

SPIN-WAVE RESONANCE (SWR)

The dynamic equation of motion for a ferromagnet may be written

$$\frac{i\omega}{\gamma}\ \vec{m} = (\vec{M}_o+\vec{m})\times\left\{\vec{H}_i + \frac{2A}{M_o^2}\nabla^2(\vec{M}_o+\vec{m})\right\} , \qquad (3)$$

where A is the exchange-stiffness constant, M_o the static component of the magnetization, m the dynamic component of the magnetization, ω the angular frequency of applied radiation, $\gamma=g(e/2mc)$, and H_i is the effective static internal magnetic field. For the case of a thin magnetic film saturated normal to the film plane (i.e., parallel to z) $\nabla^2=\partial/\partial z^2$. For circularly polarized radiation propagating along z. Eq. (1) reduces to the scalar form

$$\frac{2A}{M_o}\frac{d^2m}{dz^2} + \left\{\frac{\omega}{\gamma} - H_i(z)\right\}m = 0. \qquad (4)$$

The internal field H_i is now assumed to be composed of an applied field normal to the film plane H_a, the demagnetizing field from surface poles $4\pi M_o$, an effective anisotropy field H_u, and an effective field arising from gradients in M_o. Since $z = 0$ corresponds to the substrate/film interface, the variations in M_o are assumed to arise from this point and hence will be a maximum there. As $z \longrightarrow L$, the free surface of the film, it is assumed that the interfacial effects decrease and the film will approach the properties of bulk material. In the case of a uniform material there is no z dependence, $H_i = H_a - 4\pi M_o + H_u$, and the separation of the SWR modes in applied field away from the uniform mode obey the well-known n^2 law.[5] One recognizes that Eq. (2) is analogous to Schrödinger's equation and hence we have a one-to-one correspondence to the one-dimensional problem of a particle under the influence of a scalar potential.

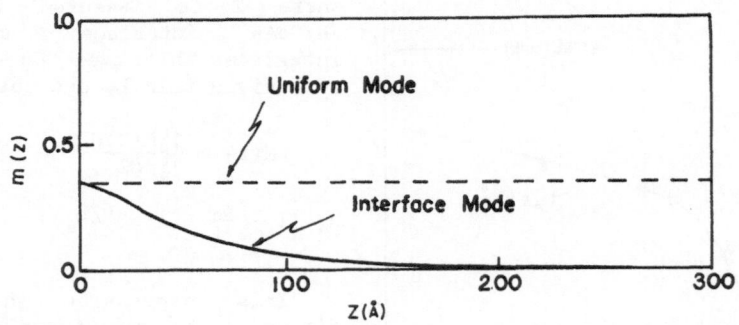

Fig. 4. Distribution of dynamic component of magnetization measured from Fe/GaAs interface (z=0). Uniform mode, when $M = M_o$; interface mode for $M(z) = M_o(1-e^{-z/10\text{Å}})$.[8]

If the internal field exhibits an exponential dependence as in Eq. (1) and appropriate boundary conditions[7] are imposed, Eq. (4) can be solved to obtain the distribution of the magnitude of the dynamic magnetization, m, within the film. This is equivalent to the distribution of the "excitation" of the spin system and from Fig. (4) it can be seen that it is largely concentrated at the Fe/GaAs interface. In the absence of the gradient in $M(z)$ one would obtain a uniform distribution of the "excitation" throughout the film. This result is very similar to the inversion layer observed due to band bending near a semiconductor interface, which provides a potential well to trap carriers. Clearly, for any device applications whose operation depends upon coupling to spin-wave modes near the magnetic/semiconductor interface, the implications of a gradient in $M(z)$ near that interface may be quite important.

FERROMAGNETIC RESONANCE (FMR)

If the applied magnetic field lies <u>in</u> the film plane and is sufficiently large to saturate the magnetization, the energy density of the system is given by

$$E = -\vec{H}\cdot\vec{M} + 2\pi M_n^2 \qquad \text{(isotropic media)}$$
$$+K_1(\alpha_1^2\alpha_2^2 + \alpha_2^2\alpha_3^2 + \alpha_3^2\alpha_1^2) \qquad \text{(cubic anisotropy)}$$
$$\left.\begin{array}{ll} +K_u \sin^2\theta & \text{(in-plane)} \\ +K_n (M_n/M_o)^2. & \text{(normal)} \end{array}\right\} \quad \text{(Surface anisotropies)} \qquad (5)$$

here $\vec{H}$ is the applied field, $\vec{M}$ is the static magnetization, M_n is the normal component of $\vec{M}$, K_1 is the usual cubic anisotropy energy arising from the bulk crystal *structure*, and the α_i's are the direction cosines. Since in this orientation, the resonance equations have not been solved analytically for the case of a gradient M(z), one assumes a uniform M throughout the film. A gradient may however give rise to the same energy terms one obtains for a termination of the magnetic film at a surface or interface, namely surface anisotropy energies. These are represented above by a uniaxial term K_u, where θ is measured from the in-plane [110] axis and another uniaxial term K_n normal to the film plane.

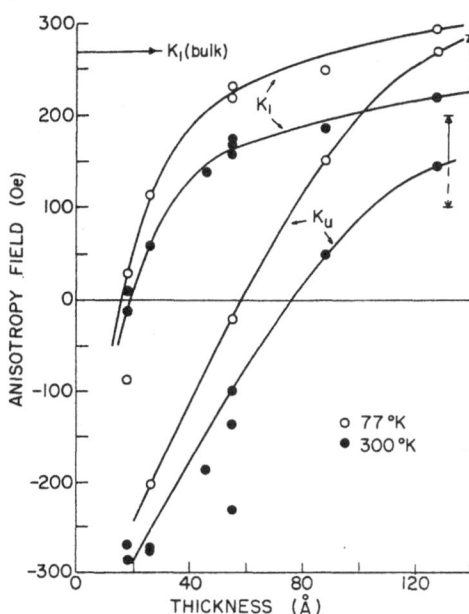

Fig. 5. Dependence of anisotropy upon film thickness. Arrows indicate calculated temperature shifts expected.[6]

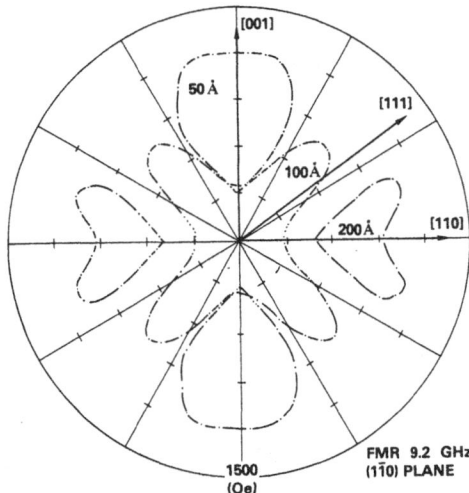

Fig. 6. Dependence of resonance field upon film thickness.[7]

When the observed FMR spectra are fitted to the modes obtained from the above energy density expression, the results shown in Fig. 5 are obtained. They show that the cubic anisotropy approaches bulk values for films >100Å, but goes to zero for a film thickness of ≈ 20Å. This indicates that the cubic symmetry of the <u>magnetic</u> structure characteristic of bulk Fe is lost near the interface. Similarly, the uniaxial term K_u introduced to account for the magnetic discontinuity at the interface becomes very large (negative) for thin films. What is surprising, and still unexplained, is the persistance of K_u for thick films. One would expect $K_u \longrightarrow 0$ when $K_1 \longrightarrow K_1$ (bulk). The observed behavior suggests a thickness independent contribution to K_u, perhaps arising from growth induced strains or dislocations. K_n was found to have a value of zero for all film thicknesses.

The implications of Fig. 5 are dramatic, since these large changes in the magnetic anisotropy are directly observed in a shift of the easy axis of magnetization as the film thickness changes. A picturesque display of this shift is shown in Fig. 6 where a polar plot of the value of the resonance field versus its orientation in the plane of the (1$\bar{1}$0) film is shown. For a "thick" film (200Å), two resonances are obtained as expected at 9.2 GHz with the applied field along [110], the "hard" axis. No resonances are seen for H along the "easy" axis [001], since 9.2 GHz is too low a frequency. For a 100 Å film, resonances are obtained in all directions in the plane showing that the two axes had now become equivalent. Finally, for a "thin" film (50Å) the roles of the two axes have completely reversed. Two resonances are observed for H||[001], showing it to be the "hard" axes, and no resonance is observed for H||[110], indicating that it is now the easy axis.

PHOTOEMISSION STUDIES

The next step in understanding the physical basis for these interface effects was a probe of electronic structure of the metal film using electron photoemission spectroscopy. In situ studies of the films as a function of deposition thickness have been performed at two different synchrotron laboratories. The first, carried out at the Tantalus ring in Wisconsin, looked at Fe deposited at room temperature on GaAs (110).[9] It yielded a very detailed picture of the Ga 3d and As 3d core level spectra as it evolved with layer thickness. The second, carried out at the BESSY ring in West Berlin, looked at single crystal epitaxial growth at 175°C.[10] The BESSY beamline is the only one in the world equipped to measure spin-polarized photo-emitted electrons and those studies were therefore specifically oriented toward questions of magnetic ordering. Nevertheless, the core level spectra obtained in the latter work agreed with that in the former.

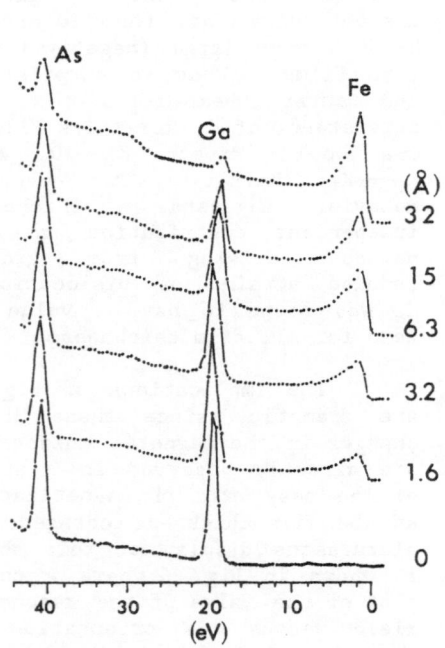

Fig. 7. Photoemission spectra obtained with 80 eV radiation from Fe on GaAs (110) at indicated coverage.[10]

If one looks at the series of unpolarized spectra displayed in Fig. 7, the principal features are from the Fe 3d valence bands and the GA-3d and As-3d core levels. With increasing deposition thickness the Fe valence bands grow in intensity, evolving into the bulk bands of α-Fe, while the Ga-3d and As-3d core level intensities decrease. There was no measurable evidence of polarization in the core level transitions, however the Fe valence bands became polarized when passing from a thickness of 6.3Å (3.1 Monolayers-ML) to 13Å (6.5ML). This corresponds very closely with the exponential decay length of 10Å found in the magnetization data discussed earlier. The detailed dependence of the intensity of the core levels is illustrated in Fig. 8.[9] The initial rise in the ratio I(Ga)/I(As) and the corresponding displacement of the Ga curve is interpreted as arising from an Fe<—>Ga interchange at the interface. The linear portion of the Ga curve corresponds to an electron escape depth of ≈5Å, as expected if the displaced Ga were then simply covered by the subsequent Fe deposition. The As core level intensity, however, reflects a decay depth corresponding to ≈20Å, which suggests As diffusing into the Fe film as well as perhaps moving to the top surface of the film. This general behavior has also been seen in a recent study of Co films on GaAs.[11]

Although the initial Fe<—>Ga interchange could yield a magnetically dead monolayer, it is the extended presence of As in the film which is the likely source of the extended diminished magnetization. While the amount of As is too small to account for the observed magnetic effects

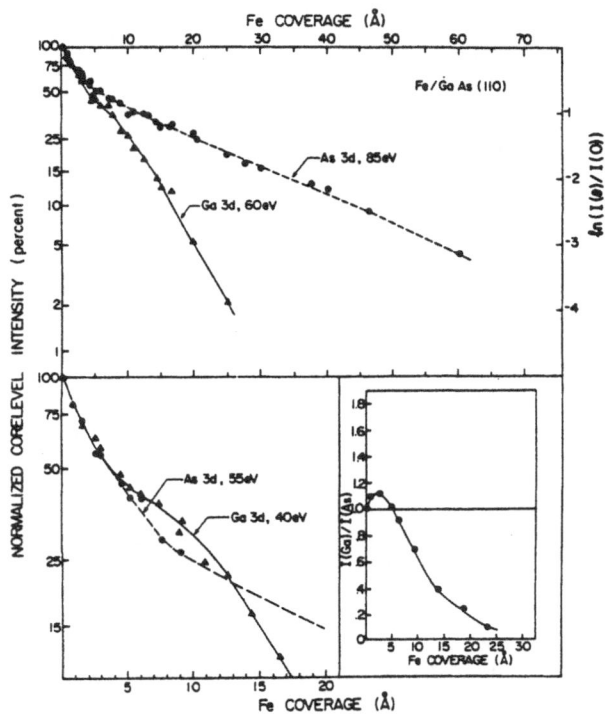

Fig. 8. Dependence of integrated intensities from As 3d core level transitions with increasing Fe coverage.[9]

if it merely acts as a dilutant, it has been pointed out[12] that an As impurity will tend to bond an Fe-Fe pair on either side of it into an antiferromagnetic alignment. In a bcc structure, therefore, a single As ion could effect up to eight Fe moments. Furthermore, the Neel temperature of such compounds can be quite high, e.g. Fe_2As (T_N=350°C). Although the exact nature of the magnetic effects of As impurities in Fe has not yet been determined, a probable cause of the decreased magnetization appears to be present. A microscopic study of the magnetic order near the interface is required to settle the issue.

One of the most interesting features of the growth of bcc-Fe on GaAs, in light of the chemical disruption of the interface, is that the growth is immediately in registry. This is indicated in Fig. 9 where the RHEED pattern for 1Å (mean coverage of 0.5 ML) is shown, along with the same orientation of the GaAs substrate just before deposition. Although the growth is obviously three-dimensional and islanded, the axes of all the Fe crystals are already aligned. This clearly would not occur if the interface were structurally disordered. The more recent results for bcc-Co grown on GaAs[11] also show that although there is chemical disruption at the interface, the near-perfect lattice match between the two structures permits immediate epitaxial growth of oriented crystals.

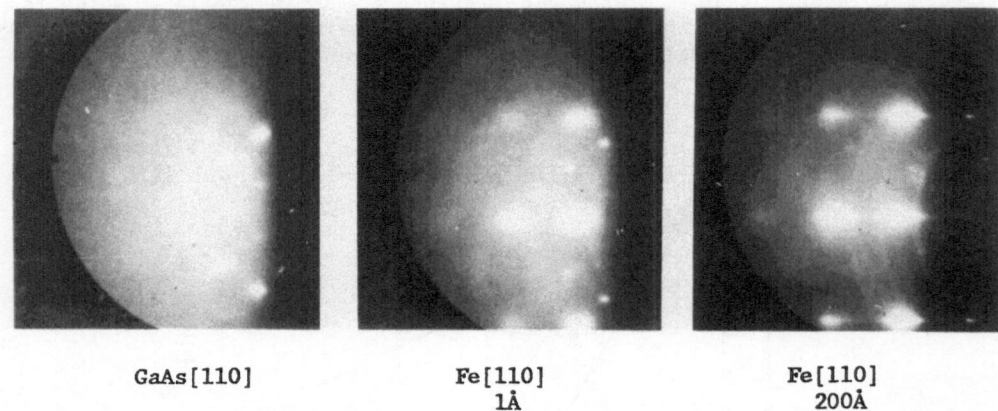

| GaAs[110] | Fe[110]
1Å | Fe[110]
200Å |

Fig. 9. RHEED patterns from (110) face of GaAs along [110] axis for Fe coverage shown.

The epitaxial growth of the magnetic transition metals bcc-Fe and bcc-Co on GaAs creates a useful bridge between the interests of two disparate elements of the solid state community. The ready growth of very high quality single crystal films of these metals makes new studies possible in metastable magnetic phases,[13] spin reorientation[14] and spin wave dynamics.[8] In the process it creates new heterostructure materials whose combined properties may be exploitable for both technological applications as well as physical studies. We have an excellent example of the latter case here, in that the interplay between measurements of the chemical structure, electronic structure and magnetic properties all contributed toward understanding the formation of the interfacial growth region of this metal/semiconductor system. This highlights the value of the metal atom possessing a magnetic moment, since the magnetic moment and the magnetic anisotropy of the atomic site can both be used as probes of the interfacial region.

References

1. A.Y. Cho, E. Kollberg, H. Zirath, W.W. Snell and M.V. Schneider, Electron. Lett. 18 (10), 424 (1982).
2. D.J. Chadi, Phys. Rev. Lett. 41, 1061 (1978).
3. J. Ihm and J.D. Joannopoulos, Phys. Rev. Lett. 47, 679 (1981).
4. T.J. McGuire, J.J. Krebs and G.A. Prinz, J. Appl. Phys. 55, 2505 (1984).
5. C. Kittel, Phys. Rev. 110, 1295 (1958).
6. J.J. Krebs, F.J. Rachford, P. Lubitz and G.A. Prinz, J. Appl. Phys. 53(11), 8058 (1982).
7. G.A. Prinz, G.T. Rado and J.J. Krebs, J. Appl. Phys. 53(3), 2087 (1982).
8. C. Vittoria, F.J. Rachford, J.J. Krebs and G.A. Prinz, Phys. Rev. B., 30, 3903 (1984).
9. M.W. Ruckman, J.J. Joyce and J.J. Weaver, Phys. Rev. B, 33, 7029 (1986).
10. C. Carbone, B.T. Jonker, K. -H. Walker, G.A. Prinz and E. Kisker, Solid State Comm. (In press).
11. F. Xu, J.J. Joyce, M.W. Ruckman, H. -W. Chen, F. Boscherini, D.M. Hill, S.A. Chambers and J.H. Weaver, Phys. Rev. B, 35, (In Press).
12. J.J. Krebs, B.T. Jonker and G.A. Prinz, J. Appl. Phys. (submitted).
13. G.A. Prinz, Phys. Rev. Lett. 54, 1051 (1985).
14. K.B. Hathaway and G.A. Prinz, Phys. Rev. Lett. 47, 1761 (1981).

METAL SEMICONDUCTOR INTERFACES:
THE ROLE OF STRUCTURE AND CHEMISTRY

R. Ludeke

IBM T.J. Watson Research Center

P.O. Box 218, Yorktown Heights N.Y.

INTRODUCTION

The objective of this work is an assessment of the role of structure and defects on the electronic properties of metal-semiconductor interfaces. It can be argued that such an undertaking is premature, since little is known about the microscopic details of the interface structure between the metal and the semiconductor and even less about structural defects and impurities. However in the spirit of the NATO workshop, which among others emphasized the problems underlying our understanding of low dimensional structures, it is of importance to identify the relevant issues, discuss their implications and address and speculate at possible approaches towards their understanding. This approach will be undertaken here.

The metal-semiconductor contact is probably the least understood interface to a semiconductor. The principal reasons for this is the sensitivity of the interfacial electronic properties to small defect and impurity concentrations at the interface[1] and the general absence of near-perfect, coherent metal-semiconductor interfaces. The former constrains the effective extent of the interface that is relevant for the full formation of the electronic interface to about a monolayer or less,[2] and the latter severely limits the identification and characterization of the structural entity responsible for the electronic interfacial properties. In addition, the problem for all interfaces prevails in that spectroscopies sensitive to minute impurity or defect concentrations in a narrow surface or interface region are not readily available as for bulk semiconductor characterizations.

Some of the relevant issues that need to be addressed for a better understanding of the interface are:
1) The role of interfacial bonding on the electronic structure; can bonding (filled) or antibonding (empty) states form in the band gap, and do these control the Fermi level?
2) The relevance of structure, defects and composition of the metal at the interface on the electronic interfacial properties.
3) The role of the metal in determining the electronic properties, in particular the Schottky barrier; or, as some theories propose, are these properties independent of the metal? [3-6]
4) The function of defects and/or impurities in producing band gap states, and their effects on the electronic structure. Are defects inherent to the semiconductor unique in determining a universal defect levels;[5] or can band gap states

be produced by deep impurity levels which vary in energy within the band gap, as recently proposed?[7]

In the following we will discuss in some detail three cases which exemplify characteristic problems related to these issues. They are: a) the Sb/GaAs(110) interface, which represents a coherent interface for which effects of disorder can be ascertained; b) the MBE-grown Al/GaAs(100) interface, for which interface chemistry and epitaxial orientation appear to play an important role in Schottky barrier heights; and c) the transition metal/compound semiconductor interface, which is a strongly reacted interface whose electronic properties are controlled by extrinsic impurity levels.

Sb/GaAs: A COHERENT MONOLAYER INTERFACE

Although Sb is a semimetal, its use in the present situation does not describe a real metal-semiconductor interface, since layers as thick as 5 nm still appear to exhibit semiconducting character.[8] However, its inclusion here underscores the necessity for understanding the formation of an interface, albeit that of a rather ideal and simple model system, in order to assess parameters relevant to the development of its electronic structure. This system was chosen because prior work [9-16] suggested that an ordered monolayer structure formed during room temperature deposition of Sb on a cleaved GaAs(110) surface. This structure consists of zigzag chains of Sb atoms overlying the troughs between the zigzag chains of alternating Ga and As atoms along the [1$\bar{1}$0] surface direction, with each Sb atom bonded alternately to a Ga or As atom below it, as well as to two adjacent Sb atoms of the chain. Theoretical studies indicated that the resulting two-dimensional band structure exhibits a band gap comparable to that of the underlying GaAs(110) surface.[13-14] A further advantage of this surface is based on the experimental observation that a well cleaved surface is devoid of band gap states, that is, the surface Fermi level is at its bulk position. Thus the measured position of the surface Fermi level is a very sensitive probe of the generation and charging of surface states produced by subsequent treatments.[17] This characteristic, which is not shared by Si or Ge, allows for an early detection of electronic surface processes long before there is any evidence for chemical or structural changes occurring at the interface.[18]

The theoretical prediction of a band gap of the two-dimensional Sb structure nearly overlying that of the GaAs(110) surface was supported by recent experimental work,[16] which indicated increased band bending with Sb coverage up to 1/2 monolayer (ML) and a decrease toward flat-band conditions near and beyond ML coverages. These results have recently been questioned based primarily on the notion that room temperature deposition may not be sufficient to create a perfect ML (the breakup of the Sb_4 molecule is one kinetic constraint) and that the measurements are incompatible with the generation and healing of surface defects in the manner described (it should be recalled that a charged defect density of less than 1 per 1000 surface atoms is sufficient to cause the observed band bending[17], and once generated, defects are unlikely to disappear with further coverage).[19]

We have undertaken annealing experiments to further understand the formation process of the ordered Sb overlayer, and correlate, if possible, the effects of structural defects to changes in the electronic structure.[19] In our experiments we deposited various thicknesses of Sb, ranging from less than a tenth ML to many ML's, onto freshly cleaved GaAs(110) surfaces held at room temperature. Photoelectron spectroscopy, excited with synchrotron radiation, was used to obtain core level and valence band spectra. The samples were subsequently annealed up to $\simeq$ 330°C and their photoelectron spectra re-measured. Fig. 1 shows the angle integrated valence band spectra of a ML of Sb on GaAs(110) as deposited (top) and subsequent to a 300°C anneal (bottom). No Sb is lost during this treatment; however a radical enhancement of spectral features is evident, which suggests either an improvement in the order or perfection of the Sb layer, and/or changes in the surface chemistry. The latter could be eliminated and the former process corroborated with results from core level spectra: no line shape changes were observed in the Ga and As 3d core levels (not shown); whereas a drastic sharpening was observed in the Sb 4d core level spectra, as shown in Fig. 2. The as-deposited spectrum shows a somewhat asymmetric spin-orbit split doublet, which could be spectrally decomposed into two components, one corresponding to Sb atoms bonded to Ga (low kinetic energy, KE, doublet) and the other to Sb atoms bonded to As. The larger amplitude of the latter suggests appreciable Sb-Sb bonding on top of the Sb adlayer, and the broader spectral width, compared to the annealed sample, indicates some disorder. The spectrum of the annealed sample (Fig. 2 top) indicates that the heat treatment increased the surface order and removed the Sb adsorbed onto the ML (when more than a ML is deposited initially, the anneal to 300°C removes all Sb in excess for the first ML).

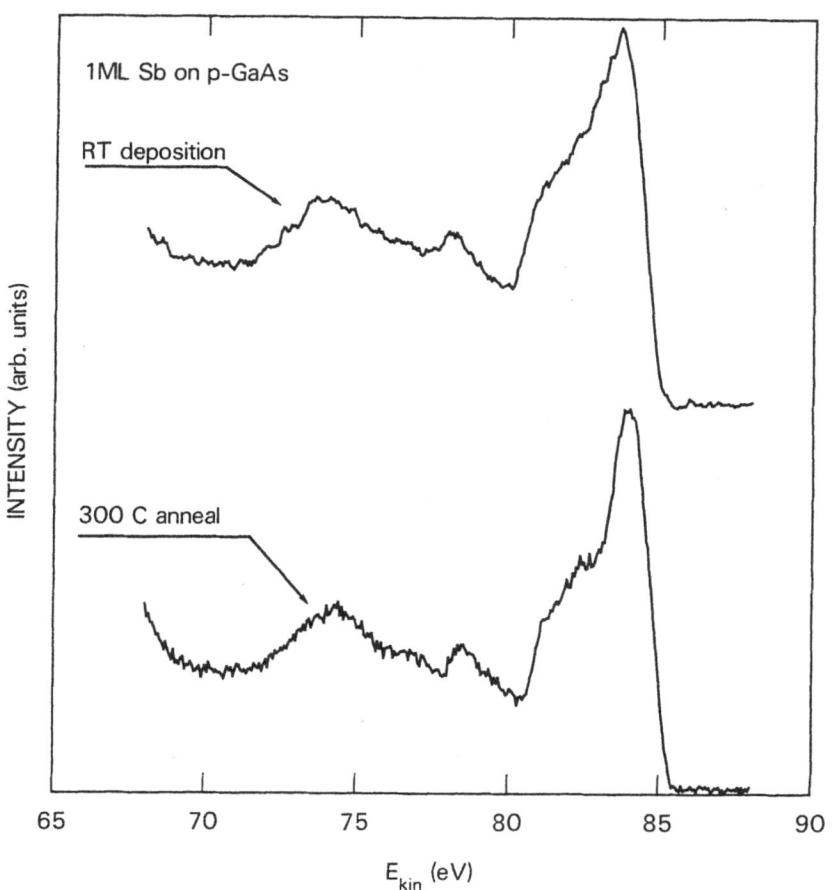

Figure 1. Valence band spectra of 1 monolayer (2.7 Å) of Sb on GaAs(110) after room temperature deposition (top) and following 300°C anneal (bottom).

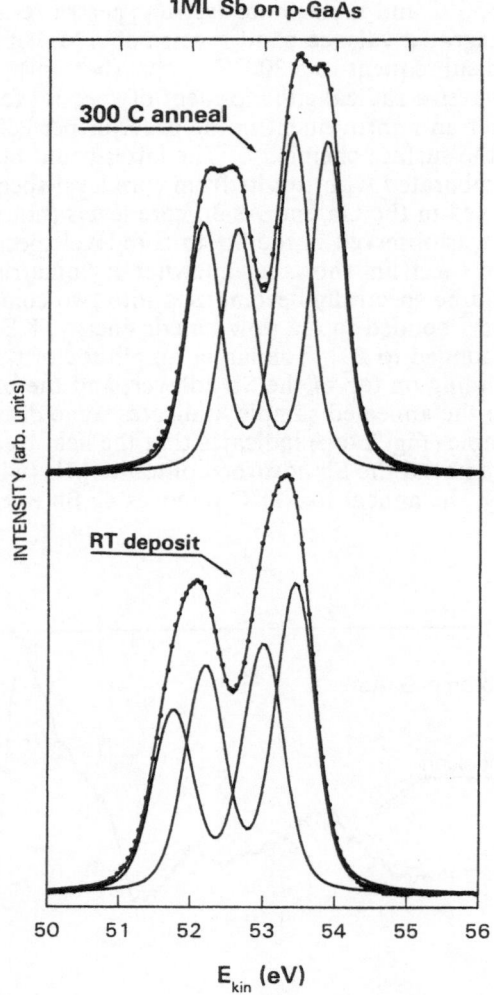

1ML Sb on p-GaAs

INTENSITY (arb. units)

300 C anneal

RT deposit

E$_{kin}$ (eV)

Figure 2. Sb 4d core level spectra of 1 monolayer of Sb on GaAs(110) after room temperature deposition (bottom) and following 300°C anneal (top). The dots represent the experimental data and the line through them the sum of a computer decomposition into two spectral components.

The heat treatment has a further effect. It may be noticed that the overall position in KE of the annealed structure in Fig. 2 lies at a higher KE than for the as-deposited sample. This effect is due to a change in band bending,[20] which in the present case for a p-type sample corresponds to a decrease towards the flat band position. Band bending following room temperature deposition is observed for all coverages studied, and is caused by the formation of surface defects which are largely removed by the heat treatment. The changes in the surface Fermi level, i.e. the band bending for p-type material, are shown in Fig. 3 as a function

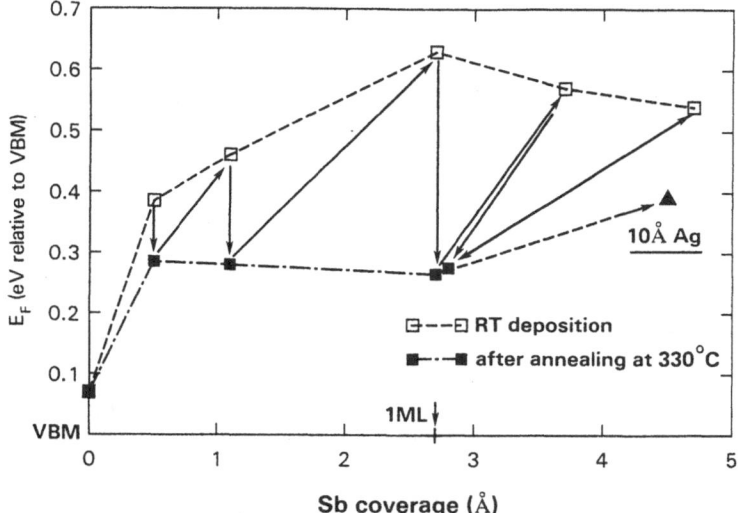

Figure 3. Fermi level shifts relative to the valence band maximum (VBM) of GaAs as a function of Sb coverage for room temperature deposition (dashed line) and following $\simeq$ 330°C anneal (dash-dotted line) after each deposition. Arrows indicate the experimental sequence. Ag was deposited subsequently; the triangle indicates the Fermi level positions for 10 Å coverage. The resulting Schottky barrier height was the same as for Ag deposited directly onto the GaAs without the intervening Sb layer.[47] (Ref. 19)

of deposit thickness and annealing cycles. The band bending was measured from the changes in KE of the Ga 3d core level component,[20] but could be obtained as well from the As 3d component. A monolayer of Sb corresponds to 2.7 Å. The results indicate that annealing is necessary to partially restore the near flat band condition and that the process furthermore produces ordered chains of Sb atoms even for submonolayer coverages. The decreases in band bending with annealing correlate well with the removal of extraneous emission and the sharpening of the Sb 4d core level discussed above, and suggest that the origin of the defects which cause the band bending can be assigned to Sb atoms on top or in between the surface chains, or possibly clusters on top of the GaAs. Thus room temperature deposition is not sufficient to totally overcome the kinetic barrier of the Sb_4 break up and the formation of an ordered surface chain. Upon annealing, these extraneous Sb atoms get merged into the ordered chain structure for coverages below a ML, but evaporate for coverages greater than one ML. It is unlikely that other defects, such as vacancies or antisites, are generated by the deposition process in the GaAs surface, and are then annealed out at a temperature which is insufficient to generate similar defects on the clean GaAs(110) surface. This is an important conclusion because the heat of condensation has been implicated in the generation of intrinsic defects necessary for the unified defect model.[5] The decrease in band bending does not reach the clean surface value because the Sb layer produces a filled surface band which lies 0.27 eV above the GaAs valence band edge at the Γ point.[13-15,19]

In summary, the Sb/GaAs(110) system provides a relatively simple scheme to assess the role of order at the monolayer level on the generation of structure-related defect states that pin the Fermi level near mid gap of the GaAs band gap. It was shown that thermal annealing was required to remove structural imperfections which induced gap states at the interface. Furthermore, it was ruled out that these levels originate from defects induced in the GaAs surface by the deposition process.

THE EPITAXIAL Al/GaAs(100) INTERFACE

In the following we will examine the role of interfacial bonding on the epitaxy of Al on GaAs(100) surfaces, and asses its influence on Schottky barrier heights. The Al films were grown epitaxially on GaAs(100) substrates covered with a GaAs buffer layer grown by molecular beam epitaxy (MBE). Details of the crystallographic relationships and growth modes will be presented next. Additional information can be found elsewhere.[20,22]

The crystallographic orientation of Al on MBE grown GaAs substrates and its dependence on growth parameters were investigated by the author and co-workers.[21,22] and subsequently confirmed by Massies, et al.[23] and Landgren, et al.[24] Following an initial 2D coverage which removed the intrinsic surface reconstruction of the GaAs surface, the Al nucleated in 3D clusters with the usual fcc symmetry. The thickness of the Al necessary for nucleation was directly dependent on the As coverage of the starting GaAs surface and varied from 1.5 to 3Å ($1-2\times10^{15}$ atoms cm^{-2}). The approximate values of As coverage for the investigated surfaces, as obtained by core level photoemission experiments,[25] were 18%, 62%, and 87% for the Ga-rich (4×6), As-stabilized $c(2 \times 8)$ and the As-rich $c(4 \times 4)$, respectively. The lack of a metallic signature in the Auger spectra for Al suggested that the Al in the prenucleation stage was not metallic, but exhibited the characteristics of covalently bonded AlAs.[22] The nuclei in nearly every instance were multi-oriented with the following crystallographic relationships: Al(110) with Al[001] ‖ GaAs[0$\bar{1}$1] [as shown in Fig. 4(A)], Al(110) R with Al[001] ‖ GaAs[011] [Fig. 4(B)], and Al(100) with Al[001] ‖ GaAs[0$\bar{1}$1] [Fig. 4(C)]. This latter structure is rotated 45° relative to the GaAs substrate and is the configuration with the lowest possible mismatch of only 1.4% .

After nucleation a particular orientation dominates the subsequent growth which exhibits a single orientation in reflection high energy electron diffraction (RHEED) after 100-200 Å of Al coverage. The final orientation for room temperature growth depended on the stoichiometry of the starting surface, being Al(110) for the As-stabilized $c(2 \times 8)$ and Al(100) for the Ga-rich (4×6) starting surfaces, although Missous, et al.[26] reported recently Al(100) growth irrespective of the reconstruction of the starting surface. For temperatures exceeding 200°C only the Al(110)R prevailed. Even though Al(100)-45° orientation is the best lattice match to the substrate, it could be consistently achieved only on the Ga stabilized surfaces. We attribute this relationship to the high density of Ga surface atoms to which the Al can make a metallic bond, which is weaker and less directional than a covalent bond. The ever present As surface atoms, even for the Ga rich surfaces, account for the formation of (110) nuclei, which appear to be stabilized by interfacial covalent bonds. The latter dominate on the more As-rich $c(2 \times 8)$ and $c(4 \times 4)$ surfaces and imparts a directional constraint to the nuclei which dominates their orientation. The directionality of the covalent bond is

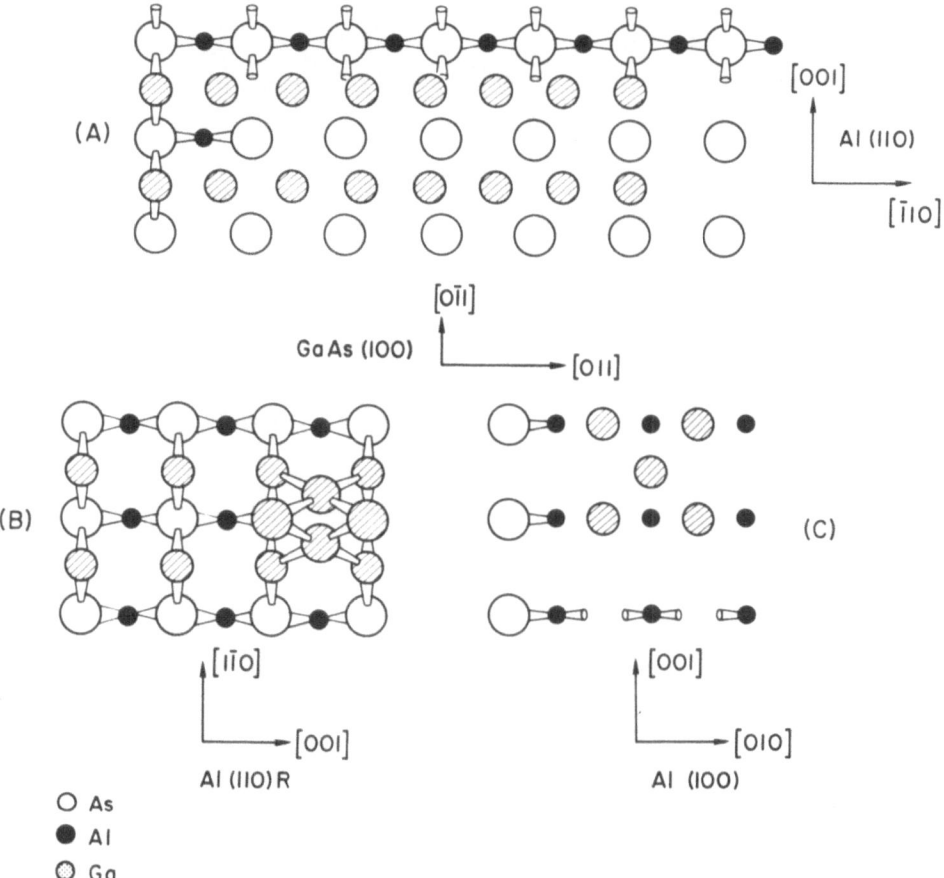

Figure 4. Model of the ideal, unreconstructed GaAs(100) surface showing the crystallographic arrangements of observed Al overgrowths: (A) Al(110), (B) Al(110)R, and (C) Al(100). Open, filled and shaded circles represent, respectively, As, Ga and Al atoms. The truncated rods in (A) and (C) indicate "broken" bonds. (Ref. 22)

suggested in Fig. 4A by the direction of the dangling As bonds. Since all three possible orientations are simultaneously present the ultimate orientation is determined by their relative surface density and growth rates. The growth rate of the (110)R orientation is particularly small at room temperature, most likely the result of severe strains necessary to accommodate this orientation [Fig. 4(B)] but becomes dominant at elevated temperature because interdiffusion between Al and the GaAs surface provides an interfacial region which can better accommodate the strains. Extensive regions of AlGaAs, which for 400°C growth were several hundred Å in thickness, were observed by cross sectional TEM, above which pyramidal Al(110) crystallites extended into the Al(110)R matrix.[27]

Details of the interfacial accommodation are not known, although high resolution cross-sectional studies are in progress to address this issue.[27] Initial studies for the Al(100) interface grown at -10°C indicate that this interface consists of predominantly Al(100) in a generally coherent alignment with the substrate. A detailed lattice image is shown in Fig. 5a. The interface is generally not atom-

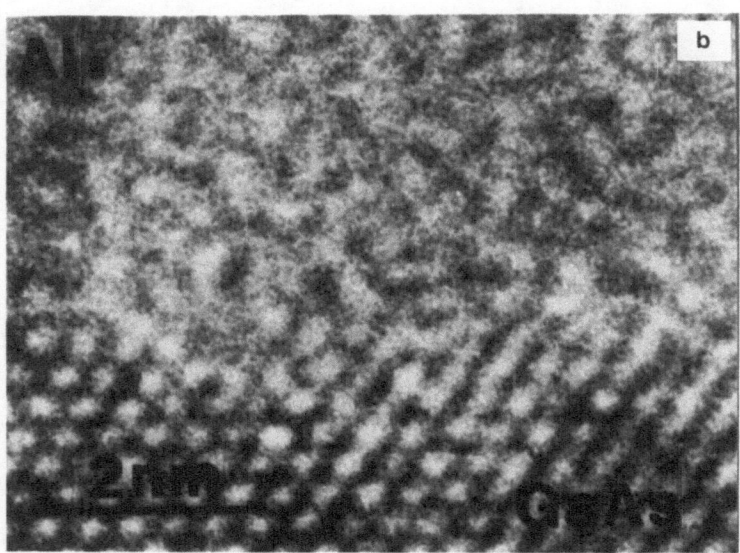

Figure 5. Cross section of the Al(100)/GaAs(100) interface viewed parallel to the GaAs[0Ī1] and Al[001] directions: a) between domains, b) at β-Ga domains (Refs. 27 and 28).

ically flat, and surface steps are frequently observed. Extraneous fringes were observed at the interface, which were tentatively assigned to an extremely thin Ga layer of the order of a monolayer.[28] Occasional clusters of predominantly β-Ga were observed at the interface as well. These domains were elongated parallel to the surface and were typically 400-700 Å long and 50-200 Å thick. The GaAs-domain interface in these regions was much rougher, as shown in Fig. 5b. The origin of the Ga is not entirely understood. That of the thin interfacial layer is probably derived from the observed Ga-Al exchange reaction which accompanies the nucleation step for Al growth on the Ga stabilized surfaces.[22] The nucleation is an exothermic process and its energy release is sufficient to trigger the exchange reaction; the exchanged Al is in an energetically more stable AlAs-like coordination.[22] The Ga in the domains may originate partially from the surface exchanged Ga and excess Ga on the surface left over from the preparation of the Ga stabilized surface. The domains may originate from the segregation of the Ga during the early growth and coalescence stage of the Al (snowplow effect). There was no evidence in the cross sectional TEM micrographs of the other oriented nuclei observed during nucleation by RHEED, suggesting that the minority-oriented nuclei and subsequent larger crystallites reoriented during coalescence and/or continuous film growth. This interpretation is supported by annealing experiments during TEM observation, which indicated already considerable regrowth of large ($\simeq 0.2$ μm) imbedded and misoriented crystallites near 100°C. The origin of these large crystallites was attributed to compositional inhomogeneities or defects in the GaAs surface.[29]

Determinations of Schottky barrier heights for differently prepared Al/GaAs(100) interfaces have been reported by several authors. Wang[30] and Svensson, et al.[31] have reported a correlation of Schottky barrier heights with As content of the surface. For the ordered surfaces: c(4×4), c(2×8) and 4×6 Wang reported a range of 0.74 to 0.80 eV for barrier heights on n-type GaAs, whereas Svensson, et al. observed a much smaller range of 0.74 to 0.76 eV. Missous, et al.[26], however, reported a constant value of 0.78 eV for all three starting surfaces. Nevertheless, excesses of Ga at the interface prior to the Al deposition increased the barrier heights to 0.83 eV, whereas excess As decreases it to $\simeq 0.65 - 0.72$ eV.[30,31]

The possible contribution to the barrier height of AlAs and/or AlGaAs at the interface needs to be considered. A barrier height of 0.85-0.94 eV has been reported for Al/AlAs(100),[30] and values of 0.86-0.91 eV have been obtained for Al deposited on GaAs(100) at 400°C,[32] under conditions where appreciable formation of an alloy of AlGaAs is observed.[22,27] In view of the small barrier heights for As-doped interfaces, the excess As does not appear to lead to the formation of AlAs at the interface for room temperature deposition. On the other hand, the more Ga-rich starting surfaces are subject to Al-Ga exchange reactions,[22] which qualitatively explains their somewhat higher barrier values. Although the workfunction of Ga exceeds that of Al and As,[33] and would account for the larger barrier height for the Ga-rich starting surface, the general failure of the Schottky model (barrier height = metal workfunction - electron affinity) makes this an unsatisfactory explanation. The same caveat needs to be applied to the crystallographic dependence of the Al on the Schottky barrier height, although the Al(100) surface exhibits the largest workfunction[33] and is consistent with some observations.[30,31] The role of impurities and defects for this interface cannot be ascertained at this time. It is not clear if these play a major role, since previous work for Ag on MBE grown n-type GaAs(100) surfaces gave Schottky barrier heights of 0.83 and 0.97 eV for the c(2×8) and (4×6) starting surfaces, respectively.[34] The larger values compared to Al barriers suggest that the metal plays an important role in the barrier heights. However, a conclusive answer must await further refinements in sample preparation, diagnostics and structural analyses.

THE TRANSITION METAL/SEMICONDUCTOR INTERFACE

In the previous case of Al we have dealt with an epitaxial system which left many basic issues unanswered, in particular the role of impurities. In the next and final system to be discussed here, we deal with a complex and reacted interface, whose structural and metallurgical quantification is beyond present capabilities. Some of the transition metals are known to grow epitaxially on GaAs, and the interested reader should refer to the article by G. Prinz in these proceedings. Even though structurally complex, the transition metal/III-V semiconductor interface offers a convenient system to study a special class of impurity states which are generated by the interaction of the transition metal with the III-V semiconductor surface. We will show that interface states responsible for the pinning of the Fermi level can be observed for small coverages of the metal (< < ML), and that these are derived from d-electrons of the transition metal atoms. Again synchrotron excited photoemission is used to study the surface chemistry and electronic properties of the developing interface. Because of the large photoionization cross sections of the d-electrons it is possible to see their effect in the photoemission valence band spectra for coverages as low as 0.01 ML. The technique of inverse photoemission[35] is also used to search for empty interface states, which are necessary to pin the Fermi level for n-type material.[36]

It is well known by now that transition metals deposited even at room temperature react strongly with GaAs and other semiconductor substrates.[7,37-39] This behavior is readily observed in the core level photoemission spectra of the semiconductor. For a representative example we show in Fig 6 the evolution of the Ga and As 3d core level spectra as a function of Y coverage.[39] The experimental spectra (dots) have been decomposed into chemically shifted individual components by computer fitting routines.[7] The spectra of the clean surface are made up of a bulk peak and a surface peak, as indicated. The latter decreases with coverage of the transition metal. For coverages near 0.1 Å an additional chemically shifted peak can be observed in the Ga 3d core level spectrum (not shown); this structure is obvious in Fig. 6 for a coverage of 0.3 Å. For some transition metals, such as Ti, Pd and V, the additional Ga peak is observable near 0.02 Å.[7] This peak is attributed to elemental Ga, which diffuses to the surface of the growing Y film. The presence of elemental Ga is strong evidence that an exchange reaction has occurred on the surface, with the transition metal (here Y) replacing the Ga. The fate of the surface As atoms is less clear since the chemically shifted As component, obvious for coverages > 1 Å, coincides with the surface peak, and becomes indistinguishable from it at the lower coverages of the transition metal. The general lack of experimental evidence for elemental As (except for Pd[7]) suggests that Ga is the predominantly replaced surface component, and that the As, both at the interface and as a diffused species, is predominantly bonded to the transition metal.

The spectral decomposition is also necessary to accurately follow the shifts in kinetic energy with coverage of the bulk component, so that the resulting band bending and final positions of the Fermi level (E_F) can be measured accurately to $\pm$ 10 meV.[40] The positions of E_F relative to the valence band edge are shown in Fig. 7 as a function of coverage for several transition metals on GaAs(110). The final position of E_F for several other metals is also indicated on the right hand ordinate. Three important conclusions can be drawn from the data: a) the final pinning position of E_F (and the Schottky barrier) varies by as much as 0.3 eV for different metals; b) the same final position of E_F is obtained independent of n or p doping; and c) the band bending is essentially complete, except for Ag, near a coverage of 0.1 Å, which is below the coverage for metallic behavior of the overlayer. These observations are generally inconsistent with previously proposed

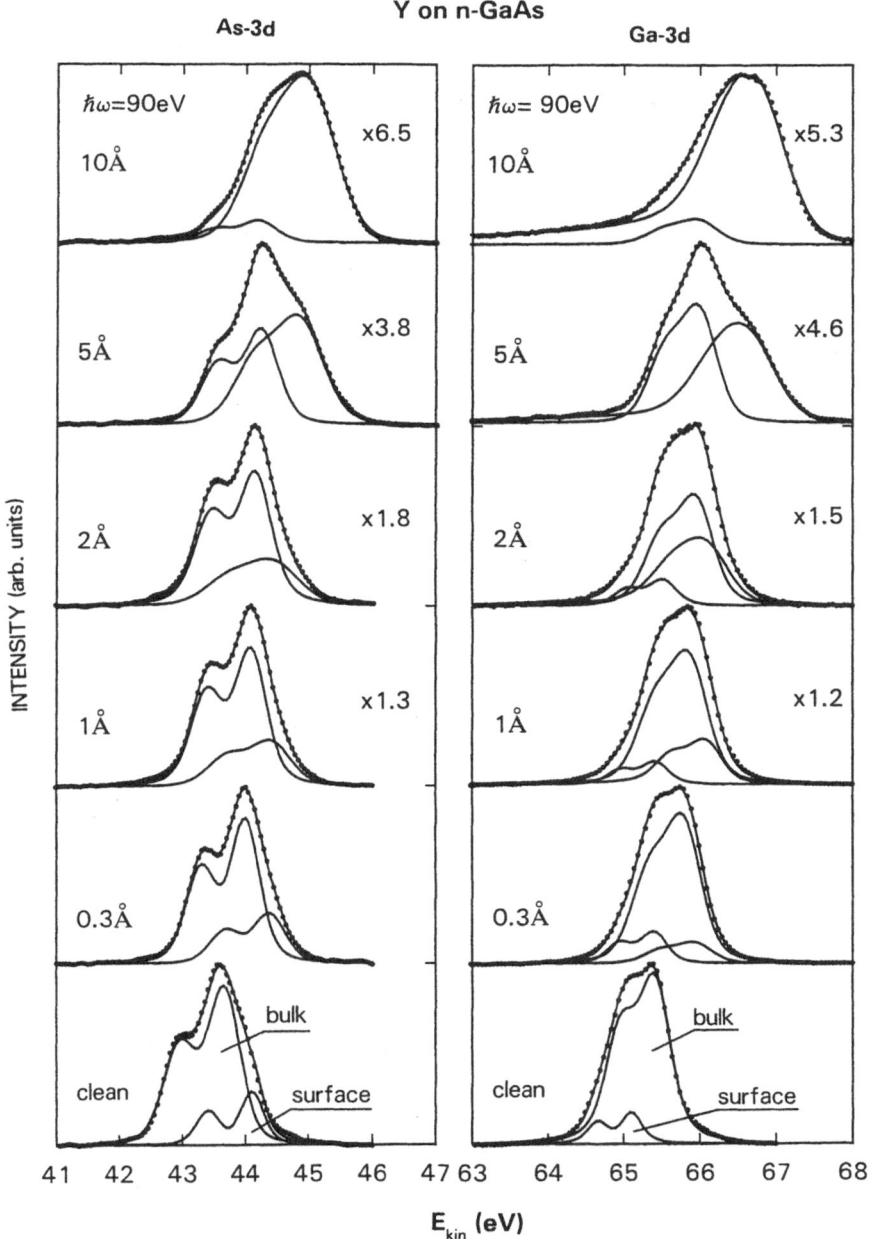

Figure 6. Spectra of the Ga 3d and As 3d core levels for increasing yttrium coverage. The data (dots) are decomposed into bulk, surface and reacted components, and their sum is represented by the line through the data points. The attenuation of the core level signals with coverage is represented by scaling factors. The shifts in KE of the bulk components is due to band bending (Ref. 39).

Schottky barrier models,[3-6] and we must look for a different origin for the interface states which determine E_F at interfaces with transition metals. The strong reactions of these metals with the semiconductor suggest states induced by a new

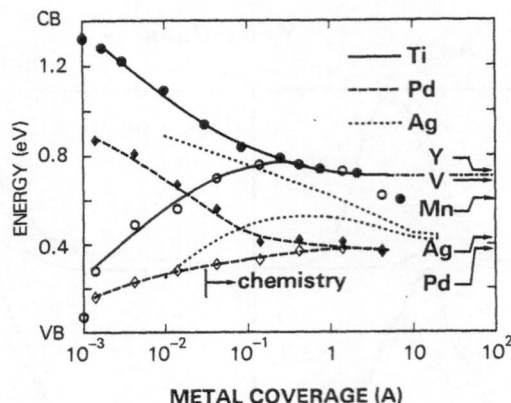

Figure 7. The position of the Fermi level relative to the valence band maximum as a function of coverage for Ti, Pd and Ag. The final positions for other metals is indicated on the right ordinate. Their large range is incompatible with most Schottky barrier models. The arrow labeled "chemistry" marks the onset of observable chemical changes in the core level spectra (Ref. 19).

chemical environment, i.e. one of extrinsic origin. These are readily observed in the valence band spectra.

The valence band spectra for clean GaAs(dotted) and for the indicated coverages of V, Mn and and Y (solid) are shown in the left panel of Fig. 8. Strong spectral features are readily observed which originate from the d-electrons of the transition metals. Difference spectra, obtained by subtracting the corresponding clean spectrum after correction for band bending shifts, are shown in the right panel of Fig. 8. The sharp spectral features are readily observable at coverages of $\simeq 0.01$ Å (not shown) and indicate a unique chemical environment of the transition metal.[7] Based on the exchange reaction deduced from the core level spectra, the Ga substitutional site is a likely site for the transition metal. In bulk semiconductors such a substitutional site often generates deep impurity levels which have been extensively investigated.[42] For the present situation a surface or near-surface analogue of this defect is being considered. Although further substantiation for a particular defect is needed, we nevertheless can draw the conclusion that the transition metals induce filled interface states which overlap the valence band of GaAs. For Ti, V, Pd and Mn these states extend into the band gap, where the emission edge coincides with the position of E_F. Complementary information on the empty bulk conduction band states and interface states is obtained from inverse photoemission (IPS).[35,36,41] A spectral composition of filled and empty states is shown in Fig. 9, which depicts the usual difference spectra (solid lines) of the valence bands (VB), obtained by phototemission, and the conduction bands (CB), obtained by IPS, after 0.2 Å of Ti, V and Pd were de-

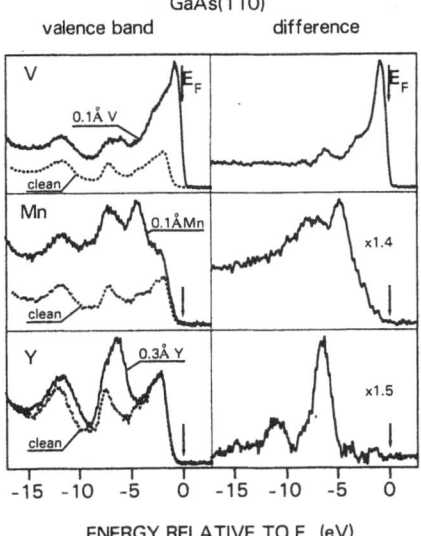

Figure 8. Valence band photoemission spectra for clean GaAs(110) surfaces (dotted) and for the indicated coverages of V, Mn and Y (left panel). The right panel shows their difference spectra. The arrows indicate the position of the Fermi level, E_F.

posited. The dashed curves show the experimental spectra for the clean surface. The IPS results clearly show the existence of empty interface states derived from empty d-states of the transition metal. These and their filled counterparts extend into the band gap and determine the position of E_F at the interface. The values of E_F are more clearly shown in Fig. 7.

The energetic positions of the emission peaks of the filled and empty interface states in the band gap for Ti and V on GaAs (Fig. 9) and V on InP[41] agree very well with the ionized donor and acceptor levels of the respective transition metal substitutional impurity for GaAs[43] and InP.[42] This correlation gives additional support for a substitutional impurity or related complex. Donor states have not been observed for Mn in the band gap, which is consistent with our observations for low coverages. Nevertheless, Mn, at coverages < 1 Å, and Y form pinning states which are not readily identifiable from our spectroscopic results. Y in a substitutional Ga site is not expected to produce donor states in the band gap, since its three valence electrons ($4d^1 5s^2$) are, like Ga, involved in covalent bonds to As. This notion is consistent with the absence of gap emission in our spectra. It was also observed that the coverage dependence of the Fermi level position for

Y is substantially slower than for the other metals, such as those in Fig. 7.[39] We speculate that for transition metals which do not generate deep impurity levels other reaction products, the exchanged Ga and structural defects for example, are the most likely source of interface states responsible for the pinning.

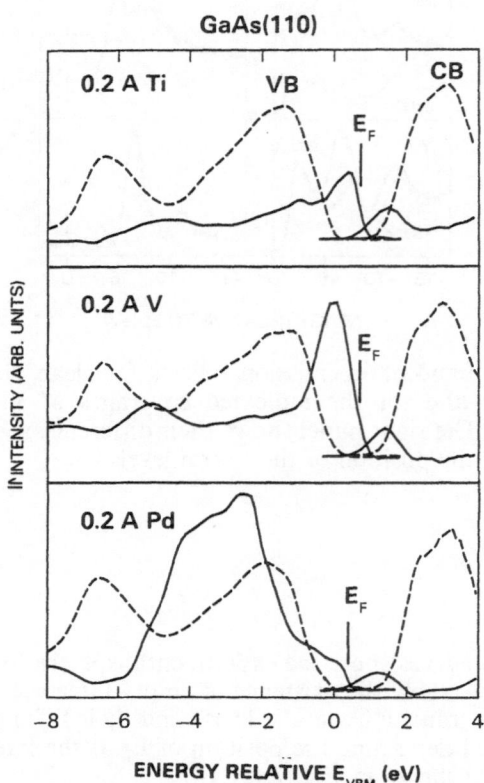

GaAs(110)

Figure 9. Valence band (VB) photoemission spectra and inverse photoemission spectra of empty states (conduction band, CB) for clean GaAs(110) surfaces (dashed lines), and the valence band and conduction band difference spectra for 0.2 Å coverages of Ti, V and Pd (solid lines). E_F marks the position of the Fermi level in the band gap, which for GaAs extends to +1.4eV (Ref. 39).

SUMMARY AND PROGNOSIS

In the last examples of transition metals on III-V semiconductors we have demonstrated that impurity levels for many transition metal atoms determine the interfacial electronic properties below submonolayer coverages and that the presence of the metallic phase appears to be superfluous. From this we may make the extension that structural details, such as phases, symmetry and orientational alignment of the transitional metal are unessential as well. The details of these impurity levels are not known, although for low metallic coverages they exhibit certain spectroscopic resemblance to substitutional impurities. At coverages for which the As chemistry becomes important as well, a simple impurity model breaks down, as defect/impurity complexes are expected to dominate the interface electronic properties. On the other hand it was demonstrated that a coherent and defect free interface is achievable for the Sb/GaAs(110) system, but that Sb adatoms create defect states that affect the position of the interface Fermi level. This system is, of course, neither metallic nor three-dimensional, but nevertheless illustrates the importance of crystalline perfection and interfacial coherence in reducing extrinsic defect levels. The importance of extrinsic defects is suggested as well for the Al/GaAs(100) interface, although some correlation was observed of the barrier heights with workfunction changes due to crystal orientation or the presence of Ga at the interface. Of the three systems discussed, this interface is the least understood.

A better understanding of the electronic properties of these interfaces requires not only novel structural and electronic diagnostic tools, but superior sample preparation and characterization as well. A promising new diagnostic tool is the scanning tuneling microscope, which recently has been used simultaneously for both structural and spectroscopic investigations with atomic scale resolution.[44-46]

Acknowledgements: The work discussed here is based on published and unpublished work in cooperation with G. Landgren, G. Hughes, F. Schäffler, D. Straub, F.J. Himpsel, W. Drube, A. Taleb-Ibrahimi, of IBM, and Z. Liliental-Weber, of the Lawrence Berkeley Lab., who is responsible for the cross sectional TEM studies. Their contributions are gratefully acknowledged.

REFERENCES

1. A. Zur, T. C. McGill and D. L. Smith, Phys. Rev. B 28:2960 (1983); C. B. Duke and C. Mailhoit, J. Vac. Sci. Technol. B 3:1170 (1985).

2. The extent of the interface is here defined by the average thickness of the overlayer beyond which the electronic properties, for example the final position of the interface Fermi level with respect to the semiconductor band edges, do not change. The final position of the Fermi level, often referred as the pinning position, is reached at coverages below a monolayer of the metal (see Fig. 7)

3. V. Heine, Phys. Rev. 138:A1689 (1965).

4. J. Tersoff, Phys. Rev. Letters 52:465 (1984).

5. W. E. Spicer, P. W. Chye, P. R. Skeath and I. Lindau, J. Vac. Sci. Technol. 16:1422 (1979).

6. J. L. Freeouf and J. M. Woodall, Appl. Phys. Lett. 39:727 (1981).

7. R. Ludeke and G. Landgren, Phys. Rev. B 33:5526 (1986); G. Hughes, R. Ludeke, F. Schäffler and D. Rieger, J. Vac. Sci. Technol. B 4:924 (1986).

8. M. Mattern-Klosson and H. Lüth, Solid State Commun. 56:1001 (1985).

9. P. Skeath, C.Y. Su, I. Lindau and W. E. Spicer, J. Vac. Sci. Technol. 17:874 (1980).

10. C.B. Duke, A. Paton and W. K. Ford, Phys. Rev. B 26:803 (1982).

11. J. Carelli and A. Kahn, Surf. Science 116:380 (1982).

12. P. Skeath, C.Y. Su, W. A. Harrison, I. Lindau and W. E. Spicer, Phys. Rev. B 27:6246 (1983).

13. C. M. Bertoni, C. Calandra, F. Manghi and E. Molinari, Phys. Rev. B 27:1251 (1983).

14. C. Mailhoit, C. B. Duke and D. J. Chadi, Phys. Rev. Letters 53:2114 (1983).

15. P. Martensson, G. V. Hanson, M. Lähdeniemi, K. O. Magnusson, S. Wiklund and J.M. Nicholls, Phys. Rev. B 33:7399 (1986).

16. K. Li and A. Kahn, J. Vac. Sci. Technol. A4:958 (1986).

17. For GaAs doped to 10^{17}cm^{-3} a surface charge of 10^{12}e/cm^2 is sufficient to cause a shift in the surface Fermi level (or band bending) of about 1 eV. See for example S. Sze, Chapter 5 in "Physics of Semiconductor Devices", John Wiley & Sons, New York. (1981).

18. R. Ludeke and G. Landgren, Phys. Rev. B 33:5526 (1986); R. Ludeke, Surf. Science 168:290 (1986).

19. F. Schäffler, R. Ludeke, A. Taleb-Ibrahimi, G. Hughes and D. Rieger, Phys. Rev. B, to be published.

20. R. Ludeke, R. M. King and E. H.C. Parker, Chap. 16 in "The Technology and Physics of Molecular Beam Epitaxy", E. H. C. Parker, ed., Plenum Press, N.Y. (1985).

21. R. Ludeke, L. L. Chang and L. Esaki, Appl. Phys. Lett. 23:201 (1973).

22. R. Ludeke and G. Landgren, J. Vac. Sci. Technol. 19:667 (1981); R. Ludeke, G. Landgren and L.L. Chang, in Proc. 8th Int. Vac. Cong., F. Abeles and M. Croset, ed., Soc. Franc. du Vide, Paris, (1980), Vol. I, p.579.

23. J. Massies and N. T. Linh, Surf. Sci. 114:147 (1982).

24. G. Landgren, S. P. Svensson and T. G. Andersson, Surf. Sci. 122:55 (1982); J. Phys. C15:6673 (1982).

25. R. Ludeke, T.-C. Chiang and D.E. Eastman, Physica 117B/118B:819 (1983).

26. M . Missous, E. H. Rhoderick and K.E . Singer, J. Appl. Phys. 60:2439 (1986).

27. Z. Liliental-Weber, C. Nelson, R. Gronsky, J. Washburn and R. Ludeke, 1986 Symposia Proc., Materials Research Society, to be published.

28. Z. Liliental-Weber presented at Electron Microscope Society of America (EMSA) meeting, Albuquerque, NM, Aug. 1986, and private communications.

29. G. Landgren, R. Ludeke and C. Serrano, J. Cryst. Growth 60:393 (1982).

30. W. Wang, J. Vac. Sci. Technol. B1:574 (1983).

31. S. P. Svensson, G. Landgren and T. G. Andersson, J. Appl. Phys. 54:4474 (1983),

334

32. M. Heiblum and R. Ludeke, unpublished results.

33. J. Hölzl and F.K. Schulte in "Springer Tracts in Modern Phys.", vol. 85, p.1, G. Höhler, ed., Springer-Verlag, Berlin (1979) vol. 85, p. 1.

34. R. Ludeke, T.-C. Chiang and D. E. Eastman, J. Vac. Sci. Technol. 21:599 (1982).

35. F. J. Himpsel and Th. Fauster, J. Vac. Sci. Technol. A2:815 (1984).

36. R. Ludeke, D. Straub, F. J. Himpsel and G. Landgren, J. Vac. Sci. Technol. A4:874 (1986).

37. M. Grioni, J. J. Joyce and J. H. Weaver, J. Vac. Sci. Technol. A3:918 (1985); J. H. Weaver, M. Grioni and J. J. Joyce, Phys. Rev. B 31:5348 (1985); M. W. Ruckman, J. J. Joyce and J.H. Weaver, Phys. Rev. B 33:7029 (1986).

38. J. Nogami, M. D. Williams, T. Kendelewicz, I. Lindau and W. E. Spicer, J. Vac. Sci. Technol. A4:808 (1986).

39. F. Schäffler, G. Hughes, W. Drube, R. Ludeke and F. J. Himpsel, Phys. Rev. B, to be published.

40. This is the accuracy for relative shifts in band bending for coverages up to $\simeq 2$ Å. For larger coverages and the absolute position of the Fermi level relative to the semiconductor band edges, the accuracy is probably no better than ± 50 meV.

41. F. Schäffler, W. Drube, G. Hughes, R. Ludeke and F. J. Himpsel, J. Vac. Sci. Technol. to be published.

42. B. Clerjaud, J. Phys. C, 18:3615 (1985).

43. C. D. Brandt, A. M. Hennel, L. M. Pawlowicz, F. P. Dabkowski, J. Lagowski and H. C. Gatos, Appl. Phys. Letters 47:607 (1985); Phys. Rev. B 33:7353 (1986).

44. J. A. Stroscio, R. M. Feenstra and A. P. Fein, Phys Rev. Letters 57:2579 (1986).

45. C. F. Quate, Physics Today, Aug. 1986; Physics Today, Jan. 1987.

46. R. M. Hamers, R. M. Tromp and J. E. Demuth, Phys. Rev. Lett. 56:1972 (1986).

47. R. Ludeke, T.-C. Chiang and T. Miller, J. Vac. Sci. Technol. B1:581, (1983), and corrections to data given in Ref. 18.

SYNTHESIS OF RARE EARTH FILMS AND SUPERLATTICES

J. Kwo

AT&T Bell Laboratories
Murray Hill, NJ 07974

ABSTRACT

The development of the metal molecular beam epitaxy technique has led to synthesis of single crystal rare earth superlattices composed of alternating magnetic rare earths (Gd, Dy) and their nonmagnetic analog Y. Long range, coherent magnetic structures modulated by the superlattice wavelengths were demonstrated for the first time in the synthetic superlattices. These include ferromagnetic, antiferromagnetic, and halimagnetic order. This article reviews the state of art growth techniques, structural characteristics, and magnetic spin structures.

INTRODUCTION

In the past decade, due to the development of advanced thin film growth techniques, enormous progress has been made in the studies of artificially structured materials.[1] The unusual material and physical properties discovered in these novel structures not only have advanced our knowledge in the fundamental phenomenon, but also have led to the development of new device technology. The original concept of "artificial superlattice" suggested by Esaki and Tsu for (III, V) semiconductors is now generally extended to systems containing lattice mismatched components with dissimilar physical or chemical properties. A notable example is the rapidly growing field of metallic multilayers in the past seven years.

The history of metallic multilayers is traced back to as early as 1940, when DuMond and Youtz first prepared Cu/Au compositionally modulated alloys.[2] Their intention was to use these multilayers as an artificial Bragg plane reflector to calibrate the X-ray wavelength. The attempt did not succeed, because the intensities of the Bragg reflections from the composition modulation diminished rapidly with time due to the interdiffusions between the Au and Cu layers. In the 60's and 70's, these compositionally modulated alloys were used by Hilliard et al. as a basis for studying the diffusion effect in an inhomogeneous system where a steep concentration gradient is present.[3] The sample preparation techniques were mostly coevaporation or cosputtering in high vacuum. Either technique is capable of producing a large quantity of samples with relative ease, and is reguarded as a useful method for exploratory work.

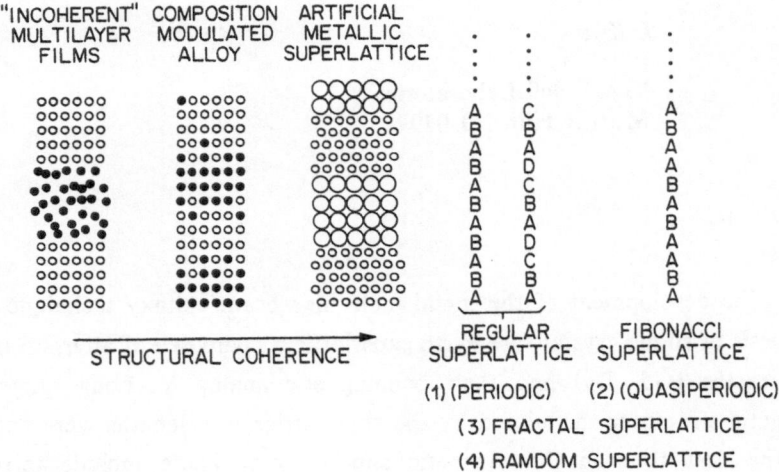

Fig. 1   Examples for illustrating the evolution of structural order and the layering sequences in metallic multilayers.

Multilayer systems reported in the literature to date span vastly different chemical and crystal structures.[4] Here the multilayers are classified primarily based on the degree of structural order.[5] Three representative examples are illustrated in Fig. 1 to demonstrate the evolution of structural oreder. The left example is an extreme case of perfect chemical order but with poor structural order. The two periodic alternating blocks can be either crystalline or amorphous. In the central case, the structural coherence is improved by choosing two components having identical crystal structures. However the chemical order is deteriorated due to interdiffusions near the interfaces. On the right, the example of perfect, coherent superlattice is shown, where the structure has now evolved from one-dimensional (1D) to three-dimensional (3D) order. At the interface, the layers are constrained to match perfectly with one another in the lateral plane, meanwhile developing coherency strain along the growth direction. In other words, a long range registry is maintained between the lateral position of a plane in one block of one material and the lateral position of a plane in a different block of the same material.

Presently, the (III, V) semiconductor superlattices produced by the molecular beam epitaxy (MBE) technique are approaching this type of ideal superlattices. In the metallic systems, we are just beginning to produce single crystal superlattices of comparable quality by the advanced metal MBE technique. The rare earth superlattice discussed in this article serves as a good demonstration for the realization of nearly ideal metallic superlattice. The diffusion related imperfections encountered in the earlier work on metallic multilayers can now be best minimized.

The most common layering sequence is following a simple periodic order by alternating two constituents repeatedly. More recently, there has been substantial interest in synthesizing quasiperiodic superlattices,[6] which in the 1D case are equivalent to the 1D quasicrystals. The quasiperiodic superlattice is produced by alternating two linearly independent periodic sequences (A and B) according to a Fibonacci series. The periods of the two sequences are incommensurate to one another with a ratio of the golden mean. Recently, we have started in our laboratory the investigation of magnetic order in quasiperiodic magnetic superlattices.[7] Studies of fractal superlattices and random superlattices were also reported by Beasley et al. at Stanford University for investigating the critical behavior near superconducting transitions.[8]

In order to modulate a specific physical property in the superlattices by artificially imposed periodicities, it is essential to match the superlattice wavelength with the length scale characteristic of the given physical phenomenon. A well known example of this approach was the (III, V) semiconducting quantum well structures in that the superlattice wavelength was introduced on a length scale less than the electron mean free path. Strong modulations on electronic transport, and band gaps were demonstrated. In the metals the well known physical phenomena are superconductivity and magnetism. The critical length scale of the former is the superconducting coherence length. By varying the thickness of the normal layer relative to the coherence length in the superconducting- normal superlattices, e.g. Nb-Ge, and Nb-Cu multilayers, a dimensional crossover behavior was observed in the upper critical field near T_c .[9]

Recent focus of our research has been on the magnetic superlattice that is produced by alternating a magnetic material with another nonmagnetic species along the growth direction. The critical length scale in magnetic order is obviously the magnetic interaction range. By varying the thickness of the magnetic region to the nonmagnetic region, magnetic superlattices can be used as a model system to investigate the interfacial magnetic behavior, two-dimensional magnetism, and interlayer exchange coupling effect. In the 3d magnetic superlattices containing itinerant magnets like Fe, Ni, and Co, because the magnetic interaction range does not extend more than a few atomic planes, the magnetic behavior depends sensitively on the chemical and crystal structures near the interfaces. Many of the studies have been directed toward the modified magnetic behavior near the interfaces or the surfaces,[10] and the possible existence of 2-D ferromagnetism in monolayer magnetic films.[11]

While these effects are interesting to pursue, our ultimate goal of study is toward the modulation effect on the overall magnetic properties of the superlattices as a whole from the imposed periodicities. So far, such effect has not been demonstrated in the research of magnetic superlattices. It is possible to observe such effect only under the condition that a long range magnetic correlation is established across the superlattice. In comparison, the magnetic interaction in the 4f magnetic rare earths is relatively long-ranged, and it is indirect through the conduction electrons in the Ruderman-Kittel-Kasuya- Yosida (RKKY) manner. The fabrication of magnetic rare earth superlattices can be made by alternating

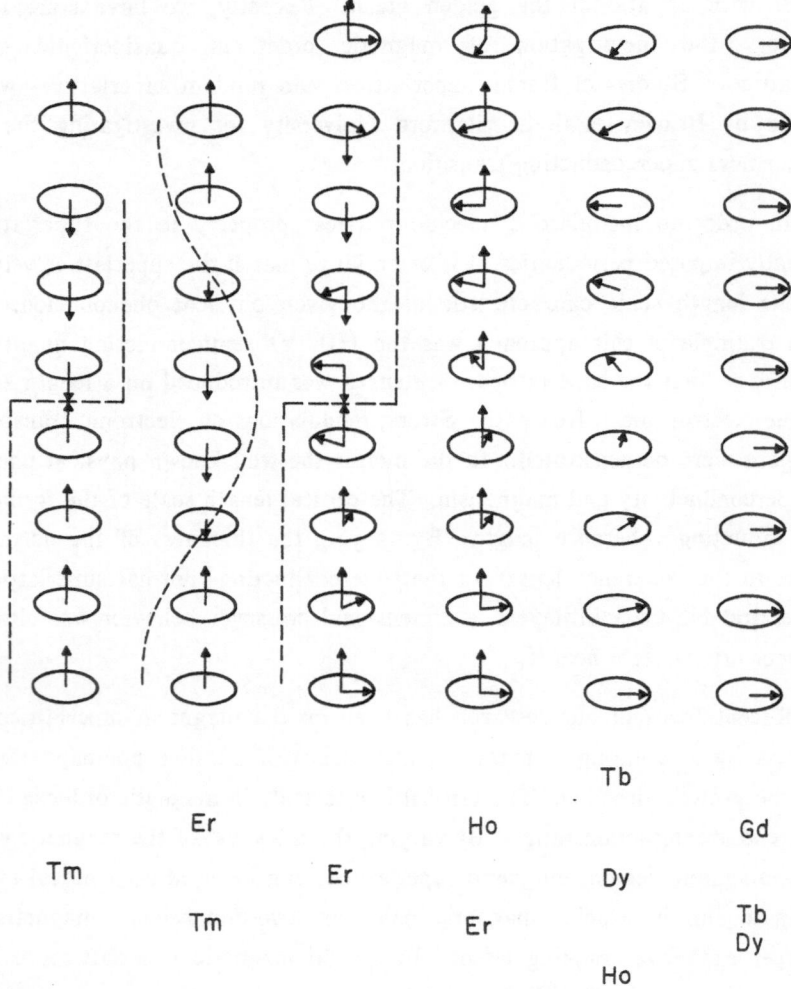

Fig. 2   Representative magnetic structures for magnetic rare earths of Tm, Er, Ho, Tb, Dy, and Gd at low temperature.

hcp magnetic rare earths and their nonmagnetic analog like Y or Lu. Therefore, it is highly probable that magnetic correlations will propagate coherently through the conduction electrons of the nonmagnetic medium, say Y, and then cause an effective coupling between separate magnetic layers.

Many magnetic rare earths exhibit long range magnetic structures modulated along the crystallographic c-axis.[12] Examples for the magnetic structures of rare earths including Gd and Dy, are shown in Fig. 2. The modulated spin structures are understood to be caused by the competition among the magnetic exchange coupling, magnetostrictions, and the crystal effect at low temperature. Among them, Gd has the simplest magnetic structure with a ferromagnetic Curie point of 292K and a spin only moment of 7/2. The main issue to address in the Gd-Y superlattices is whether the successive Gd layers are ferromagnetically or antiferromagnetically coupled to one another through the Y layer. Dy exhibits an antiferromagnetic in-plane helix at low temperatures. In the Dy-Y superlattice, the interesting question would be whether the spiral spin structure will maintain its coherency across the superlattices, and whether it is possible to tailor the spiral periodicity via commensuration with the Dy layer thickness.

In the synthesis of rare earth superlattice, a number of stringent structural requirements need to be fulfilled first, before meaningful magnetic characterizations can be made. Since the magnetic order in the elemental magnetic rare earths are modulated along the crystallographic c-axis, it is required that the rare earth films are preferentially grown in the direction of the c-axis, so that the multilayering direction of the superlattice coincides with the magnetic modulation direction. Orientation order in the plane is highly desirable, because certain magnetic rare earths like Dy exhibit in-plane anisotropy for the magnetic moment to align with an in-plane crystal axis. Furthermore, the chemical modulation should have sharp boundaries with minimum interdiffusion. This aspect is not easy to achieve considering that virtually all rare earths are mutually soluble to one another in thermal equilibrium.

In order to achieve the structural perfections and precisions required for the growth of rare earth superlattices, multiple degrees of fine controls on the growth parameters are clearly needed. The MBE technique appears to be the most viable approach to this end.

METAL MOLECULAR BEAM EPITAXY TECHNIQUE

All the crystals in this work have been prepared by the metal molecular beam epitaxy technique developed in our laboratories in the last five years. The growth was made in a versatile multisource deposition system especially designed for metal thin film work. We will first describe in length basic operations of the metal MBE system, and then the growth techniques for rare earth thin films.

To carry out the metal thin film growth in ultrahigh vacuum, the traditional MBE technique originally developed for semiconductor became inadequate. The effusion cells are operated by radiation heating, and do not yield sufficient flux for most metals of interest at the maximum operating temperature of 1300°C. Electron beam evaporation provides a solution to this need. However, at that time (1981), the electron beam deposition technology was largely used in high vacuum (10^{-7} torr) applications. Therefore, it became necessary to construct a research system by incorporating two technologies together: the MBE and e-beam evaporation techniques.

The design and the manufacture of the system were started in late 1981 with the company of Riber, ISA, France, and was in final completion after two year periods. The vacuum system comprises of two major portions: the main metal

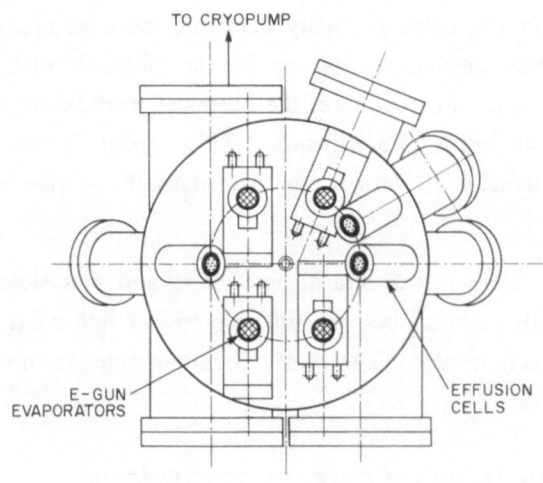

INTERNAL SOURCE CONFIGURATION

3 ELECTRON BEAM EVAPORATORS
3 EFFUSION CELLS.

Fig. 3   The internal source configuration of the metal MBE system.

growth chamber, and the sample introduction chamber for fast entries. This layout follows the standard design of the sample transfer mechanism for 2"-size substrates.

In the main growth chamber, there are totally six independent evaporation sources: 3 effusion cells and 3 electron beam heated sources. As shown schematically in Fig. 3, the centers of six sources are located in a circle with a radius of 6". The effusion cells are individually liquid nitrogen shrouded, and aiming toward the substrate with an angle of 30°. The effusion cells are suitable for evaporating metals with low melting points or with sufficiently high vapor pressures at temperature below 1300°C to yield a reasonable growth rate (~ 1.0Å/sec). The evaporation flux of the cells are checked with a nude flux gauge.

The electron beam sources are individually mounted on a 10" flange with independent power control and shutter operation. The commercial Airco Temescal 14 kW, 40cc single hearth evaporators are used. One major difference between the e-beam and the effusion cell evaporation is that the thermal configuration of the former is by far more dynamic and nonequilibrium. In order to form a uniform melting pool with stable evaporation flux, it is essential to sweep rapidly the electron beam with sufficiently large amplitude and high frequency in order to minimize the convection cooling from the surrounding water-cooled hearth. Furthermore, a fast feedback loop formed with the evaporation rate monitor is required to maintain a constant rate control. The rate monitors in use are the commercial Sentinel (200) flux monitors manufactured by Inficon, Leybold Heraeus. The flux monitoring scheme is based on the principle of the electron impact emission spectroscopy. The advantages of this type of monitor over the conventional quartz crystal thickness monitor are true flux monitoring, element selective, and insensitive to radiation heating.

The rate control for the rare earth evaporation is quite good: zero percent long term drift, and only $\pm 2\%$ for the short term fluctuations. A sensitive flux detection can be routinely made, e.g. 0.01Å/sec for Y, and 0.03Å/sec for Gd. Each e-beam evaporator is equipped with one independent set of Sentinel monitor and an adjacent crystal monitor for calibration. Another crystal monitor is located at the substrate level with an adjustable position to provide a cross check of the deposition rates precisely at the substrate location. The pneumatically driven shutter for the e-beam source operates in the open position as to expose only the substrate area to the flux; however, uninterrupted flux monitoring and rate control are maintained continuously with the Sentinel monitors.

The pumping package of the main chamber consists of a cryopump at the e-gun source level to remove the majority of gas load during evaporation, and a combination of ion pump and titanium sublimation pump at the substrate level. A base pressure less than 2×10^{-11} torr is maintained. During the evaporation, the liquid nitrogen cooled shrouds around the sources and the substrate provide the additional trapping of the gas impurities like CH_4, CO, and CO_2 generated by secondary electron desorption. The typical background pressure about 3×10^{-11} torr is achieved during the evaporation of the rare earths. The ultrahigh vacuum environment is crucial for the deposition of high-purity rare earth films at low growth rates ($< 1.0 \text{Å/sec}$).

In-situ characterization tools in the growth chamber are the residual gas analyzer based on quadruple mass spectrometry and the 10 kV reflection high energy electron diffraction (RHEED) with a grazing angle less than 1° off the substrate. The RHEED is used routinely to characterize the epitaxial interfaces, and further to optimize the growth parameters.

GROWTH OF RARE EARTH THIN FILMS AND SUPERLATTICES

Rare earth elements are well known for their chemical reactiveness. Elemental Gd, Dy, and Y exist in a stable crystalline form. High-purity rare earth sources were supplied by the Rare Earth Materials Center of Ames Research Lab, Iowa. Gd, and Y starting materials were premelted into large volume ($\sim 35cc$) ingots with a shape compatible with the e-beam hearth in order to minimize residual impurities adsorbed near the surfaces. The deposition of Dy is made by sublimation from an effusion cell source operating at about 1300°C to yield a growth rate of 1.0Å/sec. A thin wall ($\sim 0.01"$) tantalum insert is fit between the Dy source and the pyrolytic boron nitride (PBN) crucible to minimize chemical interactions between Dy and the PBN crucible, therefore keeping the PBN crucible intact for long usage.

The most severe difficulty in the growth of rare earth materials derives from the fact that no homoepitaxial substrates are available, as opposed to the field of (III, V) semiconductor and Si epitaxy. Attempts to prepare the rare earth single crystals substrates with highly polished surfaces did not succeed. Direct depositions of rare earths onto the commonly used substrates like Si, sapphire, and rare earth garnets invariably led to strong chemical interactions between the rare earths and the substrate species. For example, rare earths readily form rare earth silicide at 100°C followed by a rough islanding growth. Moreover, rare earth oxides are incorporated during the film growth by chemical interactions of rare earths with oxygen atoms supplied from sapphire (Al_2O_3).[13]

The difficulty of chemical reactions with the substrates was circumvented by intervening a chemical buffer layer which does not react either with the rare earths or with the sapphire substrates.[14,15] A Nb single crystal film is the ideal choice for this purpose based on two major reasons: The group VB and VIB elements including V, Nb, Ta, Cr, Mo, and W react very little with the rare earths. Moreover, single crystal Nb films were achieved in an earlier study in that an epitaxial growth was formed between bcc Nb(110) and sapphire $(11\bar{2}0)$.[16]

The growth of rare earth films is carried out with the following deposition sequences: After initial outgassing of the sapphire substrates at elevated temperatures of $\sim 1000°$C, the Nb (110) crystal about 2000Å thick was deposited onto sapphire with a rate of 2.0Å/sec at a growth temperature of 900°C. The in-plane epitaxial relationship for two orthogonal axes is $[\bar{1}11]$ Nb‖ [0001] Al_2O_3, and $[1\bar{1}2]$ Nb‖ $[10\bar{1}0]$ Al_2O_3. The rare earth crystals were subsequently deposited onto the Nb bcc(110) surface obeying a (0001) orientation. The in-plane orientation between hcp rare earth (0001) plane and bcc Nb (110) plane follows the Nishiyama-Wasserman relationship;[17] namely, the most densely packed row $[11\bar{2}0]$ of hcp rare earth is parallel to the densely packed row [002] of bcc Nb, and correspondingly in the orthogonal direction, the $[10\bar{1}0]$ axis of rare earth is parallel to $[1\bar{1}0]$ of Nb. The optimum growth temperature for the rare earths is about $550 \pm 50°$C at a growth rate of 1.0Å/sec. An initial Y seed layer about 600Å thick was deposited at this temperature to initiate a high-quality single crystal growth.

In order to carry out the superlattice growth with minimum interdiffusion, it is necessary to deposit alternatively Gd and Y layers at the lowest temperature while still maintaining good structural order. This is done by monotonically decreasing the substrate temperature from 550°C to about 250°C over a period of 30 minutes while continuing the Y seed layer deposition to ensure a clean surface. The optimum growth temperature is determined with the aid of RHEED. Growth at too low a substrate temperature resulted in a streaky RHEED pattern, with

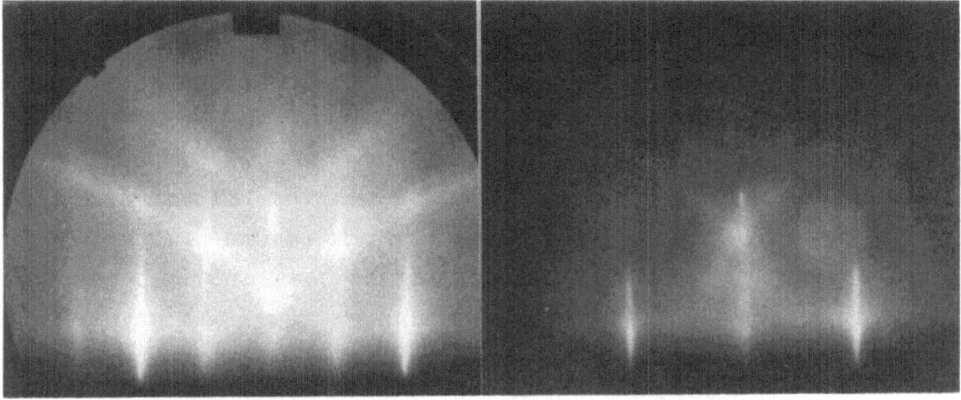

Fig. 4   The RHEED patterns along (a)$[11\bar{2}0]$, (b)$[10\bar{1}0]$ directions for a typical Gd-Y superlattice surface.

additional arrow-shaped marks occurring near the bright intersections of the streaks with Kikuchi arcs. This is, in general, indicative of facet growth. The growth temperature for the superlattices is therefore kept about 20°C higher than the faceting temperature. Typical RHEED patterns along two azimuthal directions of $[11\bar{2}0]$ and $[10\bar{1}0]$ for a (0001) Gd-Y superlattice surface are shown in Fig. 4 (a) and (b), respectively. The sharp streak pattern accompanied with distinct Kikuchi arcs suggests that a layer-by-layer growth has resulted in a highly ordered and atomically smooth surface.

A final protecting Y layer about 200-500Å thick was deposited to complete the growth. The oxidation of the single crystal rare earth films is self-limiting, and is confined to the upper 100Å. The film samples are stored in the laboratory atmosphere with no evidence of deterioration.

STRUCTURAL CHARACTERIZATION

The common structural characterization methods for the superlattices are X-ray diffraction analysis (XRD), and transmission electron microscopy (TEM). The latter method has been widely applied to semiconductor superlattices to provide a direct assessment on the epitaxial interfaces and structural perfections. However, because most metal film are deposited onto hard substrates like sapphires which are difficult for cleavage, preparation of the TEM specimen becomes a nontrivial task. Furthermore, the metallic films are subject to ion beam damage easily. Consequently TEM has not been utilized as a vital diagnostic tool in characterizing the metallic superlattices. Nevertheless, the research project of TEM studies is currently underway in our laboratory.

A more readily feasible structural probe is the standard X-ray diffraction analysis.[5] The measurements on the rare earth superlattices were made on a triple-axis diffractometer with a Cu rotating anode X-ray source from a Ge (111) monochromator. The data were taken either with a Ge analyzer (111) for high resolution studies or with an open detector for collecting integrated intensities.[18]

The $[Gd_{N_{Gd}}-Y_{N_y}]_m$ superlattice contains hcp N_{Gd} atomic planes and N_Y such planes in each bilayer period that is repeated m times. The superlattices are three-dimensional hcp single crystals. The full widths at half maximum of the rocking curves are 0.16° and 0.23° for directions perpendicular and parallel to the plane, respectively. The coherence length along the growth direction is essentially the total film thickness, and the in-plane domains are ordered on a length scale of about 2000-3000Å. The crystallinity of the rare earth (0001) films is approaching closely that of the underlying Nb (110) substrates.

Harmonics of the (0002) Bragg reflection from the chemical modulation are generally observed up to ± 5th or ± 7th order. Representative longitudinal scans for Gd–Y[18] and Dy–Y[20] superlattices are shown in Fig. 5(a) and (b), respectively. The irregularities in the bilayer periodicity are manifested in the increasing widths of the high order harmonics. In the present case, we infer that the variations of the modulation wavelength across the entire sample are as small as 1.5%, far surpassing those reported in the semiconductor superlattices. This is directly benefited from the good rate control achieved for e-beam evaporation without long term drift. The asymmetry of the amplitudes of positive and negative harmonics is caused by the coherency strain modulation along the c-axis.

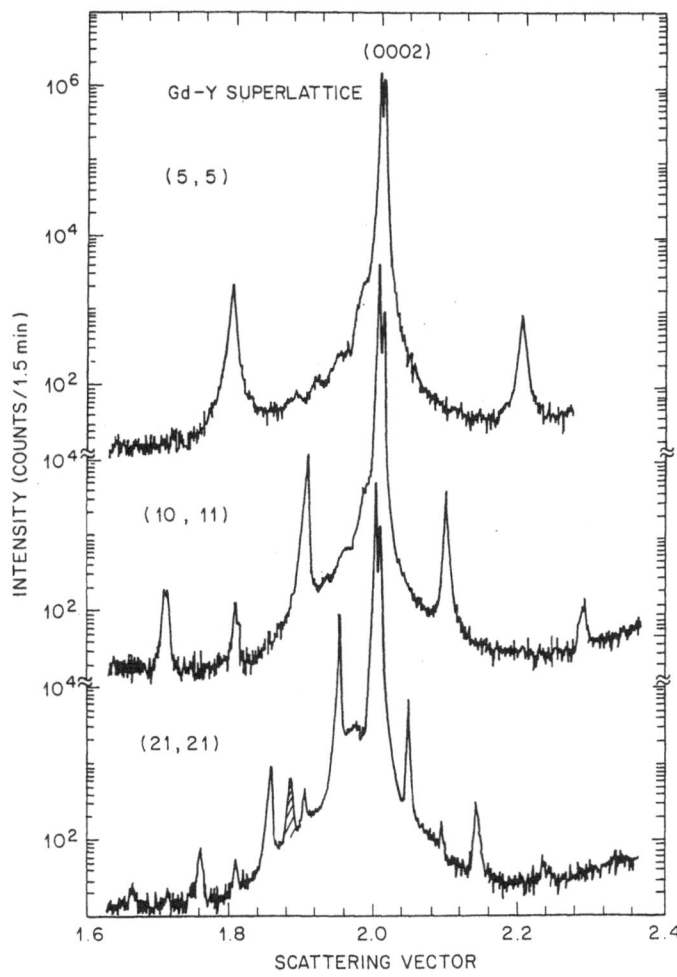

Fig. 5   Longitudinal X-ray scans about (0002) reflection for (a) Gd-Y superlattices of (21, 21), (10, 11), and (5, 5) samples.

The composition modulation was determined from a quantitative analysis on the intergrated intensities of the main Bragg reflections and their harmonics. Here the chemical modulation is approximated by an erf function to account for atomic diffusions occurring near the interfaces, and the amplitude of the strain modulation was calculated from the elastic constants assuming a coherent structure. This model analysis concluded a full interfacial width of 3.5 atomic layers thick; in other words, the compositional profile goes from 90% to 10% in about 2.0 atomic layers. The amplitude remains essentially unity except at shorter wavelength where the interfacial regions begin to overlap. Note that such sharp boundaries are approaching those achieved in the semiconductor superlattices. Although the

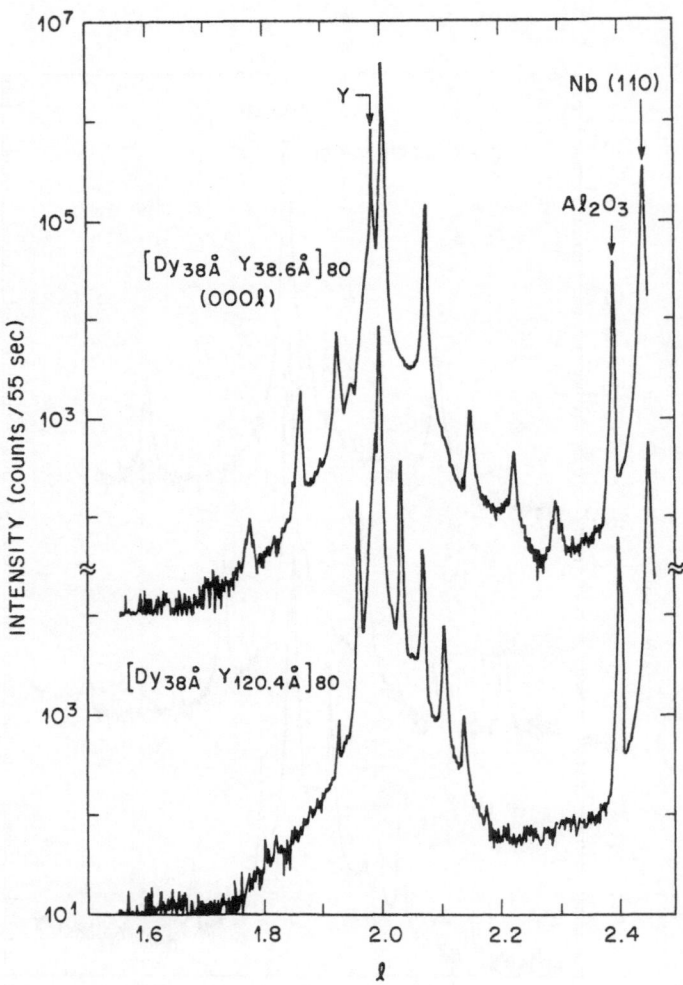

Fig. 5   (b) Dy-Y superlattices of $[Dy_{38Å} - Y_{38.6Å}]_{80}$, $[Dy_{38Å} - Y_{120.4Å}]_{80}$.

interdiffusions in the rare earth multilayers are hard to avoid, by eliminating the majority of grain boundaries and structural defects, we have shown that the interdiffusions are confined mostly to within two interfacial atomic planes.

One important aspect that was not included in this analysis is the roughening or corrugation that often occurs at highly oriented metal surfaces on a scale about a few thousand angstroms. Evidence for such roughening phenomenon was observed in an earlier cross-sectional TEM study[19] and the phase contrast optical microscopy. The observation is consistent with the smaller coherency length in the plane for metals compared to semiconductors. Present X-ray analysis cannot distinguish between an atomically smooth interface with a sinusoidally modulated composition profile, and a corrugated surface yet sustaining perfect chemical order. Further thorough analysis by including such surface roughening effect will have to be made in the future.

A typical interplanar spacing or coherency strain modulation is shown in Fig. 6(a) for a (21, 21) superlattice at T = 12K and 333K.[21] The in-plane lattice mismatch between Gd and Y at room temperature is as small as 0.3%, therefore forming a coherent interface. At temperature below T_c, the amplitude and the interfacial width of the interplanar spacing modulation increase continuously with decreasing temperature, due to the magnetostrictive effect in the ferromagnetic Gd. The presence of the coherency strain in the Gd arrays likely affects the magnetic properties, particularly for the interfacial Gd layers. This will be the main subject of discussion in the next section.

MAGNETIC PROPERTIES OF Gd-Y SUPERLATTICES

1) Interfacial Magnetism

A) Magnetic X-ray scattering Experiment

The magnetic order at the interface between a ferromagnet and a nonmagnetic substrate has long been a subject of interest. Superlattices contain regular arrays of such interfaces, thereby provide a means for studying interfacial magnetic behavior. Previously, we have shown that the X-ray diffraction yields the chemical and strain modulation information in the superlattices, where the scattering is predominantly contributed by the charge distribution. Here we will demonstrate that the magnetic moment modulation in the superlattices can be further determined, based on the X-ray scattered from regular arrays of magnetic spins.[22]

The intensity of the magnetic x-ray scattering component is given by the equation of,[23]

$$\frac{2\lambda_c}{\lambda} F''(l,m) \ (\vec{k_i} \times \vec{k_f}) \cdot \vec{S}(l,m)$$

Where F" (l,m) is the imaginary part of the charge scattering structure factor, and S(l,m) is the spin structure factor. Note that only the projection of the magnetic spin to the normal of the scattering plane contributes to the such scattering. With the aid of the high intensity synchrotron source, the cross term between the charge and the magnetic scattering can be further maximized by tuning the x-ray energy through the L_{II} and L_{III} absorption edges of Gd. Consequently, the small changes in the intensities of the Bragg reflections of the ferromagnetic Gd arrays are measurable, when the direction of the spin is reversed by an external magnetic field.[21]

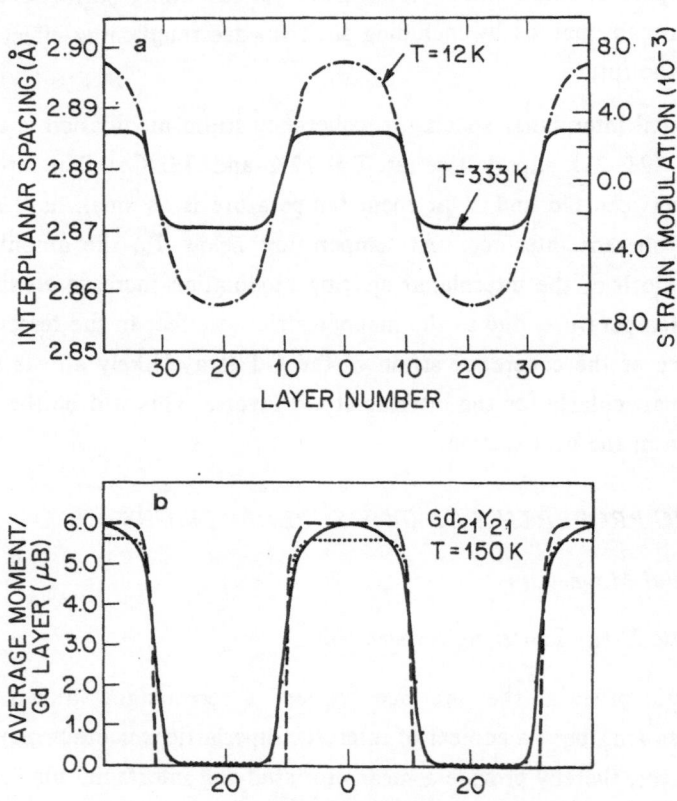

Fig. 6   (a) The interplanar spacing modulation along the c-axis at 15K and 333K for a (21, 21) superlattice. (b) Three models for the distribution of magnetic moment per Gd plane at 150K.

The flipping ratios, defined as $(I^+ - I^-)/I$, over a wide range of (l, m) were measured for the (21, 21) sample at low temperatures. The results were compared with calculations assuming three different magnetic moment distributions for the

Gd arrays as shown in Fig. 6(b). The solid curve, in that the full Gd moment is maintained at the center of the Gd array with a smooth reduction of the moment in approaching the interfaces, gives the best agreement with the measurements.

B) Magnetization Measurements

The reductions of the magnetic moments near the interfaces were observed directly in the magnetic measurements.[18] The interfacial magnetic behavior can be probed systematically with a series of superlattices with decreasing N_{Gd}, hence with increasing contributions from the interfaces. The magnetic measurements were carried out at low temperature with vibrating sample magnetometer (VSM) (0-15 kOe), and the Farady force method (1.5-12.8 kOe).[18]

The magnetization curves for most superlattice samples (if not all) closely follow the behavior expected for an ideal thin film with weak anisotropy; namely, a square loop for H in the plane of the film ($H_\perp$), and a linear dependence of $\sigma_\perp$ on H for $H_\perp < 4\pi M$ applied perpendicular to the plane ($H_\perp$). A coercive field is typically less than 100 Oe. An example of a σ–H loop is shown in Fig. 7 for a (5, 5) superlattice. The small remanence for $H_\perp$ seen in Fig. 7 was actually an exceptional case, and was not observed in other samples.

The σ–H loop can be further examined in details by defining the following characteristic parameters: (1) the zero field remanence, σ_r, according to the standard definition, (2) the saturation field, H_s, as defined to be the field at which $\sigma_\parallel$ no longer exhibits a rapid rise with field, and $\Delta\sigma_\parallel/\Delta H$ becomes constant, (3) the

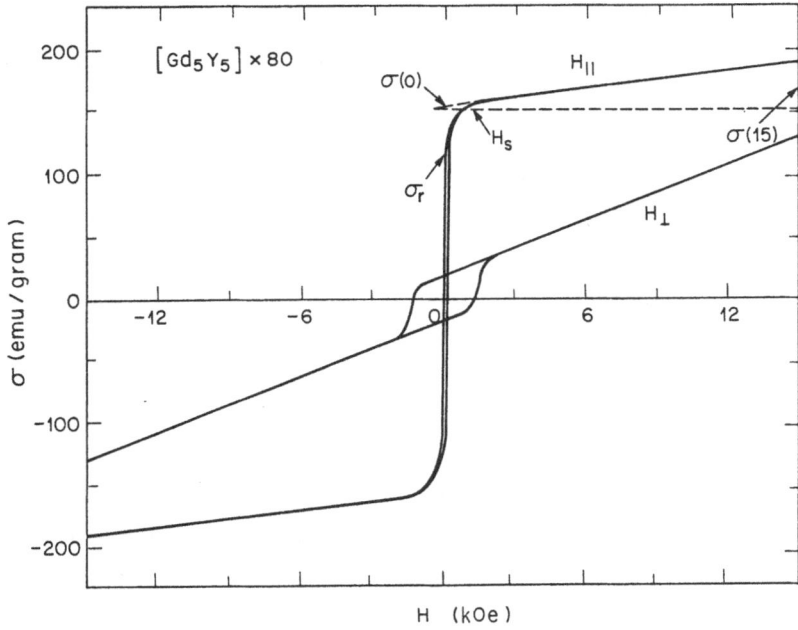

Fig. 7   The σ–H loop for $H_\parallel$ and $H_\perp$ of a (5, 5) superlattice.

spontaneous ferromagnetic moment, $\sigma(0)$, as the extrapolation of $\sigma_{||}$ at high field ($10 \text{ kOe} \leq \text{H} \leq 15 \text{ kOe}$) back to zero field, (4) $\Delta\sigma$ (15), as the difference between $\sigma(15)$ and $\sigma(0)$, resulting form the linear rise of $\sigma_{||}$ with field. (5) H_K, the effective anisotropy field, as the $H_\perp$ field at which $\sigma_\perp$ equals $\sigma_{||}$ obtained by extrapolation to high fields.

As to be discussed in the following section, the saturation field H_s is the field needed to overcome the interlayer exchange coupling between the adjacent Gd arrays. For the present discussions, we focus on the magnetization behavior at $H_{||} > H_s$. The magnetic properties of each Gd array in a superlattice are regarded essentially identical for $H > H_s$. As N_{Gd} decreases from 21 to 5, the systematic trend is that $\sigma(0)$ decreases and σ (15) increases linearly with $1/N_{Gd}$. Such linear dependence was accounted for by a simple two-layer model, in that the magnetization of each Gd array is contributed from two separate regions: the central ferromagnetic Gd region of (N_{Gd}-2) atomic layers thick, and the two interfacial Gd atomic planes which do not order ferromagnetically, and can be characterized by a susceptibility value independent of N_{Gd}. The central ferromagnetic Gd region retains the ideal magnetic thin film behavior even for N_{Gd} as small as 5. This is supported by the observation that the anisotropy field H_K remains nearly constant at 24 ± 2 kOe which is close to the demagnetization field $4\pi M$ for thin magnetic films.

The two-layer model is a simple approximation of the real moment distribution where the exchange coupling between two regions is clearly present. However, the most important aspect derived from this model is the rapid depression of the magnetic moment near the interfaces, which, at least, is in qualitative agreement with the magnetic x-ray scattering results.[21] The absence of the ferromagnetic order in the interfacial Gd plane is closely related to the finite interfacial width of the chemical modulation, and presumably the coherency strain effect as well. For example, the bulk alloys of Gd-Y with the Y concentration over 33 at.% are antiferromagnetically ordered in an in-plane helical structure.[24] These effects, together with exchange coupling with the neighboring Y layer, strongly favor a nonferromagnetic order in the interfacial plane. It is conceivable that in-plane spins are arranged in a relatively disordered manner which leads to the cancellation of the total moment. Further analyses on the interfacial spin structures by neutron diffraction are underway.

2) Interlayer Exchange Coupling Effects

As discussed in the Introduction section, our ultimate goal is to investigate the long range order in the superlattice derived from an interlayer exchange coupling. Magnetic superlattice, again, provides a convenient means for studying such effect by varying N_{Gd}.[25]

The systematic variations of the magnetic saturation behavior with the intervening Y layer thickness are shown in Fig. 8 for two series of superlattices: $N_{Gd} = 4$, and 10. Here the data show the dependence on N_y of the two characteristic magnetization parameters, the ratio of $\sigma_r/\sigma(0)$ and the saturation field H_s. The continuous oscillatory dependence of the data is evident, with a periodicity of 7 Y atomic layers over a range of 20 Y atomic layers. Furthermore, such modulation on the magnetization behavior depends strongly solely on N_Y, and is not affected by varying N_{Gd}.

With the aid of polarized neutron diffraction,[26] a clear picture for the magnetically ordered spin structures has now emerged. The magnetic order of the

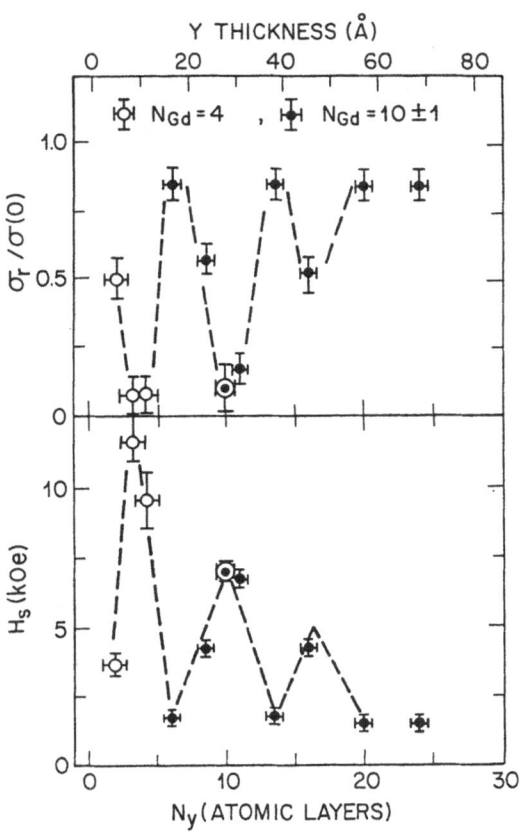

Fig. 8   The oscillatory dependence of (a) $\sigma_r/\sigma(0)$ and (b) H_s on N_Y for two series of superlattices with $N_{Gd} = 4$, and 10.

superlattices oscillates in two distinctly different states depending on the exact intervening Y layer thickness. One magnetic order is the simple ferromagnetic order where the spins in the plane of all Gd arrays are in parallel alignment relative to one another at zero field. The other magnetic order is the so called antiphase domain structure, where in zero field, the adjacent Gd arrays are 180° antiparallel to one another. Both magnetic structures are of long-range with a coherence length over 1000Å. In a small applied field, the moments rotate freely in the plane because of the weak anisotropy of Gd. The final energy minimum state was reached where the adjacent spin arrays are arranged symmetrically with respect to the external field. This is analogous to the spin flop configuration. For example, the angle between the spins and a field of 150 Oe at T = 150K is determined to be 79°, as depicted in Fig. 9,[26]. A full alignment of the spins with the field is reached at about 6 kOe at 77K, and the antiphase domain structure is replaced with the simple ferromagnetic order. The magnetic antiphase domain structure accounts for the low remanence and high H_s seen in the magnetization measurements, and an anomalous temperature dependence of moment at low field.[25]

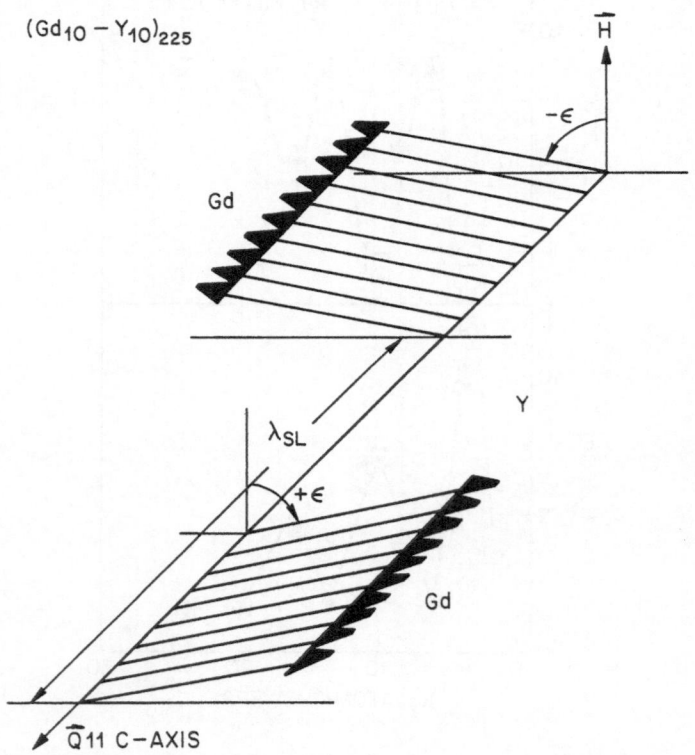

Fig. 9   The magnetic antiphase domain structure for a (10, 10) superlattices determined by the neutron diffraction.

The fundamental mechanism governing the long rang magnetic order and hence the modulation on the magnetic properties is originated from the long range RKKY interaction across the nonmagnetic Y medium. The conduction electrons of the Y layer are responsible for propagating the exchange coupling indirectly. The conduction electrons of Y respond to the local exchange fields penetrating from the neighboring Gd layers in a manner described by the generalized susceptibility function $\chi_Y(q)$ of Y. It is well known both experimentally[27] and theoretically[28] that a maximum occurs in $\chi_Y(q)$ at a wavevector of $q_{max} = 0.28 \times 2\pi/c$. This is directly resulted from the nesting feature of parallel sheets of the Y Fermi surfaces along such direction.[28] This feature gives rise to an enhanced exchange coupling strength between the two adjacent Gd arrays to produce coherence in the magnetic order.[29] Moreover, the sign of this exchange interaction oscillates with a periodicity corresponding to $2\pi/q_{max}$, hence leads to the oscillatory occurrence of the long range ferromagnetic order and antiferromagnetic order. Furthermore, this RKKY interaction in the present one-dimensional configuration depends on the intervening Y thickness as $1/r$-like rather than $1/r^3$ as in the 3-dimensional case. The exchange coupling extends over quite long range, similar to the observed dependence of H_s on N_y over 20 Y atomic planes.

MAGNETIC PROPERTIES OF Dy-Y SUPERLATTICES

Dy is another magnetic rare earth exhibiting long range modulated magnetic order along the c-axis. Below the Neel temperature of 178K, Dy orders antiferromagnetically with an in-plane helix. The turn angle between the successive spirals decreases continuously from 43.2° at T_N to 26.5° at $T = 90K$ before collapsing to the ferromagnetic state through the first order transition. Magnetostriction is known to affect strongly on the progressive changes of the spiral wavevector with temperature.

In the artificially layered structures of Dy epitaxially grown on the Y substrate, the presence of coherency strain leads to an modified magnetostrictive effect on the magnetic order. Specifically, the lattice mismatch (1.6%) between Dy and Y is significantly larger than the Gd-Y case. In the basal plane, the Dy lattice has to expand in order to match with the Y substrate, meanwhile the interplanar spacings along the c-axis will contract correspondingly. The situation is analogous to the case of applying an uniaxial stress along the c-axis direction to the Dy crystal. The systematic suppression of the 1st order transition temperature (T_c) with the uniaxial pressure was observed in a previous study on a bulk Dy crystal.[30]

Indeed, the suppression of the T_c in the presence of coherency strain was found in a recent magnetization study of the single layer Dy films epitaxially grown on the Y substrate,[20] and the results of M vs T are shown in Figure 10. The T_c of a 5000Å thick Dy film is nearly the same as in the bulk. In reducing the single layer thickness systematically to 200Å, the T_c decreases accordingly to as low as 25K, and by the thickness of 76Å, the first order transition is no longer present.

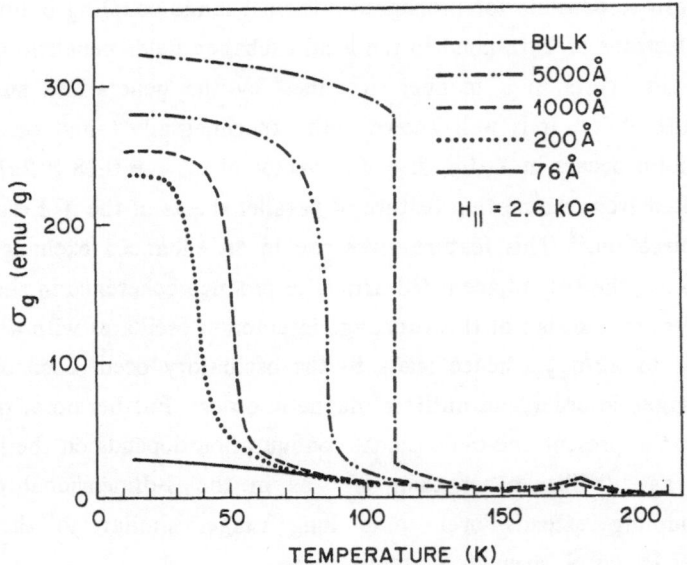

Fig. 10   Moment vs temperature at H = 2.6 kOe for single layer Dy films of 200Å, 1000Å, 5000Å thick and a bulk Dy crystal.

In the superlattice configuration, both Dy and Y layers are accommodating the coherency strains. The absence of the first order ferromagnetic transition in the Dy-Y superlattices is accounted for by the similar mechanism. Furthermore, as determined by the neutron diffraction studies,[20] the spiral wavevectors of the individual Dy layers in the Dy-Y superlattices show a more modest temperature dependence than in the bulk Dy, and are shown in Fig. 11.[20,31] In a $[Dy_{38Å} - Y_{38.6Å}]_{80}$ superlattice, the spiral wavevector exhibits a lock-in transition at T ≈ 45K to a constant value corresponding to a turn angle of 36.2° per layer. For a $[Dy_{38Å} - Y_{120.4Å}]_{80}$ superlattice, the spiral wavevector becomes essentially temperature independent for temperature less than 90K. These results suggest that the constraining of the D_Y in-plane lattice by the adjacent Y layers has reduced the tendency of developing the magnetoelastic strain associated with ferromagnetic order. The detailed mechanism is still a subject of further investigation.

The neutron diffraction studies by the group at University of Illinois and National Bureau of Standards have shown that the propagation of the Dy spiral is coherent through the Y layer.[32] Our recent investigations on Dy-Y superlattices support this conclusion as well. The relative pitch angle between the successive

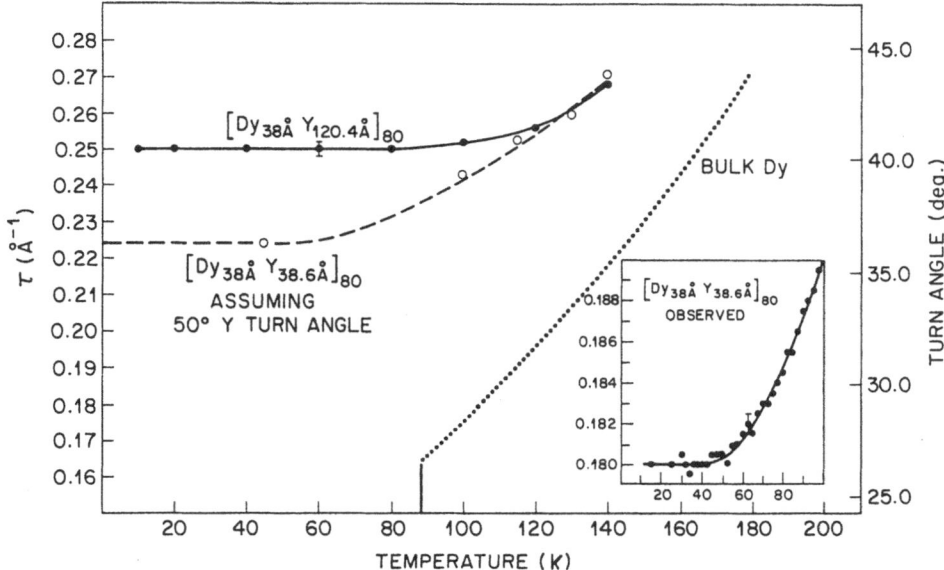

Fig. 11 The temperature dependence of the Dy spiral wavevector in (a) a bulk Dy crystal, [b] a $[Dy_{38Å}-Y_{38.6Å}]_{80}$ superlattice (c) a $[Dy_{38Å}-Y_{120.4Å}]_{80}$ superlattice.

spirals is largely determined by the wavevector of q_{max} in $\chi_{Y(q)}$, and the Y layer thickness. However, the coherence of the spirals deteriorates with thicker intervening Y layer. Recent experiments on Dy-Y superlattices with the Y layer as thick as 42 atomic planes showed that spiral coherence is confined to be about the bilayer thickness, and the pitch angles between the adjacent spirals are essentially random.[20]

Therefore, both studies on Gd-Y and Dy-Y superlattices have demonstrated that the magnetic correlation in the superlattices is predominantly the RKKY coupling with a finite range of approximately 20 atomic layers. Speculations of possible existence of a spin density wave[33] stabilized in the conduction band of Y is apparently not supported by our experimental observations, since one expects the strength and the range of a magnetic correlation based on the spin density wave mechanism to exceed substantially those based on the RKKY coupling. The existence of a spin density wave in the real laboratory materials requires a conduction electron mean free path of a few hundred angstroms extending over several bilayer periods. The magnitude of this critical parameter still awaits further experimental determinations by transport measurements along the multilayering direction.

SUMMARY AND CONCLUSIONS

Advances in the synthesis of magnetic rare earth superlattices have been made with the development of metal molecular beam epitaxy technique. The chemically reactive rare earths can now be prepared as single crystal films for the first time. What is most remarkable is the long range magnetic order tailored by the superlattice periodicities. The overall magnetic properties of the superlattices as a whole are modulated by the artificially layering sequence. This approach has opened up a wide spectrum of exciting opportunities in the fundamental studies of thin film magnetism. Novel magnetic properties and structures are expected to be discovered continually in the future as the research can now be easily extended to include the 3d itinerant magnets. As a close analogy to the "band gap engineering" in the semiconductor superlattices, our work has demonstrated that the "spin structure engineering" in the magnetic superlattices is becoming feasible. No doubt some of the new findings will have good potential for real device applications. The knowledge that is gained by this basic research will eventually be served as a basis for future technology.

ACKNOWLEDGEMENT

The work is done in collaboration with F. J. DiSalvo, L. C. Feldman, R. M. Fleming, E. M. Gyorgy, M. Hong, A. R. Kortan, D. B. McWhan, S. Nakahara, L. F. Schneemeyer, C. Vettier, J. V. Waszczak, and Y. Yafet at AT&T Bell Labs; J. Bohr, L. D. Gibbs, H. Grimms, and C. F. Majkrzak at Brookhaven National Laboratory; and J. W. Cable at Oak Ridge National Laboratory. It is a pleasure to acknowledge valuable contributions from R. Superfine, and J. P. Mannaerts in constructing the metal MBE facility. The author also thanks the continuing encouragement and support for this research from J. M. Rowell, A. Y. Cho, T. H. Geballe, G. Y. Chin, and C. K. N. Patel.

REFERENCES

1. For a complete review, see *Synthetic Modulated Structures*, edited by L. L. Chang, and B. C. Giessen, (Academic, New York, 1985).

2. J. DuMond and J. P. Youtz, J. Appl. Phys. **11**, 357, (1940).

3. J. E. Hilliard, in *Phase Transformations*, Ed. H. I. Aaronson, ASM (1970), p. 557.

4. C. M. Falco, I. K. Schuller, in reference 1, p. 339.

5. D. B. McWhan, in reference 1, p. 43.

6. R. Merlin, K. Bajema, R. Clarke, F. Y. Juang, and P. K. Bhattacharya, Phys. Rev. Lett. **55**, 1768, (1986).

7. A. R. Kortan, A. I. Goldman, C. F. Majkrzak, M. Hong, and J. Kwo, to be published.

8. V. Matijasevic and M. R. Beaseley, Procceedings of Materials Research Society Meeting, Boston, Dec. 1985, and to appear in Phys. Rev. B.

9. S. T. Ruggiero, and M. R. Beaseley, in reference 1, p. 365.

10. T. Shinjo, N. Hosoito, K. Kawaguchi, N. Nakayama, T. Takada, and Y. Endoh, J. of Mag. Mag. Mat. **54-57**, 737 (1986), and references therein.

11. H. K. Wong, H. Q. Yang, J. E. Hilliard, and J. B. Ketterson, J. Appl. Phys. **57**, 3660, (1985).

12. *Magnetic Properties of Rare Earth Metals*, edited by R. J. Elliot, Plenum Press, London and New York, (1972).

13. L. H. Greene, W. L. Feldmann, J. M. Rowell, B. Batlogg, E. M. Gyorgy, W. P. Lowe, and D. B. McWhan, Superlattices and Microstructures, **1**, 407, (1985).

14. J. Kwo, D. B. McWhan, M. Hong, E. M. Gyorgy, L. C. Feldman, and J. E. Cunningham, in *Layered Structures, Epitaxy, and Interfaces*, Eds. Gibson, and Dawson, Materials Research Society Symposia Proceedings, vol 37, p. 509, 1985.

15. J. Kwo, M. Hong, S. Nakahara, Appl. Phys. Lett. **49**, 319, (1996).

16. S. M. Durbin, J. E. Cunningham, M. E. Mochel, and C. P. Flynn, J. Phys. **F11**, L223 (1981).

17. L. A. Bruce and H. Jaegger, Philos. Mag. **A38**, 223 (1978) and references therein.

18. J. Kwo, E. M. Gyorgy, D. B. McWhan, M. Hong, F. J. DiSalvo, C. Vettier, and J. E. Bower, Phys. Rev. Lett. **55**, 1402, (1985).

19. S. Nakahara, R. Schutz, and L. R. Tetardi, Thin Solid Films, **72**, 277, (1980).

20. M. Hong, R. M. Fleming, J. Kwo, J. V. Waszczak, L. F. Schneemeyer, J. P. Mannaerts, C. F. Majkrzak, Doon Gibbs, and J. Bohr, in the 31st Annual Conf. on Mag. and Mag. Mat., Baltimore, Nov, 1986, to appear on J. Appl. Phys.

21. C. Vettier, D. B. McWhan, E. M. Gyorgy, J. Kwo, B. M. Buntschuh, and B. W. Batterman, Phy. Rev. Lett. **56**, 557, (1986).

22. Doon Gibbs, D. E. Moncton, K. L. D 'Amico, J. Bohr, B. H. Grier, Phy. Rev. Lett. **55**, 234, (1985).

23. F. DeBergevin and M. Brunel, Acta. Cryst. **A37**, 314 (1981).

24. W. C. Thoburn, S. Legvold, and F. H. Spedding, Phys. Rev. **110**, 1298, (1958).

25. J. Kwo, M. Hong, F. J. DiSalvo, J. V. Waszczak, and C. F. Majkrzak, to be published.

26. C. F. Majkrzak, J. W. Cable, J. Kwo, M. Hong, D. B. McWhan, Y. Yafet, and J. V. Waszczak, Phys. Rev. Lett. **56**, 2700, (1986).

27. N. Wakabayashi and R. M. Nicklow, Phy. Rev. B. **10**, 2049, (1974).

28. R. P. Gupta, and A. J. Freeman. Phy. Rev. B. **13**, 4376 (1976).

29. Y. Yafet, in the 31st Annual Conf.on Mag. and Mag. Mat. Baltimore, Nov, (1986), to appear in J. Appl. Phys.

30. H. Bartholin, J. Beilif, D. Bloch, P. Boutron, and J. L. Feron, J. Appl. Phys. **42**, 1679, (1971).

31. Doon Gibbs, J. Bohr, and C. F. Majkrzak to be published.

32. M. B. Salamon, S. Sinha, J. J. Rhyne, J. E. Cunningham, R. W. Erwin, J. Borchers, and C. P. Flynn, Phys. Rev. Lett. **56**, 259 (1986).

33. J. J. Rhyne, R. W. Erwin, M. B. Salamon, S. Sinha, J. Borchers, J. E. Cunnnigham, and C. P. Flynn, to appear in the Less Common Metals (Proc. 17th Rare Earth Research Conference).

FERROMAGNETIC METALLIC MULTILAYERS : FROM ELEMENTARY SANDWICHES TO SUPERLATTICES

J.P. Renard

Institut d'Electronique Fondamentale
CNRS UA 022, Bâtiment 220, Université Paris-Sud
91405 - Orsay, Cédex, France.

ABSTRACT

The elaboration and the structural characterization of ultrathin films and multilayers of Ni and Co on Au (111) are described. From *in situ* resistivity measurements, and *ex situ* grazing X-ray interference and transmission electron microscopy, the samples show sharp interfaces and good crystalline coherence perpendicular to the layer. The magnetic properties of the films are reviewed. In particular, the variation of the Curie temperature of Ni films versus thickness and the one of Ni bilayers versus gold spacing layer thickness, is studied. The hcp Co films show a strong perpendicular anisotropy evidenced by SQUID magnetometry and ferromagnetic resonance.

1. INTRODUCTION

The ultrathin films of ferromagnetic metals show new magnetic properties, different from the bulk ones [1, 2, 3, 4]. Besides that, the development of thin film technology allows now to prepare metallic multilayers of good quality, keeping on a macroscopic scale the interesting properties of the thin films. This opens a way towards new magnetic materials, potentially useful for practical applications.

When a film of a ferromagnetic metal M is deposited on a metallic substrate M', its magnetic properties can be modified by the following mechanisms :
i - the reduction of the space dimensionality from 3 to 2, which favors the magnetic fluctuations.
ii - the variation of the conduction electron state density with the thickness of the film as it becomes of the order of the Fermi wavelength.
iii - the local changes of the energy band structure at the interface.
    These last effects are very important, as shown in the self-consistent band calculations [2, 3, 4, 5, 6,]. In

particular the reduced coordination, the symmetry breaking and the band hybridization lead to important differences with respect to the bulk system, which in certain cases, such as Cr could consist in a change from antiferromagnetism to ferromagnetism [2, 4].

An other important feature is the possibility of obtaining by epitaxy metastable phases [7] with uncommon magnetic properties.

The experimental study of the effects mentionned before requires samples as perfect as possible with good crystallinity and sharp interfaces. The simplest system allowing *ex situ* measurements is the sandwich M'/M/M' which presents only two interfaces (more or less identical) and a good stability in time, with a proper choice of M', avoiding oxidation and interdiffusion. The elementary sandwich is the first step towards magnetic metallic multilayers or superlattices $(M'/M)_n/M'$ which, for large n values, display the interface magnetic properties on a macroscopic scale. These multilayer systems allow to study the interactions between magnetic layers through metallic spacing layers [8, 9].

Both sandwiches and multilayers of ferromagnetic 3d transition metals and non magnetic metals have yet been prepared and studied [10, 11,]. To date, many of them concern Fe [12,13, 14, 15] because of its large magnetic moment moment and the opportunity of Mössbauer studies [12,16].

Several experimental studies were also devoted to Cu/Ni/Cu sandwiches [17] and $(Ni/Cu)_n$ superlattices [18, 19, 20] because of a good matching of the fcc lattices of Ni and Cu, and of the existence of theoretical predictions for these systems [5, 6, 21, 22, 23]. The present paper concerns the ultrathin films of Ni and Co on Au (111), prepared in sandwiches Au/M/Au and in multilayers $(Au/M)_n/Au$ with n = 2 - 5.

2. SAMPLE PREPARATION

The metal films of gold, cobalt and nickel are deposited by slow evaporation in ultrahigh vacuum (starting pressure 10^{-10} torr), on a polished glass substrate with very small surface roughness ($\simeq$ 5 A).The substrate shown in Fig.1 is equipped of Cr Au electrical contacts and gold wires which allow *in situ* resistivity measurements during the growing process of the film. The metals are evaporated either by Joule heating of tungsten boats or by electron bombardment (for Ni and Co). The deposition rates are of about 1 A/sec for Au and 1 A/mn for Ni and Co. During deposition, the film thickness is monitored by a quartz oscillator, which is calibrated by X-ray techniques (see section 3.2).

Before growing the first magnetic layer, a buffer gold film of about 250 A thickness, is deposited on the glass substrate at room temperature and annealed during 1 hour at 150°C. The textured structure of the Au buffer film after annealing is shown in Fig. 2. It consists of crystals of mean lateral size of about 2000 A with random orientations in the plane of the film but with the surface oriented (111). The film surface is atomically flat, made of terraces (compact planes) of size around 250 A, separated by monoatomic steps. This high planeity is revealed both by transmission electron microscopy (TEM) [24] and by the small value of the electrical resistance which is not far from the bulk value (see section 3).

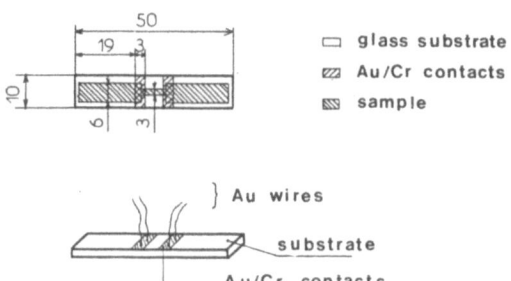

Fig. 1.   View of the substrate and geometry of the thin films. The lengths are in mm.

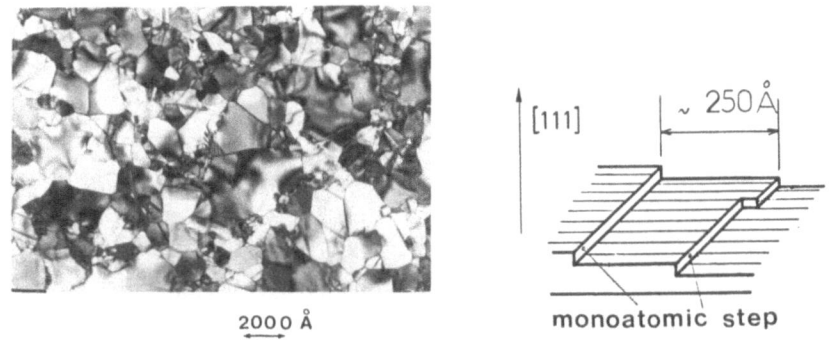

Fig. 2.   TEM bright field image of the gold buffer layer and schematic view of its surface.

3. STRUCTURAL CHARACTERIZATION

3.1. *In situ* electrical resistivity

It is well known that the surface of a metal film contributes to the electrical resistivity when the thickness t is comparable to the electron mean free path λ .

The resistivity increase, $\delta\rho$, with respect to the bulk value depends on the coefficient p of specular reflection as shown in the early works of Fuchs [25] and Sondheimer [26] which give for $t \gg \lambda$:

$$\delta\rho / \rho_{oo} = \frac{3}{8} \frac{\lambda}{t} (1 - p) \tag{1}$$

For high quality metal films with smooth surfaces, such as our Au buffer layer, p is close to 1 (p $\simeq$ 0.85) [27], and the resistivity is very sensitive to atoms deposited on the surface which increase the roughness and thus the diffusivity and the resistivity [28, 29]. The resistivity variation during the deposition of the metal overlayer gives informations on the growing process [30]. In particular for a monolayer by monolayer growing process (Franck-van der Merve [31]), a periodic oscillation of resistivity related to the one of the surface roughness can be observed, as exemplified in Fig.3 which shows the variation of resistance during the homoepitaxy of indium at low temperature [32,33], Similar oscillations were also observed in heteroepitaxy of In on Au, whose minima correspond perfectly to the breaks of the slope of the Auger intensity of In and Au [33,34].

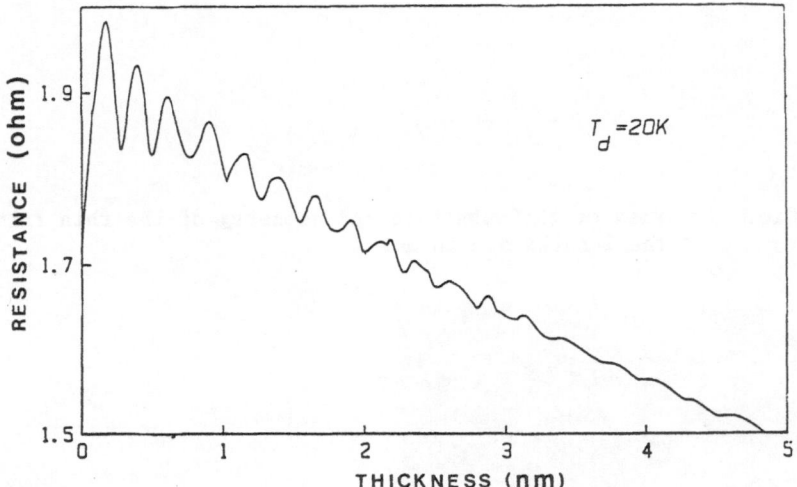

Fig. 3. Variation of resistance versus the thickness of an indium overlayer deposited on an annealed In film.The period of the oscillations corresponds exactly to the distance between In(111) planes. (from thesis of C. Marlière, Orsay 1985)

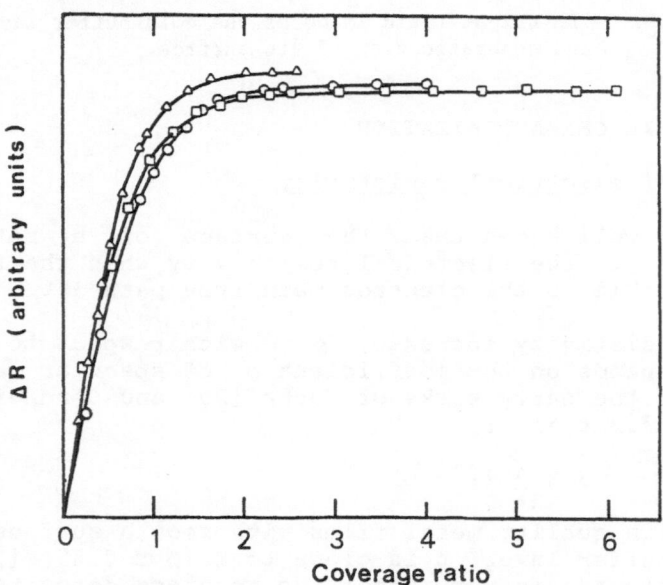

Fig. 4. Variation of the resistance of the Au buffer layer versus the Ni coverage ratio (equivalent monolayers) for three samples.

In the case of deposition of Ni and Co at room temperature, the resistivity exhibits a monotonous increase and reaches a plateau for a coverage equivalent to about two monolayers (Fig.4). It can be concluded that the growth mechanism is not a F-vdM one but that the (n+1)th atomic layer begins to grow before the nth one is complete. However the fact that a plateau is attained shows that the whole Au surface is perturbed after having deposited an average thickness of about 5 A. The resistance variation versus time, during the growth of a superlattice of (Au/Ni)₅/Au with respective Ni and Au thickness of 8 A and 30 A is shown in Fig.5. The resistance increases during each Ni layer deposition, while it decreases during each Au layer deposition. This can be interpreted by corresponding successive increases and decreases of the surface roughness. Even after having grown several metal layers, the coverage of the Au surface by Ni leads to a sizable resistivity increase which shows that the Ni/Au interface keeps a good quality.

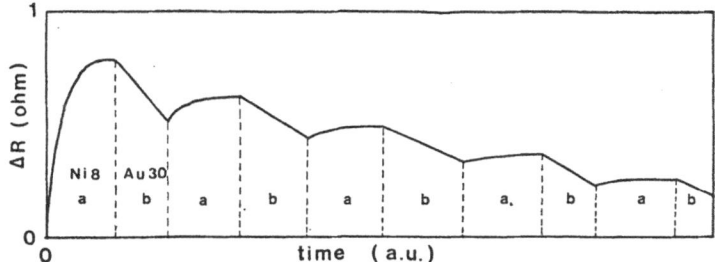

Fig. 5.   Variation of the resistance during the growth of a (Au/Ni)₅/Au superlattice. The thickness of the Ni and Au layers are respectively 8 A and 30 A.

3.2. X-ray studies

The samples were studied out of the vacuum chamber by two different X-ray techniques :

3.2.1. Au (111) Bragg diffraction.

The diffraction profile of Au(111), for Au/M/Au sandwiches with same Au thickness t, shows fringes with a spacing proportional to 1/t. This provides a control of the film quality and a precise measurement of the Au thickness allowing the calibration of the quartz oscillator. The presence of the magnetic metal M can be deduced from the apparition of secondary peaks with a weak intensity. For thicker layers, the (111) peak of M is observed. Besides the systematic calibration of the quartz, this technique is interesting since it shows the absence of interdiffusion between M and Au. Indeed, in the case of efficient interdiffusion, the fringe system would be thinner and would correspond to an homogeneous film with a thickness about twice than that of the individual gold layers [35].

3.2.2. Interference of grazing X-rays

The reflectivity of a multilayer film for a monochromatic parallel X-ray beam at a small angle Θ with the surface, exhibits a complex variation with Θ, because of interferences between the beams reflected and refracted by the several surfaces or interfaces.

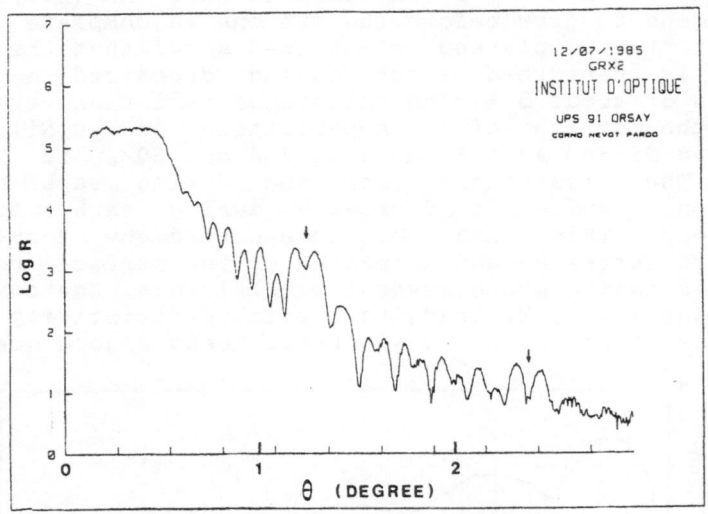

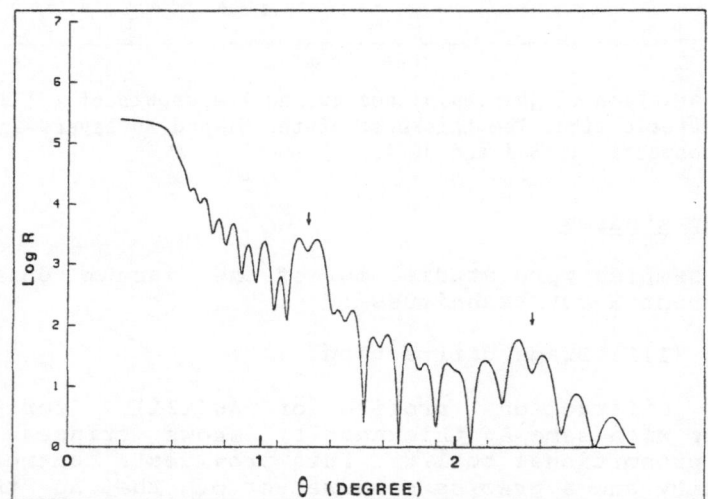

Fig. 6.   Grazing X-ray pattern of a (Au/Co)₅/Au superlattice. The thicknesses of the Co and Au layers are respectively 8 A and 30 A. The small arrows indicate the Bragg peaks of the superlattice with a period of 38 A. The experimental curve is at the top of the figure and the computed one at the bottom.

For a single film, a simple variation is observed : the reflectivity shows Kiessig fringes [36], whose maxima occur at angles Θ_k given by the following relation :

$$2t (\sin^2 \Theta_k - \sin^2 \Theta_c)^{1/2} = k.\lambda \qquad (2)$$

where t is the film thickness, λ the X-ray wavelength, k an integer number and Θ_c the angle of total reflection related to the index of refraction $1-\alpha$ by $\Theta_c = (2\alpha)^{1/2}$. The Kiessig fringes provide a precise measurement of the film thickness and valuable information on the roughness of the film surface and of the interface with the substrate [37].

For the elementary sandwich Au/M/Au, the interference pattern is not too complicated and can be fitted by calculations done on simple models [38]. By this way, the thickness, the roughness and the average density of the different layers can be obtained. These studies show that the first M/Au interface can be considered as atomically flat while the second one extends on about 2-3 atomic planes.

An example of diffraction pattern for a superlattice, (Au/Co)₈/Au is shown in Fig.6. In this case, the angular variation of the reflectivity is rather complex and its detailed analysis is almost impossible because of a too large number of fit parameters. Nevertheless, the computer simulation done on an idealized structure : low roughness, equal thickness of the layers of same metal, reproduces satisfactorily the experimental data (Fig.6) showing, at least qualitatively, that the sample is no too far from the model.

3.3. Transmission Electron Microscope (TEM) studies

The Au/M/Au sandwiches were peeled off from the glass substrate, mounted in copper grids and studied by TEM. The bright field images are very similar to the ones obtained for single gold films (Fig.2) with a thickness equal to the total thickness (typically about 500 A). In particular, the grain boundaries are as neat as for the single layer, showing that they extend through the whole sample thickness.

Electron diffraction diagrams of good quality are obtained by focusing the electron beam on carefully selected large single crystals (Fig.7). For both sandwiches with Ni and Co, bright spots corresponding to the (2 $\bar{2}$ 0) reflections of Au are observed. Each of these spots is surrounded by 6 satellites which are due to the magnetic metal M. One of the 6 satellites is a component of the reciprocal lattice of M. while the 5 other ones are due to double diffraction between Au and M. These diagrams show that the observed Au and M crystals are oriented with their [111] axis roughly parallel to the incident electron beam. The spots due to M do not appear for M thickness below 12 A. As soon as they appear, above 12 A, they correspond to crystal parameters which are identical to the bulk ones.

While the Au spots have the usual circular shape, the M ones have an elongated shape along a circle centered on the transmitted beam. This confirms that the M parameters are well defined but show that their direction has small fluctuations. The rows of M atoms in the compact planes are not strictly parallel to the corresponding ones for Au; indeed, the former have a slightly variable direction, in order to adjust different metals, Ni or Co to Au, which have a large lattice mismatch of about - 0.14.

In addition, the diagrams show the existence of Au layer spots (1/3(422) in the fcc notation). In the case of Au/Co/Au samples, each Au layer spot is surrounded by 3 satellites which cannot be detected for Ni. A careful comparison of the measured intensities of these satellites with calculation for fcc and hcp structures [24] demonstrates that the thin cobalt layers on Au(111) have the hcp structure, while the Ni films have the fcc structure of bulk Ni.

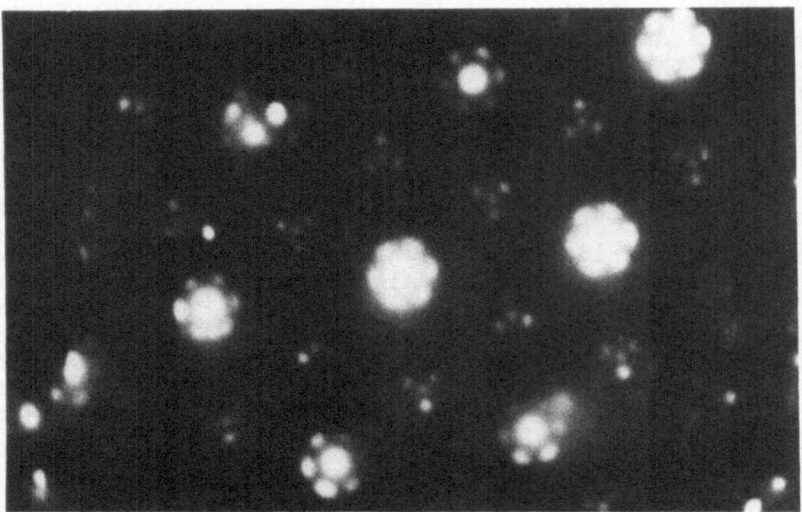

Fig. 7.   Electron beam diffraction pattern of a monocrystalline Au/Co/Au sandwich with Co thickness of 80 A. Notice the six satellites round the intense (2$\bar{2}$0) Au spots and the weak Au layer spots surrounded by three satellites due to hcp Co.

Indirect lattice images of Au/Co/Au obtained from the central spot and its 6 satellites, show that the Co crystals consist of perfectly ordered small regions of 60 to 100 Å lateral size, with slight local changes (a few degrees) in their orientation. Many dislocations can be seen on the images. Most of them arise from the magnetic material. These results are confirmed by high resolution TEM using the 6 Au layer spots with their 3 satellites [39]. In conclusion, the TEM study of the Au/Co/Au sandwiches reveals a quasi epitaxy of hcp Co on Au (111). In spite of slight local changes in Co crystal orientation, the top Au layer has the same orientation as the Au buffer layer; this would permit to grow superlattices with a long range coherence in the direction perpendicular to the layer.

3.4. ^{59}Co Nuclear Magnetic Resonance

As TEM, Nuclear Magnetic Resonance (NMR) is a technique which provides informations on the local structure. In a ferromagnet, the nuclear spins experience a local hyperfine field which is proportional to the average magnetization and depends on the structure. Due to the high permeability of ferromagnetic samples, the NMR signal is strongly enhanced and can be observed, even in very thin films, at sufficiently low temperature. In the Au/Co/Au sandwiches with an area of about 1 cm^2, the ^{59}Co NMR signal at 1.8 K was easily observed by spin echo technique for Co thicknesses down to 20 A [40]. The NMR line profile for a Co thickness of 80 A is shown in Fig.8. The NMR line is shifted to upper frequencies with

368

respect to the Co frequency in bulk fcc Co. The maximum corresponds to the NMR frequency of nuclei located in the Bloch walls of bulk hcp Co. Thus the NMR data are fairly consistent with the Co hcp structure observed by TEM.

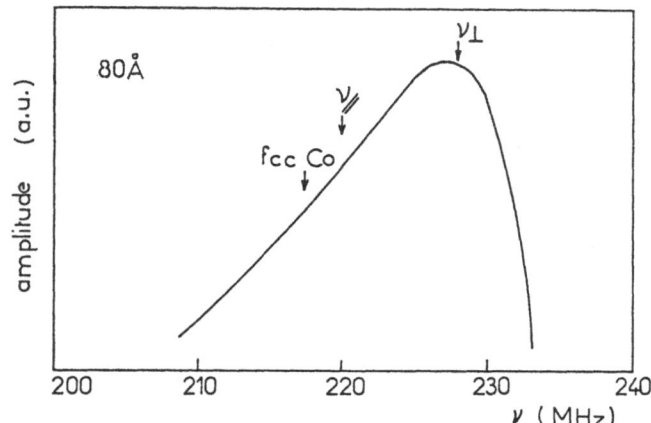

Fig. 8. ^{59}Co spin-echo spectrum at 1.8 K in zero external field for Au/Co/Au with Co thickness of 80 Å. The arrows indicate the NMR frequencies of bulk hcp Co in domains ($\nu_{//}$) and at the center of 180° Bloch walls ($\nu_{\perp}$) and that of fcc Co.

The potentialities of NMR for characterization of thin films are also revealed by the study of $(Co/Mn)_n$ superlattices [41]. In the latter case, the observation of coexisting broad Co and Mn NMR lines in superlattices of short period shows the existence of intermixing.

4. MAGNETIC PROPERTIES

4.1. Curie temperature of the ultrathin Ni films

The low field magnetization of the Au/Ni/Au sandwiches has been measured by SQUID magnetometry [42] as a function of the temperature. For the orientation perpendicular to the film, the magnetization is vanishingly small while it is easily measured for field orientation parallel to the film. For all Ni thicknesses, the field cooled magnetization (FCM), the zero field cooled magnetization (ZFCM) and the remanent magnetization (RM) show similar behaviours versus temperature which are exemplified in Fig.9. These magnetization curves have been interpreted by a simple model of magnetic domains following the polycrystalline structure of the film, and assuming that the magnetization of each domain is frozen along one of its easy axes at low temperature [43]. The Curie temperature of the film is obtained either for the slope break of the FCM measured in very low field, or as the temperature at which the RM vanishes.

The measured Curie temperatures for several Au/Ni/Au sandwiches versus the Ni thickness, are reported in Fig.10 and compared to available experimental data on Ni overlayers on rhenium [44], Cu/Ni/Cu (100) sandwiches [17], $(Ni/Mo)_n$ [45] and $(Ni/Cu)_n$ superlattices [18]. For a given thickness, the experimental data of different sources are widely dispersed.

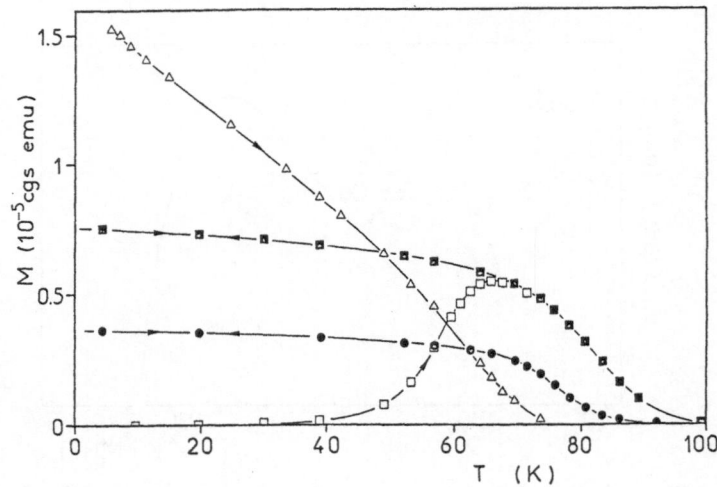

Fig. 9. Low field magnetization of a Au/Ni/Au sandwich of Ni thickness 8.8 Å elaborated by Joule heating evaporation versus temperature: FCM in 0.1 Oe (full circles); FCM in 1 Oe (full squares); ZFCM in 1 Oe (open squares); Remanent magnetization in zero field after applying 5 kOe (triangles).

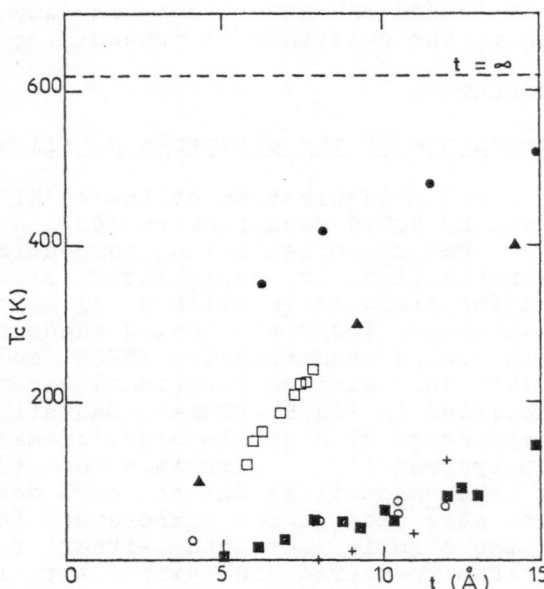

Fig. 10. Experimental data for the Curie temperature of ultrathin nickel films : Au/Ni/Au, Joule Heating evaporation (full squares), electron bombardment (open squares); Ni (111) on Rhenium [44] (black circles); Cu/Ni/Cu (100) [17] (full triangles); Ni-Cu superlattices [18] (open circles); Ni-Mo superlattices [45] (crosses).

Even for the presently studied Au/Ni/Au sandwiches, two different sets of data are obtained, depending on the method of preparation. The Curie temperatures of the samples prepared by Joule heating of Ni are systematically lower than the ones obtained by electron bombardment. They show a weak non monotonous variation versus thickness which was previously interpreted as a quantum size effect [46]. The Ni samples obtained by electron bombardment show a fast monotonous T_c increase with thickness in very good agreement with the experimental data in Cu/Ni/Cu [17]. The presence in the vacuum jar of hydrogen released by the tungsten boat during the Joule heating process might produce Ni hydride giving the *"exotic"* T_c behaviour reported before. In the case of superlattices, the low T_c values are likely due to an intermixing extending over a few atomic layers.

In fact, the wide dispersion of the T_c values in Ni ultrathin films reflects the fragility of the itinerant ferromagnetism of Ni, to magnetic dilution and disorder. Nevertheless, a common feature to all data is the disappearence of ferromagnetism at thicknesses below 1.5 - 2.5 monolayers. These apparent dead layers are predicted in a recent mean field calculation of the T_c variation versus thickness in Cu/Ni/Cu sandwiches [23] within the framework of the itinerant electron model [47]. A comparison of these numerical calculations with the relevant experimental data is shown in Fig.11. The predicted T_c increase versus thickness

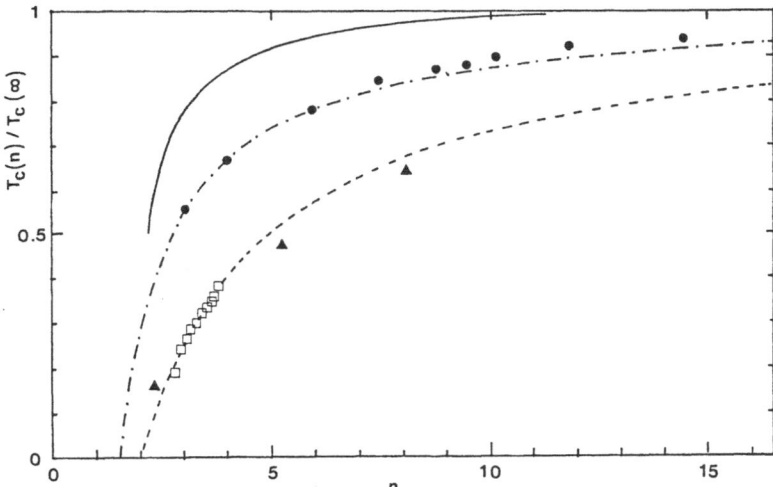

Fig. 11.   Reduced Curie temperature of ultrathin Ni films versus the number n of atomic layers. The experimental data relative to Ni/Rh (black circles); Au/Ni/Au (open squares) and Cu/Ni/Cu (black triangles) are compared to the theoretical prediction of Hasegawa (full curve). The dashed curve and dashed-dotted curve correspond to relation (3) for respective parameter values n_{DL} = 2, γ = 0.25 and n_{DL} = 1.5, γ = 0.35.

is more important than the observed one. This discrepancy could be due to an insufficient quality of the studied samples, but also to the underestimation, in the mean field model, of the large two-dimensional (2D) fluctuations. Indeed, the shape of the experimental $T_c(t)$ curves are well reproduced by the following simple approximation deduced from the theory of 2D spin waves in a weakly anisotropic ferromagnet with localised spins [48].

$$T_c(n^*)/T_c(\infty) = \gamma n^*/[1 + \gamma(n^* - 1)] \qquad (3)$$

in which n^* is the number of living magnetic layers, $n^* = n - n_{DL}$ and $\gamma = T_c(1)/T_c(\infty)$.

4.2. Interaction between Ni films through Au

The fast increase of the Curie temperature of Ni versus thickness, allows the straightforward experimental study of the interaction between two Ni layers of same thickness t through a spacing Au layer of thickness t_s, from $T_c(t_s)$. Indeed, for large t_s values for which the interlayer coupling is negligible, T_c has the value of the single layer : $T_c(t)$, while with t_s reducing to zero, T_c would tend towards $T_c(2t)$ >> $T_c(t)$. In the Ni bilayer systems Au/Ni/Au/Ni/Au, T_c is practically independent on the thickness of the Au spacing layer down to t_s = 7.5 A and shows an abrupt increase at lower values (Fig.12)[49].

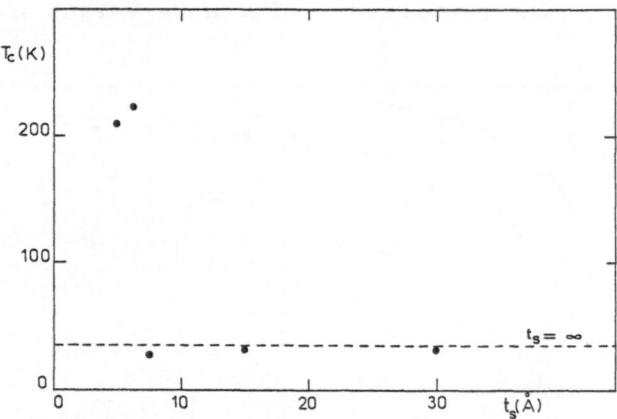

Fig. 12. Ferromagnetic transition temperature T_c of Ni bilayers Au/Ni/Au/Ni/Au versus the thickness t_s of the gold spacing layer. The Ni layers elaborated by Joule Heating have the same thickness 7.5 A.

This experiment first confirms the good quality of the samples. In particular, the interfaces are sharp enough as Ni-Ni contacts are suppressed by an Au film of 7.5 A thickness.

Secondly, the interaction between Ni films through Au has a very short range (< 7.5 A). This could be related either to the small value of the Fermi wavelength of Au : λ_F = 5.2 A in the (111) direction, or, according to P. Grünberg [9], to the efficient effect of dilution by Au, for destroying the ferromagnetism of Ni.

Finally, we cannot conclude about the fast T_c variation

around t_s = 6 - 7 A; it can indicate, either the range of the interaction, or the occurence of shorts connecting the Ni layers. This latter effect might be responsible for the T_c increase observed in the Cu/Ni multilayer films [18] at relatively large Cu spacing ($\simeq$ 20 A), in contrast with the present study.

4.3. Magnetic anisotropy of the hcp Co ultrathin films

4.3.1. Au/Co/Au sandwiches

Contrarily to the Ni samples, the Au/Co/Au sandwiches remain ferromagnetic down to one Co monolayer [46] and their magnetization exhibits a perpendicular easy axis at low Co thicknesses [46,50]. This can be deduced from the dependence of the parallel and perpendicular remanent magnetizations, $M_{R//}$ and $M_{R\perp}$ versus the cobalt thickness (Fig.13). For small

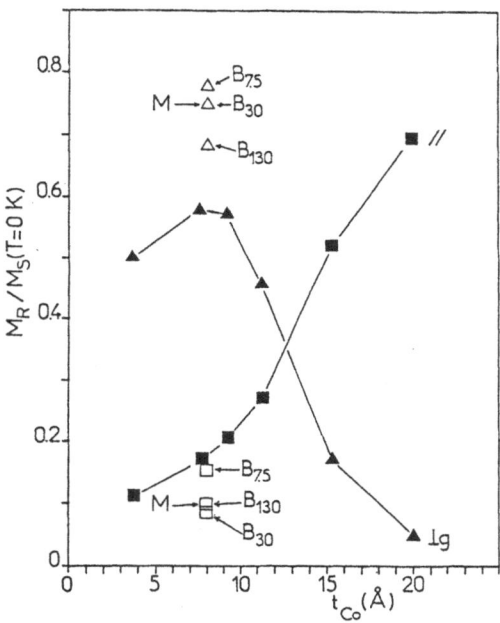

Fig. 13.   Perpendicular (triangles) and parallel (squares) relative remanent magnetization after applying a 5 kOe field, of Au/Co/Au layered structures versus the cobalt thickness. The black symbols are relative to the single Co systems whereas the letters B and M correspond respectively, to Co bilayers and to the multilayer film.

thicknesses, $M_{R\perp}$ is much stronger than $M_{R//}$, then it decreases rapidly around $t_{Co} \simeq$ 12 A to vanish at large thicknesses, while $M_{R//}$ increases and tends towards the saturation value of bulk Co, M_s . The existence of a large interface contribution to the hcp Co magnetic anisotropy, which favours a perpendicular orientation of the magnetization is confirmed by Ferromagnetic Resonance (FMR) experiments [50,51]. The variation of the resonance fields versus the angle Θ_H between the magnetic field and the plane of the film (Fig.14) is well interpreted by the classical FMR theory using the proper expression of the magnetic anisotropy energy for a system with hexagonal structure :

$$E_{an} = K_1 \sin^2 \alpha + K_2 \sin^4 \alpha \qquad (4)$$

where α is the angle between the magnetization M and the c axis. The values of K_1 and K_2 for different Co thicknesses, are obtained from the fits to the FMR data [50,51] assuming that the magnetization is equal to the bulk value ($4\pi M_s = 17.69$ kOe at room temperature).

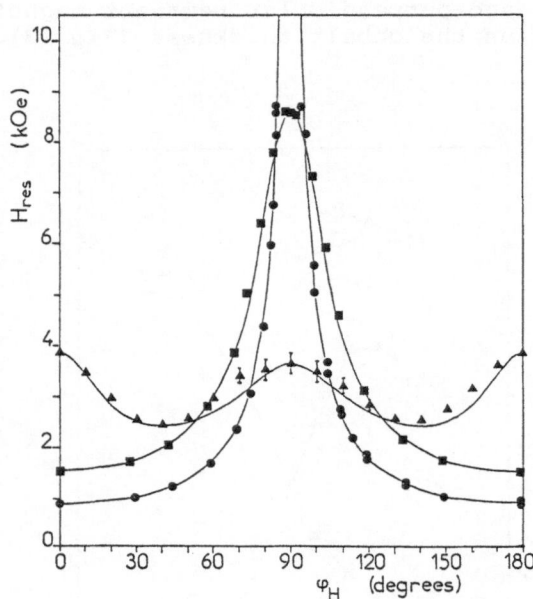

Fig. 14.   Resonance field H_{res} at T = 291 K, versus the orientation of the magnetic field (angle φ_H). $\varphi_H = 0$ corresponds to an orientation parallel to the film. $t_{Co} = 80$ A (full circles); $t_{Co} = 20$ A (black squares); $t_{Co} = 11.3$ A (full triangles). The solid lines for H_{res} are calculated curves as described in ref. [51].

The dependence of the anisotropy constants versus thickness t_{Co}, is well described by the simple relation :

$$K = K_v + 2K_s/t_{Co} \qquad (5)$$

where K_v is an homogeneous volume contribution and K_s an interface contribution.

From such a plot in Fig.15 for K_1 (t_{Co}), the value of the interface anisotropy constant characteristic of the Co/Au (111) interface is obtained : $K_s = 0.53 + 0.03$ erg.cm^{-2} at room temperature. To date, there are few available experimental data on interface anisotropy [1]. In the case of the interfaces of Ni (111) with various non magnetic metals, the data are consistent with relation (5), and the corresponding values of K_s have the same order of magnitude than for Co/Au (111) but with opposite sign [52]. A recent band calculation [53] of the anisotropy constants for a Fe(100) monolayer yields an axial contribution of the same sign and order of magnitude than the present data for Co/Au (111). However, recent experiments performed on a Co monolayer on Cu(100) show a much lower surface anisotropy [54]. Further theoretical and experimental works are clearly needed on this important topic .

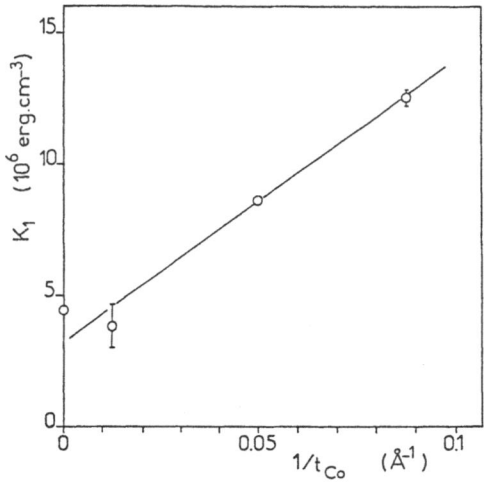

Fig. 15.   Plot of the K_1 anisotropy constant measured at room tempe-rature, on Au/Co/Au sandwiches, versus the inverse of the cobalt thickness. The point on the vertical axis is the tabulated value for hcp cobalt.

4.3.2. Multilayers (Au/Co)$_n$/Au.

In the three bilayers with t_{Co} = 7.8 A and gold spacing thickness 7.5, 30 and 130 A, the easy axis is perpendicular. The ratio $M_{R\perp}/M_{R//}$ is even sensibly higher than for sandwiches of same Co thickness (Fig.13). As for the Ni bilayers, an interlayer Au film of 7.5 A thickness provides an efficient separation of the magnetic films. Indeed, contacts between the Co films would reduce interface anisotropy and favour an in plane equilibrium orientation of the magnetization with respect to the out of plane one.

In these bilayers the FMR signals are too weak and broad to allow a relevant study (It should be noticed that for sandwiches, FMR cannot be detected below t_{Co} < 10 A). On the contrary, the sample with five Co layers with same Co thickness, 7.8 A and Au spacing of 30 A, shows a fairly intense FMR signal [50]. The angular dependence of the FMR resonance field cannot be explained in the frame of the simple model used for the sandwiches. This peculiar behaviour suggests a dynamical coupling between the five Co layers which deserves further experimental studies.

5. CONCLUSION

Ultrathin ferromagnetic films of Ni and Co have been grown on polycrystalline textured Au(111) films with atomically flat surface. The magnetic films do not grow layer by layer, but nevertheless exhibit sharp interfaces extending on 2-3 atomic layers. This sharpness is shown from *in situ* resistivity and grazing X-ray interference and indirectly, from the magnetic behaviour of bilayer films.

In spite of a large lattice mismatch, a quasi-epitaxy of fcc Ni and hcp Co on Au (111) is observed by TEM, leading to a good crystalline coherence through the whole depth of the sandwiches and related multilayer systems.

The fcc Ni (111) films show ferromagnetism only for thicknesses above about 2 monolayers. The temperature dependence of the Curie temperature versus thickness, for the best samples is consistent with the predictions of a 2D spin-waves model. In bilayers, no significant interlayer coupling is observed at Au spacing layer thicknesses down to 7.5 A.

The hcp Co films show an interface magnetic anisotropy which favours a perpendicular orientation of the magnetization. The competition with shape anisotropy produces a rotation from the out of plane orientation to the in plane one with increasing Co thickness. The interface anisotropy is obtained from FMR measurements. The FMR results in a five layers sample arise the possibility of a dynamic interlayer coupling.

To date, both theoretical and experimental data are insufficient to provide a comprehensive view of the ferromagnetism of ultrathin Ni and Co films. In particular, the present experiments deserve to be completed by high field magnetization measurements on the same samples and by their extension to other Au orientations, as well as to other metallic substrates such as Ag and Cu.

ACKNOWLEDGMENTS

This work is the result of a collaboration between three laboratories of the Université Paris Sud (UPS), Orsay and a laboratory of the Université de Toulon (UT) I am greatly indebted to my coworkers : C. Marlière and D. Renard from Institut d'Optique, UPS, Orsay; P. Beauvillain, C. Chappert, K. Le Dang, J. Seiden and P. Veillet from Institut d'Electronique Fondamentale, UPS, Orsay. H. Hurdequint from Laboratoire de Physique des Solides, UPS, Orsay; C. Césari and G. Nihoul from GMET and CNRS UA 797, UT. I would also like to thank J.P. Chauvineau for helpful suggestions and valuable discussions, J. Corno, L. Nevot and B. Pardo for experiments on their grazing X-ray apparatus, R. Mégy for a photographic reproduction and Mrs. F. Genet for the typewriting. A financial support was supplied by the ATP *"couches minces magnétiques"* of PIRMAT-CNRS.

REFERENCES

[1] U. Gradmann, J. Mag. Mag. Mater., <u>54-57</u>, 733 (1986)

[2] C.L. Fu, A.J. Freeman and T. Oguchi, Phys. Rev. Lett. <u>54</u>, 2700 (1985)

[3] R. Richter, J. G. Gay and J. R. Smith, Phys. Rev. Lett., <u>54</u>, 2704 (1985)

[4] R. H. Victora and L. M. Falicov, Phys. Rev. B,<u>31</u>, 7335 (1985)

[5] A. J. Freeman, J. Mag. Mag. Mater., <u>35</u>, 31 (1983)

[6] J. Tersoff and L. M. Falicov, Phys. Rev. B <u>25</u>, 4937 (1982); Phys. Rev. B <u>26</u>, 6186 (1982)

[7] See for instance the contributions of R. F. Willis, and B. Heinrich et al in the present book, and G. Prinz, Phys. Rev. Lett. <u>54</u>, 1051 (1985).

[8] Contribution of J. Kwo in this book, and J. Kwo, D. B. Mc Whan, E. M. Gyorgy and F. J. Di Salvo, Mat. Res. Soc. Symp. Proc., <u>56</u>, 211 (1986)

[9] P. Grünberg, J. Appl. Phys.,<u>57</u>, 3673 (1985) and contribution in this book.

[10] M. B. Brodsky, J. Mag. Mag. Mater., <u>35</u>, 99 (1983)

[11] C. M. Falco and I. K. Schuller, Synthetic Modulated Structures, edit. L. Chang and B. C. Giessen, Jr. Academic Press, New York, 339 (1985)

[12] G. Bayreuther and G. Lugert, J. Mag. Mag. Mater., <u>35</u>, 50 (1983)

[13] T. Shinjo, N. Hosoito, K. Kawaguchi, N. Nakayama, T. Takada and Y. Endoh, J. Mag Mag. Mater, <u>54-57</u>, 737 (1986)

[14] N. Hosoito, K. Kawaguchi, T. Sinjo and T. Takada, J. Phys. Soc. Japan, <u>51</u>, 2701 (1982)

[15] H. K. Wong, H. Q. Yang, J. E. Hilliard and J. B. Ketterson, J. Appl. Phys. <u>57</u>, 3660 (1985)

[16] Contribution of U. Gradmann in this book.

[17] L. R. Sill, M.B. Brodsky, S. Bowen and H. C. Hamaker, J. Appl. Phys. <u>57</u>, 3663 (1985)

[18] W. S. Zhou, H. K. Wong, J. R. Overs - Bradley and W.P. Halperin, Physica, <u>108B</u>, 953 (1981)

[19] J. Q. Zheng, J.B. Ketterson, C. M. Falco and I. K. Schuller, J. Appl. Phys., <u>53</u>, 3150 (1982)

[20] E. M. Gyorgy, D. B. Mc Whan, J. F. Dillon, L. R. Walker and J. Waszczak, Phys. Rev.B <u>25</u>, 6739 (1982)

[21] X. Y. Zhu, H. Hong and J. Hermanson, Phys. Rev. B <u>29</u> 3009 (1984); H. Hong, X.Y. Zhu and J. Hermanson, Phys. Rev. B <u>29</u>, 2270 (1984)

[22] A. J. Freeman, J. H. Xu and T. Jarlborg, J. Mag. Mag. Mater. <u>31-34,</u> 909 (1983)

[23] H. Hasegawa, Surface Sci., to be published

[24] D. Renard and G. Nihoul, Phil. Mag., to be published

[25] K. Fuchs, Proc. Camb. Philos. Soc., <u>34</u>, 100 (1938)

[26] E. H. Sondheimer, Adv. in Phys., <u>1</u>, 1 (1952)

[27] J. P. Chauvineau and C. Marlière, Thin Solid Films <u>125</u>, 25 (1985)

[28] K. L. Chopra and M. R. Randlett, J. Appl. Phys. <u>38</u>, 3144 (1967)

[29] M. S. P. Lucas, Thin Solid Films, <u>2</u>, 337 (1968)

[30] J. P. Chauvineau and C. Pariset, Surf.Sci, <u>36</u>, 55 (1973); J. de Physique, <u>37</u>, 1325 (1976)

[31] F.C. Franck and J.H. van der Merwe, Proc. R. Soc. London, Ser A, <u>198</u>, 205 (1949)

[32] D. Schumacher and D. Stark, Thin Solid Films, <u>120</u>, 15 (1984)

[33] C. Marlière, Thesis, Université Paris-Sud, Orsay (1985) unpublished.

[34] C. Marlière, Thin Solid Films, <u>136</u>, 181 (1986)

[35] J.P. Chauvineau and C. Pariset, Acta Cryst., <u>A30</u>, 246 (1974)

[36] H. Kiessig, Naturwissenschaften <u>18</u>, 847 (1930); Ann. Phys.[5], <u>10</u>, 715 (1931)

[37] P. Croce and L. Névot, Rev. Physique Appliquée, <u>11</u>, 113 (1976)

[38] L. Névot and P. Croce, J. Appl. Cryst. <u>8</u>, 304 (1975)

[39] C. Césari, G. Nihoul and D. Renard, J. de Physique, to be published.

[40] K. Le Dang, P. Veillet, C. Chappert, P. Beauvillain and D. Renard, J. Phys. F : Met. Phys., <u>16</u>, L109 (1986)

[41] K. Le Dang, P. Veillet, H. Sakajima and R. Krishnan, J. Phys. F: Met. Phys. <u>16</u>, 93 (1986)

[42] P. Beauvillain, C. Chappert and J.P. Renard, J. Phys. E, Sci. Instr., <u>18</u>, 839 (1985)

[43] C. Chappert, D. Renard, P. Beauvillain and J. P.Renard, J. Physique. Lett. <u>46</u>, L59 (1985)

[44] R. Bergholz and U. Gradmann, J. Mag. Mag. Mater <u>45</u>, 389 (1984)

[45] I.K. Schuller and M. Grimsditch, J. Appl. Phys. <u>55</u>, 2491 (1984)

[46] C. Chappert, D. Renard, P. Beauvillain, J.P. Renard and J. Seiden, J. Mag. Mag. Mater., <u>54-57</u>, 795 (1986); C. Chappert, thesis, Université Paris-Sud, Orsay n° 3088 (1985), unpublished.

[47] H. Hasegawa, J. Phys Soc. Japan, <u>46</u>, 1504 (1979); <u>49</u>, 178 (1980); <u>49</u>, 963 (1980)

[48] J. Seiden, to be published

[49] P. Beauvillain, C. Chappert, J.P. Renard, C. Marlière and D. Renard, MRS European Division Meeting, Strasbourg, june 86, Proceedings to be published in J. de Physique.

[50] C. Chappert, K. Le Dang, P. Beauvillain, H. Hurdequint, D. Renard and C. Marlière, MRS European Division Meeting, Strasbourg, June 86, Proceedings to be published in J. de Physique.

[51] C. Chappert, K. Le Dang, P. Beauvillain, H. Hurdequint and D. Renard, Phys. Rev. B, <u>34</u>, 3192 (1986)

[52] U. Gradman, R. Bergholz and E. Bergter, IEEE Trans. Magn., <u>20</u>, 1840 (1984); E. Bergter, U. Gradman and R.Bergholz, Solid State Commun., <u>53</u>, 565 (1985)

[53] J. G. Gay and R. Richter, Phys. Rev. Lett, <u>56</u>, 2728 (1986)

[54] Contribution of D. Pescia in this book.

THE CHARACTERIZATION OF MODULATED

METALLIC STRUCTURES BY X-RAY DIFFRACTION

Roy Clarke

Department of Physics
The University of Michigan
Ann Arbor, MI 48109-1120 USA

1. INTRODUCTION

Artificial heterostructures provide fascinating examples of solid state systems in which structure and composition can be controlled on an atomic scale in order to achieve extraordinary physical properties [1,2]. This microscopic control of physical properties, for long a dream of solid state physicists and materials scientists, is made possible by recent progress in thin film deposition techniques and new methods for their characterization. Many of these advances are described in accompanying papers in these proceedings.

The vast majority of research in this area has focussed on the technologically important III-V semiconductor heterostructures. Of this subset of materials, most of the effort so far has been concentrated on lattice matched heterostructures such as GaAs-Al$_x$Ga$_{1-x}$As [3,4]. In these cases, the resulting thin film, typically in the form of a quantum well structure [5] or a superlattice [1,6] grown on a GaAs substrate, is ideally a single crystal in the sense that long-range order extends over the whole sample. The high degree of coherence here is, of course, a requirement for the operation of new types of high-speed electronic and optical devices based on such high-mobility materials.

As yet it has not been possible to make metallic heterostructures with structural perfection matching that of the best semiconductor systems. Several factors influence the structural quality including, interdiffusion, miscibility of the constituents, degree of lattice matching and compatibility with appropriate refractory substrate materials. The latter are crucially important for epitaxial growth of thin films. The interplay of these various factors leads to a wide variety of structural forms; it cannot be overstated that the problems associated with growth morphology are the main challenge in the development of high quality heterostructures.

In order to get a feel for the variety of structural arrangements, it is worthwhile here to summarize the different possibilities. Some multilayer systems have been grown in which one or both of the constituents does not have regular crystal structure, yet in which very precise layering can be achieved with atomically abrupt interfaces. An example is the tungsten-carbon multilayer system [7], where the carbon layers are amorphous. In this case there is no structural coherence in atomic positions between neighboring layers, and such structures are often referred to as "incoherent multilayer films" (see Fig. 1).

For other combinations of materials it may be possible to achieve a high degree of coherence in the crystal lattice but the chemical composition modulation may not be abrupt at the interface. This case has been referred to as a "composition modulated alloy" in the literature.

The highest state of coherence in both chemical composition and lattice structure is usually referred to as a "superlattice". Very few examples of true superlattices exist. In classifying heterostructures it is therefore important to distinguish between the coherence of the chemical modulation and the coherence of the crystal lattice. Both depend critically on the constituents being used to form the heterostructure.

In general, it will not be possible to attain lattice matching of the constituents of a heterostructure. The important point is that only in a very few carefully chosen cases will sufficiently small differences exist between the lattice parameters of the substrate and the film deposited on it, and further, between the individual strata of which the heterostructure is composed. In most systems, then, lattice parameter mismatches of greater than a fraction of a percent cannot be accommodated by elastic strain [8], unless the film is very thin (~ few monolayers). If there is significant mismatch the resulting structure will exhibit extended defects such as misfit dislocations [9]. In equilibrium, such defects are predicted [9, 10] to form an ordered array, preserving the overall coherence of the structure. However, since the growth conditions in most methods of thin film deposition (e.g., sputtering, MBE, MOCVD) are far from equilibrium, structural defects are determined by growth

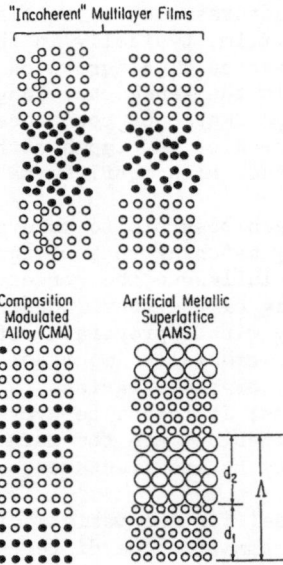

Fig. 1. Schematic classification of multilayer structures. From C.M. Falco [11].

kinetics [12, 13] rather than equilibrium thermodynamics. Thus it is to be expected that even a small degree of mismatch in lattice parameters will result in some structural disorder.

In this paper, I will survey the different kinds of information that can be obtained, non-destructively, by modern X-ray scattering techniques. While X-ray diffraction is the traditional probe for investigating crystal structures, the full power of the technique has not as yet been applied to thin film heterostructures.

The focus of this contribution is on the metallic heterostructures. This class of materials was chosen for several reasons: firstly, the metallic systems represent a much less well developed research area than that of III-V epitaxy and only a few materials have been studied in any detail. Secondly, the interesting metallic heterostructures that one would like to fabricate will not, in general, be blessed with good lattice matching. In many cases not even the point group symmetries of the constituents will be related to each other. Thus, it is to be expected that in many metallic heterostructures the structure will be far from that of a perfect crystal. Nevertheless, by careful control of the rate of deposition, it is still possible in many cases to obtain coherent layering [14,15] so that the film retains some semblance of long-range order, at least normal to the layers. Fortunately the consequences of structural defects are less severe in many of the metallic systems of interest, (e.g., magnetic and superconducting structures) than in the semiconductor heterostructures.

When discussing materials of this type it is important to note that the total X-ray scattering function consists of both Bragg diffraction and diffuse scattering contributions [16], the former relating to long-range order (LRO) and the latter to departures from LRO caused by imperfections and disorder in the heterostructures. This aspect represents one of the great advantages of the X-ray scattering technique.

An additional motivation for selecting the metallic heterostructures for study is that the nature of the bonding and interactions is very different than in the semiconductor systems such as the III-V's. For example, imperfectly screened interatomic forces across bimetallic interfaces can give rise to new kinds of structural ordering which are peculiar to this class of heterostructure. Specific examples will be discussed in Section 5. In this vein it should be stated that, since the physical properties are expected to depend strongly on the microscopic structure, the principal motivation for all of these studies is to better understand the physical behavior, particularly the electronic transport.

The layout of this presentation is as follows: In Section 2 some general aspects of X-ray characterization will be reviewed in the context of two central issues, namely atomic interdiffusion and interfacial quality. In Section 3, I will describe some experimental methods for X-ray characterization of layering, epitaxial ordering, layer stacking, mosaic texture, etc. Sections 4 and 5 are concerned with some specific examples of experimental results illustrating the structural behavior of two representative classes of metallic heterostructure: textured multilayers and bicrystal superlattices. Finally, in Section 6, I outline some future trends in structural measurements and characterization utilizing X-ray synchrotron radiation techniques. In this section I also include some recent developments in aperiodic heterostructures, i.e., multilayer systems in which the layers are arranged intentionally in some definite sequence yet lacking the usual translational symmetry.

2. INTERDIFFUSION AND INTERFACE QUALITY

One of the major problems encountered in heterostructure fabrication is the tendancy of species on either side of the heterointerface to diffuse into each other. This may take place during growth, at the elevated temperatures necessary to achieve good epitaxial ordering, or in some cases during prolonged storage at ambient temperature. This factor has been recognized for many years; in fact, one of the first applications of artificially modulated materials was to measure diffusion constants in the solid state. It is worth recounting the very early work of DuMond and Youtz [17] in this area since it sets the stage for one of the most important uses of X-ray characterization: as a probe of interface quality.

DuMond and Youtz's studies were a continuation of earlier work [18, 19] aimed at fabricating artificial gratings for absolute X-ray wavelength measurements. In these pioneering experiments DuMond and Youtz carried out the deposition of a modulated Cu/Au structure on a glass substrate. Their method consisted of alternately raising and lowering the temperature of a trough of molten Au in order to produce a periodically varying flux of Au atoms. Simultaneously, a constant flux of evaporated Cu atoms is allowed to impinge on the substrate. The resulting structure had an approximately sinusiodal concentration profile with a spatial periodicity, L, of ~100Å. The small angle X-ray diffraction pattern obtained for this structure is reproduced in Fig. 2 and shows essentially a single Fourier component in the scattering, confirming the harmonic nature of the concentration profile. More interestingly, it was observed that over a period of several days the relative intensity of this single diffraction peak decreased dramatically with a "half-life" of roughly 2 days.

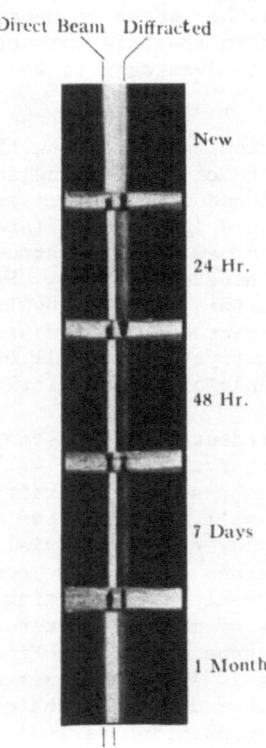

Fig. 2.   Decay of diffraction peak in a Cu/Au modulated
          structure (After DuMond and Youtz [17].

These results were interpreted in terms of a simple diffusion
equation, from which an estimate of the interdiffusion constant could be
extracted. Although more sophisticated developments have since appeared
in the literature, for example taking into account the effects of
coherency strains and higher order gradient energy terms [20, 21], a
simple analysis suffices to illustrate the basic ideas involved.

It is normal to represent the concentration profile, $c(z,t)$, of
the diffusing atoms (Au in this example) as a Fourier sum:

$$c(z,t) = \sum_{\ell} Q_{\ell}(t)e^{i\ell kz} \quad , \qquad k = \frac{2\pi}{L} \qquad (1)$$

Since X-ray diffraction pattern can also be represented as a Fourier
transform of the electron density with coefficients evaluated at
reciprocal lattice points, ℓ, the intensity of the ℓ'th order diffraction
peak is proportional to $Q_{\ell}^2(t)$.

Taking for example a concentration profile which is initially
atomically abrupt, i.e., an alternating superlattice of equal thickness
layers of atom types A and B,

$$Q_{\ell} = \frac{4}{\ell\pi} \sin(\frac{\ell\pi}{2}) \qquad (2)$$

and assuming that a simple diffusion equation holds in the small amplitude
limit:

$$\dot{c} = D\nabla^2 c \ , \qquad (3)$$

the Fourier coefficients are predicted to decay exponentially with time:

$$Q_{\ell}(t) = Q_{\ell}(0)\exp[-(\ell k)^2 Dt] \qquad (4)$$

In this way the interdiffusion constant D can be determined; because the
atomic displacements involved are microscopic, values of D can be measured
that are many orders of magnitude smaller than is possible from standard
tracer techniques (see Fig. 3). Note also that the anisotropy of the
interdiffusion coefficient [22], can be determined conveniently by
choosing appropriate satellite reflections (e.g., in the <100>, <111>....
directions).

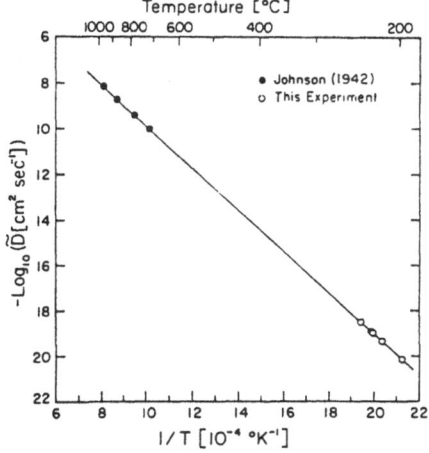

Fig. 3. Arrhenius plot of diffusion coefficient $\tilde{D}$ for a Au/Ag
alloy. From Cook and Hilliard [20].

Characterization of Interface Quality

The above analysis suggests that detailed microscopic information can be gained from an analysis of the intensities of X-ray diffraction satellites where the diffraction vector is normal to the layers. In practice it is necessary to model the interface structure in order to extract reliable quantitative information about the chemical composition, strain and amount of disorder at the heterointerface. In general, each of these parameters has a different effect on the X-ray diffraction profile. These effects are briefly summarized as follows:

i) The **relative intensities** of the satellite peaks are very sensitive to chemical composition of individual layers and also to the lattice spacings (strain) between atomic layers. Although both chemical composition and strain affect the satellite intensities, the presence of strain will typically show itself as an asymmetry in the intensities of equivalent-order peaks on either side of the main peak. An example of this is shown in Fig. 4.

ii) The sharpness of the interface, on the other hand, determines the overall **envelope** of the diffraction profile; generally, higher order satellites are suppressed more by interface broadening than those closer in to the main peak. This fact allows a rough estimate of the abruptness of the interface to be made based on the number of satellite peaks (or Fourier components in Eq. 1) that are observed. Alternatively, if one wishes to probe the chemical composition of layers near the interface, the intensities of low-order satellites are the most useful since they will be relatively insensitive to effects from interface smearing.

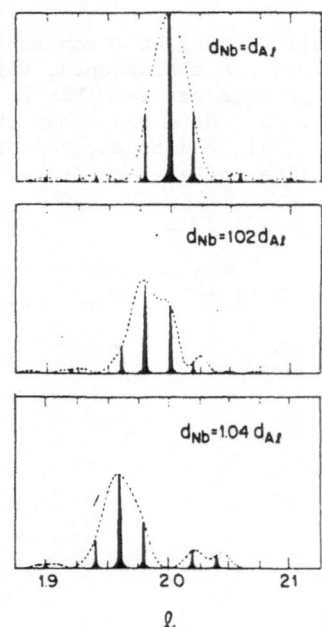

Fig. 4. Calculated superlattice peaks from Nb-Aℓ multilayers showing the effects of strain. The graphs show the effects on the (00ℓ) pattern for 0, 2% and 4% uniaxial compression of the Aℓ interplanar spacing. (After McWhan et al. [15]).

iii) Inhomogeneities due to grainy texture and island formation will tend to make the interfaces rough on a microscopic scale. There is little effect on the satellite peaks, apart from an overall Debye-Waller-like damping [23] factor. As far as the 00ℓ) profile is concerned roughening is hard to distinguish from interface smearing, e.g., due to interdiffusion.

Chrzan and Dutta [24] have pointed out that **rocking scans** (around the 00ℓ peaks) are sensitive to interface roughness. These authors have calculated a two-dimensional structure factor for a superlattice containing steps at the interface (see Fig. 5) such that $\alpha(= \sigma/x)$ is the variance in layer thickness per unit distance, x, parallel to the layers:

$$S(\sigma, k_x, k_z) = \sum_{\ell = -\infty}^{\infty} \exp(ik_x \ell d_x) S_1(k_z, \alpha|\ell|d_x) \tag{5}$$

where d_x is the lattice spacing parallel to the layers, and:

$$S_1(k_z, \sigma) = \exp(ik_z L) \exp(-\sigma M)$$
$$\times \left[\frac{(1 - \exp(ik_z NL)\exp(-\sigma NL)}{1 - \exp(ik_z L)\exp(-\sigma M)} \right]$$
$$\times \{ f_a(k_z, \sigma)\exp(-ik_z N_b d_b)$$
$$\exp[-\sigma \sin^2(k_z d_b/2)] + f_b(k_z, \sigma) \} \tag{6}$$

is the full structure factor for a one-dimensional superlattice [24].

In Eq. (6), $M = \sin^2\left(\frac{k_z d_a}{2}\right) + \sin^2\left(\frac{k_z d_b}{2}\right)$ and,

$$f_{a,b}(k_z, \sigma) = g_{a,b} \frac{1 - \exp(ik_z N_a d_{a,b})\exp[-\sigma\sin^2\left(\frac{k_z d_{a,b}}{2}\right)]}{1 - \exp(ik_z d_{a,b})}$$

f_a and f_b are the average form factors of layers of type A and B, respectively, taking into account the random distribution of steps; g_a and g_b are atomic form factors; N_a and N_b are the total number of atoms of type A and B respectively and d_a and d_b, the interplanar spacings of these

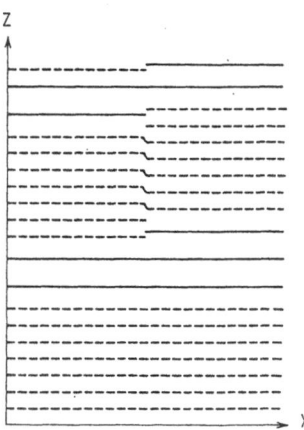

Fig. 5. Schematic of a superlattice with rough interfaces [24].

layers. Note that, in the limit of no interface roughness ($\sigma = 0$), the structure factor given by Eq. 6 reduces to that of a perfect binary superlattice. The term in square brackets in Eq. 6 gives rise to peaks at the reciprocal lattice positions $k_z = 2\pi/L$ (superlattice satellites) and the terms f_a and f_b modulate the intensities so that the strongest satellites are found closest to the Bragg peaks for the bulk structures of A and B. The latter would occur at wavevectors given by $k_z = 2\pi/d_a$ and $2\pi/d_b$ respectively. In superlattices with thin strata (d_a, $d_b \lesssim 100\text{Å}$), there is one prominent peak between these wavevectors which is referred to the as "main" or "central" peak while the sidebands are called "satellites". Note that $k_z = 0$ is a reciprocal lattice point of the structure so that strong satellite peaks should also appear at low angles.

Interface Roughening

As mentioned above in point (iii), plots of the (00ℓ) profile from an expression such as Eq. 6 reveal only a progressive decrease in satellite intensities with increasing roughness, α. However, the rocking curves, corresponding to a trajectory in reciprocal space $k_x^2 + k_z^2 = $ const., do reveal interesting information about the interface roughness [24].

Fig. 6 shows the calculated full widths at half maximum (FWHM) of the rocking curves plotted as a function of α. It can be seen that the central peak is practically insensitive to variations in α compared to the satellite peaks. Interestingly, Chrzan and Dutta [24] find that the satellite at smaller wavevector, k_z, is more sensitive to roughness than the upper one. The important point to make here is that the relative (00ℓ) satellite intensities are found to be sensitive to interface roughness only up to a certain level of $\alpha \approx 0.5 \times 10^{-3}\text{Å}^{-1}$, i.e., in rather homogeneously layered samples. Conversely, the rocking curve widths continue to depend strongly on α up to $\alpha \approx 10^{-2}\text{Å}^{-1}$ (see Fig. 6). Thus rocking curve measurements would seem to be useful in artificially modulated structures that are far from perfect crystals.

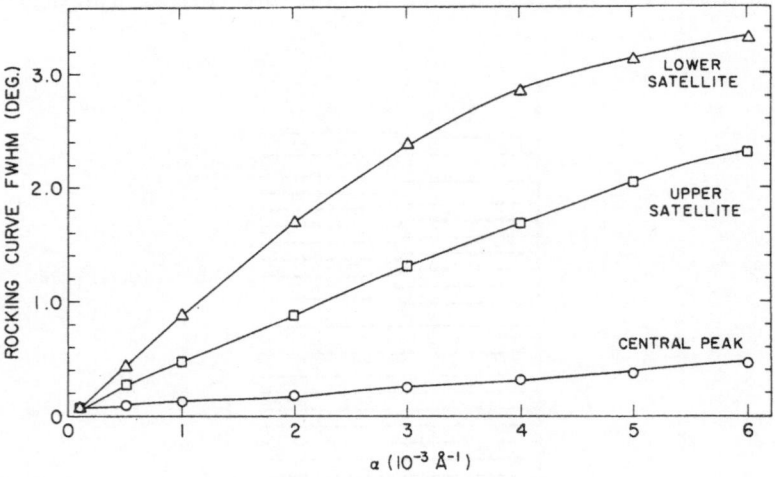

Fig. 6. FWHM of rocking curves as a function of roughness, α [24].

Interfacial Composition Grading

As we have seen there is a tendency for intermixing of the constituents across the heterointerface either microscopically, via a diffusion process, or by means of step formation at a rough interface. Methods to characterize the chemical composition at interfaces have been developed by several authors including Segmüller and Blakeslee [25] and McWhan et al. [15]. One example [15] models the interface with a trapezoidal composition variation. In this case the satellite intensities are determined by the Fourier coefficients:

$$Q_\ell = (\frac{L}{t_o+t_1} -1)^{-1} \; [\; \frac{\sin(\pi\ell t_1/L}{\pi\ell t_1/L}] \, [\; \frac{\sin[\pi\ell(t_o+t_1)/L]}{\pi\ell(t_o+t_1)/L} \;] \qquad (7)$$

where t_1 is the thickness of the interface and t_o is the thickness of the pure A or pure B strata.

In another model considered by McWhan et al. [15], the interface was considered to be an abrupt step but with intermetallic compound formation, such as $NbA\ell_3$ in the Nb-Aℓ multilayer films.

Plainly, there are a multitude of variable parameters that can be included in such models of the interface. However, the analytical expressions for the structure factors become hopelessly unwieldy when more than one or two aspects of the real structure are treated (e.g., strain, chemical composition, interface steps, variable stratum thickness, etc.). Although expressions such as Eqs. 6 and 7 serve as useful intuitive guides to understanding the microstructure, it eventually becomes necessary to employ computer simulation methods in order to handle more realistic structure factor modeling. This is particularly true when a significant amount of randomness is present in the heterostructure.

The approach we have taken recently is to construct a model of the superlattice, layer-by-layer on a VAX computer. Parameters of the structure can be changed conveniently and the method has the added advantage that very large numbers of layers can be dealt with, sometimes as many as are in the actual sample. Moreover, disorder can be introduced as a natural part of the algorithm that is used to construct the computer model of the structure. The structure factor is calculated by fast Fourier transform techniques and the calculated diffraction profile is compared with the one measured from X-ray scattering.

As an example of this approach I take the superlattice Ru-Ir. This very well lattice matched system can be prepared as an excellent single-crystal structure in the form of alternating layers of close-packed ruthenium and iridium atoms [26]. The material is discussed in greater detail in section 5. In Fig. 7 we show the (00ℓ)-scans for two samples, with L = 41Å and L = 62Å, in order to probe the nature of the interfaces between Ru and Ir strata.

Immediately one can judge from the many Fourier components that the interfaces in this system are very sharp. Detailed structure factor modeling has been performed for both samples using computer simulation of the actual superlattice. We find that the envelope of the (00ℓ) profiles is quite sensitive to small amounts of intermixing at the interfaces. On the other hand, the intensity ratios of individual superlattice satellites are very sensitive to the average number of layers of each constituent within the unit cell. We find the best fit to the observed profile with Ru-Ir model sequences containing on average 12 atomic planes

of Ru and 16 of Ir for the 62Å sample, and 11 atomic planes of Ru and 9 of
Ir for the 41Å sample. These figures are to within ±1 monolayer of the
intended composition ratios. The best fit envelope is obtained with
intermixing confined to a pair of neighboring monolayers with an
approximately 15±5% content of Ru in Ir in one of the interface layers,
and the same amount of intermixing of Ir in Ru in the other interface
layer. This intermixing ratio is found to be somewhat dependent on the
type of strain relaxation that is used in the model of the interface. For
simplicity we assumed a trapezoidal variation of the interplanar spacing
at the interface. Although the fits shown in Fig. 7(b) are very
promising, some additional improvement should be gained by including
disorder in the form of a slightly variable number of layers per stratum.

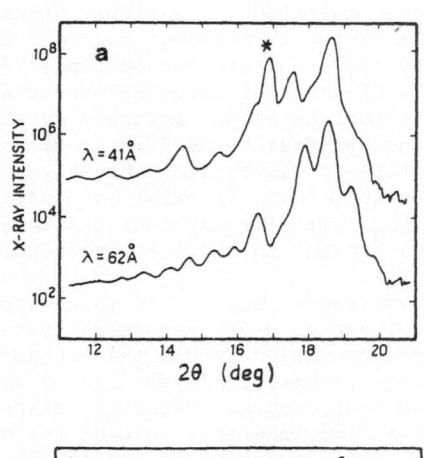

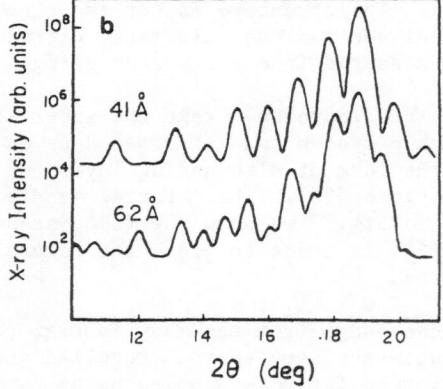

Fig. 7. (00ℓ) diffraction profiles for two Ru-Ir superlattice
samples: (a) measured; (b) calculated according to the
model discussed in the text including instrumental
broadening. The peak marked (*) is the sapphire (11$\bar{2}$0)
substrate peak. The upper trace is displaced vertically
by two orders of magnitude [27].

3. EXPERIMENTAL METHODS

X-ray Source

The X-ray source used in most of our measurements is a 12 kW Rigaku RU-200 generator fitted with a rotating anode. Typical power settings are 58 kV and 180 mA.

The X-ray spectrum produced is a superposition of wide-band Bremsstrahlung radiation and the characteristic monochromatic lines of the metal anode target, in this case molybdenum. This particular target is chosen because the photon energy is relatively large (~17.5 keV) so that a highly penetrating beam is obtained. This turns out to be important for thin-film characterization since it allows measurements to be made in transmission. We will describe some experiments below in which thin film heterostructures were studied by X-ray scattering parallel to the layers without removing the film from the thick substrates.

The X-ray beam is monochromated by a slab of highly oriented pyrolytic graphite (HOPG) which isolates the $K\alpha_1$ and $K\alpha_2$ doublet, of weighted photon wavelength, $\lambda = 0.7107\text{Å}$. HOPG has a large mosaic spread and high reflectivity producing a high X-ray flux at the sample but with sacrificed resolution. Typical longitudinal resolution (FWHM) is $\Delta k \approx 0.03\text{Å}^{-1}$. An alternative would be to use the (111) reflection of Ge which provides an order of magnitude better resolution with a corresponding decrease in available X-ray flux. For most scans on metallic thin films, HOPG monochromatization is sufficient and the extra intensity of the incident beam outweighs the disadvantage of the lower resolution. A typical beam size is ~0.5x0.5 mm^2 in these experiments.

X-ray Diffractometer and Detector

A diffractometer that we find to be most versatile and convenient for diffraction scans on the thin-film heterostructures consists of a Huber 420 turntable on which is mounted a Huber type 512 Eulerian cradle. This combination is sometimes referred to as a '4-circle' diffractometer (see Fig. 8). The ω, ϕ, χ and 2θ (detector arm) movements are driven by digital stepping motors controlled by an LSI-11 microprocessor. The advantage of 4-circle geometry here is that any plane of atoms in the sample can be brought into the diffracting condition by a combination of Euler rotations. Moreover, any trajectory in reciprocal space can be programmed by the computer in order to probe various aspects of the structure (see Fig. 9). The X-ray detector consists of a standard Th-NaI scintillator crystal/photomultiplier combination mounted on the 2θ arm of the diffractometer. The scattered photons are counted by means of a single channel pulse height analyzer, the output of which goes to a scalar interfaced with the LSI-11.

X-ray Scans for Layered Structures

Fig. 9 illustrates the various types of reciprocal space scans that can be performed with the 4-circle diffractometer. Two representative reciprocal lattice planes are shown normal to the layering axis (conventionally designated as c^*). A scan in which the diffraction vector lies parallel to c^* and passes through the origin (k=0) is referred to as a (00ℓ) scan. Note that any such scan along a radial direction through k=0 is also referred to as a θ-2θ scan since the sample and detector rotations are maintained in a 1:2 ratio in order to obey the Bragg reflection condition. The (00ℓ) scan probes the regular **layering** of the atoms in the thin film and is sensitive to atomic displacements normal to the layers. Such scans are often used, therefore, to measure the modulation period and to probe interface quality and abruptness (see section 2).

Fig. 8. Photograph of 4-circle X-ray diffractometer.
(1) Rotating-anode X-ray generator; (2) Monochromator;
(3) Slits; (4) Huber goniometer; (5) Detector;
(6) Sample holder.

A second type of scan is when the diffraction vector passes through a reciprocal lattice point [e.g., (100)] but the locus in reciprocal space does not intersect k=0. An example might be that shown in Fig. 9 where the scan is along (10ℓ) axis. This is sometimes referred to as a "c^*-scan" since the tip of the diffraction vector traces out a line parallel to the c^*-axis, through the (100) point, during this type of scan. The structural information so obtained relates to the correlations of atomic planes in a direction cutting across the stack of planes at some angle to the c-axis, i.e., the spacing and intensities of (10ℓ) peaks contain information about the **stacking** of atomic layers in the structure. In particular, if the (10ℓ) scans exhibit broadening (streaks) the existence of stacking faults can be inferred.

Yet another type of scan, referred to below as a χ-scan, is a constant-2θ diffraction profile with the diffraction vector sweeping out a circular path exactly in the plane of the sample. In this case the fixed length of the diffraction vector is chosen to coincide with one or more in-plane reciprocal lattice vectors. In this way the orientational order or **epitaxy** of the layered structure can be probed. For example, if the strata of the heterostructure are in good epitaxial alignment, the χ-scan will show sharp peaks in the plane of the sample. On the other hand, if the film is polycrystalline in the plane, the χ-scan will show a uniform Debye ring of scattering. In order to carry out such a scan, the diffraction condition is achieved **in transmission** with the diffraction vector lying exactly in the plane of the sample, and the c-axis (normal to the layers) exactly parallel to the horizontal rotation axis of the χ-circle.

The χ-scan is really an example of a rocking-curve type of scan. Another rocking curve scan, at right angles to the χ-scan, was described in section 2 in connection with interface roughness. This was the 00ℓ rocking curve. Note that the "two-crystal" diffractometer described by Bartels [28] is specifically designed to make rocking curve scans of this type with high resolution (~ few seconds of arc FWHM). However, the two-crystal geometry is not as versatile as the 4-circle goniometer described here. Examples of our results from the various kinds of reciprocal-space scans discussed above will be given in later sections.

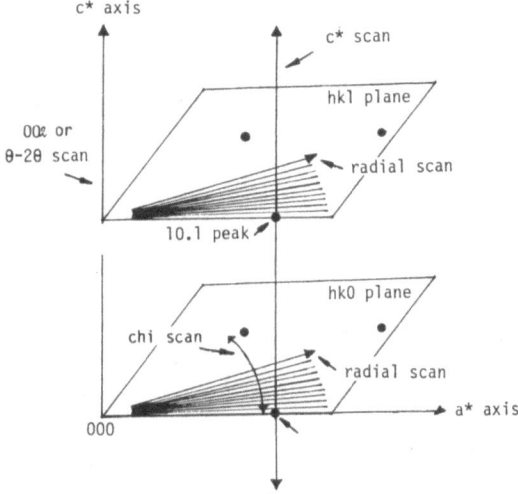

Fig. 9. Schematic of various diffractometer scans used to probe: layering (00ℓ), stacking (10ℓ) and epitaxy (chi-scan).

4. TEXTURED MULTILAYERS

To date, very few examples of metallic heterostructures exist which are true superlattices in the sense that long range order extends over the whole sample in three dimensions. Some well-known examples are Nb-Ta, which was the first metallic superlattice to be constructed [29], and the Y-Gd system [30]. All of these systems have very close lattice matching (<<1%) and the paired components are isostructural.

By far the most common situation in the metallic heterostructures is where no close lattice matching exists and numerous examples can be found in the literature. Ruggiero [31] has recently compiled a summary of structural and transport studies in metallic multilayer systems. In this section I will review some of our recent results on the Mo-Ni system. This serves as an excellent example of a metallic multilayer system with coherent layering but which is polycrystalline in the direction parallel to the layers. This type of structure, in which the crystallites (~100Å in size) have a preferred orientation (or 'texture') relative to the c-axis, is very common in the field of metallic heterostructures.

In spite of the random orientation of crystallites within the plane of such samples it is still possible to have some well-defined contiguity of atomic planes in the interior of the strata, i.e., well away from the interface boundary. The interesting question is, what happens at the interfaces of such a badly mismatched structure? Using some of the ideas discussed in section 2 it should be possible to make some general statements.

First, consider the (00ℓ) diffraction profiles of a rather badly mismatched textured multilayer system, Mo/Ni, prepared by sequential sputtering [32]. When the Ni or Mo layers are thicker than about 10Å (~4 monolayers) there is relatively good coherent layering as evidenced by the central peak flanked by two satellites (see Fig. 10). The fact that never more than two or three orders of satellites are observed is indicative not of chemical interdiffusion, since Mo and Ni are immiscible, but probably of island morphology. As we have seen above, this roughens the interface and leads to suppression of satellite intensities and to broad rocking curves. Inset (a) in Fig. 10 shows the (00ℓ) rocking curve for an L $\approx$ 23Å sample. The width (~15° FWHM) suggests an interface step density in excess of 0.05Å^{-1}. When the Mo or Ni strata are made even thinner than 10Å (see inset (b) of Fig. 10) then the structure loses its layering morphology and more resembles a metallic glass. Clearly, what is happening here is that when the layers become very thin the rough interface regions begin to overlap and the structure loses coherence normal to layers.

The increasing dominance of the interfaces, when the strata are made thinner and thinner, is also shown up in the in-plane electrical resistivity. At low temperatures the resistivity of a metal is determined primarily by scattering from imperfections. In Mo/Ni multilayer films, the in-plane resistivity is found to be inversely proportional to the modulation wavelength [33] a clear sign that it is the disordered interface regions that give rise to the electron scattering. Very low temperature studies [34] of the in-plane resistivity show evidence of three-dimensional localization again suggesting highly disordered interface regions (see Fig. 11).

The presence of interface roughening coupled with disordered structure in textured monolayers leads one to suspect that anomalous behavior such as the 'supermodulus' effect [35], may be an artifact of the

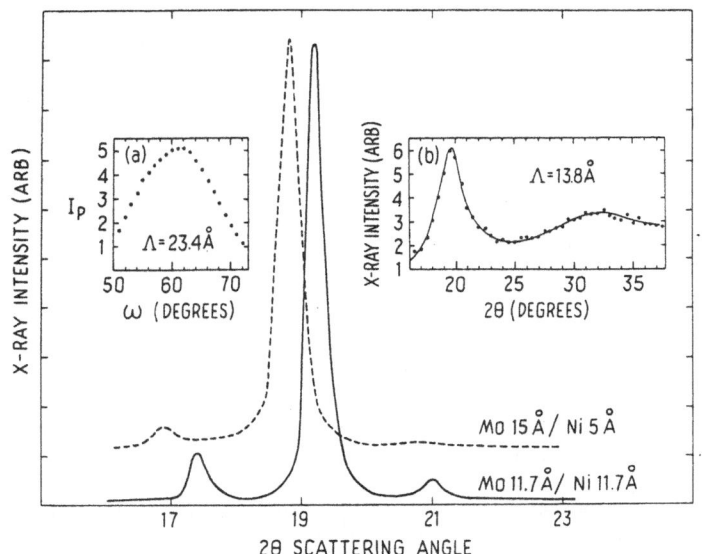

Fig. 10. 00ℓ X-ray diffraction profiles for two samples showing coherent layering. Inset (a): rocking curve for Mo 11.7Å/Ni 11.7Å periodic multilayer film. I_p is the intensity of the main peak near 2θ=19°. Inset (b): Glassy behavior of Mo 7Å/Ni 7Å sample [34].

structure when the layers become very thin ($\lesssim$ 30Å). In order to identify this as a fundamental effect it will be necessary to make measurements on samples in which the interfaces are characterized to monolayer precision. In principal this is possible with current techniques; it would be easiest to perform these experiments on systems which have excellent lattice matching, i.e., superlattices.

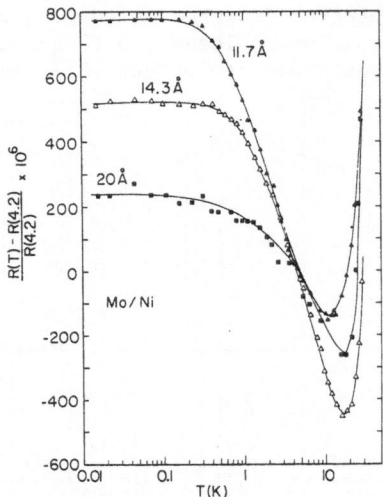

Fig. 11. Temperature dependence of normalized resistance for
non-superconducting Mo/Ni structures. The labels
refer to 1/2 L [34].

5.   BICRYSTAL SUPERLATTICES

Recently it has been demonstrated that under some circumstances high
quality superlattice growth is possible even when the crystal structures
of the two components of the superlattice differ from each other [26]. In
particular, fcc-hcp "bicrystal" superlattices were recently fabricated at
the University of Illinois by sequential molecular beam epitaxy deposition
of iridium and ruthenium, respectively. Several aspects make this
particular combination of metals favorable for growth of a
three-dimensionally ordered superlattice; most important is the
fcc(111) ÷ hcp(001) epitaxial relationship [36,37]. This factor,
together with the fairly close (0.3%) matching of nearest neighbor
distances in Ru and Ir, promotes the coherent growth of close-packed
layers to form an essentially single-crystal three-dimensional
heterostructure.

In this section I will use this intriguing, if somewhat exotic,
superlattice to illustrate some effects which are peculiar to metallic
systems. Particular reference is made to the observation [26] that
when the Ir strata are very thin ($\lesssim$10 monolayers) the stacking within
the Ir component takes an hcp configuration in contrast to the bulk fcc
structure. Using the X-ray techniques described in section 3 we were able
to identify very long-range coherent stacking sequences near the fcc-hcp
crossover and to probe stacking faults associated with the fcc-hcp
transition.

The Ru-Ir superlattices described here were electron-beam evaporated
on (11$\overline{2}$0) sapphire substrates held at 920°C in an ultrahigh vacuum MBE
chamber [26]. This particular orientation was preferred because the
rectangular symmetry of the (11$\overline{2}$0) face was found to inhibit the
nucleation of twin domains with vectors rotated by ±30°. Two samples of
modulation wavelengths L = 41Å and 62Å were studied in detail. In both
samples the Ru strata were of constant thickness (11 atomic planes or 24Å
thickness), whereas the Ir strata in the L = 41Å sample were 8 atomic
planes thick, and were 15 atomic planes thick in the L = 62Å sample. The
total thickness of the superlattice films was ~2000Å.

Fig. 12 shows an in-plane diffraction pattern of the X-ray scattering intensity from the L = 41Å Ru-Ir superlattice. The pattern is actually an intensity contour map of the (hk0) reciprocal lattice plane obtained by making a 60° fan of θ-2θ scans each incremented in χ by an azimuthal angle of 1°, as described in section 2. The hexagonal symmetry of the close-packed metal layers is clearly shown, confirming a high degree of epitaxial ordering. There are some weak additional spots in Fig. 12 which are rotated ±30° from the {100} reflections and are due to the twinning problem noted above.

Stacking of the atomic planes was studied by making (10ℓ) diffractometer scans [27]. Fig. 13(a) shows one such scan for L = 62Å. The sharp, strong peaks at $k_c^* = 2\pi(2\bar{d})^{-1} = 1.45\text{Å}^{-1}$ and $2\pi(3\bar{d})^{-1} = 0.96\text{Å}^{-1}$ where $\bar{d}$ is the mean interplane spacing, are associated with very coherent hcp (ABAB....) and fcc (ABCABC....) stacking of Ru and Ir layers respectively. In addition, we observe sidebands on all the principal peaks; the separation of these peaks of 0.10Å^{-1} is consistent with the superlattice modulation wavelength of 62Å.

A similar (10ℓ) scan for the L = 41Å sample [see Fig. 13(b)] reveals some interesting and unexpected differences in stacking order compared to the larger wavelength sample in which the Ru strata are the same thickness but the Ir strata have 15 monolayers instead of 8 monolayers. Firstly, the diffraction peaks at $k_c^* = 0.96\text{Å}^{-1}$ (fcc stacking peaks) are absent. Secondly, the hcp peak at 1.45Å^{-1} is substantially broader than in the L = 62Å sample.

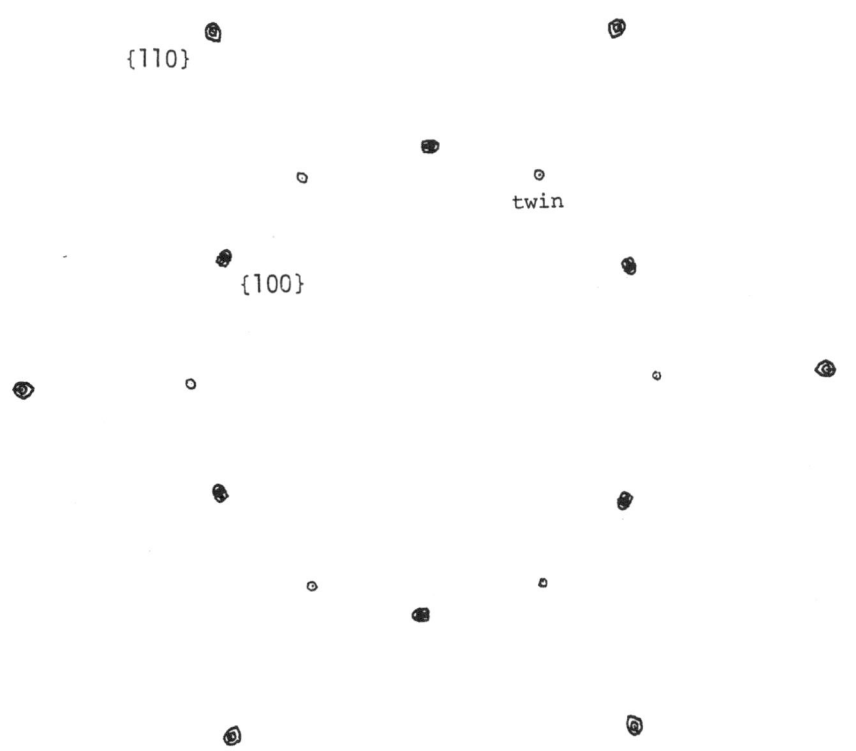

Fig. 12. In-plane diffraction pattern measured on Ru-Ir superlattice with λ = 41Å modulation wavelength [27].

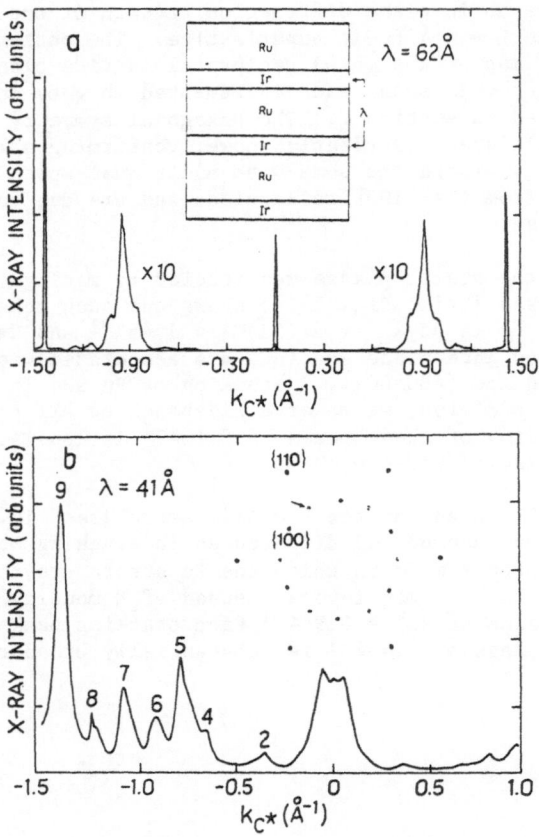

Fig. 13. (10ℓ) scans for two Ru-Ir superlattices. The labels
in (b) denote the ℓ-values of superlattice peaks.

The diffraction data for the L = 41Å sample can be interpreted in
terms of a stacking arrangement in which the Ru and Ir components both
exist in a predominantly hcp conformation. Thus, when the Ir strata are
thin enough, $\leq$ 10 monolayers, there is a crossover to Ir hcp stacking from
the bulk fcc structure. Some information about the coherence of the
stacking can be obtained from the widths of the (10ℓ) peaks in the
c^*-scans shown in Figs. 13(a) and 13(b). We estimate a stacking coherence
length of > 750Å in the L = 62Å sample, but about a factor of 5 less in
the L = 41Å sample. We interpret this reduction in coherence as being
due to the presence of stacking faults in the Ir strata which may still
retain some vestige of the fcc stacking. Thus a typical Ir stratum in
this sample may have the following stacking sequence: ABAB//CBCB, where
the double slashes denote a stacking fault resulting from a remnant fcc
package (ABC). In an alternative notation [38], this sequence can be
described by hhhchhhh, where h and c, refer respectively to hexagonal and
fcc atomic plane coordinations.

One additional piece of information in Fig. 13(b) relates to the
triplet of peaks around $k_c^* = 0$. The triplet is composed of a
principal {100} peak flanked by two side peaks at $\pm 0.051Å^{-1}$. Higher
order peaks at $\pm 0.102Å^{-1}$ etc., are also observed. This set of Fourier
components corresponds precisely to a new periodicity of 3L = 123Å

superimposed on the normal modulation wavelength, L. We tentatively
suggest that the origin of this unexpectedly large commensurate
superstructure is a polytype structure containing an ordered array of
stacking faults. If a small amount of fcc stacking still persists in the
L = 41Å sample then one stacking fault (or fcc package) per Ir layer would
introduce a phase shift of $2\pi/3$ in the stacking sequence. Thus, the full
stacking sequence would repeat after 3 cycles, i.e., in a distance of 3L.
We envisage a complete stacking sequence such as
....hhchhhhhhhhhchhhhhhhhhhchhhhh.... etc. It should be noted that the
stacking coherence of 150Å noted above is not that much larger than the
true repeat distance (123Å). Thus there is some randomness as to the
exact placement of the stacking fault within each Ir layer.

A theoretical argument for the appearance of such long-period
stacking-fault configurations has recently been proposed by Redfield and
Zangwill [39]. In their analysis, Friedel-like oscillatory interactions
are associated with the interfaces of the metallic superlattice and these
can stabilize polytype structures which are intermediate between hcp and
fcc. Fig. 14 shows the stacking fault chemical potential calculated by
Redfield and Zangwill [39] for a single bimetallic interface. To
first-order the near-interface region (I) will exhibit stacking faults at
every atomic plane where $\mu(d)<0$. Farther away from the compositional
boundary, the modulation amplitude drops off and a region (II) is
encountered where the influence of the interface is comparable to the bulk
defect-defect interaction. In this region a repulsive interaction between
stacking faults may stabilize a single-defect 'compromise' structure,
ccchccchcc..., as shown in Fig. 14(c). A bulk-dominated region (III)
extends at still greater distances from the interface. For short-period
superlattices, the oscillatory contributions from neighboring interfaces
must be superposed. If the two modulations are exactly in phase, a
stacking structure may be observed that is quite different than in the
bulk. In particular, if the compositional layer thickness is less than a
Friedel wavelength (i.e., region I dominates) then a stacking fault occurs
on every atomic plane. In this way it is energetically favorable to
convert an entire stratum from fcc to hcp, or vice versa. This could

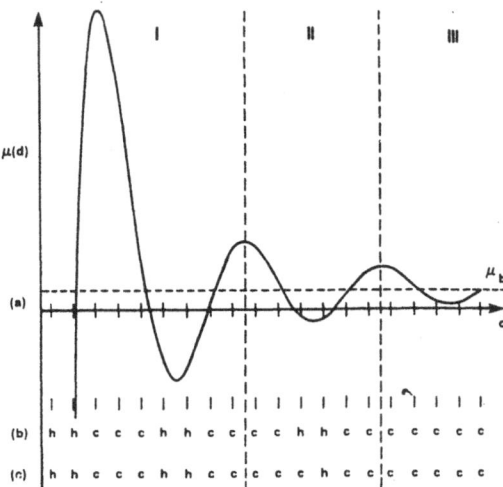

Fig. 14. Redfield and Zangwill's calculation of the chemical potential
for an h stacking fault, $\mu(d)$, as a function of distance from
a bimetallic interface. The horizontal axis is marked in
units of interplanar distance [39].

explain the stacking crossover described above. Redfield and Zangwill
have also pointed out that, in addition to the above oscillatory
interactions, other effects such as elastic distortions [40] and charge
transfer can influence the stacking structure. The Ru-Ir system may be
particularly sensitive to charge transfer effects since Ru and Ir straddle
a structural phase boundary in the periodic table [41].

6. FUTURE TRENDS

    Research on the growth and characterization of thin film
heterostructures of metals and metal-semiconductor composites is still in
its infancy. It is difficult, therefore, to predict in which directions the
field will evolve. However, there are two fairly recent developments which
should have an impact on both the fabrication of thin films and on the
precision with which one can probe the atomic structure. In this section I
will touch briefly on each of these developments, pointing out some
interesting opportunities for future research.

Synchrotron Radiation

    The routine availability of very bright, highly collimated, X-ray
beams derived from synchrotron storage rings presents a wealth of exciting
possibilities for measurements on thin film heterostructures. Kwo [42] has
described one such application in an accompanying paper on magnetic X-ray
scattering. This particular measurement utilized the high degree of
polarization of synchrotron X-ray radiation and exploits the small
differences in total scattering depending on the relative orientation of
the magnetic spin vector and the polarization of the X-ray beam [43].

    Another useful characteristic of synchrotron radiation is that,
perpendicular to the plane of the storage ring, the radiation has very
good collimation. For example, for typical relativistic storage ring
energy of $E \approx 2$ GeV, the opening angle of the radiation can be as low as
10 μrad.

    This high degree of collimation, coupled with the intense brightness
of the X-ray beam permits scattering experiments on surfaces and
interfaces that are impractical with normal laboratory sources [44]. In
one such technique, developed by Marra, Eisenberger and Cho [45], the
X-ray beam is incident at some small grazing angle relative to the plane
of the surface (see Fig. 15). Because the refractive index for X-rays in
the sample is less than unity by a few ppm, the X-ray beam undergoes total
external reflection. Thus it is possible to satisfy the Bragg diffraction
condition with the diffraction vector lying precisely in the plane of the
sample. Moreover, since the X-ray beam is close to the critical angle for
total external reflection, it is only the surface layers of the sample
that participate in the diffraction process. Of course this is where the
collimation of the incident beam becomes very important.

    One of the first applications of this technique was to study the
interface structure of an epitaxial layer of Aℓ deposited on GaAs [45,46].
In this case the grazing incidence X-ray beam is used as a means to probe
the structure as a function of depth beneath the surface. The
depth-profiling capability made possible by highly collimated polarized
synchrotron radiation shows great promise for in-situ studies of
interfaces in a large variety of thin-film heterostructures. The
advantage over more conventional high resolution electron microscopy is

398

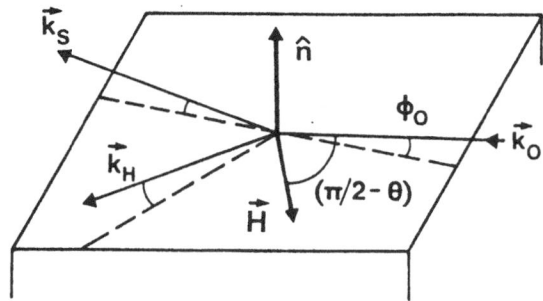

Fig. 15. Schematic of a surface diffraction experiment [49]. $\vec{k}_s$ is the specularly reflected beam and $\vec{k}_H$ the surface diffracted wave.

that preparation of the sample, i.e., thinning by ion-milling, is in general not required so that the interfaces could be studied in the undisturbed, as-grown, condition.

Another potentially important application of X-ray synchrotron radiation to diffraction experiments on artificial heterostructures is based on standing wave effects. The idea here, pioneered by Batterman [47] is that a standing-wave field is established external to the crystal surface via interference between the incident and Bragg reflected X-ray beams. Near the total external reflection condition the phase of the standing wave is very sensitive to the angle of incidence. This dynamical effect has been used recently to determine the positions of ordered impurity atoms deposited on a crystal surface [48].

As yet very little of this kind of work has been done using synchrotron sources and the techniques are only just beginning to be applied to heterostructures; one aspect of synchrotron radiation that has not been fully exploited in this respect is the **tunability** of the source. By selecting an X-ray energy close to an absorption edge of a particular constituent of the structure, it should be possible in principle to enhance or supress the X-ray fluorescence from these particular atoms. The interference of the fluorescence and the standing-wave fields in the Batterman method can be used to determine the relative phases of atoms.

Aperiodic Heterostructures

A second interesting development that is just beginning to come into prominence in the context of artificial heterostructures concerns **aperiodic** multilayers; that is, thin films in which the strata are deposited so that there is no translational symmetry normal to the layers. There is an infinite range of possibilities, extending from random deposition of layers to "almost periodic" (quasiperiodic) sequences of layers which conform to some predetermined, yet non-periodic, mathematical sequence.

Physically, such aperiodic systems are of interest because they offer the potential to realize novel properties that are quite unlike either crystalline (periodic) materials or bulk amorphous solids [50]. One example of recent interest is the so-called "quasiperiodic superlattice" recently reported by Merlin, et al. [51]. In this new kind of heterostructure, layers of different materials, A and B, are deposited in a Fibonacci sequence: ABAABABA...[51,52]. An interesting property of this system is that its Fourier spectrum is a dense set of sharp peaks, i.e., there are Fourier components at every wavevector. Thus, the reflectivity of the quasiperiodic superlattice is finite at all wavevectors, and at all angles of incidence greater than the critical angle. This situation is in contrast to the usual periodic multilayer systems where Bragg's law is obeyed only at discrete diffraction vectors given by $2\pi/c_0$, where c_0 is the layering period. Fig. 16 shows a transmission electron micrograph of a metallic Fibonacci superlattice grown at the University of Michigan. The layers A (Nb) and B(Ta) are chosen to be of thickness 55Å and 34Å, respectively. These dimensions are not critical in order to obtain the quasiperiodic structure: it is the sequencing of the layers that is important. This fact allows for a great deal of flexibility in making a superlattice with desired properties. Nb and Ta were chosen in this case because the superconducting behavior should show some interesting features [53,54].

The physical properties of this new class of quasiperiodic materials are just beginning to be studied. Many new developments are expected in this area in the near future.

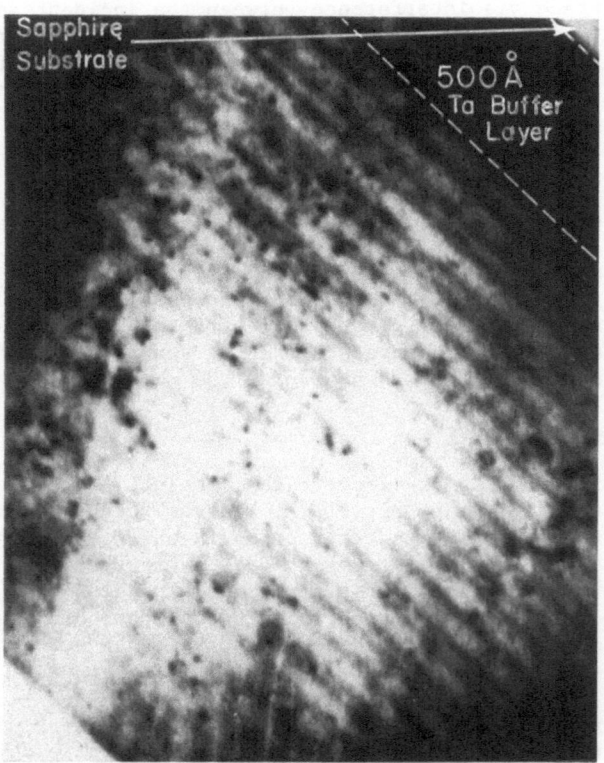

Fig. 16. Transmission electron micrograph of quasiperiodic Nb/Ta superlattice based on a Fibonacci series. The darker bands are Ta, of nominal thickness 34Å. (Electron microscopy by H. He) [54].

400

7. CONCLUSIONS

Preliminary research into modulated metallic structures has revealed a wealth of new phenomena and promises further improvements of our understanding of bulk metallic systems. The field is potentially as exciting as that of III-V epitaxy. However, given the rather undeveloped state of UHV deposition techniques for metallic systems, particularly using electron-beam evaporation, much development work remains to be done in order to achieve the control of growth conditions that are necessary for making high-quality epitaxial heterostructures. Although significant progress has been made on a few specific materials, metal-MBE is still far from becoming routine.

Along with continuing development of growth techniques a parallel effort must be made in methods for structural characterization. Clearly, no single technique can provide all the information required; X-ray scattering should be viewed as but one of a battery of powerful tools for probing the microstructure. However, since it is exceedingly difficult to prepare metallic structures for high-resolution TEM, X-ray probes, particularly synchrotron radiation, will undoubtedly play a major role in unraveling the complex nature of bimetallic and metal-semiconductor interfaces.

ACKNOWLEDGEMENTS

It is a pleasure to acknowledge the contributions of my co-workers during the course of this work: firstly, my students Frank Lamelas and Hui He, for their perseverance in growing refractory metal superlattices by MBE; secondly, my colleagues Ctirad Uher and Roberto Merlin for stimulating research collaborations. Much of the X-ray work described in this review was made possible by access to a wide range of excellent samples of metallic heterostructures. For this I am indebted to Jack Cunningham, Peter Flynn and his students, and to Ivan Schuller.

Support for my research is provided in part by NSF Materials Research Group Grant DMR8602675, by NSF Low Temperature Physics Grant DMR840975 and by the US Army Research Office under Grant DAAG-29-83-K-0131.

REFERENCES

1. L. Esaki and R. Tsu, IBM J. Res. Dev. 14, 61 (1970).
2. For a comprehensive review, see Synthetic Modulated Structures, eds. L.L. Chang and B.C. Giessen (Academic, New York, 1985); see also papers in Surface Science 142, (1984).
3. L. Esaki, Proceedings of the 17th International Conference on the Physics of Semiconductors, eds. J.D. Chadi and W.A. Harrison (Springer-Verlag, New York, 1985) p. 473.
4. C. Weisbuch, R. Dingle, A.C. Gossard and W. Weigmann, in Proceedings of the 8th International Symposium on Gallium Arsenide and Related Compounds, ed. H.W. Thim (Institute of Physics, London, 1981) Inst. Phys. Conf. Ser. No. 56, p. 711.
5. For a recent review, see L. Esaki, IEEE J. Quantum Electron 22, 1611 (1986).
6. G.H. Dohler, J. Vac. Sci. Technol. 16, 851 (1979).
7. T.W. Barbee, Jr., Superlattices and Microstructures 1, 311, 1985).
8. J.W. Matthews, Epitaxial Growth (Academic, New York, 1975).
9. F.C. Frank and J.H. van der Merwe, Proc. Roy. Soc. A 198, 216 (1949); E. Bauer and J.H. van der Merwe, Phys. Rev B33, 3657 (1986).

10. J.C. Bean, L.C. Feldman, A.T. Fiory, S. Nakahara and I.K. Robinson, J. Vac. Sci. Technol. A2, 436 (1984); A.T. Fiory, J.C. Bean, L.C. Feldman and I.K. Robinson, J. Appl. Phys. 56, 1227 (1984).

11. C.M. Falco, W.R. Bennett and A. Boufelfel, in Dynamical Phenomena at Surfaces, Interfaces and Superlattices, eds., F. Nizzoli, K.H. Rieder and R.F. Willis (Springer-Verlag, Berlin, 1985) p.35.

12. R. Bruinsma and A. Zangwill (to be published).

13. See, A. Madhukar in this proceedings; J. Singh and K.K. Bajaj, Superlattices and Microstructures, 2, 185 (1986).

14. I.K. Schuller, Phys. Rev. Lett. 44, 1597 (1980).

15. D.B. McWhan, M. Gurvitch, J.M. Rowell and L.R. Walker, J. Appl. Phys. 54, 3886 (1983), and references therein.

16. A. Guinier, X-ray Diffraction in Crystals, Imperfect Crystals and Amorphous Bodies, Transl. P. Lorrain and D. Sainte-Marie Lorrain (San Francisco: Freeman, 1963).

17. J. DuMond and J.P. Youtz, J. Appl. Phys. 11, 357 (1940).

18. J. DuMond and J.P. Youtz, Phys. Rev. 48, 703 (1935).

19. W. Deubner, Ann. der Physik 5, 261 (1930).

20. H.E. Cook and J.E. Hilliard, J. Appl. Phys. 40, 2191 (1969).

21. J.W. Cahn, Acta Met. 9, 525 (1961).

22. S.M. Durbin, J.E. Cunningham and C.P. Flynn, (to be published).

23. B. Warren, X-ray Diffraction (Addison-Wesley, Reading, 1969).

24. D. Chrzan and P. Dutta, J. Appl. Phys. 59, 1504 (1986).

25. A. Segmuller and A.E. Blakeslee, J. Appl. Cryst. 6, 19 (1973).

26. J.E. Cunningham and C.P. Flynn, J. Phys. F 15, L221 (1985).

27. R. Clarke, F. Lamelas, C. Uher, C.P. Flynn, and J.E. Cunningham, Phys. Rev. B 34, 2022 (1986).

28. See paper by W.J. Bartels in this proceedings.

29. S.M. Durbin, J.E. Cunningham, M.E. Mochel, and C.P. Flynn, J. Phys. F: Met. Phys. 11, L223 (1981).

30. J.R. Kwo, E.M. Gyorgy, D.B. McWhan, M. Hong, F.J. DiSalvo, C.Vettier and J.E. Bower, Phys. Rev. Lett. 55, 1402 (1985).

31. S.T. Ruggiero, Superlattices and Microstructures, 1, 441 (1985).

32. I.K. Schuller and C.M. Falco, in Inhomogeneous Superconductivity-1979 eds. D.U. Gubser, J.L. Francavilla, J.R. Leibowitz and S.A. Wolf (American Institute of Physics, New York,1980) p. 197.

33. M.R. Khan, C.S.L. Chun, G.P. Felcher, M. Grimsditch, A. Kueny, C.M. Falco and I.K. Schuller, Phys. Rev. B27, 7186 (1983).

34. R. Clarke, D. Morelli, C. Uher, H. Homma and I.K. Schuller, Superlattices and Microstructures 1, 125 (1985).

35. W.M.C. Yang, T. Tsakalakos and J.E. Hilliard, J. Appl. Phys. 48, 876 (1977); A. Jankowski and T. Tsakalakos, J. Appl. Phys. 57, 1835 (1985); Scripta Met. 19, 625 (1985).

36. J.Q. Zheng, J.B. Ketterson and G.P. Felcher, J. Appl. Phys. 53, 3624 (1982).

37. W. B. Pearson, Handbook of Lattice Spacings and Structures of Metals and Alloys, (Pergamon, Oxford, 1969).

38. H. Jagodzinski, Acta Crystallogr. 2, 201 (1949).

39. A.C. Redfield and A.M. Zangwill, Phys. Rev. B 34, 1378 (1986).

40. R. Bruinsma and A. Zangwill, Phys. Rev. Lett. 55, 214 (1985).

41. H.L. Skriver, Phys. Rev. B 31, 1909 (1985).

42. see, J.R. Kwo, in this proceedings; and C. Vettier, D.B. McWhan, E.M. Gyorgy, J.R. Kwo, B.M. Buntshuh and B.W. Batterman, Phys. Rev. Lett. 56, 757 (1986).

43. P.M. Platzman and N. Tzoar, Phys. Rev. B2, 3556 (1970); F. DeBergevin and M. Brunel, Acta Crystallogr. Sect. A 37, 314 (1981).

44. H. Winick, S. Doniach eds., Synchrotron Radiation Research (Plenum, New York, 1980).

45. W.C. Marra, P. Eisenberger and A.Y. Cho, J. Appl. Phys. 50, 6927 (1979).

46. A.Y. Cho and P.D. Dernier, J. Appl. Phys. 49, 3328 (1978).

47. B.W. Batterman, Phys. Rev. 133, A759 (1964); Phys. Rev. Lett. 22, 703 (1969).

48. J.A. Golovchenko, J.R. Patel, D.R. Kaplan, P.L. Cowan and M.J. Bedzyk, Phys. Rev. Lett. 49 (1982); P.L. Cowan, S. Brennan, T. Jach, M.J. Bedzyk and G. Materlik, Phys. Rev. Lett. 57, 2399 (1986), and references therein.

49. P.L. Cowan, Phys. Rev. B 32, 5437 (1985).

50. J.B. Sokoloff, Phys. Rev. B 22, 5823 (1980); S. Ostlund and R. Pandit, Phys. Rev. B 29, 1394 (1984); S. DasSarma, A. Kobayashi and R.E. Prange, Phys. Rev. Lett. 56, 1280 (1986).

51. R. Merlin, K. Bajema, R. Clarke, F.-Y. Juang and P.K. Bhattacharya, Phys. Rev. Lett. 55, 1768 (1985).

52. J.P. Lu, T. Odagaki and J.L. Birman, Phys. Rev. B 33, 4809 (1986).

53. F. Nori and J.P. Rodriguez, (to be published).

54. J.J. Lin, J. Cohn, F. Lamelas, He Hui, R. Merlin, R. Clarke and C. Uher (to be published).

SPIN-POLARIZED NEUTRON REFLECTION FROM METASTABLE

MAGNETIC THIN FILMS

J.A.C.Bland   R.F.Willis

Clarendon Laboratory   Cavendish Laboratory
University of Oxford   University of Cambridge
Oxford OX1 3PU, U.K.   Cambridge CB3 0HE, U.K.

INTRODUCTION

Many optical phenomena have been demonstrated using a neutron source of radiation [1]. For example, at the boundary between two media with different neutron refractive indices, a collimated neutron beam may, like light, be totally reflected when the incident angle is less than the critical glancing angle θ_c. Under conditions close to critical reflection, the neutron reflectivity is sensitive to the refractive index profile $n(z)$ normal to the interface boundary [2]. The measurement of the reflected intensity for different angles of incidence $\theta_i > \theta_c$ gives the depth profile of the scattering density. Variations in the neutron interaction potential, arising from inhomogeneities in the interface region, produces diffuse scattering around the specularly reflected beam which may be analyzed for interface roughness and strain effects. It is possible to extract separately the spatial dependence of the magnetisation (magnetisation profile) from the density profile at a ferromagnetic surface. The magnetic interaction potential is comparable in magnitude with the total interaction potential, which is a distinct advantage over grazing-incidence X-ray scattering where the magnetic part of the scattering cross-section is 10^{-5} weaker than the total interaction potential. Thus, the specular reflectivity of neutrons as a function of momentum transfer normal to the interface provides information on long-range (spin) ordering in magnetic films while the parallel momentum transfer component relates to the short-range order associated with, for example, critical fluctuations during a phase transition [3]. Similarly, the technique can be extended to magnetic superlattices, Bragg diffraction from which provides information on the layer spacing and structure [4].

The sensitivity of both X-rays and neutrons to surface and thin film phenomena is greatly enhanced in the grazing angle of incidence geometry. Even so, the scattered intensities are weak since the number of scattering centres in the surface layer is so small compared with the bulk. In the case of X-ray scattering, this has led to the recent development of powerful synchrotron radiation sources [5]. The fluxes from current thermal neutron sources are many orders of magnitude less so that it has proved difficult to date to study ultrathin films down to a few monolayers thickness. This is unfortunate since it is precisely at the ultrathin thickness level that there is considerable interest in the properties of ferromagnetic materials - see the article by U. Gradmann in this volume [6]. For example, it is possible to grow metastable ferromagnetic phases, which are not normally stable at room temperature; and which possess unusual magnetic properties absent in the bulk equilibrium phase.

In this article, we demonstrate that it is possible to study the magnetic properties of ultrathin magnetic films of thickness down to a single monolayer using neutron glancing-angle reflectance. This greatly enhanced sensitivity over previous measurements is achieved by interference enhancement in the neutron wavefield in a thin-film sandwich structure consisting of the magnetic layer sandwiched between a non-magnetic metallic matrix. The magnetic films are epitaxially grown on single crystal substrates and coated with an epitaxial film of the same material.

After some experimental considerations, section 2, we discuss the thin-film interference method in section 3. In section 4, we extend the discussion to theoretical considerations of spin-polarized neutron reflection from a 1-dimensional potential, which effectively models the optical interference. In section 5, we illustrate the method with some results for f.c.c. cobalt and iron films epitaxially grown on Cu (001) substrates. Finally, secton 5, we address the usefulness of this method for characterising low-dimensional magnetic thin-film structures and properties.

SPIN-POLARIZED NEUTRON REFLECTIVITY MEASUREMENTS

Magnetic thin films epitaxed to single crystal substrates of surface area ~1 cm^2 are routinely amenable to characterisation by surface science techniques. However, the small surface area imposes severe requirements on the experimetnal determination of polarized neutron reflectivities.

The usable beam width "a" for momentum transfer $k_i - k_f = 2q$, with $|k_i| = |k_f|$ is given by $(|q| l / |k_i|) = a$, where l is the sample length illuminated (Fig. 1).

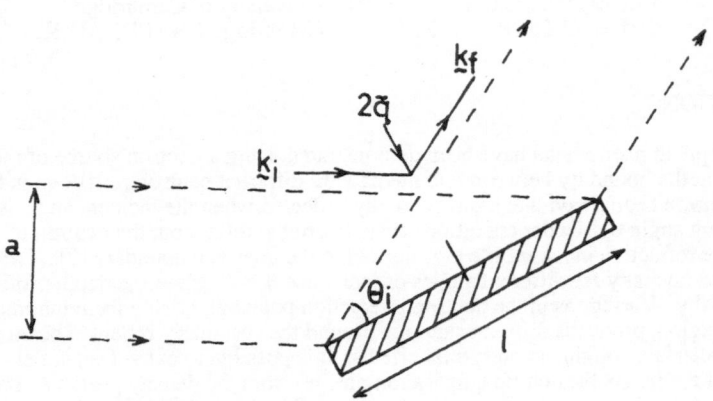

Fig. 1. Neutron reflection geometry for a beam of cross-section, a = l sin θ_i for a sample of length l and angle of incidence θ_i: momentum transfer, $k_i - k_f = 2q$.

For an incident neutron flux $n_i (\lambda)$ and neutron intensity reflectance R(q), the reflected intensity is:

$$
\begin{aligned}
I_R &= n_i (\lambda) \, a \, R(q) \, \Delta q \\
&= n_i (\lambda) \, R(q) \, \lambda \, l \, |q| \, \Delta q \, / 2\pi
\end{aligned}
\tag{1}
$$

i.e. for a given resolution $\Delta q/q$, the measured reflected intensity is larger at longer incident wavelength (as are the apertures needed for the appropriate collimation). The resolution Δq must be sufficiently large that I_R is measurable but, also, small enough that structure in R(q) can be resolved. Eqn (1) represents the reflected intensity per unit depth perpendicular to the interface so that, for ultrathin films, the effect of the film on the reflectivity is felt at large q values i.e. large $\theta_i > \theta_c$; under these circumstances, R(q) is slowly varying, permitting some relaxation in Δq. Δq values ~10^{-3} Å^{-1} are particularly suitable for thin film measurements. However, R(q) falls off rapidly with q, so that signal to noise limitations are the important criterion.

Thus, the requirements are: an intense source ($\geq 10^6$ n cm^{-2} Å^{-1}) of monochromatic, long wavelength ($\lambda \geq 10$ Å) highly collimated beam of neutrons and some means of varying the

406

perpendicular component q of the incident neutron momentum. Two methods are currently available: either one can vary θ_i while keeping λ fixed [7]; or one fixes θ_i and employs a range of λ, as provided by a pulsed neutron source together with time-of-flight analysis [8].

In figure 2, we illustrate schematically one such arrangement [7] in which q is varied by mechanical rotation of the crystal in a fixed λ beam.

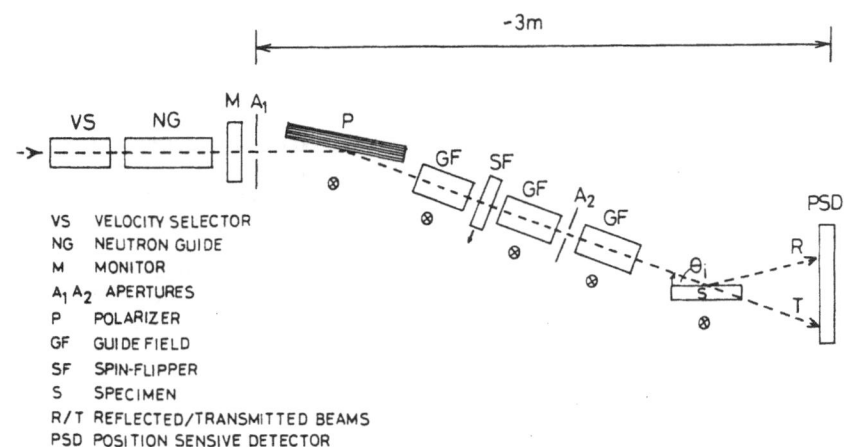

VS	VELOCITY SELECTOR
NG	NEUTRON GUIDE
M	MONITOR
$A_1 A_2$	APERTURES
P	POLARIZER
GF	GUIDE FIELD
SF	SPIN-FLIPPER
S	SPECIMEN
R/T	REFLECTED/TRANSMITTED BEAMS
PSD	POSITION SENSITIVE DETECTOR

Fig. 2. Spin-polarized neutron reflectance beam line showing the perpendicular ($\otimes$) and parallel ($\downarrow$) configuration of the H fields with respect to the scattering plane.

A highly collimated incident beam of neutrons whose incident intensity distribution peaks at $\lambda = 12$ Å is defined to within $\Delta\lambda = 10\%$ by a mechanical velocity selector and the beam cross-section a by cadmium aperture A_1. The beam is efficiently polarized by reflection off a Co-Ti 'supermirror' (itself a ferromagnetic superlattice thin film structure [9]) and then guided through a region of homogeneous magnetic field (H ~100 Oe) onto the sample via a spin-flipping electro-magnetic coils arrangement consisting of two orthogonally wrapped coils which produce a $\pi/2$ Larmor precession of the neutron spin [10]. Aperture A_2 limits the beam divergence at the sample ($\delta\theta$ ~10^{-3} rad.). The latter is aligned optically, and held rigidly in a magnetic assembly (H $\leq$800 Oe) inside a liquid He cryostat which can be rotated about the vertical axis. A second supermirror is installed just in front of the sample position to measure the polarization efficiency of the spin-flipping coils; P_o as high as 90% is readily achievable. The neutron spin is either parallel (+) or antiparallel (−) to the magnetic field axis H ($\otimes$, Fig. 2). The transmitted and reflected beams off the sample are collected by a large area, position-sensitive-detector (PSD) of sufficient spatial resolution to resolve the profile of the reflected beam [7].

In an arrangement where time structure in the incident beam or variable λ is employed, a non-adiabatic spin flipper must be employed [11].

Having maximized the polarized neutron intensity reflected from the sample, we must now consider the optimisation of the reflectivity off the thin film sample structures.

NEUTRON INTERFERENCE IN THIN FILMS

A simple application of Snell's law of optics illustrates the optical interference of neutrons reflected at grazing incidence from thin metallic films.

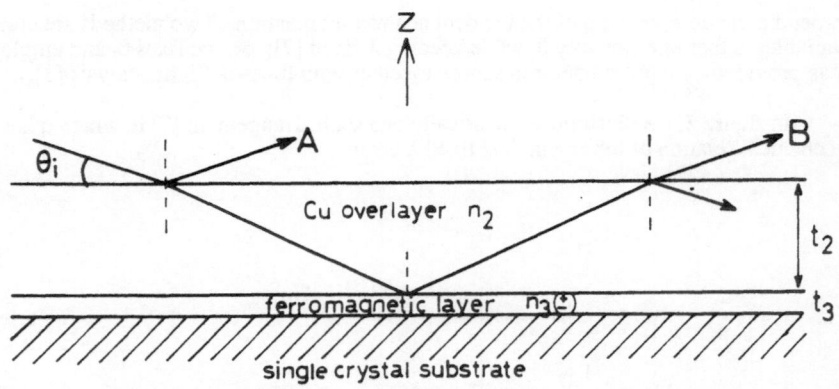

Fig. 3. Optical interference between rays A,B produced by neutrons reflected at grazing incidence angle θ_i from a non-metallic copper film
(thickness t_2 and and refractive index n_2) deposited on top of a thinner, ferromagnetic layer ($t_3 n_3(\pm)$) epitaxed to a single crystal substrate.

Neutrons of wavelength λ incident at grazing angle θ_i on a film of refractive index n_2, characterized
by a mean coherent scattering length b and a density of scattering centres per unit volume N,
undergo total reflection when $\theta_i < \text{Re}\,(\theta_c)$ - absorption is small and can be neglected:

$$\theta_c = \lambda\,(Nb/\pi)^{1/2} \qquad\qquad\qquad (2)$$

$$n_2 = 1 - \theta_c^2/2 \qquad\qquad\qquad (3)$$

i.e. b may take positive or negative values so that n_2 may be smaller or larger than 1. For positive
values of b (the case here), interference fringes arising from the optical path difference between
beams A and B (Fig. 3) are only observed for $\theta_i > \theta_c$. (For negative b, interference is possible for
all grazing angles [12]). For a magnetised thin-film sample, b in eqn. (2) is effectively replaced by
$(b \pm p)$ with signs representing the two neutron spin states and p is the magnetic scattering length,
$b \sim p$. The fall-off in reflectivity R(q) observed for $\theta_i > \theta_c \le 1°$ depends on the depth dependent
refractive index variation n(z) which, in turn, takes different values for spin (+) and spin (−)
neutron beam polarization. This gives rise to a spin-asymmetry, $S = P_0^{-1}\,[R(+) - R(-)]/[R(+) +
R(-)]$ in the measured reflectivities $R(\pm)$ for a neutron beam spin polarized to a degree P_0. Alter-
natively, this behaviour is often referred to simply in terms of a "Flipping Ratio", $F = R(+)/R(-)$
[13] such that $S(q) = P_0^{-1}\,(F(q) - 1)/(F(q) + 1)$.

The reflectivity from the three medium systems, Fig. 3, is given by

$$R^{(\pm)}(q) = \left| \frac{r_{12}^{(\pm)} + r_{23}^{(\pm)}\,e\,(2iq_2^{\pm}\,t_2)}{1 + r_{12}^{(\pm)}\,r_{23}^{(\pm)}\,e\,(2iq_2^{\pm}\,t_2)} \right|^2 \qquad\qquad (4)$$

where t_2 is the thickness of a non-magnetic film overlayer on top of a magnetic film of thickness t_2.
Note that the phase variation which introduces interference in the neutron wavefield within the
sandwich-film structure is controlled principally by the thickness t_2 and perpendicular component
of the wavevector q_2 in the overlayer film.

For a system where the change in refractive index at a boundary is not abrupt, the boundary
region may be broken up into successive thin layers with slightly differing refractive indices such
that standard methods of multilayer film optics may be applied to the reflection and transmission at
each boundary. The intensity of the reflected and transmitted neutrons follow approximately the

same laws as for electromagnetic radiation with the electric vector perpendicular to the plane of incidence [1].

However, although the optical method is well-suited to computations involving complicated refractive index profiles normal to the interfaces, an alternative method which gives a better physical insight, is to treat the neutron as a particle penetrating a potential barrier and solve Schrödinger's equation for the system.

SPIN-POLARIZED NEUTRON REFLECTION FROM A 1-D POTENTIAL

A surface or interface represents a break in symmetry normal to the interface. If we treat the refracting medium as possessing translational invariance in the (x, y) plane, then we can model the neutron-solid interaction as scattering from a suitable one-dimensional potential. The neutron wavefield on both sides of the interface can be calculated by solving Schrödinger's equation for a model potential $V(z)$ representing the wave-solid interaction as a function of depth.

The simplest such potential is the step potential shown schematically in Fig. 4 (a)

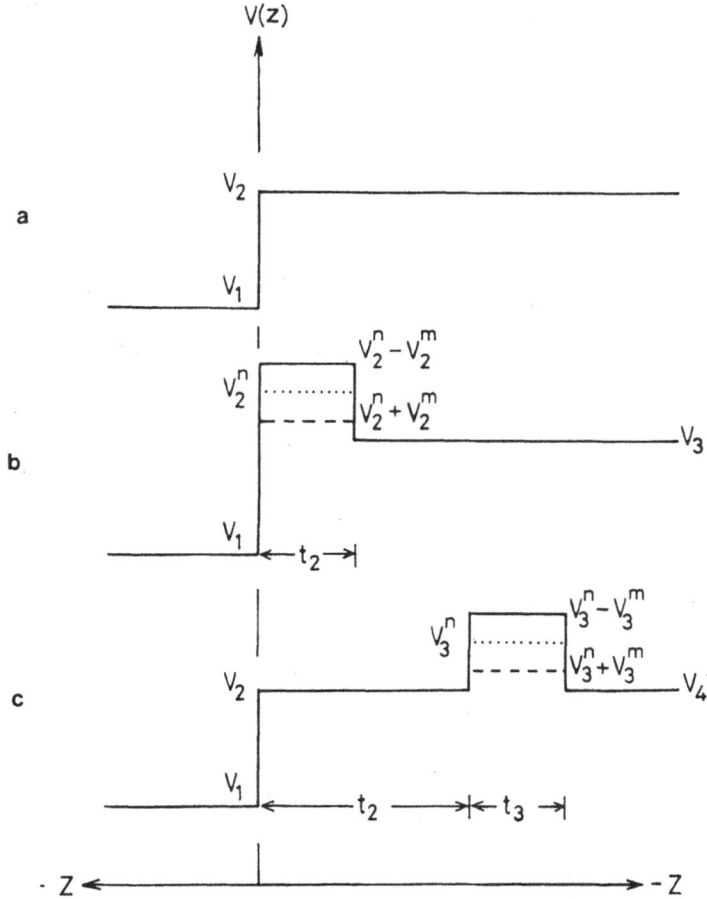

Fig. 4. The potential $V(z)$ for (a) a uniform non-magnetic medium, (b) non-magnetic medium overcoated with a ferromagnetic film, thickness t_3, and (c) system b overcoated with a non-magnetic layer of the substrate material of thickness t_2: $V(\pm) = V^n \pm V^m$; V^n, V^m, are the nuclear and magnetic components of the interaction potential $V(z)$ for spin-up (+) and spin-down (-) polarized neutrons.

The perpendicular wavevector with the lth medium can be written in terms of its refractive index $n_l^{(\pm)}$ as:

$$q_l^{(\pm)2} = k_i^2 \, (n_l^{(\pm)2} - \cos^2 \theta_i) \tag{5}$$

$$\text{and} \qquad q_c^{(\pm)} = k_i \, (1 - n_s^{(\pm)2})^{1/2} \tag{6}$$

where q_c is the critical wavevector perpendicular to a totally externally reflecting interface. The critical wavevector associated with medium l is related to the interaction potential:

$$q_{cl}^{(\pm)} = (2mV_l^{(\pm)})^{1/2}/\hbar \tag{7}$$

Considering the case of a step potential variation due to a uniform non-magnetic surface (Fig. 4a), the reflectivity is given as:

$$r_{12}^{(\pm)} = (q_1^{(\pm)} - q_2^{(\pm)})/(q_1^{(\pm)} + q_2^{(\pm)}) \tag{8}$$

which is a function only of the perpendicular component of the wavevector. We can extend this analysis to more complicated potentials - for example, solving Schrödinger's equation for the three medium potential (Fig. 3) yields:

$$r_{123}^{(\pm)} = \frac{[r_{12}^{(\pm)} + r_{23}^{(\pm)} \, e(2iq_2^{(\pm)}t_2)]}{1 + r_{12}^{(\pm)} \, r_{23}^{(\pm)} \, e(2iq_2^{(\pm)}t_2)} \tag{9}$$

which is the 'optical' result arrived at earlier, equn. (4). We can regard any layered system composed of uniform regions of thickness t_l, each with associated interaction potential $V_l^{(\pm)}$, as approximating the overall spatially variant potential $V^{\pm}(z)$, given that $q_l^{(\pm)}t_l << \pi$.

For the case of a magnetic surface layer (Fig. 4b), the interaction potential can be written as a linear combination of nuclear and magnetic terms:

$$V_l^{(\pm)} = V_l^{nuc} \pm V_l^{mag} \tag{10}$$

$V^{nuc}(z) = [2\pi\hbar^2 b\rho(z)]/m$ for ions with nuclear scattering length b and density $\rho(z)$. The magnetic term, V_l^{mag} is given by $-\mu_N . B(z)$, with $B(z)$ the magnetic induction within the layer and μ_N is the neutron magnetic moment. In the presence of an applied magnetic field H, the magnetic induction $B(z)$ arises from the ordering of the domains in a spin-ordered ferromagnetic film: $B(z) \approx \mu_0 M$, where M is the saturation magnetisation. As we shall see, this simple relationship is not always valid for thin films showing strong magnetic anisotropy. The classical boundary condition $\nabla B = 0$ requires that we apply the external field H parallel to the interface since it is the difference in $V(z)$ either side of side of an interface which gives rise to refraction.

In Fig. 3c, we show the form of the potential for a magnetic film epitaxed to a single crystal non-magnetic substrate and coated with a thicker film of the substrate metal. Appropriate choice of the overlayer film thickness t_2 induces oscillations in $R^{(\pm)}(q)$ sufficient to enhance $R^+(q)/R^-(q)$ in a region of q where the reflectivity is falling-off sharply [14]. This is illustrated in Fig. 5 showing the variation in flipping ration $F(q)$ for the specific case of a single magnetic film layer $t_3 = 18$ Å and $\mu = 1.7 \, \mu_B$ sandwiched below a non-magnetic film of thickness $t_2 = 42$ Å.

Note that for an uncoated film (curve A, Fig. 5), one observes only a small deviation from unity $\Delta F \approx 0.1$ at $2q_c$ (the negative slope of the curve arising from the specific nature of the phase factor, eqns (4,9) for the single film system). Oscillatory interference behaviour in $R(q)$ is not resolved for film thicknesses less than a few 100 Å [12]. In contrast, if we repeat the calculation for $S(q)$ in the presence of an overlayer coating (Fig. 4c), we observe resonance enhancement $\Delta F(q) \approx 50$ at $2q_c$ (curve B, Fig. 5). That is, the 'sandwich-structure' greatly enhances the spin dependence of the reflectivity, the position of the maximum in $F(q)$ being determined solely by the overlayer film thickness, t_2 [15]. The important point to emerge, however, is that the magnitude of the departure from unity $\Delta F(q)$ is entirely due to the presence of the magnetic layer and provides a direct measure of the absolute magnitude of the magnetic moment per atom μ of the magnetic phase [16].

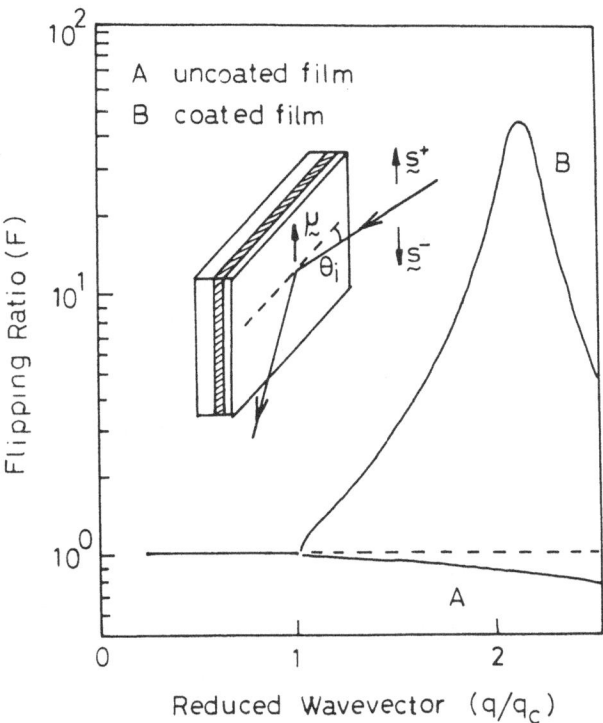

Fig. 5. The "flipping ratio" $F(q) = R^+(q)/R^-(q)$ calculated for: curve A, an 18 Å f.c.c. cobalt film epitaxed to a Cu(001) substrate; curve B, showing the effect of overcoating the 18 Å f.c.c. Co(001) film with 42 Å of f.c.c. Cu(001), assuming $\mu_{bulk} = 1.7 \, \mu_B$. Dashed line shows $F(q) = 1$ and inset diagram the scattering geometry.

ULTRA-THIN f.c.c. Fe and Co FILMS

Fig. 6 shows the spin averaged reflectivity obtained at 4 K for an 18 Å f.c.c. cobalt, film epitaxed to Cu(001) and overcoated with 42 Å of an epitaxially grown Cu(001) film. Correction was made for any background scattering and the incident perpendicular wavevector component was normalised with respect to the critical wavevector for the substrate. Correction was also made for the change is surface reflectivity area with q (proportional to $\sin \theta_i$). The solid and dotted curves were calculated employing the 1-D potential approximation (inset diagram); the dotted curve includes the effect of angular divergence in the neutron beam, $\Delta q \sim 2 \times 10^{-2} q_c$ in Fig. 2. The error bars represent the statistical error for counting times, 30-60 mins. per q value.

The experimental points show small deviations from the theoretical curve indicative of interfacial roughness and film inhomogeneities. Interface roughness produces variations in the interaction potential with the consequence that the spatial average of $\bar{V}(z)$ may be lowered. Nérot and Croce [17] have shown that the reduction in reflectivity may be expressed in terms of the mean square amplitude variations $\langle\zeta\rangle^2$ at an interface: $\zeta = z - \langle z \rangle$, and $r^{obs} = e^{-4q_1q_2\langle\zeta\rangle^2} r^{ideal}$, where r^{ideal} corresponds to the ideal perfectly flat interface (i.e. the spatially averaged step-potential, Fig. 4 and 6). If we imagine the interface to exhibit long wavelength height variations [18] (i.e. macroscopic waveness) in addition to the microscopic roughness ζ, then Δq is effectively increased by scattering from the sample. It is possible that the undulation of the experimental data about the

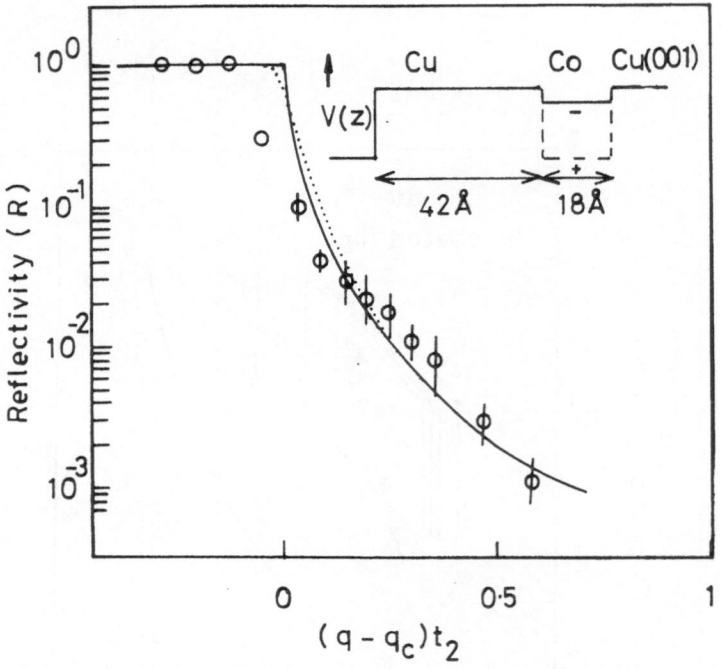

Fig. 6.   Spin-averaged reflectivity R measured at 4 K for 10 ML f.c.c. Co(001)
film incorporated into the heterostructure (inset). Solid and dotted
curves are calculated for $t_2 = 42$ Å and $\Delta q = 0$ (solid), $\Delta q \sim 10^{-3}$ Å^{-1}
(dotted). Increasing statistical error is shown for $q > q_c$.

theoretical curves in the region $q_c < q < 2q_c$ (Fig. 6) is due to variations in V(z) parallel to the
interface boundary [19]. Finally, at small q, the collimation of the beam itself introduces a finite Δq
which causes the rounding of the curve calculated for $\Delta q = 0$ around q_c.

These small variations in R(q) introduced by microroughness factors become second order
when we come to consider the resonance-enhanced spin asymmetry behaviour (Fig. 7). Fig. 7
shows S(q) vs q curves for curves for cobalt films epitaxed to Cu(001) substrates, normalized to
take into account differing magnetic film thicknesses, 1 to 10 monolayers (ML).

The 10 ML (18 Å) thick Co(001) film overcoated with 42 Å of Cu(001) (Fig. 7a) shows data
in good agreement with the 1-D potential calculations (solid and dotted curves, as in Fig. 6),
assuming a moment μ per Co atom of 1.8 ± 0.25 μ_B. This value is in accord with ground state
energy spin-density-functional calculations of bulk phase f.c.c. cobalt [20] assuming that the
Cu(001) substrate strains the Co lattice parameter by ~2.9%. The observatiion of the resonance
enchancement peak structure confirms the validity of the spatially averaged 1-D potential model to
describe the S(q) behaviour.

Fig. 7b shows similar data for a 4 ML (7.2 Å) Co(001) film overcoated with 120 Å of
Cu(001) and epitaxed to a Cu(001) single crystal substrate. The 1-D potential calculation gives best
agreement with the data for an average magnetic moment per Co atom throughout the film,
$\mu = 2.3 \pm 0.3$ μ_B. This value is considerably higher than the bulk phase moment ($\mu_{bulk} = 1.69\mu_B$
[20]) and also the value derived from the 10 ML film. This may be an indication of either d-band
narrowing with reduced thickness or interfacial enhancement at the Cu interface, which is more
manifest in the thinner Co film [21]. Similar behaviour has been observed for 2 ML and 6 ML
thick Co films [16].

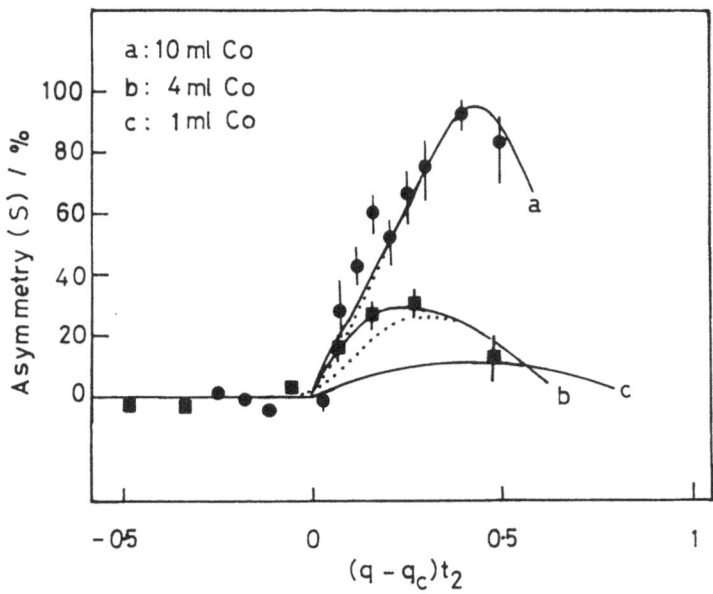

Fig. 7. Spin-asymmetry ratio S observed for f.c.c. Co(001) ferromagnetic films of thickness (a) 10 ML (0), (b) 4 ML ( ), and (c) projection over 1 ML film with overcoating thicknesses $t_2 = 42$ Å;
130 Å, and 75 Å of Cu(001) respectively. Solid and dotted
curves show the effect of angular divergence of the neutron beam,
calculated assuming (a) $\mu = 1.8\ \mu_B$, (b) $\mu = 2.3\ \mu_B$, and (c) $\mu =$
$1.7\ \mu_B$. Any interfacial enhancement in the 1 ML film, $\mu > 1.7\ \mu_B$,
will further enhance the S(q) peak value.

Although measurements have yet to be made for a single ML film, Fig. 7c demonstrates the
feasibility of such a measurement for a Cu(001) overlayer, $t_2 = 75$ Å. Signal-to-noise can be
improved to obtain resonance curve (c) and so derive μ per atom. The principle difficulty is mainly
in the preparation of sufficiently homogeneous films.

The sample surfaces have typically an area of 5 x 10 mm² with a mirror surface finish. They
are prepared by careful epitaxial growth onto single crystal substrates, which have first been characterized using surface analytical techniques, using low-flux (MBE) evaporation sources of the
pure metals. We have conducted an extensive investigation of the growth and lattice structure
characteristics using LEED, Auger and photoemission analysis [22]. The principle results are:
(i) films of cobalt up to 50 ML thickness grow by a layer-by-layer growth mechanism at substrate
temperatures of 300 K; (ii) the films appear to be quite stable and no interdiffusion is observed
below 450 K; (iii) the Cu(001) overlayer films epitax quite readily to the f.c.c. magnetic Co films
up to ~100 Å; (iv) the films investigated to date (2 ML, 4 ML, 6ML and 10 ML) show no fall off
in μ with temperature over a range 4 to 450 K.

This last result is unexpected since conventional spin-wave theory [23] predicts a fall-off in
$|\mu|$ to zero as the film thickness approaches 1 or 2 ML. Also, recent spin-polarized photoemission
measurements [24] on these films conducted in a varying magnetic field up to 1 Tesla [25] have
confirmed that: a) the preferred direction of M_s lies in the plane of the films, and b) similar results
to the spin-polarized neutron reflection results are obtained from cobalt films grown *in situ* in uhv
without a Cu(001) overlayer. That is, the spin-polarised photoemission measurements confirm the
view that the neutron reflection method accurately measures the absolute magnitude of μ in these
ferromagnetic metastable Co films.

The situation is much more complex in the case of epitaxially grown f.c.c. Fe films [25].

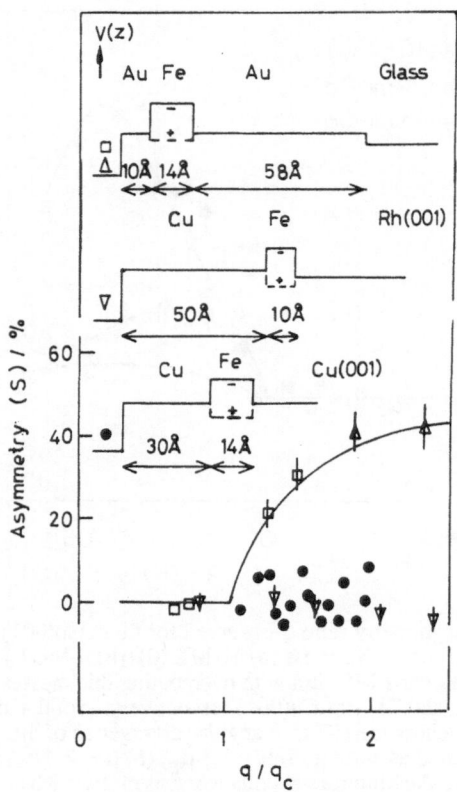

Fig. 8. Spin-asymmetry ratio S observed for thin
films of polycrystalline b.c.c. iron (μ_B =
2.2 μ_B) epitaxed to Au(□,△), and metastable
f.c.c. Fe(001) epitaxed to Cu(001) (0) and
Rh(001) (∇). The respective film thicknesses in relation to V(z) are shown inset.

In Fig. 8 we show data obtained at 300 K for a 12 Å polycrystalline b.c.c. Fe film sandwiched
between polycrystalline Au layers, supported on an optically flat glass substrate. The actual peak
of the resonance curve occurs at a q-value outside the range of our measurements (due to very thin
Au overlayer thickness-inset figure). Nevertheless, the theoretical curve (full line) shows good
agreement for a spatially averaged magnetic moment per Fe atom μ = 2.2 μ_B, the bulk b.c.c.
value, despite the considerable roughness of these polycrystalline films. This behaviour is to be
compared with MBE grown f.c.c. films of similar thickness of Fe on Cu(001) and Rh(001)
overcoated with Cu(001) films of 30 Å and 50 Å respectively (ref. inset diagrams). The
significance of these data is the absence of a spin-dependent reflectivity for the f.c.c. Fe films.

When the f.c.c. Fe(001) films are cooled to 4 K, we still observe negligible magnetism.
However, when the magnetic field strength is increased in the surface plane, then a small S(q)
resonance response appears [26]. This suggests that the f.c.c. Fe(001) films are indeed
ferromagnetic but now with a preferred magnetic anisotropy normal to the film interface.

Recent confirmation of this view has been obtained using spin-polarized photoemission with
H oriented and variable normal to the film plane [27]. Measurements on uncoated f.c.c. Fe(001)
films of thickness 1 ML, 2 ML, 3 ML and 5 ML show quite complex magnetic hysteresis
behaviour dependent on both temperature T and film thickness: firstly, the f.c.c. Fe(001) films are

ferromagnetic for all thicknesses; secondly, the Curie temperature is lowered (<300 K for 1 ML; <400 K for 3 and 5 ML); thirdly, the degree of magnetic remanence is dependent on both the film thickness and temperature; fourthly, while films of thickness greater than 2 ML produce a remanence curve indicative of strong magnetic anisotropy normal to the film plane, the 1 ML films show no remanence, behaving like the Co films with the easy direction of magnetization in the film's plane.

CONCLUDING REMARKS

The important points to emerge to date from these preliminary results are: (i) the neutron reflectivity measurements accurately reflect the magnetic properties of ultrathin films down to 1 ML, unhindered by the non-magnetic metal overlayer, (ii) the magnetic properties show, in some cases, complex behaviour related to magnetic anisotropies which have yet to be understood and do not scale according to simple spin-wave theory, and (iii) surface roughness clearly manifests itself in diffuse scattering but a spatially averaged 1-D potential model accurately predicts the absolute magnitude of the ferromagnetic moments per atom in films of only a few monolayers thickness. The complementary nature of the spin-polarized photoemission results on the 'free standing' films suggests that the neutron method will be particularly useful for characterizing the magnetic properties of multilayer structures. Clearly, preparation technique is very important. In many cases, and particularly with the metastable f.c.c. Fe(001) films, $|\mu|$ is sensitive to lattice strain [20,25]. This implies that controlled MBE techniques in UHV must be used in the future to guarantee reproducibility. The converse of this is that, along with grazing incidence X-ray scattering, the spin-polarized neutron reflectance method will provide a sensitive technique for characterizing the microstructure and strain-related effects of magnetic films and superlattices with new and unusual properties.

ACKNOWLEDGEMENTS

We wish to thank the staff of the I.L.L. Laboratory, Grenoble, France - particularly Prof. O. Schaerpf and Dr H Lauter, for their helpful advice with our measurements. RFW is grateful to Prof. H. C. Siegmann and Dr. F. Meier at ETH, Zurich, for collaborating with them on the spin polarized photoemission measurements. Also, the Science and Engineering Council's Rutherford-Appleton Laboratory for financial support.

REFERENCES

[1]   M. L. Goldberger and F. Seitz, Phys. Rev., 71:294 (1947).
[2]   J. B. Hayter, J. Penfold, and W. G. Williams, J. Phys. E (Sci. Instr.), 11:454 (1978).
[3]   S. Dietrich and H. Wagner, Phys. Rev. Lett., 51:1469 (1983); S. Dietrich and R. Schack, Phys. Rev. Lett., 58:140 (1987).
[4]   C. F. Majkrzak, J. W. Cable, J. Kwo, M. Hong, D. B. McWhan, Y. Yafet, J. V. Waszczak, and C. Vettier, Phys. Rev. Lett., 56:2700 (1986).
[5]   C. Vettier, D. B. McWhan, E. M. Gyorgy, J. Kwo, and B. M. Buntschuh, Phys. Rev. Lett., 56:757 (1986).
[6]   U. Gradman, Review article, this volume.
[7]   Small-angle diffractometer D17, I.L.L., Grenoble, France (1987).
[8]   CRISP instrument at ISIS source, Rutherford-Appleton Lab, U.K. (1987).
[9]   O. Schaerpf, J. Phys. E (Sci. Instr.), 8:269 (1975), and [2].
[10] F. Mezei, in: "Neutron Spin Echo", F. Mezei, ed., Lecture Notes in Physics Vol.128, pp.3-26, Springer-Verlag, Heidelberg (1980).
[11] T. J. L Jones and W. G. Williams, Rutherford-Appleton Lab, Internal Report RL-77-079; G. P. Felcher, R. Felici, R. T. Kampwirth, and K. E. Gray, J. Appl. Phys., 57:3789 (1985).
[12] J. B. Hayter, J. Penfold, and W. G. Williams, Nature, 262:569 (1976).
[13] G. P. Felcher, Phys. Rev. B, 24:1595 (1981).
[14] G. P. Felcher, K. E. Gray, R. T. Kampwirth, and M. B. Brodsky, Physica, 136B:59 (1986).
[15] J. A. C. Bland, D. Pescia, R. F. Willis, and O. Schaerpf, Physica Scripta, in press (1986).
[16] J. A. C. Bland, D. Pescia, and R. F. Willis, Europhysics Letters, in press (1986).
[17] L. Nerot and D. Croce, Rev. Phys. Appl., 15:761 (1980).

[18] A. Steyerl, Z. Physik, 254:169 (1972).
[19] J. A. C. Bland and R. F. Willis, to be published.
[20] P. M. Marcus and N. L. Moruzzi, Sol. State Commun., 55:97 (1985).
[21] C. L. Fu, A. J. Freeman, and T. Oguchi, Phys. Rev. Lett., 54:2700 (1985).
[22] A. Clarke, G. Jennings, and R. F. Willis, to be published.
[23] For recent theoretical work see: H. Hasegawa, J. Phys. F, 16:347 (1986); experimental
         work see: L. R. Sill, M. B. Brodsky, S. Bowen, and H. C. Hamaker, J. Appl.
         Phys., 57:3663 (1985).
[24] D. Pescia, G. Zampieri, M. Stampanoni, G. L. Bona, R. F. Willis, and F. Meier,
         Phys. Rev. Lett., in press (1987).
[25] J. A. C. Bland, D. Pescia, and R. F. Willis, Phys. Rev. Lett., in press (1987).
[26] J. A. C. Bland, W. Schwarzacher, and R. F. Willis, to be published.
[27] D. Pescia, G. Zampieri, M. Stampanoni, G. L. Bona, R. F. Willis, and F. Meier,
         Phys. Rev. Lett., in press (1987).

PROBING SEMICONDUCTOR MQW STRUCTURES BY X-RAY DIFFRACTION

Paul F. Fewster

Philips Research Laboratories
Cross Oak Lane, Redhill, Surrey, U.K.

1. INTRODUCTION

This paper presents a systematic approach to the analysis of Multiple Quantum Well (MQW) structures, most of which can be undertaken on simple X-ray diffraction equipment. The structural parameters of interest are the layer thicknesses and their alloy compositions, and any deviations in these parameters. In an MQW structure, the quantum well width and the composition in the barriers, which gives the barrier height, determine the confined particle energy-states, which can be modified by compositional grading at the interfaces between the wells and the barriers. Therefore knowledge of the alloy composition in the barriers compared to that in the wells is an important parameter. Careful analysis of the X-ray diffraction profiles will give the well width and the compositional grading at the interfaces. To relate these results to that of the required parameters, for example the exciton associated with these transitions between the confined particle states we must first define the probe size. This can be considered as the region over which the X-rays are coherently diffracted. If the X-ray source size, and the slits, etc., are not too large the coherent region parallel to the diffracting planes will be large and the diffraction features will be the sum of the intensities from these regions over the incident beam area projected on the sample.

The structures reported in this paper were all grown by Molecular Beam Epitaxy (MBE), and based on the AlGaAs alloy system:

- $Al_xGa_{(1-x)}As/GaAs$ MQWs

- AlAs/GaAs MQWs

- $AlAs/Al_xGa_{(1-x)}As/GaAs$ 3-layer repeat structure

- $(AlAs)_n (GaAs)_n$ superlattice structures, where n is small

A distinction is made here between Multiple Quantum Well structures and superlattices in that the barrier widths of the latter are thin enough to allow coupling between the wells from overlapping wave-functions, but not in the former.

2. OBTAINING THE BASIC PARAMETERS

The method for obtaining the basic structural parameters will only be given in outline here, for a more detailed account of the method see Fewster[1]. For other closely related approaches see Kervarec et al.[2] and Kervarec[3], and different but comparable approaches see for example[4] and [5]. The method employed involves the determination of:

- The average composition of the MQW.

- Other compositions and thicknesses in the structure.

- The average period.

- The well width, the barrier width, the barrier alloy composition.

- Period fluctuations.

- Interfacial quality

The latter two parameters, although routine in our laboratory, will be discussed in section 4.

The apparatus used to determine these parameters are a powder diffractometer and a double crystal diffractometer, or other high resolution diffractometer, all operated under microcomputer control. The former instrument is used purely as a $\omega-2\theta$ diffractometer to give high intensity for obtaining the high order Fourier coefficients which determine the interface profile. This diffractometer has Bragg–Brentano geometry with a curved graphite analyser crystal. The double–crystal diffractometer operated in the (+ , -) arrangement[6,7], on the other hand is used to obtain an accurate value for the mismatch and hence the alloy composition.

2.1. The Average Composition in the MQW and other Compositions and Thicknesses in the Structures

The geometrical arrangement of the double-crystal diffractometer in the (+ , -) mode reduces wavelength dispersion (i.e. $d\lambda/d\omega \sim 0$), which in turn results in narrow diffraction peaks thus allowing high resolution for accurate measurement of the lattice mismatch. An example of a typical diffraction profile from the MQW structure KLB269 is given in Fig. 1. This profile was obtained with a Ge 1st crystal, <111> orientation and 115 reflection, and a 115 sample reflection from the <001> orientated sample. The MQW appears as a single peak in the profile and so all we can determine is the average alloy composition. The mismatch is related to the angular separation, $\Delta\omega$, of the layer peaks with that of the substrate, by the equation

$$\frac{\Delta a}{a} = \frac{-\Delta\omega\,(1-\nu)}{\cos^2\phi(\tan\theta + \tan\phi)(1+\nu)} \qquad \ldots \quad \ldots \quad \ldots \;(1)$$

where ϕ is the angle between the diffracting and surface planes (for planes tilted towards the incident beam), θ the Bragg angle, and ν is Poisson's ratio to account for the tetragonally distorted layer. This simplistic approach has been shown to be inadequate for determining the mismatch in thin layer structures[1,8], and this will be covered in more detail in section 3. The alloy composition can be derived from the mismatch by Vegard's Law since the ternary $Al_xGa_{(1-x)}As$ exists for the full range of x[9], hence

$$x = \frac{a_{GaAs} \frac{\Delta a}{a}}{a_{AlAs} - a_{GaAs}} \qquad \qquad \text{... ... ... (2)}$$

The compositions in the various layers can be determined if they exhibit well defined peaks in the diffraction profile, although diffraction effects can give misleading results, and for full interpretation of the profile, computer simulations are used not only to confirm the derived compositions but also to determine the various layer thicknesses[1]. The method for simulating the diffraction profile is given in section 3. The measured average composition in the MQW for the sample KLB269 was found to be 0.302.

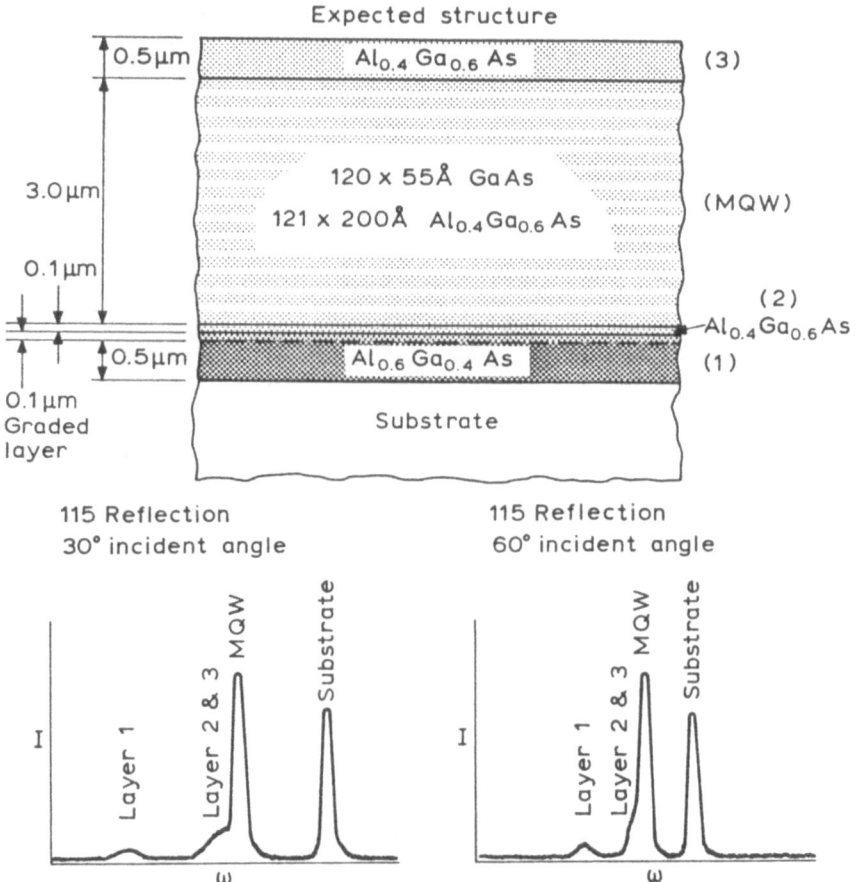

Fig. 1.   The structure and 115 high resolution diffraction profiles for
sample KLB 269

2.2. The Average Period

A structure that has an additional periodicity apart from the normal lattice repeat will result in a series of subsidiary maxima closely associated with the main Bragg maxima in the diffraction pattern and will pass through these maxima on a line parallel to the modulation direction. This has been shown most simply by Guinier[10], by transforming the distribution and scattering power of the atoms to obtain the diffracted amplitudes, and by de Fontaine[11], using Bessel functions. The result of this is that the

diffraction pattern now has a series of satellite maxima that are separated
by the inverse of the modulation wavelength in reciprocal space and by

$$\bar{\Lambda} = \frac{(L_1 - L_2)\,\lambda}{2(\sin\theta_1 - \sin\theta_2)} \qquad \dots \ \dots \ \dots \ (3)$$

in diffraction space, where λ is the wavelength of the radiation and θ_1
and θ_2 are the angles at which the L_1 and L_2 satellite orders diffract.
We can consider then, that satellites associated with the Bragg maxima are
directly related to the Fourier coefficients of the modulation waveform
resulting from this additional periodicity. For example a sinusoidal
wave will have just the first order coefficients and therefore only the
1st order satellites. If however the interface between the layers is
abrupt, resulting in a rectangular waveform, then there should be a high
number of satellites with the highest order satellites giving the interface
detail and the lower order satellites giving the overall shape of the
waveform.

We must remember that in measuring the period that the diffracting
volume is large, and we are averaging over this region. Because of the
large lateral coherence length the X-ray beam determines the averge
interface positions for determining the period. The double-crystal
diffractometer or the powder diffractometer can be used for measuring the
satellite positions, the choice we make depends on the ease of locating
the peak positions and the number of observable satellites, and generally
the powder diffractometer is favoured. A diffraction profile showing the
satellites around the 002 reflection for the structure KLB269 is given in
Fig. 2. The full dynamic range is not apparent in this figure, but 17
satellites were measured and the period was determined as 276.3Å, with a
standard deviation of 0.57Å, from equation 3.

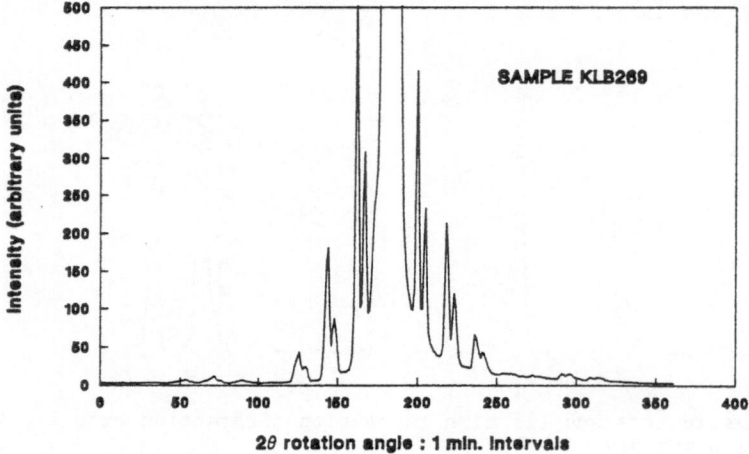

Fig. 2. A diffraction profile obtained with a powder diffractometer in
the vicinity of the 002 Bragg reflection for sample KLB 269

2.3. Well and Barrier Widths and the Composition in the Latter

To obtain this additional information the integrated intensities of
the satellites need to be known, and the powder diffractometer is more than
adequate for this purpose. As with most X-ray diffraction experiments the
phase information is not directly obtainable and we need to propose a model
for the structure and compare calculation with experiment. To do this

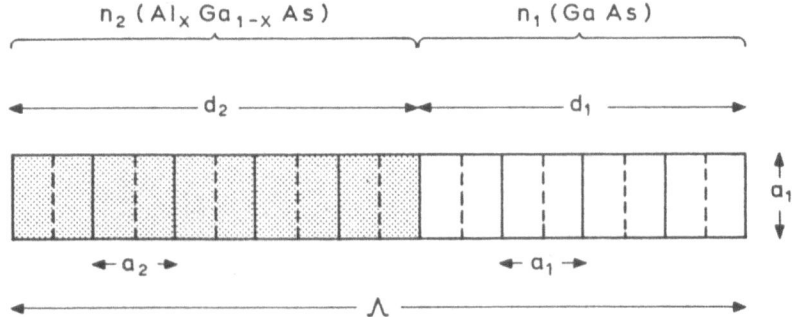

Fig. 3. The large unit cell (corresponding to the period) for calculating the structure factors of the satellites

consider the unit cell for the MQW structure to be the period, Fig. 3. We approximate in the first instance that the period is commensurate with the lattice periodicity, and that the lattice parameters in the plane of the interface are that of the substrate, that is the interfaces are all coherent and the density of misfit dislocations is small, then

$$\Lambda = d_1 + d_2 \qquad \text{... ... ... (4)}$$

and

$$x = \frac{\bar{x}\Lambda}{d_2} \qquad \text{... ... ... (5)}$$

Clearly since Λ and $\bar{x}$ are known then only one parameter need be determined to obtain the well width, barrier width and composition in the barrier. The structure factor of the Lth subsidiary maxima for this large unit cell is given by

$$F_c = \sum_{j=1}^{n_1+n_2} \left[f_{GaAl} \exp 2\pi iLz_j + f_{As} \exp 2\pi iLz_j \right] \qquad \text{... ... (6)}$$

where f and z_j represent the scattering factor and fractional position of the jth atom in this large unit cell. It is now a simple matter of calculating the structure factor for various well widths, and hence barrier widths and barrier compositions and comparing with the value, F_O, derived from the integrated intensities. The "observed" structure factor, F_O, is equal to the square-root of the integrated intensity after systematic effects have been taken into account. This assumption is perfectly valid since the satellite intensities have very low reflectivities. The satellite F_O values are scaled to the F_c values by the ratio of their sums, and the main 002 Bragg peak is not included in the calculation since the observed intensity saturates the counter and is not easily separated from the substrate and cladding diffraction peaks. The agreement of the calculated to the "observed" structure factors can be monitored as the well width is changed and for the structure KLB269, the variation is given in Fig. 4. This also gives an indication of the sensitivity of the method. The best-fit values, that give the minimum R-factor (which relates to the level of agreement) are listed in Table 1.

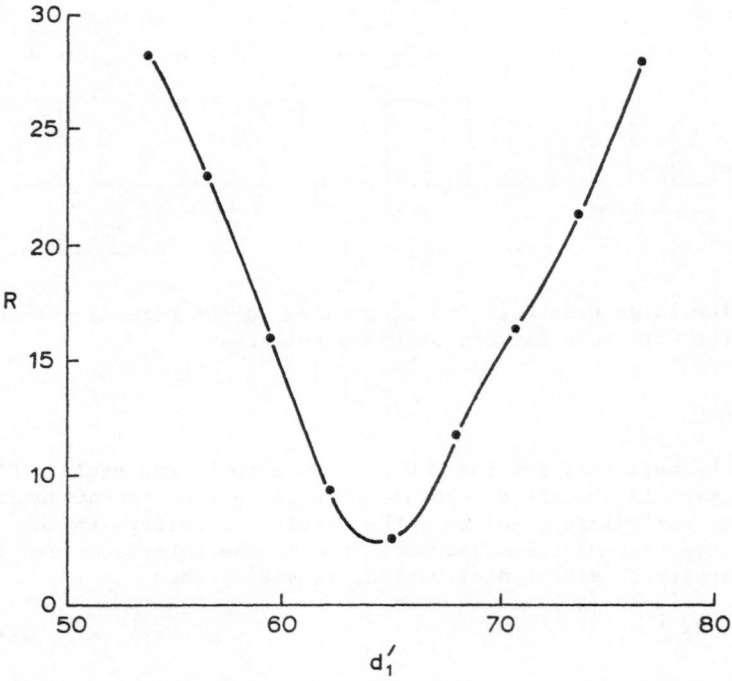

Plot R-factor versus d_1

Fig. 4. The variation of the R-factor with well width, plotted at 1-monolayer intervals

TABLE 1

h	k	l	F_O	F_C
0	0	90	51	38
0	0	91	100	117
0	0	92	139	151
0	0	93	103	105
0	0	94	31	32
0	0	95	258	239
0	0	96	483	471
0	0	97	681	688
0	0	98	(not measured)	(002) main peak
0	0	99	535	551
0	0	100	468	444
0	0	101	269	253
0	0	102	74	57
0	0	103	64	85
0	0	104	155	144
0	0	105	120	122
0	0	106	24	50
0	0	107	44	29

$$R = \frac{\sum \left| |F_o| - |F_c| \right|}{\sum |F_o|} = 6.4\%$$

Up to now we have assumed that the period is commensurate with the lattice parameter, and to work in average thicknesses, which are not integer numbers of monolayers, we need to rescale the well width to the minimum in the R-factor curve, d_1, and to the average period with the following equations

$$\bar{d}_1 = d_1' \left(1 + \frac{\bar{\Lambda} - \Lambda}{\Lambda} \right) \qquad \ldots \quad \ldots \quad \ldots \quad (7)$$

hence

$$\bar{d}_2 = \bar{\Lambda} - \bar{d}_1 \quad \text{and} \quad x = \frac{\bar{x}\bar{\Lambda}}{\bar{d}_2} \qquad \ldots \quad \ldots \quad \ldots \quad (8)$$

The final values are all given in Table 2.

TABLE 2

Final parameters for KLB 269

$\bar{\Lambda} = 276.3\ (0.57)\text{Å} \qquad \bar{d}_1 = 63.9\text{Å} \qquad \bar{d}_2 = 212.4\text{Å}$

x (in barriers)   = 0.395

x (in inner cladding)   = 0.38

x (in outer cladding)   = 0.69

2.4. Five-period MQW Structure

This approach clearly works for this relatively thick structure. To illustrate that thinner structures can easily be accommodated, an example of a 5-period MQW is given which has been analysed whose well widths were nominally 25Å and the total MQW thickness was just 1000Å. A total of 10 satellites could be observed with the powder diffractometer and a good fit to the experimental data was found, Fig. 5. To indicate the usefulness of a powder diffractometer for this work consider the −5 satellite, then assuming that the structure is perfect, the calculated peak intensity is 1 part in 100,000,000 of the incident intensity. With this experimental method we can measure these weak satellites and its profile using 5 seconds count-time per step!

2.5. Comparison with Photoluminescence Excitation Spectroscopy

Many of the structures analysed in this way have also had their well widths determined by Photo-Luminescence Excitation spectroscopy (PLE), Dawson et al[12]. Considering that the PLE probe is the exciton (whose diameter is approximately 150Å), the agreement with the well width determined by X-ray diffraction (whose probe size is several orders of magnitude larger) is remarkable. A list of results is given in Table 3 and a more detailed analysis of these results will be given elsewhere, Orton et al[13].

TABLE 3:   Comparison of Well Width by XRD and PLE

Sample	PLE	XRD
G50	24.5	25.7
G51	74.5	73.4
G52	146	149.1
G53	54	53
G54	95	91.8
G55	112	112.5
G56	25	33.9
G67	54	56
G126	76	76.3
G127	80	82.9
G143	58	64.8
G144	90	93.2
KLB219	66	70.6
KLB222	56	56.5
KLB262	51	56.5
KLB269	59	63.9

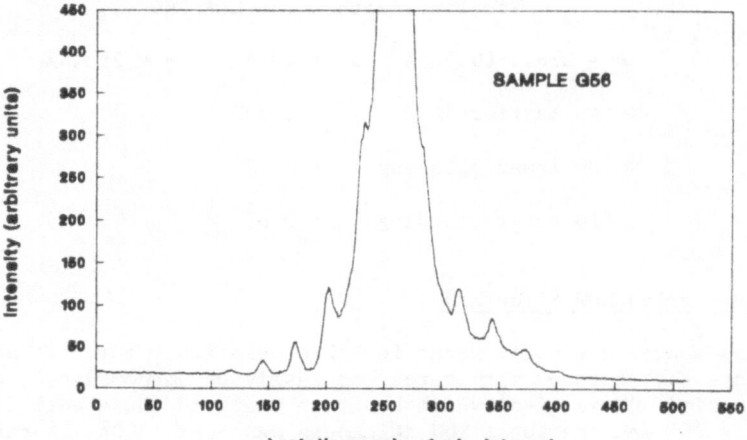

rotation angle : 1 min. intervals

STRUCTURE FACTOR LISTING
AND DERIVED PARAMETERS

L	Fo	Fc
60	377	151
61	733	630
62	1171	1154
63	1636	1642
64	1961	2049
65	not meas.	11833
66	1761	1775
67	1359	1560
68	1116	1147
69	640	657
70	199	189

$d(GaAs) = 33.9 \text{Å}$ $d(AlGaAs) = 150.0 \text{Å}$ $x = 0.329$ $R = 6.5\%$

Fig. 5.   The structure and the diffraction pattern in the vicinity of the
002 reflection, and the satellite structure factor listing with
final derived parameters for G56

424

3. DETERMINATION OF THE MISMATCH

So far we have assumed that the mismatch can be determined from the separation of the peaks in the high resolution diffraction profile. This is only reliable for relatively thick layers (> 1 micron for the 004 reflection), therefore the diffraction profile should be fully interpreted by profile fitting. The method used here is the solution of the Takagi-Taupin equations, (Takagi[14]; Taupin[15]), which are two differential equations relating the change in diffracted and incident amplitudes (D_H and D_O) with depth (z) into the crystal. This is a full dynamical scattering theory in the 2-beam approximation, taking into account crystal distortions. The 2-beam approximation is valid because the radius of the Ewald sphere is of the order of the reciprocal lattice vectors and would require a determined effort to achieve multiple-beam conditions.

$$\frac{i\lambda\gamma_H}{\pi}\frac{dD_H}{dz} = \psi_O D_H + C\psi_H D_O - \alpha_H D_H \qquad \dots \quad \dots \quad \dots \quad (9)$$

and

$$\frac{i\lambda\gamma_O}{\pi}\frac{dD_O}{dz} = \psi_O D_O + C\psi_{\bar{H}} D_H \qquad \dots \quad \dots \quad \dots \quad (10)$$

where $\alpha(\omega)$ is a distortion parameter for each layer ($= -2\lambda(\theta-\theta_O)\cos\theta_O/d$, where θ_O is the Bragg angle and d the interatomic plane spacing), C the polarisation factor (1 or $|\cos2\theta|$) and ψ is related to the structure factor F, electron radius r_e and unit cell volume V ($= -\lambda r_e F/(\pi V)$).

Equations 5 and 6 can now be combined by defining an amplitude ratio X ($= D_H/D_O$), and by integrating this expression (this is a standard integral), we obtain the amplitude ratio X(Z) with respect to an amplitude ratio at a different depth into the crystal X(z)

$$X(Z,\omega) = \frac{SX(z,\omega) + i(E + BX(z,\omega))\tan GS(z-Z)}{S - i(AX(z,\omega) + B)\tan GS(z-Z)} \qquad \dots \quad (11)$$

where
$$A = -\psi_{\bar{H}}|\gamma_H|/\gamma_O$$
$$B = 0.5\left[\psi_O\left(1 - |\gamma_H|/\gamma_O\right) - \alpha_H(\omega)\right]$$
$$G = -\pi/\left(\lambda|\gamma_H|\right)$$
$$E = \psi_H, \quad \gamma_O = \underline{n}\cdot\underline{s}_O = \sin(\theta-\phi), \quad \gamma_H = \underline{n}\cdot\underline{s}_H = \sin(\theta+\phi)$$

To actually calculate a diffraction profile we only need to establish a starting amplitude ratio, and progress up through the crystal layers (or regions of constant F and lattice parameter, a), and use the amplitude ratio at the top of one layer as the start value for the bottom of the next. Halliwell, Juler and Norman[16] have used this approach to successfully predict diffraction profiles from graded layers. The initial value for the amplitude ratio is taken as zero, for a region deep inside the substrate, (i.e. we assume that a negligible proportion of the X-ray beam is reflected out from deep inside the crystal), from this we can obtain the amplitude ratio at the substrate surface which is the same as that for the bottom of the first layer.

This procedure is clearly repetitive and requires calculation for each step in the crystal rotation angle, ω, for the angular range of the experiment. The required amplitude ratio is that for the top surface $X_O(\omega)$ and to obtain the reflectivity ratio $P(\omega)$

$$P(\omega)_{calc} = X_o(\omega) \, X_o^*(\omega) \, \frac{|\gamma_H|}{|\gamma_o|} \qquad \ldots \quad \ldots \quad \ldots \quad (12)$$

since we are dealing in complex quantities. Having now obtained this expression for the reflectivity, and convoluted this profile with the instrument function (e.g.: first crystal reflectivity profile in the double-crystal case), we can compare this directly with the experimental reflectivity, I(diffracted at ω)/I(incident).

To illustrate the importance of simulating diffraction profiles, three examples are given.

3.1. A Single Thin Layer on a Substrate

The actual sample studied was G56, described in section 2.4, which included 5 narrow quantum wells, but the effect is similar to modelling a thin single layer 0.38 microns thick. The experimental diffraction profile is given in Fig. 6 and taking the simple approach (equation 1) the alloy composition appears to be x = 0.30. If now the profile is simulated assuming the alloy composition to be x = 0.30, then the profile will appear as in Fig. 6 (dotted line) and the agreement in peak positions is poor. The experimentally determined substrate profile is broadened because the substrate is bent. The best fit to the experimental profile was obtained when the alloy composition was equal to 0.33, Fig. 6, (dashed line). The diffraction peak for a thin layer on a substrate is therefore "pulled" towards the substrate peak as the layer is reduced in thickness.

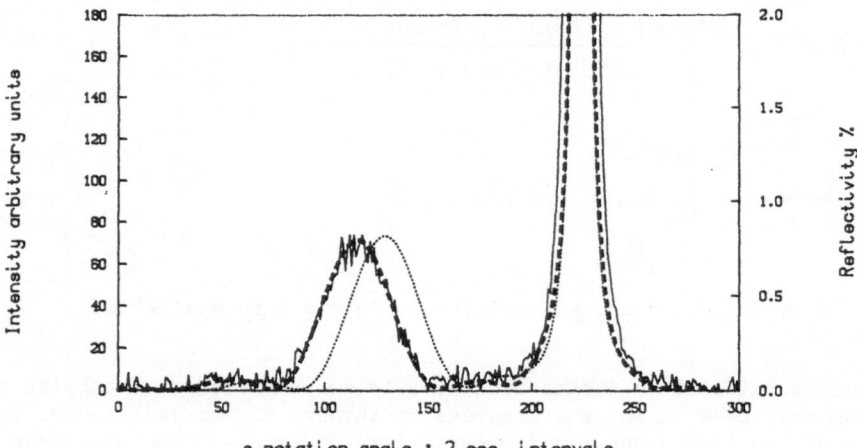

Fig. 6. The experimental diffraction profile, obtained with a double-crystal diffractometer with an asymmetric 115 reflection (solid line), compared with a simulated profile with the composition derived from the experimental profile (dotted line), and the best fit profile with a 10% increase in the composition value (broken line)

3.2. A Thin Buried Layer

Sample D355, a AlAs/GaAs superlattice structure 0.17 microns thick and cladded in Al(0.67) Ga(0.33)As 0.1 microns thick top and bottom, described in section 7, gave a 004 diffraction profile shown in Fig. 7. Again if we deduced the mismatch and hence the alloy composition according to equation 1, then we obtain x = 0.51. Using this value and simulating the profile we have a poor fit, and this time we have overestimated the composition and the best fit value is for x = 0.45, Fig. 7. This time the layer moves away from the susbtrate peak as its thickness is reduced, another example of this was given by Fewster[1], in a more complicated structure.

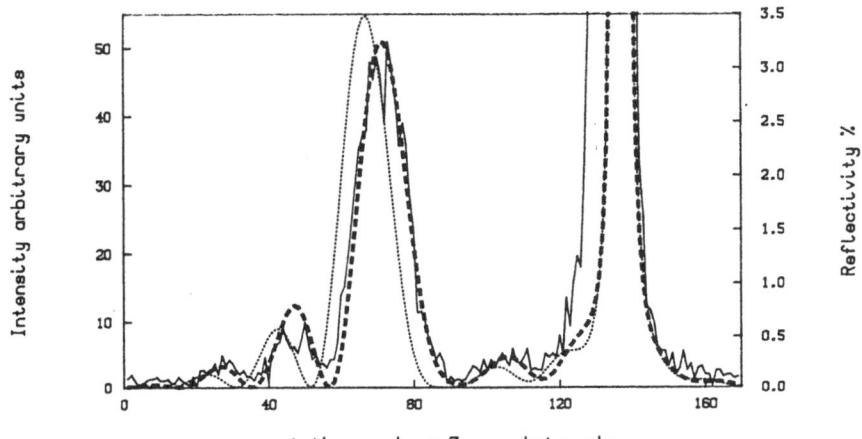

Fig. 7. The diffraction profile for a thin buried layer cladded with very thin layers, 5-crystal diffractometer and 004 reflection, (solid line), compared with simulated profiles using the derived composition (dotted line), and a composition 13% less than this value which gave the best fit (broken line)

3.3. Cladding Regions with a Thin Layer between

Tanner and Hill[17] and Fewster[1] have given examples of the peak splitting by separating two regions of the same composition by a thin layer, and this can lead to incorrect interpretation for the unwary. An example, which was given in the latter article, is given in Fig. 8, and the sample analysis would suggest that the two cladding regions have different compositions which is clearly the wrong interpretation.

These few examples should convince all those working with similar structures that profile simulation is the only way to ensure correct interpetation. Further examples are given by: Halliwell, Lyons and Hill[18], Fukuhara and Takano[19], and a more detailed investigation of the former two structural types is given by Fewster and Curling[8].

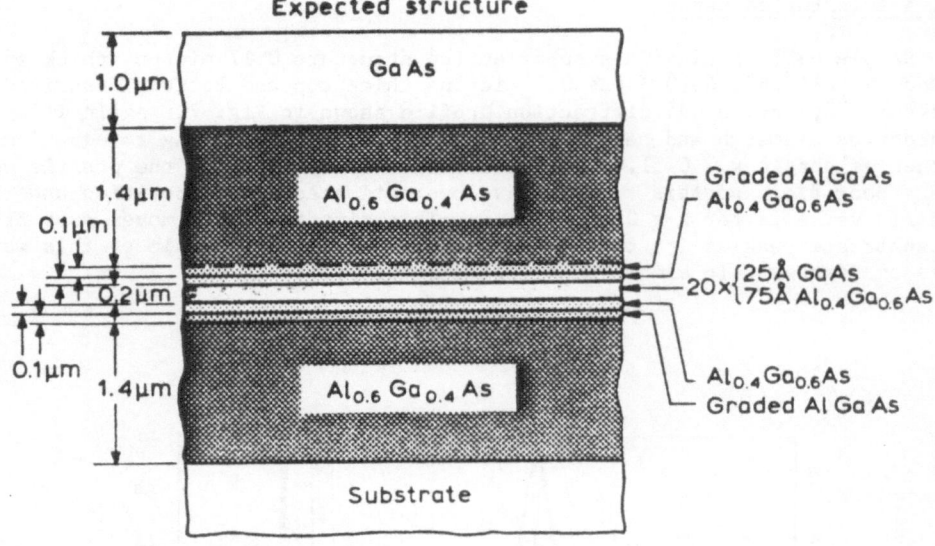

Expected structure

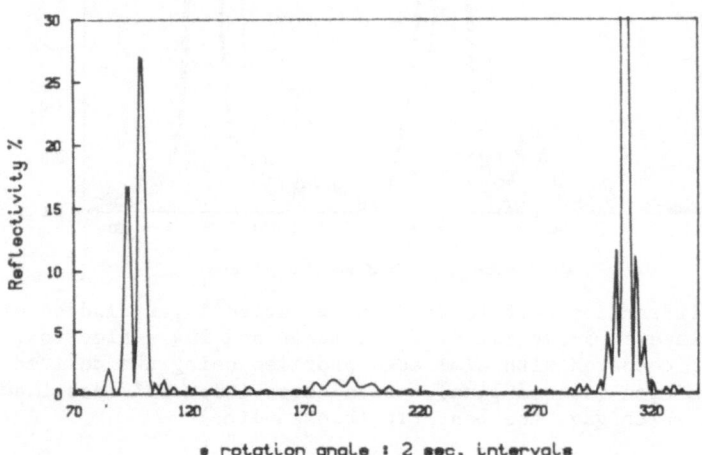

• rotation angle : 2 sec. intervals

Fig. 8. The structure and its 115 diffraction profile showing the split peak
arising from the interference between the two separated cladding
regions

4. STRUCTURAL IMPERFECTIONS

4.1. The Method

Suppose the discrepancy index, or R-factor is larger than we would
expect experimentally, then the simple model for determining the thicknesses
and compositions in the MQW is not describing the full structure. The
experimental R-factor relates to the intensity of the measured satellite
reflections:

$$R_{ex} = \frac{\sum (I_{meas})^{1/4}}{\sum (I_{meas})^{1/2}} \qquad \dots \quad \dots \quad \dots \quad (13)$$

Up until now we have assumed the interfaces to be abrupt, and the period to be constant throughout the structure. If we relax these conditions then the agreement improves dramatically. Interfacial roughness is likely to be present during growth since the shutters of the MBE growth equipment would have to be operated exactly at the point when sufficient material for a monolayer coverage has been delivered, and also the growth must be such that there can be no material carried over from one monolayer to the next. These stringent conditions are very unlikely to be fulfilled, and these effects we shall call interface roughness, and structurally will appear as island growth or incomplete regions in the layer. Interfacial grading on the other hand arises from intermixing of group III atoms across the interface, either during or after growth, but is clearly on a more microscopic scale.

The influence of these two effects on the diffraction pattern, are identical since the lateral coherence length for the X-ray beam will be larger than the growth islands or incomplete regions of the layer. Within the coherence length of the X-ray beam, the diffracted beam will be that from the average structure. We can now consider it in terms of a degrading of the interface with the higher order Fourier coefficients becoming smaller, that is the higher order satellite intensities in general becoming weaker.

Fluctuations in the period (the combined well and barrier thicknesses), will broaden the satellites with increasing order[1,20]. This broadening does not arise from interface grading as suggested by[21]. This can be considered in a simple way, Fewster[1], by assuming that each region of the structure with different periods contribute to different parts of each satellite. The longer the period the shorter the distance between the satellites, and the shorter the period the larger the distance between the satellites, therefore since the satellites are associated with each Bragg peak for the average structure this period dispersion will progressively broaden the satellites further from the Bragg peaks. Again the X-rays average laterally and therefore this period variation measured by X-ray diffraction is the period fluctuation along the modulation direction.

To model the interfaces we follow this general procedure:

- Introduce minimum roughness, < monolayer, to account for the incommensurability in the structure.

- Include period variations.

- Vary the grading of both interfaces.

- Monitor

$$R = \frac{\sum ||F_o| - |F_c||}{\sum |F_o|}$$ ... ... (14)

Having finally found a good fitting model for the structure we could do a Fourier transform, but in general the final resolution of this method is limited from series termination effects and in the assumptions that both interfaces have the same amount of roughness/grading. The Fourier transform approach for obtaining the interface shape directly, Clarke[22], Fleming et al[23], would rely heavily on the correct assignment of phases, and the extent to which the predicted model would tolerate deviations would be unknown. In the case of a non-centrosymmetric distribution arising from two different interfaces, additional phase information for the sine as well as the cosine functions in the Fourier expansion would have to be determined.

4.2. An Example of the Method

An example of this procedure will now be given. The structure and the diffraction profile close to the 002 reflection for the sample G55 is given in Fig. 9. The R-factor remained high at 18.7%, assuming the interfaces to be abrupt. Each satellite was isolated in the diffraction profile, and after the background intensity was removed, the $K\alpha_2$ was stripped using a standard algorithm, Rachinger[24], then the satellite profiles were fitted to a Gaussian, Fig. 10(a). The Full Width at Half Maximum (FWHM) of the Gaussian profiles, for each of the satellites were then plotted against the satellite order, Fig. 10(b), and the best-fit lines were determined by linear regression for the positive and negative orders to determine the average extrapolated breadth of the "zeroth" order profile. This "zeroth" order breadth corresponds to a convolution of the intrinsic crystal

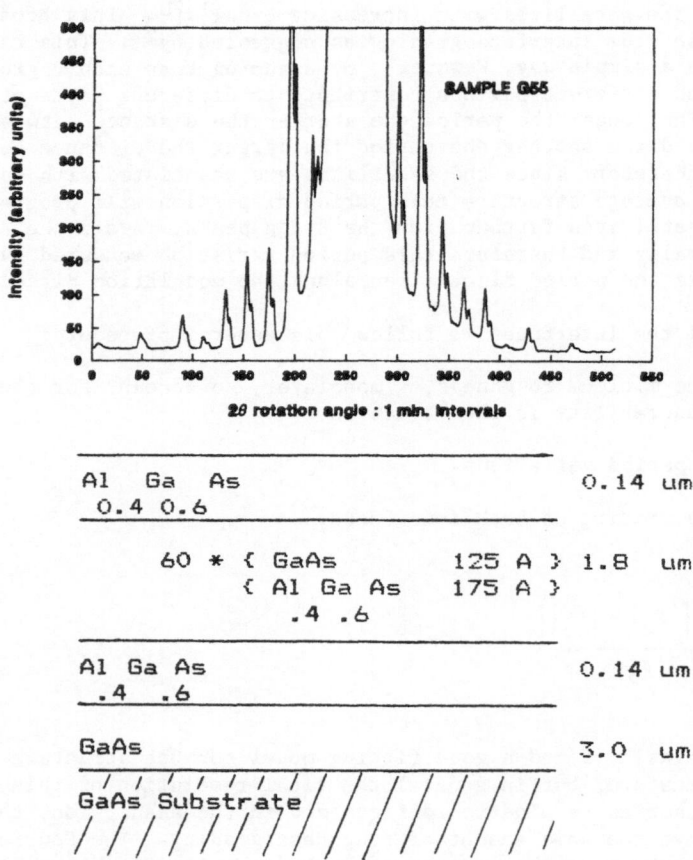

Fig. 9. The structure and diffraction profile close to the 002 Bragg reflection for sample G55

430

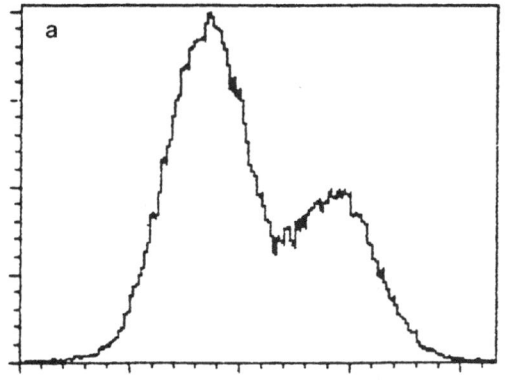
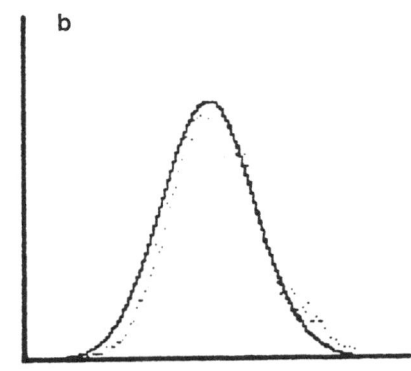

SATELLITE WIDTH VERSUS ORDER

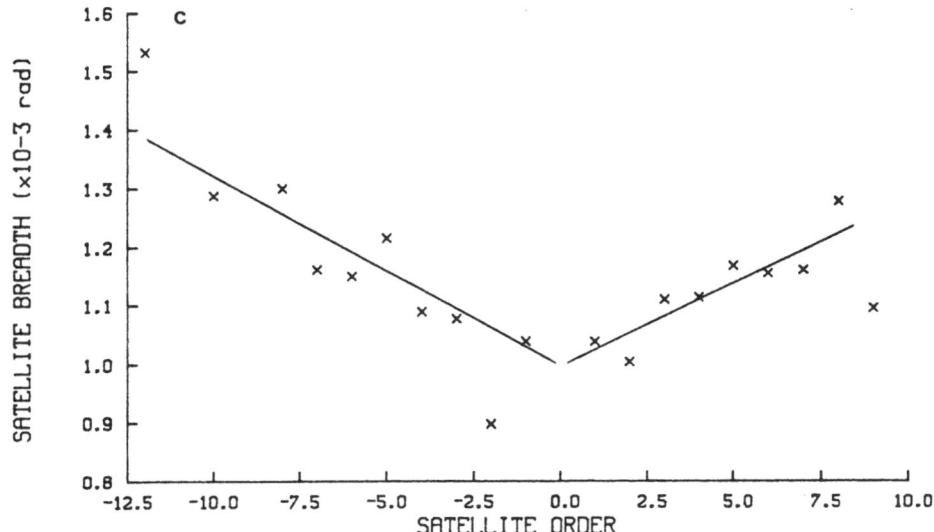

Fig. 10. a) The satellite profile from the powder diffractometer after background removal showing the $K\alpha_1$-$K\alpha_2$ doublet, and b) after α_2 stripping and fitting to a Gaussian distribution, c) gives the plot of the satellite breadth as a function of order for deriving the period variations in the MQW structure of G55

reflection breadth and the instrumental function, and is deconvoluted from each of the satellite breadths to give the real satellite integral breadths with order. The period variation can the be calculated from any satellite since their profiles should now represent a distribution of periods from the relation, Fewster[25].

$$\Delta \Lambda = \frac{\pi \lambda \Delta \omega_{1/2}}{\cos\theta (\Delta\theta)^2} \qquad \dots \quad \dots \quad \dots \quad (15)$$

where $\Delta\theta$ is the angular separation of the satellite from the Bragg peak and $\Delta\omega_{\frac{1}{2}}$ the FWHM of the satellite. The period variation in terms of the FWHM

431

of the Gaussian distribution of periods, Fewster[1] for the sample G55 was found to be 0.0172Å with an estimated standard deviation of 0.0003Å.

To model the interfaces which have some roughness we need only have two well width sizes, both with an integer number of monolayers straddling the average value. Then for each possible period, in this case there are only two, we introduce a similar procedure, resulting in the calculation of the satellite structure factors for 4 of the large unit cells and adding their values coherently in porportion to their distribution. If the period variation was larger (>1 monolayer), then more than two components of the Gaussian distribution would have to be included.

The interface grade is now introduced to the calculation, and the R-factor monitored. The interface is defined as being on an Arsenic plane. The variation in R-factor with interface grade, varying both the ternary to binary and binary to ternary interfaces independently is given in Fig. 11. The minimum in the three-dimensional plot corresponds to the width of the two interfaces, and these are given in Table 4, with the structure factor listings for the observed and calculated values, assuming abrupt and graded interfaces. The first thing to note is that the R-factor has dropped dramatically by introducing graded interfaces, and that by introducing this grade the higher order satellite structure factors are on average reduced. Also that the two interfaces have different grades. The interface which has the largest grade cannot simply be determined because the original choice of the large unit cell origin was arbitarily chosen, but this can be inferred from the following section.

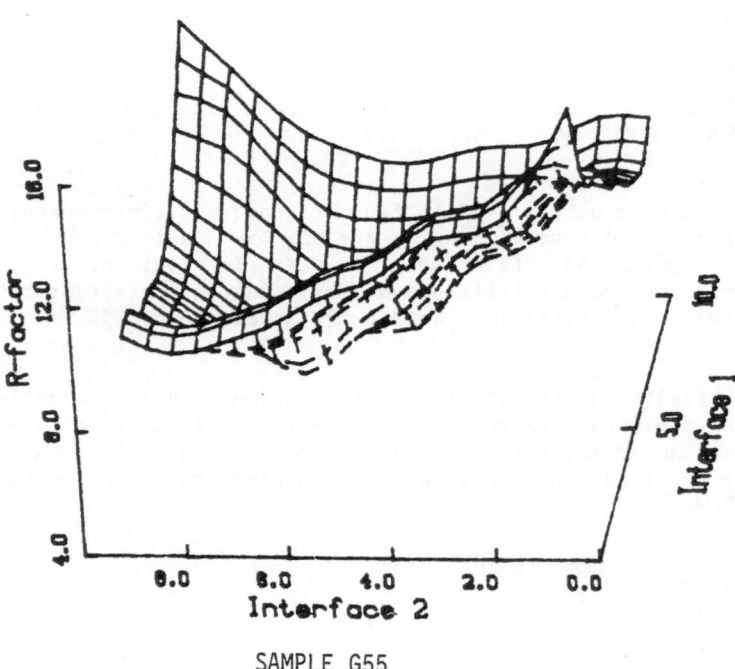

SAMPLE G55

Fig. 11. The 3-dimensional plot of the R-factor (in %) as a function of the two interface widths (in monolayers)

TABLE 4

Satellite Order	Observed F_O	Abrupt Interface F_C	Graded Interface F_C
-12	116	304	103
-11	23	7	34
-10	213	361	191
- 9	23	121	83
- 8	290	410	294
- 7	163	293	238
- 6	392	443	394
- 5	475	566	544
- 4	484	455	465
- 3	1168	1127	1209
- 2	522	415	460
- 1	4507	3773	4269
0	not measured	9779	11140
1	3659	3193	3614
2	571	605	671
3	1016	989	1061
4	567	527	539
5	395	471	452
6	405	468	415
7	147	217	177
8	302	406	292
9	40	62	50
10	193	340	180
11	65	41	37
12	128	272	92

R = 18.7%   6.58%

PERIOD VARIATION   = 0.02Å

INTERFACE 1   = 4.28 Monolayers

INTERFACE 2   = 6.49 Monolayers

5. GRADING IN AlAs/GaAs MQW STRUCTURES

5.1. Results on AlAs/GaAs MQW Structures

A series of 5 MQW structures with 60 GaAs wells and AlAs barriers have been studied with the aforemention method. The AlAs barriers were all approximately 65Å wide and the well width in each was different, and the MQW was cladded top (0.1 microns) and bottom (1.01 microns) with GaAs. The R-factor plots for the different interface grades all show a principal minimum, indicating the best fit to the graded interfaces. The quantitative information is given in Table. 5.

The well widths range from 16 to 65Å and the periods are all about 100Å. The period variation, the FWHM of the Gaussian distribution, was in all cases less than 0.5Å indicating that once set very good control over the growth rate was maintained. The R-factors calculated for an abrupt interface were larger than would be expected from uncertainties in the data, but on introduction of the grading the R-factors all reduced to values comparable to the experimentally determined R-factors, equation 13.

TABLE 5

	Angstroms			Monolayers			
Sample	well width	period	$\Delta\omega_{1/2}$	Interf1 width	Interf2 width	R abrupt	R grade
G167	46.9	112.5	0.07	4.51	4.47	14.6%	3.1%
G168	16.0	80.1	0.29	2.79	4.58	34.3%	9.6%
G169	27.9	95.7	0.43	4.41	4.37	32.9%	7.1%
G170	23.2	92.9	0.42	3.08	4.64	29.5%	8.2%
G171	65.1	138.6	0.31	4.55	3.09	31.9%	4.9%

In all the samples listed the interfacial grading is of the order of 4 monolayers for both interfaces, what is meant by this we shall define in the next subsection.

5.2 Definition of Interface Grade

We have established that grading and roughness are indistinguishable, in this subsection the actual grade included in the calculation will be defined. This is best described in Fig. 12, which is the example for an interface with 4 monolayers of grade. This is assuming a linear variation in Al concentration with depth, and the interface extent from pure Ga monolayers to pure Al monolayers takes 4 monolayer steps. We can see from this that there are only 2 monolayers with significant intermixing and the monolayers adjacent to these have less than 13% of the other atomic species.

GRADING DEFINITION

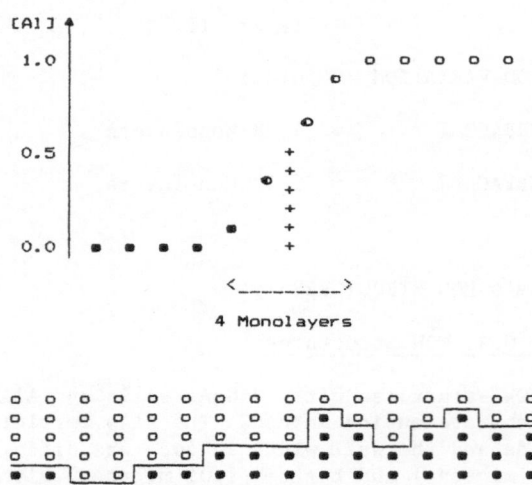

4 Monolayers

Fig. 12. The representation of a 4-monolayer interface grade, showing that there is only significant mixing of the group III species on two atomic planes

Therefore an interface of 4 monolayers extent represents a fairly sharp interface, considering we are averaging a large sample area! To equate this grading to a distribution, FWHM value, we need to halve these values (i.e. 2 monolayers) since the grade is linear and the subsequent distribution is a triangular function. To emphasise the meaning of the quoted figures, it is difficult to visualise how these interfaces can be improved unless the "molecular beam" is switched when an exact monolayer is delivered to a previously perfect surface, which would only be possible with in-situ RHEED monitoring. However with sample rotation, which is essential for good layer uniformity, this is not possible.

5.3. Comparison of AlAs/GaAs and AlGaAs/GaAs MQW Structures

It is interesting at this stage to compare these results with those for the AlGaAs/GaAs MQW structure G55. This analysis as mentioned previously cannot distinguish between the two interfaces in the structure, but we can perhaps obtain information by comparing the results quoted here. We have established that for the binary AlAs-binary GaAs the interfaces are very similar regardless of which of these layers was grown first, and the value of the grade was approximately 4 monolayers. But for the ternary AlGaAs-binary GaAs interface there is a difference, suggesting that the roughness/ grading is sensitive to which layer is grown first, and the value of these grades were approximately 4 and 6 monolayers. Since the roughness/grading is determined by the underlying layer and that sample G55 is typical, then the GaAs (binary)-AlGaAs (ternary) interface corresponds to the 4 monolayer grade, and the AlGaAs (ternary)-GaAs (binary) interface corresponds to the 6 monolayer grade. This idea that the top of the AlGaAs is rougher than that of the GaAs would make sense on the assumption that a ternary has more degrees of freedom during growth and therefore likely to have a rougher surface.

This comparison of binary/binary and ternary/binary interface widths is being studied and collated on a range of samples to ascertain the above observation, Fewster and Curling[26].

6. A 3-LAYER REPEAT STRUCTURE

To illustrate further the applicability of the method for determining thicknesses and compositions in periodic structures, an example of a 3-layer repeat MQW, shown in Fig. 13, will be given. With this particular structure there are 3 thicknesses and 1 composition to be determined. The analysis is very similar to that of section 2.3, in that the following relations can be written:

$$\bar{\Lambda} = d_1 + d_2 + d_3$$

and

$$\bar{x} = \frac{d_1 + d_2 x}{\bar{\Lambda}}$$

As described in previous sections, $\bar{\Lambda}$ and $\bar{x}$ can be determined, therefore there are 4 unknowns and 2 equations. This simply means that there are two parameters to be determined, with the added knowledge that $0 < x < 1$. The calculation for this has been carried out by monitoring the R-factor as these two parameters are varied, Fig. 14. The final determined dimensions are given in Fig. 13.

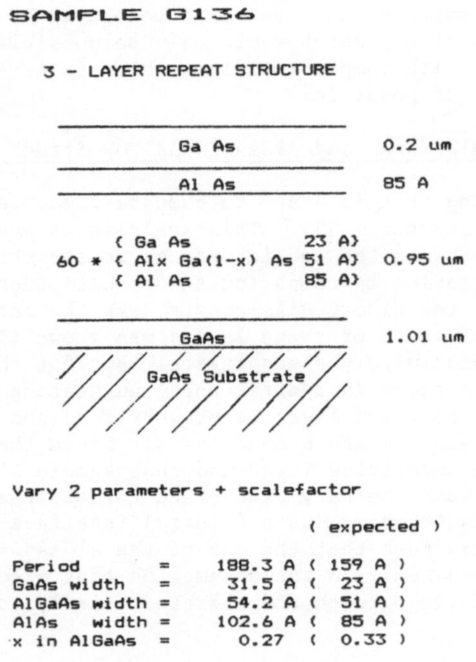

SAMPLE G136
3 - LAYER REPEAT STRUCTURE
Ga As 0.2 um
Al As 85 A
(Ga As 23 A)
60 * (Alx Ga(1-x) As 51 A) 0.95 um
(Al As 85 A)
GaAs 1.01 um
GaAs Substrate
Vary 2 parameters + scalefactor
(expected)
Period = 188.3 A (159 A)
GaAs width = 31.5 A (23 A)
AlGaAs width = 54.2 A (51 A)
AlAs width = 102.6 A (85 A)
x in AlGaAs = 0.27 (0.33)
R = 15.1 %

Fig. 13. The 3-layer repeat structure with its final derived parameters

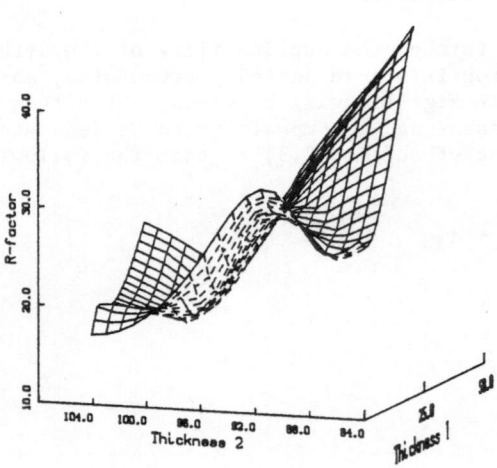

R-factor
10.0
20.0
30.0
40.0
104.0
100.0
96.0
92.0
88.0
84.0
Thickness 2
Thickness 1
SAMPLE G136

Fig. 14. The 3-dimensional plot of R-factor as a function of the GaAs thickness (1) and AlAs thickness (2) for the 3-layer repeat structure

7. SUPERLATTICE STRUCTURES

In this section, 3 examples of superlattice structures will be given, which illustrates samples close to perfection and samples with imperfect structures where incorrect interpretation of the diffractograms could arise.

7.1. Incommensurate Structure

Consider a superlattice with the nominal structure $(GaAs)_3$ $(AlAs)_3$ that is not perfectly commensurate, sample D355, since the measured modulation wavelength is not an integer number of lattice periods. The diffractogram from this sample is given in Fig. 15(a), and besides the main Bragg peaks there are satellites associated with them from which we can measure the period to be 6.12 monolayers. This means that there is interfacial rough-ness, since the minimum growth step is one monolayer, and therefore this fractional coverage represents a rough interface. By determining the average alloy composition in the superlattice the individual thicknesses have been determined from the equation

$$\bar{d}_1 = \bar{\Lambda} (1 - \bar{x})$$

and

$$\bar{d}_2 = \bar{\Lambda} \bar{x}$$

where $\bar{d}_1$ and $\bar{d}_2$ are the GaAs and AlAs thicknesses respectively. With these short periods, minimal roughness will result in a near sine waveform, leading to one Fourier coefficient and only first-order satellites.

7.2. Commensurate Structure

This is a much more complex structure, which is essentially a combination $(GaAs)_3(AlAs)_6$ and $(GaAs)_6(AlAs)_3$ superlattices, G166. The diffractogram is given in Fig. 15(b), and all expected satellites up to the 006 main Bragg peak could be observed. Above this they gradually became less visible. This clearly means that the period is, or is very close to, an exact number of monolayers and therefore constitutes a commensurate superlattice. The growth rate in this structure was judged very accurately before growth, indicated by this result, but the interfaces could still be rough. This latter statement arises from what can be deduced from this cursory glance, that is we know the period is equal to 9 monolayers, but we do not know its relative position with respect to the completion of a monolayer. The period is only the difference between every other interface and therefore independent of the interface shape. The interface shape can be obtained from the analysis given in section 4, although the slow intensity fall-off of satellites is indicative of fairly abrupt interfaces.

7.3. Imperfect Structure

This structure, D365, was grown with the intention of being $(GaAs)_2$-$(AlAs)_2$ and the diffractogram is given in Fig. 15(c). The measured period was 3.6 monolayers, which makes the superlattice incommensurate. Also many of the diffraction features cannot simply be related to the expected structure, but the first-order satellites are consistent with the 3.6 monolayer period. If now the same analysis was performed as for D355 (section 7.1), then the GaAs and AlAs widths are unrealistic compared with those expected from the growth. The analysis therefore may be inappropriate for a structure that is very imperfect, since the average x-value would be correctly determined but the period can only be determined from the areas that are purely periodic, and hence the diffracting volumes for the two experments are different.

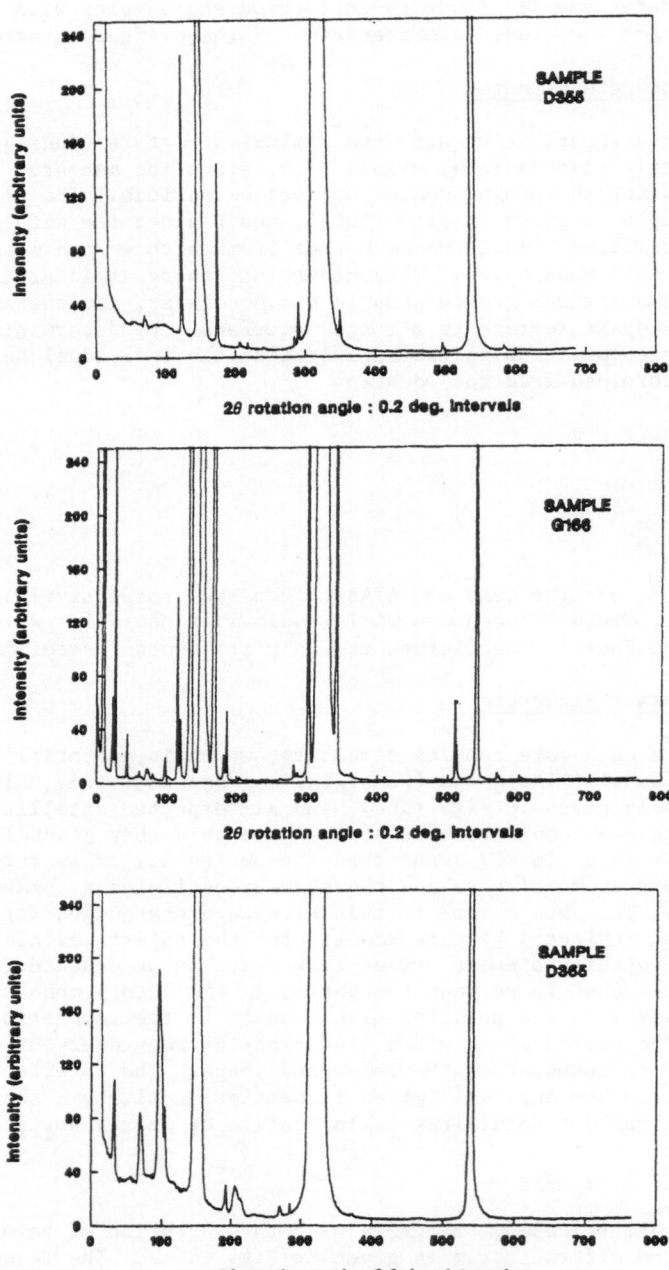

Fig. 15. Diffraction profiles obtained on the powder diffractometer showing
          the 002, 004, 006 Bragg peaks and their associated satellites for
          a) an incommensurate superlattice, b) a commensurate lattice and
          c) an imperfect superlattice

It is interesting to note at this stage that the background level in the imperfect structure is very high, indicative of imperfection, and this diminishes for D355 and G166 as the structural quality increases. Certainly the example of G166 does show the control that is possible by MBE growth of these very thin layers. The period of two of these superlattice structures (D355, D365) have also been determined by transmission electron microscopy, and despite averaging different sample areas, the periods are in very close agreement[27,28].

8. CONCLUSIONS

This paper has endeavoured to show how information can be obtained on MQW structures with X-ray diffraction methods. Much of the work can be done using simple apparatus, but careful analysis is necessary for correct interpretation. The probe size in these experiments is large and therefore in comparing with other techniques this must be born in mind. The measurement of alloy composition is a well established X-ray diffraction technique, but this paper points out how the simple standard technique is inadequate for thin layers, and suggests that full dynamical theory profile fitting is the only way to arrive at the correct value. The well and barrier thicknesses and the interface widths on the other hand, have been obtained using kinematical theory modelling, and the well widths agree very closely with those obtained by PLE. The width of an interface when the binary is the underlayer has also been found to be more abrupt than when the ternary is the underlayer, although the results from a larger number of samples are being analysed to see if this is a general observation.

ACKNOWLEDGEMENTS

The author is grateful to Miss Catherine Curling for assisting with the diffraction experiments, Drs Tom Foxon and Karl Woodbridge and Mr David Hilton for growing the epitaxial layers, and other members at Philips Research Laboratories for many stimulating discussions.

REFERENCES

1. Fewster, P.F. Philips J. Res. 41, 268 (1986).
2. Kervarec, J., Baudet, M., Caulet, J., Auvray, P., Emery, J.Y. and Regreny, A. J. Appl. Cryst. 17, 196 (1984).
3. Keravec, J. Doctor-Ingenieur thesis (1983).
4. Palatnik, L.S., Koz'ma, A.A., Mikhailov, I.F. and Maslov, V.N. Kristallografiya 23, 570 (1977).
5. Segmuller, A. and Blakeslee, A.E. J. Appl. Cryst. 6, 19 (1973).
6. Compton, A.H. and Allison, S.K. "X-rays in Theory and Experiment", 2nd Ed. Van Nostrand Reinhold, New York (1935).
7. Fewster, P.F. J. Appl. Cryst. 18, 334 (1985).
8. Fewster, P.F. and Curling, C.J. To be published.
9. Estop, E., Izrael, A. and Sauvage, M. Acta Cryst. A32, 627 (1976).
10. Guinier, "X-ray Diffraction in Crystals, Imperfect Crystals and Amorphous Bodies", W.H. Freeman and Company, San Francisco (1963).
11. de Fontaine, D. "Local Atomic Arrangements Studied by X-ray Diffraction", 36, 51 (1966).
12. Dawson, P., Duggan, G., Ralph, H.I., Woodbridge, K. and 't Hooft, G.W. "Superlattices and Microstructures", Vol.1, 3, 231 (1985).
13. Orton, J.W., Dawson, P., Duggan, G., Fewster, P.F., Foxon, C.T., Gowers, J.P., Moore, K.J., Curling, C.J., Dobson, P.J., Ralph, H.I. and Woodbridge, K. To be published.

14. Takagi, S. Acta Cryst. 15, 1311 (1962), and J. Phys. Soc. Japan 26, 1239 (1969).

15. Taupin, D. Bull. Soc. Franc. Mineral Crist. 87, 469 (1964).

16. Halliwell, M.A.G., Juler, J. and Norman, A.G. Inst. Phys. Conf. Ser. No. 67, 365 (1983).

17. Tanner, B.K. and Hill, M.J. "Advances in X-ray Analysis", 29, 337 (1986).

18. Halliwell, M.A.G., Lyons, M.H. and Hill, M.J. J. Cryst. Growth, 68, 523 (1984).

19. Fukuhara, A. and Takano, Y. Acta Cryst. A33, 137 (1977).

20. Hill, M.J., Tanner, B.K. and Halliwell, M.A.G., Mat. Res. Soc. Symp. Proc. 37, 53 (1985).

21. Tapfer, L. and Ploog, K., Phys. Rev. B33, 5565, (1986).

22. Clarke, R. This volume.

23. Fleming, R.M., McWhan, D.B., Gossard, A.C., Wiegmann, W. and Logan, R.A. J. Appl. Phys. 51, 357, (1980).

24. Rachinger, W.A. J. Sci. Instrum. 25, 254 (1948).

25. Fewster, P.F. To be published.

26. Fewster, P.F. and Curling, C.J. To be published.

27. Fewster, P.F., Gowers, J.P., Hilton, D. and Foxon, C.T. Fourth International Conference on MBE, York. To be published in J. Cryst. Growth (1986).

28. Gowers, J.P. In this volume.

CHARACTERIZATION OF SUPERLATTICES BY X-RAY DIFFRACTION

W. J. Bartels

Philips Research Laboratories, P.O. Box 80.000
5600 JA Eindhoven, The Netherlands

1. INTRODUCTION

X-ray diffraction line profiles from layered structures grown epitaxially on perfect single crystal substrates contain a lot of information which can be correlated with the concentration depth profile in the grown structure (Bartels and Nijman, 1978). The diffraction profiles (rocking curves) of perfect crystals like silicon and gallium arsenide have a very narrow intrinsic half-width down to 2", so that it is possible to detect the small changes in lattice constant typically related with processes like epitaxy, diffusion and ion-implantation. For this purpose a high-resolution X-ray diffractometer has been designed, where the germanium four-crystal monochromator results in an almost parallel and monochromatic incident beam for investigating the specimen (Bartels, 1983; Bartels, 1983/84). The actual concentration depth profile in a given layered structure can only be obtained after a detailed comparison of observed and calculated diffraction profiles.

GaAs-AlAs superlattices grown by molecular beam epitaxy (MBE) have been studied by X-ray diffraction, but only intensities of diffraction satellites were compared with theory (Segmüller et al., 1977; Fleming et .al., 1980; Kervarec et al., 1984; Terauchi et al., 1985). Diffraction profiles of superlattices were calculated by Speriosu and Vreeland (1984), who used a geometric series within the framework of kinematical theory. This theory makes the assumption that multiple reflections can be neglected, which is only allowed when the reflectivity is below 10%. Above this limit the dynamical theory of X-ray diffraction must always be applied in order to describe diffraction profiles accurately. Vardanyan et al. (1985) described the dynamical diffraction of an ideal superlattice with Chebyshev polynomials of the second kind. However, the computation of these polynomials is rather complicated and requires more time as compared with the use of a geometric series. The most important limitation of both types of calculations is that they can only describe ideal superlattices where all periods are identical in thickness and concentration profile. In the case of a drift of period or statistical variations of layer thicknesses or a changing concentration profile we have a non-ideal superlattice for which a different approach is necessary.

The dynamical theory for crystals with a strain gradient perpendicular to the surface was developed by Takagi (1969) and Taupin (1964). They derived a set of differential equations that describe the dynamical diffraction of X-rays of distorted crystals in the neighbourhood of a Bragg reflection. The latter restriction is known as the so-called two-beam case. The Runge-Kutta method of numerical integration was applied for calculating rocking curves of ion-implanted and diffused silicon (Larson and Barhorst, 1980; Fukuhara and Takano, 1977). The first iteration of Taupin's equation for the Bragg case was used as a semikinematical approximation for calculating diffraction profiles of thin GaAs-AlAs superlattices grow by MBE (Tapfer and Ploog, 1986). For epitaxial layers an integrated solution of the differential equation (Halliwell et al., 1984; Wie et al., 1986; Bartels et al., 1986) is most useful, since numerical integration requires much more computing time and becomes unstable for larger step size. The integrated solution of Taupin's differential equation allows one to calculate the X-ray diffraction profiles of non-ideal superlattices, which were shown to be in good agreement with experimentally observed profiles (Bartels et al., 1986). The effect of random variations in the thickness of the layers in superlattices was studied by Chrzan and Dutta (1986).

An important restriction of the dynamical theory as formulated above is the two-beam case condition so that diffraction profiles can only be calculated in the neighbourhood of a Bragg reflection. However, when the superlattice period decreases, the angular separation of diffraction satellites will increase. In the kinematical theory it is possible to describe the diffraction of superlattices without using the departure from Bragg condition as an input parameter of the calculation (Segmüller and Blakeslee, 1973; McWhan et al., 1983; McWhan, 1985). However, improvement of the theory is needed in order to make detailed comparison with observed diffraction profiles possible. Thus, it is necessary to include refraction and absorption of X-rays and the intensity is required on a reflectivity scale.

A recursion formula from the optical theory of thin films was used successfully to calculate the low order Bragg reflections corresponding to the deposition period of multilayers (Chang et al., 1976; Segmüller, 1979; Underwood and Barbee, 1981; Spiller and Rosenbluth, 1985; Bartels et al., 1986). The simple description with the modulation of the electron density and Fresnel reflection coefficients is only allowed for amorphous multilayers. In the case of strictly periodic multilayers a solution with Chebyshev polynomials of the second kind was obtained by Lee (1981) based on the characteristic matrix method of optical theory (Born and Wolf, 1980). A kinematical theory for multilayers was given by Saxena and Schoenborn (1977).

In this paper a recursion formula is derived from the Takagi-Taupin differential equations, which describe the dynamical diffraction of X-rays in strained crystals. Calculated diffraction profiles of symmetric and asymmetric reflections of AlAs-GaAs superlattices will be compared with rocking curves observed with a high-resolution X-ray diffractometer. In section 3 a kinematical theory for superlattices is developed which includes refraction and absorption of X-rays and gives intensity on a reflectivity scale. The kinematical theory will be compared with the dynamical theory for calculating the profiles of diffraction satellites of superlattices. Comparison with experiment will illustrate that high-resolution X-ray diffraction is very sensitive to variations in the growth period of superlattices. In section 5 the kinematical theory will be used in a macroscopic description of the X-ray diffraction of periodic multilayers and comparison is made with the optical theory of thin films.

2. DYNAMICAL THEORY

In the dynamical theory the change of the amplitudes D_0 and D_H of the incident and the diffracted beam with the depth coordinate is described with a set of differential equations derived independently by Takagi (1969) and Taupin (1964). Taupin has combined the two equations for the Bragg case to give one differential equation for the amplitude ratio X, but has discussed only centrosymmetrical reflections. The differential equation for the polar Bragg case can be written as

$$- i \frac{dX}{dT} = X^2 - 2\eta X + 1 \ , \tag{1}$$

where X, η and T are complex quantities given by

$$X = \left(\frac{F_{\bar{H}}}{F_H} \right)^{\frac{1}{2}} \left| \frac{\gamma_H}{\gamma_0} \right|^{\frac{1}{2}} \frac{D_H}{D_0} \ , \tag{2}$$

$$\eta = \frac{- b \, (\theta - \theta_B) \sin 2\theta_B - \frac{1}{2}\Gamma F_0(1 - b)}{\sqrt{|b|} \ C \ \Gamma \ \sqrt{F_H F_{\bar{H}}}} \ , \tag{3}$$

$$T = \frac{\pi \ C \ \Gamma \ \sqrt{F_H F_{\bar{H}}}}{\lambda \ \sqrt{|\gamma_0 \gamma_H|}} \ t \ , \tag{4}$$

$$\Gamma = \frac{r_e \lambda^2}{\pi V} \ , \quad r_e = \frac{e^2}{4\pi\epsilon_0 m c^2} \ , \quad b = \frac{\gamma_0}{\gamma_H} \ . \tag{5}$$

T is determined by the crystal thickness t and the structure factor F_H of the reflection. The departure from the Bragg angle θ_B determines the deviation parameter η. The second part of the numerator of η corresponds to the refraction and absorption of the X-rays. In the Bragg case the direction cosines γ_0 and γ_H of the incident and the diffracted beam with respect to the surface normal are opposite in sign so that the asymmetry factor b is negative. The classical electron radius r_e is equal to 2.818×10^{-5} Å, λ is the X-ray wavelength and V is the volume of the unit cell. C=1 for perpendicular (σ) polarization and C=$|\cos 2\theta_B|$ for parallel (π) polarization of the incident beam.

The differential equation can be solved for layers of constant η and arbitrary thickness. This solution can also be used for sections for which η can be considered to be constant. The following recursion equation gives the relation between the amplitude ratio X_0 at the bottom of the layer and X_t at its top (Bartels et al., 1986)

$$X_t = \eta + \sqrt{\eta^2 - 1} \left[\frac{S_1 + S_2}{S_1 - S_2} \right] \ , \tag{6}$$

where

$$S_1 = (X_0 - \eta + \sqrt{\eta^2 - 1}) \exp(- iT\sqrt{\eta^2 - 1}) \ , \tag{7}$$

$$S_2 = (X_0 - \eta - \sqrt{\eta^2 - 1}) \exp(iT\sqrt{\eta^2 - 1}) \ . \tag{8}$$

443

The recursion formula we have obtained is the general solution for the dynamical reflection of an epitaxial layer of arbitrary thickness. The recursion process allows one to calculate rocking curves of complicated layered structures such as non-ideal superlattices on perfect crystals. For an infinitely thick crystal the equation reduces to the well known Darwin-Prins formula (Fingerland, 1971; James, 1963; Pinsker, 1978; Zachariasen, 1945)

$$X_\infty = \eta \pm \sqrt{\eta^2 - 1} \, , \qquad (9)$$

where the sign to be selected is opposite to the sign of $Re(\eta)$. The rocking curve of the crystal is given by the reflectivity P_H as a function of the deviation parameter η. For an asymmetric reflection we must take into account the change in beam cross section, so that the reflectivity P_H is given by

$$P_H = \frac{1}{|b|} \frac{I_H}{I_0} = \left| \frac{\gamma_H}{\gamma_0} \right| \left| \frac{D_H}{D_0} \right|^2 = \left| \frac{F_H}{F_{\bar{H}}} \right| \left| X \right|^2 . \qquad (10)$$

Ideal superlattices were grown by metalorganic vapour-phase epitaxy on (001) GaAs substrates. A high-resolution X-ray diffractometer (Bartels, 1983; Bartels, 1983/84) was used for measuring rocking curves. The germanium four-crystal monochromator of this instrument produces a parallel and monochromatic incident X-ray beam of sufficient intensity. In this way, almost intrinsic diffraction profiles can be measured at any Bragg angle, whereas double-crystal diffractometers have the severe restriction of equal Bragg angle for monochromator and specimen. Observed and calculated diffraction profiles of the (002) $CuK\alpha_1$ reflection of an ideal superlattice are shown in Fig. 1. The superlattice is composed of 47 periods of 107 Å AlAs and 810 Å GaAs, where the layer thicknesses have been deduced from trial and error fitting of the profiles. The finite divergence of 5" of the (σ) polarized incident beam was taken into account by applying a corresponding smoothing function to the calculated profile obtained with the dynamical recursion formula Eq.(6). The computing time was 20 s on the IBM 3081 computer for an angular resolution of 1" in the given angular range. We used a value of 2.2 μm for the extinction depth (=t/T) of AlAs, which is slightly larger than the theoretical value of 1.8 μm deduced from "The International Tables for X-Ray Crystallography", Vol. IV (1974).

Fig. 2 gives a comparison between observed and calculated diffraction profiles of the (115) $CuK\alpha_1$ reflection of the same specimen, where the same layer thicknesses were used in the calculation. The correspondence between the profiles is excellent. The asymmetric (115) reflection is sensitive to the tetragonal lattice deformation of the AlAs epitaxial layers, which is described by elasticity theory (Hornstra and Bartels, 1978; Bartels and Nijman, 1978). The zeroth order reflection of the superlattice is well separated from the GaAs substrate reflection, which was taken as origin of the angular scale. The width of the diffraction satellites visible in Figs. 1 and 2 is independent of the satellite order, which is characteristic for an ideal superlattice where all periods are equal. In the case of a non-ideal superlattice the width of the satellites increases progressively with the satellite order (Bartels et al., 1986). Computer simulation of diffraction profiles is very useful for studying how different structural parameters act on the fine structure of diffraction satellites. In this way were studied the influence of statistical variations in layer thicknesses, drift of the superlattice period and variations in the concentration profile.

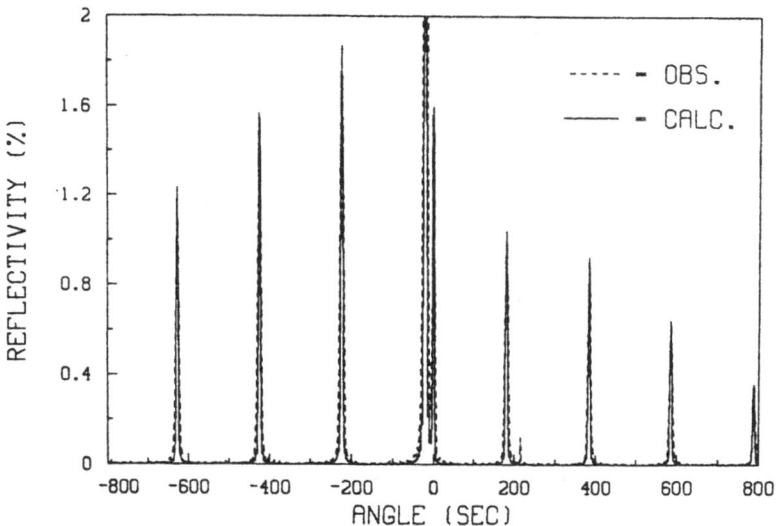

Fig. 1. Observed and calculated (002) CuKα_1 X-ray diffraction
profiles of an AlAs-GaAs ideal superlattice with a
period of 817 Å on a (001) GaAs substrate.

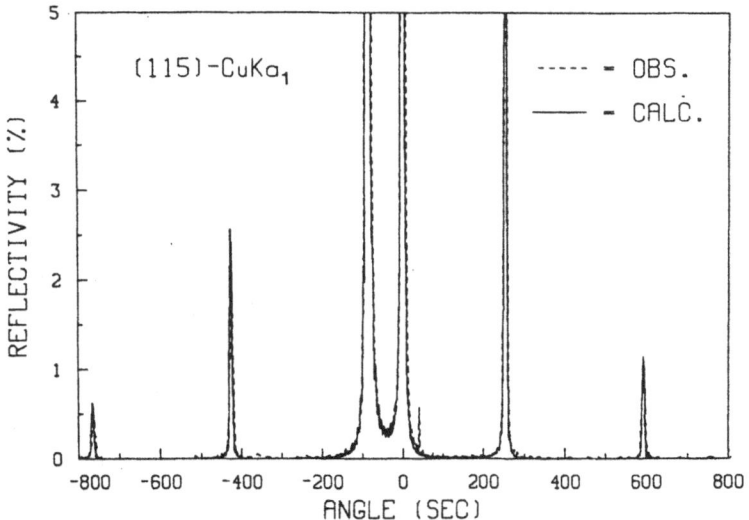

Fig. 2. Observed and calculated asymmetric (115) CuKα_1 reflection
of the superlattice shown in Fig. 1. The reflection of
the GaAs substrate is taken as origin of the angle scale.

3. KINEMATICAL THEORY

In the kinematical theory the mutual exchange of energy in the diffracted and incident beam directions is neglected. This theory assumes a simple addition of scattered waves taking the differences in phase and absorption into account. When the reflectivity is below 10% we may neglect multiple reflections. For the differential equation of Taupin this implies loss of the quadratic term. Integration of the equation results then in the well known kinematical formulae (Bartels et al., 1986). The simple addition of scattered waves in the kinematical approximation is an advantage for calculating the diffraction profile of an ideal superlattice. In this case a geometric series is obtained, which can save computing time as compared with the use of the dynamical recursion formula. The differential equation of Taupin is only valid for a two-beam case. Thus, it is assumed that we are always in the neighbourhood of a Bragg reflection. However, the angular separation between the diffraction satellites and the zeroth order peak increases with decreasing superlattice period. For very small periods it is essential to use a description without the deviation parameter. In the kinematical theory we can put all phase information in the structure factor F by using the reflection index l as a continuous variable. We confine the case to symmetrical ($00l$) reflections of perpendicular polarized X-rays. The reflected amplitude ratio X of N identical cubic cells with lattice constant a is then given by (James, 1967)

$$ X = \frac{i \pi \Gamma F a}{\lambda \sin \theta} \left(\frac{1 - R^N}{1 - R} \right) , \quad \Gamma = \frac{r_e \lambda^2}{\pi V} , \tag{11} $$

$$ R = \exp(- a \mu / \sin \theta - 2\pi i l) , \tag{12} $$

$$ l = \frac{2a \sin(\theta - 2\delta / \sin 2\theta)}{\lambda} , \tag{13} $$

$$ F = \Sigma f_j \exp(2\pi i l z_j) \exp \left(- B_j \frac{\sin^2\theta}{\lambda^2} \right) . \tag{14} $$

The second exponential term in Eq.(14) is the Debye-Waller factor, which takes into account the thermal vibration of atoms. Further μ is the linear absorption coefficient and δ is the real part of the deviation of the refractive index from unity. The angle of incidence θ at the crystal surface must be corrected for refraction before calculating the phase-lag across the unit cell. The factor R in the geometric series takes into account the change in phase and absorption across the unit cell.

The kinematical formulae given above can be applied to a superlattice unit cell with period p as presented in Fig.3. The number of monolayers in the given two-layer sequence are n1 and n2. The thickness of a monolayer of GaAs is equal to half of the lattice constant a. The composition x and the perpendicular lattice mismatch of AlAs and GaAs must be taken into account for calculating the thickness of a monolayer of the ternary compound. The atomic scattering factor of the group III element in each layer depends on the composition x. Thus, the superlattice unit cell can be described by

$$p = n_1 \, p_1 + n_2 \, p_2 \, , \quad z_0 = n_1 \, p_1 \, / \, p \, , \tag{15}$$

$$p_1 = \left[1 + x \left(\frac{\Delta a}{a} \right)_{\perp} \right] \frac{a}{2} \, , \quad p_2 = \frac{a}{2} \, , \tag{16}$$

$$Alg = x \, f_{AL} + (1 - x) \, f_{Ga} \, . \tag{17}$$

The reflection index L is used for describing the phase-lag across the superlattice unit cell and for refraction and absorption we must take average values for this cell. The index L varies continuously with the angle of incidence θ and for integer values of L we obtain diffraction satellites. The structure factor F of the superlattice unit cell with n1 and n2 as the number of monolayers in the given two-layer sequence is then given by (Kervarec et al., 1984)

$$F_{SL} = 2(As + Alg \, R_1^{\frac{1}{2}}) \, \frac{1 - R_1^{n1}}{1 - R_1} + 2(As + Ga \, R_2^{\frac{1}{2}}) \, \frac{R_1^{n1}(1 - R_2^{n2})}{1 - R_2} \, , \tag{18}$$

$$L = \frac{2p \, \sin(\theta - 2\bar{\delta} \, / \, \sin 2\theta)}{\lambda} \, , \tag{19}$$

$$R_1 = \exp(2\pi i L p_1 \, / \, p) \, , \quad R_2 = \exp(2\pi i L p_2 \, / \, p) \, . \tag{20}$$

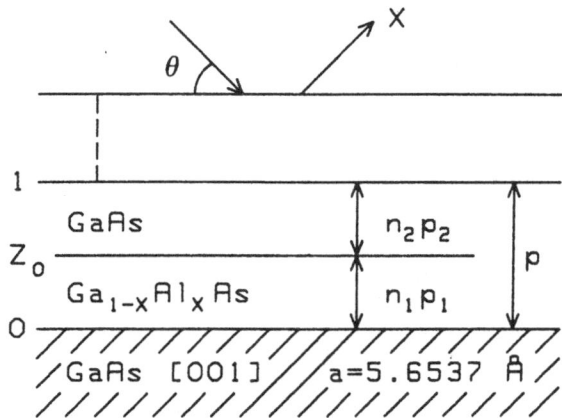

Fig. 3. Superlattice unit cell with period p and n1 and n2 as number of monolayers in the two-layer sequence. Theta is the angle of incidence of X-rays with the crystal surface.

The atomic scattering factors have been abbreviated in Eq. (18) by writing chemical symbols of the elements.

The contribution of the substrate to the reflected amplitude ratio of a specimen can be described with an infinite geometric series and we must take into account the change in phase and absorption by the N unit cells of the superlattice. The reflected amplitude ratio X of the superlattice specimen is derived from Eq. (11) - (14) so that

$$X = \frac{i\, r_e\, \lambda}{a^2 \sin\theta} \left[\frac{F_S\, R_{SL}^N}{1 - R_S} \exp(-2\pi i l) + F_{SL} \frac{1 - R_{SL}^N}{1 - R_{SL}} \exp(-2\pi i L) \right], \quad (21)$$

$$R_{SL} = \exp(-p\, \bar{\mu}\, /\, \sin\theta - 2\pi i L) \cdot \quad (22)$$

R_S and F_S are given by Eq. (12) and Eq. (14) respectively.
The exponentials in Eq. (21) have been added in order to relate the contributions of the substrate and the superlattice to the same physical surface. This is necessary since the reflection index varies continuously with the angle of incidence. The kinematical theory for superlattices given above contain a geometric series and consequently it is in this form only valid for ideal superlattices. However, in the case of variations from period to period it is possible to apply the same principles. Therefore the equation has to be rewritten in a recursion formula so that the change in the reflected amplitude ratio X can be calculated for each period of the superlattice.

The calculations presented above provide X-ray intensity on a reflectivity scale and they contain absorption and refraction of X-rays. Thus, it is possible to make direct comparison with observed diffraction profiles or with calculations based on the dynamical theory. Fig.4 shows the diffraction profile calculated for a superlattice with 120 periods of 25 monolayers of AlAs and 15 monolayers of GaAs (n1=25, n2=15). Diffraction satellites are visible in the neighbourhood of the (002) and (004) reflection of the GaAs substrate. Besides, several reflection orders of the superlattice period occur at low angle of incidence. These low order reflections have indices L=1, 2, ..., whereas the zeroth order reflection of the superlattice close to the (002) GaAs substrate reflection has as reflection index L=n1+n2=40. The calculated reflection profiles are too narrow to be displayed as diffraction peaks on this large angular scale.

A comparison of the kinematical theory and the dynamical theory is shown in Fig. 5 for the diffraction satellites close to the (002) reflection. The zeroth order of the superlattice corresponds to L=40 and is well separated from the extremely narrow substrate reflection denoted by S. The (002) reflection of GaAs is very weak and the large extinction depth of 18 μm explains the narrow profile. Both theories result in equal peak positions for the diffraction satellites and it is difficult to detect any difference between the diffraction profiles.
Thus, the kinematical theory as formulated above results in reliable diffraction information when the reflectivity is well below 10%. This kinematical theory allows one to calculate X-ray diffraction profiles within a large angular range so that the limitation of the deviation parameter used in the dynamical theory has been overcome.

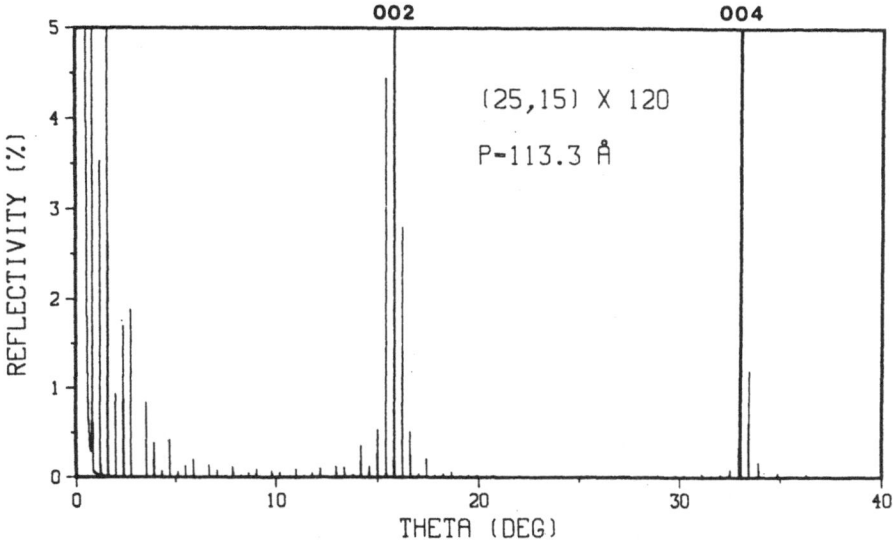

Fig. 4. X-ray diffraction profile calculated with kinematical
theory for an AlAs-GaAs superlattice with 120 periods
with n1=25 and n2=15 as number of monolayers.

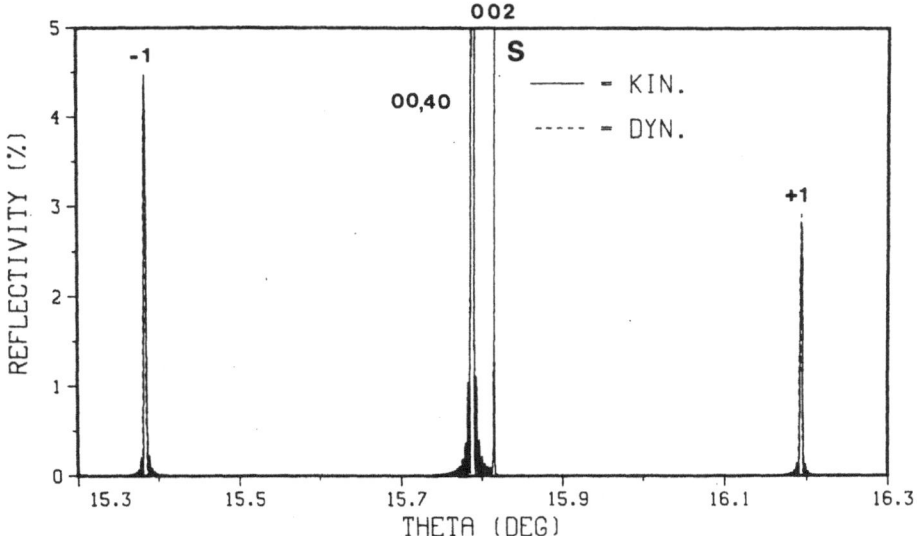

Fig. 5. Comparison of the kinematical theory with the dynamical
theory for the superlattice diffraction satellites of Fig. 4
close to the (002) reflection of the GaAs substrate (S).

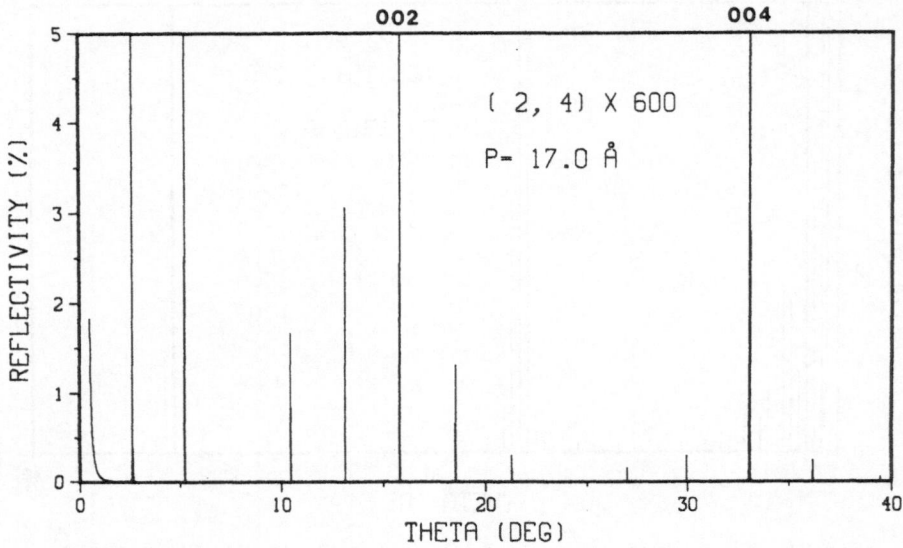

Fig. 6. X-ray diffraction profile of superlattice with 600 periods
of 2 monolayers of AlAs and 4 monolayers of GaAs on a
(001) GaAs substrate calculated with kinematical theory.

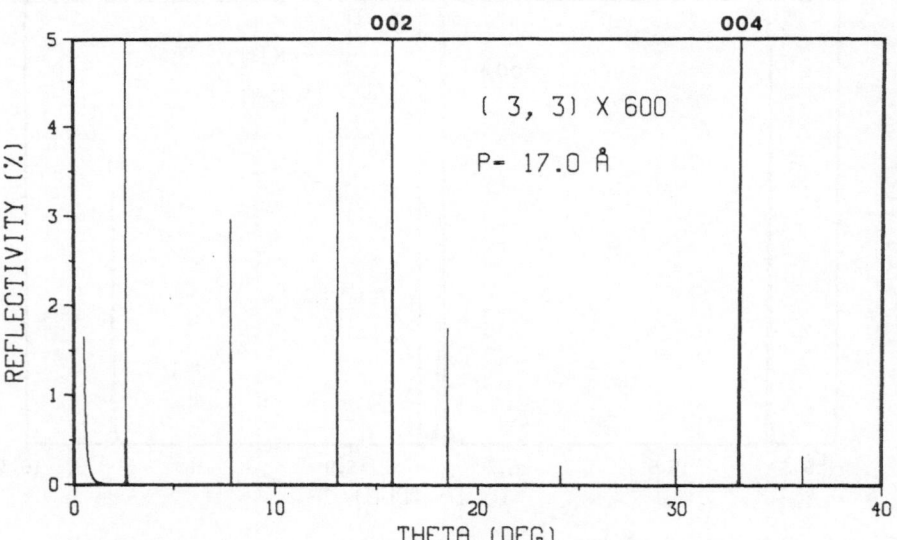

Fig. 7. X-ray diffraction profile around the (002) and (004)
reflection of a (001) GaAs substrate covered with a
superlattice as in Fig. 6, but with n1=3 and n2=3.

4. SUPERLATTICES

The dynamical theory as formulated by Takagi (1969) and Taupin (1964) is only valid in the neighbourhood of a Bragg reflection so that the deviation parameter occurring in their equations is relatively small. This so-called two-beam case condition is a severe limitation for calculating the diffraction profile of superlattices with extremely small periods. In this case the angular distance between diffraction satellites is so large that it is better not to speak of satellites. However, the kinematical theory as formulated in the preceding section allows one to calculate accurately and over a wide angular range the diffraction profile of superlattices with small periods. Figs. 6 and 7 provide examples for superlattices with a period of 17 Å. When the number of monolayers of AlAs and GaAs is equal, all even orders vanish except those corresponding to reflections of the average lattice which occur close to the (002) and (004) substrate reflections. Similar rules apply to other ratios of numbers of monolayers. Thus, some orders of superlattice diffraction will be very sensitive to interface roughness, in which case the number of monolayers in each layer is less well defined.

The diffraction profile of a superlattice of 200 periods of 49 Å of AlAs and 18 Å of GaAs is shown in Fig. 8. The superlattice was grown by metalorganic vapour-phase epitaxy on a (001) GaAs substrate. The first and the second order satellite are compared in Fig. 9 with a profile calculated by kinematical theory for the (002) CuKα_1 reflection. The differences between observed and calculated profiles can be explained by the presence of a very small drift of the superlattice period as is illustrated in Figs. 10 and 11. A linear variation of the period from 67 Å to 66 Å results in a rather good fit for the profiles, whereas a small increase of the drift of period is dramatically unfavourable. At the same time the decrease of the average period is visible as a displacement of the diffraction satellites. These figures illustrate that for superlattices the sensitivity of high-resolution X-ray diffraction can reach the Ångström level, which is less than the thickness of a monolayer.

A non-ideal superlattice is characterized by unequal periods. A drift of the period or statistical variations in layer thicknesses or a changing concentration profile all result in non-ideal superlattices. Computer simulation of X-ray diffraction profiles is a very useful tool for studying how different structural parameters act on the fine structure of diffraction satellites. Thus, it was shown that in the case of a drift of period, the width of diffraction satellites increases progressively with the satellite order, whereas ideal superlattices are characterized by a constant width of the diffraction satellites (Bartels et al., 1986). It is very important to distinguish interface roughness from statistical variations in layer thicknesses. Ideal superlattices can be grown with rough interfaces and computer simulation of diffraction profiles indicates that interface roughness only reduces the peak height of diffraction satellites, the more the higher the order of the satellite. This effect is comparable to the Debye-Waller factor describing thermal vibration of atoms. Statistical variations in layer thicknesses always results in non-ideal superlattices where the shape and width of high order satellites have changed. With X-ray diffraction it seems impossible to distinguish interface roughness from diffusion or intentionally graded interfaces. All three cases result in the same diffraction profile and the difference is merely in the size of the area over which is averaged in the description. Chrzan and Dutta (1986) have recently discussed the influence of statistical variations of layer thicknesses on a lateral two-dimensional scale. It is important to observe the width of high-order diffraction satellites of rocking curves from asymmetric reflections.

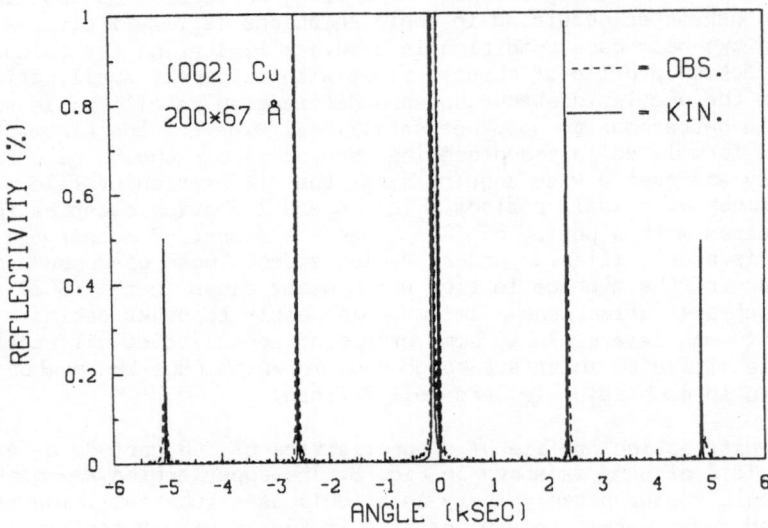

Fig. 8. X-ray diffraction satellites of an AlAs–GaAs superlattice
with 200 periods of 67 Å around the (002) CuKα_1 reflection of
the (001) GaAs substrate.

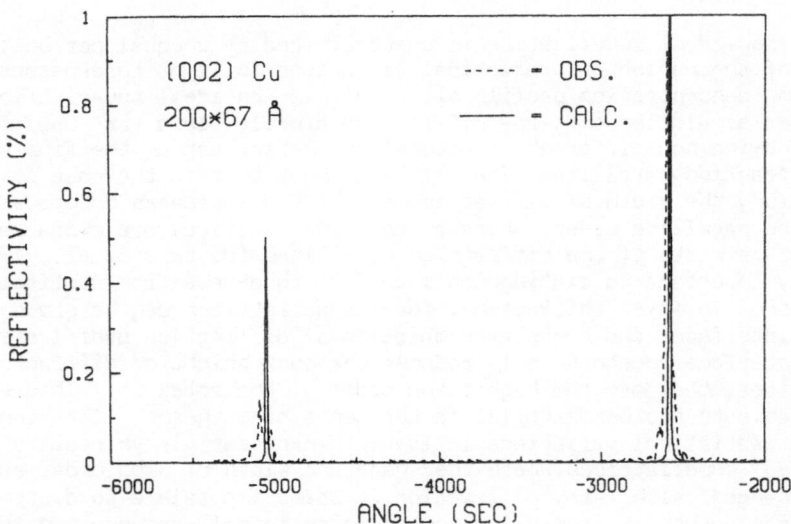

Fig. 9. Comparison of observed and calculated profiles of the
first and second order satellites of the superlattice
shown in Fig. 8.

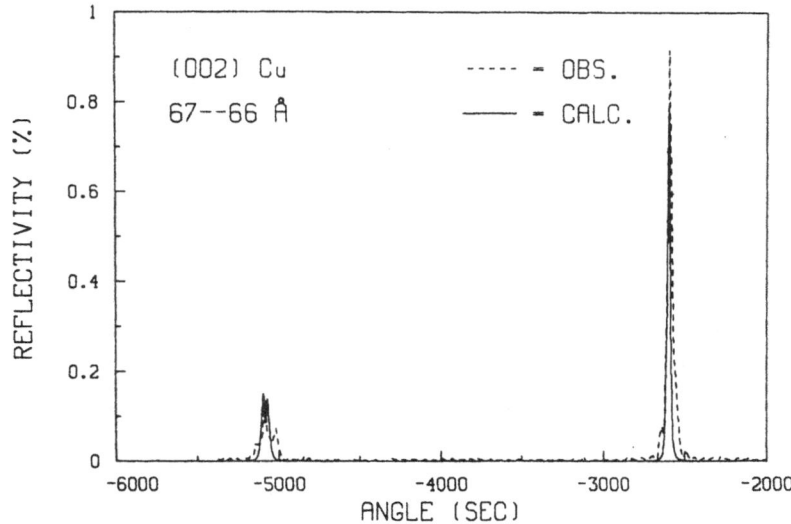

Fig. 10. The same satellites as shown in Fig. 9, but in the
calculation a linear drift of the superlattice period
from 67 Å to 66 Å was assumed.

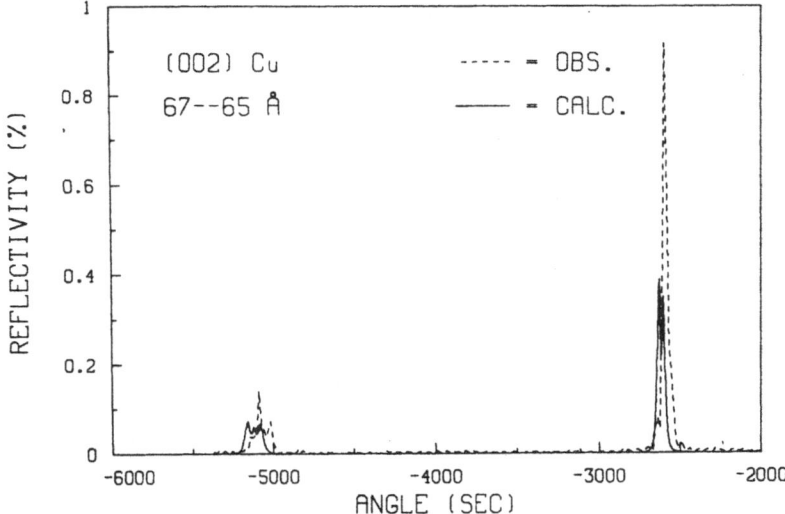

Fig. 11. First and second order diffraction satellites close to the
(002) CuKα_1 reflection of the same superlattice of Fig. 10,
but with an assumed drift of the period from 67 to 65 Å.

5. PERIODIC MULTILAYERS

Sputter-deposited periodic multilayers of tungsten and carbon are used as soft X-ray monochromators. Insight into their reflecting properties was obtained with $CuK\alpha_1$ radiation where the Bragg angle is about 1 degree for an artificial period of 50 Å. Lee (1981) used the characteristic matrix method from the optical theory of thin films (Born and Wolf, 1980) as a macroscopic description of the Bragg reflection of periodic multilayers. A solution with Chebyshev polynomials of the second kind was obtained, but this can only be applied to strictly periodic multilayers. In the case of non-periodic multilayers the optical recursion formula for thin films is very useful (Chang et al., 1976; Segmüller, 1979; Underwood and Barbee, 1981; Spiller and Rosenbluth, 1985; Bartels et al., 1986). The details of the atomic structure of the layers is lost in the optical theory. The simple macroscopic description with the modulation of the electron density and Fresnel reflection coefficients at the interfaces is only allowed for amorphous multilayers. The optical recursion formula for multilayers is given by

$$X_t = \frac{r + X_0 \exp(-i2\phi)}{1 + r\, X_0 \exp(-i2\phi)} \; , \tag{23}$$

where r is the Fresnel reflection coefficient for σ or π polarization at the top of a layer (Born and Wolf, 1980) and X_0 and X_t are the reflected amplitude ratios at the bottom and the top of this layer. The phase difference φ across layer j of thickness t can be related to the angle of incidence θ with the surface of the multilayer, when we make use of Snell's law, so that

$$\phi = \frac{2\pi}{\lambda}\, t\, n_j\, \sin\theta_j = \frac{2\pi}{\lambda}\, t\, \sqrt{n_j^2 - \cos^2\theta} \; , \tag{24}$$

where n_j is the complex index of refraction whose value for X-rays is close to unity and given by

$$n = 1 - \delta - i\beta \quad , \qquad \beta = \frac{\mu\lambda}{4\pi} \cdot \tag{25}$$

The imaginary part of n is related to the linear absorption coefficient of the layer. The first medium in the recursion process is the infinitely thick substrate where at the bottom $X_0 = 0$. Recursion proceeds from bottom to top for each layer taking the double phase difference into account. The final value of the amplitude ratio X_t at the surface of the multilayer gives the reflectivity as $P = |X_t|^2$. Non-correlated and correlated interface roughness can be taken into account by using a Debye-Waller multiplication factor with the Fresnel reflection coefficient r at every interface or with the final value of X_t at the surface of the multilayer respectively.

When a multilayer is exactly periodic we can consider the structure as an artificial crystal. The parameters δ and β are related to the real and imaginary parts of atomic scattering factors used in calculations of structure factors. The relation is given by (James, 1967; Zachariasen, 1945)

$$\Gamma F_0 = -\psi_0 = 2(\delta + i\beta) \cdot \tag{26}$$

It is interesting to compare the macroscopic optical description given above with the kinematical diffraction theory developed in section 3.

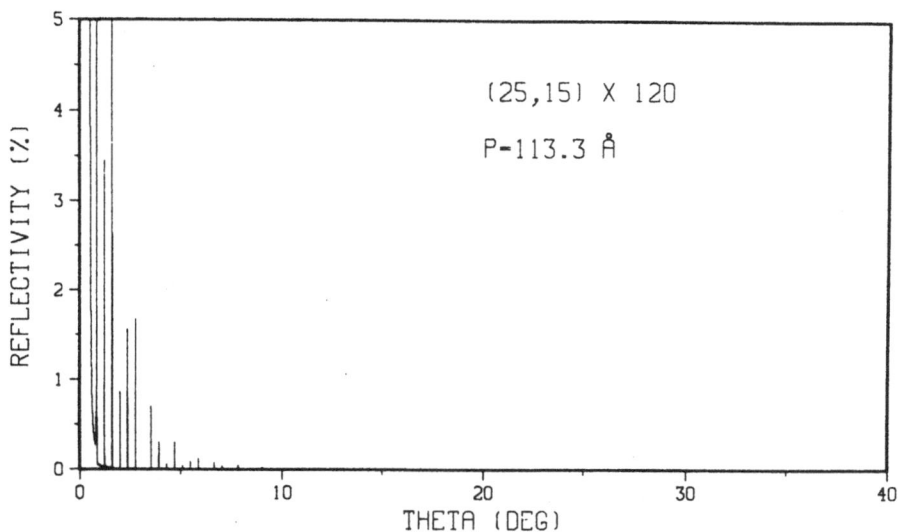

Fig. 12. X-ray diffraction profile of a multilayer with 120 periods
of 25 monolayers of AlAs and 15 monolayers of GaAs
calculated with a macroscopic kinematical theory.

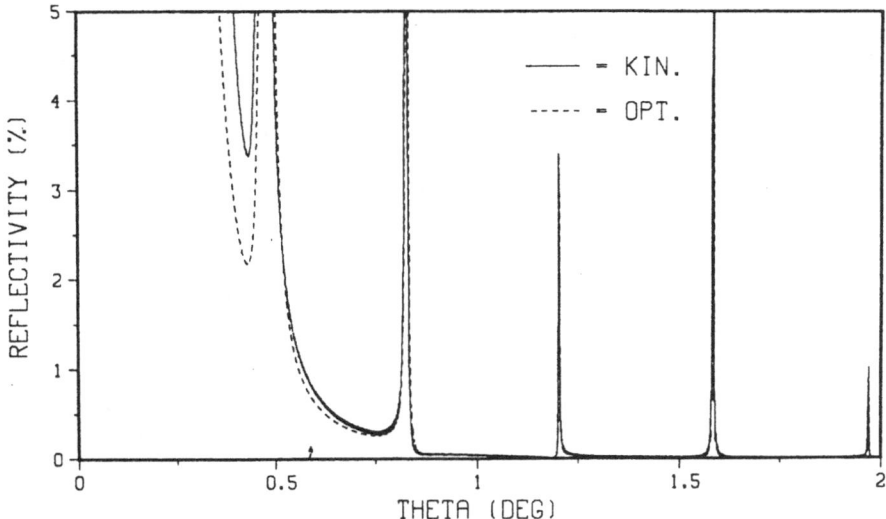

Fig. 13. Comparison of optical theory and kinematical theory for
calculating X-ray diffraction profiles of periodic
multilayers for the same structure as in Fig. 12.

The structure factor for the multilayer unit cell (see Fig. 3) with the continuously varying reflection index L is given by

$$\Gamma F_{ML} = -\int_0^1 \psi \exp(2\pi iLz)dz \; , \qquad (27)$$

$$\Gamma F_{ML} = -\frac{(\psi_1 - \psi_2)\exp(2\pi iLz_0) - \psi_1 + \psi_2\exp(2\pi iL)}{2\pi iL} \; . \qquad (28)$$

In the macroscopic description of a multilayer only the modulation of the electron density is taken into account for calculating the structure factor of a Bragg reflection. Thus, when considering AlAs-GaAs multilayers the information of atomic structure is lost. This is illustrated in Fig. 12, which presents the multilayer diffraction profile of the same structure as used for calculating Fig. 4. As expected, the (002) and (004) reflections and their satellites have disappeared in the calculated profile. The diffraction of this multilayer is compared in Fig. 13 with calculations based on the macroscopic optical description (Eq. 23). At very low glancing angle there is some difference between the calculated profiles, but in general the agreement of kinematical and optical theory is rather good. The diffraction peaks occur at the same angle and are equal in height and shape.

6. CONCLUSIONS

The diffraction profiles of superlattices calculated with the kinematical theory developed in section 3 are in excellent agreement with results obtained with the dynamical recursion formula given in section 2. However, the kinematical theory allows one to calculate diffraction profiles over a large angular range, whereas the dynamical theory requires a small departure from a Bragg reflection owing to the two-beam case restriction. Good agreement between calculated and observed diffraction profiles was obtained for asymmetric and symmetric Bragg reflections of ideal superlattices of AlAs-GaAs grown by metalorganic vapour-phase epitaxy. It was shown that high-resolution X-ray diffraction is capable of detecting variations in the growth period of a superlattice with a sensitivity reaching the Ångström level.

The low order Bragg reflections of periodic multilayers were calculated with a macroscopic description of the structure in terms of the modulation of the electron density. The atomic scattering factors used in structure factor calculations can be related to the complex index of refraction of the optical theory. The diffraction profiles calculated within the framework of the kinematical diffraction theory were shown to be in good agreement with results obtained with the macroscopic optical description in terms of the complex index of refraction and Fresnel reflection coefficients.

ACKNOWLEDGMENTS

I am grateful to J. Hornstra for helpful discussions throughout this work and to D.J.W. Lobeek for measuring X-ray rocking curves. The author thanks H.F.J. van 't Blik for the growth of superlattices by metalorganic vapour-phase epitaxy.

REFERENCES

Bartels, W.J., 1983, Characterization of thin layers on perfect crystals with a multipurpose high resolution X-ray diffractometer, J. Vac. Sci. Technol., B1:338.

Bartels, W.J., 1983/84, High-resolution X-ray diffractometer, Philips Tech. Rev., 41:183.

Bartels, W.J., Hornstra, J., and Lobeek, D.J.W. 1986, X-ray diffraction of multilayers and superlattices, Acta Crystallogr., A42:xx.

Bartels, W.J., and Nijman, W., 1978, X-ray double-crystal diffractometry of Ga(1-x)Al(x)As epitaxial layers, J. Cryst. Growth, 44:518.

Born, M., and Wolf, E., 1980, "Principles of Optics," Pergamon Press, Oxford.

Chang, L.L., Segmüller, A., and Esaki, L., 1976, Smooth and coherent layers of GaAs and AlAs grown by molecular beam epitaxy, Appl. Phys. Lett., 28:39.

Chrzan, D., and Dutta, P., 1986, The effect of interface roughness on the intensity profiles of Bragg peaks from superlattices, J. Appl. Phys., 59:1504.

Fingerland, A., 1971, Some properties of the single-crystal rocking curve in the Bragg case, Acta Crystallogr., A27:280.

Fleming, R.M., McWhan, D.B., Gossard, A.C., Wiegmann, W., and Logan, R.A., 1980, X-ray diffraction study of interdiffusion and growth in (GaAs)n(AlAs)m multilayers, J. Appl. Phys., 51:357.

Fukuhara, A., and Takano, Y., 1977, Determination of strain distributions from X-ray Bragg reflexion by silicon single crystals, Acta Crystallogr., A33:137.

Halliwell, M.A.G., and Lyons, M.H., and Hill, M.J., 1984, The interpretation of X-ray rocking curves from III-V semiconductor device structures, J. Cryst. Growth, 68:523.

Hornstra, J., and Bartels, W.J., 1978, Determination of the lattice constant of epitaxial layers of III-V compounds, J. Cryst. Growth, 44:513.

International Tables for X-Ray Crystallography, Vol. IV, 1974, J.A. Ibers and W.C. Hamilton, eds., The Kynoch Press, Birmingham.

James, R.W., 1963, The dynamical theory of X-ray diffraction, in: "Solid State Physics Vol. 15," F. Seitz and D. Turnbull, eds., Academic Press, New York.

James, R.W., 1967, "The Optical Principles of the Diffraction of X-Rays," Bell, London.

Kervarec, J., Baudet, M., Caulet, J., Auvray, P., Emery, J.Y., and Regreny, A., 1984, Some aspects of the X-ray structural characterization of (GaAlAs)n1(GaAs)n2/GaAs(001) superlattices, J. Appl. Cryst., 17:196.

Larson, B.C., and Barhorst, J.F., 1980, X-ray study of lattice strain in boron implanted laser annealed silicon, J. Appl. Phys., 51:3181.

Lee, P., 1981, X-ray diffraction in multilayers, Opt. Commun., 37:159.

McWhan, D.B., 1985, Structure of chemically modulated films, in: "Synthetic Modulated Structures," L.L. Chang and B.C. Giessen, eds., Academic Press, New York.

McWhan, D.B., Gurvitch, M., Rowell, J.M., and Walker, L.R., 1983, Structure and coherence of NbAl multilayer films, J. Appl. Phys., 54:3886.

Pinsker, Z.G., 1978, "Dynamical Scattering of X-Rays in Crystals," Springer-Verlag, Berlin.

Saxena, A.M., and Schoenborn, B.P., 1977, Multilayer neutron monochromators, Acta Crystallogr., A33:805.

Segmüller, A., 1979, Small-angle interferences of X-rays reflected from periodic and near-periodic multilayers, in: "AIP Conference Proceedings No. 53," J.M. Cowley, J.B. Cohen, M.B. Salamon, and B.J. Wuensch, eds., American Institute of Physics, New York.

Segmüller, A., and Blakeslee, A.E., 1973, X-ray diffraction from one-dimensional superlattices in GaAs(1-x)P(x) crystals, <u>J. Appl. Cryst.</u>, 6:19.

Segmüller, A., Krishna, P., and Esaki, L., 1977, X-ray diffraction study of a one-dimensional GaAs-AlAs superlattice, <u>J. Appl. Cryst.</u>, 10:1.

Speriosu, V.S., and Vreeland Jr., T., 1984, X-ray rocking curve analysis of superlattices, <u>J. Appl. Phys.</u>, 56:1591.

Spiller, E., and Rosenbluth, A.E., 1985, Determination of thickness errors and boundary roughness from the measured performance of a multilayer coating, <u>in</u>: "Applications of Thin-Film Multilayered Structures to Figured X-Ray Optics," SPIE Vol. 563:221, G.F. Marshall, ed., SPIE, Washington.

Takagi, S., 1969, A dynamical theory of diffraction for a distorted crystal, <u>J. Phys. Soc. Jpn</u>, 26:1239.

Tapfer, L., and Ploog, K., 1986, Improved assessment of structural properties of Al(x)Ga(1-x)As/GaAs heterostructures and superlattices by double-crystal X-ray diffraction, <u>Phys. Rev. B</u>, 33:5565.

Taupin, D., 1964, Théorie dynamique de la diffraction des rayons X par les cristaux déformés, <u>Bull. Soc. Fr. Minéral. Crystallogr.</u>, 87:469.

Terauchi, H., Sekimoto, S., Kamigaki, K., Sakashita, H., Sano, N., Kato, H., and Nakayama, M., 1985, X-ray studies of semiconductor superlattices grown by molecular beam epitaxy, <u>J. Phys. Soc. Jpn</u>, 54:4576.

Underwood, J.H., and Barbee Jr., T.W., 1981, Layered synthetic microstructures as Bragg diffractors for X-rays and extreme ultraviolet: theory and predicted performance, <u>Appl. Opt.</u>, 20:3027.

Vardanyan, D.M., and Manoukyan, H.M., and Petrosyan, H.M., 1985, The dynamic theory of X-ray diffraction by the one-dimensional ideal superlattice. I. Diffraction by the arbitrary superlattice, <u>Acta Crystallogr.</u>, A41:212.

Wie, C.R., Tombrella, T.A., and Vreeland Jr., T., 1986, Dynamical X-ray diffraction from nonunifrom crystalline films: Application to X-ray rocking curve analysis, <u>J. Appl. Phys.</u>, 59:3743.

Zachariasen, W.H., 1945, "Theory of X-ray Diffraction in Crystals," Wiley, New York.

HIGH RESOLUTION ELECTRON MICROSCOPY AND CONVERGENT BEAM ELECTRON DIFFRACTION OF SEMICONDUCTOR QUANTUM WELL STRUCTURES

Colin J. Humphreys

Department of Materials Science and Engineering
University of Liverpool
P.O. Box 147, Liverpool L69 3BX, England, U.K.

INTRODUCTION

For the structural characterisation of Low-Dimensional Structures the parameters we need to know include the following,: (i) the precise thickness of each layer, (ii) the presence of interface steps, (iii) the positions of atoms in the layers and at the interfaces, (iv) whether or not crystallographic defects such as dislocations and planar faults are present, (v) the local chemical composition on a nanometre scale, (vi) the electron configuration of the atoms (particularly the d-band occupancy of transition metals in magnetic superlattices) and (vii) local strains with nanometre scale spatial resolution. In addition it would be very useful to have a 'strain map' superimposed on the image.

Transmission electron microscopy (TEM) can perform all of the above, and hence it is a very powerful technique for studying and characterising low dimensional structures. TEM methods are in many ways complementary to X-ray diffraction methods, described in this book by Wim Bartels and Paul Fewster. For example, X-ray diffraction normally gives results which are averaged over relatively large areas, corresponding to the cross-sectional area of the incident X-ray beam, whereas TEM is one of the few techniques capable of probing materials with a lateral resolution better than the layer thickness of quantum well structures. On the other hand, X-ray methods are normally non-destructive, whereas TEM requires the preparation of thin specimens, thereby destroying part of the original specimen. The chapter by John Gowers on TEM concentrates on the more conventional uses of TEM with special reference to case studies. In this chapter I will focus particularly on high resolution electron microscopy (HREM) and convergent beam electron diffraction (CBED), overlapping with John Gowers where appropriate for continuity.

MEASUREMENT OF LAYER THICKNESS USING DIFFRACTION CONTRAST

The ability of dark-field images to reveal interfaces and layers in heterostructures is well established (see for example, Petroff, 1977). Fig. 1 shows a typical dark-field image of a multiple quantum well (MQW) of $GaAs/Ga_{0.7}Al_{0.3}As$ in (110) cross section using a 002 reflection (for further details see Hetherington et al, 1985). It is important to use a composition sensitive reflection, such as 002, to obtain high contrast

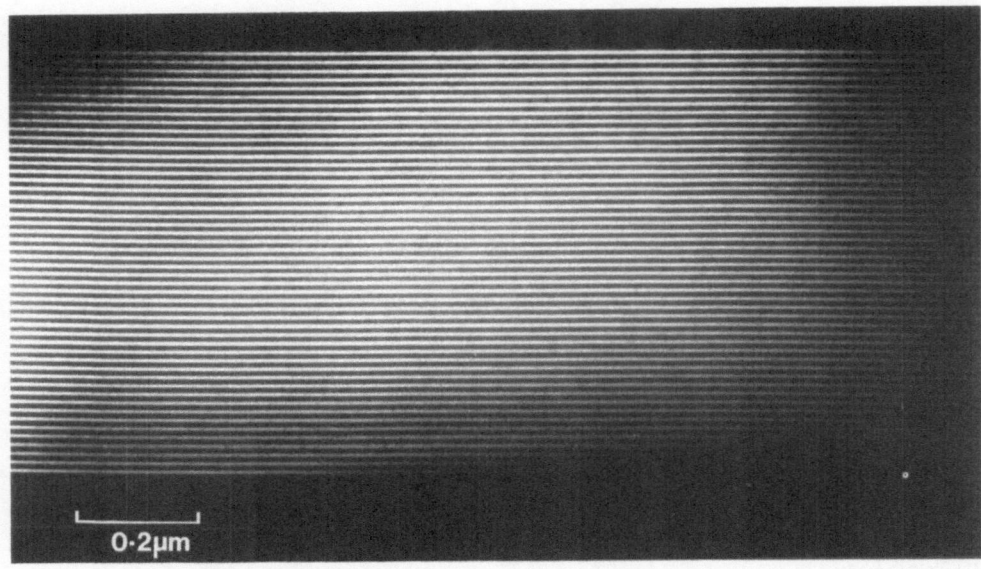

Fig. 1. Dark field (002) image of a multiple quantum well structure. Light bands are GaAlAs, dark bands are GaAs.

between the GaAs and GaAlAs layers.

The high interlayer contrast of such images arises because the structure amplitude F_{002} is equal to the difference between the atomic scattering amplitudes (f) for the group III and the group V atoms. For example, for $Ga_{1-x}Al_xAs$:

$$F_{002} = \sum_j f_j \exp(2\pi i\underline{g}\cdot\underline{r}_j) = 4\{(1-x)f_{Ga} + xf_{Al} - f_{As}\} \tag{1}$$

Since Ga and As are near neighbours in the Periodic Table, $f_{Ga} \simeq f_{As}$ and hence (1) can be written as

$$F_{002} \simeq 4x(f_{Al} - f_{Ga})^2 \tag{2}$$

The kinematical theory intensity of the 002 reflection is

$$I \propto |F_{002}|^2 \propto 16x^2(f_{Al} - f_{Ga})^2 \tag{3}$$

Thus the intensity (on kinematical theory) of the (002) reflection is proportional to the <u>square</u> of x. An accurate treatment of diffracted intensities requires dynamical electron diffraction theory (see, for example, Humphreys, 1979). However, equation (3) provides a rough indication of the sensitivity of the 002 reflection to the composition of GaAlAs and related materials. In practice, layers with 5% variations in Al composition from one layer of GaAlAs to the next have been imaged (Leys et al, 1984). Reflections such as 111 are relatively insensitive to composition because, unlike 002, the structure amplitude depends upon the sum of atomic scattering amplitudes for group III and group V atoms.

The power of the dark field 002 imaging method of cross-sectional specimens lies in the fact that <u>individual</u> layer thicknesses can be measured. Fig. 2 shows an example of GaAs/AlGaAs multilayers in which the

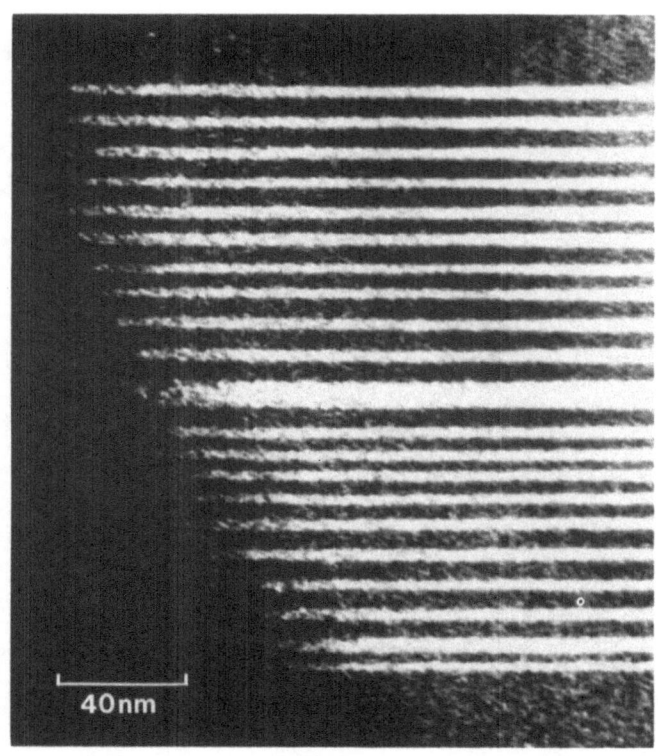

Fig. 2. Dark field (002) image of multilayers of GaAs/Ga$_{0.75}$Al$_{0.25}$As with the middle GaAlAs layer twice the width of the others, producing a strong negative differential mobility. Light bands GaAlAs, dark bands GaAs.

middle AlGaAs layer is twice the width of the others. It would have been extremely difficult to deduce this using X-ray diffraction methods. The specimen has interesting properties, exhibiting strong negative differential mobility (NDM) and acting as a very fast switch (Davies et al, 1985).

The resolution of the dark field method is at best about 5Å, irrespective of the instrumental resolution of the microscope. This is because only one diffracted beam is allowed to pass through the objective aperture in diffraction contrast imaging. Thus the maximum range of spatial frequencies that can pass through the objective aperture to form an 002 image is from 0 to 0.5 g_{002}, corresponding to a real space reso- lution of $(0.5\ g_{002})^{-1}$ = 2d$_{002}$ ≃ 5Å. If higher resolution than this is required, so that atomic height interface steps can be detected, then a larger objective aperture including more than one diffracted beam must be used, as in lattice imaging.

LATTICE IMAGING

Lattice imaging, often called high resolution electron microscopy (HREM), can utilise the full instrumental resolution capability of an electron microscope. The highest resolution microscopes currently avail- able have a point resolution of about 1.7Å, hence atomic height interface steps can in principle be imaged, provided the contrast is sufficient.

461

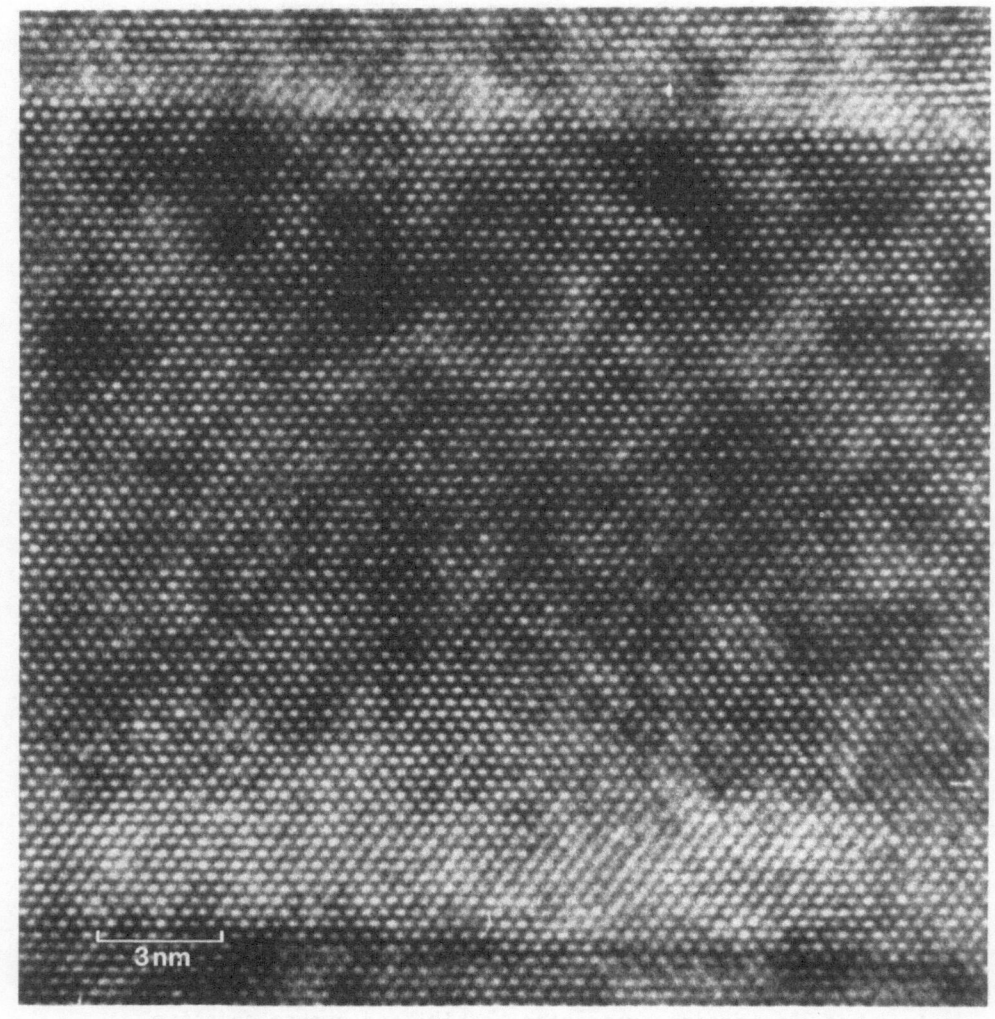

Fig. 3. Axial [110] lattice image of GaAlAs graded layers. Note the terrace of interface steps at the interface between GaAs (darker region) and AlAs (lighter region). Image formed at Scherzer defocus, 200 keV electrons, courtesy Dr. J. Hutchison.

Fig. 3 shows a lattice image of GaAlAs graded layers: with AlAs at the bottom of each layer and GaAs at the top. HREM images such as this may be used to measure accurately and absolutely the widths of each layer simply by counting lattice fringes of known separations (in fig. 3, the 2 sets of inclined fringes correspond to 111 planes and the horizontal fringes are 002 fringes, of separation d(002) = 2.8Å). The layers of fig. 3 were grown on a GaAs substrate cut slightly off (001) so that a terrace of interface steps is formed at each interface, as revealed directly in fig. 3 (Humphreys, 1986). HREM is unique in its ability to image atomic scale interface steps such as these.

It is straightforward to image terrace type steps in which the step runs through the thickness of the specimen. It is very much harder to detect isolated small area steps. However careful microdensitometry can probably be used on lattice images to detect interface roughness and to quantify Fourier coefficients of interface roughness.

462

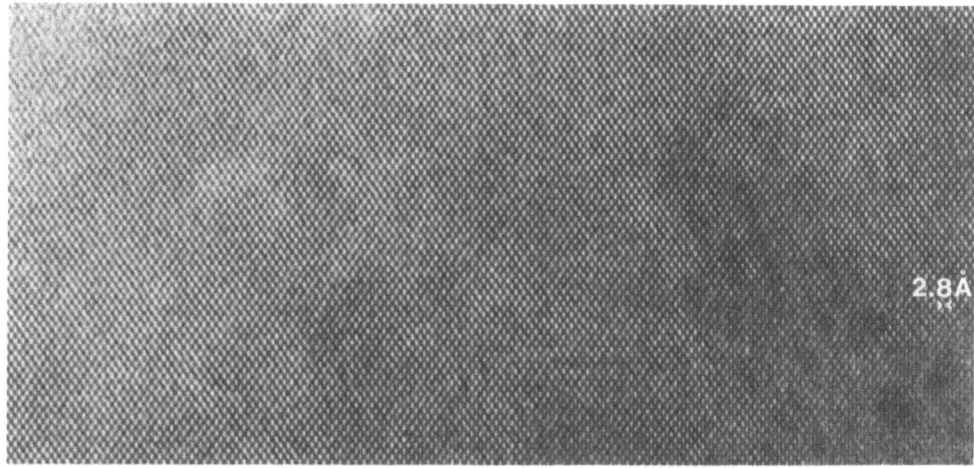

Fig. 4. Axial [110] lattice image of GaAs/$Ga_{0.7}Al_{0.3}$As layers. The interfaces are invisible owing to the lack of contrast between the layers.

A particular problem arises in HREM images when the layers are composed of very similar materials, for example GaAs and $Ga_{1-x}Al_xAs$ with x < 0.3. In such cases, as shown in fig. 4, axial [110] lattice images reveal little or no contrast between the layers, because the image intensity is dominated by the composition insensitive (111) reflections, so that the interfaces between layers are almost invisible. However, Hetherington et al, 1985, have shown that suitably tilting off [110] increases the relative contribution of the composition sensitive {002} reflections and this increases the interface contrast (see fig. 5). Alternatively, axial [100] images containing four {002} type reflections exhibit good interface contrast (Hetherington et al, 1985; Suzuki and Okamoto, 1985), as shown in fig. 6. Thus HREM may be used to image, at atomic resolution, interfaces between similar materials if the appropriate diffraction conditions are chosen.

Fig. 5. As for fig. 4, but an off-axis [110] lattice image. The interfaces are now apparent. The fringes correspond to (002) lattice planes, spacing 2.8A, and the GaAs layers are lighter than the GaAlAs layers. The diffraction pattern is shown inset.

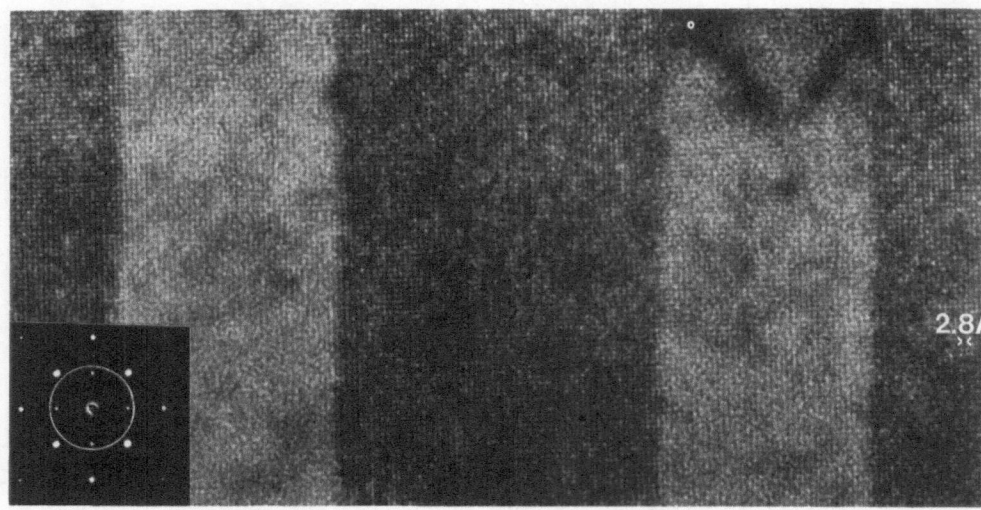

Fig. 6. Axial [100] lattice image of GaAs/Ga$_{0.7}$Al$_{0.3}$As. Note the strong interface contrast. GaAs layers are dark (note the changed contrast from Fig. 5). In the diffraction pattern shown inset, the objective aperture includes the direct beam plus four {002} type reflections.

STRUCTURE OF INTERFACES

Since high resolution electron microscopy can image interfaces at atomic, or near-atomic, resolution, the determination of the atomic structure of interfaces is possible. At this level of interpretation it is essential to perform detailed image simulation calculations. For example, the atomic coordinates of the NiSi$_2$-(111) Si interface and of the NiSi$_2$-(100) Si interface have been determined for the first time by high resolution electron microscopy (Cherns et al, 1982, 1984), and these coordinates are now being used by theoretical physicists in attempts to explain the electrical properties of silicon-silicide interfaces (e.g. Robertson, 1985).

The procedure used to determine atomic structure from HREM images is to record images for known specimen thicknesses at known defocus. Atomic models are proposed for the interface structure, and an image for each model is computed, for the required specimen thickness, microscope defocus, etc. It is usually found that all proposed models except the correct one do not fit the experimental micrographs, although insufficient image contrast or resolution can render this computer matching technique difficult.

MEASURING LOCAL COMPOSITIONS WITH HIGH SPATIAL RESOLUTION

Conventional methods of microanalysis often do not have the spatial resolution required for measuring composition changes in low dimensional structures. For example, the spatial resolution of Auger spectroscopy is at best about 500A. Energy, or wavelength, dispersive X-ray analysis (EDX or WDX) using an electron probe has a spatial resolution of about 1μm in a bulk specimen due to spreading of the electron beam. With a finely focussed probe and a thin specimen, EDX and WDX can have a spatial resolution of about 40A, but peak overlap problems can make quantitative measurements difficult (e.g. in GaAlAs, the AlKα peak overlaps the AsLα peak). For the above reasons new microanalytical methods, with very high

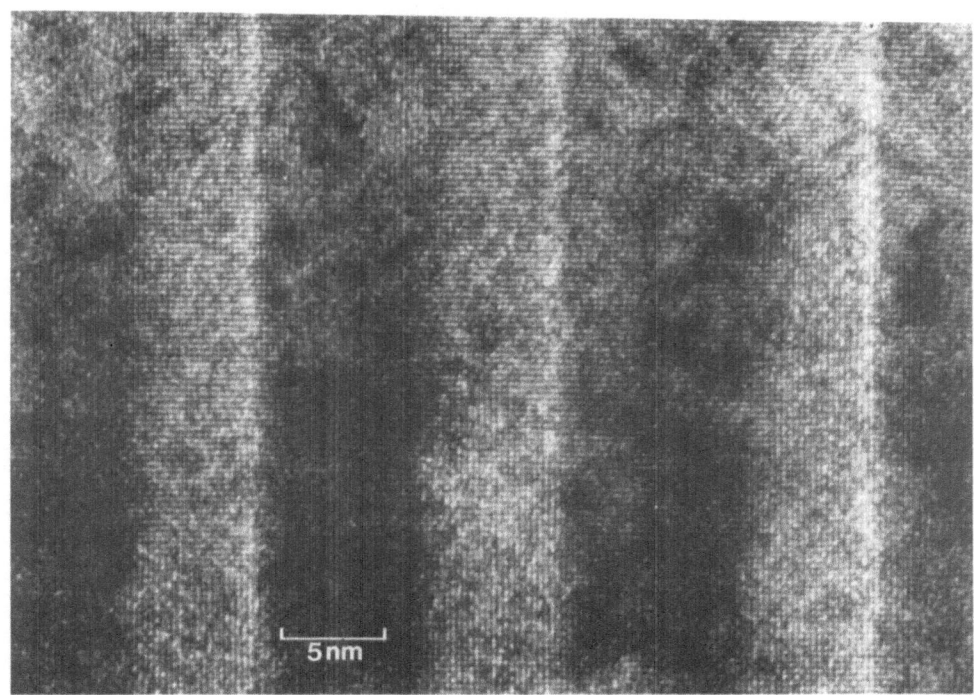

Fig. 7. Axial [100] image of MOCVD grown GaAs/GaAlAs multilayers. Bright layers are GaAlAs, dark layers are GaAs. The thin bright stripes about 8Å wide are due to a sudden increase in the Al concentration in the GaAlAs layers resulting from poor growth control. The growth direction is from left to right across the figure. The crossed fringes correspond to the (020) and (002) planes of 2.8Å spacing.

spatial resolution, are required to probe the detailed compositional variations which may occur in low dimensional structures. Three recently proposed methods are discussed below.

Measuring Local Compositions Using High Resolution Electron Microscopy

Fig. 7 shows a high resolution [100] axial lattice image of a MQW structure of GaAs/AlGaAs, of period 160Å, grown by MOCVD, with nominal composition in the AlGaAs layers of $Ga_{0.73}Al_{0.27}As$. A high intensity band about 8Å wide (three 002 fringes) occurs consistently on one side of all the GaAlAs layers in the specimen, due to a compositional change. Thus HREM is able to detect compositional changes on an Å scale. (In the limit, one lattice fringe of differing contrast could be detected, corresponding to a spatial resolution of about 2Å). Note that the resulting quantum well is highly asymmetric, and it would be extremely difficult to deduce this using X-ray diffraction methods.

Thus HREM can detect compositional variations, but can these be quantified? In the case of the specimen of fig. 7, a scanning transmission electron microscope (STEM) with a field emission gun and EDX detector was used to show that the high intensity bands of fig. 7 corresponded to an increase in Al concentration (Bullock et al, 1986). However the STEM could not be used to quantify this increase because the beam broadened probe size (∿40Å) was significantly greater than the width of the Al spike (∿8Å from the HREM image). The knowledge that the 'spike' is definitely due to Al, however, enables HREM image calculations to be

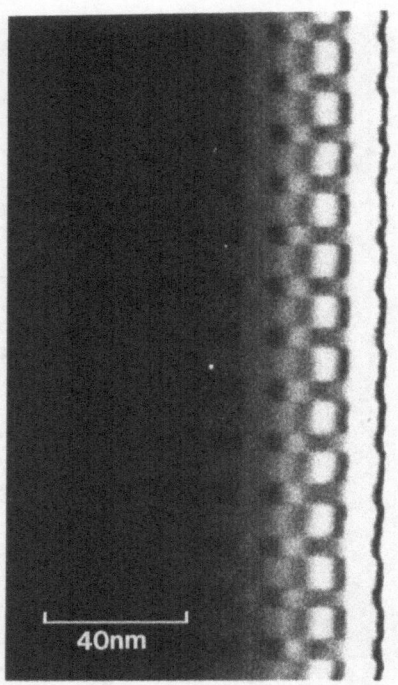

Fig. 8. Bright field micrograph of a
MQW GaAs/AlGaAs wedge specimen. The
alternating GaAs and AlGaAs layers are
horizontal. The wedge tapers towards
the right hand side so that the
thickness fringes are vertical.

performed for a range of Al concentrations and matched with experimental
micrographs. This procedure shows that the Al concentration rises from
its nominal value of 0.27 to about 0.5 in the spike.

Thus HREM can directly and visually detect composition changes. If
the elements involved are known from other information (e.g. EDX or EELS)
then image simulation and matching can be performed to yield quantitative
microanalysis on a scale of a few Å.

Measuring Local Compositions Using Thickness Fringes

Kakibayashi and Nagata (1985) have proposed a novel method of compositional analysis using cleaved specimens and thickness fringes. The
method entails cleaving the specimen along two {011} planes with normals
perpendicular to the [100] growth direction, and imaging the crystal in
bright field, with the electron beam along the [010] direction perpendicular to the 90 degree wedge. An example of such an image of an AlGaAs
layer on GaAs is shown in fig. 8. Kakibayashi and Nagata noticed that
the spacing of the thickness fringes seemed to be proportional to the
alloy composition, thus allowing x in $Al_xGa_{1-x}As$ to be measured from
the thickness fringes, with a very high spatial resolution of a few Å.

Further analysis of the situation (Eaglesham et al, 1986) shows that
the interpretation is more complex than originally assumed by Kakibayashi
and Nagata, in particular the alloy composition x is not simply
proportional to the spacing of the thickness fringes, due to many-beam

466

diffraction effects. However x can be determined by detailed matching of experiment with calculated thickness fringe profiles, including the effects of electron absorption (Eaglesham et al, 1986). The technique has the advantage of rapid specimen preparation, it is immediately visual - an Al (say) spike looks like a spike in the image, and values of x to an accuracy of 5-10%, with a spatial resolution of a few A, can be obtained using image matching with calculations.

Measuring Local Compositions Using Convergent Beam Electron Diffraction. (CBED)

The third technique recently proposed for microanalysis is a diffraction method rather than an imaging method, and it may be explained in terms of the localisation of the Bloch wave states of the fast incident electrons within the crystal. For the case of 100 keV electrons incident along [010] in AlGaAs, only the first, second and fifth states contribute significantly. The first Bloch wave state has maxima on columns of As atoms, the second on mixed Ga/Al atomic columns and the fifth has weak maxima on all atomic columns. Since the total energy of each Bloch state is equal to the incident electron energy, the kinetic energy of the Bloch states is determined by the strengths of the potentials they experience, so all these states have different wave vectors. These wave vectors are directly displayed in the higher order Laue zone (HOLZ) lines of CBED patterns. Thus by measuring the separation of the appropriate HOLZ lines on a CBED pattern with a ruler we can directly measure the strength of the crystal potential and hence deduce the composition (Eaglesham and Humphreys, 1986). This method has an accuracy of about 3% and a spatial resolution that of the incident beam probe diameter, broadened only by elastic scattering effects, (typically 50A). It is of interest to note that this method of microanalysis essentially measures the eigenstates of the fast electrons within the crystal, whereas EDX and EELS measures core-electron eigenstates.

MEASURING ELECTRON CONFIGURATIONS ON A NANOMETRE SCALE

A knowledge of, for example, the d-state occupancy of transition metals in metallic multilayers as a function of layer thickness could be very revealing. Recent work by Waddington et al (1986) shows that this is possible in principle using electron energy loss spectroscopy (EELS) which can be performed using an electron probe only 1 nm in diameter, in a STEM.

The EELS spectra for 2p electron excitation in the 3d transition metals (and their compounds) consist of 2 sharp peaks, due to transitions from the 2p states which are split by spin orbit coupling, to empty 3d states. The relative intensities of these 2 sharp peaks have been unexplained for many years. However, Waddington et al (1986), using multiconfigurational Dirac-Fock theory, have recently obtained good agreement with experiment. It is clear that the peak intensities are sensitive to the d state occupations of the atoms. These can therefore now be studied as a function of the layer thickness in metallic multilayers.

MEASURING LOCAL STRAINS IN SUPERLATTICES

Until recently, all semiconductor superlattices have been grown from materials that are closely lattice matched (to within about 0.1%) e.g. GaAs/AlGaAs. In the last few years it has become possible to grow high-quality strained layer superlattices (SLS's) from a wide variety of lattice mismatched semiconductors. A knowledge of the local strain in layers and around interfaces is essential to understand and quantify properties such as strain induced band gap variations and mobilities. This knowledge can also be used deliberately to 'fine-tune' band gaps: the so-called "strain engineering" of band gaps.

Fraser et al (1985) and Maher et al (1985, 1986) have used convergent beam electron diffraction with a 40A diameter probe to detect and measure local strains in a Si/GeSi SLS. When GeSi is grown epitaxially upon Si, the GeSi layers distort tetragonally in the growth direction. When cross-sectional electron microscope specimens are prepared, surface relaxation effects must be taken into account, and this leads to the GeSi layers in Si/Ge$_{0.05}$Si$_{0.95}$ becoming orthorhombic. In Si/Ge$_{0.1}$Si$_{0.9}$ layers, the GeSi distorts to become monoclinic. These strain-induced new phases of material which is cubic when unstrained will have new electrical properties, and CBED is an essential technique to quantify the local strains. Surface strain relaxation effects are also expected to be important in thin film devices, and again CBED using small probes will be an important analytical technique.

The angular resolution of CBED is about 10^{-4} rad., and all three lattice parameters can be measured to 1 part in 10^4, with a spatial resolution of the probe size (typically 50A). The local symmetry of the crystal is immediately apparent from the symmetry of the HOLZ lines in the CBED pattern. For example in [001] CBED patterns from a thin layer of GeSi on Si it is immediately evident that the pattern from the Si substrate has 4-fold symmetry, hence the substrate is cubic, whereas the pattern from GeSi has only 2-fold symmetry (Maher et al, 1985, 1986). In order to obtain quantitative values for lattice parameters, and hence strain values, it is necessary to fit experimental patterns of HOLZ lines with computer simulations.

MAPPING LOCAL DISTORTIONS ACROSS INTERFACES

Recently a new TEM technique has been developed for detecting, mapping and measuring small crystalline distortions with an accuracy approaching that attainable using X-ray diffraction, but with far higher spatial resolution (Humphreys et al, 1987; Maher et al, 1985). With this technique, called convergent beam imaging (CBIM), strains, lattice parameter variations and structural distortions are revealed as displacements of higher-order Laue zone (HOLZ) lines which are superimposed on the image.

The CBIM technique uses a convergent beam focussed either above or below the specimen. The specimen should be reasonably thick and the visibility of HOLZ lines in the CBIM image is significantly improved by cooling the specimen. The varying curvature of HOLZ lines across the image of an interface maps the rate of strain variation across the interface and immediately reveals whether or not strain changes abruptly at an interface. The CBIM technique has general applications to the investigation of multilayers where localised variations in strain, lattice parameter, crystal symmetry and crystallographic rotation are of interest.

CONCLUSIONS

Transmission electron microscopy is a powerful technique for providing local information, with high spatial resolution, on low dimensional structures. It can measure layer thicknesses, interface steps, local compositions and local strains. TEM using a defocussed probe can also provide a "strain map" superimposed on an image.

REFERENCES

Bullock, J.B., Huxford, N.P., Titchmarsh, J.M. and Humphreys, C.J., 1986,
    Proc. XIth Int. Cong. on Electron Microscopy, Kyoto, p.1473.
Cherns, D., Hetherington, C.J.D., and Humphreys, C.J.,1984, Phil. Mag.
    A49, 165.
Cherns, D., Spence, J.C.H., Anstis, G.R. and Hutchison, J.L., 1982,
    Phil. Mag. A46, 849.
Davies, R.A., Kelly, M.J., Kerr, T.M., Hetherington, C.J.D. and
    Humphreys, C.J., 1985, Nature, 317, 418.
Eaglesham, D.J., Hetherington, C.J.D. and Humphreys, C.J., 1986.
    Proc. Mat. Res. Soc. Symp., to be published.
Fraser, H.L., Maher, D.M., Humphreys, C.J., Hetherington, C.J.D.,
    Knoell, R.V. and Bean, J.C., 1985, Microscopy of Semiconductor or
    Materials. Inst. Phys. Conf. Ser. No.76, p.307.
Hetherington, C.J.D., Barry, J.C., Bi, J.M., Humphreys, C.J., Grange, J.
    and Wood, C., 1985, Mat. Res. Soc. Symp. Proc., 37, 41.
Humphreys, C.J., 1979, Rep. Prog. Phys., 42, 1825.
Humphreys, C.J., 1986, Proc. XIth Int. Cong. on Electron Microscopy,
    Kyoto, p.105.
Humphreys, C.J., Maher, D.M., Fraser, H.L. and Eaglesham, D.J., 1987,
    Submitted to Phil. Mag.
Kakibayashi, H. and Nagata, F., 1985, Japan. J. Appl. Phys., 24, 1.905.
Leys, M.R., Van Opdorp, C., Viegers, M.P.A. and Tulan-Van Der Mheen, H.J.,
    1984, J. Cryst. Growth, 68, 431.
Maher, D.M., Fraser, H.L., Humphreys, C.J., Knoell, R.V., Field, R.D.,
    Woodhouse, J.B. and Bean, J.C., 1985, Electron Microscopy and
    Analysis 1985. Inst. Phys. Conf. Ser. No. 78, p.49.
Maher, D.M., Fraser, H.L., Humphreys, C.J., Field, R.D. and Bean, J.C.,
    1986, Appl. Phys. Lett. (in press).
Petroff, P.M., 1977, J. Vac. Sci. Technol., 14, 973.
Robertson, J., 1985, J. Phys. C., 18, 947.
Suzuki, Y. and Okamoto, J., 1985, J. Appl. Phys., 58, 3456.
Waddington, W.G., Rez, P., Grant, I.P. and Humphreys, C.J., 1986,
    Phys. Rev. B, 34, 1467.

THE TEM CHARACTERISATION OF LOW-DIMENSIONAL STRUCTURES IN EPITAXIAL

SEMICONDUCTOR THIN FILMS

J.P. Gowers

Philips Research Laboratories
Cross Oak Lane, Redhill, Surrey, U.K.

1. INTRODUCTION

Many types of low-dimensional structures can now be grown in epitaxial semiconductor thin films using growth techniques such as molecular beam epitaxy (MBE) and metal-organic chemical vapour deposition (MOCVD)[1,2]. Recent advances in the understanding of the fundamental growth processes and in the control technology, now enable layer thickness, composition and interface abruptness to be controlled at an atomic level of precision[3-6].

The powerful range of techniques that transmission electron microscopy (TEM) comprises allows the resultant structural properties to be studied in detail, with under some circumstances, near atomic resolution[7,8]. This unique ability of analytical electron microscope techniques to provide local nanometre scale structural information complements X-ray diffraction techniques which are able to provide often high precision parameters but averaged over regions many orders of magnitude larger[9]. The main purpose of this article is to illustrate some of the many types of information available from the wide range of TEM techniques with which it is possible to study the structural, compositional and dimensional properties of low-dimensional structures. The microscope techniques fall broadly into the three categories corresponding to the imaging, diffraction and probe modes of operation of the TEM, here assumed to be a modern ~ 100 keV conventional analytical machine.

The types of information that can be obtained by TEM characterisation are illustrated by a necessarily limited range of examples, although others are demonstrated in Prof. Humphrey's article and more are described in the recent literature[10]. The examples in the next section are grouped under three broad headings according to whether the principal interest is dimensional, compositional or structural, although in reality all three aspects of a given low-dimensional structure may be of importance. The next section contains examples of the use of TEM techniques in the study of the dimensional properties of some structures including the measurement of local layer thicknesses, uniformity, growth rate and hence the group III element flux calibration for MBE growth. Local compositional fluctuations can occur in III-V alloy structures[11,12] and may arise during growth from either kinetic of thermodynamic effects[13,14], their presence can influence the electronic properties of resultant structures[15]. One aspect involving TEM studies of these fluctuations is discussed in section 2.

Extended structural defects are nearly always present in epitaxial films and the literature contains numerous examples of detailed TEM analyses of their type, density, location and origin for a number of different combinations of materials and growth methods[10]. For the purposes of this article two structural topics are discussed in section 3, namely the imaging of antisite domains on thin GaAs films grown on (001)Ge[16], and diffusion induced disordering effects in GaAs-AlAs superlattices, both grown by MBE[17].

2. DIMENSIONAL STUDIES

The growth of III-V multiple quantum well (MQW) and superlattice (SL) structures for device and physics applications has generated a need for measurements of the thickness of individual well and barrier layers. Often some sort of average values of these parameters are reasonably well known (say better than ± 10%) simply from the calibration of the growth conditions. For example in the case of GaAs-AlAs structures grown by MBE, the individual layer thickness grown is in the absence of re-evaporation, proportional to the product of the group III flux, sticking coefficient and shutter-time. Also, the use of insitu RHEED oscillations during MBE growth may be used to monitor the growth of successive monolayers[4]. While techniques such as photoluminescence[18] and X-ray diffraction[9] are also able to measure the MQW well thickness and well and barrier thicknesses respectively, the measurements are essentially averages of a number of periods over an area some hundred or more micrometers across.

More detailed results may be obtained from cross-sectional TEM images which are particularly valuable for revealing local variations of layer thickness. Either (100) or (110) cross-sectional specimens are made by conventional polishing and argon ion beam milling operations[19,20] and dark field (DF) images are made using an objective aperture around the 002 or 00$\bar{2}$ spot in the diffraction pattern[21]. The aperture will necessarily mean that the resolution in the image cannot be better than about 5Å. The 002-type dark field images are useful because the image intensity is sensitive to the structure factor F_{002} which is proportional to the difference in group V and III element scattering factors ($f_V - f_{III}$). Good contrast occurs between III-V binary layers e.g. GaAs and AlAs, but the intensity is only directly proportional to F^2_{002} when the specimen thickness is less than a few nanometres and where kinematical (single) scattering conditions may dominate.

The use of the TEM as a nanometre ruler is illustrated by way of the 002 dark field micrographs of relatively uniform and non-uniform MQW structures in Fig. 1. The GaAs well (W) and $Ga_{0.7}Al_{0.3}As$ barrier (B) thicknesses can readily be obtained as a function of period number. The variations of W and B for the two structures are shown in Fig. 2 where it can be seen that the structure in Fig. 2(b) – the first MQW that was grown by MBE at PRL Redhill – is relatively non-uniform compared with that in Fig. 2(a). The thickness values of W and B are of course not unique but correspond to the local regions of specimen imaged in Fig. 1 and are averages through several hundred angstroms of the (110) cross-sectional specimen.

The precision with which the values of B and W can be measured from micrographs such as Fig. 1 can be a few percent, corresponding to a few angstroms. However the accuracy of the values obtained will in general be lower because the images are not simply related to the projected atomic potential of the specimen crystal lattice[22]. Several effects combine to limit the accuracy of measurement of well and barrier thicknesses from 002 dark field micrographs. They include (a) the ~ 5Å resolution limit imposed

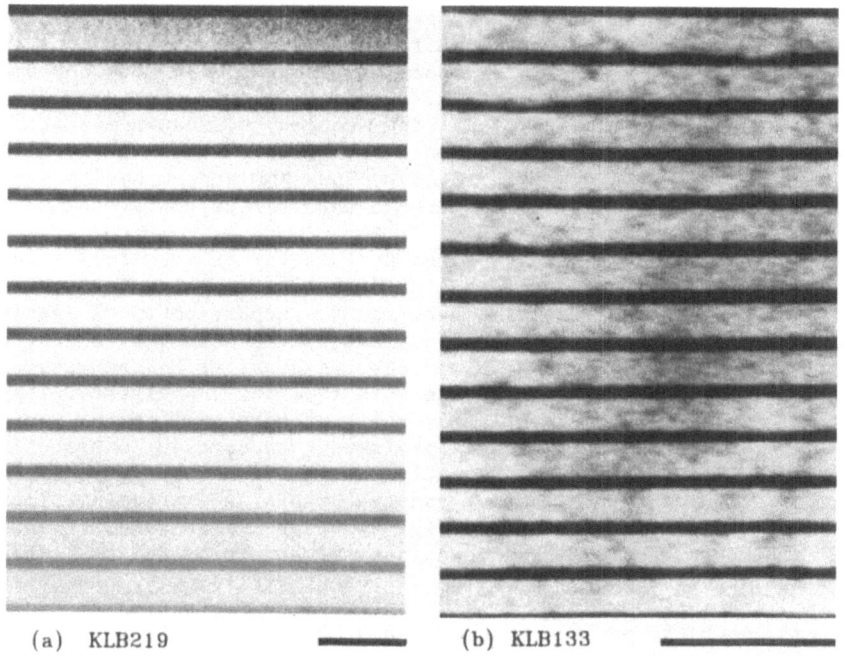

(a) KLB219   (b) KLB133

Fig. 1.   Cross-sectional (110) TEM micrographs taken in 002 dark field
          conditions showing (a) part of a uniform 15 period MBE grown MQW
          structure having GaAs wells (dark) and (Al,Ga)As barriers (light),
          and (b) part of a relatively non-uniform 60 period MQW structure.
          The magnification bars are 500Å.

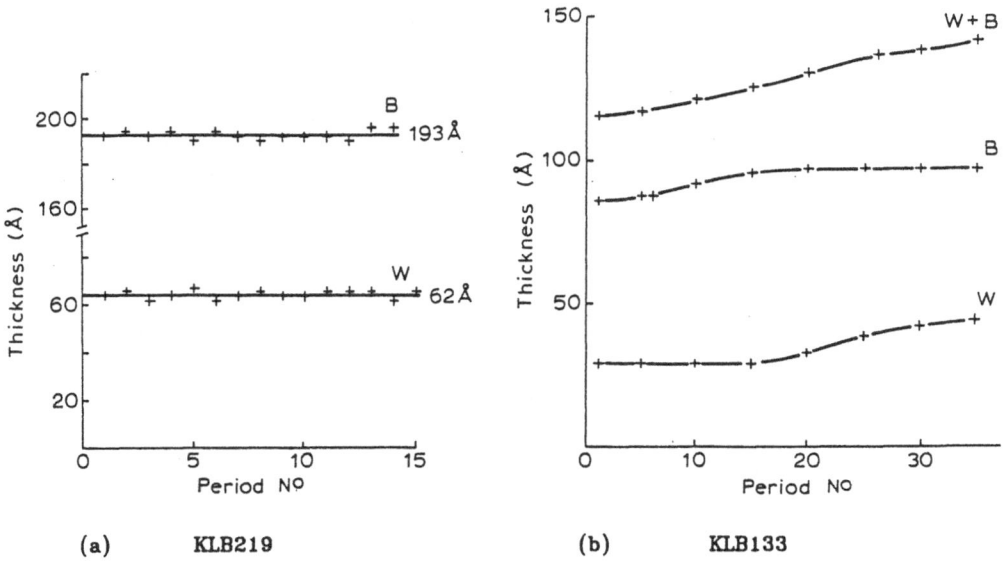

(a)   KLB219   (b)   KLB133

Fig. 2.   Plots of the GaAs well thickness (W), (Al,Ga)As barrier thickness
          (B) as a function of period number for the two structures shown in
          Fig. 1.

by the use of an objective aperture to make the image, (b) the projection effect arising from the specimen tilt to the 002 Bragg angle of 0.34° which broadens the interface contrast, (c) monolayer steps in GaAs-AlAs interfaces, (d) ion beam milling damage of the specimens surfaces, (e) multiple beam dynamical diffraction effects, (f) phase grating contrast which may occur when MQW satellite spots on either side of the 002 spot are transmitted by the objective aperture, and finally (g) associated Fresnel fringe contrast. Although in general the accuracy of measurements will be reduced by these effects they may often be relatively small and results of independent measurements of well widths by TEM, XRD and PL have suggested the accuracy of 002 dark field TEM measurements can be better than ± 5Å[23].

An alternative way of studying the periodicity of a uniform structure uses the coherent diffraction of the incident electron beam to produce satellite spots in the diffraction pattern of a cross-sectional specimen. The satellites occur about the usual crystal lattice spots by means of multiple diffraction and lie on a line that is parallel to the growth plane normal. The first order satellite distance from its associated lattice spot in the diffraction pattern is inversely proportional to the average MQW or SL period in the part of the specimen cross-section contributing to the diffraction pattern[21]. Consequently the period can be determined directly using the 002 spot spacing and the GaAs-AlAs average lattice parameter of 5.657Å. In Fig. 3 and Fig. 4 002 DF images and diffraction patterns are shown of (110) cross-sections of MBE grown $(GaAs)_m-(AlAs)_n$ superlattice structures having design values of period (m+n) monomolecular layers of 6 and 4 respectively[24]. The diffraction patterns give measured average values of the periods of 6.12 and 3.62 respectively in good agreement (2%) with X-ray diffraction measurements over a much larger area. The individual average values of m, n are more difficult to obtain, but values with a precision of about ± 5% depending on the number of periods probed by the beam selected area of the structure may be obtained by energy dispersive X-ray analysis (EDX). In effect the MQW or SL structure is treated as an alloy and the average composition determined by standard thin specimen analysis techniques[25].

These illustrations of the use of the TEM as a nanometre ruler for obtaining dimensional measurements have been valuable in complementing other measuring techniques which often have much larger effective probe sizes, e.g. X-ray diffraction, photoluminescence and Raman spectroscopy. In addition to the direct measurement of layer thickness values, other growth parameters such as group III element MBE fluxes, sticking coefficients and growth rates characteristic of the growth experiment may also be derived.

3. COMPOSITIONAL STUDIES

The measurement of average layer composition and the detection of local compositional changes are important in low dimensional structures comprised of III-V binaries and their alloys. Not only are the resultant electronic and optical properties directly determined by these parameters but they can also influence the formation and behaviour of extended crystallographic defects. The measurement of composition in the TEM may be achieved using what are now standard thin specimen EDX (energy dispersive X-ray) analysis techniques[25]. The electron beam may be focussed so as to probe regions of the specimen only a few hundred angstroms across, or defocussed to analyse much larger volumes.

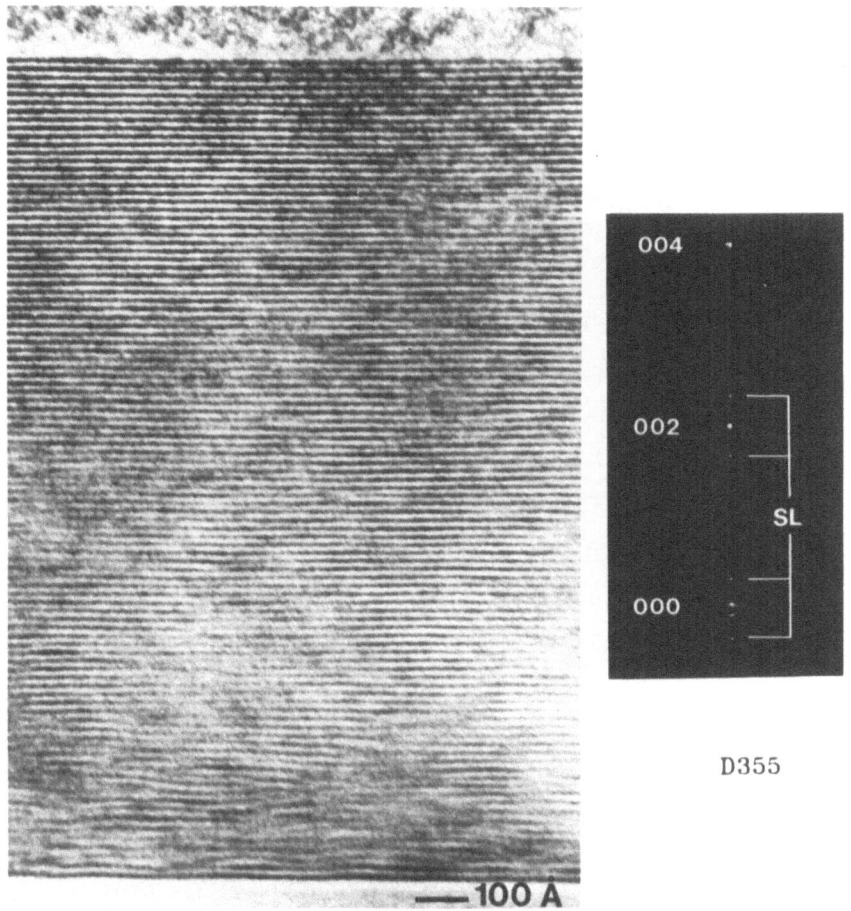

D355

Fig. 3. Cross-sectional (110) TEM micrograph (a) taken in 002 dark
conditions of an MBE grown superlattice made of 100 alternate
layers of GaAs and AlAs each designed to be three monolayers
thick. The superlattice spots (SL) in the diffraction pattern
have spacings corresponding to an average period of 6.12
monolayers.

Compositional information may also be obtained from diffraction
patterns in those cases where the lattice parameter changes in a known way
with alloy composition. The precise interpretation of spot spacings in
diffraction patterns in terms of local composition can be complicated by
strain relaxation effects which occur in thin specimens[26]. In the near
lattice-matched GaAs-AlAs system composition measurements using electron
diffraction are more difficult although the convergent beam technique
described in Prof. Humphrey's article is clearly a useful alternative.

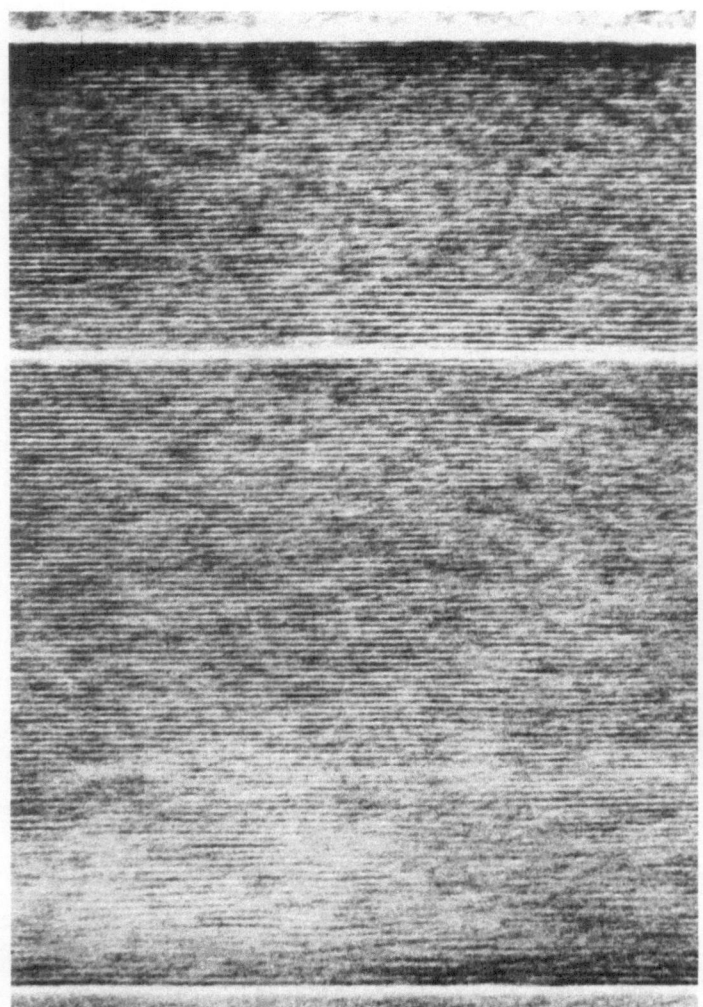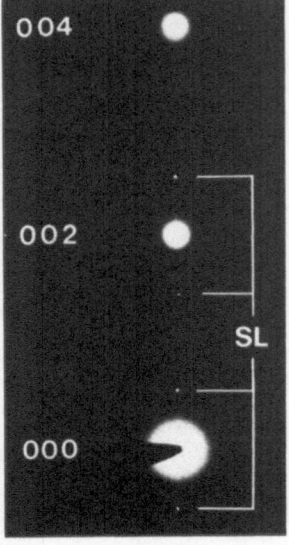

D365

Fig. 4.   Cross-sectional (110) TEM micrograph (a) and diffraction pattern
(b) of an MBE grown superlattice having 100 plus 50 alternating
layers of GaAs and AlAs of designed thickness 2 monolayers each.
The spacing of the superlattice spots (SL) indicates an average
period of 3.62 monolayers.

The determination of $Al_xGa_{1-x}As$ alloy composition is possible by two
other techniques which are mentioned for completeness, although not
illustrated here. In one it is assumed that the kinematical approximation
can be used to determine the scattered intensity in the 002 reflection from
a uniformly thin specimen. The intensity I_{002} in the 002 DF image is then
related to the composition by way of the electron scattering factors[27]:-

$$I_{002} = 16 \left(x\, f_{Al} + (1-x)\, f_{Ga} - f_{As} \right)^2 \qquad \ldots \quad \ldots \quad \ldots \quad (1)$$

where f_{Al}, f_{Ga} and f_{As} are the elemental electron scattering factors. This method may be of some use with a TEM having a scanning attachment to measure the value of I_{002} in the region of interest, but it is not accurate because of the effects of multiple scattering which are strong, even in thin specimens. In the other method proposed by Kakibayashi and Nagata[28] the spacing of thickness fringes due to a 90° {110} cleaved wedge was used to determine the composition of MBE and MOCVD grown $Al_xGa_{1-x}As$ multilayers. This method has been further refine by Eaglesham and Humphreys, as reported elsewhere in this volume.

The microscopic structure and composition of semiconductor III-V alloys has been the subject of considerable interest in recent years. Qualitatively the effects of local fluctuations in composition, attributed variously to clustering, ordering or spinodal decomposition, have been observed in TEM images, diffraction patterns and also by EDX analysis using a small electron probe[11,29,30]. In two-beam 220 TEM images the fluctuations can give rise to a grainy contrast on the scale of 10 nanometres, and from the contrast amplitude it is possible to make an approximate estimate of the amplitude of the lattice distortions arising from the fluctuations in composition[29]. Vegard's law is assumed to relate the local lattice parameter to the local composition and in addition strain relief mechanisms which can be important in thin TEM specimens have been neglected in the first order treatment. Examples of the contrast are shown in Fig. 5 for GaAs, $Ga_{0.46}In_{0.54}As$, $Ga_{0.52}In_{0.48}P$ and $Ga_{0.27}In_{0.73}As_{0.63}P_{0.37}$. It is strong for the Ga-In containing alloys and arises by dynamical diffraction between the incident and diffracted beams when the lattice parameter varies in an approximately periodic way through the crystal. The contrast is particularly strong when the compositional periodicity L in the beam direction is comparable with the extinction distance ξ_g. This can be seen by solving the coupled differential equations of the two-beam dynamical theory in the column approximation[31] using an oscillatory function for the local displacement vector of the diffracting planes down the column. When the displacement vector $\bar{R}$ is represented by a sine wave:

$$\bar{R} = (\bar{b}/2) \sin \left(\theta + \frac{2\pi z}{L} \right) \qquad \ldots \quad \ldots \quad \ldots \quad (2)$$

where $\bar{b}$ is the maximum difference in the 220 interplanar distance for Ga-rich and In-rich clusters say, z is the distance through the crystal in the beam direction, and θ is a phase angle describing the local displacement at the upper crystal surface where z is zero, then local variations in θ will give rise to image contrast. It is this type of contrast which is visible in the micrographs of the III-V alloys in Fig. 5(b)-(d) and which is absent from the GaAs binary in Fig.5(a). Although variations in z and L are also important, it is mainly $|b|$ that determines the magnitude of the contrast. For the ternary alloy $Ga_{0.52}In_{0.48}P$ the estimated contrast in the negatives of the micrograph in Fig. 5(c) is a few tens of a percent, and the cluster size is about 100Å, consistent with compositional deviations $\Delta x/x$ of $\pm 10\%$[29].

4. STRUCTURAL STUDIES

Transmission electron microscopy has been extensively used to study structural properties of low-dimensional epitaxial semiconductor thin films. The properties cover a wide range of types of deviation from an ideal perfect crystal film including extended crystallographic defects such as dislocations, stacking faults, microtwins, antisite boundaries and morphological features such as interfacial steps, facets, rigid body displacements, strain fields and so on. In this section two of many

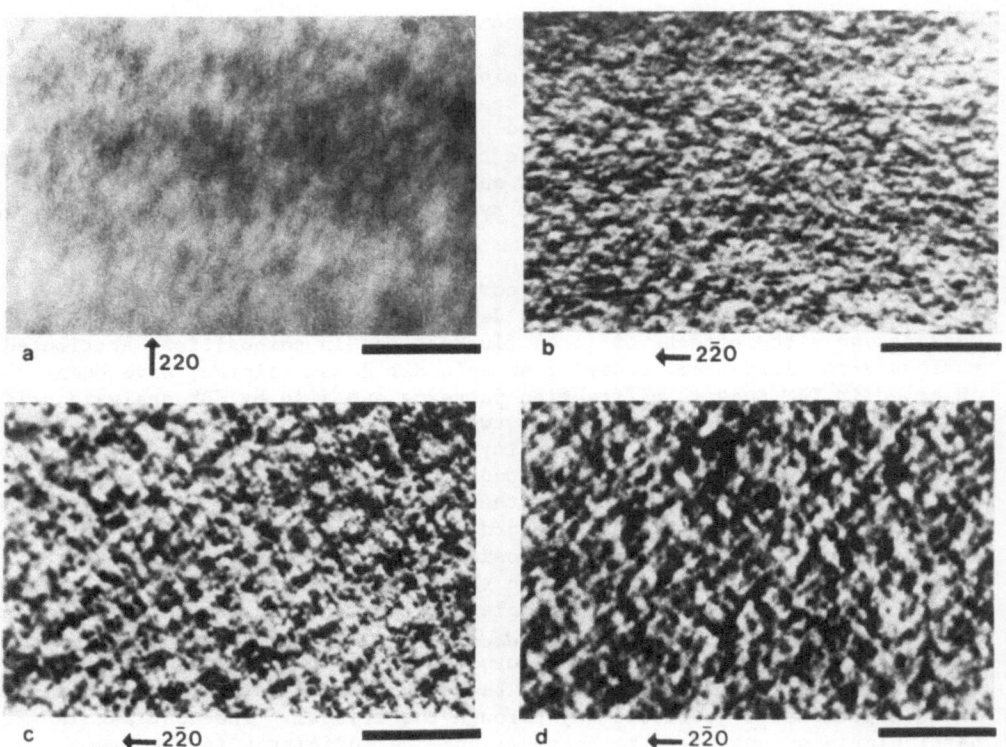

Fig. 5(a)-(d) TEM bright field images of [001] epitaxial layers of
(a) GaAs, (b) $Ga_{0.46}In_{0.54}As$, (c) $Ga_{0.52}In_{0.48}P$ and
(d) $Ga_{0.27}In_{0.73}As_{0.63}P_{0.37}$. The deviation parameter
$S=0$, the magnification bar is 1000Å.

possible examples illustrate TEM techniques that have application in the
study of a pair of material systems of considerable current interest. The
first concerns the growth of polar III-V semiconductors on non-polar Si or
Ge substrates, and the second the disordering of III-V superlattices and
MQW structures by dopant diffusion.

The growth of GaAs on Ge has been studied both theoretically and also
experimentally using MBE[16]. The system has a very small lattice mismatch
which should result in relatively low densities of extended defects being
generated at the interface. However the possibility of antisite domains
being generated in the polar GaAs layer for some orientations of the Ge
substrate has been recognised for sometime[32]. More recently a general
treatment of the possible crystallographic relationships between the
substrate & layer has been developed by Pond and co-workers based on the
principle of symmetry compensation[33,34]. The principle allows for the
two-fold reduction in the number of symmetry operations that distinguish
the GaAs lattice from the Ge lattice by predicting the possible existence
of two equivalent structural variants for particular substrate
orientations.   In this case the variants correspond to antisite domains
(also widely referred to by the misleading name of antiphase domains). The
presence of antisite domains in thick GaAs films was originally confirmed
by etching studies[35], although they have only recently been unambiguously
identified by transmission electron microscopy in GaAs films a few hundred
angstoms thick grown by MBE on Ge substrates. The TEM observations used a
new imaging technique based on an asymmetry in the 002 convergent beam

478

diffraction disc intensity arising from multiple diffraction effects[36] in the two domain types.

The origin of the domain contrast may be explained by considering the phase relationships between the amplitudes of the diffracted beams which may be written after Taftφ and Spence[37] as

$$\omega = -n\pi/2 + \sum_{1}^{n} \omega_i \qquad \ldots \quad \ldots \quad \ldots \quad (3)$$

where $-\pi/2$ is the phase retardation of the incident beam for each of n elastic scattering events at the Bragg angle, and ω_i are the structure factor phase changes for each of these n reflections. Knowing the number of reflections n and their structure factor phase angles ω_i allows ω to be calculated from equation (3). For the non-centrosymmetric GaAs sphalerite structure an arbitrarily chosen unit cell has Ga at the origin and at the three other positions related by face-centring operations (0 $\frac{1}{2}$ $\frac{1}{2}$, $\frac{1}{2}$ 0 $\frac{1}{2}$, $\frac{1}{2}$ $\frac{1}{2}$ 0) and As at $\frac{1}{4}$ $\frac{1}{4}$ $\frac{1}{4}$ and at three other positions related by face-centring operations ($\frac{1}{4}$ $\frac{3}{4}$ $\frac{3}{4}$, $\frac{3}{4}$ $\frac{1}{4}$ $\frac{3}{4}$, $\frac{3}{4}$ $\frac{3}{4}$ $\frac{1}{4}$). Although the arbitrary choice of origin will determine the structure factor phase angle ω_i for a given reflection, the phase difference between reflections will be independent of the choice of origin. In the usual way the structure factor phase angle is calculated from the complex scattering factor and is the angle whose tangent is the ratio of the imaginary divided by the real component. For the chosen unit cell of GaAs the scattering factor F_{hkl} for plane (hkl) is given by:

$$F_{hkl} = \text{CONSTANT} \left\{ f_{Ga} + f_{As} \exp\left[i(h+k+l)/2 \right] \right\} \qquad \ldots \quad \ldots \quad \ldots \quad (4)$$

where the value of the CONSTANT is either 4 for unmixed (hkl all odd or all even) and zero for mixed reflections. Consequently four types of reflection can be identified depending on whether the value of (h+k+l) in equation (4) is 4m, (4m+2) or (4m±1), where the integer m = ...-2, -1, 0, 1, 2...., and these are shown in Table 1 below.

From equation (3) and Table 1 the overall phase relationships between singly and doubly elastically scattered beams may be readily calculated for specific beam paths (i.e. hkl values).

Table 1. The structure factor F_{hkl} and approximate structure factor phase angle ω_i for the four types of reflection hkl in GaAs. The integer m enables the type and phase of a given reflection to be readily deduced.

hkl	example	(h+k+l)	F_{hkl}	ω_i (radians)
all even	220,$\bar{4}$00	4m	$4\{f_{Ga}+f_{As}\}$	0
all even	$\bar{2}$00,222	4m+2	$4\{f_{Ga}-f_{As}\}$	π
all odd	1$\bar{1}$1,311	4m+1	$4\{f_{Ga}+if_{As}\}$	$\approx \pi/4$
all odd	111,331	4m-1	$4\{f_{Ga}-if_{As}\}$	$\approx -\pi/4$

In the case of an antisite domain in which the Ga atoms on one sublattice and the As atoms on the other have been interchanged, then As is now at the origin causing f_{Ga} and f_{As} to be exchanged in equation (4). The effect of this on ω_1 in Table 1 is to change only the phase of (4m±2) reflections from π to zero[33]. Consequently the combination of a (4m+2)-type singly diffracted beam (e.g. 002) with a doubly diffracted beam of another type, say (4m+1), can result in the in-phase addition of amplitudes in one domain and the out-of-phase subtraction of amplitudes in another[37]. In the 002 dark field image the domains appear in light and dark contrast. This can be seen in Fig. 6 which shows convergent beam diffraction patterns and images of a ~ 200Å GaAs film on 500Å of Ge grown by MBE. The images are similar to 002 dark field images but the GaAs film has been carefully oriented with respect to the incident electron beam so that a pair of doubly diffracted beams from the first order Laue zone also contribute to the image[36]. These beams are visible as a cross in the 002 convergent beam diffraction disc. In Fig. 6(a) the beams involved are $\overline{9}31$, $\overline{2}00$ and 731, in Fig. 6(b) they are $\overline{7}31$, 200, and 931 and the corresponding $\overline{2}00$ and 200 images in Fig. 6(c) and Fig. 6(d) show the contrast reversal of the domains which can be seen in more detail in Fig. 7. The phase relationships are summarised in Table 2 for the relevant beam paths and also for the two domain types. It can be seen that the Ge film contributes uniformly to the overall image intensity since the doubly diffracted beams are nearly exactly in phase with their counterparts in the GaAs film. The effect of the Ge film is to reduce the contrast between antisite domains.

Additional observations made using 220 weak beam images and energy dispersive X-ray analysis with a ~ 1000Å probe confirmed that the GaAs layer was in the form of a continuous film and not islands.

Table 2. The approximate phase change of singly and doubly diffracted beams following the indicated paths in the two types of GaAs domain and also in Ge.

Dark field imaging reflection	Path	Bragg reflections	Relative phase change (approx)		
			GaAs		
			Ga(000)	As(000)	Ge
$\overline{2}00$	I	$000 \rightarrow 731 \rightarrow \overline{2}00$	$\pi/2$	$\pi/2$	$\pi/2$
	II	$000 \rightarrow \overline{9}31 \rightarrow \overline{2}00$	$\pi/2$	$\pi/2$	$\pi/2$
	III	$000 \rightarrow \overline{2}00$	$\pi/2$	$-\pi/2$	
200	I	$000 \rightarrow 931 \rightarrow 200$	$-\pi/2$	$-\pi/2$	$-\pi/2$
	II	$000 \rightarrow \overline{7}31 \rightarrow 200$	$-\pi/2$	$-\pi/2$	$-\pi/2$
	III	$000 \rightarrow 200$	$\pi/2$	$-\pi/2$	

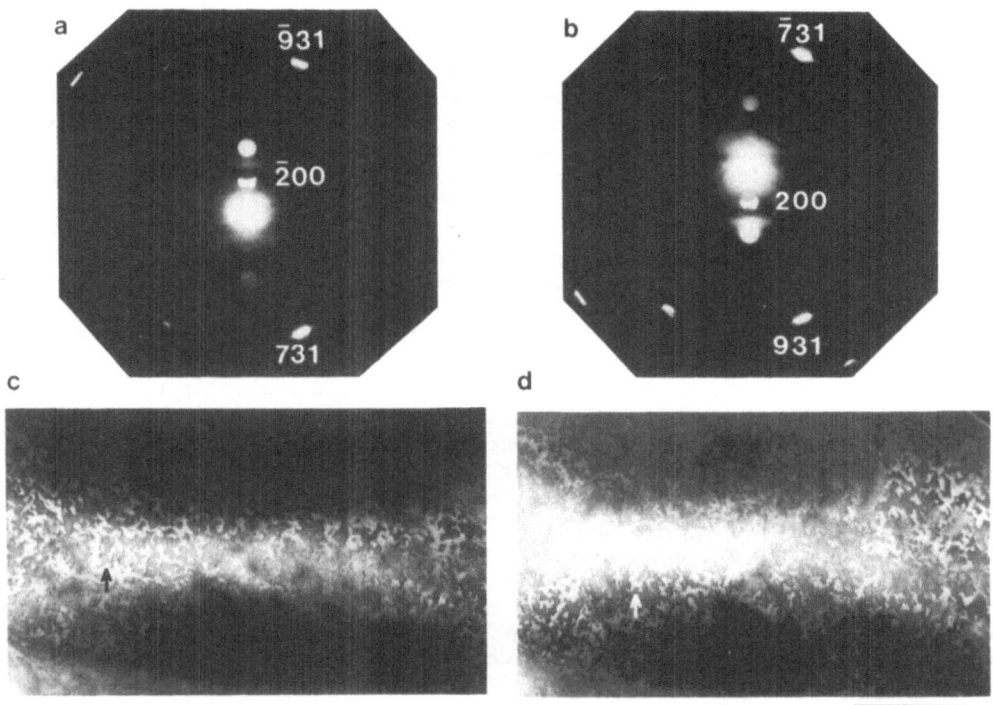

Fig. 6(a)-(d) Convergent beam diffraction patterns of MBE grown GaAs on Ge(001) for (a) $\bar{2}00$, $\bar{9}31$ and 731 excited, and (b) 200, $\bar{7}31$ and 931 excited. Below are the corresponding (c) $\bar{2}00$ and (d) 200 dark field images; the arrows mark a reference feature. The magnification bar is 0.5 μm.

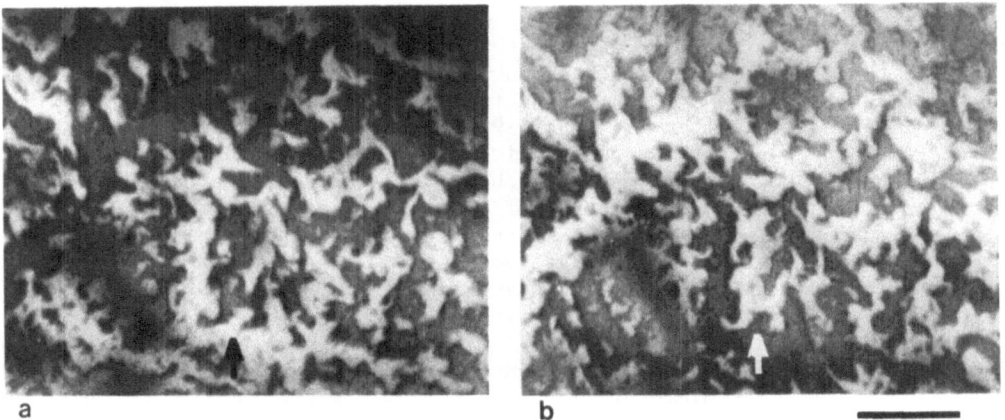

Fig. 7 Higher magnification images of Fig. 6(c) and 6(d), showing the domain contrast in the $\bar{2}00$ dark field image (a) which is clearly reversed in the 200 image (b). The reference feature is arrowed; the magnification bar is 1000Å.

Finally the origin of the antisite domains has already been discussed in[16] in terms of two possible processes. First, roughly equal probabilities of nucleating either Ga or As atoms onto an atomically flat Ge(001) surface would result in domains of approximately equal areas overall. Secondly, the nucleation of As first onto $a_0/4$ steps would also give domains with a lateral size characteristic of the step interval. Although the experimental evidence available does not allow the dominant process to be determined with certainty, the latter process is most probable for two reasons. High resolution core level studies suggest the interface bonding is predominantly Ge-As rather than Ge-Ga and also the growth of the GaAs was initiated in the presence of an arsenic flux[16].

The second example in this section relates to the use of deliberately induced local disordering of superlattices and MQW structures and the concomitant modification of the local electronic material properties which has obvious applications in the fabrication of devices. Most work has involved the ion implantation and subsequent annealing of periodic structures[39]. Closely related to this aspect of disordering but perhaps of more fundamental interest are studies of the disordering of III-V superlattices and MQW structures by diffusion alone[17,40].

Fig. 8 is a 002 dark field cross-sectional TEM micrograph showing the disordering effect that occurs during growth due to the diffusion of silicon from a heavily doped $Al_{0.3}Ga_{0.7}As$ alloy layer (A) into adjacent superlattice layers (SL) on either side. The structure was grown by MBE at 700°C on (001) GaAs and the 0.5 μm thick alloy layer doped with Si to about 10^{20} cm^{-3} is sandwiched between a pair of undoped 50 period GaAs-AlAs superlattices with individual well and barrier thicknesses of about 20Å. The lower disordered region (D) is wider than the upper one because it experienced a greater time at the growth temperature. In combination with SIMS measurements of the Si concentration profiles in the same specimens, the TEM measurements of the extent of superlattice disordering in Fig. 8 provide additional insight into the mechanisms of incorporation and migration of Si during MBE growth. The results of this study show that the Si migrates by diffusion alone and that surface segregation effects during growth are small since (a) there is no significant displacement of the Si concentration profile on the upper side of the alloy layer, and (b) the widths of the disordered regions on either side of the alloy layer are accounted for by a single value of the diffusion coefficient of disordering of 4×10^{-15} cm^{-2} s^{-1} and the respective times of growth from (i) the lower and (ii) the upper alloy interfaces to the final surface. Furthermore the disordering effect sheds light on the transport mechanism of the Si at doping levels greater than about 5×10^{18} cm^{-3}. It seems reasonable to assume that the superlattice disordering must occur via a diffusion process in which a Si atom jump causes the exchange of nearest neighbour group III atoms and therefore the interchange of Ga with Al and vice versa. Thus a high Si concentration on acceptor sites would be necessary, as occurs above about 5×10^{18} cm^{-3}, and the principal species involved could be Si-Si pairs which move substitutionally by exchange with either (Ga,Al) or As vacancies[41].

Superlattice disordering has also been observed by the diffusion of Be from AlGaAs into GaAs-AlAs superlattices during MBE growth[42] as can be seen in Fig. 9 although here the Be diffusion mechanism is essentially differentfrom that of Si. The diffusion of Be occurs by an interstitial-substitutional mechanism and it seems probable that the observed superlattice disordering involves the "kick-out" mechanism[43] whereby the incorporation of interstitial Be into the lattice generates Ga and Al self-interstitials in local supersaturation which allows enhanced intermixing[42].

482

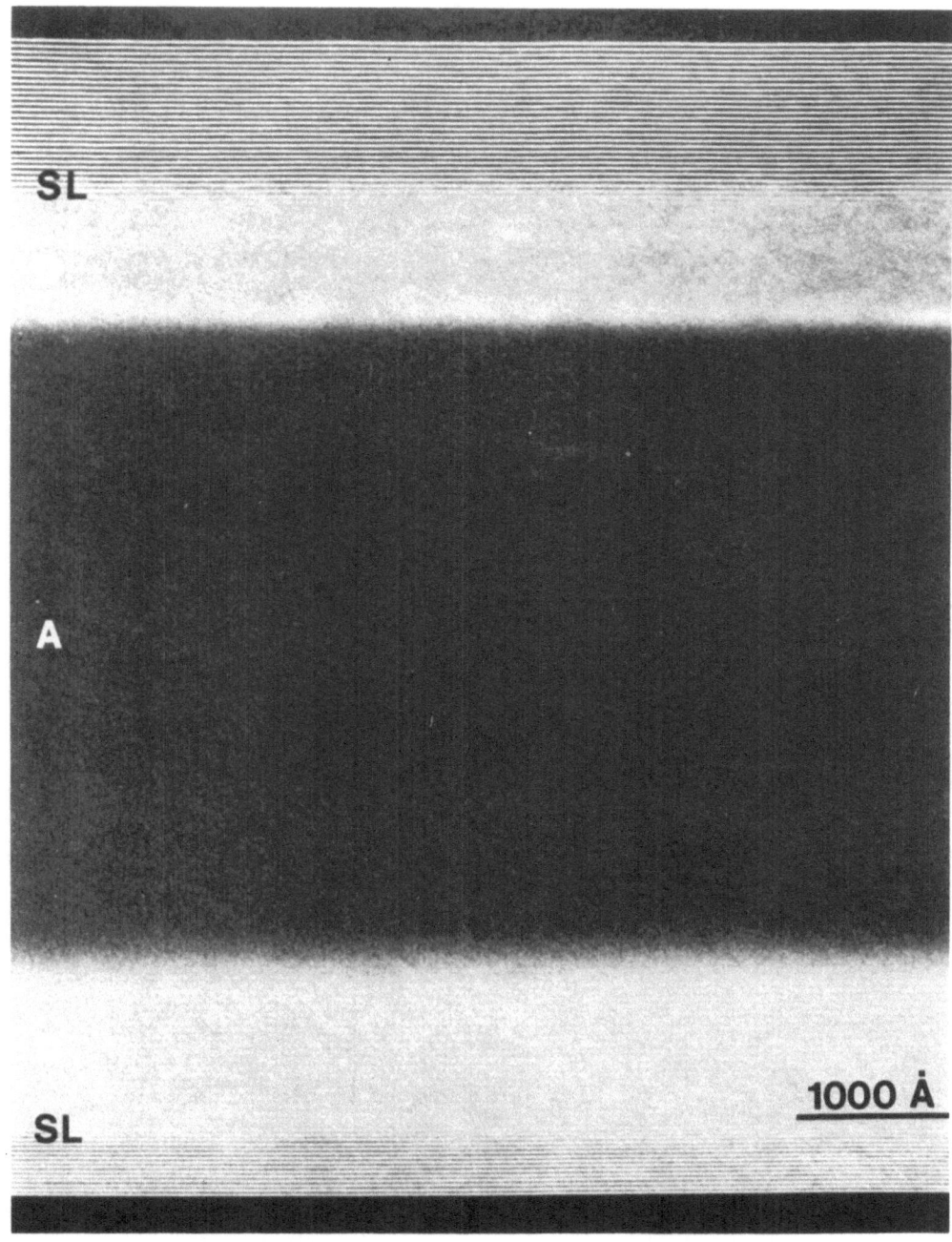

Fig. 8. A (110) cross-sectional TEM micrograph taken in 002 dark field conditions of an MBE grown structure showing the formation during growth of disordered regions (D) in the original superlattices (SL) grown on either side of a heavily Si-doped (Al,Ga)As alloy layer (A).

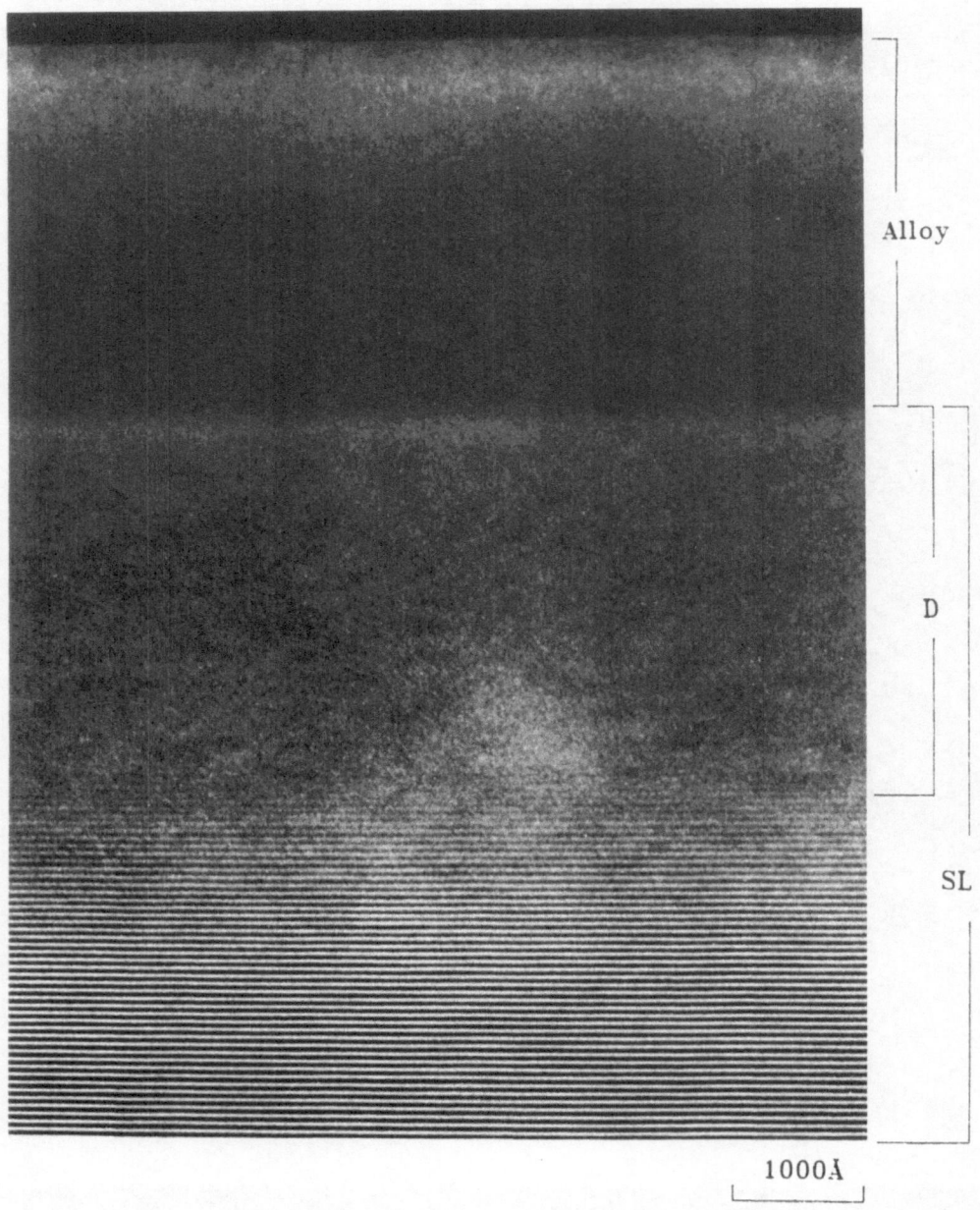

Alloy

D

SL

1000Å

Fig. 9. A (110) cross-sectional TEM micrograph taken in 002 dark field
conditions showing disordering in the region (D) of the original
GaAs-AlAs superlattice (SL) due to Be diffusion from the (Ga,Al)As
alloy cap during MBE growth.

5. CONCLUSIONS

Recent advances both in the understanding of the growth mechanisms and also in the degree of control over them have been achieved for the growth by MBE and MOCVD of low-dimensional semiconductor films. This has involved the use of a range of experimental techniques both during and after growth; many are described in this volume. The most successful techniques exploit the use of X-rays, light or electron beams to probe the structures and rely on good models of the resultant interactions to enable the interpretation of the observed processes. The application of transmission electron microscopy has been mainly directed at resolving problems of a dimensional, compositional or structural nature and has been of most value where high spatial resolution information is required. Rather than attempt to cover the complete range of TEM techniques which have been applied to those areas of the characterisation of low-dimensional structures in epitaxial semiconductor films, this article covers some recent examples related to the author's own activities.

Looking to the future, it must be mentioned that although considerable improvements have been achieved in the performance of electron microscopes and in the simulation of images, we are still hampered by the major limitation of not being able to tell which atom is where at interfaces. Finally two other limitations that might be more readily overcome include the need for a full quantitative interpretation of relatively low resolution 002 dark field images such as those of the 2 monolayer super-lattice, and also further development of specimen thinning techniques to provide thin, artefact-free specimens.

6. ACKNOWLEDGEMENTS

It is a pleasure to thank the many colleagues who have made direct contributions to the work described here by growing structures, preparing specimens, and providing results of other measurements, and also those who have helped by freely sharing ideas in discussions. They include: J.B. Clegg, P.J. Dobson, P.F. Fewster, C.T. Foxon, L. Gonzalez (Universidad Autónoma de Madrid), D. Hilton, B.A. Joyce, P.K. Larsen (Philips Research Laboratories, Eindhoven), C.D. Mayne, J.A. Morice, J.H. Neave, R.C. Pond (University of Liverpool), K. Woodbridge.

7. REFERENCES

1. K. Ploog and G.H. Dohler, Advances in Physics, 32, 285 (1983).
2. A. Ishibashi, Y. Mori, F. Nakamura and N. Watanabe, J. Appl. Phys., 59, 2503 (1986).
3. A.C. Gossard, P.M. Petroff, W. Weigmann, R. Dingle and A. Savage, Appl. Phys. Lett., 29, 323 (1976).
4. J.H. Neave, B.A. Joyce, P.J. Dobson and N. Norton, Appl. Phys. Lett. A31, 1 (1983).
5. J.H. Neave, P.J. Dobson and B.A. Joyce, Appl. Phys. Lett. 47, 100 (1986).
6. T. Sakamoto, H. Funabashi, K. Ohta, T. Nakagawa, N.J. Kawai and T. Kojima, Jpn. J. Appl. Phys. 23, L657 (1984).
7. P.M. Petroff, Gallium Arsenide and Related Compounds 1984, Institute of Physics Conference Series No. 74, p.259, Adam Hilger Ltd, Bristol and Boston.
8. J.M. Gibson, Ultramicroscopy 14, 1 (1984).
9. See for example the articles by W.J. Bartels and P.F. Fewster in this volume.

10. See for example, Microscopy of Semiconducting Materials 1985, Institute of Physics Conference Series No. 76, Adam Hilger Ltd, Bristol and Boston.

11. P. Henoc, A. Izrael, M. Quillec and H. Launois, Appl. Phys. Lett. 40, 963 (1982).

12. G.B. Stringfellow, J. Cryst. Growth, 58, 194 (1982).

13. S.N.G. Chu, S. Nakahara, K.E. Strege and W.D. Johnston, J. Appl. Phys. 57, 4610 (1985).

14. J. Singh, S. Dudley, B. Davies and K.K.Bajaj, J. Appl. Phys. 60, 3167 (1986).

15. P. Blood and A.D.C. Grassie, J. Appl. Phys. 56, 1866 (1984).

16. J.H. Neave, P.K. Larsen, B.A. Joyce, J.P. Gowers and J.F. van der Veen, J. Vac. Sci. Technol. B1(3), 668 (1983).

17. Luisa Gonzalez, J.B. Clegg, D. Hilton, J.P. Gowers, C.T. Foxon and B.A. Joyce, Appl. Phys. A41, 237 (1986).

18. P. Dawson, G. Duggan, H.I. Ralph and K. Woodbridge, Superlattices Microstructure 1, 231 (1985).

19. H.R. Pettit and G.R. Booker, Proc. of the 25th Anniversary Meeting of EMAG, Inst. of Physics Conf. Series 10, 290 (1971).

20. M.S. Abrahams and C.J. Buiocchi, J. Appl. Phys. 45, 3315 (1974).

21. P.M. Petroff, A.C. Gossard, W. Weigmann and A. Savage, J. Cryst. Growth, 44, 5 (1978).

22. D.J. Smith, Helvitica Physica Acta 56, 463 (1983).

23. J.P. Gowers, P.F. Fewster, P. Dawson, G. Duggan (unpublished).

24. J.P. Gowers, P.F. Fewster, D. Hilton and C.T. Foxon, to be published.

25. J.I. Goldstein, Introduction to Analytical Electron Microscopy eds. J.J. Hren, J.I. Goldstein and D.C. Joy, Plenum Press, New York and London, p.83 (1979).

26. J.M. Gibson, R. Hull, J.C. Bean and M.J. Treacy, Appl. Phys. Lett. 46, 649 (1985).

27. P.A. Doyle and P.S. Turner, Acta Cryst. A24, 390 (1968).

28. H. Kakibayashi and F. Nagata, Japan, J. Appl. Phys. 24, L905 (1985).

29. J.P. Gowers, Appl. Phys. A31, 23 (1983).

30. A.G. Norman and G.R. Booker, J. Appl. Phys. 57, 4715 (1985).

31. See for example, Electron Microscopy of Thin Crystals, by P.B. Hirsch, A. Howie, R.B. Nicholson, D.W. Pashley and M.J. Whelan, Butterworths, London (1965).

32. L.C. Bobb, H. Holloway, K.H. Maxwell and E. Zimmerman, J. Appl. Phys. 37, 4687 (1966).

33. R.C. Pond, J.P. Gowers, D.B. Holt, B.A. Joyce, J.H. Neave and P.K. Larsen, Mat. Res. Soc. Symp. Proc. 25, 273 (1984).

34. R.C. Pond, J.P. Gowers and B.A. Joyce, Sur. Science 152/153, 1191 (1985).

35. K. Morizane, J. Cryst. Growth, 38, 249 (1977).

36. J.P. Gowers, Appl. Phys. A34, 231 (1984).

37. J. Taftø and J.C.H. Spence, J. Appl. Cryst. 15, 60 (1982).

38. D.B. Holt, J. Phys. Chem. Solids 30, 1297 (1969).

39. Y. Hirayama, Y. Suzuki and H. Okamoto, Japan, J. Appl. Phys. 24, 1498 (1985).

40. K. Ishida, T. Ohta, S. Semura and H. Nakashima, Japan J. Appl. Phys. 24, L620 (1985).

41. M.E. Greiner and F.J. Gibbons, Appl. Phys. Lett. 44, 750 (1984).

42. R. Devine, B.A. Joyce, C.T. Foxon, J.B. Clegg and J.P. Gowers, (Submitted to Applied Physics).

43. U. Gösele and F. Morehead, J. Appl. Phys. 52, 4617 (1981).

MAGNETO-OPTIC KERR EFFECT AND LIGHTSCATTERING FROM SPINWAVES: PROBES OF

LAYERED MAGNETIC STRUCTURES

P. Grünberg

KFA-Jülich, IFF
5170 Jülich, W. Germany

ABSTRACT

Magnetooptic Kerr effect (MOKE) as well as lightscattering (LS) from spinwaves have now emerged as powerfull tools to measure some properties of thin magnetic films and multilayered structures. MOKE responds to the static-, LS to the dynamic part of the magnetization. The magnetic properties in turn can be changed due to effects caused by the layering. The aim of this article is to show how MOKE and LS can be used to explore the nature and size of dipolar and exchange coupling of ferromagnetic films across nonmagnetic or antiferromagnetic intermediate films. The interlayer exchange can be ferro- or antiferromagnetic, resulting in unique differences in the static- and dynamic magnetic properties of such structures.

list of abbreviations: MO = magnetooptic, MOKE = magnetooptic Kerr effect, MA = microwave absorption, LS = lightscattering, DE = Damon Eshbach, S = Stokes, aS = anti Stokes.

1. INTRODUCTION

Experimental methods for the investigation of surfaces or thin films are generally characterized either by high instrumental sensitivity or by strong interaction between probe and matter. Optical methods have been developed for many different purposes for a long time and generally have now reached a high degree of sensitivity. If applied to metals then the interaction with the material is also relatively strong. For magnetic materials in addition there is the magnetooptic interaction which is also strongest in the case of the ferromagnetic metals. The purpose of this article is to demonstrate how the magnetooptic (MO) effects can be used to obtain valuable information on some properties of layered magnetic structures. We focus on the question of magnetic interlayer coupling. This can be due to the dipole-dipole interaction or interlayer exchange. The nature and strength of this coupling has an effect on the static and dynamic magnetic properties of such layered structures. These in turn can be detected via the magnetooptic interactions. Fig. 1 illustrates how the magnetooptic Kerr rotation can be used to measure the hysteresis curve of a ferromagnetic sample. Basically this is very simple because one needs only to measure the Kerr rotation of polarization of a reflected light beam which depends on the direction of the magnetization. Ellipticity very often can be neglected. It is convenient to use a laser as light source. Since the laser beam can be focussed or defocussed a large or a small area can be probed. The latter is of particular interest if the magnetization reversal is via domain wall motion. This will produce a jump in the Kerr rotation when the domain wall moves through the laser focus. Hence one feature of this method is high spatial resolution. Another feature of particular interest for metallic layered structures, is due to the strong absorption of the light. If the top layer is thicker than the penetration depth of light then only its hysteresis is seen in the experiment and can be compared with a hysteresis of the whole sample obtained by other methods. Also we can replace the detector by a telescope or a TV camera and observe domains in a defocussed laser beam or a conventional light source. Recently it has been shown that the sensitivity of the magnetooptic hysteresis measurement can be increased to such an extent that the probing of monolayers of Fe became possible [1]. This was achieved by computer assisted multiscanning.

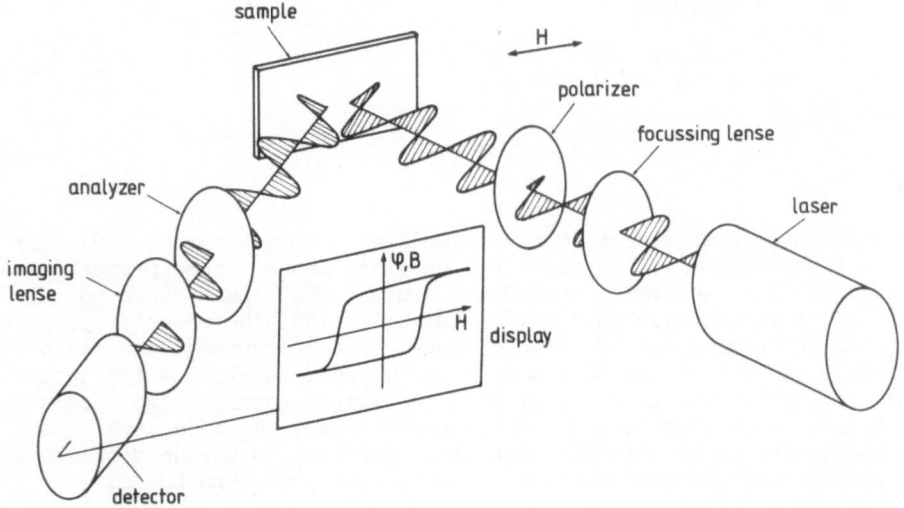

Fig. 1. Experimental arrangement for the measurement of hysteresis curves via the magnetooptic Kerr rotation φ. The plane of polarization of incident light is rotated in opposite sense for opposite magnetization directions.

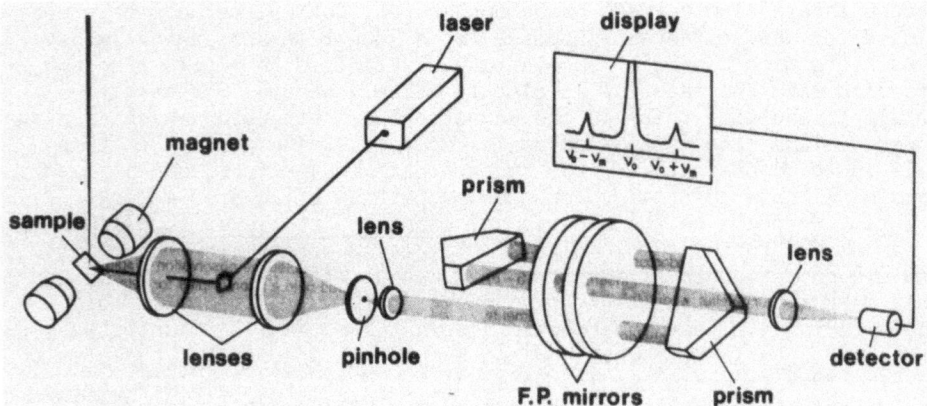

Fig. 2. Experimental arrangement for the meausrement of lightscattering from spinwaves using a three passed single interferometer.

Hysteresis curves of monolayers of Fe could be measured within a few minutes. With the possibility of magnetooptic recording there is also renewed interest in this method for applications.

We turn now to the lightscattering method. The magnetic fluctuations of a sample can be decomposed into its harmonic components, the spinwaves. These modulate the optical constants of a material via the magnetooptic interaction and hence can be observed by means of inelastic lightscattering. This method has made a lot of progress during approximately the last decade through the invention of the multipassed tandem interferometer by J. Sandercock. Fig. 2 illustrates the basic idea for a single interferometer. Light scattered from the sample passes an interferometer different times, so finally only a small frequency band is transmitted with very good contrast. The interferometer can be scanned and hence a spectrum of the scattered light can be obtained. Further improvement is achieved by using two interferometers of the kind shown in fig. 2 in tandem.

Summing up the successfull application of the magnetooptic techniques to layered magnetic structures is mainly due to an improvement in sensitivity and suppression of background or noise. As was mentioned we detect by the MO effects the status of the static magnetization and the spinwaves. Here we want to study how these are modified by the interlayer coupling. We restrict ourselves to double layers because these are the simplest systems which show already all the important effects. Such a double layer is illustrated in fig. 3. In the subsequent chapters we will see how the spinwaves in such a double layer are modified by the magnetic coupling. This can be dipolar or via interlayer exchange and the latter can be positive or negative. At the end of this article we will see that the static properties can also be modified, in particular when the interlayer exchange is negative or antiferromagnetic.

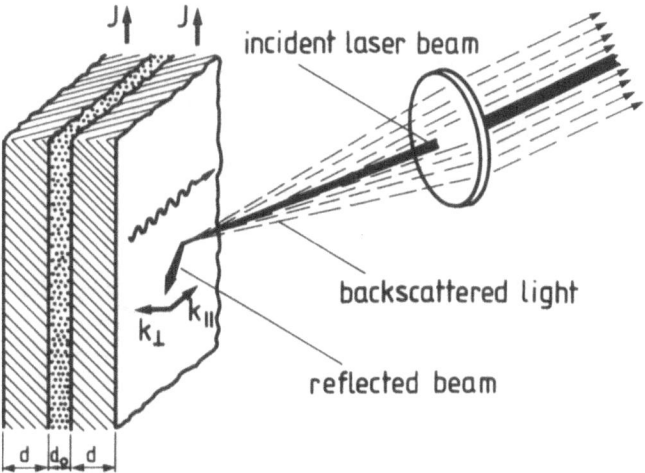

Fig. 3. The magnetic double layers considered here consist of two magnetic films (hatched) of equal thickness d and magnetization J separated by a nonmagnetic or an antiferromagnetic film with thickness d_0. The geometrical arrangement for the LS-experiments is also indicated.

2. SPINWAVES IN SINGLE FILMS

Spinwaves in single films have been studied for a long time by means of microwave absorption (MA) and more recently also by means of inelastic lightscattering (LS)[3]. Here we repeat only the most important features which serve as a basis for the following disucssion.

We restrict ourselves to modes travelling perpendicular to the direction of the static magnetization J which is assumed to lie in the sample plane. We neglect surface pinning and bulk anisotropy. The wavevector can be decomposed into its component in the sample plane, $k_{\parallel}$, and perpendicular to it, $k_{\perp}$. The nature of these modes is displayed in fig. 4a. They can be classified by the mode number n, which is the number of nodes across the film. The n = 0 mode is a uniform mode when $k_{\parallel} = 0$ but has an increasingly decaying amplitude profile for increasing $k_{\parallel}$. In this case it is also called the Damon Eshbach (DE) surface mode. Fig. 4b displays the mode frequencies if only the dipole-dipole interaction acts as a restoring force on the precessing moments. All the modes with n = 0 are now degenerate and form the horizontal line labelled "bulk modes" in fig. 4b. Their frequency ω_b is given by

$$\omega_b = \gamma \left[B_o (B_o + J) \right]^{1/2} \tag{1}$$

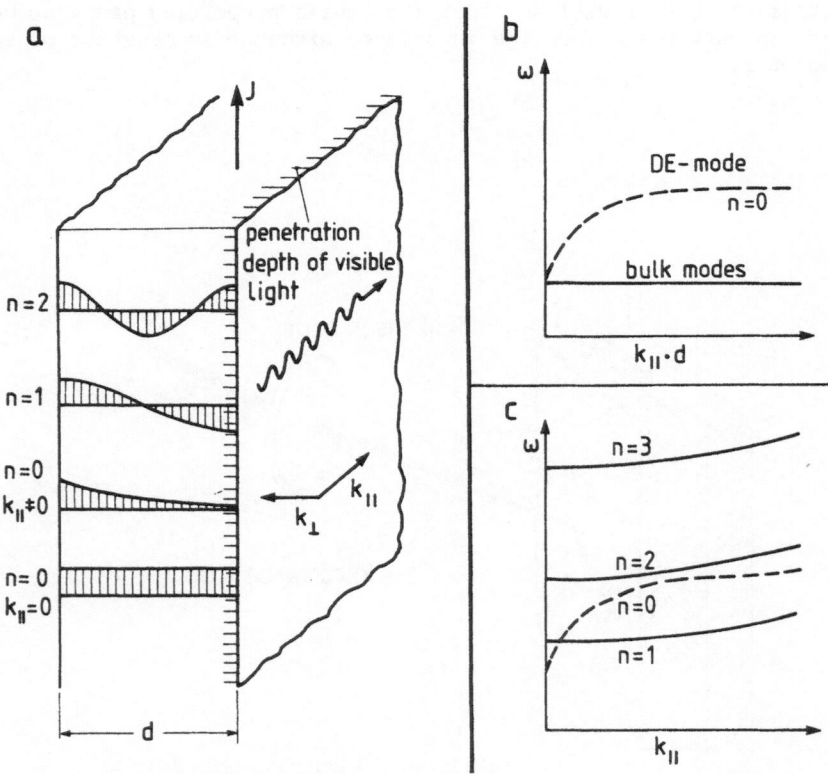

Fig. 4. Part a) displays amplitude profiles of spinwaves in thin films. Penetration depth of light (hatched zone) in metals is $\approx 100\ \overset{\circ}{A}$. Part b) shows mode frequencies schematically in the dipolar approximation. In part c) exchange has been included.

where γ is the gyromagnetic ratio, B_o the external field and J the samples magnetization. The n = 0 mode, which is uniform for $k_{\parallel} = 0$ and exponentially decaying otherwise is given by

$$\omega_{DE} = \gamma \left[(1-e^{-2k_{\parallel} \cdot d})(J/2)^2 + B_o^2 + B_o \cdot J \right]^{1/2} \qquad (2)$$

where d is the film thickness.

The effect of exchange on these modes is displayed in fig. 4c. The degeneracy of the bulk modes with respect to n is lifted and equ. (1) is replaced by

$$\omega_n = \gamma \left\{ (B_o + \left[2\mu_o A/J \right] k_n^2) \cdot (B_o + \left[2\mu_o A/J \right] \cdot k_n^2 + J) \right\}^{1/2} \qquad (3)$$

here A is the exchange constant and the total wavevector k_n is given by

$$k_n^2 = k_{\parallel}^2 + k_{\perp}^2, \qquad k_{\perp} = n \cdot \frac{\pi}{d} \qquad (4)$$

The effect of exchange on the DE-mode is in practical cases negligible.

From the foregoing follows that it is the dipolar coupling which lifts the n = 0 mode above the bulk modes for $k_{\parallel} \neq 0$ whereas the exchange has its main effect on the bulk modes and leaves the n = 0 mode essentially unchanged. The explanation for this unique difference lies in the amplitude patterns of fig. 4a. The n = 0 mode carries a resulting net dipole moment which precesses around the direction of static magnetization. For the standing modes with n = 0 this moment cancels out. The standing modes on the other hand have a strong spin variation across the film which makes them sensitive to the exchange interaction. This makes it plausible why the n = 0 mode reacts to the exchange only for large $k_{\parallel}$. The differences which have just been described persist when we proceed to interlayer dipolar- or exchange coupling.

Before we do this another feature which follows also from the amplitude patterns should be mentioned. The MA- and LS-techniques are both characterized by the fact that the wavelength of the probe is usually much larger than the film thickness. Both the light- and the microwave interact with the net precessing moment which they "see". For microwaves which penetrate thin films entirely the observation of standing modes therefore is only possible if the net moment through the presence of in-homogeneities is not cancelled completely. For the lightwaves in the case of metals the small penetration of ≈ 100 Å is quite helpfull. This is illustrated in fig. 4a by the hatched area within which the net moments of the standing modes do not cancel. Hence the observation of LS from these modes becomes allowed by the optical absorption which is a natural material property whereas the interaction with microwaves becomes allowed by inhomogeneities which is mostly an unwanted material property.

3. DIPOLAR INTERLAYER COUPLING

As was just mentioned standing modes do not develope a net precessing moment (under the ideal conditions considered here). This is the main reason why we can neglect dipolar coupling of standing modes on different films in layered structures. This comes also as a result of a more quantitative approach [4]. It turns out that even for n = 0 modes this type of coupling dissappears in the limit of $k_{\parallel} = 0$ (MA experiments). Hence we have to consider only the dipolar coupling of DE-modes with $k_{\parallel} \neq 0$. Theoretically this is achieved by looking for solutions of the equation of motion which at the interfaces fullfill the boundary conditions following from Maxwells's equations. The result of such a calculation

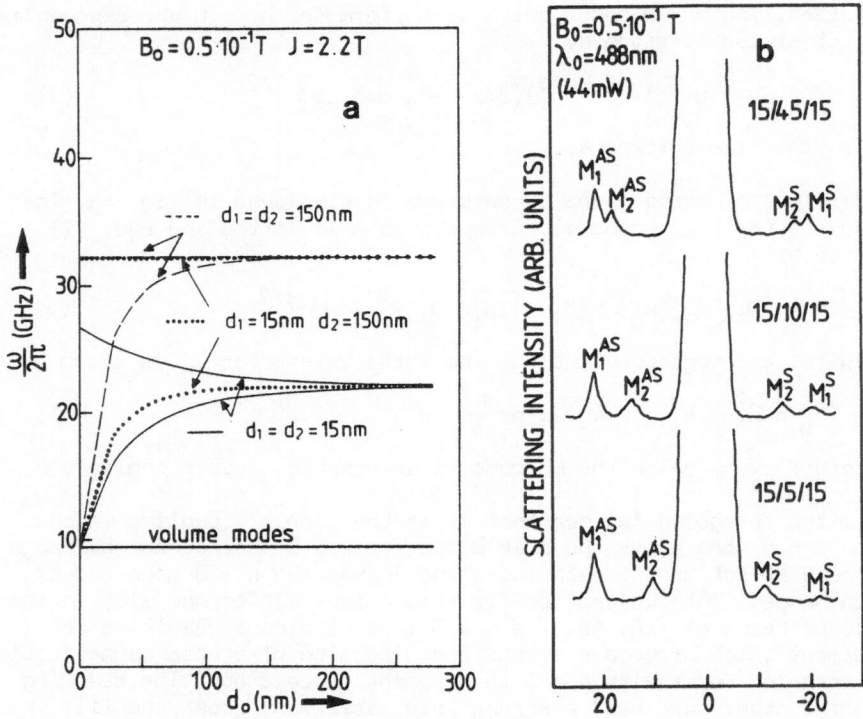

Fig. 5. Part a) displays how the DE-modes of single films are modified by the dipolar coupling in double layers. Various sets of parameters have been used. Part b) shows the LS-experiments for $d_1 = d_2 = 15$ nm corresponding to the solid lines of part a. Thicknesses in units of nm are given in the notation $d_1/d_0/d_2$. For large d_0 these modes would be degenerate.

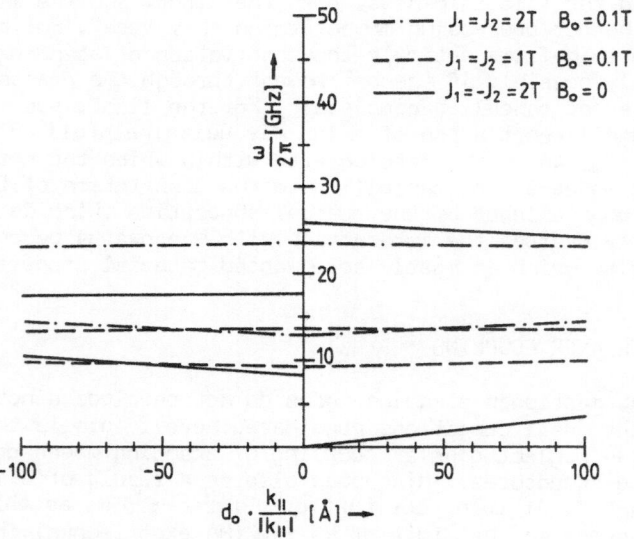

Fig. 6. Mode frequencies of dipolar coupled DE-modes in double layers for small d_0. Various sets of parameters, as indicated. The case $J_1 = -J_2$ corresponds to antiferromagnetic coupling, discussed in sec. 6.

and comparison with experiment [5] is displayed in fig. 5. In fig. 5a the mode frequencies as a function of interlayer thickness d_0 are plotted for various sets of parameters. For large values of d_0 where any coupling vanishes we are dealing with the DE modes of the single films. These are degenerate if $d_1 = d_2$. Below $d_0 = 1000$ Å appreciable coupling sets in which lifts this degeneracy. This is also seen in the experimental example of fig. 5b. For any set of parameters the lower branches of fig. 5a for $d_0 = 0$ end up at the position of the volume modes, given by equ. (1). The other branches in this limit become the DE modes of the resulting single films with thickness $d = d_1 + d_2$. Their frequency is given by equ. (2).

In the following the range of d_0 will be of particular interest where in addition to the dipolar – we have also interlayer exchange coupling. This will be for $d_0 \lesssim 50$ Å. It is therefore important to know what the effect of the dipolar coupling is in that d_0-range. This is displayed in fig. 6 for two sets of parameters which correspond to the experiments on Fe and on $Ni_{0.8} Fe_{0.2}$ (permalloy) to be discussed later. Also shown is the case $J_1 = -J_2$ which will be of interest in the context of negative interlayer exchange coupling. We see, that the resulting mode shifts are on the order of 1 GHz or less which is close to the resolution of the LS experiments. This is due to the long range nature of the dipolar coupling effects. In the thickness range displayed they are practically constant, close to saturation.

4. INTERLAYER EXCHANGE COUPLING

The problem of interlayer exchange coupling has been treated in a phenomenological approach using continuum theory by Hoffmann et al.[6]. In a similar way but in much more detail it has recently been considered by Vayhinger et al [4]. Also recent work by Hinchey et al.[7] treats the problem with a microscopic approach. Since none of these calculations can be direcly compared with the experiments presented in the following we will describe now a coupling scheme [8] which will be sufficient for the evaluation of the experiments.

To make the application of symmetry considerations possible we restrict ourselves to the symmetrical case of fig. 3 where the two magnetic films have equal thickness and magnetization. We describe the interlayer exchange by a coupling parameter A_{12} and denote the exchange constant of the material by A. For simplicity we first neglect the interlayer dipolar coupling hence restrict ourselves to $k_{\parallel} = 0$. In the decoupled limit where $A_{12} = 0$ we are dealing with two single films and we can calculate the mode positions from equ. (2), (3) by setting $k_{\parallel} = 0$. In the fully coupled limit of $d_0 = 0$ we have $A_{12} = A$ and we can again calculate the mode positions from equ.s (2), (3) only that d is replaced by 2d. The question then is how the modes evolve from one extrem case to the other. To find this out, we note that the general procedure to find the modes (or eigenstates) of a coupled system is, to form symmetry adapeted linear combinations of the modes in the single systems. Since the midplane between the two films is a mirror plane any coupled mode has to be either symmetric or antisymmetric with respect to that symmetry operation.

For the uniform– and the first standing mode on the single films this is displayed in fig. 7a. Since the symmetry character has to be conserved all along the way from the fully decoupled to the fully coupled limit, it is clear that the symmetric combination of uniform modes produces again a uniform mode and the antisymmetric one the first standing mode. In the same way we obtain from the coupling of the two first standing modes the second and third mode of the coupled system. It is straightforward to extend this receipe to other higher modes. Fig. 7b repeats the result of fig. 7a in an energy level scheme. For simplicity here only the exchange part $\sim k_n^2$ of equ. (3) has been considered and we have set $k_n^2 = k_\perp^2$. To describe the coupling effects for $k_n = 0$ we can take equ.s (1) and (3) to calculate the frequencies in the fully decoupled and coupled cases and obtain the inter-mediate cases by interconnecting according to fig. 7b.

When $k_\parallel \neq 0$ in addition we have to take into account interlayer dipolar coupling. For standing modes due to the cancellation of the net amplitude this effect is weak. For the $n = 0$ mode which for $k_\parallel \neq 0$ is now the DE-mode, the effect of the dipolar coupling has already been described in sec. 3. In anticipating experimental results we have to go below $d_0 \approx 50$ Å in order to obtain $A_{12} \neq 0$. Here according to fig. 6 the dipolar coupling produces a practically constant, thickness independent, splitting. Hence, for $k_\parallel \neq 0$ the dipolar effect is simply considered by introducing this splitting for the $n = 0$ mode on the right hand side of fig. 7b. This is achieved by replacing in fig. 7b the lowest horizontal line by the dashed line. It should be mentioned that even in the limit of small d_0 the position of the dashed line will still depend on the thickness d and magnetization J of the magnetic films. A position as in fig. 7b results approximately for the parameters of fig. 6 when $J_1 = J_2$. The modes corresponding to the lowest solid inclined line is in the position of the uniform mode when $A_{12} = 0$ (right hand side of fig. 7b). When $A_{12} = A$ (left hand side) it is in the positon of the first standing mode of the single film which results for $d_0 = 0$. Since its shift is a measure for the effective interlayer exchange we want to call it "exchange mode" in the following.

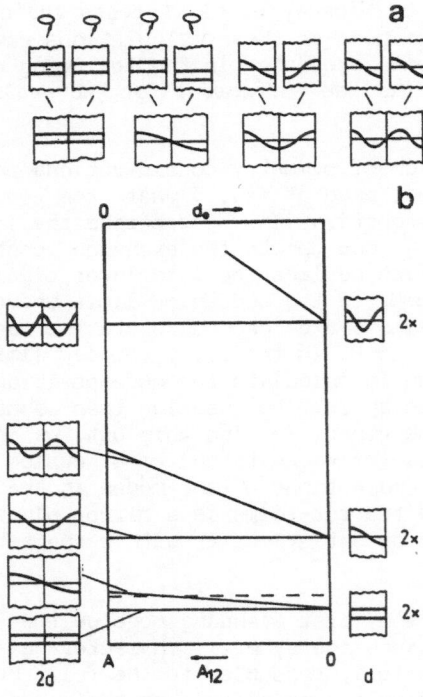

Fig. 7. Coupling scheme for modes in double layers. From the modes on the single layers we form symmetric and antisymmetric combinations as displayed in fig. 7a. Part b) shows the energy shifts which follow.

494

5. EXPERIMENTAL RESULTS ON FERROMAGNETIC INTERLAYER EXCHANGE

In the previous section we saw how interlayer exchange coupling modifies the spinwave mode spectra of magnetic double layers. Two experimental

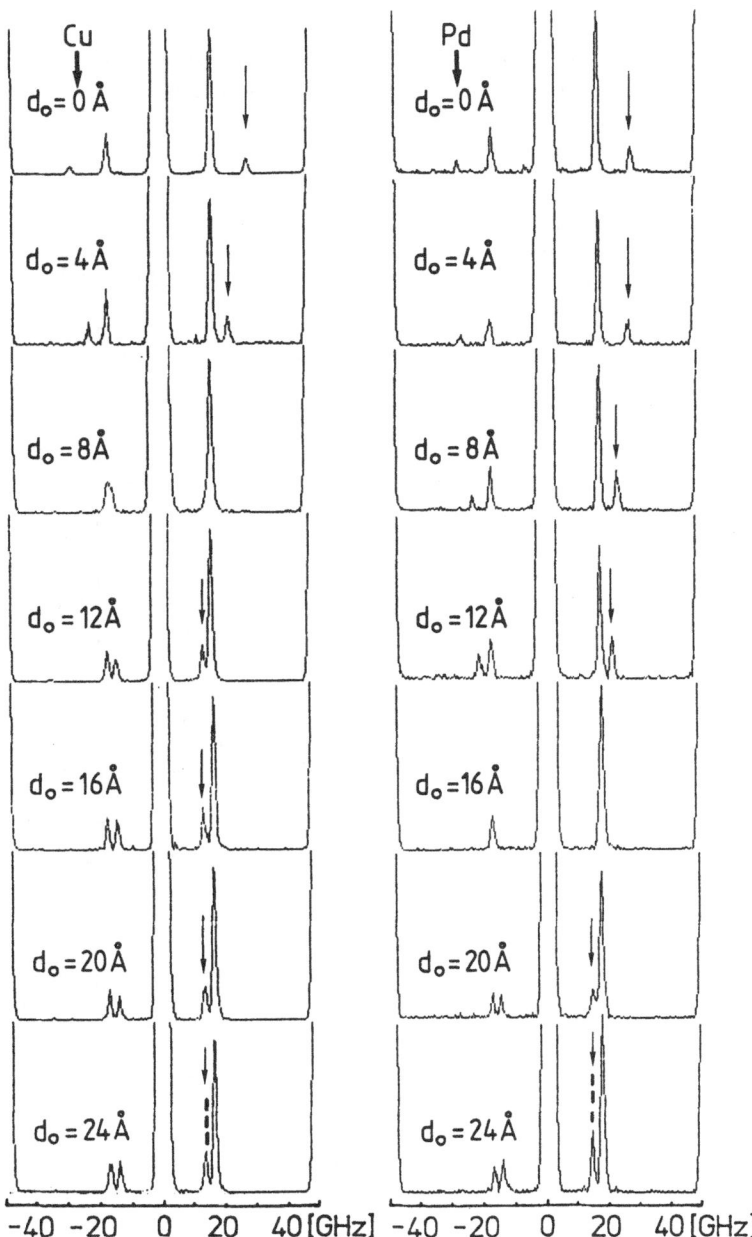

Fig. 8. Lightscattering spectra from spinwaves in $Ni_{0.8}Fe_{0.2}$ double layers with Cu and Pd interlayers of thickness d_o as indicated. Exchange mode on aS side marked by arrow. Line positions for $d_o = 0$ and $d_o \geqslant 20$ Å can be fitted with $J = 1.02$ T, $A = 7.0 \cdot 10^{-12}$ J/m, $B_o = 0.2$ T, $k_{\parallel} = 1.73 \cdot 10^5 cm^{-1}$, $\gamma = 1.77 \cdot 10^{11}$ $T^{-1}s^{-1}$. Position of uniform mode marked by dashed line in the lowest traces. Downshift of exchange mode for increasing d_o is more rapid for Cu interlayers.

examples[9] are displayed in fig. 8. These spectra were obtained from double
layers of permalloy (Ni$_{0.8}$Fe$_{0.2}$) with an individual thickness of d = 100 Å
separated by interlayers of Pd or Cu. The value of d was chosen such, that
standing modes on the individual films are outside the spectral range of the
instrument and only the modes corresponding to the lowest two branches of
fig. 7b are observed. The value of $k_\parallel$ of the observed spinwaves is determined by the laser light used for excitation and by the scattering geometry.
It is $k_\parallel = 1.73 \times 10^5$ cm^{-1} (green laser light impinging under 45° to the surface, backscattering). Since $k_\parallel \neq 0$ the modes are subject to dipolar coupling
which replaces the lowest horizontal line of fig. 7b by the dashed line.
The strong peak in the spectra is due to scattering from the DE-mode (dashed
line in fig. 7b) the other weaker one, is due to the exchange mode. When
d_0 increases the DE mode stays in the same position whereas the exchange
mode shows a downshift, all as expected from fig. 7b. In the limit of
$A_{12} = 0$ but still full dipolar coupling the latter mode is as expected at
the position of the uniform resonance, given by equ. (1). In the spectra
this position is marked by a dashed line. For given d_0 we can always confirm the criterion of full dipolar coupling by inspection of fig. 4. Hence
the modeshift just described can be taken as a measure for the strength
of the effective interlayer exchange A_{12}. In the following we assume
that the scale on the abscissa of fig. 7b is linear. This establishes

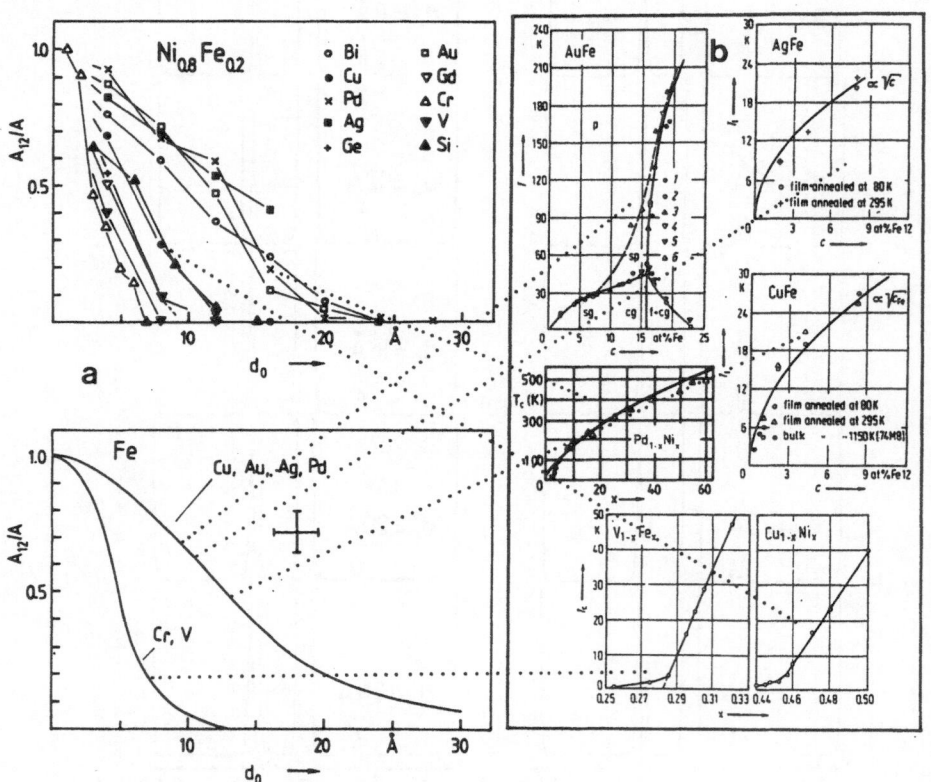

Fig. 9. Part a) summarizes the results for A_{12} from Ni$_{0.8}$Fe$_{0.2}$ - and Fe
        double layers with various interlayer materials as indicated. Slow
        decrease of A_{12} with increasing d_0 represents strong exchange of
        the magnetic films across the interlayers. Part b) displays Curie
        temperature T_c or spinglass freezing temperature T_f for various
        alloy systems. Rapid increase of T_f or T_c with concentration
        of magnetic atoms represents strong exchange interaction across
        host material. Dotted lines connect layered structures with corres-
        ponding alloy systems.

a linear relationship between shift and value of A_{12}. Since we know that $A_{12} = A$ in the limit of $d_0 = 0$ we can also normalize A_{12} to A. This has been done in fig. 9a where A_{12}/A has been plotted as a function of d_0 for double layers of permalloy and of Fe for the interlayer materials indicated. The double layers analyzed here have all been obtained by e-gun evaporation in a vacuum of $\approx 10^{-9}$ torr onto substrates of oxydized silicon or sapphire held at room temperature. Rates were at ≈ 1 Å/sec. As expected and confirmed by X-ray diffractometry the films were polycrystalline. The relatively low substrate temperature was chosen to avoid diffusion as much as possible. It has also the advantage that there will be less tendency to island formation, which for very thin interlayers would be particularly critical.

The basic idea behind fig. 9a is, that the exchange interaction of the magnetic films across the nonmagnetic ones is stronger in those cases where A_{12} as a function of increasing d_0 decreases slowly. Comparison of these results with the corresponding alloy systems [10] seemingly show that looking for a correspondence between the interlayer couplings in film systems and interatomic exchange in diluted alloys might be good working concept [9]. We see this from comparison of fig. 9a with fig. 9b. In fig. 9b data for the magnetic ordering temperature T_c or spinglass freezing temperature T_f in various alloy systems have been compiled. Here rapid increase of T_f or T_c with concentration of magnetic component suggests strong interaction across the nonmagnetic one. Since data for alloys with $Ni_{0.8}Fe_{0.2}$ which could be compared are not available we imply here, that the difference between $Ni_{0.8}Fe_{0.2}$ and Ni is only small and compare our $Ni_{0.8}Fe_{0.2}$ double layers with Ni-alloys. The comparison then shows, that whenever in the alloy system T_f or T_c increases rapidly with magnetic concentration, in the corresponding double layer A_{12} decreases slowly as a function of d_0. We note in particular the different behavior of alloys of Fe and Ni with Cu. It is correctly reproduced in the behavior of the double layers. Furthermore we note that, whenever in the double layers the interlayer is nonmetallic there is also a rapid decrease of A_{12} upon increasing d_0. It is therefore suggestive to make the conduction electrons responsible for the long range interaction seen here. This is for example in agreement with the well known RKKY interaction which in addition to the direct exchange predicts an exchange interaction via the electrons at the Fermi energy, which are also responsible for metallic conduction.

6. ANTIFERROMAGNETIC EXCHANGE BETWEEN FERROMAGNETIC LAYERS

The fact, that Cr is an antiferromagnet at room temperature makes it very attractive to use it as an interlayer and see whether antiferromagnetic coupling can be achieved. The results from the polycrystalline films displayed in fig. 9a, however indicate only a weakening of ferromagnetic coupling as d_0 increases even for Cr interlayers. With the possibility of making [100] oriented epitaxial Fe-films, antiferromagnetic coupling of Fe across Cr has recently indeed been found [11].

The epitaxy, which in this case seems to be of crucial importance was achieved in the following way. We exploit the fact, that the (100) planes of bcc Fe match the (100) plances of fcc Au almost perfectly after a 45° rotation. Furthermore Cr and Fe have both bcc structure and also practically the same lattice constant ($a_0(Fe) = 2.87$ Å, $a_0(Cr) = 2.88$ Å). We continue with a perfect match by making the films on LiF substrates which has the same lattice constant as Au (4.08 Å). Fig. 10 illustrates the good match of the quoted fcc and bcc systems. To obtain the atomic positions during growth one has to shift the atoms of the fcc lattice to the midpoints of the bcc unit cells as indicated by the arrow. Epitaxial

Au-films with [100] -orientation can also be made on NaCl substrates which in the same way as LiF cleaves along (100)-planes. The Au-films are used as buffer layers for the subsequent growth of the Fe double layers. To achieve good growth the Au-films were evaporated onto LiF or NaCl at substrate temperatures $T_S \approx 300°$ C. Here it is not advisable to use freshly cleaved NaCl surfaces. Also exposure to humidity before use seems to be favorable and surprisingly the vacuum during deposition should not be too good ($\approx 10^{-6}$ torr). There is an extended literature on the growth of Au on NaCl [12]. It is known that at $T_S = 300°C$ the initial growth of Au on NaCl is in the form of islandes which coalesce to a continuous film at a thickness of $d \approx 500$ Å. In the case described here the Au-films were made 1000 Å thick at rates of 1 Å/sec. The [100] -growth of Au was confirmed by X-ray diffractometry. Fig. 10b displays a Laue pattern obtained in transmission from such a film. For the subsequent preparation of the Fe double layers separated by Cr- or Au-interlayers the substrate temperature was decreased to $T_S = 200°C$. This was done mainly in order to avoid diffusion as much as possible. Rates were at 0.1 Å/sec for Fe and Au and at 0.5 Å/sec for Cr. The somewhat higher rate for Cr was chosen because Cr is known to be a strong getter hence at small rates the possibility of contamination increases. During preparation of the double layers pressure never rose higher than 10^{-9} torr. The epitaxy of the Fe double layers can be confirmed by exploiting the magnetic anisotropy of the Fe as will be shown below.

We return now to the determination of the effective interlayer exchange as described in the foregoing sections. Including the anisotropy of Fe,

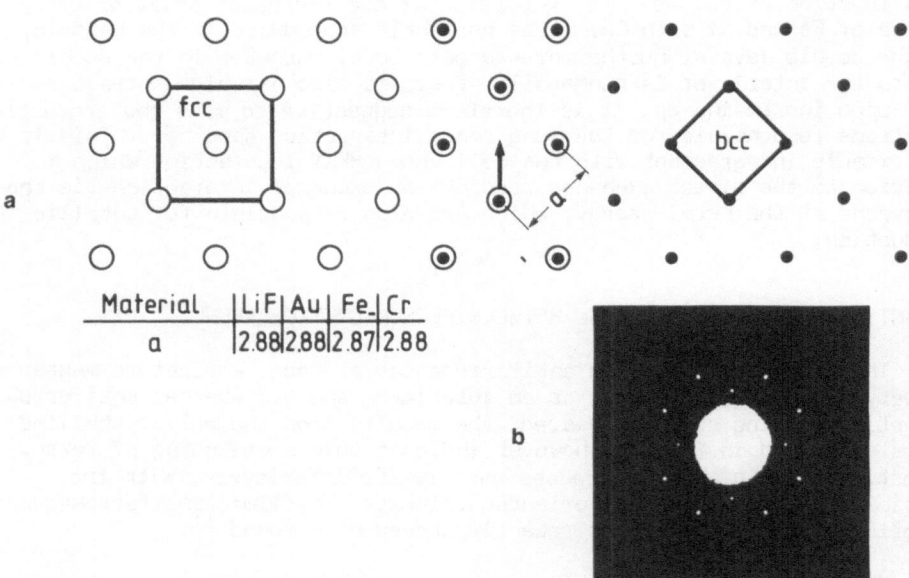

Material	LiF	Au	Fe	Cr
a	2.88	2.88	2.87	2.88

Fig. 10. Part a) displays the perfect match of some fcc and bcc lattices in (100) planes. To the left we see the fcc to the right the bcc lattices allone, in the middle part they overlap. Nearest neighbor distance "a" is given for the different materials. The arrow indicates how one of the lattices in the fig. has to be shifted to obtain the situation during growth. Part b) shows a Laue pattern from an Au-film in transmission grown on NaCl. It proves epitaxy with [100] oriented growth.

498

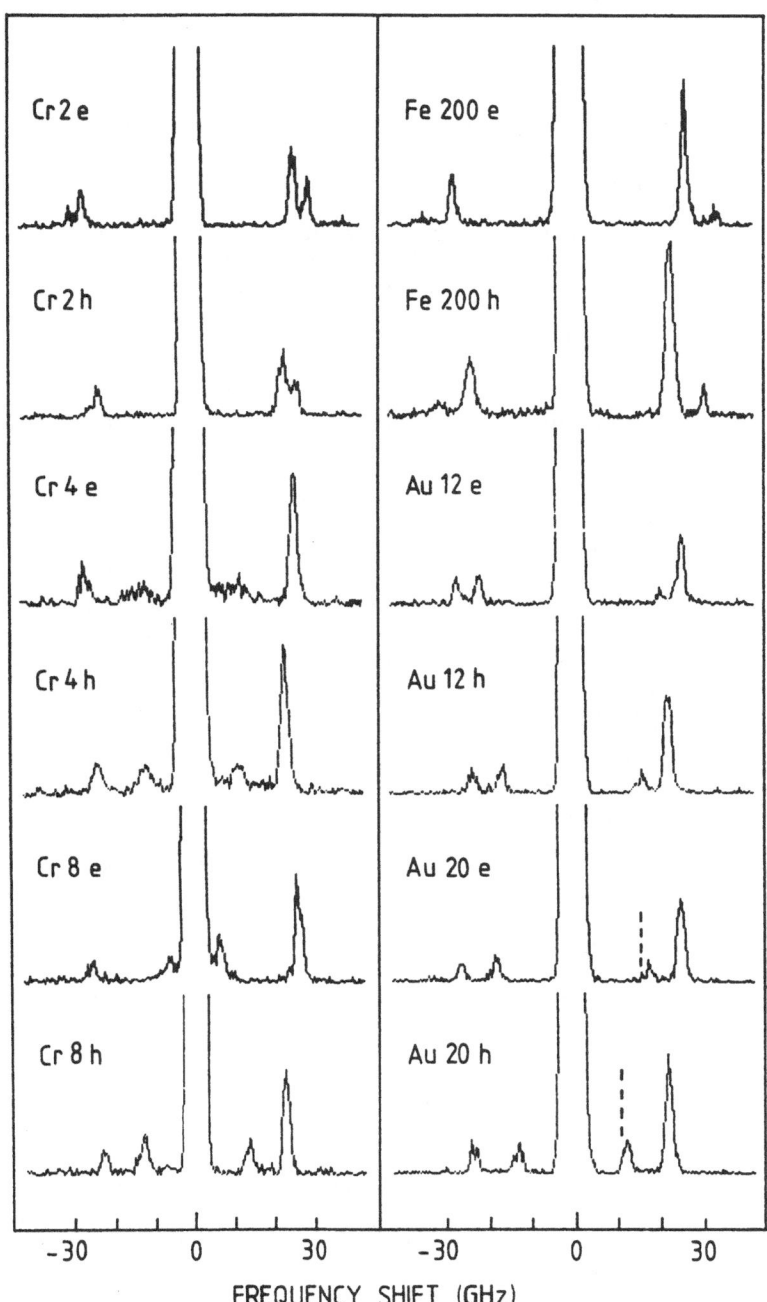

FREQUENCY SHIFT (GHz)

Fig. 11. Lightscattering spectra from spinwaves in 200 Å thick single
layers of Fe (upper two traces on left hand side) and Fe double
layers where the interlayer material and thickness is marked
on the various traces (e and h indicate direction of external
field along easy and hard axis of Fe). From traces Fe 200e and
Fe 200h we evaluate $J = 2.0$ T and $B_{an} = 0.035$ T which yields
the position of the uniform mode marked by dashed lines in traces
Au 20e and Au 20h. $B_o = 0.1$ T in all cases.

equ.s (1) - (3) have to be somewhat modified. The frequency of the uniform mode is now given by

$$\omega = \gamma \left[(B_o + B_{an})(B_o + B_{an} + J) \right]^{1/2} \tag{5}$$

for B_o along the easy direction which for Fe is along [100] and

$$\omega = \gamma \left[(B_o + B_{an}/2 + J)(B_o - B_{an}) \right]^{1/2} \tag{6}$$

for B_o along [110]. Here B_{an} is the anisotropy field. We chose $\gamma = 1.85 \cdot 10^{11} \ T^{-1} \ s^{-1}$, equivalent to a g-factor of $g = 2.1$ for Fe. For $B_o \parallel [100]$ equ. (2) for the DE-mode is now replaced by

$$\omega = \gamma \left[(1 - e^{-2k_\parallel d})(J/2)^2 + (B_o + B_{an})^2 + (B_o + B_{an})J \right]^{1/2} \tag{7}$$

and as mentioned before is independent of the interlayer exchange. For $B_o \parallel [110]$ the frequency of that mode has also recently been worked out [13] but for the present purpose it is sufficient to know that its shift from $B_o \parallel [100]$ to $B_o \parallel [110]$ is only about half the value of the shift diplayed by the uniform mode. Hence for the two extreme cases of full and zero interlayer exchange coupling and including the Fe-anisotropy we can predict the mode positions as just described.

We discuss first results obtained from Fe-Au-Fe double layers which were used as a reference. Spectra with increasing Au thickness are displayed on the righ hand side of fig. 11. For each thickness two spectra were taken - one with B_o along [100] (e = easy) and one with B_o along [110] (h = hard). B_o is always in the plane of the film. Clearly for B_o along 100 both modes shift to higher frequencies which is in agreement with the fact that here the anisotropy field assists the external field. The observation that these films show the expected magnetic anisotropy of Fe confirms their epitaxial growth. From the position of the DE-mode and the formalism described above we evaluate magnetization J and anisotropy field B_{an}. We find $J = 2.0$ T and $B_{an} = 0.035$ T. Next we use these parameters to predict the position of the uniform mode including anisotropy. In the lower traces at the righthand side of fig. 11 these positions are indicated by dashed lines. They agree well with the actual line positions. The slight disagreement is of unknown origin. We conclude that in double layers with Au interlayers of thickness $d_{Au} = 20 \text{Å}$ the Fe layers are exchange decoupled. This is also confirmed by the observation that a further increase of the Au thickness (not shown) slowly shifts the mode to ever higher freuquencies, in agreement with the dipolar theory as displayed in fig. 6. Next we turn to the spectra from the Fe-Cr-Fe double layers shown on the left hand side of fig. 11. Again for each film we confirm the expected magnetic anisotropy and hence epitaxial growth. The DE mode as before shows only small dependence on the interlayer thickness. The exchange mode however now behaves quite different than in the Fe-Au-Fe reference. Let us first discuss the spectra of $B_o \parallel [100]$ (e). For $d_o(Cr) = 4$ Å we see that the exchange mode has already dropped below the value expected for the exchange decoupled case (compare to the lowest traces on the righthand side). At the same time there is appreciable line broadening. When the Cr-thickness is increased the mode stays at a low frequency but becomes more narrow. The fact that this mode is observed at a frequency which is lower than that of the uniform mode can only be interpreted such that the effective coupling has now become antiferromagnetic. This is also in agreement with the phenomenological theory [4]. The linebroadening at $d_o(Cr) = 4$ Åcould for example be due to a nonuniform Cr-thickness, which would cause strong variations in the effective interlayer exchange.

A remarkable behavior is also observed when B_o is now brought over to

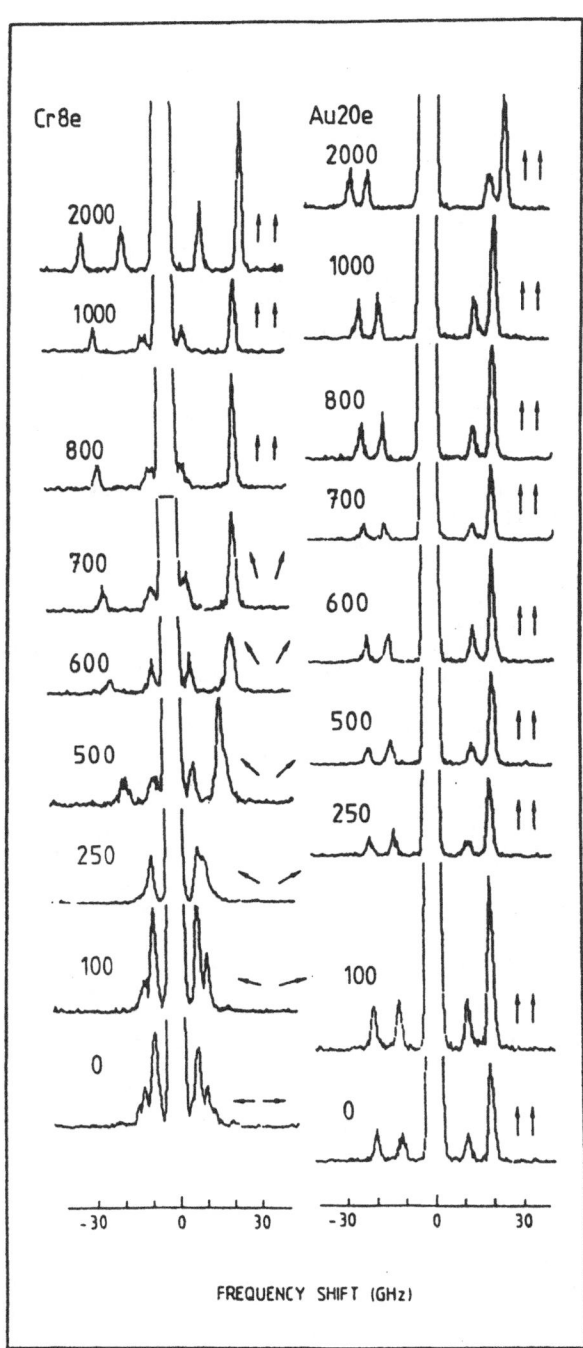

Fig. 12. Spectra from samples Cr8 and Au20 (see fig. 11) with B_0 along the easy axis of Fe. Numbers on the traces mark values of external field in units of 0.1 mT $\hat{=}$ Oe. The arrows indicate the suggested magnetization directions on the two Fe layers where B_0 is supposed to point up. Observed spinwave propagation then is along a horizontal line.

the [110] direction (h). There is a clear frequency upshift for the exchange mode – opposite to what is expected from the Fe-anisotropy (note that the DE-mode shifts according to the Fe anisotropy). This could be due to an anisotropy within the Cr-film. If the Cr-moments want to stay aligned along a $\pm$ 100 direction then the coupling to the adjacent Fe layers becomes weaker when we force the Fe-moments into a 110 direction. Hence the overall antiferromagnetic coupling of the two Fe films also decreases which now means a frequency upshift.

The low frequency of the exchange mode as displayed e.g. in trace Cr 8e of fig. 11 suggestes that it might go soft when we decrease the external field by which it should shift to even lower values. This has also been tested. The result is shwon in fig. 12. Again we discuss first the behaviour of our Fe-Au-Fe reference system displayed on the right hand side of fig. 12. Here a sample with an Au thickness of 20 Å has been chosen. When the external field B_0 is decreased all modes shift gradually downwards but stay in the same position for $B_0 \lesssim 0.05$ T (≈ 500 Oe). In agreement with equ. (5) – (7) this is due to the effect of the anisotropy field $B_{an} = 0.035$ T (350 Oe) of these Fe-films. At the lowest applied fields shown, there is still enough remanence to hold the magnetization in the same position as in higher fields otherwise the modes would shift. Opposite to this for the Fe-Cr-Fe system as B_0 is decreased the exchange mode goes through a minimum at $B_0 = 0.08$ T (800 Oe) and comes slightly up again as B_0 is further decreased. At the same time the DE-mode starts to downshift which indicates a change in the direction of the samples magnetization. It is very likely that this has to do with a softening of the exchange mode which makes the previous magnetization configuration unstable. However due to this change the soft mode again increases its frequency. The interpretation is also possible that the soft mode disappears from the spectrum and another mode comes up instead [14]. To gain further information on the behavior of the static magnetization the net moment of the sample was measured with a Faraday balance as a function of external field. It was found that for decreasing B_0 below 0.05 T (500 Oe) the net moment gradually disappeared. The obvious interpreation is, that in small external fields the moments on the two films due to the antiferromagnetic coupling by the interlayer align themselves antiparallel. More evidence for this comes again from spinwave mode spectra.

We repeat the low field experiment of fig. 12 on the Fe-Cr-Fe sample but change the lightscattering geometry such, that we observe now modes which propagate in the direction of the external field. The result is displayed in fig. 13b. From comparison with theory [15] we find, that this unique pattern occurs for a double layer with antiparallel orientation of the magnetization in a small field and modes which travel transverse to the magnetization. Note that opposite to the spectra shown in fig. 11 and 12 modes occuring on the Stokes- and the Antistokes side now have different frequencies. This comes from the fact that an antiparallel layer by symmetry is a truly nonreciprocal system [15]. The frequencies as a function of d_0 are qualitatively displayed in fig. 13a and the correspondence to the spectrum in fig. 13b is indicated by the numbers. The curves of fig. 6 with $J_1 = - J_2$ give a more quantitative description using dipolar theory and the parameters appropriate for the Fe-films. However since in these calculations the interlayer exchange was not included a qualitative comparison is more adequate. The spectrum in fig. 13b should be seen as a qualitative fingerprint of the situation displayed on the right hand side of fig. 13a. The remarkable aspect remains that not only can we prove by this that the magnetization of the two Fe-films is antiparallel but also that it is perpendicular to the small external field. This is in complete analogy to the spin flop phase of an antiferromagnet.

Such antiferromagnetically coupled double layers show also a quite remarkable behavior as far as the reversal of the magnetization in an

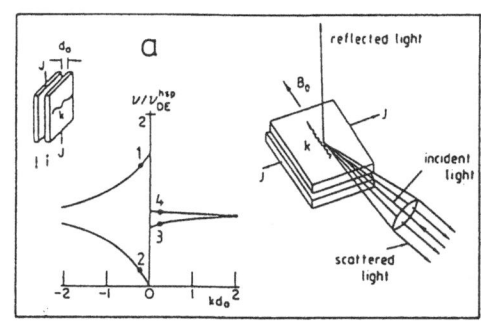

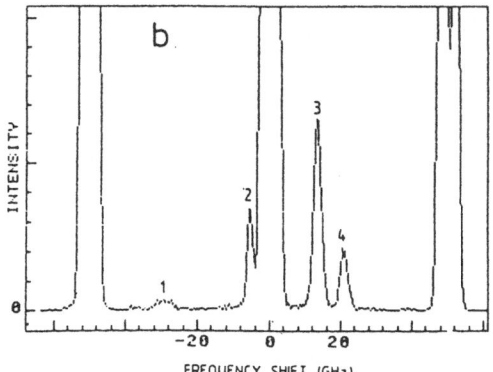

Fig. 13. Part a) shows dispersion curves of spinwaves in antiparallel layers (compare fig. 6 for $J_1 = -J_2$) and the employed scattering geometry. Part b) shows the resulting lightscattering spectrum. The numbers indicate the correspondence to the branches in part a). The spectrum proves that the magnetizations J in Cr8e for small external fields are aligned as proposed in fig. 12.

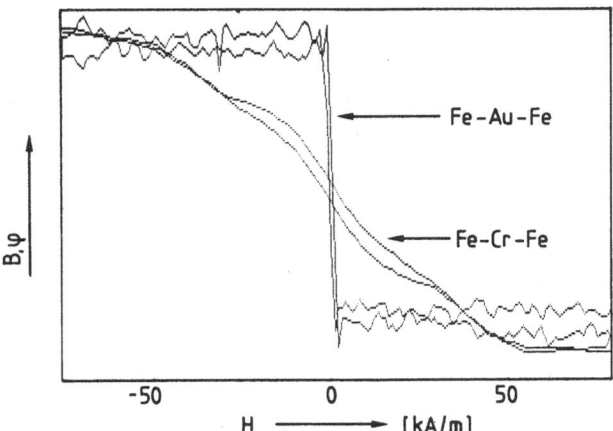

Fig. 14. B-H loops from the same samples as used for fig. 12. B is assumed to be proportional to the Kerr rotation φ which was measured using an arrangement as displayed in fig. 1.

503

external field is concerned. Fig. 14 displays B-H loops for the two samples used also for fig. 12. The reference sample with an Au-interlayer yields an essentially square hysteresis loop as do also single epitaxial Fe-films. By visual observation it was confirmed that here the magnetization reversal is via nucleation of domains and domain wall motion. Opposite to this for the sample with the antiferromagnetic interlayer coupling the B-H loop is strongly sheared and saturation is reached only in relatively high fields of H = 50 kA/m ($\approx$ 500 Oe). At a first glance the interpretation seems straightforward if we assume that by decreasing H below 50 kA/m the magnetization in the two Fe films rotates in opposite sense until for H→0 the situation displayed in fig. 13a is reached. By increasing H in the opposite direction the rotation would continue until parallel alignment is reached again. The fact that in such samples we need much higher fields to saturate, certainly is caused by the fact that the external field has to overcome the antiferromagnetic exchange between the films. However under the assumption of a rotation of the magnetization one can easily show that it requires much higher fields than 50 kA/m to get over the barrier along the [110] direction imposed by the Fe-anisotropy [16]. Domains on the other hand so far have not been observed or have escaped detection. This puzzle is presently investigated [17].

7. FINAL REMARKS

In recent years layered magnetic structures have been of increasing interest. There is some hope to obtain new magnetic materials this way. Since for applications anisotropy very often is of great importance one might for example try to exploit interface anisotropy by the multiplication of layers and interfaces. Also as described in other contributions to this book it is sometimes possible to force materials in sufficiently thin films into another structure than in the bulk which also results in changes of the magnetic properties. Here the question of interlayer exchange coupling has been addressed which is thought to be also an important and fundamental one. Two cases of antiferromagnetic interlayer coupling have recently been discovered: the Gd-Y-Gd [18] case described in this volume by Kwo, and the Fe-Cr-Fe [11] case described here. A systematic search which includes also ferromagnetic coupling certainly will continue. In this article techniques are described which allow a quantitative determiation of the interlayer exchange coupling. They rely on the magnetooptic interactions and have the great advantage that probably the least amount of material for such investigation is neccessary. Opposite to other techniques we can suffice ourselves with double layers and the individual layers are also fairly thin. Of great advantage is also the fact the LS probe (opposite to the conventional MA experiments) provides a frequency scan of the spin waves. Using a field scan as in the conventional MA-experiment it would not have been possible to obtain for example the result displayed in fig. 13b. The question as to how many other magnetic materials beyound the 3d-metals Fe, Ni, Co can be investigated will depend on the size of the magnetooptic interaction and also the exchange constant of the material. One important criterion for example is that at least one standing spinwave in a single film of that material has to be observable. Nevertheless even if we concentrate on the 3d-metals than considering the freedom with respect to the interlayer materials and including temperature as a parameter it can not be expected that the number of possible or even interesting combinations can be exhausted quickly.

I want to thank my colleguas from Argonne National Laboratory in particular S.D. Bader, M. Brodsky, M. Grimsditch, L. Moog, H. Sowers, who have contributed in many ways to results presented here. Also I want to thank W. Zinn of IFF-KFA Jülich for continuous support and many helpful discussions and R. Schreiber whose valuable technical assistance in particular in preparing the samples I have enjoyed now for many years. Illuminating discussions with L.Hinchey, K. Vayhinger and D.L. Mills concerning the theoretical background are also acknowledged. I also thank W. Vach, P. Swiatek, F. Saurenbach and U. Walz who have performed many of the experiments during their ph.D. works.

REFERENCES

1.   S.D. Bader, E.R. Moog, P. Grünberg, J. Magn. Magn. Mat. 53, L295 (1986).
2.   J.R. Sandercock in Topics in Appl. Phys. 51, p. 200, Springer 1982.
3.   for a review see e.g. P. Grünberg, Progr. in Surf. Sci., Vol. 18, no. 1 (1985).
4.   K. Vayhinger, H. Kronmüller, submitted and K. Vayhinger diploma thesis, Univ. of Stuttgart (1986).
5.   P. Grünberg, M.G. Cottam, W. Vach, C. Mayr, R. Camley, J. Appl. Phys. 53, 2078 (1982).
6.   F. Hoffmann and H. Pascard, AIP Conf. Proc. No. 5, 1103 (1971) and references therein.
7.   L. Hinchey and D.L. Mills, Phys. Rev. B33, 3329 (1986).
8.   P. Grünberg, J. Appl. Phys. 57, 3673 (1985).
9.   P. Swiatek, F. Saurenbach, Y. Pang, P. Grünberg, W. Zinn, Proc. of the 3. ICPMM, 1986 Szczyrk, Poland to be published by world Sci. Publ. Cy., Singapore and MMM Conf. Nov. 1986, Baltimore MD.
10.   K.H. Fischer in "Landolt-Börnstein" New Series Vol. 15a (1982). Data for $Pd_{1-x} Ni_x$ are from J. Beille and R. Tournier, J. Phys. F6, 621 (1979).
11.   P. Grünberg, R. Schreiber, Y. Pang, M.B. Brodsky, H. Sowers, Phys. Rev. Lett. 57, 2442 (1986) and P. Grünberg, R. Schreiber, Y. Pang, U. Walz, M.B. Brodsky, H. Sowers, MMM Conf. Nov. 1986, Baltimore MD.
12.   see e.g. R.W. Adam, Z. Naturforsch. 23a, 1526 (1986).
13.   G. Rupp, W. Wettling to be pulbished, and G. Rupp private communication.
14.   L. Hinchey, private communication.
15.   K. Mika, P. Grünberg, Phys. Rev. B31, 4465 (1985).
16.   W. Zinn, private communication.
17.   U. Walz, S. Mantl, P. Grünberg, W. Zinn, in preparation.
18.   C.F. Maykrzak, J.W. Cable, J. Kwo, M. Hong, D.B. McWham, Y. Yafet, J.W. Waszczak, C. Vettier, Phys. Rev. Lett. 56, 2700 (1986).

MAGNETISM AT SURFACES AND SPIN POLARIZED ELECTRON SPECTROSCOPY

H.C. Siegmann
Swiss Federal Institute of Technology
8093 Zürich, Switzerland

INTRODUCTION

Magnetic properties are often very different at the surface compared to the bulk. The inherent breaking of the symmetry at the surface, compositional and structural changes as well as surface adsorbates are possible causes of surface induced magnetic changes. The complex phenomena associated with magnetism at surfaces pose a great challenge to the experimentalist and require new techniques of measurement.

Spin polarized electron beam techniques provide unique possibilities to measure the magnetization at surfaces. If one neglects the orbital moment which is possible in most cases because it is quenched in the crystal field, the magnetization is given by $M = n\uparrow - n\downarrow$ where $n\uparrow(n\downarrow)$ is the density of up spin (down spin) electrons in the solid. The spin polarization of electrons is $P = (n\uparrow - n\downarrow)/(n\uparrow + n\downarrow)$. In other words, instead of measuring the magnetization, one can equally well measure the spin polarization. At first glance, this does not appear to be an advantage, because it is not straight forward to measure P of electrons in a solid. There are only indirect ways to do this, for instance over the contact terms to the nuclear magnetic moment, neutron-scattering, μ-meson depolarization or similar techniques. However, one of the major achievements of the past decade is that one has learned how to extract electrons from a solid and emit them into vacuum. Generally, in electron emission the energy is supplied in the form of electric fields. Electric fields do not couple to the electron spin in the nonrelativistic limit, and therefore P is conserved in the process of electron emission. Once the electrons are in vacuo, one can form an electron beam and measure the spin polarization P in a scattering experiment. This is the basis of magnetometry with electrons. The following advantages are achieved with this new type of magnetometry:

1. It is very fast. With lasers, one can generate short pulses of ultraviolet light. If such a pulse of say 10^{-12} sec duration strikes the surface of a solid, enough electrons are emitted to perform a very accurate measurement of P. No other magnetometer exists that can measure as fast as this.

2. It has very high spatial resolution: A primary electron beam can be focused into a small spot, say of 100 Å diameter. If it strikes a magnetic solid, secondary electrons will be emitted from the region of the

focus. The spin polarization P of the secondary electrons yields the magnetization of the spot from which they were emitted. No other magnetometer can measure with so little magnetic material, it actually takes only 10^6 magnetic atoms to do a measurement.

3. It is element specific: If one excites electrons from specific atomic shells, one will obtain a measure of the local magnetization, around the atom from which the electron was emitted. That is, one can obtain element specific magnetization in an alloy or an epitaxial structure.

4. It can measure the energy density of magnetization. This is a new magnetic quantity that describes how much each electronic state contributes to the total magnetization. It has helped significantly to understand metallic magnetism. Even if two different excitations nearly coincide in energy, they can be separated if they have different P. In this way, one also has identified new excitations in ferromagnets; one example are the Stoner excitations.

5. It can provide nondestructive depth profiles of magnetization. Due to the energy dependence of the escape depth of electrons, one can observe for the first time how the magnetization varies with distance from a surface or interface. It is important that this is done at the scale of the interlayer spacing in crystals, since magnetism is a short range interaction.

Altogether we can say that spin polarized electrons appear to be the key to the newly developing fields of surface- and epitaxial magnetism, and to magnetic material design on an atomic scale. This paper will not cover all aspects of this. It will mainly demonstrate some applications of material specific depth profiling techniques.

SURFACE INDUCED MAGNETIC STRUCTURES

The first layer of a magnetic material normally has a magnetization M_S very different from the bulk magnetization M_B; M_S also has a different dependence on temperature T. Furthermore, the exchange coupling of the first layer to the underlying bulk or substrate in case of a monolayer is important as well as the magnetic anisotropy K_S. K_S is vastly different from the bulk anisotropy K_B, both in direction and magnitude. The crystalline anisotropy arises from spin orbit coupling. With 3d-magnetic materials, the orbital moments are almost completely quenched in the highly symmetric crystal field of the bulk. The lack of cubic symmetry at the surface means that the orbital moment is not completely quenched and furthermore, that terms of lower order can enter into the power series expansion of the spin-orbit coupling. Both effects lead to the above mentioned large changes of K_S. However, it is very difficult to calculate K_S; theorists disagree by orders of magnitude on their estimates[1,2]. The exchange coupling in single layers and surface layers is perhaps even more difficult to predict. Already in bulk material, the exchange shiftness $A = 2 J \cdot S^2/a$ is basically a parameter that can only be determined in experiments; furthermore each type of experiment yields a different value for A. Near surfaces, A becomes anisotropic. $2 J \cdot S^2$ is the exchange energy of a pair of neighboring atoms with spin S and a is the lattice parameter. Hence A, the exchange energy per unit length, depends on whether one chooses a path within the surface layer or perpendicular to it.

These considerations lead to one of the key problems in the measurement of surface magnetism: the surface may induce special magnetic structures very different from the bulk structures. These surface induced magnetic structures (SMS) are also expected to critically depend on temperature. Therefore, one has to try to predict under which conditions SMS may occur in order to interpret surface magnetic measurements properly.

As a model case we consider first the 100-surface of an Fe-single crystal remanently saturated in the (100)-easy direction. Fig. 1a) illustrates the model employed to investigate the conditions under which SMS occur. Only the first layer with magnetization M_S and uniaxial surface anisotropy K_S is assumed to be different from the bulk. M_S may include an angle $\theta \neq \pi/2$ with the surface normal (z - axis). The first layer is coupled to the 2nd layer by an exchange interaction $2 JS_1S_2 = \xi \cdot A_{12}$, where S_1 and S_2 are neighboring magnetic moments in the 1st and 2nd layer respectively, ξ is the interlayer distance and A_{12} the exchange stiffness in the z - direction.

We assume that the properties are bulklike in all subsequent layers. The 2nd layer magnetization may enclose an angle $\psi \neq \pi/2$ with the z - axis. This surface induced distortion leads to the formation of a tail of a domain wall extending into the bulk. The total energy δ per unit area of the resulting SMS is given by:

$$\delta = \sqrt{AK} \, (1 - \sin \psi) + A_{21}/2\xi(1 - \cos(\psi - \theta)) + K_S \cdot \xi \sin^2\theta + \left(M_S^2/2\mu_0\right) \cdot \xi \cos^2\theta \qquad (1)$$

The first term represents the energy of the tail of the domain wall into the bulk according to Zijlstra[3], the second is the familiar energy of the exchange coupling between surface and bulk, and the two remaining terms are the magnetostatic and structural anisotropy respectively. A is the exchange stiffness and K stands for the anisotropies opposing the formation of a 90°-domain wall of the Néel type with energy $\sqrt{AK}$ per unit area. Dropping a term that does not depend on θ and ψ, one obtains the surface energy density δ^* in units of $\sqrt{AK}$:

$$\delta^* = (1 - \sin \psi) + \lambda(1 - \cos(\theta - \psi)) + \mu \cos^2\theta \qquad (2)$$

λ may be positive or negative depending on whether the bulk couples anti- or ferromagnetic to the surface, and $\mu > 0$ or $\mu < 0$ depending on whether the difference between magnetostatic and structural anisotropy favors $M_S \perp$ or $\parallel$ to z respectively.

If δ^* is at a minimum for $\theta = \psi = \pi/2$, we will say that no SMS occurs; if, however, at least $\theta \neq \pm \pi/2$ for minimal δ^*, SMS do exist. Fig. 2 shows the values of μ and λ for which non-trivial SMS are predicted. With $\lambda < 0$ the trivial structure occurs for M_S antiparallel to the bulk. The case of $\lambda < 0$ must be included, since the exchange can change sign when the atomic positions are changed as may be the case on a surface. The change from ferromagnetic to antiferromagnetic coupling has indeed been observed in the case of the Gd(0001) surface by D. Weller et al.[4].

Fig. 2 shows that SMS exist for negative μ only; hence the easy direction of M_S must be perpendicular to the surface and $K_S > M_S^2/2\mu_0$, the magnetostatic shape anisotropy. Assuming for $M_S = 2.1$ Tesla (the bulk value) leads to the conclusion that SMS can occur on Fe(100) only if $K_S > 1.75 \cdot 10^6$ Wattsec/m^3. Since the crystalline anisotropy of bulk Fe is $K_B = 5.3 \cdot 10^4$ Wattsec/m^3, $K_S > 33 \, K_B$ is the precondition that SMS may occur at T = 0.

Gay and Richter[2] were the first to show that such high values of K_S are possible. They calculated $K_S = 100 \, K_B$ for an isolated layer of Fe(100) on an Ag-substrate. Assuming that this holds for Fe on Fe(100) as well, we can estimate whether or not SMS occur at finite temperatures. The magnetization near the surface has frequently been reported to decrease linearly with T. This yields $M_S = 1.41$ Tesla and $AK = 2.56 \cdot 10^{-3}$ Wattsec/m^2 at room temperature T_R, resulting in $\mu = -0.25$. From Fig. 2 we see that eq. (2) predicts SMS in this case if the bulk exchange $\lambda_B = 3.85$ at T = 0 is reduced by at least a factor ~ 4 at the surface. This reduction is easily conceivable because the number of nearest neighbors that the atoms in the first layer have is reduced by 50% with Fe(100). However, due to the lack of reliable theory for λ, the actual evidence for SMS on Fe(100) must come from measurements. The experimentalist faces the enormous task to do a non-destructive magnetic depth profiling on a scale of the lattice parameter a with an atomically clean and possibly defect free surface at various T.

Equation (2) describes of course the simplest possible model. It underestimates the likelyhood of the occurrence of SMS because subdivision into SMS with a magnetization component out of the surface and SMS into the surface could occur reducing the magnetostatic energy of SMS. These SMS-waves depend on the exchange stiffness in the surface plane. However, there are cases where the model can be rigorously tested, even when a external magnetic field H is applied. These include a very thin ferromagnetic film on a thick antiferromagnetic substrate as employed in magnetic recording heads to produce transferred exchange fields[5] or a very thin ferromagnetic film on a ferrimagnetic substrate. A thin ferromagnetic film occurs naturally at the surface of 3d-4f intermetallic compounds such as the permanent magnet $Nd_2Fe_{14}B$ or the surface of 3d-4f amorphous alloys such as the magneto-optic recording material FeTb. Fig. 1b) shows the model applicable to these 2 cases. Uniaxial anisotropy is present in the bulk of these materials, and Fig. 1b) depicts a surface perpendicular to this anisotropy. Due to the naturally occurring surface segregation and oxidation of the R.E., a ferromagnetic layer consisting mainly of the 3d-element is formed in the subsurface with the nonmagnetic R.E. oxide film at the outermost surface[6]. The subsurface layer of thickness d is exchange coupled to the bulk, exhibits large M_S and low structural anisotropy. Both effects produce $\mu > 0$ making the preferential direction of M_S in-plane. This naturally leads to SMS. Equation (2) has to be modified as follows to account for the different direction of the bulk remanent magnetization:

$$\delta^* = (1 - \cos\psi) + \lambda(1 - \cos(\theta - \psi)) + \mu\cos^2\theta \tag{3}$$

Non-trivial SMS occur now only if $\mu > 0$. The values of $|\mu|$ and λ for which SMS exist is otherwise the same as in the case of (2), see Fig. 2. It is also evident from Fig. 1b) that SMS induce fractional domain walls and hence lead to a reduction of coercivity. This mechanism is responsible for the deterioration of $Nd_2Fe_{14}B$ and $SmCo_5$ permanent magnets on oxidation[7].

In the case of e.g. amorphous $Fe_{72}Tb_{28}$ one can go one step further. This composition has a small bulk magnetic moment since the dominant Tb-subnetwork magnetization is almost compensated by the antiparallel Fe-subnetwork magnetization[8]. Therefore, an external magnetic field H will predominantly interact with the Fe-rich film at the surface. One can then calculate magnetization curves (MC) by finding the values of θ (and ψ) for which

$$\delta^* = (1 - \cos\psi) + \lambda(1 - \cos(\theta - \psi)) + \mu\cos^2\theta + \kappa(1 - \cos\theta) \tag{4}$$

has a minimum; the reduced magnetic field is given by $\kappa = H \cdot M_S \cdot d/\sqrt{AK}$. We will show below that nondestructive magnetic depth profiling using spin polarized photoemission in combination with magneto-optic measurements yields detailed evidence that equation (4) can describe the general situation of a ferro- or ferrimagnetic layer which is exchange coupled to an underlying bulk of zero or weak magnetization.

Experimental evidence will also be presented showing that SMS definitely exist in many cases. It is important to identify SMS in order to understand surface magnetic measurements properly; it is also very helpful

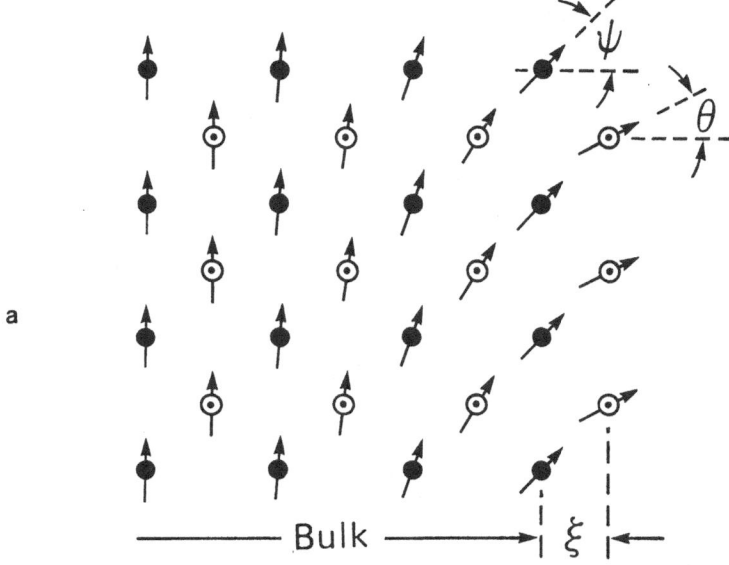

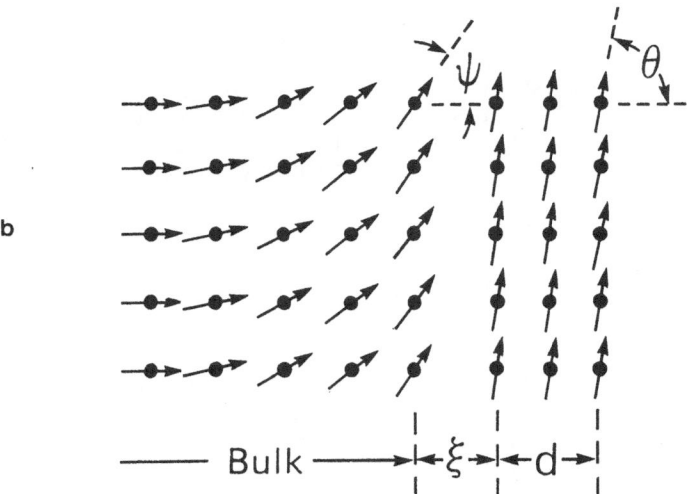

Fig. 1. Models for Surface Magnetic Structures in the remanence.
a) Model for the Fe(100) surface. The bulk is remanently
magnetized upwards, the surface anisotropy, K_S, is perpendicular
to the surface. It is assumed that only the 1st layer is different from the bulk. Note that atoms at circled positions are
out of plane of the drawing. b) Model for 3d-4f alloys. The bulk
is remanently magnetized perpendicular to the surface, whereas
the surface layer of thickness d has the easy direction in-plane.
Since 3d-3d-interactions are dominant, the 3d-magnetic moments are
depicted only.

to characterize the quality of surfaces and interfaces by the M_S, λ and μ parameters which are of course dependent on contamination and surface defects. In this way, one might hope to make possible the intercomparison of different experiments[9].

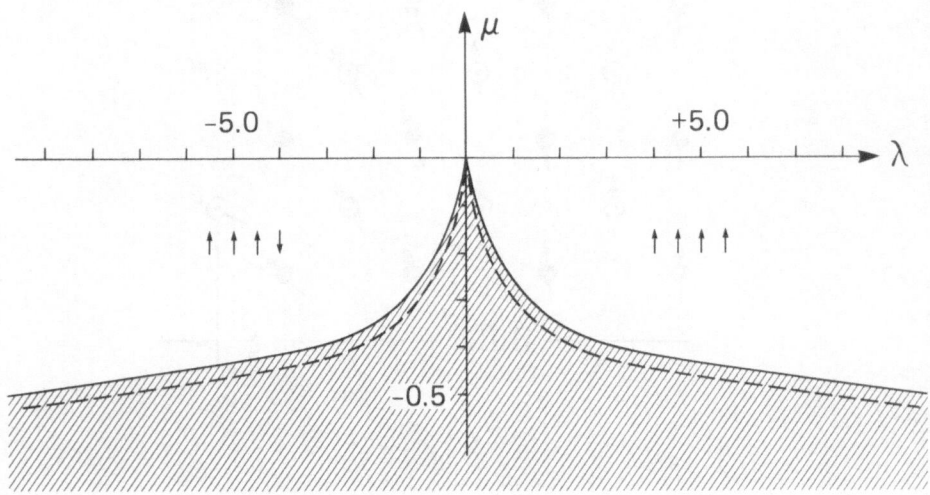

Fig. 2. Phase diagram of SMS. SMS exist in the hatched area. The full line indicates values of μ and λ for which θ deviates by more than 2° from ±90° (or 0° in the case of model Fig. 1b). The dashed line indicates values of μ and λ for which the occurrence of SMS reduces the energy by 1% compared to the trivial structure, giving a picture of the soft boundary between SMS and no SMS.

DEPTH DEPENDENT INFORMATION ON MAGNETIZATION FROM SPIN POLARIZED SECONDARY ELECTRONS

Depth dependent information on magnetization has been obtained by elastic scattering of spin polarized electrons, particularly in low energy electron diffraction[10]. The problem of multiple electron scattering can be difficult to handle and this is why it is not straightforward to extract the desired information from the data. The spin polarization of the secondary electrons provides magnetic depth profiles in the 5-50 Å range by variation of 2 parameters: the energy E_p of the primary (unpolarized) electrons and the energy E_S of the secondary (spin polarized) electrons. This can best be understood in a three step model for emission of secondary electrons: 1. production of hot electrons by electron - electron collisions, 2. transport of electrons to the surface, 3. emission of electrons over the surface barrier potentials. The depth distribution of hot electron production is given by the energy dissipation profile of the primary electrons impinging onto the solid. Inelastic scattering of the hot electrons on their way to the surface then leads to an attenuation which is energy dependent via the energy dependence of the inelastic mean free path (IMFP). Step 3 eliminates the very low energy electrons. The separation into steps l and 2 is somewhat artificial since the cascade process of hot electron production includes IMFP-effects. To estimate the escape depth of secondary electrons one can take the calculated energy dissipation profiles and multiply them by an exponential attenuation $I = I_0 \exp(-z/JMFP(E_S))$. This neglects multiple scattering during transport to the surface. The result of this simplified procedure is qualitatively correct to within a factor of 2. The maximum probing depth is $z \cong 50$ Å at $E_p = 2500$ eV and $E_S = 1$ eV, and the minimum probing depth is $z \cong 5$ Å at $E_p = 100$ eV and $E_S = 50$ eV. Of course these z - values represent the most probable depth of origin only; the actual distribution shows a steep increase at low z and a large tail towards higher z.

As an example, Fig. 3 shows hysteresis loops obtained with an Fe single crystal and with secondary electrons emitted from the Fe(100) surface[13]. The reason for choosing this surface is that it is an allowed plane of a 180° domain wall in bulk Fe. Hence, one expects that a domain wall nucleates at this surface when an external field is applied opposite to the remanent magnetization. The reason why one wants to observe this process is Browns paradoxon according to which the theoretical coercive field is orders of magnitude higher than the experimental coercive field. It has been proposed that rudimentary traces of domain walls existing at planar defects like grain boundaries, interfaces or surfaces can resolve this problem[3]. The embryonic domain walls only have to be further developed and torn free from the planar defects to reverse the magnetization. In this way, the coercive field is much lower since the domain walls do not have to nucleate from a uniformly magnetized state. However, as one has better crystals and epitaxial thin films of high quality one actually would like to see these embryos.

In the preceding chapter it was shown how the surface might induce special magnetic structures (SMS) which in turn can be equated in some, but not all cases, with domain wall embryos. Fig. 3 gives clear evidence for SMS. The upper panel shows a hysteresis loop taken with secondary electrons that have a most probable depth of origin of $z = 10$ Å. This loop turns out to be identical to the magneto-optic loop representing material at a depth of ∼ 200 Å; it shows no anomalies and it is a typical bulk loop of a single crystal magnetized in the easy direction. However, as the most probable depth of origin of the secondary electrons is reduced to $z = 5$ Å, the loop changes; it has rounded edges instead of rectangular ones. The special feature is that the component M_x of the magnetization in the direction of the remanent magnetization changes at field strength below the coercive field H_c. Further, the observation was made that the change of M_x is reversible for $|H| < .8 H_c$. The fact that M can be changed in the first few layers only with very weak external field means that the first layer spins are frustrated: they are under the opposing influence of the surface anisotropy K_S and the exchange coupling λ to the bulk, and hence may be influenced by very weak external fields. This is very strong evidence for SMS on Fe(100).

The saturation value of the spin polarization P in the direction of the remanence is also different for the 2 probing depth shown in Fig. 3. This has 2 reasons: 1. the surface may have a lower magnetization compared to the bulk due to thermal excitation of surface magnons. This T-induced decrease of P must be distinguished from the T-dependent decrease of P if SMS are present. 2. P depends on the energy of the secondary electrons; P decreases as E_S increases. This phenomenon is not understood in detail and therefore no conclusions can be drawn yet from the observed decrease of the saturation value of P with decreasing probing depth.

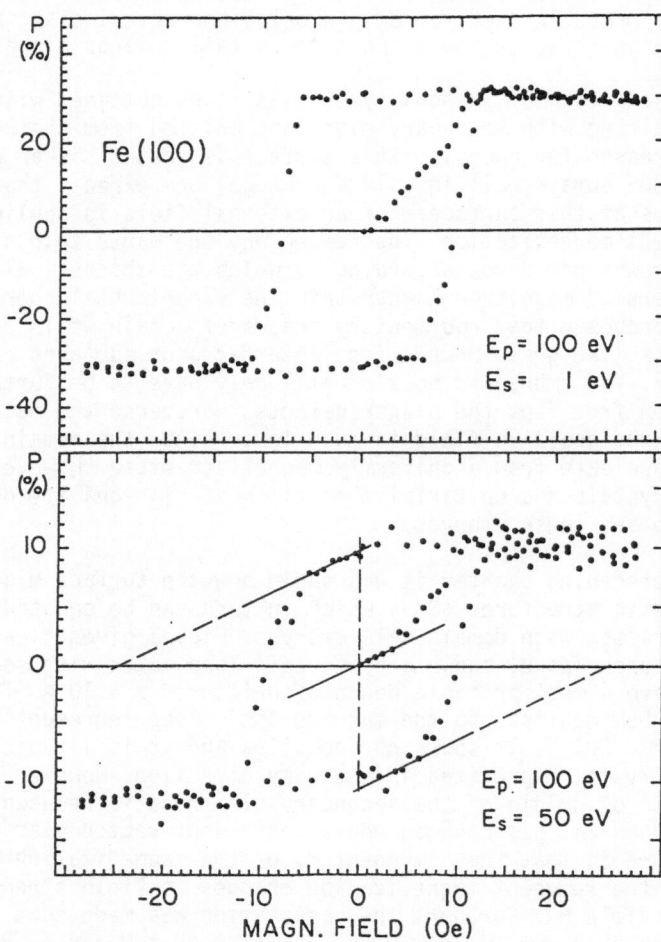

Fig. 3. Magnetic hysteresis loops and the virgin magnetization curve recorded via the spin polarization P of secondary electrons at room temperature. The 2 curves are obtained with the same primary energy $E_p = 100$ eV but with different secondary electron energies E_S. The values of the external magnetic field are somewhat uncertain. The full lines are the predictions of the model for $\lambda \ll 1$ and $H \ll H_c$.

There is also a difference in the virgin magnetization curve at the surface compared to the bulk. In the bulk, the initial reduced slope of the virgin curve indicates reversible domain wall motion in potential minima created by internal tensions; the subsequent higher slope is due to irreversible Barkhausen jumps over the potential maxima. At the very surface, it appears that the virgin magnetization curve closely resembles to the magnetization curve obtained when the remanent magnetization decreases upon reversing the external field.

All these features are quantitatively predicted by the model for SMS if $\lambda \ll 1$, and for $\psi = \pm 90°$. The latter applies at $H \ll H_c$ in the remanence and the demagnetized state of the sample; the exchange term in equation (2) reads then $\lambda(1 - \sin\theta)$. We also have no tail of a domain wall, and it becomes possible to study the surface layer in an external field H; the field energy is $\kappa(1 - \sin\theta)$ with $\kappa = H M_S \cdot d/\sqrt{AK}$, and $d =$ lattice parameter. The surface energy density δ^* becomes

$$\delta^* = (\kappa + \lambda)(1 - \sin\theta) + \mu\cos^2\theta \tag{2'}$$

We see that the exchange coupling of the surface layer can be simulated by a magnetic field acting parallely to κ. According to Stoner and Wohlfarth[15] (2') yields a hysteresis free linear increase of the magnetization with a slope $\chi = M_S/(2\mu - \lambda)$, and saturation occurs at $K_S = 2\mu - \lambda$. All these features are observed as Fig. 3 proves; in particular, the virgin curve starts with the same slope as M_S decreases from the remanence when H is inverted. The experiment shows $H_S = 25$ Oe yielding $K_S \cong 3 \cdot 10^{-4}$; this leads to $2\mu \cong \lambda$. In this way one can determine λ from the experiment if one knows μ; with $\mu = .25$ from the work of Gay and Richter[2] we obtain $\lambda \cong .5$, which is ~7.5 times smaller than the bulk value at $T = 0$, indeed a very reasonable result. One also notes that the line $2\mu = \lambda$ is located completely in the region where SMS exist in Fig. 2.

There is further evidence for SMS on the Fe(100) surface from thin film experiments. Jouker et al.[14] have built an Fe(100) film on a Ag substrate by depositing layer by layer and measuring the development of the spin polarized band structure. The first layer had no remanent magnetization in the plane of the film yet exchange split energy bands. Remanent magnetization in the film plane developed from 2.5 layers on only. This also yields evidence for a shallow SMS on that surface. The existence of SMS depends of course crucially on cleanliness and structural defects. However, in the case of Fe(100), SMS can probably not directly be related to domain wall embryos reducing coercivity. This arises because the coercivity of an Fe single crystal does not crucially depend on the state of the very surface. D. Pescia et al.[16] have shown that $\mu > 0$ for fcc Co-films on Cu(100); hence SMS do not exist in this case.

However, if SMS exist as it is very likely the case with Fe(100), they will be T-dependent. This means that T-dependences observed in a surface sensitive experiment like inverse or normal spin polarized photoemission do not have a simple interpretation as was previously assumed[17].

A 2nd case where SMS may exist is the surface of $\gamma - Fe_2O_3$ sputtered films. It has been reported that the remanent magnetization in the film plane as observed with secondary or Auger-electrons is zero[18], whereas the spin polarization of photoelectrons with external field perpendicular to the film surface is large[19]. In both configurations, one sees however magneto-optic signals with good samples. One obvious explanation covering all different observations is that SMS with a larger depth, of the order of 30 Å, exist at the surface of $\gamma - Fe_2O_3$ and magnetite, Fe_3O_4, which exhibits similar behavior.

DEPTH DEPENDENT INFORMATION ON MAGNETIZATION FROM THRESHOLD PHOTO-EMISSION

Alloys or intermetallic compounds of 3d-transition metals with rare earth, in short 3d-4f materials, have found many interesting applications e.g. as magneto-optic recording media and permanent magnets. Famous examples are FeTb amorphous thin films and $SmCo_5$ and NdFeB permanent magnets. The magnetic anisotropy and the coercivity have been reported in numerous papers to crucially depend on the processing conditions and particularly on the state of the surface. The surface scientist working in well defined ultrahigh vacuum conditions faces the problem that the outermost surface layers are ill defined because of the spontaneous segregation of the rare earth (R.E.) to the surface, leaving behind a subsurface layer enriched in the 3d component. The reactive R.E. usually forms a nonmagnetic oxide in a short time protecting the subsurface layer which remains unoxidized 3d-metal even when the speciment is exposed to atmospheric pressure. Therefore, the surface tools with a low probing depth like Auger-analysis invariantly detect mainly the uninteresting R.E. oxide only. One then first needs a surface analytical tool with a large probing depth; in conjunction with measurement of electron spin polarization, threshold photoemission (TP) is ideal for investigating magnetic phenomena in 3d-4f materials because it has a very large probing depth. TP is achieved by irradiating the sample surface with photons of energy $h\nu$ close to photoelectric threshold Φ. Φ can be lowered from typically 4 - 5 eV to less than 2 eV by depositing a fraction of a monolayer of Cs onto the sample surface thereby increasing the probing depth from $\approx$ 25 to $\approx$ 80 Å. The Cs layer does not affect the spin polarization P of the photoelectrons. P represents the spin polarization at the Fermi energy E_F in the case of TP. $P(E_F)$ is highly specific for the chemical composition of a given material; for instance, it is $\sim$ + 50% in γ - Fe_2O_3 and $\sim$ - 70% in Fe_3O_4 or $\sim$ - 30% in polycristalline Ni and + 60% in polycristalline Fe.

Magnetism in compositionally modulated structures of 3d-4f - materials is one of the most complex phenomena; one reason is that the compensation point in materials like FeTb changes by 20°C if the composition changes by 1%, another reason is the exchange coupling between compositionally modulated layers. The 3d-3d - exchange is strong and positive, the 3d-4f - exchange is weaker and negative, and 4f-4f - exchange is weakest and has unpredictable sign. Because of this hierarchy, the exchange coupling between layers favors a state in which the 3d-subnetwork magnetizations are parallel. The exchange coupling manifests itself in the most anomalous hysteresis loops[20] and/or magnetic anisotropies.

As noted above, the naturally occurring surface segregation and subsequent oxidation of the R.E. leaves behind a ferromagnetic layer consisting mainly of the 3d - element in the subsurface. The subsurface layer of thickness d is exchange coupled to the bulk and exhibits large magnetization because it has been depleted of the compensating R.E. Furthermore, it has low structural anisotropy because the process of diffusion of the R.E. is of course random. Both effects produce $\mu > 0$, which means that M_S of the subsurface has the lowest energy when in-plane.

It turns out that TP has just the right probing depth to obtain magnetization curves (MC) of the subsurface in-plane film. The deeper lying "bulk" material is detected by measuring the Kerr elliplicity (K) with a probing depth of $\sim$ 160 Å (in FeTb). Thus combining TP and K is ideal to disentangle the various contributions to the total magnetization and to study the exchange coupling λ at the interface between subsurface and bulk.

As an example we show results obtained with the magneto-optic material $Fe_{72}Tb_{28}$. In this composition the Tb subnetwork magnetization is dominant at all T, yet almost compensated by the Fe - subnetwork magnetization. The Néel point is 130°C. Hence at T ≥ 130°C, one can observe the 3d - rich subsurface alone ($\lambda = 0$). As T is lowered, the bulk becomes magnetic and the exchange coupling is switched on.

Let us first consider on the basis of the model what we expect in this case. Fig. 4 shows MC's obtained from eq. (4) at $\lambda = 0$ and at various $\lambda \neq 0$ and μ values. At $\lambda = 0$, we see the Stoner - Wohlfarth[15] hysteresis-free MC of a film magnetized perpendicular to its easy axis of magnetization. As λ increases, the MC shifts to the right and tilts at high λ/μ ratios. The shift is equivalent to the notion of an exchange field transferred from the bulk. The tilting arises if the exchange energy λ is large enough to induce a domain wall in the bulk. This is possible at $\lambda \gtrsim 1$. The arrow in Fig. 4 indicates the external field at which ψ is changed by 90°.

Turning to the actual measurements shown in Fig. 5 we see a remarkable agreement with the predictions of the model. The full line are MC's calculated from eq. (4) with the following assumptions:

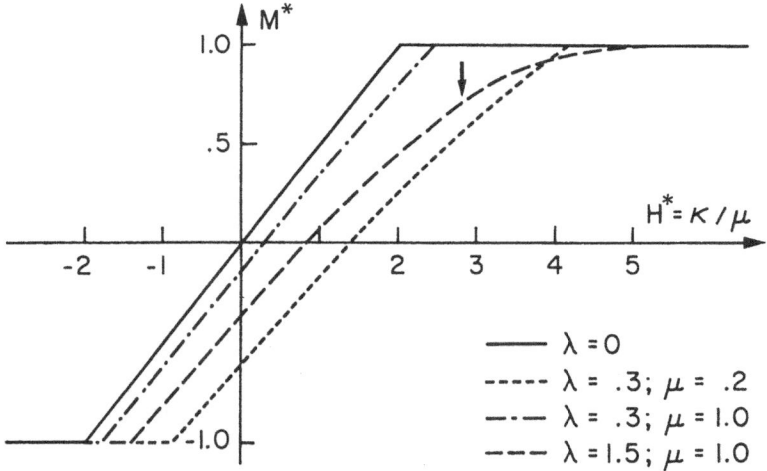

$$\lambda = 0$$
$$\lambda = .3; \mu = .2$$
$$\lambda = .3; \mu = 1.0$$
$$\lambda = 1.5; \mu = 1.0$$

Fig. 4. Reduced magnetization $M^* = M(H)/M(H \to \infty)$ vs reduced magnetic field $H^* = \kappa/\mu$ as calculated from eq. (4). The anisotropy μ, exchange coupling λ of surface layer to the bulk, and the magnetic field κ are dimensionless parameters defined in the text. The arrow indicates the field at which the bulk magnetic moment closest to the surface deviates by 90° from its equilibrium easy direction. We see how the Stoner-Wohlfarth[15] magnetization curve at $\lambda = 0$ is shifted and distorted on switching on the exchange coupling.

1. The Fe - rich in plane film has shape anisotropy only, i.e.
$\mu = M_S^2 \cdot d/2\mu_0$. 2. This film is d = 25 Å thick, which is consistent
with depth dependent chemical analysis by sputter etching and with the
escape depth of threshold photoelectrons. From the MC at T = 130°C
(λ = 0) one obtains then μ = .14 and M_S = 0.38 Tesla. Cooling increases
M_S as is evident from the increase of the saturation value of P from 15%
to 18%. This yields μ = .20 at T = -60°C. With λ = + .35 one obtains a
good fit to the data, indicating ferromagnetic coupling between bulk and
subsurface as expected from the dominant Fe-Fe - interaction. λ is roughly 10 times weaker than in bulk FeTb; this weakening arises possibly from
the spongy structure of the subsurface film produced by diffusing Tb to
the surface.

Fig. 5 also shows the Kerr loop representing the "bulk" magnetization. It is inverted since the magneto-optically active electrons are the
3d-electrons and since the Fe subnetwork is antiparallel to the dominant
Tb magnetization. One notes that little evidence of SMS is seen with the
conventional Kerr - technique except that the steep drop of the magneti-

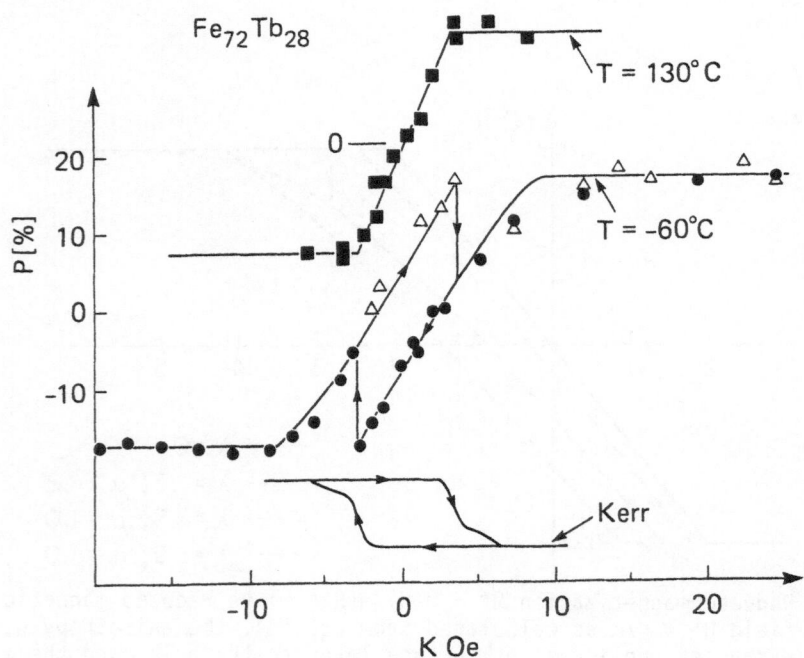

Fig. 5. Spin polarization P of photoelectrons in % excited near photoelectric threshold from $Fe_{72}Tb_{28}$ vs. magnetic field strength,
from Ref. 22. The field is applied perpendicular to the surface. The full lines are magnetization curves obtained from (4)
with μ = .14; λ = 0 at T = 130°C and μ = .20; λ = .35 at
T = -60°C. Kerr ellipicity vs. field strength is also shown as
measured at T = -60°C.

zation at H_c is replaced by smooth transitions. The smooth transitions are caused by SMS in the subsurface. In this case, as evident also from Fig. 2, the SMS can be equated with embryonic domain walls at the surface causing a reduction of the coercivity.

It should be mentioned that excellent results on the microscopic aspects of 3d - 4f - coupling have been obtained with spin polarized Auger - spectroscopy on the system Fe-Gd[21]. Auger lines at various energies provide depth dependent analysis on an atomic scale which is also element specific. This arises because the spin polarization of Auger electrons reflects the magnetization around the atom from which the Auger - electron was emitted.

CONCLUSIONS

Non destructive depth profiling with spin polarized electrons provides important insight into magnetic phenomena at surfaces. Surface anisotropy and surface bulk exchange interaction may induce special surface magnetic structures (SMS) that must be identified and properly taken into account when interpreting surface magnetic measurements. Various spin polarized electron beam techniques are available to study magnetism at surfaces. The range of the depth profiling capabilities must be adapted to the specific material. Threshold photoemission is unique in that it is very simple, as an adjustable probing depth, and allows the application of large magnetic fields perpendicular to the surface. Spin polarized secondary and Auger electrons provide non-destructive depth resolution on a scale comparable to the lattice parameter and in the case of the Auger-electrons are also element specific. As model cases demonstrating these techniques we show results obtained with Fe(100) and FeTb. Fe(100) very likely exhibits SMS. In 3d-4f - alloys, SMS occur by surface segregation and can be characterized with few parameters.

The author wishes to thank Daniele Mauri, Paul Bagus, Eric Kay and Alan Bell at the IBM Almaden Research Center and Martin Aeschlimann, Gianni Bona and Mauro Taborelli at the Swiss Federal Institute of Technology for contributions and support of this work in many areas.

REFERENCES

1. U. Gradmann, J. Mag. and Mag. Mat. 54-57:733 (1986)
2. J.G. Gay and R. Richter, Phys. Rev. Lett. 56:2728 (1986)
3. H. Zijlstra, IEEE Trans. on Mag., Vol. Mag. 15:1246 (1979)
4. D. Weller, S.F. Alvarado, W. Goudat, K. Schröder and M. Campagna, Phys. Rev. Lett. 54:1555 (1985)
5. D. Mauri, H.C. Siegmann, P.S. Bagus and E. Kay, to be published
6. G.L. Bona, F. Meier, M. Taborelli, H.C. Siegmann, A.E. Bell, R.J. Gambino and E. Kay, J. Mag. and Mag. Mat. 54-57:1403 (1986)
7. L. Schlapbach, J. of the Less Common Met. III:291 (1985)
8. G.A.N. Conell, R. Allen and M. Mansuripur, J. Appl. Phys. 53:7759 (1982)
9. H.C. Siegmann, P.S. Bagus and E. Kay, to be publ.
10. For a review see H.C. Siegmann, F. Meier, M. Erbudak and M. Landolt, Adv. El. and El. Phys. 62:1 (1984)
11. D. Mauri, R. Allenspach and M. Landolt, J. Appl. Phys. 58:906 (1985)
12. J.P. Ganachaud and M. Cailler, Surf. Sci. 83:519 (1979)
13. R. Allenspach, M. Taborelli, M. Landolt and H.C. Siegmann, Phys. Rev. Lett. 56:953 (1986)
14. B.T. Jonker, K.H. Walker, E. Kisker, G.A. Prinz and C. Carbone, Phys. Rev. Lett. 57:142 (1986)

15. E.C. Stoner and E.P. Wohlfarth, Phil. Trans. Roy. Soc. (London) A240:599 (1948)
16. D. Pescia, G. Zampieri, M. Stampanoni, G.L. Bona, R.F. Willis and F. Meier, to be publ.
17. E. Kisker, K. Schroeder, W. Goudat and M. Campagna, Phys. Rev. B31:329 (1985), and ref. cited
18. M. Taborelli and R. Allenspach, priv. comm.
19. E. Kay, R.A. Sigsbee, G.L. Bona, M. Taborelli and H.C. Siegmann, Appl. Phys. Lett. 47:533 (1985)
20. S. Esho, Suppl. Japan J. Appl. Phys. 15:93 (1976)
21. M. Taborelli, R. Allenspach, G. Boffa and M. Landolt, Phys. Rev. Lett. 56:2869 (1986)
22. M. Aeschlimann, Diploma ETH Zurich 1985, unpubl.

EPITAXIAL GROWTHS AND SURFACE SCIENCE TECHNIQUES APPLIED

TO THE CASE OF NI OVERLAYERS ON SINGLE CRYSTAL FE(001)

B. Heinrich, A.S. Arrott, J.F. Cochran, S.T. Purcell,
K.B. Urquhart, N. Alberding, and C. Liu[*]

Physics Department
Simon Fraser University
Burnaby, B.C. Canada V5A-1S6

INTRODUCTION

The rapidly increasing interest and activity in the study of epitaxially deposited magnetic films on single crystal substrates stem both from the ability to stabilize metastable crystalline structures which do not exist otherwise in nature and from theoretical predictions of enhanced magnetic moments and crystalline anisotropies in low dimensional systems. For example, recent spectacular experimental results[1,2] and theoretical calculations[3] show that the crystalline anisotropy field in ultrathin Fe films is capable of overcoming the demagnetizing field perpendicular to its surface, making such films an ideal building block for multilayered permanent supermagnets. This is an example of the creation of new magnetic materials by means of atomic engineering. It should be pointed out that such recent advances and future progress in atomic engineering would not be possible without Molecular Beam Epitaxy (MBE) techniques using controlled atomic beams in Ultra High Vacuum (UHV) and using state of the art surface science techniques such as Reflection High Energy Electron Diffraction (RHEED), spin polarized or unpolarized Auger Electron Spectroscopy (AES) and X-Ray Photoelectron Spectroscopy (XPS).

The SFU group has concentrated on the investigation of ultrathin metal overlayers deposited by MBE on single crystal metallic substrates. Mn overlayers were deposited on a single crystal of Ru(0001). Ni, Mn, MnO, and Cr overlayers were grown on a single crystal Fe(001) substrate, Ni was epitaxially grown on Au, and Au was grown on Fe(001) and Ni(001). Our studies have been carried out with several objectives in mind:

(a) The growth and characterization of metastable structures stabilized by the substrate,

(b) The study of magnetic moment formation and magnetic crystalline anisotropy in a different atomic environment,

(c) The investigation of magnetic properties and magnetic interface interactions using microwave ferromagnetic resonance (FMR) measurements.

---

[*]Present address: Physics Department, Rice University, Houston, TX 77251, USA

In this paper we highlight the results of our studies of Ni overlayers grown on a single crystal Fe(001) substrate.

1. PREPARATION OF THE FILMS

1.1 Substrate Preparation

The Fe crystal substrate used in this work was cut from a single crystal plate which was misoriented two degrees from the (001) plane. The plate was prepared by Takeuchi and Ikeda[4] using a strain anneal technique. The substrate was spark cut in the form of a 20 mm diameter disk and then mechanically polished on a lapping wheel to a thickness of 0.5 mm. Final polishing was carried out using a commercial technique developed for polishing Si wafers: Rodel Products' NALCO 2350 colloidal silica slurry (50-70 nm particle size) was applied to adhesive-backed NALCO Suba IV pads mounted on the lapping wheel. After a short polish (2-4 minutes) the sur- face of the Fe substrate was thoroughly washed in distilled water and then acetone. The finished Fe substrate showed no signs of mechanical scratching when observed under high optical magnification and illuminated at oblique incidence.

The Fe substrate surface was sputter-etched using an Ar ion beam at 2 keV and ~8 µa over 1 cm^2 for ~45 minutes. The first anneal in UHV was carried out at 700 °C leading to a well recrystallized surface. However, the Auger spectra show that at this temperature the elements S and N are segregated at the surface. The sample surface was resputtered for 30 minutes and the second anneal cycle was carried out at 500 °C. This temperature was high enough for removal of the surface damage caused by the ion sputtering but remained sufficiently low to preclude further S and N surface segregation.

At 500 °C the RHEED patterns are respectable and the Fe surface shows no C impurities. However, at ambient temperature the XPS line intensities reveal that O, C, and Fe are in the ratio 1:2:40. Presumably the interstitial potential wells on the surface for carbon are lower than in the bulk resulting in carbon surface segregation upon cooling. At ambient temperature a small amount of surface reconstruction c(2x2) is visible; see Fig. 1a. Such reconstruction streaks rapidly disappear after evaporating a small fraction (~1/10) of the first monolayer of various materials used as overlayers.

1.2 Growth and Experimental Facilities

We use a Physical Electronics Model 400 Molecular Beam Epitaxy system for the growth of new epitaxial ferromagnetic structures. The layout of the PHI 400 MBE machine, equipped with RHEED and cylindrical mirror analyser (CMA) for Auger and XPS, is shown in Fig. 2. Film growth is monitored *in situ* by means of RHEED pattern observations. The sample surface can be rotated 360° with respect to the RHEED electron gun allowing one to study the full azimuthal dependence of the RHEED patterns.

Both the chemical composition and overlayer growth are studied in the analysis chamber by means of AES and XPS using an angular resolved double pass cylindrical mirror analyser (PHI model 15-110A CMA).

Magnetic behaviour in UHV is investigated by means of a 10 GHz microwave spectrometer placed in the analysis chamber[5]. In the PHI-MBE-400 machine, the substrate is attached to a sample holder at the end of the long sample arm which allows one to move the substrate from the intro-chamber, through the analysis chamber, into the growth chamber.

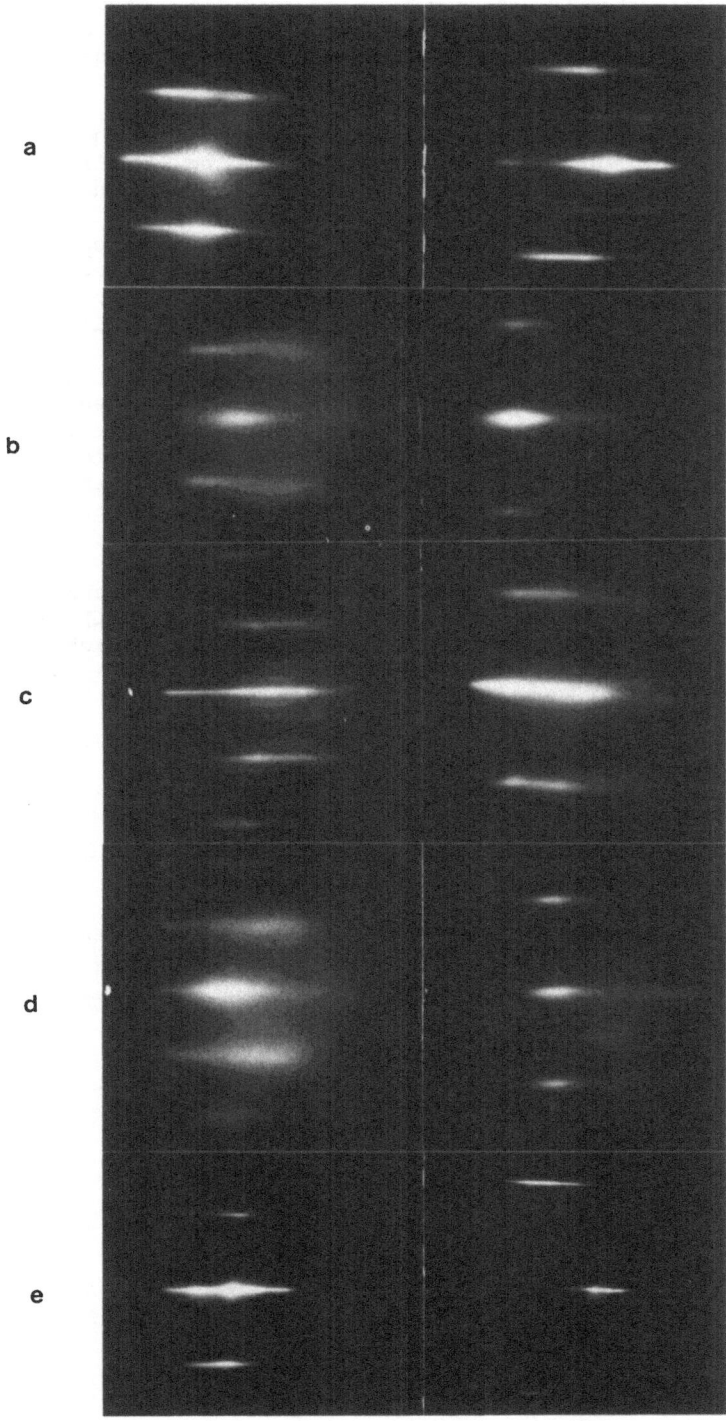

Fig. 1. Reflection High Energy Electron Diffraction (RHEED) patterns measured for Ni and Au overlayers deposited on an Fe(001) surface. The photographs on the left were obtained for the electron beam oriented along the [10] direction; those on the right for the beam oriented along the [11] direction: (a) the starting iron substrate; (b) Ni on Fe(001); (c) Au on Fe(001); (d) Ni on Au on Fe(001); (e) the surface of the fcc Ni(001) crystal.

When the FMR spectrometer is extended into the analysis chamber, a 3 mm hole in its gold-plated microwave cavity faces the surface of the Fe single crystal substrate. The substrate is pushed against the 3 mm hole to become part of the wall of the cavity.

In this paper we report the results of measurements on the following epitaxial layers: 3.0 nm of Au on Fe(001), 6.0 nm Ni on Fe(001), 3.0 nm of Au on 6.0 nm of Ni on Fe(001). We have also epitaxially grown 3.0 nm of Au on 3.0 nm of Ni on 2.6 nm of Au on Fe(001). The Au epitaxial layers were used to protect the sample surface when exposed to atmospheric conditions and to magnetically decouple the Ni overlayers from the Fe substrate. In order to minimize interfacial diffusion, the overlayers were prepared with the Fe substrate held at ambient temperature. During deposition the vacuum in the growth chamber did not exceed 5.0×10^{-10} Torr, compared with the general background of 6.0×10^{-11} Torr. The Ni was evaporated from a pure Ni wire (99.999%) wrapped around a W filament; Au was deposited from high purity shot (99.999%) contained in a pyrolytic boron nitride crucible and heated by means of an external coil of Ta wire. The evaporation rate of the deposited materials was monitored using the residual gas analyser (PHI model 1400 RGA) mounted above the substrate. The RGA mass intensity measurements were calibrated using the thickness dependence of AES and XPS intensities from the substrate and overlayer. The growth rate could be also determined from oscillations of RHEED or AES and XPS intensities. The rate of deposition was held approximately at 0.7 monolayer (ML) per minute. The uniformity of growth was checked by comparing apparent thicknesses obtained at different AES and XPS electron energies.

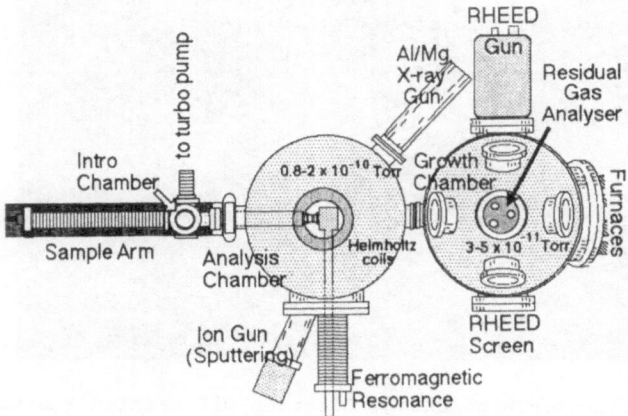

Fig. 2. Schematic diagram of the Molecular Beam Epitaxy system used to prepare and to characterize the growth modes of overlayers on an Fe(001) substrate.

The FMR measurements at 10 GHz were carried out with the dc magnetic field applied parallel to the specimen plane (parallel configuration) which required fields up to 3 kOe. These fields were provided by a pair of Helmholtz coils mounted in two interconnected, stainless steel dewar sections which were separated by a 3.5 cm gap. The coils were cooled by immersion in liquid nitrogen. A smaller pair of coaxially mounted coils were used to generate a 10 kHz modulation field. The double pass CMA was replaced by the unit containing the Helmoltz coils in order to carry out the FMR measurements. This meant that AES and XPS measurements could not be made when the apparatus was configured to perform *in situ* FMR studies. The growth conditions and physical characteristics of the overlayer under investigation had to be thoroughly investigated using the CMA prior to mounting the FMR apparatus. Since the final preparation of the specimen used for FMR studies had to be carried out without the CMA in place, it was necessary to rely on the RHEED patterns for characterization of the overlayer growth.

2. RHEED STUDIES

2.1 RHEED Patterns

In situ RHEED (10keV) with a variable glancing angle of incidence (1 - 3 degrees) was used as a major instrument for the characterization of overlayer growth.

RHEED patterns of prepared overlayers and substrates are shown in Fig. 1. Intensity variations along the streaks are caused by the presence of steps and terraces on the Fe substrate vicinal surface (the sample surface a few degrees away from a low index plane). The vicinal surface introduces unidirectional distortions of the RHEED patterns; for example, the electron beam directed down the staircase (the electrons striking the terrace faces) forms a simpler RHEED pattern than when the beam is directed up the staircase (the electrons striking the step faces). Thickening of RHEED streaks observed for thick overlayers is very likely due to surface roughness caused, for example, by a variable terrace width and step size.

The four-fold hollow hole sites of the Fe(001) plane create favorable conditions for epitaxial growth.

It is not surprising that Au grows epitaxially on iron because the 2-d square net formed by Au atoms on the fcc(001) plane forms a very good match with the 2-d square net formed by Fe atoms on the bcc(001) plane if the axis of the gold net is rotated by 45° with respect to the axis of the iron net. However, the results obtained for nickel overlayers were completely unexpected. RHEED patterns showed that Ni grows epitaxially with the lattice spacing and symmetry of Fe(001); see Figs. 1b and 1c. As the usual lattice spacing of fcc Ni (a = 3.59 Å) would have to be expanded by 14 percent (a volume change of 45 percent) to match the bcc lattice of iron, it is most unlikely that the Ni overlayers grow with the fcc structure.

From RHEED patterns we see only the structure of the top layer of atoms. In order to determine the lattice spacing perpendicular to the surface, an additional *in situ* technique is required. A method based upon the Reflection Electron Energy Loss Fine Structure (REELFS) is described below. Results of REELFS measurements prove that the Ni overlayer does not grow in an expanded, face-centered form. Z.Q. Wang et al.[6] have also recently grown Ni on a bulk Fe(001) substrate. They measured the energy and angular dependence of LEED spot intensities. Using the results of the full dynamic diffraction theory they concluded that at least the first six monolayers grow in a pure bcc structure having the lattice spacing of the Fe. After 6 monolayers, a different LEED structure c(2x2) was observed which has not yet been deciphered.

2.2 RHEED Pattern Intensity Oscillations

We have observed oscillations in the intensity of the RHEED for Ni overlayers grown on our Fe(001) substrate; see Fig. 3. Intensity oscillations give information about the molecular beam growth process and have been extensively studied in connection with the MBE growth of GaAs and SiGe. However, no observations of RHEED oscillations have been reported for metal epitaxy.

It is useful at this point to introduce the mechanism which has been found to play a major role in the RHEED intensity oscillations observed for epitaxial growth of semiconductors. A more detailed discussion, and the controversies involved, can be found in the contributions to this book by B.A. Joyce, J.H. Neave, P.J. Dobson and P.I. Cohen. We lean towards P.I. Cohen's point of view that the origin of the RHEED intensity oscillations lies in the interference between electron waves diffracted from the growing layer and those diffracted from the layer upon which the growth is taking place. Let us assume that the substrate surface consists of terraces of equal width separated by evenly spaced step edges. If RHEED intensity is a maximum for a particular configuration then the intensity should initially decrease during the deposition of the next monolayer. This decrease in intensity is caused by the interference of waves diffracted from randomly distributed islands with waves diffracted from the terraces. Upon completion of the monolayer, the surface recovers its original configuration and consequently the RHEED pattern intensity recovers as well. A minimum

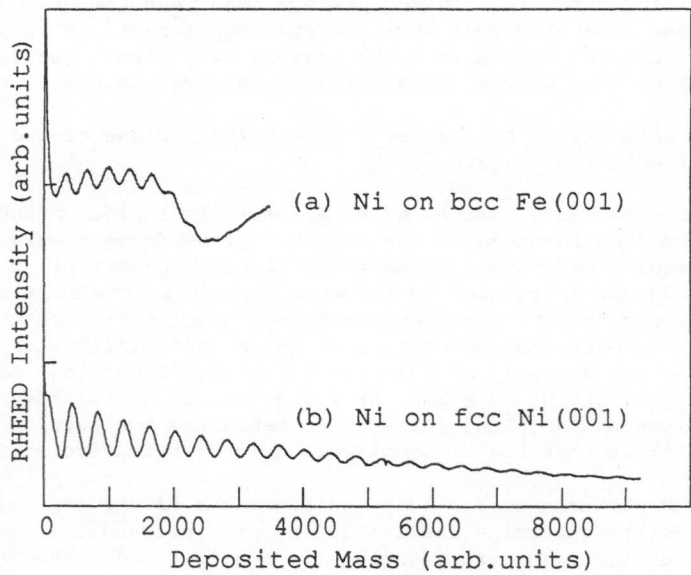

Fig. 3. RHEED intensity measured as a function of deposited mass:
(a) Nickel deposited upon an Fe(001) substrate; (b) Nickel deposited upon a Ni(001) substrate. The electron beam was oriented along the [11] azimuth in each case.

RHEED intensity is reached when the terraces are half covered by randomly distributed islands of the next monolayer. For RHEED-intensity oscillation to occur it is sufficient that something close to the original surface condition is recovered upon completion of the monolayer formation. The extent to which the initial condition is degraded leads to damping of the oscillations. The most common source of RHEED-intensity oscillation damping is the growth of upper layers on top of an unfinished overlayer.

It was only during our recent studies of the bcc Ni overlayers on Fe(001) that we decided to monitor RHEED-pattern intensities during growth[7]. The light emitted from the RHEED fluorescent screen is focussed onto a slit in front of a photomultiplier tube. The slit is wide enough to integrate over the structure (due to a stepped surface) across a particular streak and short enough along the streak to measure only the region having maximum intensity. The photomultiplier slit was located along the zeroth streak at the mirror reflection point. No special precautions for the Bragg or anti-Bragg condition were taken. The sample was oriented along the [110] azimuth with the glancing incidence electron beam (~3°) going down the staircase. Results of RHEED-intensity measurements are shown in Fig. 3, curve (a). After 6 monolayers of Ni on Fe(001) the amplitude of the oscillations sharply decreases, while the overall intensity increases.

In order to clarify some aspects of RHEED oscillations, and at the same time to avoid the complexities associated with the bcc Ni, we studied RHEED oscillations of Ni overlayers deposited on a single crystal substrate of fcc Ni(001). The Ni substrate was cut from a Ni crystal boule in the form of a disk with a surface misorientation of 1/2° from the (001) plane. The sample was fine polished using 3 μm diamond paste on a felt lap and then electropolished using a solution of 60% (V/V) H_2SO_4 and 40% (V/V) H_2O. The sample was sputtered for 45 minutes at 2 keV and ~ 8 μa over 1 cm^2, annealed to 650 °C for 30 minutes, resputtered for 45 minutes and then annealed to 350 °C for 30 minutes. AES showed $C/O_2/Ni$ to be ~ 3/1/40. The RHEED streak patterns were very respectable (Fig. 1e) and comparable with those of Fe whiskers grown by chemical vapor deposition (CVD). As shown in Fig. 3, curve (b), RHEED oscillations associated with the Ni overlayers grown on the Ni substrate at room temperature are well defined.

After the deposition of 6 monolayers of Ni on Fe (001), the intensity decreases but the oscillation amplitude is still observable and the periodicity is measurable. At higher substrate temperatures, the amplitude of the RHEED oscillations decreases gradually. For a substrate temperature of 100 °C these amplitudes are smaller by a factor of 10 than those associated with a substrate held at room temperature and no oscillations in the RHEED intensity with growth were observable for a substrate temperature of 200 °C.

Similar results were reported by J.H. Neave et al.[8] on GaAs. They observed that the oscillations are well developed for a substrate temperature of 540 °C but become weaker and disappear as the temperature approaches 590 °C. They argue that well developed oscillations are observed only when a layer-by-layer growth is established otherwise the RHEED oscillations are suppressed by the rough surface which forms as a consequence of multilayer growth. On the other hand, when the diffusion radius of a condensing atom becomes comparable to an average terrace width, growth occurs mainly at the step edges. A mere translation of step edges across the sample surface does not affect the RHEED intensity which again results in a suppression of RHEED oscillations. Our results on Ni overlayers clearly show that the mechanism of growth is very similar to that for GaAs epitaxy.

The observation of RHEED oscillations opens new avenues for exploration. For example, one can investigate phase-locked growth, widely used in the epitaxy of III-V elements, in which the deposition is periodically interrupted after the completion of each monolayer, corresponding to a maximum in the RHEED oscillations. The sample is then annealed at a suitable temperature. Such a procedure results in a significant improvement of overlayer crystal structure, surface smoothness and superior superlattice interfaces. The cycle can be repeated until the desired thickness has been achieved.

The observation of RHEED-intensity oscillations in Ni overlayers proves that the Fe substrate held at ambient temperature is suitable for layer-by-layer epitaxial growth and that temperatures of 200 °C are more than adequate for thermal annealing of the deposited overlayers. The periodicity of the RHEED oscillations should reflect the surface coverage of the structures grown. Indeed, the period of oscillations observed for bcc Ni is 17 percent smaller than the period observed for fcc Ni on the Ni(001) substrate. In bcc Ni the number of atomic sites on the (001) plane is 25 percent smaller than for fcc Ni, which is in good agreement with the measured periods. The 8 percent discrepancy is well within the accuracy of deposition rate measurements performed several weeks apart.

2.3 AES and XPS Intensity Oscillations

Ni overlayer growth was also studied using the dependence of AES and XPS peak intensities of substrate and overlayer atoms on the added film thickness. The results for the AES KLL Fe line (45 eV) peak intensity as a function of Ni mass is shown in Fig. 4. Our results do not reveal a typical sequence of linear segments terminating upon the completion of each layer. However, the averages of the AES and XPS peak intensities of both the substrate and overlayer atoms follow an exponential dependence as is expected for a layer-by-layer growth. The deviations from the exponential

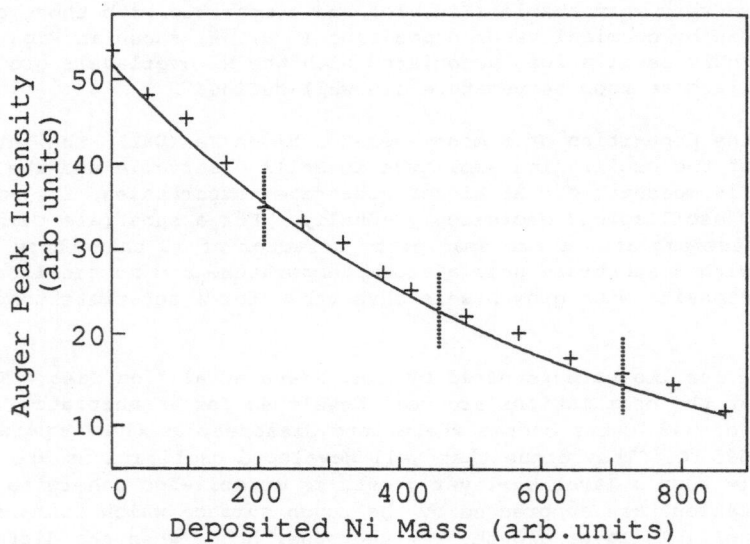

Fig. 4. The dependence of the intensity of the Auger KLL Fe substrate line on the mass of nickel deposited upon the substrate. The solid line is an exponential curve constructed so as to be parallel, but displaced downwards by one unit for clarity, with those data points which correspond to the completion of a monolayer (indicated by the vertical lines).

528

dependence displays an oscillatory behaviour. It is important to note that the periods of the AES and XPS oscillations are in good agreement (within 10 percent) with those measured for the RHEED oscillations.

Let us illustrate how these oscillations could arise as a result of the growth process. The growth on the substrate held at ambient temperature originates from small patches of atoms which nucleate randomly across the terrace surfaces. The small diffusion radius of condensing atoms permits 2-d growth of one atom thick, flat, pancake-like islands which eventually form a continuous monolayer due to their growth and coalescence. Islands of deposited material significantly increase the relative amount of edge area compared with flat area. The presence of additional step edges modifies the AES and XPS peak intensities. The CMA accepts Auger electrons which are approximately 45° away from the surface normal. The presence of step edges very likely increases the AES signal as a result of a decreased effective path followed by electrons escaping from the sample through step edges; see Fig. 5. One can contrast the variation of Auger peak intensity expected for growth due to patches of atoms with that expected if the density of steps remained constant - as would happen if the growth originated from step edges. In the latter case, the AES intensity from the substrate should decrease linearly with time until the new monolayer has been completed. In the former case, the AES intensity should decrease more slowly with coverage due to the presence of the many newly formed edges. Eventually the islands coalesce and the effectiveness of the faces decreases. In either case, upon completion of the monolayer one must have the same AES intensity. Hence, a patchy growth mode can be expected to lead to the arc-like dependence of AES intensity with time as shown by the data of Fig. 4.

Oscillations were observed in both Auger peak intensities and the XPS integrated peak intensities, and for both the substrate and the overlayer atoms. These oscillations, therefore, appear to be associated with the mode of growth and can be understood on the basis of the mechanism proposed above. One would expect to see such oscillations for any materials whenever growth occurs by the formation of many islands. In this respect it differs from oscillations in Auger splitting reported for Cu and Ag by R.W. Vook et al.[9] and by Y. Namba et al.[10] which are observable only for overlayer atoms having an Auger line whose splitting is sensitive to the proximity of step faces.

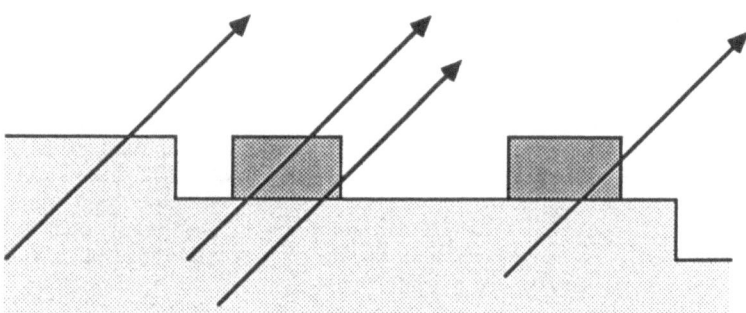

Fig. 5. Stylized representation of the substrate surface
showing typical Auger electron trajectories during
a "patchy" growth emphasizing the role of "step-faces"
in decreasing the Auger electron paths relative to a
growth mode having the same percentage of surface
coverage but few steps.

The dependence of the Fe and Ni XPS peak intensities on the thickness of the bcc Ni overlayer clearly shows an exponential decrease for an overlayer coverage well in excess of six monolayers; see Fig. 6. Therefore the decrease of the RHEED oscillation amplitude after the first six monolayers (Fig. 3a) does not signify a dramatic change in the growth mechanism; it merely reflects the sensitivity of the RHEED oscillations to the surface smoothness on an atomic scale.

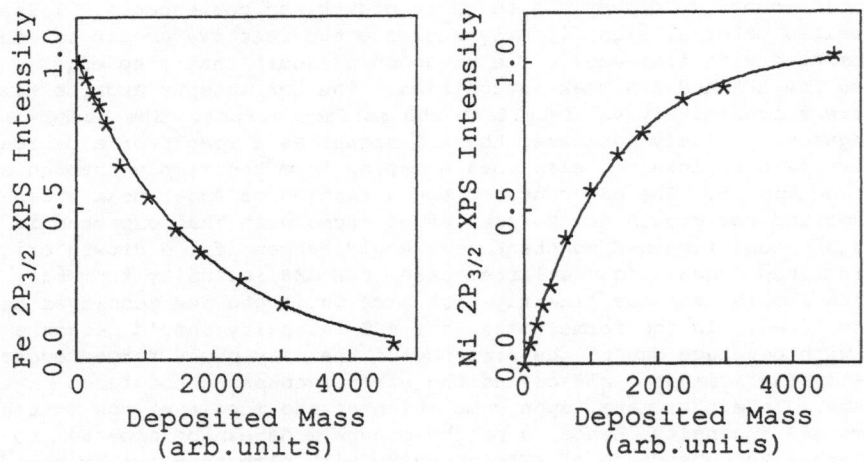

Fig. 6. XPS signal intensities for Ni(001) deposited on Fe(001) as a function of deposited mass showing (a) the decrease of the principal Fe XPS speak signal and (b) the growth of the Ni XPS signal with thickness.

3. REELFS OF NI(001) AND NI LAYERS ON FE(001)

3.1 Theoretical Basis and Experimental Method

REELFS spectroscopy and RHEED studies are both *in situ* methods which give structural information. They are complementary in that while the RHEED pattern indicates the lattice structure of the surface layers, REELFS spectra reveal the local structure around individual atoms.[11,12]

Experimentally, the REELFS spectrum is obtained by directing an electron beam with a sharply peaked kinetic energy distribution onto the sample at nearly normal incidence. The flux of inelastically scattered electrons is then measured as a function of energy E below the incident energy E_p. The electron flux (or current), dN/dE, as a function of energy loss, $E_L = E_p - E$, is the electron energy loss spectrum. The studies described in this paper used E_p = 2400 keV. The incident beam and the CMA are coaxial; the diameter of the CMA is such that the electrons are collected at an angle of 44° with the incident beam direction. The resolution of the CMA, $\Delta E/E$, is 0.006.

The interpretation of REELFS spectra proceeds analogously to that of extended x-ray absorption fine structure spectra (EXAFS) which has undergone extensive development.[13] In the case of REELFS one is measuring the relative probability of an incident electron suffering a certain energy loss, E_L. Similarly, in EXAFS one measures the absorption probability of an x-ray with energy E_v. Both processes depend on being able to excite a core electron to an energy, E_L or E_v above its quiescent state. For REELFS this excitation is governed by the matrix element $\langle f|\exp(i\mathbf{q}\cdot\mathbf{r})|i\rangle$ and in EXAFS the relevant matrix element is $\langle f|\varepsilon_k\cdot\mathbf{r}|i\rangle$ where $\mathbf{q}$ is the momentum transfer of the inelastically scattered electron, $\mathbf{r}$ is the interatomic distance vector, and ε_k is a unit vector along the photoelectron wavevector.[14] Because $\mathbf{q}$ is the momentum transfer relative to a reciprocal lattice vector, the dot product $\mathbf{q}\cdot\mathbf{r}$ is sufficiently small that the dipole approximation holds. Thus REELFS spectra are formally similar to EXAFS spectra.

The final state wave function of the ejected electron is responsible for the structural information in REELFS and EXAFS spectra. The initial state wave function is localized near the atomic core. For our REELFS studies it is a 3p state. If the electron in this initial state does not obtain sufficient energy to reach an unoccupied level the spectrum has a small value (low absorption or small reflected electron flux). When the energy imparted to the electron is above E_0, the Fermi energy (the threshold necessary to reach unoccupied levels) the spectrum value increases. For a small interval of energy there is structure in the spectrum which is related to the energy and symmetry of unfilled, bound states. At higher energies, the electron reaches unbound levels and the spectrum is modulated by a fine structure that is small relative to the threshold edge jump. This fine structure carries information about distances to nearest neighbour atoms. The fine structure occurs because the final state wave function of the ejected electron is composed of an outgoing wave and the reflection of this wave from neighbouring atoms. The final state wave function at the central atom position is larger or smaller depending upon the phase change $2kr$ from the central atom to the reflecting atom and back, where k is the electron wavevector magnitude. There is an additional phase change, slowly varying with k, due to the complex backscattering amplitude and the atomic potential near the central atom. Thus the modulation of the REELFS spectrum due to a neighboring atom, $\chi(k)$, is a sum of waves of the form $\sin[2kr + \delta(k)]$. A Fourier transform of $\chi(k)$ is a function of r showing peaks corresponding to the various near neighbour atomic shells around the central atom. The positions of the peaks are slightly shifted because of δ. This shift can be calibrated from measurements on known structures or, less accurately, calculated theoretically.

The data are accumulated in the derivative mode and are analysed by first passing a smooth, splined background through dN/dE which is then integrated to get $N(E)$. The fine structure spectrum, $\chi(k)$, is the variation around the smooth background, N_o, and expressed as a function of k instead of E_L: $\chi(k) = (N-N_o)/N_o$, $k=[0.263(E_L - E_0)]^{1/2}$. It is usual to plot $k^3\chi(k)$ rather than $\chi(k)$ in order to emphasize features at large k; see Fig. 7. The analysis then proceeds as for EXAFS with Fourier analysis using an apodization window to avoid spurious sidelobes. The principle conclusions regarding nearest neighbour positions can be drawn from examining the Fourier transform magnitude (Fig. 8).

The existence of an atom aligned colinearly with the backscattering atom greatly enhances the signal from the backscattering atom. This can be understood because the forward scattering of the intermediate atom, $F(0,k)$, is larger than unity. Thus in passing through the middle atom twice, the signal may be amplified several times. Such multiple scattering arises in the fcc structure along the diagonal of the face, and in the bcc structure along the diagonal of the cube. In both cases, the signal from the atom at twice the nearest neighbour distance is greatly enhanced. The Fourier transform of a spectrum is dominated by the nearest neighbour peak and the peak at twice the nearest neighbour distance. If the middle atom is displaced from a colinear position, the magnitude of this latter peak decreases rapidly with the amount of displacement. Thus a decrease in the magnitude of the "multiple-scattering" peak could indicate a deformation from a pure bcc or fcc structure.

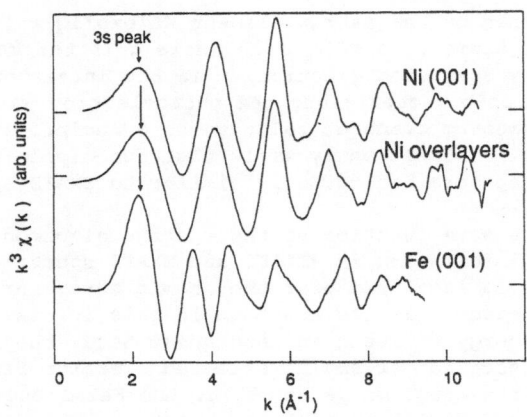

Fig. 7. The REELFS spectra of Ni(001), Ni overlayers on Fe(001)
and Fe(001).. The $\chi(k)$ were obtained from the measured
first derivative .electron flux as described in the text.
The spectra are plotted multiplied by k^3 to emphasize
the high k features. The 3p absorption edge lies slightly
to the left of the k-axis origin.

Despite the formal similarity of REELFS and EXAFS spectroscopy, REELFS
has certain practical limitations. In REELFS the value of E_p used limits
the maximum energy loss that can be measured; therefore, the initial state
must be at most only several hundred eV below the threshold of unfilled
levels (the Fermi energy). While EXAFS usually measures the Ni K edge
(initial 1s state) REELFS measures the M edge (initial 3p state). As the M
edges of Ni and Fe are only about 12 eV apart, measurements of Ni on
Fe(001) must be made on surface layers that are thicker than the incident
electron penetration depth in order to avoid overlap of the Ni and Fe
spectra. For this reason we only carried out REELFS measurements on samples
covered by approximately·30 monolayers. It would have been preferable to
have made measurements on films 6 or less monolayers thick: such films are
a pure bcc epitaxial structure. The k-space range of REELFS spectra is
more limited than is normally the case for EXAFS: (1) at low k because of
the existence of a small 3s transition 43 eV above the large 3p edge and
(2) at high k because of lower signal to noise ratio. This will limit the
distance resolution of. the analysis. Furthermore, because the resolution
of the electron energy analyser is lower than that of typical x-ray
monochromators, higher frequencies of the fine structure, corresponding to
more distant neighbouring shells, are attenuated.

3.2 Experimental Results of REELFS Measurements

The size of an atom determines the nearest neighbour distance (n.n.d.)
in an atomic crystalline structure. The n.n.d. in bcc Fe (2.48 Å) and in
fcc Ni (2.49 Å) are almost identical. Let us point out what constitutes
the difference between the bcc and fcc structures when they are formed on
the (001) plane. Since both lattices possess only one set of 4-fold hollow

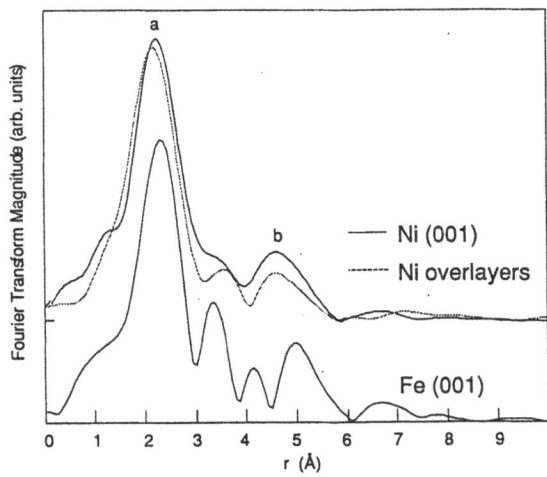

Fig. 8. The Fourier transforms of the spectra of Fig 7. The peaks
in the Fourier transforms are related to neighbouring
atomic positions around a central atom. The peaks
labelled (a) and (b) correspond to the nearest neighbour
distance and to twice the nearest neighbour distance in
Ni: see the text. The transforms are of $k^3 \chi$ (k) over
a range of 3.5-9.5 $Å^{-1}$.

hole sites, the body and face centered cubic lattices can differ only by
their interplanar spacing. The interplanar distance is given by (1/2)a in
bcc and by $(1/\sqrt{2})a$ in fcc structures, where a is the size of the 2-d square
lattice; see Fig. 9. The face centered cell is $\sqrt{2}$ larger than the body
centered cell because the fcc lattice is rotated by 45° in the (001) plane
with respect to its bcc counterpart. The first Ni monolayer is formed by
filling the strongly attractive potential wells of the 4-fold hollow holes
of the Fe(001) plane. The Ni atoms of the next Ni monolayer again see the
4-fold hollow hole sites. The Ni atoms which are trapped by the attractive
potential of the 4-fold hollow hole sites adjust the interplanar distance
to accommodate the Ni atomic size. The size of the Ni atoms is consistent
with the Fe bcc interplanar spacing because the distance between the corner
and center atoms of the cube is equal to the Ni atomic diameter, and it is
therefore not surprising that the Ni overlayer on the Fe(001) substrate
grows in the bcc structure with the Fe lattice spacing. One can speculate
that the formation of bcc Ni is a consequence of a more common rule: a
metastable lattice structure is likely to be formed if the atoms can fit
into the n.n.d. of the lattice on which it is formed.

The value of the nearest neighbour distance can be determined from the
REELFS measurements. Results of the REELFS measurements in the bcc Fe, fcc
Ni and 6.0 nm thick Ni overlayer on Fe are shown in Figs. 7 and 8. The

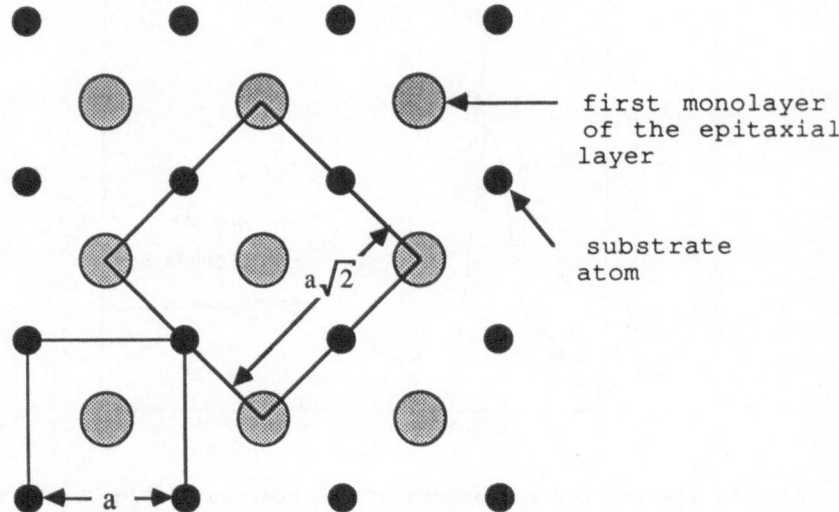

Fig. 9. Schematic diagram showing the surface atom positions
for a bcc substrate and epitaxially deposited atoms
in the first monolayer.

REELFS signal is dominated by the first and second harmonics, see Fig. 10.
The second harmonic originates from an enhanced forward scattering along
the direction of nearest neighbour atoms. Within the experimental error
(0.05 Å) the n.n.d. in the Ni overlayers is identical to that of ordinary
fcc Ni which is very nearly the same as the n.n.d. of bcc Fe; see the
Fourier transform in Fig. 8. It therefore follows that the nickel
overlayers on Fe(001) cannot be simply an expanded form of the fcc lattice.

There are two possible interpretations of the decrease in the height
of the "multiple-scattering" peak (peak (b) in Fig. 8) for Ni on Fe(001) as
compared with Ni(001): (1) The position of the atom at twice the n.n.d. may
be much more disordered than that of the nearest neighbour atom, or (2) the
multiple scattering arrangement may deviate from colinearity, implying a
buckling of the surface layer. In either case there must be a substantial
deviation from the pure bcc or fcc structure.

The decrease in RHEED intensity oscillations after the first six monolayers is accompanied by the presence of superlattice streaks. They occur
mostly with the c(2x2) symmetry, but sometimes we observe another reconstruction whose symmetry has not been yet identified due to a complicated
contribution from the surface steps. The superlattice streaks come with a
varying degree of intensities ranging from almost invisible to more
pronounced, but always significantly weaker than the main streaks along the
[100] and [110] azimuths.

Measurements of RHEED streak separations carried out using a
photomultiplier indicate that the thick overlayers are 2 percent expanded
in the lateral direction; see Fig. 11. This result is reflected in the
periodicity of the RHEED oscillations. For films thicker than six monolayers the period of the RHEED oscillations is between 5 and 6 percent
lower than the period in the first six monolayers. This observation is in
good agreement with the decrease in the separation of the RHEED streaks.

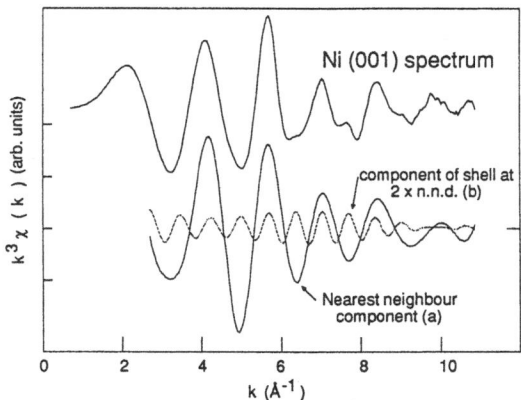

Fig. 10. The components of the Ni(001) REELFS spectra corresponding
to the first nearest neighbour and the shell at twice the
first nearest neighbour distance which is enhanced by multiple
scattering. The components were obtained by filtering peaks
(a) and (b) of Fig. 8 and back-transforming.

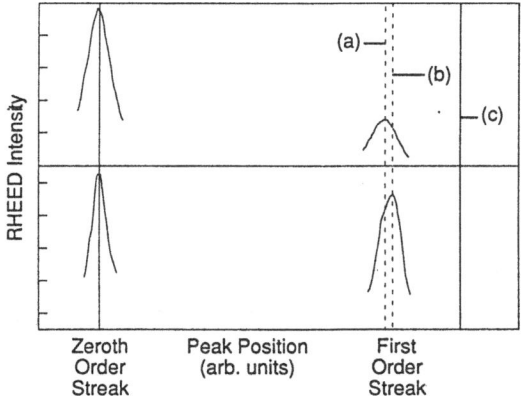

Fig. 11. The positions of the zeroth and first order RHEED streaks
for (a) Ni grown on the Fe(001) substrate and (b) for the
bare Fe substrate. The line (c) indicates the position of
the first order RHEED streak for a fcc Ni(001) surface.

It remains to be seen whether the observed superlattice streaks, and 2 percent lateral expansion, represent only a modulation of the bcc lattice or a new phase of Ni which forms beyond the sixth layer. Because the main features of the RHEED patterns are very similar to those obtained from bcc Fe, and for the sake of convenience, we shall from now on refer to Ni overlayers thicker than six monolayers as "bcc Ni".

4. FERROMAGNETIC RESONANCE

4.1 Theoretical Outline

Ferromagnetic resonance (FMR) is an extremely useful tool in the study of magnetic materials, and in particular has played a crucial role in the characterization of magnetic thin films. The virtues of the FMR method can be demonstrated from the Landau-Lifshitz (L-L) equation of motion which describes the response of the magnetization to a driving radio frequency (RF) magnetic field:

$$[1] \qquad -\frac{1}{\gamma}\frac{\partial M}{\partial t} = M \times H_{eff}$$

where $\qquad M = M_S + m \; ; \; H_{eff} = H_0 + H_{dmag} + H_k + H_G + h,$

M_S is the saturation magnetization vector and m is the RF component of M. H_0 is the externally applied magnetic field taken to be parallel to the plane of a disc-shaped specimen of thickness d whose [001] direction is perpendicular to the plane. $H_{dmag} = - N_x M_S - 4\pi m_\perp$ is the demagnetizing field; N_x is the dc demagnetizing factor in the plane of the specimen and $4\pi m_\perp$ is the RF demagnetizing field ($m_\perp$ is the component of m perpendicular to the specimen plane). The form of the RF term in H_{dmag} arises as follows: at microwave frequencies the high refractive index of the metal causes all RF fields in the metal to propagate along the specimen normal. For a thick specimen ($d \gg$ RF skin depth) the Maxwell equation $\nabla \cdot b = 0$ requires $b_\perp = 0$ resulting in a RF demagnetizing field along the direction of the normal whose magnitude is given by $h_\perp = -4\pi m_\perp$.

The anisotropy field H_k depends on the direction of the magnetization with respect to the crystal axes. The actual form of the anisotropy field[15], H_k, is derived from magnetic anisotropy energy for cubic symmetry for which the dominant term is given by

$$[2] \qquad E_k = -K_1(\alpha_1^2\alpha_2^2 + \alpha_2^2\alpha_3^2 + \alpha_1^2\alpha_3^2),$$

where α_1, α_2, α_3 are the direction cosines of the magnetization vector M.

The exchange field

$$[3] \qquad H_{ex} = \frac{2A}{M_S}\left(\frac{\nabla^2 m}{M_S}\right)$$

arises from the excess of the exchange energy density, E_{ex}, due to a spatial variation of m (caused by eddy currents)

$$[4] \qquad E_{ex} = \frac{A}{M_S^2}\left\{(\nabla\cdot m)^2 + (\nabla\times m)^2\right\}$$

where A is the exchange stiffness constant which, for the nearest neighbour interaction $-2J\mathbf{S}_i \cdot \mathbf{S}_j$, is given in the bcc lattice by $A = (\sqrt{3}) JS^2 / (\text{n.n.d.})$. $\mathbf{H}_G = -G/(\gamma M_S)^2 (\partial \mathbf{m}/\partial t)$ is the Gilbert form of the intrinsic damping for the precessing magnetization $\mathbf{M}$. The field $\mathbf{h}$ is the internal RF field which satisfies Maxwell equations which, for a metal at microwave frequencies, are written in Gaussian units as:

[5]

$$\nabla \times \mathbf{h} = \frac{4\pi\sigma}{c} \mathbf{e}$$

$$\nabla \times \mathbf{e} + \frac{1}{c}\frac{\partial \mathbf{b}}{\partial t} = 0$$

where σ is the dc conductivity. The Landau-Lifshitz equation and Maxwell's equations have been applied to the case of two magnetic metal slabs in which the magnetizations are coupled together by an exchange interaction at the interface[16]. The specimen geometry and configuration of the RF driving fields are shown in Fig. 12. Parameters which refer to the overlayer are labeled with the subscript A; those which refer to the substrate are labeled with the subscript B. In order to solve the coupled Landau-Lifshitz and Maxwell's equations it is necessary to specify boundary conditions for the magnetization components of the overlayer at the surface z = 0 (Fig. 12), and for the magnetization components at the interface (z = d) for both the overlayer and the bulk. These boundary conditions depend upon whether or not there are surface torque densities acting upon the magnetization[17]. Such torques are the consequence of surface terms in the free energy of the magnetic system (surface anisotropy energies).

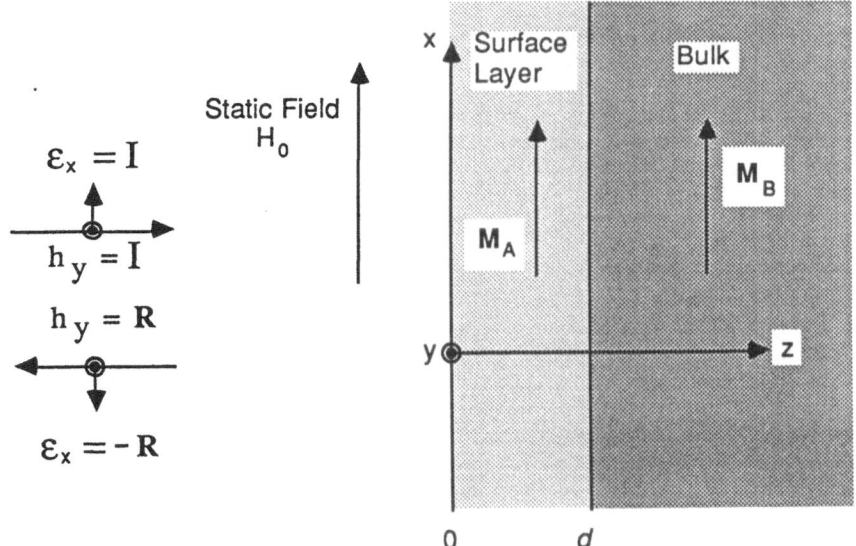

Fig. 12. The geometry used to calculate the response to incident microwave radiation of a bulk ferromagnet covered by a thin film of a second ferromagnet. The magnetizations in the two metals, $\mathbf{M}_A$ and $\mathbf{M}_B$, are assumed to be uniform in equilibrium and parallel with the external field, $\mathbf{H}_0$. $\mathbf{M}_A$ and $\mathbf{M}_B$ are assumed to be coupled at their interface by an exchange energy of the form $E_{ex} = -J\,\mathbf{M}_A \cdot \mathbf{M}_B$.

As there is no strong experimental evidence[18,19] for surface anisotropy (in a well prepared specimen), and as the Nèel anisotropy[20] should be zero in the (001) plane, we will assume that intrinsic surface torques are absent in our specimens. Therefore the magnetization of the overlayer must satisfy the condition[17]

[6]
$$\left.\frac{\partial m_A}{\partial z}\right|_{z=0} = 0$$

The overlayer/substrate interface boundary conditions were constructed using the model of F. Hoffmann et al.[21] based on nearest neighbour exchange coupling which results in the surface energy

[7]
$$E_{interface} = -J \, \mathbf{M}_A \cdot \mathbf{M}_B \, \delta(z-d)$$

Detailed balance of torques in the interface leads to the interface boundary conditions:

[8]
$$\frac{2A_A}{M_A}\left.\frac{\partial m_A^y}{\partial z}\right|_d = J M_A M_B \left[\frac{m_B^y(d)}{M_B} - \frac{m_A^y(d)}{M_A}\right] = \frac{2A_B}{M_B}\left.\frac{\partial m_B^y}{\partial z}\right|_d$$

$$\frac{2A_A}{M_A}\left.\frac{\partial m_A^z}{\partial z}\right|_d = J M_A M_B \left[\frac{m_B^z(d)}{M_B} - \frac{m_A^z(d)}{M_A}\right] = \frac{2A_B}{M_B}\left.\frac{\partial m_B^z}{\partial z}\right|_d$$

The equations of motion for the substrate and overlayer magnetizations coupled with the Maxwell equations [5] give a system of linear differential equations. When solved, together with the magnetic boundary conditions and continuity of the electromagnetic fields $e_{||}$ and $h_{||}$ across the substrate/overlayer and overlayer/vacuum interfaces, the equations lead to a unique solution which may be written in the form[17,22]:

[9]
$$\mathbf{e}, \mathbf{h}, \mathbf{m} \sim e^{-kz + i\omega t}$$

This plane wave approximation leads to a secular equation for the wavevector k. The eigenvalues of the secular equation describe the modes of the system. At any value of the external field a conducting ferromagnetic sample supports three waves. The wavevector k_1 (exchange wave) corresponds to a non-propagating wave whose magnitude is relatively independent of the value of the magnetic field. This wave has a very small amplitude and affects the microwave absorption very little. The wavevector k_2 corresponds, for fields below resonance, to a propagating spin wave and becomes a purely non-propagating spin wave (surface mode) for fields larger than the resonance field. The wavevector k_3 corresponds to an electromagnetic wave. The solution of the problem involves 9 amplitudes of wavevector modes (3 for the substrate and 6 for the overlayer), and the amplitude of the electric field component at z = 0, $e_x(0)$. These 10 unknown amplitudes are matched by 10 equations; 4 equations arise from continuity of $e_{||}$ and $h_{||}$ on both interfaces and 6 equations arise from magnetic boundary conditions [6] and [8].

The degree to which each of the wavector modes is excited at any given frequency and applied field is determined by the boundary conditions. Such a problem is ideally suited for computer calculation and some of the computer results will be demonstrated below. In order to understand the main features of our experimental results we will now apply this theory to the problem of a thin magnetic overlayer deposited on a magnetic substrate.

For a very thin overlayer one can assume that

$$m_A^y(0) = m_A^y(d) = M_y$$

[10]
$$m_A^z(0) = m_A^z(d) = M_z$$

$$\frac{1}{d} \int_0^d h_y(z)\,dz = h_y(A)$$

Integrating the equation of motion for the overlayer and replacing the first derivative terms by eqs. [6] and [8] leads to

$$i\left(\frac{\omega}{\gamma_A}\right) M_y + \left[B_A + \frac{JM_B}{d} + i\omega\frac{G_A}{\gamma_A^2 M_A}\right] M_z = \frac{JM_A}{d} m_B^y(d)$$

[11]

$$\left[H_A + \frac{JM_B}{d} + i\omega\frac{G_A}{\gamma_A^2 M_A}\right] M_y - i\left(\frac{\omega}{\gamma_A}\right) M_z = h_y(A)M_A + \frac{JM_A}{d} m_B^y(d)$$

which are just the equations of motion for a uniform magnetization with an additional driving term $(J\,M_A/d)\mathbf{m}_B(d)$ and an additional effective field, $(J\,\mathbf{M}_B/d)$, which adds to the applied field $\mathbf{H}_0$. In eq. [11], $H_A = H_0 + H_{K1}$ and $B_A = H_0 + 4\pi M_A + H_{K2}$, where H_{K1} and H_{K2} are static effective magneto-crystalline anisotropy fields[23].

4.2 Strong coupling

For a clean interface between two metals one would expect the magnetic coupling to be comparable with the bulk coupling. This is so strong that the angular deviations from equilibrium are the same on both sides of the substrate/overlayer interface. For strong coupling

$$\frac{m_B^y(d)}{M_B} = \frac{m_A^y(d)}{M_A} = \frac{M_y}{M_A}$$

[12]

$$\frac{m_B^z(d)}{M_B} = \frac{m_A^z(d)}{M_A} = \frac{M_z}{M_A}.$$

Inserting eqs. [12] into eqs. [11], and using the substrate/bulk equations of motion for $h_y(d)$, leads, after the application of simplifying assumptions[16], to the approximate boundary equations for the thick substrate:

[13a]
$$\frac{2A_B}{M_B} \frac{dm_B^y}{dz}\bigg|_d = -\frac{d}{M_B} M_A [H_B - H_A] m_B^y(d)$$

[13b]
$$\frac{2A_B}{M_B} \frac{dm_B^z}{dz}\bigg|_d = -\frac{d}{M_B} M_A [B_B - B_A] m_B^z(d)$$

where

$$(H_B - H_A) = \frac{2K_B}{M_B} - \frac{2K_A}{M_A}$$
$$(B_B - B_A) = 4\pi(M_B - M_A) + \frac{2K_B}{M_B} - \frac{2K_A}{M_A}$$

$\left.\right\}$ for $\mathbf{M_A}$, $\mathbf{M_B}$ along [100]

and

$$(H_B - H_A) = -\left(\frac{2K_B}{M_B} - \frac{2K_A}{M_A}\right)$$
$$(B_B - B_A) = 4\pi(M_B - M_A) + \frac{K_B}{M_B} - \frac{K_A}{M_A}$$

$\left.\right\}$ for $\mathbf{M_A}$, $\mathbf{M_B}$ along [110]

The form of the boundary conditions [13] is characteristic of a partially pinned surface[17], with the exception that the pinning parameters are different for the y and z components of the magnetization.

The differences in the volume anisotropies and saturation magnetizations which occur in the boundary conditions [13] will have a profound effect on the substrate microwave absorbed power. For an Fe substrate supporting a Ni overlayer $4\pi M_A < 4\pi M_B$ and one can neglect the anisotropy terms in the boundary condition for the magnetization component perpendicular to the surface $(2K/M \ll 4\pi M_s)$. If $(2K_B/M_B - 2K_A/M_A) < 0$ then the in-plane boundary condition for M_A, M_B parallel to [110] reinforces the out-of-plane boundary condition in producing an upward shift of the resonance field and can cause eventually a splitting of the FMR line[22]. The upward shift and, if strong enough, a splitting come about because the negative surface torques enhance the amplitude of the non-propagating spinwave. This results in a positive shift in the resonance field due to the negative value of the exchange field $(2A_B/M_B)k_2^2$. The rotation of magnetizations M_A, M_B into the easy axis [100] changes the sign of the in-plane boundary condition and results in a downward shift of the resonance field[22]. The overall FMR response is given by an interplay between the in-plane and out-of-plane boundary conditions.

540

<u>4.3 Experimental results</u>

The dynamic response of Fe is unique among the 3d transition elements because the damping of the magnetization is small. The intrinsic Gilbert damping in Fe[18,24] is $G < 5 \times 10^7$ sec^{-1} while that of Ni[19,25] is $G = 2.5 \times 10^8$ sec^{-1}. Consequently the intrinsic linewidth in Fe is dominated by the exchange interactions caused by a spatial inhomogeneity of the RF magnetization in the presence of eddy currents. Therefore the FMR linewidth in Fe is given mostly by the exchange-conductivity mechanism[16] and the position of the FMR line is strongly dependent on the magnetic boundary conditions. Hence a single crystal of Fe is an ideal substrate for the study of magnetic coupling and properties of deposited overlayers.

In this paper we will concentrate on "bcc Ni" grown on an Fe(001) substrate.

In order to carry out FMR measurements at higher microwave frequencies we grew fcc Au epitaxially on the "bcc Ni". 10 GHz measurements were carried out *in situ* in UHV. However, it proved to be desirable to work at frequencies up to 73 GHz for which the specimen would have to be removed from the vacuum system. We used *in situ* FMR to investigate the temperature dependence of the Ni on Fe system in the absence of Au. The resonance measured at room temperature was unchanged by heating to 200 °C. That the Fe FMR at 200 °C was still affected by the Ni magnetism puts a lower bound on the Curie temperature of "bcc Ni". We also verified, by using *in situ* 10 GHz FMR measurements, that the effect of Ni on the Fe resonance was not changed by covering the Ni with Au. The Au covered sample was then exposed to the air and it was verified that the FMR results were not changed by a prolonged exposure to the air. This gives us confidence that our samples could be removed from the UHV system for measurements in several FMR spectrometers without affecting their properties. The angular variations of FMR lineshapes as the external field was rotated in the specimen plane indicated that the FMR measurements carried out at lower frequencies were affected by a severe lag of the Ni saturation magnetization behind the dc applied field. Presumably the applied fields (1 kOe for 10 Ghz and 2.5 kOe for 24 GHz) were not sufficient to saturate the Ni overlayers.

The FMR measurements carried out in high magnetic fields (~18 kOe) at 73 GHz provided the most interesting results. The 73 Ghz cavity operating in the cylindrical TE$_{012}$ mode is particularly useful for angular FMR studies. The sample is mounted flush with a 3mm diameter hole in the center of the end plate. The microwave magnetic field is radially directed and the microwave electric field circulates about the cavity axis. As the applied field is rotated about the cavity axis the components of the microwave cavity field perpendicular to the dc field do not change as they would do for a rectangular cavity. For the cylindrical cavity the applied dc magnetic field can be rotated in the sample plane without producing any major changes in the magnitude of the active microwave magnetic field component.

The FMR linewidth $\Delta H = 170$ Oe observed for the Fe(001) substrate at 73.55 GHz is quite narrow and its value compares favorably with the results obtained on Fe platelets grown by chemical vapor deposition (CVD)[18]. The comparision of the 73.55 GHz linewidth with a theoretical calculation results in a value for the Gilbert damping of $G = 5 \times 10^7$ sec^{-1}. The rest of the Fe parameters are listed in the caption of Fig. 15.

The angular dependence of the Fe(001) FMR peak position obeys perfectly the calculated dependence using the crystalline anisotropy constants K_1, K_2 quoted by Wohlfarth[26], see Fig. 13. This result demonstrates that there are no noticeable strains in our Fe crystal and

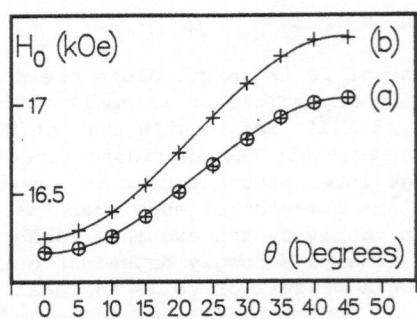

Fig. 13. The applied magnetic fields at which the microwave
absorption is maximum are shown for 5° steps in angle as
the applied field is rotated from the easy [10] to the
planar hard [11] axis. Maxima are found from the zero
crossings of the measurements of the derivative of the
Ferromagnetic Resonance response at 73.55 GHz. Curve (a) is
for the Fe(001) with a 3 nm Au coverlayer with the +'s for
the data points and the O's for the theory points. Curve (b)
is for the 6 nm Ni(bcc) overlayer on Fe(001) with a 3 nm Au
coverlayer, see Fig. 14. The solid lines are of the form
A + B cos(4θ) which are, by symmetry, the leading terms in
the angular variation.

that it is magnetically well-behaved. The effect of a Ni overlayer on the
FMR of Fe is shown in Fig. 14. The Ni overlayer preserves the 90° symmetry
of the (001) plane, see Fig. 13b. The shift in the Fe FMR line position
and splitting of the FMR line around the semi-hard axis [110] shows clearly
that the "bcc Ni" is indeed ferromagnetic.

V.L. Moruzzi and P.M. Marcus[27] recently reported the calculated
magnetic state diagram for bcc Ni. Using the local density approximation
for electron exchange and correlation, they predict that "bcc Ni" would be
on the edge of a first order transition from the non-magnetic to the
ferromagnetic state. The non-magnetic phase corresponds to the absolute
energy minimum, but it would require only a 1.5 percent increase in lattice
spacing to obtain a ferromagnetic state. If bcc Ni had the lattice spacing
of Fe, it would be ferromagnetic with a moment smaller than, but close to,
that of fcc Ni. According to V.L. Moruzzi and P.M. Marcus one should expect
bcc Ni grown epitaxially on Fe to be ferromagnetic.

The calculations shown in Fig. 15 are for Fe(001) with and without a
6.0 nm fcc Ni overlayer and for the dc field applied along the easy and
semi-hard axes of the Fe. An ordinary fcc Ni overlayer which has a small
anisotropy 2K/M = -200 Oe would create splitting of the FMR line along both
axes caused by the presence of the strongly negative pinning term
$4\pi(M_{Fe}-M_{Ni})$ in the boundary condition [13b]. The crystalline anisotropy term

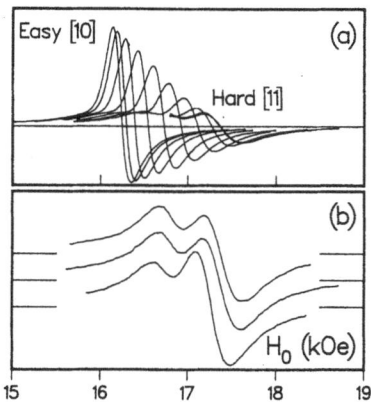

Fig. 14. Ferromagnetic Resonance response for 6 nm Ni(bcc) overlayer on Fe(001) with a 3 nm Au coverlayer at 73.55 GHz
in 5° steps in angle from left to right as the applied
field is rotated from the easy [10] to the planar hard
[11] axis. Note the splitting that occurs in the hard
direction. The variation of the zero crossings with
angle are shown in Fig. 13. Note that only part of the
last three curves are shown in (a). The full curves
for 35°, 40° and 45° angles are shown in (b) from bottom
to top. The displaced zeros are indicated.

(2K/M) is too weak and has the wrong sign to cause any anisotropic
behaviour of the FMR lineshape. However the "bcc Ni" FMR line shows no
tendency for a splitting when the magnetization lies along the Fe easy axis
[100]. This can only mean that the crystalline anisotropy field of the Ni
overlayer in the in-plane boundary equation [13a] is sufficiently large to
overcome the out-of-plane negative surface pinning due to the difference
in magnetizations between the overlayer and the Fe substrate. This
strongly suggests that the anisotropy of "bcc Ni" has the same sign as that
for Fe but is larger. An explanation based upon an appropriate uniaxial

anisotropy normal to the plane would remove the splitting of the FMR line for the external field in any direction in the (001) plane and therefore cannot be used in the interpretation of the angular dependence of our measured FMR lineshapes.

The interpretation of our experimental results based on the simplified boundary conditions [13] reflects the main features but describes only crudely the FMR response of the samples.

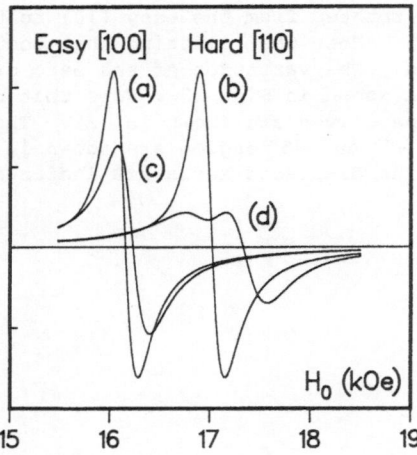

Fig. 15. Calculated field derivative of the microwave absorption for the (100) surface of Fe at 73.55 GHz with and without an epitaxial 6.0 nm bcc Ni overlayer for both the easy [10] and planar hard [11] directions. The curves (a) and (b) are for the easy and hard axis with no Ni overlayer. The curves (c) and (d) are for the easy and hard axes with the Ni overlayer. To achieve the agreement between Fig. 14 and Fig. 15, it is essential that the damping constant G take a value near 6.0×10^{8} sec^{-1} compared to 2.5×10^{8} sec^{-1} found in fcc Ni at 300 K. The parameters used are: frequency f = 73.55 GHz; thickness of the overlayer d = 6.0 nm; surface exchange constant $J = 10^{-4}$ cm; ω/γ = 24.01 kOe for Ni(bcc) and 25.03 kOe for Fe; $4\pi M_{s}$ = 4.5 kOe for Ni(bcc) and 21.55 kOe for Fe; anisotropy, $K_{1} = 6 \times 10^{5}$ ergs/cc for Ni(bcc) and 4.81×10^{5} ergs/cc for Fe; K_{2} = 0 for Ni(bcc) and 1.2×10^{3} ergs/cc for Fe; resistivity ρ = 7.2 $\mu\Omega$-cm for Ni(bcc) and 10 $\mu\Omega$-cm for Fe; the exchange constant is $A = 10^{-6}$ ergs/cm for Ni and 2×10^{-6} ergs/cm for Fe.

Computer calculations, using the exact theory outlined in section 4.1, shown in Fig. 15, were carried out using the fcc nickel parameters $A = 1.0 \times 10^{-6}$ ergs/cm, $g = 2.187$ and resistivity $\rho = 7.5 \times 10^{-7}$ Ω-cm. The parameters used for Fe are listed in the Fig. 15 caption. The saturation induction $4\pi M_S = 4.5$ kG was taken from the V.L. Moruzzi and P.M. Marcus[27] magnetic moment calculations corresponding to the lattice spacing of the bcc Fe. The crystalline anisotropy constant K_1 for the "bcc Ni" was chosen to obtain the best agreement between the experimentally determined and the theoretically calculated peak positions along the [100] and [110] crystalline directions. Note that this choice of the crystalline anisotropy, $K_1 = 6.0 \times 10^5$ ergs/cc, removes the splitting along the easy axis [100]. The intrinsic Gilbert damping parameter $G = 6.0 \times 10^8$ sec^{-1} in the "bcc Ni" was chosen in order to achieve agreement between the experimental and calculated FMR lineshape for the dc magnetic field applied along the semi-hard [110] axis.

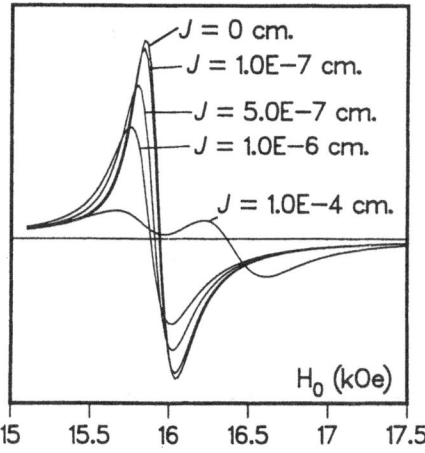

Fig. 16. Calculated absorption derivatives for a 6.0 nm thick epitaxial bcc Ni overlayer grown on an Fe(001) surface for various values of the interface exchange parameter J : the nickel layer is assumed to be coupled to the iron substrate by an interface surface energy of the form $E_{interface} = - J\, M_A \cdot M_B \delta(z-d)$. The position of the derivative zero crossing is extremely sensitive to the value of the interface exchange coupling parameter, and shifts corresponding to J as small as 10^{-7} cm would be detectable. The calculations were carried out using the parameters listed in Fig. 15 and for the applied field, H_0, parallel to the [11] planar hard axis.

Clearly, a change in crystalline structure results in a drastic change of the magnetic crystalline anisotropy: the crystalline anisotropy constant K_1 in the "bcc Ni" has the same sign as in bcc Fe, and the effective field $(2K_1/M)_{bcc} \approx 3$ kOe significantly exceeds the value $|(2K/M)_{fcc}| = 0.20$ kOe in the Ni face-centered structure. The large crystalline anisotropy in "bcc Ni" overlayers explains why the magnetization and the applied field are not colinear at low frequencies when the field is oriented along the [110] direction.

The ability to grow Au epitaxially on Fe and Ni on Au allows us to study the role of non-magnetic interlayers on the strength of the magnetic coupling between two different and well defined materials. This study can very likely be extended to all materials which have a lattice matched to the Fe substrate such as Cr(001), Ag(001) and Al(001).

In Fig. 16, we see the sensitivity of FMR measurements to the interface exchange J. Even the choice of $J \approx 1000\times$ smaller than its fully saturated value would result in a detectable effect on the substrate resonance. This makes FMR a particularly sensitive tool for studying the interactions between the substrate and magnetic layers separated by a non-magnetic metal which could carry the exchange interaction from one side to the other. Recent light scattering results by P.Grünberg et al.[28] provide an alternative approach in the thick ($d > 150$ Å) film limit. A wide range of experimental studies and exchange of ideas should result in a better understanding of one of the most important aspects of atomic engineering - the magnetic interactions in magnetic superlattices.

ACKNOWLEDGEMENT

The authors would like to thank the Natural Sciences and Engineering Research Council of Canada for grants which provided support for this work. The authors are grateful to Mr. K. Myrtle for invaluable technical assistance and to Prof. E.D. Crozier for his original suggestion for doing the REELFS measurements in our MBE system.

REFERENCES

1. B.T. Jonker, K.H. Walker, E. Kisker, G.A. Prinz, and C. Carbone, Spin-polarized photoemission study of epitaxial Fe(001) films on Ag(001), Phys. Rev. Lett. 57:142 (1986).
2. S. Shultz, D. Youm, A.F. Starr, and J.P. Armstrong, Direct dc magnetization measurements of Fe on Ag, J. Appl. Phys. (to be published).
3. J.G. Gay and Roy Richter, Spin anisotropy of ferromagnetic films, Phys. Rev. Lett. 56:2728 (1986).
4. T. Takeuchi and S. Ikeda, Studies on iron single crystals, Trans. ISJI 9:484 (1969).
5. B. Heinrich, A.S. Arrott, J.F. Cochran, C. Liu, and K. Myrtle, Ferromagnetic resonance in ultrahigh vacuum: Effect of epitaxial overlayers, J. Vac. Sci. Technol. A4(3):1376 (1986).
6. Z.Q. Wang, Y.S. Li, F. Jona, and P.M. Marcus, Ultra-Thin Epitaxial Films of bcc Nickel, to be published in Mat. Res. Soc. (1986).
7. S.T. Purcell, B. Heinrich, and A.S. Arrott, RHEED oscillations during the epitaxial growth of metals on metals, (submitted for publication).
8. J.H. Neave, B.A. Joyce, P.J. Dobson, and N. Norton, Dynamics of film growth of GaAs by MBE from RHEED oscillations, Appl. Phys. A31:1 (1983).

9. R.W. Vook and Y. Namba, Auger line shape analyses for epitaxial growth in the Cu/Cu, Ag/Ag AND Ag/Cu Systems, Appl. of Surf. Sci. 11/12:400 (1982).

10. Y. Namba, R.W. Vook and S.S. Chao, Thickness periodicity in the Auger line shape from epitaxial (111)Cu Films, Surf. Sci. 109:320 (1981).

11. M. De Crescenzi, L. Papagno, G. Chiarello, R. Scarnozzino, E. Colavita, R. Rosei and S. Mobilio, Extended ELS fine structures above the $M_{2,3}$ edges of Cu and Ni, Sol. St. Comm. 40:613 (1981) and L. Papagno, M. De Crescenzi, G. Chiavello, E. Colavita, R. Scarnozzino, L.S. Caputi and R. Rosei, Radial distribution functions of Cu and Ni by reflection energy loss spectroscopy, Surface Sci. 117:525 (1982).

12. A.P. Hitchcock and C.H. Teng, Extended energy loss fine structure in reflection electron energy loss spectra of Cu and Ni, Surface Sci. 149:558 (1984) and C.H. Teng and A.P. Hitchcock, EXELFS in the reflection electron energy loss spectra of Cu and Ni, J. Vac. Sci. Tech. A1:1209 (1983).

13. F.W. Lytle, D.E. Sayers and E.A. Stern, Extended x-ray absorption fine structure technique: II. Experimental technique and selected results, Phys. Rev. B11:4825 (1975) and E.A. Stern, D.E. Sayers and F.W. Lytle, Extended x-ray absorption fine structure technique: III. Determination of physical parameters, Phys. Rev. B11:4836 (1975).

14. Mitio Inokuti, Inelastic collisions of fast charged particles with atoms and molecules - the Bethe theory revisited, Rev. Mod. Phys. 43:297 (1971).

15. J.R. Macdonald, Ferromagnetic resonance and the internal field in ferromagnetic materials, Proc. Phys. Soc. London A64:968 (1951).

16. J.F. Cochran, B. Heinrich, and A.S. Arrott, Ferromagnetic resonance in a system composed of a ferromagnetic substrate and an exchange coupled thin ferromagnetic overlayer, to be published in Phys. Rev. B34, #11.

17. G.T. Rado and J.R. Weertman, Spin-wave resonance in a ferromagnetic metal, J. Phys. Chem. Solids 11:315 (1959) and W.S. Ament and G.T. Rado, Electromagnetic effects of spin wave resonance in ferromagnetic metals, Phys. Rev. 97:1558 (1955).

18. Z. Frait, D. Fraitová, M. Kotrbová, Z. Hauptmann, Ferromagnetic resonance in thin single crystal platelets of Iron, Czech. J. Phys. B16:837 (1966).

19. D.S. Rodbell, Magnetic resonance of high quality ferromagnetic metal single crystals, Physics 1:279 (1965).

20. L. Neél, Anisotropie magnétique superficielle et surstructures d'orientation, J. de. Physique et Radium 15:225 (1954).

21. F. Hoffmann, A. Stankoff, and H. Pascard, Evidence for an Exchange coupling at the interface between two ferromagnetic films, J. Appl. Phys. 41:1022 (1970). and F. Hoffmann, Dynamic pinning induced by Nickel layers on Permalloy films, Phys. Stat. Sol. 41:807 (1970).

22. B. Heinrich, J.F. Cochran, and R. Baartman, Ferromagnetic resonance absorption in Supermalloy at 9, 24, and 38 GHz., Can. J. Phys. 55:806 (1977).

23. J.F. Cochran and B. Heinrich, Microwave transmission through ferromagnetic metals, IEEE Trans. Magn. MAG-16:660 (1980).

24. B. Heinrich, D. Fraitová, and V. Kambersky, The influence or s-d exchange on relaxation of magnons in metal, Phys. Stat.Sol. 23:501 (1967).

25. B. Heinrich, D.J. Meredith, and J.F. Cochran, Wave number and temperature dependent Landau-Lifshitz damping in Nickel, J. Appl. Phys. 50:7726 (1979).

26. E.P. Wohlfarth, Iron, Cobalt and Nickel, in: "Ferromagnetic Materials, Vol. 1," E.P. Wohlfarth, ed., North-Holland, Amsterdam (1980).

27. V.L. Moruzzi, P.M. Marcus, K. Schwarz and P. Mohn, Ferromagnetic phases of bcc and fcc Fe, Co, and Ni, Phys. Rev. B. 34:1784 (1986).

28. P. Grünberg, Y. Pang, R. Schreiber, M.B. Brodsky, and H. Sowers,
     Layered magnetic structures: Evidence for antiferromagnetic
     coupling of Fe layers across Cr interlayers, to be published in
     J. Appl. Phys., see also P. Grünberg's chapter in this book.

INDEX